The Art of ANALOG LAYOUT
Second Edition

For My Father

Contents

Preface to the Second Edition xvii
Preface to the First Edition xix
Acknowledgments xxi

1 *Device Physics 1*
1.1 Semiconductors 1
1.1.1. Generation and Recombination 4
1.1.2. Extrinsic Semiconductors 6
1.1.3. Diffusion and Drift 9
1.2 PN Junctions 11
1.2.1. Depletion Regions 11
1.2.2. PN Diodes 13
1.2.3. Schottky Diodes 16
1.2.4. Zener Diodes 18
1.2.5. Ohmic Contacts 19
1.3 Bipolar Junction Transistors 21
1.3.1. Beta 23
1.3.2. I-V Characteristics 24
1.4 MOS Transistors 25
1.4.1. Threshold Voltage 27
1.4.2. I-V Characteristics 29
1.5 JFET Transistors 32
1.6 Summary 34
1.7 Exercises 35

2 *Semiconductor Fabrication 37*
2.1 Silicon Manufacture 37
2.1.1. Crystal Growth 38
2.1.2. Wafer Manufacturing 39
2.1.3. The Crystal Structure of Silicon 39
2.2 Photolithography 41
2.2.1. Photoresists 41
2.2.2. Photomasks and Reticles 42
2.2.3. Patterning 43
2.3 Oxide Growth and Removal 43
2.3.1. Oxide Growth and Deposition 44
2.3.2. Oxide Removal 45
2.3.3. Other Effects of Oxide Growth and Removal 47
2.3.4. Local Oxidation of Silicon (LOCOS) 49
2.4 Diffusion and Ion Implantation 50
2.4.1. Diffusion 51
2.4.2. Other Effects of Diffusion 53
2.4.3. Ion Implantation 55

2.5 Silicon Deposition and Etching 57
2.5.1. Epitaxy 57
2.5.2. Polysilicon Deposition 59
2.5.3. Dielectric Isolation 60
2.6 Metallization 62
2.6.1. Deposition and Removal of Aluminum 63
2.6.2. Refractory Barrier Metal 65
2.6.3. Silicidation 67
2.6.4. Interlevel Oxide, Interlevel Nitride, and Protective Overcoat 69
2.6.5. Copper Metallization 71
2.7 Assembly 73
2.7.1. Mount and Bond 74
2.7.2. Packaging 77
2.8 Summary 78
2.9 Exercises 78

3 *Representative Processes 80*
3.1 Standard Bipolar 81
3.1.1. Essential Features 81
3.1.2. Fabrication Sequence 82
Starting Material 82
N-Buried Layer 82
Epitaxial Growth 83
Isolation Diffusion 83
Deep-N+ 83
Base Implant 84
Emitter Diffusion 84
Contact 85
Metallization 85
Protective Overcoat 86
3.1.3. Available Devices 86
NPN Transistors 86
PNP Transistors 88
Resistors 90
Capacitors 92
3.1.4. Process Extensions 93
Up-Down Isolation 93
Double-Level Metal 94
Schottky Diodes 94
High-Sheet Resistors 94
Super-Beta Transistors 96
3.2 Polysilicon-Gate CMOS 96
3.2.1. Essential Features 97
3.2.2. Fabrication Sequence 98
Starting Material 98
Epitaxial Growth 98
N-Well Diffusion 98
Inverse Moat 99
Channel Stop Implants 100
LOCOS Processing and Dummy Gate Oxidation 100
Threshold Adjust 101

Polysilicon Deposition and Patterning 102
Source/Drain Implants 102
Contacts 103
Metallization 103
Protective Overcoat 103
3.2.3. Available Devices 104
NMOS Transistors 104
PMOS Transistors 106
Substrate PNP Transistors 107
Resistors 107
Capacitors 109
3.2.4. Process Extensions 109
Double-Level Metal 110
Shallow Trench Isolation 110
Silicidation 111
Lightly Doped Drain (LDD) Transistors 112
Extended-Drain, High-Voltage Transistors 113
3.3 Analog BiCMOS 114
3.3.1. Essential Features 115
3.3.2. Fabrication Sequence 116
Starting Material 116
N-Buried Layer 116
Epitaxial Growth 117
N-Well Diffusion and Deep-N+ 117
Base Implant 118
Inverse Moat 118
Channel Stop Implants 119
LOCOS Processing and Dummy Gate Oxidation 119
Threshold Adjust 119
Polysilicon Deposition and Pattern 120
Source/Drain Implants 120
Metallization and Protective Overcoat 120
Process Comparison 121
3.3.3. Available Devices 121
NPN Transistors 121
PNP Transistors 123
Resistors 125
3.3.4. Process Extensions 125
Advanced Metal Systems 126
Dielectric Isolation 126
3.4 Summary 130
3.5 Exercises 131

4 *Failure Mechanisms 133*
4.1 Electrical Overstress 133
4.1.1. Electrostatic Discharge (ESD) 134
Effects 135
Preventative Measures 135
4.1.2. Electromigration 136
Effects 136
Preventative Measures 137

4.1.3. Dielectric Breakdown 138
Effects 138
Preventative Measures 139
4.1.4. The Antenna Effect 141
Effects 141
Preventative Measures 142
4.2 Contamination 143
4.2.1. Dry Corrosion 144
Effects 144
Preventative Measures 145
4.2.2. Mobile Ion Contamination 145
Effects 145
Preventative Measures 146
4.3 Surface Effects 148
4.3.1. Hot Carrier Injection 148
Effects 148
Preventative Measures 150
4.3.2. Zener Walkout 151
Effects 151
Preventative Measures 152
4.3.3. Avalanche-Induced Beta Degradation 153
Effects 153
Preventative Measures 154
4.3.4. Negative Bias Temperature Instability 154
Effects 155
Preventative Measures 155
4.3.5. Parasitic Channels and Charge Spreading 156
Effects 156
Preventative Measures (Standard Bipolar) 159
Preventative Measures (CMOS and BiCMOS) 162
4.4 Parasitics 164
4.4.1. Substrate Debiasing 165
Effects 166
Preventative Measures 167
4.4.2. Minority-Carrier Injection 169
Effects 169
Preventative Measures (Substrate Injection) 172
Preventative Measures (Cross-Injection) 178
4.4.3. Substrate Influence 180
Effects 180
Preventative Measures 180
4.5 Summary 183
4.6 Exercises 183

5 *Resistors 185*
5.1 Resistivity and Sheet Resistance 185
5.2 Resistor Layout 187
5.3 Resistor Variability 191
5.3.1. Process Variation 191
5.3.2. Temperature Variation 192

5.3.3. Nonlinearity 193
5.3.4. Contact Resistance 196
5.4 Resistor Parasitics 197
5.5 Comparison of Available Resistors 200
5.5.1. Base Resistors 200
5.5.2. Emitter Resistors 201
5.5.3. Base Pinch Resistors 202
5.5.4. High-Sheet Resistors 202
5.5.5. Epi Pinch Resistors 205
5.5.6. Metal Resistors 206
5.5.7. Poly Resistors 208
5.5.8. NSD and PSD Resistors 211
5.5.9. N-Well Resistors 211
5.5.10. Thin-Film Resistors 212
5.6 Adjusting Resistor Values 213
5.6.1. Tweaking Resistors 213
Sliding Contacts 214
Sliding Heads 215
Trombone Slides 215
Metal Options 215
5.6.2. Trimming Resistors 216
Fuses 216
Zener Zaps 219
EPROM Trims 221
Laser Trims 222
5.7 Summary 223
5.8 Exercises 224

6 *Capacitors and Inductors 226*
6.1 Capacitance 226
6.1.1. Capacitor Variability 232
Process Variation 232
Voltage Modulation and Temperature Variation 233
6.1.2. Capacitor Parasitics 235
6.1.3. Comparison of Available Capacitors 237
Base-Emitter Junction Capacitors 237
MOS Capacitors 239
Poly-Poly Capacitors 241
Stack Capacitors 243
Lateral Flux Capacitors 245
High-Permittivity Capacitors 246
6.2 Inductance 246
6.2.1. Inductor Parasitics 248
6.2.2. Inductor Construction 250
Guidelines for Integrating Inductors 251
6.3 Summary 252
6.4 Exercises 253

7 *Matching of Resistors and Capacitors 254*
7.1 Measuring Mismatch 254

7.2 Causes of Mismatch 257
7.2.1. Random *Variation* 257
Capacitors 258
Resistors 258
7.2.2. Process Biases 260
7.2.3. Interconnection Parasitics 261
7.2.4. Pattern Shift 263
7.2.5. Etch Rate Variations 265
7.2.6. Photolithographic Effects 267
7.2.7. Diffusion Interactions 268
7.2.8. Hydrogenation 270
7.2.9. Mechanical Stress and Package Shift 271
7.2.10. Stress Gradients 274
Piezoresistivity 274
Gradients and Centroids 275
Common-Centroid Layout 277
Location and Orientation 281
7.2.11. Temperature Gradients and Thermoelectrics 283
Thermal Gradients 285
Thermoelectric Effects 287
7.2.12. Electrostatic Interactions 288
Voltage Modulation 288
Charge Spreading 292
Dielectric Polarization 293
Dielectric Relaxation 294
7.3 Rules for Device Matching 295
7.3.1. Rules for Resistor Matching 296
7.3.2. Rules for Capacitor Matching 300
7.4 Summary 303
7.5 Exercises 304

8 *Bipolar Transistors 306*
8.1 Topics in Bipolar Transistor Operation 306
8.1.1. Beta Rolloff 308
8.1.2. Avalanche Breakdown 308
8.1.3. Thermal Runaway and Secondary Breakdown 310
8.1.4. Saturation in NPN Transistors 312
8.1.5. Saturation in Lateral PNP Transistors 315
8.1.6. Parasitics of Bipolar Transistors 318
8.2 Standard Bipolar Small-Signal Transistors 320
8.2.1. The Standard Bipolar NPN Transistor 320
Construction of Small-Signal NPN Transistors 322
8.2.2. The Standard Bipolar Substrate PNP Transistor 326
Construction of Small-Signal Substrate PNP Transistors 328
8.2.3. The Standard Bipolar Lateral PNP Transistor 330
Construction of Small-Signal Lateral PNP Transistors 332
8.2.4. High-Voltage Bipolar Transistors 337
8.2.5. Super-Beta NPN Transistors 340
8.3 CMOS and BiCMOS Small-Signal Bipolar Transistors 341
8.3.1. CMOS PNP Transistors 341
8.3.2. Shallow-Well Transistors 345

8.3.3. Analog BiCMOS Bipolar Transistors 347
8.3.4. Fast Bipolar Transistors 349
8.3.5. Polysilicon-Emitter Transistors 351
8.3.6. Oxide-Isolated Transistors 354
8.3.7. Silicon-Germanium Transistors 356
8.4 Summary 358
8.5 Exercises 358

9 *Applications of Bipolar Transistors 360*
9.1 Power Bipolar Transistors 361
9.1.1. Failure Mechanisms of NPN Power Transistors 362
Emitter Debiasing 362
Thermal Runaway and Secondary Breakdown 364
Kirk Effect 366
9.1.2. Layout of Power NPN Transistors 368
The Interdigitated-Emitter Transistor 369
The Wide-Emitter Narrow-Contact Transistor 371
The Christmas-Tree Device 372
The Cruciform-Emitter Transistor 373
Power Transistor Layout in Analog BiCMOS 374
Selecting a Power Transistor Layout 376
9.1.3. Power PNP Transistors 376
9.1.4. Saturation Detection and Limiting 378
9.2 Matching Bipolar Transistors 381
9.2.1. Random Variations 382
9.2.2. Emitter Degeneration 384
9.2.3. NBL Shadow 386
9.2.4. Thermal Gradients 387
9.2.5. Stress Gradients 391
9.2.6. Filler-Induced Stress 393
9.2.7. Other Causes of Systomatic Mismatch 395
9.3 Rules for Bipolar Transistor Matching 396
9.3.1. Rules for Matching Vertical Transistors 397
9.3.2. Rules for Matching Lateral Transistors 402
9.4 Summary 402
9.5 Exercises 403

10 *Diodes 406*
10.1 Diodes in Standard Bipolar 406
10.1.1. Diode-Connected Transistors 406
10.1.2. Zener Diodes 409
Surface Zener Diodes 410
Buried Zeners 412
10.1.3. Schottky Diodes 415
10.1.4. Power Diodes 420
10.2 Diodes in CMOS and BiCMOS Processes 422
10.2.1. CMOS Junction Diodes 422
10.2.2. CMOS and BiCMOS Schottky Diodes 423
10.3 Matching Diodes 425
10.3.1. Matching PN Junction Diodes 425

10.3.2. Matching Zener Diodes 426
10.3.3. Matching Schottky Diodes 428
10.4 Summary 428
10.5 Exercises 429

11 *Field-Effect Transistors 430*
11.1 Topics in MOS Transistor Operation 431
11.1.1. Modeling the MOS Transistor 431
Device Transconductance 432
Threshold Voltage 434
11.1.2. Parasitics of MOS Transistors 438
Breakdown Mechanisms 440
CMOS Latchup 442
Leakage Mechanisms 443
11.2 Constructing CMOS Transistors 446
11.2.1. Coding the MOS Transistor 447
Width and Length 448
11.2.2. N-Well and P-Well Processes 449
11.2.3. Channel Stop Implants 452
11.2.4. Threshold Adjust Implants 453
11.2.5. Scaling the Transistor 456
11.2.6. Variant Structures 459
Serpentine Transistors 461
Annular Transistors 462
11.2.7. Backgate Contacts 464
11.3 Floating-Gate Transistors 467
11.3.1. Principles of Floating-Gate Transistor Operation 469
11.3.2. Single-Poly EEPROM Memory 472
11.4 The JFET Transistor 474
11.4.1. Modeling the JFET 474
11.4.2. JFET Layout 476
11.5 Summary 479
11.6 Exercises 479

12 *Applications of MOS Transistors 482*
12.1 Extended-Voltage Transistors 482
12.1.1. LDD and DDD Transistors 483
12.1.2. Extended-Drain Transistors 486
Extended-Drain NMOS Transistors 487
Extended-Drain PMOS Transistors 488
12.1.3. Multiple Gate Oxides 489
12.2 Power MOS Transistors 491
12.2.1. MOS Safe Operating Area 492
Electrical SOA 493
Electrothermal SOA 496
Rapid Transient Overload 497
12.2.2. Conventional MOS Power Transistors 498
The Rectangular Device 499
The Diagonal Device 500
Computation of R_M 501

Other Considerations 502
Nonconventional Structures 503
12.2.3. DMOS Transistors 505
The Lateral DMOS Transistor 506
RESURF Transistors 508
The DMOS NPN 510
12.3 MOS Transistor Matching 511
12.3.1. Geometric Effects 513
Gate Area 513
Gate Oxide Thickness 514
Channel Length Modulation 515
Orientation 515
12.3.2. Diffusion and Etch Effects 516
Polysilicon Etch Rate Variations 516
Diffusion Penetration of Polysilicon 517
Contacts Over Active Gate 518
Diffusions Near the Channel 518
PMOS versus NMOS Transistors 519
12.3.3. Hydrogenation 520
Fill Metal and MOS Matching 521
12.3.4. Thermal and Stress Effects 521
Oxide Thickness Gradients 522
Stress Gradients 522
Thermal Gradients 522
12.3.5. Common-Centroid Layout of MOS Transistors 523
12.4 Rules for MOS Transistor Matching 528
12.5 Summary 531
12.6 Exercises 531

13 *Special Topics 534*
13.1 Merged Devices 534
13.1.1. Flawed Device Mergers 535
13.1.2. Successful Device Mergers 539
13.1.3. Low-Risk Merged Devices 541
13.1.4. Medium-Risk Merged Devices 542
13.1.5. Devising New Merged Devices 544
13.1.6. The Role of Merged Devices in Analog BiCMOS 544
13.2 Guard Rings 545
13.2.1. Standard Bipolar Electron Guard Rings 546
13.2.2. Standard Bipolar Hole Guard Rings 547
13.2.3. Guard Rings in CMOS and BiCMOS Designs 548
13.3 Single-level Interconnection 551
13.3.1. Mock Layouts and Stick Diagrams 551
13.3.2. Techniques for Crossing Leads 553
13.3.3. Types of Tunnels 555
13.4 Constructing the Padring 557
13.4.1. Scribe Streets and Alignment Markers 557
13.4.2. Bondpads, Trimpads, and Testpads 558
13.5 ESD Structures 562
13.5.1. Zener Clamp 563
13.5.2. Two-Stage Zener Clamps 565

13.5.3. Buffered Zener Clamp 566
13.5.4. V_{CES} Clamp 568
13.5.5. V_{ECS} Clamp 569
13.5.6. Antiparallel Diode Clamps 570
13.5.7. Grounded-Gate NMOS Clamps 570
13.5.8. CDM Clamps 572
13.5.9. Lateral SCR Clamps 573
13.5.10. Selecting ESD Structures 575
13.6 Exercises 578

14 *Assembling the Die 581*
14.1 Die Planning 581
14.1.1. Cell Area Estimation 582
Resistors 582
Capacitors 582
Vertical Bipolar Transistors 583
Lateral PNP Transistors 583
MOS Transistors 583
MOS Power Transistors 584
Computing Cell Area 584
14.1.2. Die Area Estimation 584
14.1.3. Gross Profit Margin 587
14.2 Floorplanning 588
14.3 Top-Level Interconnection 594
14.3.1. Principles of Channel Routing 594
14.3.2. Special Routing Techniques 596
Kelvin Connections 597
Noisy Signals and Sensitive Signals 598
14.3.3. Electromigration 600
14.3.4. Minimizing Stress Effects 603
14.4 Conclusion 604
14.5 Exercises 605

Appendices
A. Table of Acronyms Used in the Text 607
B. The Miller Indices of a Cubic Crystal 611
C. Sample Layout Rules 614
D. Mathematical Derivations 622
E. Sources for Layout Editor Software 627

Index 628

Preface to the Second Edition

I originally wrote *The Art of Analog Layout* as a companion volume to a series of lectures. Many people encouraged me to publish it. At first I was reluctant to do so, for I thought that it would find a rather limited audience. Publication has proven my concerns quite unfounded. To my astonishment, *The Art of Analog Layout* has even been translated into Chinese!

The passage of several years has alerted me to the limitations of the first edition and prompted an extensive revision. Every chapter has been examined and corrected. Many new passages have been added, along with some 50 new illustrations to accompany them. New topics introduced in the second edition include the following:

- Advanced metallization systems
- Dielectric isolation
- Failure mechanisms of MOS transistors
- Integrated inductors
- MOS safe operating area
- Nonvolatile memory

In preparing this edition, I have drawn extensively upon the experience and wisdom of my colleagues at Texas Instruments. I have also made constant reference to the resources available upon the IEEE Xplore website, most particularly those contained in the *IEEE Journal of Electron Devices*. I thank all the many people who have contributed to my own understanding or who have corrected my many mistakes. A work of this length and magnitude will never prove perfect, but the second edition greatly improves upon the first.

ALAN HASTINGS

Preface to the First Edition

An integrated circuit reveals its true appearance only under high magnification. The intricate tangle of microscopic wires covering its surface and the equally intricate patterns of doped silicon beneath it, all follow a set of blueprints called a *layout*. The process of constructing layouts for analog and mixed-signal integrated circuits has stubbornly defied all attempts at automation. The shape and placement of every polygon requires a thorough understanding of the principles of device physics, semiconductor fabrication, and circuit theory. Despite 30 years of research, much remains uncertain. What information there is lies buried in obscure journal articles and unpublished manuscripts. This textbook assembles that information between a single set of covers. While primarily intended for use by practicing layout designers, it should also prove valuable to circuit designers who desire a better understanding of the relationship between circuits and layouts.

The text has been written for a broad audience, some of whom have had only limited exposure to higher mathematics and solid-state physics. The amount of mathematics has been kept to an absolute minimum, and care has been taken to identify all variables and to use the most accessible units. The reader need only have a familiarity with basic algebra and elementary electronics. Many of the exercises assume that the reader also has access to layout editing software; but those who lack such resources can complete many of the exercises with pencil and paper.

The text consists of 14 chapters and five appendices. The first two chapters provide an overview of device physics and semiconductor processing. These chapters avoid mathematical derivations and instead emphasize simple verbal explanations and visual models. The third chapter presents three archetypal processes: standard bipolar, silicon-gate CMOS, and analog BiCMOS. The presentation focuses upon development of cross sections and the correlation of these cross sections to conventional layout views of sample devices. The fourth chapter covers common failure mechanisms and emphasizes the role of layout in determining reliability. Chapters 5 and 6 cover the layout of resistors and capacitors. Chapter 7 presents the principles of matching, using resistors and capacitors as examples. Chapters 8 through 10 cover the layout of bipolar devices, while Chapters 11 and 12 cover the layout and matching of field-effect transistors. Chapters 13 and 14 cover a variety of advanced topics, including device mergers, guard rings, ESD protection structures, and floorplanning. The appendices include a list of acronyms, a discussion of Miller indices, sample layout rules for use in working the exercises, and the derivation of formulas used in the text.

ALAN HASTINGS

Acknowledgments

The information contained in this text has been gathered through the hard work of many scientists, engineers, and technicians, the vast majority of whom must remain unacknowledged because their work has not been published. I have included references to as many fundamental discoveries and principles as I could, but in many cases I have been unable to determine original sources.

I thank my colleagues at Texas Instruments for numerous suggestions. I am especially grateful to Ken Bell, Walter Bucksch, Taylor Efland, Lou Hutter, Clif Jones, Alec Morton, Jeff Smith, Fred Trafton, and Joe Trogolo, all of whom have provided important information for this text. I am also grateful for the encouragement of Bob Borden, Nicolas Salamina, and Ming Chiang, without which this book would never have been written.

1 Device Physics

Before 1960, most electronic circuits depended upon vacuum tubes to perform the critical tasks of amplification and rectification. An ordinary mass-produced AM radio required five tubes, while a color television needed no fewer than twenty. Vacuum tubes were large, fragile, and expensive. They dissipated a lot of heat and were not very reliable. So long as electronics depended upon them, it was nearly impossible to construct systems requiring thousands or millions of active devices.

The appearance of the bipolar junction transistor in 1947 marked the beginning of the solid-state revolution. These new devices were small, cheap, rugged, and reliable. Solid-state circuitry made possible the development of pocket transistor radios and hearing aids, quartz watches and touch-tone phones, compact disc players and personal computers.

A *solid-state device* consists of a crystal with regions of impurities incorporated into its surface. These impurities modify the electrical properties of the crystal, allowing it to amplify or modulate electrical signals. A working knowledge of device physics is necessary to understand how this occurs. This chapter covers not only elementary device physics but also the operation of three of the most important solid-state devices: the junction diode, the bipolar transistor, and the field-effect transistor. Chapter 2 explains the manufacturing processes used to construct these and other solid-state devices.

1.1 SEMICONDUCTORS

The inside front cover of the book depicts a long-form periodic table. The elements are arranged so those with similar properties group together to form rows and columns. The elements on the left-hand side of the periodic table are called *metals,* while those on the right-hand side are called *nonmetals*. Metals are usually good conductors of heat and electricity. They are also malleable and display a characteristic metallic luster. Nonmetals are poor conductors of heat and electricity, and those that are solid are brittle and lack the shiny luster of metals. A few elements in the middle of the periodic table, such as silicon and germanium, have electrical properties that lie midway between those of metals and nonmetals. These elements are called *semiconductors*. The differences between metals, semiconductors, and nonmetals result from differences in the electronic structure of their respective atoms.

Every atom consists of a positively charged nucleus surrounded by a cloud of electrons. The number of electrons in this cloud equals the number of protons in the nucleus, which also equals the atomic number of the element. Therefore, a carbon atom has six electrons, because carbon has an atomic number of six. These electrons occupy a series of *shells* that are somewhat analogous to the layers of an onion. As electrons are added, the shells fill in order from innermost outward. The outermost or *valence* shell may remain unfilled. The electrons occupying this outermost shell are called *valence electrons*. The number of valence electrons possessed by an element determines most of its chemical and electronic properties.

Each row of the periodic table corresponds to the filling of one shell. The leftmost element in the row has one valence electron, while the rightmost element has a full valence shell. Atoms with filled valence shells possess a particularly favored configuration. Those with unfilled valence shells will trade or share electrons so that each can claim a full shell. Electrostatic attraction forms a chemical bond between atoms that trade or share electrons. Depending upon the strategy adopted to fill the valence shell, one of three types of bonding will occur.

Metallic bonding occurs between atoms of metallic elements, such as sodium. Consider a group of sodium atoms in close proximity. Each atom has one valence electron orbiting around a filled inner shell. Imagine that the sodium atoms all discard their valence electrons. The discarded electrons are still attracted to the positively charged sodium atoms, but, since each atom now has a full valence shell, none accepts them. Figure 1.1A shows a simplified representation of a sodium crystal. Electrostatic forces hold the sodium atoms in a regular lattice. The discarded valence electrons wander freely through the resulting crystal. Sodium metal is an excellent electrical conductor due to the presence of numerous free electrons. These same electrons are also responsible for the metallic luster of the element and its high thermal conductivity. Other metals form similar crystal structures, all of which are held together by metallic bonding between a sea of free valence electrons and a rigid lattice of charged atomic cores.[1]

Ionic bonding occurs between atoms of metals and nonmetals. Consider a sodium atom in close proximity to a chlorine atom. The sodium atom has one valence electron, while the chlorine atom is one electron short of a full valence shell. The sodium atom can donate an electron to the chlorine atom, and by this means both can achieve filled outer shells. After the exchange, the sodium atom has a net positive

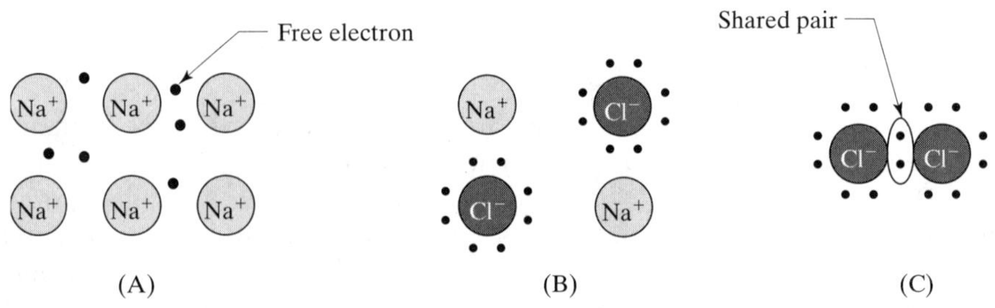

FIGURE 1.1 Simplified illustrations of various types of chemical bonding: a small part of a metallically bonded sodium crystal (A), a small part of an ionically bonded sodium chloride crystal (B), and a covalently bonded chlorine molecule (C).

[1] Some metals conduct by means of holes rather than electrons, but the general observations made here still apply.

charge and the chlorine atom, a net negative charge. The two charged atoms (or *ions*) attract one another. Solid sodium chloride thus consists of sodium and chlorine ions arranged in a regular lattice, forming a crystal (Figure 1.1B). Crystalline sodium chloride is a poor conductor of electricity, since all of its electrons are held in the shells of the various atoms.

Covalent bonding occurs between atoms of nonmetals. Consider two chlorine atoms in close proximity. Each atom has only seven valence electrons, while each needs eight to fill its valence shell. Suppose that each of the two atoms contributes one valence electron to a common pair shared by both. Now each chlorine atom can claim eight valence electrons: six of its own, plus the two shared electrons. The two chlorine atoms link to form a molecule that is held together by the electron pair shared between them (Figure 1.1C). The shared pair of electrons forms a *covalent bond*. The lack of free valence electrons explains why nonmetallic elements do not conduct electricity and why they lack metallic luster. Many nonmetals are gases at room temperature because the electrically neutral molecules exhibit no strong attraction to one another and thus do not condense to form a liquid or a solid.

The atoms of a semiconductor also form covalent bonds. Consider atoms of silicon, a representative semiconductor. Each atom has four valence electrons and needs four more to complete its valence shell. Two silicon atoms could theoretically attempt to pool their valence electrons to achieve filled shells. In practice this does not occur because eight electrons packed tightly together strongly repel one another. Instead, each silicon atom shares one electron pair with each of four surrounding atoms. In this way, the valence electrons are spread around to four separate locations and their mutual repulsion is minimized.

Figure 1.2 shows a simplified two-dimensional representation of a silicon crystal. Each of the small circles represents a silicon atom. Each of the lines between the circles represents a covalent bond consisting of a shared pair of valence electrons. Each silicon atom can claim eight electrons (four shared electron pairs), so all of the atoms have full valence shells. These atoms are linked together in a molecular network by the covalent bonds formed between them. This infinite lattice represents the structure of the silicon crystal. The entire crystal is literally a single molecule, so crystalline silicon is strong and hard, and it melts at a very high temperature. Silicon is a poor conductor of electricity because all of its valence electrons are used to form the crystal lattice.

A similar macromolecular crystal can theoretically be formed by any group-IV element,[2] including carbon, silicon, germanium, tin, and lead. Carbon, in the form of

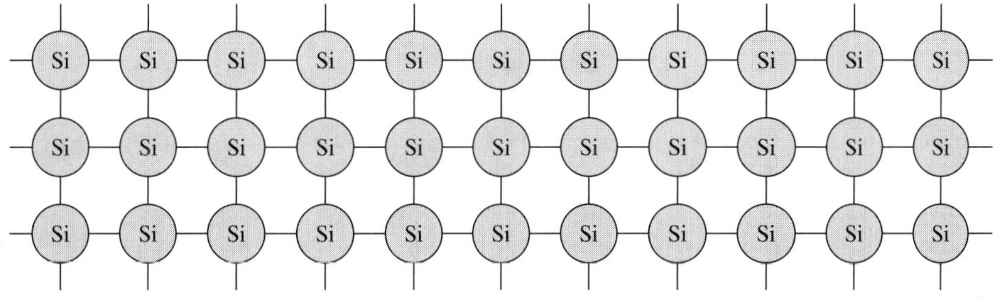

FIGURE 1.2 Simplified two-dimensional representation of a silicon crystal lattice.

[2] The group-III, IV, V, and VI elements reside in columns III-B, IV-B, V-B, and VI-B of the long-form periodic table. The group-II elements may fall into either columns II-A or II-B. The A/B numbering system is a historical curiosity and the International Union of Pure and Applied Chemists (IUPAC) has recommended its abandonment; see J. Hudson, *The History of Chemistry* (New York: Chapman and Hall, 1992), pp. 122–137.

diamond, has the strongest bonds of any group-IV element. Diamond crystals are justly famed for their strength and hardness. Silicon and germanium have somewhat weaker bonds due to the presence of filled inner shells that partially shield the valence electrons from the nucleus. Tin and lead have weak bonds because of numerous inner shells; they typically form metallically bonded crystals instead of covalently bonded macromolecules. Of the group-IV elements, only silicon and germanium have bonds of an intermediate degree of strength. These two elements act as true semiconductors, while carbon is a nonmetal, and tin and lead are both metals.

1.1.1. Generation and Recombination

The electrical conductivity of group-IV elements increases with atomic number. Carbon, in the form of diamond, is a true insulator. Silicon and germanium have much higher conductivities, but these are still far less than those of metals such as tin and lead. Because of their intermediate conductivities, silicon and germanium are termed *semiconductors*.

Conduction implies the presence of free electrons. At least a few of the valence electrons of a semiconductor must somehow escape the lattice to support conduction. Experiments do indeed detect small but measurable concentrations of free electrons in pure silicon and germanium. The presence of these free electrons implies that some mechanism provides the energy needed to break the covalent bonds. The statistical theory of thermodynamics suggests that the source of this energy lies in the random thermal vibrations that agitate the crystal lattice. Even though the average thermal energy of an electron is relatively small (roughly 0.04 electron-volt at 25°C), these energies are randomly distributed, and a few electrons possess much larger energies. The energy required to free a valence electron from the crystal lattice is called the *bandgap energy.* A material with a large bandgap energy possesses strong covalent bonds and therefore contains few free electrons. Materials with lower bandgap energies contain more free electrons and possess correspondingly greater conductivities (Table 1.1).

TABLE 1.1 Selected properties

Element	Atomic Number	Melting Point, °C	Electrical Conductivity, $(\Omega \cdot \text{cm})^{-1}$	Bandgap Energy, eV
Carbon (diamond)	6	3550	$\sim 10^{-16}$	5.2
Silicon	14	1410	$4 \cdot 10^{-6}$	1.1
Germanium	32	937	0.02	0.7
White tin	50	232	$9 \cdot 10^{4}$	0.1

A vacancy occurs whenever an electron leaves the lattice. One of the atoms that formerly possessed a full outer shell now lacks a valence electron and therefore has a net positive charge. This situation is depicted in a simplified fashion in Figure 1.3. The ionized atom can regain a full valence shell if it appropriates an electron from a neighboring atom. This is easily accomplished since it still shares electrons with three adjacent atoms. The electron vacancy is not eliminated; it merely shifts to the

[3] Bandgap energies for Si, Ge: B. G. Streetman, *Solid State Electronic Devices*, 2d ed. (Englewood Cliffs, NJ: Prentice-Hall, 1980), p. 443. Bandgap for C: N. B. Hanny, ed., *Semiconductors* (New York: Reinhold Publishing, 1959), p. 52. Conductivity for Sn: R. C. Weast, ed., *CRC Handbook of Chemistry and Physics*, 62d ed. (Boca Raton, FL: CRC Press, 1981), pp. F135–F136. Other values computed. Melting points: Weast, pp. B4–B48.

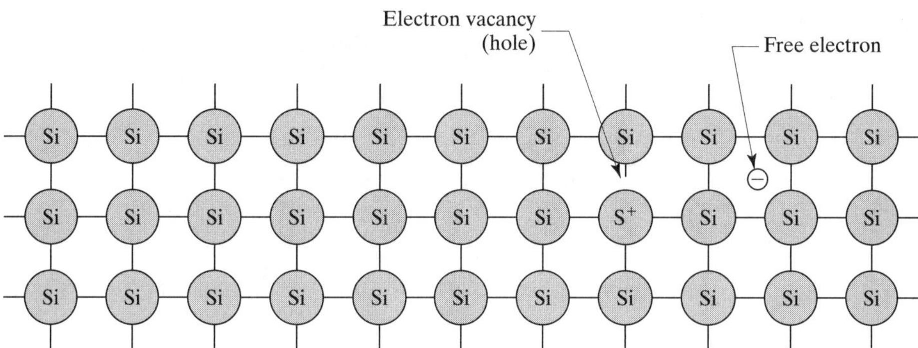

FIGURE 1.3 Simplified diagram of thermal generation in intrinsic silicon.

adjacent atom. As the vacancy is handed from atom to atom, it moves through the lattice. This moving electron vacancy is called a *hole*.

Suppose an electric field is placed across the crystal. The negatively charged free electrons move toward the positive end of the crystal. The holes behave as if they were positively charged particles and move toward the negative end of the crystal. Holes in a crystal lattice are like bubbles in a liquid. Just as a bubble is a location devoid of fluid, a hole is a location devoid of valence electrons. Bubbles move upward because the fluid around them sinks downward. Holes shift toward the negative end of the crystal because the surrounding electrons shift toward the positive end.

Holes are usually treated as if they were actual subatomic particles. The movement of a hole toward the negative end of the crystal is explained by assuming that holes are positively charged. Similarly, their rate of movement through the crystal is measured by a quantity called *mobility*. Holes have lower mobilities than electrons; typical values in bulk silicon are 480 $cm^2V^{-1}s^{-1}$ for holes and 1350 $cm^2V^{-1}s^{-1}$ for electrons.[4] The lower mobility of holes makes them less efficient charge carriers. The behavior of a device therefore depends upon whether its operation involves holes or electrons.

A free electron and a hole are formed whenever a valence electron is removed from the lattice. Both particles are electrically charged and move under the influence of electric fields. Electrons move toward positive potentials, producing an electron current. Holes move toward negative potentials, producing a hole current. The total current equals the sum of the electron and the hole currents. Holes and electrons are both called *carriers* because of their role in transporting electric charge.

Carriers are always generated in pairs since the removal of a valence electron from the lattice simultaneously forms a hole. The generation of electron–hole pairs can occur whenever energy is absorbed by the lattice. Thermal vibration produces carriers, as do light, nuclear radiation, electron bombardment, rapid heating, mechanical friction, and any number of other processes. To consider only one example, light of a sufficiently short wavelength can generate electron–hole pairs. When a lattice atom absorbs a photon, the resulting energy transfer can break a covalent bond to produce a free electron and a free hole. Optical generation will occur only if the photons have enough energy to break bonds, and this in turn requires light of a sufficiently short wavelength. Visible light has enough energy to produce electron–hole pairs in most semiconductors. Solar cells make use of this phenomenon to convert sunlight into electrical current. Photocells and solid-state camera detectors also employ optical generation.

---

[4] Streetman, p. 443.

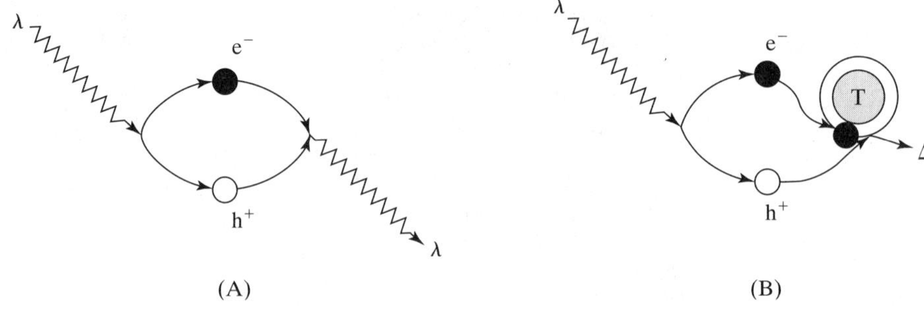

FIGURE 1.4 Schematic representations of recombination processes: (A) direct recombination, in which a photon, λ, generates a hole, h^+, and an electron, e^- that collide and re-emit a photon; and (B) indirect recombination, in which one of the carriers is caught by a trap, T, and recombination takes place at the trap site with the liberation of heat, Δ.

Just as carriers are generated in pairs, they also recombine in pairs. The exact mechanism of carrier recombination depends on the nature of the semiconductor. Recombination is particularly simple in the case of a *direct-bandgap semiconductor.* When an electron and a hole collide, the electron falls into the hole and repairs the broken covalent bond. The energy gained by the electron is radiated away as a photon (Figure 1.4A). Direct-bandgap semiconductors can, when properly stimulated, emit light. A *light-emitting diode* (LED) produces light by electron–hole recombination. The color of light emitted by the LED depends on the bandgap energy of the semiconductor used to manufacture it. Similarly, the so-called *phosphors* used in manufacturing glow-in-the-dark paints and plastics also contain direct-bandgap semiconductors. Electron–hole pairs form whenever the phosphor is exposed to light. A large number of electrons and holes gradually accumulate in the phosphor. The slow recombination of these carriers causes the emission of light.

Silicon and germanium are *indirect-bandgap semiconductors.* In these semiconductors, the collision of a hole and an electron will not cause the two carriers to recombine. The electron may momentarily fall into the hole, but quantum mechanical considerations prevent the generation of a photon. Since the electron cannot shed excess energy, it is quickly ejected from the lattice and the electron–hole pair reforms. In the case of an indirect-bandgap semiconductor, recombination can only occur at specific sites in the lattice, called *traps,* where flaws or foreign atoms distort the lattice (Figure 1.4B). A trap can momentarily capture a passing carrier. The trapped carrier becomes vulnerable to recombination because the trap can absorb the liberated energy.

Traps that aid the recombination of carriers are called *recombination centers.* The more recombination centers a semiconductor contains, the shorter the average time between the generation of a carrier and its recombination. This quantity, called the *carrier lifetime,* limits how rapidly a semiconductor device can switch on and off. Recombination centers are sometimes deliberately added to semiconductors to increase switching speeds. Gold atoms form highly efficient recombination centers in silicon, so high-speed diodes and transistors are sometimes made from silicon containing a small amount of gold. Gold is not the only substance that can form recombination centers. Many transition metals such as iron and nickel have a similar (if less potent) effect. Some types of crystal defects can also serve as recombination centers. Solid-state devices must be fabricated from extremely pure single-crystal materials in order to ensure adequate carrier lifetimes for proper device operation.

1.1.2. Extrinsic Semiconductors

The conductivity of semiconductors depends upon their purity. Absolutely pure, or *intrinsic,* semiconductors have low conductivities because they contain only a few thermally generated carriers. The addition of certain impurities greatly increases the number of available carriers. These *doped,* or *extrinsic,* semiconductors can approach the conductivity of a metal. A lightly doped semiconductor may contain only a few parts per billion of dopant. Even a heavily doped semiconductor contains only a few hundred parts per million due to the limited solid solubility of dopants in silicon. The extreme sensitivity of semiconductors to the presence of dopants makes it nearly impossible to manufacture truly intrinsic material. Practical semiconductor devices are, therefore, fabricated almost exclusively from extrinsic material.

Phosphorus-doped silicon is an example of an extrinsic semiconductor. Suppose a small quantity of phosphorus is added to a silicon crystal. The phosphorus atoms are incorporated into the crystal lattice in positions that would otherwise have been occupied by silicon atoms (Figure 1.5). Phosphorus, a group-V element, has five valence electrons. The phosphorus atom shares four of these with its four neighboring atoms. Four bonding electron pairs give the phosphorus atom a total of eight shared electrons. These, combined with the one remaining unshared electron, result in a total of nine valence electrons. Since eight electrons entirely fill the valence shell, no room remains for the ninth electron. This electron is expelled from the phosphorus atom and wanders freely through the crystal lattice. Each phosphorus atom added to the silicon lattice thus generates one free electron.

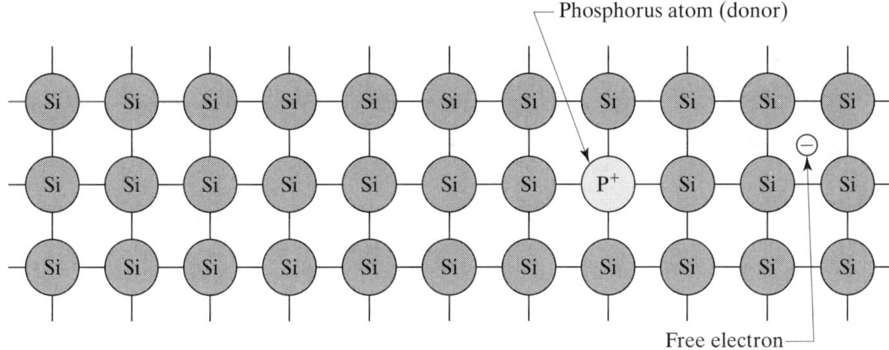

FIGURE 1.5 Simplified crystal structure of phosphorus-doped silicon.

The loss of the ninth electron leaves the phosphorus atom with a net positive charge. Although this atom is ionized, it does not constitute a hole. Holes are electron vacancies created by the removal of electrons from a filled valence shell. The phosphorus atom has a full valence shell despite its positive charge. The charge associated with the ionized phosphorus atom is therefore immobile.

Other group-V elements will have the same effect as phosphorus. Each atom of a group-V element that is added to the lattice will produce one additional free electron. Elements that donate electrons to a semiconductor in this manner are called *donors.* Arsenic, antimony, and phosphorus are all used in semiconductor processing as donors for silicon.

A semiconductor doped with a large number of donors has a preponderance of electrons as carriers. A few thermally generated holes still exist, but their numbers actually diminish in the presence of extra electrons. This occurs because the extra electrons increase the probability that the hole will find an electron and recombine. The large number of free electrons in N-type silicon greatly increases its conductivity (and greatly reduces its resistance).

A semiconductor doped with donors is said to be *N-type.* Heavily doped N-type silicon is sometimes marked N+, lightly doped N-type silicon N−. The plus and minus symbols denote the relative numbers of donors, not electrical charges. Electrons are considered the *majority carriers* in N-type silicon due to their large numbers. Similarly, holes are considered the *minority carriers* in N-type silicon. Strictly speaking, intrinsic silicon has neither majority nor minority carriers because both types are present in equal numbers.

Boron-doped silicon forms another type of extrinsic semiconductor. Suppose a small number of boron atoms are added to the silicon lattice (Figure 1.6). Boron, a group-III element, has three valence electrons. The boron atom attempts to share its valence electrons with its four neighboring atoms, but, because it has only three, it cannot complete the fourth bond. As a result, there are only seven valence electrons around the boron atom. The electron vacancy thus formed constitutes a hole. This hole is mobile and soon moves away from the boron atom. Once the hole departs, the boron atom is left with a negative charge caused by the presence of an extra electron in its valence shell. As in the case of phosphorus, this charge is immobile and does not contribute to conduction. Each atom of boron added to the silicon contributes one mobile hole.

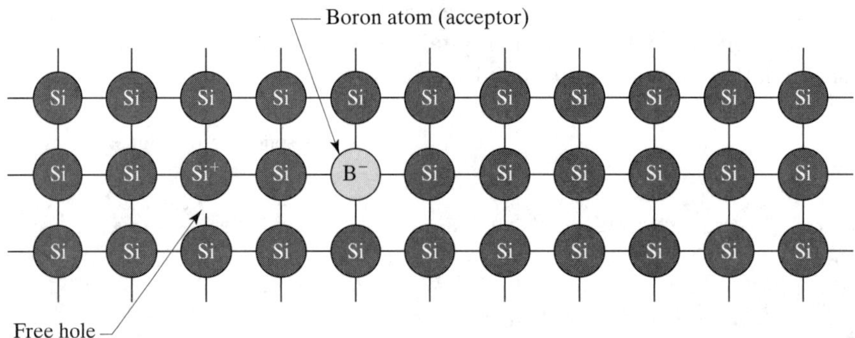

FIGURE 1.6 Simplified crystal structure of boron-doped silicon.

Other group-III elements can also accept electrons and generate holes. Technical difficulties prevent the use of any other group-III elements in silicon fabrication, but indium is sometimes used to dope germanium.[5] Any group-III element used as a dopant will *accept* electrons from adjoining atoms, so these elements are called *acceptors.* A semiconductor doped with acceptors is said to be *P-type.* Heavily doped P-type silicon is sometimes marked P+, and lightly doped P-type silicon P−. Holes are the majority carriers and electrons are the minority carriers in P-type silicon. Table 1.2 summarizes some of the terminology used to describe extrinsic semiconductors.

A semiconductor can be doped with both acceptors and donors. The dopant present in excess determines the type of the silicon and the concentration of the carriers. It is thus possible to invert P-type silicon to N-type by adding an excess of donors. Similarly, it is possible to invert N-type silicon to P-type by adding an excess of acceptors. The deliberate addition of an opposite-polarity dopant to invert the type of a semiconductor is called *counterdoping.* Most modern semiconductors are made by selectively counterdoping silicon to form a series of P- and N-type regions. Much more will be said about this practice in the next chapter.

---

[5] These difficulties include limited solid solubility and incomplete dopant ionization. Indium-doped silicon suffers the latter problem. A certain amount of energy is required to ionize a dopant atom so as to generate a free carrier. Random thermal vibrations suffice to ionize boron, phosphorus, arsenic, and antimony in silicon, but indium becomes fully ionized only at elevated temperatures. This effect has recently been exploited to produce what is in effect very lightly doped silicon. See H. Tian, J. Hayden, B. Taylor, L. Wu, and P. Rehmann, "A Comparative Study of Indium and Boron Implanted Silicon Bipolar Transistors," *IEEE Trans. on Electron Devices*, Vol. 48, #11, 2001, pp. 2520–2524.

Semiconductor Type[6]	Dopant Type	Typical Dopants for Silicon	Majority Carriers	Minority Carriers
N-type	Donors	Phosphorus, arsenic, and antimony	Electrons	Holes
P-type	Acceptors	Boron	Holes	Electrons

TABLE 1.2 Extrinsic semiconductor terminology.

If counterdoping were taken to extremes, the entire crystal lattice would consist of an equal ratio of acceptor and donor atoms. The two types of atoms would be present in exactly equal numbers. The resulting crystal would have very few free carriers and would appear to be an intrinsic semiconductor. Such *compound semiconductors* actually exist. The most familiar example is *gallium arsenide,* a compound of gallium (a group-III element) and arsenic (a group-V element). Materials of this sort are called III-V compound semiconductors. They include not only gallium arsenide but also gallium phosphide, indium antimonide, and many others. Many III-V compounds are direct-bandgap semiconductors, and some are used in constructing light-emitting diodes and semiconductor lasers. Gallium arsenide is also employed to a limited extent for manufacturing very high-speed solid-state devices, including integrated circuits. II-VI compound semiconductors are composed of equal mixtures of group-II and group-VI elements. Cadmium sulfide is a typical II-VI compound used to construct photosensors. Other II-VI compounds are used as phosphors in cathode ray tubes. A final class of semiconductors includes IV-IV compounds such as silicon carbide.

Compound semiconductors offer an opportunity to tailor the physical properties of the material to the requirements of the application. They have proven to be of tremendous value in the development of optical devices such as LEDs and lasers. Unfortunately, manufacturing difficulties have hampered their use in integrated circuits. Of all the possible compound materials, only germanium-doped silicon seems to be compatible with high-volume low-cost integrated circuit manufacture.

1.1.3. Diffusion and Drift

The motion of carriers through a silicon crystal results from two separate processes. *Diffusion* is a random process caused by thermal agitation, while *drift* is a unidirectional movement of carriers caused by electric fields.

The carriers within a semiconductor are constantly in motion. Each carrier moves in a random direction until it strikes an atom. The carrier bounces off this atom and travels in a different direction until it strikes another atom. This process occurs over and over again, causing the carrier to leap and jerk about in an unpredictable fashion, like a drunkard staggering about in a darkened room. Just as carriers are in constant motion, so is the semiconductor crystal lattice. The bonds that hold the lattice together act as little springs. A carrier striking an atom may lose energy, causing the carrier to move more slowly and the atom to begin vibrating. A carrier colliding with a vibrating atom may actually gain energy, causing the carrier to move more quickly and the atom to vibrate more slowly. The constant agitation of the crystal lattice, as

[6] Most device physicists use the terms *n-type* and *p-type* rather than *N-type* and *P-type*. Many circuit designers, including the author, favor the latter forms because of a preference for capitalizing such terms as *NPN transistor* and *PN junction*.

well as the motion of the carriers within it, creates the macroscopic phenomenon we call heat. Higher temperatures correspond to faster carriers and more vigorous lattice vibrations, while lower temperatures correspond to slower carriers and less vigorous vibrations.

So long as the carriers are uniformly distributed, thermal agitation produces no net current flow. If a certain number of carriers happen to move leftwards, on average an equal number move rightwards. However, thermal agitation can result in a net current flow if the carriers are not uniformly distributed. Imagine a situation where 20 carriers lie to the right of a certain point, but only 10 carriers lie to the left. Of the 20 carriers on the right, about half (say, 10) move leftwards. Of the 10 carriers on the left, about half (say, 5) move rightwards. The difference between these two constitutes a net current flow of five carriers to the right. This is an example of a *diffusion current.*

Diffusion currents always flow from a region of high carrier concentration to a region of low carrier concentration. Unless some mechanism replenishes the supply of carriers, the diffusion current will eventually redistribute the carriers uniformly throughout the crystal and then subside.

An electric field exerts a force on any carrier within it. Electrons are pulled toward positive potentials, while holes are pulled toward negative potentials. These forces accelerate the carriers. If they did not collide with anything, the carriers would soon be traveling at enormous speeds. This doesn't happen, because the carriers constantly collide with the atoms that form the crystal lattice. Because lattice collisions occur so frequently, only intense electric fields produce any perceptable increase in instantaneous carrier velocities.

Although weak electric fields have negligible effects upon instantaneous carrier velocities, over a long period of time they can displace carriers and thus cause electric currents to flow. Even the weakest electric field inexorably nudges electrons toward positive potentials and holes toward negative ones. After a sufficient length of time, a gradual shift in the positions of carriers becomes evident (Figure 1.7B). The gradual movement of carriers under an electric field is called *drift.* Electrons drift toward positive potentials and holes drift toward negative ones. The drift of carriers results in a *drift current.*

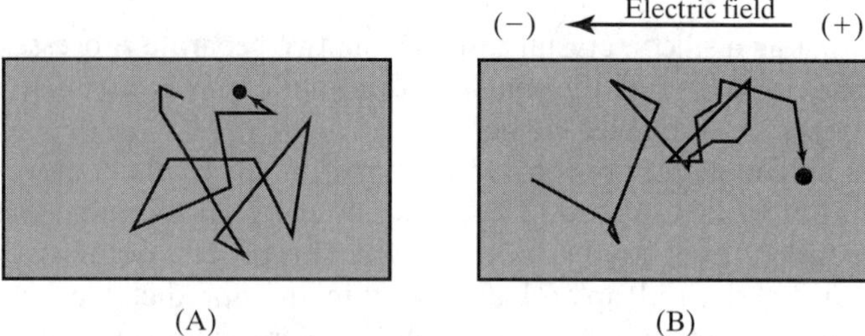

FIGURE 1.7 Comparison of conduction mechanisms for an electron: diffusion (A) and drift superimposed on diffusion (B). Notice the gradual motion of the electron toward the positive potential.

The rate at which carriers drift through a conductive material depends upon the electric field. Under low-to-moderate electric fields, this relationship is linear. This in turn gives rise to the linear relationship between current and voltage called *Ohm's law.*

Uniformly doped silicon behaves as an Ohmic conductor so long as the electric field does not exceed about 5000 V/cm. Larger electric fields produce noticeable increases in instantaneous carrier velocities. Since faster carriers correspond to higher temperatures, they are called *hot carriers.* These fast-moving carriers interact more strongly with the crystal lattice and therefore experience more frequent lattice collisions. The resulting increase in scattering causes the drift velocity to asymptotically approach a limiting value at high electric fields. This mechanism, called *velocity saturation*, causes resistors to deviate from Ohm's law under large applied voltages.

1.2 PN JUNCTIONS

Uniformly doped semiconductors have few applications. Almost all solid-state devices contain a combination of multiple P- and N-type regions. The interface between a P-type region and an N-type region is called a *PN junction,* or simply a *junction.*

Figure 1.8A shows two pieces of silicon. On the left is a bar of P-type silicon, and on the right is a bar of N-type silicon. No junction exists as long as the two do not contact one another. Each piece of silicon contains a uniform distribution of carriers. The P-type silicon has a large majority of holes and a few electrons; the N-type silicon has a large majority of electrons and a few holes.

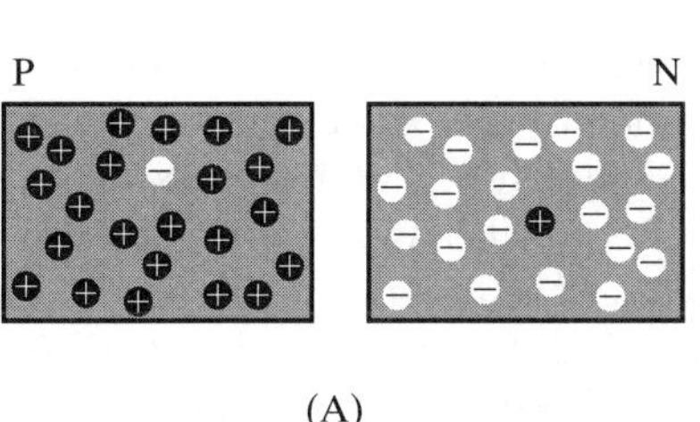

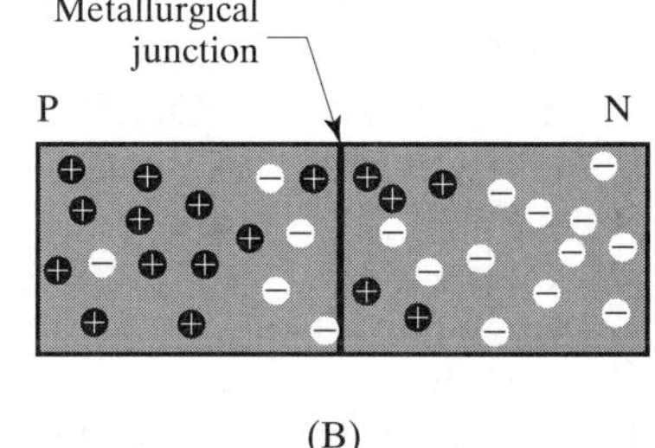

FIGURE 1.8 Carrier populations in silicon before the junction is assembled (A), and afterward (B).

Now, imagine the two pieces of silicon are brought into contact with one another to form a junction. The point where the silicon transitions from an excess of donors to an excess of acceptors is called the *metallurgical junction*. This metallurgical junction presents no barrier to the flow of carriers. There is a great excess of holes in the P-type silicon, and a great excess of electrons in the N-type silicon. Some of the holes diffuse from the P-type silicon to the N-type. Likewise, some of the electrons diffuse from the N-type silicon to the P-type. Figure 1.8B shows the result. A number of carriers have diffused across the junction in both directions. The concentration of minority carriers on either side of the junction has risen above that which would be produced by doping alone. The excess of minority carriers produced by diffusion across a junction is called the *excess minority carrier concentration.*

1.2.1. Depletion Regions

The presence of excess minority carriers on either side of a junction has two effects. First, the carriers produce an electric field. The extra holes in the N-type silicon represent a positive charge, while the excess electrons in the P-type silicon represent a negative charge. Thus, a voltage difference develops across the PN junction that biases the N-side of the junction positive with respect to the P-side.

When carriers diffuse across the junction, they leave equal numbers of ionized dopant atoms behind. These atoms are rigidly fixed in the crystal lattice and cannot move. On the P-side of the junction lie ionized acceptors that produce a negative charge. On the N-side of the junction lie ionized donors that produce a positive charge. A voltage difference again develops that biases the N-side of the junction positive with respect to the P-side. This voltage adds to the one produced by the separation of the charged carriers. The presence of this voltage difference implies the presence of an electric field across the junction.

Carriers drift in the presence of an electric field. Holes are attracted to the negative potential on the P-side of the junction. Similarly, electrons are attracted to the positive potential on the N-side of the junction. The drift of carriers thus tends to oppose their diffusion. Holes diffuse from the P-side of the junction to the N-side and drift back. Electrons diffuse from the N-side of the junction to the P-side and drift back. Equilibrium occurs when the drift and diffusion currents are equal and opposite. The excess minority carrier concentrations on either side of the junction also reach equilibrium values, as does the voltage across the junction.

The voltage difference across a PN junction in equilibrium is called its *built-in potential,* or its *contact potential.* In a typical silicon PN junction, the built-in potential can range from a few tenths of a volt to as much as a volt. Heavily doped junctions have larger built-in potentials than lightly doped ones. Because of the higher doping levels, more carriers diffuse across the heavily doped junction, and thus a larger diffusion current flows. In order to restore equilibrium, a larger drift current is also needed; hence, a stronger electric field develops. Heavily doped junctions therefore have larger built-in potentials than lightly doped ones.

Although the built-in potential is quite real, it cannot be measured with a voltmeter. This apparent paradox can be explained by a closer examination of a circuit containing a PN junction and a voltmeter (Figure 1.9). The two probes of the meter are made of metal, not silicon. The points of contact between the metal probes and the silicon also form junctions, each of which has a contact potential of its own. Because the silicon beneath the two probes has different doping levels, the two contact potentials of the probe points are unequal. The difference between these two contact potentials exactly cancels the built-in potential of the PN junction, and no current flows in the external circuit. This situation must occur because any current flow would constitute a free energy source, allowing the construction of a perpetual motion machine. The cancellation of the built-in potentials ensures that energy cannot be extracted from a PN junction in equilibrium and thus prevents a violation of the laws of thermodynamics.

The built-in potential has two causes: the separation of ionized dopant atoms and the separation of charged carriers. The carriers are free to move, but the dopant atoms are rigidly fixed in the crystal lattice. The region occupied by these charged atoms is subject to a strong electric field. Any carrier that enters this region must move quickly or it will be swept out again by the field. As a result, this region contains very few carriers at any given instant in time. This region is sometimes called a *space charge layer* because of the presence of the charged dopant

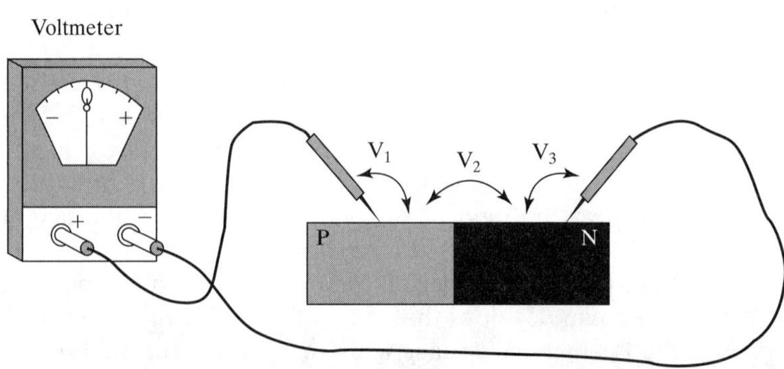

FIGURE 1.9 Demonstration of the impossibility of directly measuring a built-in potential. Contact potentials V_1 and V_3 exactly cancel built-in potential V_2.

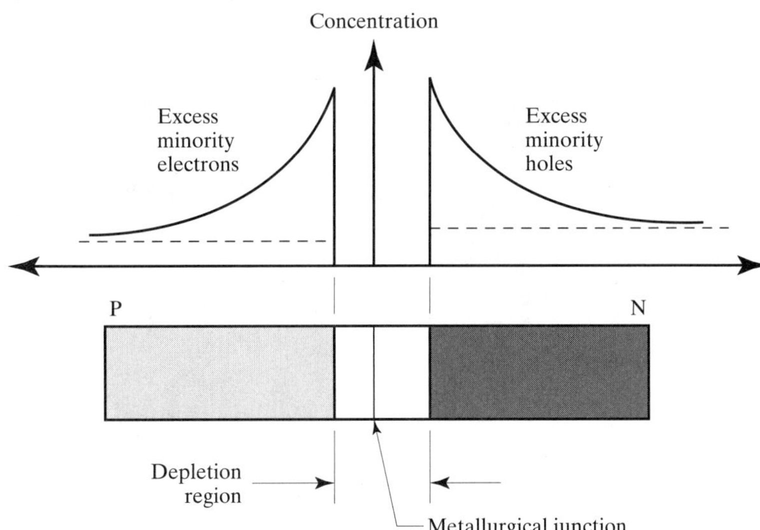

FIGURE 1.10 Diagram of excess minority carrier concentrations on either side of a PN junction in equilibrium.

atoms. More commonly, it is called a *depletion region* because of the relatively low concentration of carriers found there.

If the depletion region contains few carriers, then the excess minority carriers must pile up on either side of it. Figure 1.10 graphically shows the resulting distributions of excess minority carriers. The concentration gradients cause these carriers to diffuse away from the junction. The electric field produced by the separation of these charged carriers pulls them back toward the junction. An equilibrium is soon established, resulting in steady-state distributions of minority carriers resembling those shown in Figure 1.10.

The behavior of a PN junction can be summarized as follows: The diffusion of carriers across the junction produces excess minority carrier concentrations on either side of a depletion region. The separation of ionized dopant atoms causes an electric field to form across the depletion region. This field prevents most of the majority carriers from crossing the depletion region, and the few that do are eventually swept back to the other side.

The thickness of a depletion region depends on the doping of the two sides of the junction. If both sides are lightly doped, then a substantial thickness of silicon must deplete in order to uncover enough dopant atoms to support the built-in potential. If both sides are heavily doped, then only a very thin depletion region need be uncovered to produce the necessary charges. Therefore, heavily doped junctions have thin depletion regions and lightly doped junctions have thick ones. If one side of the junction is more heavily doped than the other, then the depletion region will extend further into the lightly doped side. In this case, a substantial thickness of lightly doped silicon must be uncovered to yield enough ionized dopants. Only a thin layer of heavily doped silicon need be uncovered to yield a counterbalancing charge. Figure 1.10 illustrates this case, since the N-side of the junction is more lightly doped than the P-side.

1.2.2. PN Diodes

A PN junction forms a very useful solid-state device called a *diode.* Figure 1.11 shows a simplified diagram of the structure of a PN diode. The diode has, as its name suggests, two terminals. One terminal, called the *anode,* connects to the P-side of the

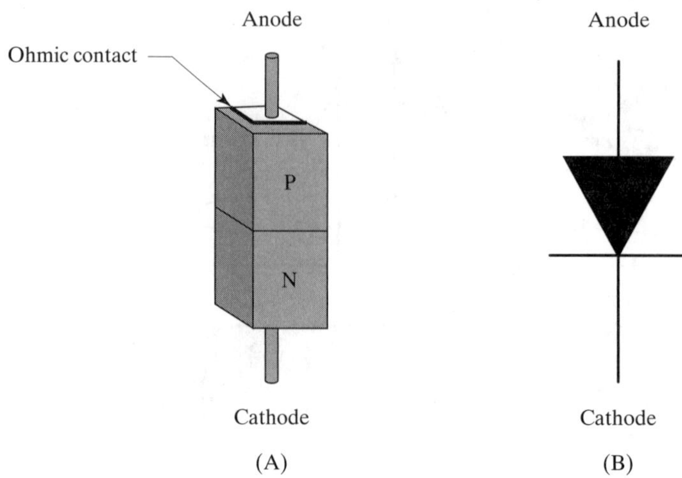

FIGURE 1.11 PN junction diode: simplified structure (A) and standard schematic symbol (B).

junction. The other terminal, called the *cathode,* connects to the N-side. These two terminals are used to connect the diode into an electrical circuit. The schematic symbol for a diode consists of an arrowhead representing the anode and a perpendicular line representing the cathode. Diodes conduct current preferentially in one direction—that indicated by the arrowhead.

To illustrate how the diode operates, imagine that an adjustable voltage source has been connected across it. If the voltage source is set to zero volts, then the diode is under *zero bias.* No current will flow through a zero-biased diode. If the voltage source is set to bias the anode negative with respect to the cathode, then the diode is *reverse biased.* Very little current flows through a reverse-biased diode. If the voltage source is set to bias the anode positive with respect to the cathode, then the diode is *forward biased* and a large current flows. This accords with the following simple mnemonic: *Current flows with the arrow, not against it.* Devices that conduct in only one direction are called *rectifiers*. They find frequent application in power supplies, radio receivers, and signal processing circuits.

Diode rectification depends upon the presence of a junction. Each of the three bias conditions can be explained by an appropriate analysis of carrier flows across this junction. The case of the zero-bias diode is particularly simple since it is identical to the case of the equilibrium junction already discussed. The only potential present across the junction is the built-in potential. When the diode is connected in a circuit, the contact potentials of the leads touching the silicon balance the built-in potential of the junction. Thus, no current flows in the circuit.

The behavior of a reverse-biased diode is also simple to explain. The reverse bias makes the N-side of the junction even more positive with respect to the P-side. The voltage seen across the junction increases, so excess minority carriers continue to be swept back across it and majority carriers continue to be held on their respective sides of the junction. The increased voltage across the junction causes the ionization of additional dopant atoms on either side, so the depletion region widens as the reverse bias increases.

The behavior of a forward-biased junction is somewhat more complex. The voltage applied to the terminals opposes the built-in potential. The voltage across the junction therefore lessens and the depletion region thins. The drift currents caused by the electric field are simultaneously reduced. More and more majority carriers make the transit across the depletion region without being swept back by the electric field. Figure 1.12 shows graphically the overall flow of carriers: Holes are injected across

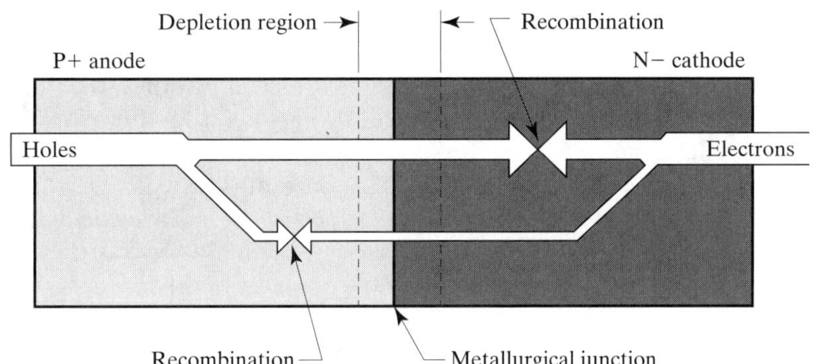

FIGURE 1.12 Carrier flow in a forward-biased PN junction.

the junction from anode to cathode (left to right), while electrons are injected across the junction from cathode to anode (right to left). In the illustrated diode, the hole current across the junction outweighs the electron current because the anode is more heavily doped than the cathode and there are more majority holes available in the anode than there are majority electrons in the cathode. Once these carriers have been injected across the junction, they become minority carriers and recombine with majority carriers present on the other side. Currents are drawn in from the terminals in order to replenish the supply of majority carriers in the neutral silicon. This illustration is somewhat simplified, since it only shows the general flow of carriers through the diode. Some of the carriers injected across the junction are swept back by the electric field before they can recombine. Such carriers do not contribute to the net current flow through the diode, so they are not illustrated. Likewise, the tiny numbers of thermally generated minority carriers that cross the junction are not shown since they form an insignificant portion of the overall current flow through a forward-biased diode.

The current through a forward-biased diode depends exponentially upon the applied voltage (Figure 1.13). About 0.6 V suffices to produce substantial forward conduction in a silicon PN junction at room temperature.[7] Because diffusion is caused

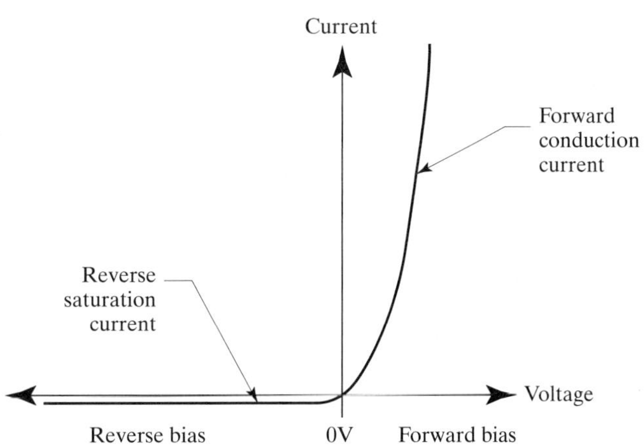

FIGURE 1.13 Diode conduction characteristics. The current scale is greatly magnified to show the reverse saturation current, which typically equals no more than a few picoamps at 25°C.

[7] Many elementary texts quote a typical forward-bias voltage of 0.7 V, but an integrated circuit base-emitter junction under microamp-level bias at 25°C actually exhibits a typical forward bias voltage of 0.6–0.65 V.

by the thermal motions of carriers, higher temperatures cause an increase in diffusion currents. The forward current through a PN junction actually increases exponentially as temperature increases. Expressed another way, the forward bias required to sustain a constant current in a silicon PN junction decreases by approximately 2 mV/°C.

Figure 1.13 also shows a low level of current flow when the diode is reverse-biased. This current flow is called *reverse conduction* or *leakage.* Leakage currents are produced by the few minority carriers thermally generated in the depletion region. The electric field opposes the flow of majority carriers across a reverse-biased junction, but it aids the flow of minority carriers. The application of a reverse bias sweeps these minority carriers across the junction. Because low-to-moderate electric fields have little effect upon the rate of generation of minority carriers in the bulk silicon, the leakage currents do not vary much with reverse bias. Thermal generation increases with temperature, and leakage currents are therefore temperature-dependent. In silicon, leakage currents double approximately every eight degrees Celsius. At high temperatures, the leakage currents begin to approach the operating currents of the circuit. The maximum operating temperature of a semiconductor device is therefore limited by leakage current. A maximum junction temperature of 150°C is widely accepted for silicon-integrated circuits.[8]

1.2.3. Schottky Diodes

Rectifying junctions can also form between a semiconductor and a metal. Such junctions are called *Schottky barriers,* and they form the basis of the semiconductor devices called *Schottky diodes*. Schottky barriers share a number of similiarites with PN junctions, but they also exhibit some significant differences that are best illustrated by presenting a concrete example.

Consider the case of a Schottky barrier between aluminum metal and lightly doped N-type silicon (Figure 1.14B). The aluminum metal is filled with electrons, while the N-type silicon contains far fewer electrons. One might guess that electrons would diffuse from the aluminum (where they are numerous) to the silicon (where they are few). However, due to differences in the electronic structures of aluminum and silicon, the electrons in the aluminum have less potential energy than those in the silicon. In order for an electron to cross from the aluminum to the silicon, it must gain energy from some source (such as thermal excitation). The potential energy difference between aluminum and silicon represents a barrier that holds the electrons in the aluminum back, preventing them from diffusing into the silicon. At the same time, this energy difference actually encourages electrons to migrate from the silicon into the aluminum. As the electrons exit the N-type silicon, they leave behind ionized dopant atoms that form a depletion region. The electrons arriving in the aluminum pile up in a thin sheet along the surface (Figure 1.14A). Eventually, an equilibrium is reached in which the diffusion of electrons from the metal to the silicon exactly counterbalances the drift of electrons from the silicon to the metal. The excess of electrons in the aluminum generates a voltage difference that biases the aluminum negative with respect to the silicon. This voltage difference is called the *contact potential* of the Schottky barrier, and it is in most ways analogous to the built-in potential of a PN junction.

---

[8] Discrete silicon transistors and diodes sometimes quote maximum junction temperatures of as much as 250°C. Integrated circuits typically operate at much lower currents than discrete devices, so they are more sensitive to the presence of small leakage currents. Few integrated circuits are designed to operate at junction temperatures above 175°C.

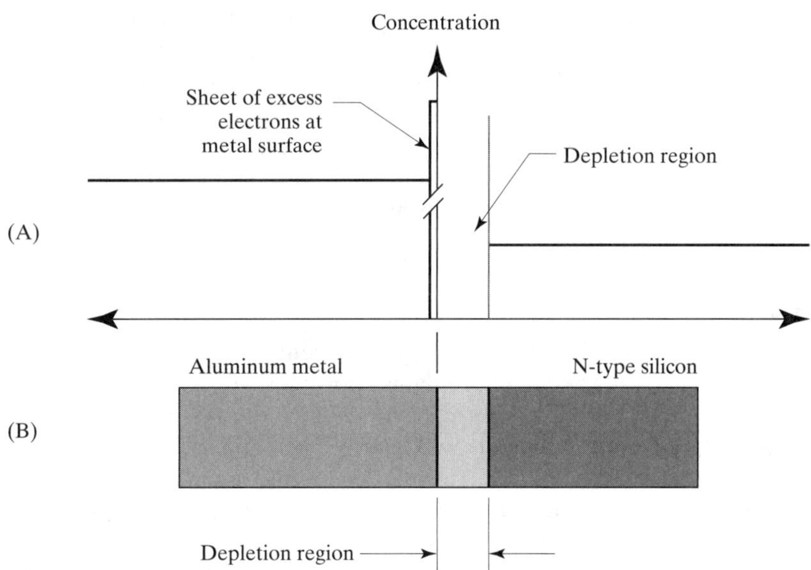

FIGURE 1.14 Diagram of excess carrier concentrations on either side of the Schottky barrier (A), and cross section of the corresponding Schottky structure (B).

Comparing an aluminum/N-silicon Schottky diode with a PN diode reveals one crucial difference. The PN diode depends upon populations of excess minority carriers for its operation, so it is called a *minority-carrier device*. The Schottky diode depends upon a population of excess majority carriers for its operation, so it is called a *majority-carrier device*. The switching speed of a minority carrier device depends upon how quickly the minority carriers recombine. The switching speed of a majority carrier device has no such limitation. Therefore, majority carrier devices such as Schottky diodes can operate at substantially higher switching speeds than minority carrier devices like PN diodes.

The behavior of a Schottky diode under bias resembles that of a PN diode. The N-type silicon forms the *cathode* of the Schottky diode, while the metal plate forms the *anode*. The case of a zero-biased Schottky diode is identical to the case of the equilibrium Schottky barrier analyzed above. A reverse-biased Schottky has an external voltage connected in order to bias the semiconductor positively with respect to the metal. The resulting voltage difference adds to the contact potential. The depletion region widens to counterbalance the increased voltage difference, equilibrium is restored, and very little current flows through the diode.

A forward-biased Schottky diode has an external voltage connected in order to bias the metal positively with respect to the semiconductor. The resulting voltage difference across the junction opposes the contact potential, and the width of the depletion region shrinks. Eventually, the contact potential is entirely offset, and a depletion region attempts to form on the metal side of the junction. The metal, being a conductor, cannot support an electric field, and no depletion region can form to oppose the externally applied potential. This potential begins to sweep electrons across the junction from the semiconductor into the metal, and a current flows through the diode.

Schottky diodes exhibit current-voltage characteristics similar to those of a PN diode (Figure 1.13). Schottky diodes also exhibit leakage currents caused by low levels of minority carrier injection from the metal into the semiconductor. These conduction mechanisms are accelerated by high temperatures, producing temperature dependencies similar to those of a PN diode.

Schottky diodes can also be formed to P-type silicon, but the forward biases required for conduction are usually quite low. This renders P-type Schottky diodes rather leaky, and they are therefore rarely used.[9] Most practical Schottky diodes result from the union between lightly doped N-type silicon and a class of materials called *silicides*. These substances are definite compounds of silicon and certain metals—for example, platinum and palladium. Silicides exhibit very stable electronic properties and therefore form Schottky diodes that have consistent and repeatable characteristics.

1.2.4. Zener Diodes

Under normal conditions, only a small current flows through a reverse-biased PN junction. This leakage current remains approximately constant until the reverse bias exceeds a certain critical voltage, beyond which the PN junction suddenly begins to conduct large amounts of current (Figure 1.15). The sudden onset of significant reverse conduction is called *reverse breakdown,* and it can lead to device destruction if the current flow is not limited by some external means. Reverse breakdown often sets the maximum operating voltage of a solid-state device. However, if appropriate precautions are taken to limit the current flow, a junction in reverse breakdown can provide a fairly stable voltage reference.

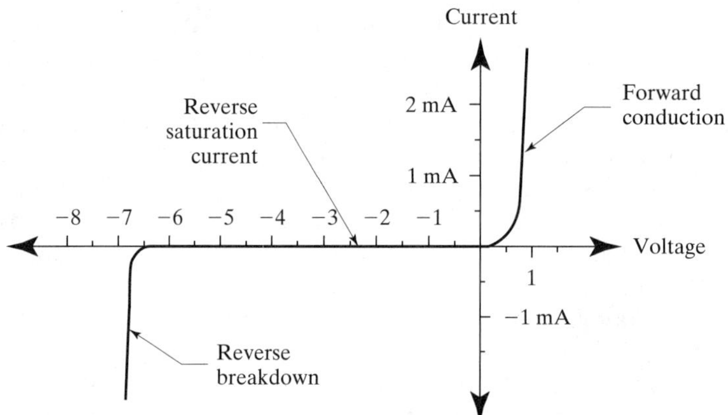

FIGURE 1.15 Reverse breakdown in a PN junction diode.

One of the mechanisms responsible for reverse breakdown is called *avalanche multiplication.* Consider a PN junction under reverse bias. The width of the depletion region increases with bias, but not fast enough to prevent the electric field from intensifying. The intense electric field accelerates the few carriers crossing the depletion region to extremely high velocities. When these hot carriers collide with lattice atoms, they knock loose valence electrons and generate additional carriers, a process called *impact ionization*. The resulting sharp increase in current flow is aptly called avalanche multiplication because a single carrier can spawn literally thousands of additional carriers through collisions, just as a single snowball can start an avalanche.

The other mechanism behind reverse breakdown is called *tunneling.* Tunneling is a quantum-mechanical process that allows particles to move short distances regardless of any apparent obstacles. If the depletion region is thin enough, then carriers can leap across it by tunneling. The tunneling current depends strongly on

---

[9] For example, compare the differences in work functions for platinum with respect to N-type silicon (0.85 V) and P-type silicon (0.25 V): R. S. Muller and T. I. Kamins, *Device Electronics for Integrated Circuits,* 2d ed. (New York: John Wiley and Sons, 1986), p. 157.

both the depletion region width and the voltage difference across the junction. Reverse breakdown caused by tunneling is called *Zener breakdown.*

The reverse breakdown voltage of a junction depends on the width of its depletion region. Wider depletion regions produce higher breakdown voltages. As previously explained, the more lightly doped side of a junction sets its depletion region width and therefore its breakdown voltage. When the breakdown voltage is less than five volts, the depletion region is so thin that Zener breakdown predominates. When the breakdown voltage exceeds five volts, avalanche breakdown predominates. A PN diode designed to operate in reverse conduction is called either a *Zener diode* or an *avalanche diode,* depending on which of these two mechanisms predominates. Zener diodes have breakdown voltages of less than five volts, while avalanche diodes have breakdown voltages of more than five volts. Engineers traditionally call all breakdown diodes Zeners regardless of what mechanism underlies their operation. This can lead to confusion because a 7 V Zener conducts primarily by avalanche breakdown.

In practice, the breakdown voltage of a junction depends on its geometry as well as its doping profile. The preceding discussion analyzed a *planar junction* consisting of two uniformly doped semiconductor regions intersecting in a planar surface. Although some real junctions approximate this ideal, most have curved sidewalls. The curvature intensifies the electric field and reduces the breakdown voltage. The smaller the radius of curvature, the higher is the electric field and the lower is the breakdown voltage. This effect can have a dramatic impact on the breakdown voltages of shallow junctions. Most Schottky diodes have particularly sharp discontinuities at the edge of the metal-silicon interface. Electric field intensification can drastically reduce the measured breakdown voltage of a Schottky diode unless special precautions are taken to relieve the electric field at the edges of the Schottky barrier.

Figure 1.16 shows schematic symbols for all of the diodes we have discussed. The PN junction diode uses a straight line to denote the cathode, while the Schottky diode and Zener diode are indicated by modifications to the cathode bar. In all cases, the arrow indicates the direction of conventional current flow through the forward-biased diode. In the case of the Zener diode, this arrow can be somewhat misleading because Zeners are normally operated in reverse bias. To the casual observer, the symbol may thus appear to be inserted "the wrong way around."

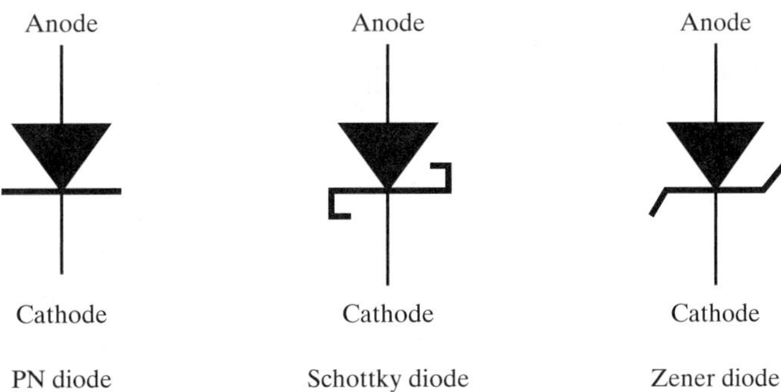

FIGURE 1.16 Schematic symbols for PN junction, Schottky, and Zener diodes. Some schematics show the arrowheads unfilled or show only half the arrowheads.

1.2.5. Ohmic Contacts

Contacts must be made between metals and semiconductors in order to connect solid-state devices into a circuit. These contacts would ideally be perfect conductors, but in practice they are *Ohmic contacts* that exhibit a small amount of resistance. Unlike rectifying contacts, these Ohmic contacts conduct current equally well in either direction.

Schottky barriers can exhibit Ohmic conduction if the semiconductor material is doped heavily enough. The high concentration of dopant atoms thins the depletion region to the point where carriers can easily tunnel across it. Unlike normal Zener diodes, Ohmic contacts can support tunneling at very low voltages. Rectification does not occur since the carriers can effectively bypass the Schottky barrier by tunneling through it.

An Ohmic contact can also form if the potential energy difference across the Schottky barrier causes surface accumulation rather than surface depletion. In accumulation, a thin layer of majority carriers forms at the semiconductor surface. In the case of an N-type semiconductor, this layer consists of excess electrons. The metal is a conductor and therefore cannot support a depletion region. A thin film of charge thus appears at the surface of the metal to counterbalance the accumulated carriers in the silicon. The lack of a depletion region on either side of the barrier prevents the contact from supporting a voltage differential, and any externally applied voltage will sweep carriers across the junction. Carriers can flow in either direction, so this type of Schottky barrier forms an Ohmic contact rather than a rectifying one.

In practice, rectifying contacts form to lightly doped silicon, and Ohmic contacts form to heavily doped silicon. The exact mechanism behind Ohmic conduction is unimportant since all Ohmic contacts behave in essentially the same manner. A lightly doped silicon region can be Ohmically contacted only if a thin layer of more heavily doped silicon is placed beneath the contact. Contact resistances of less than 50 $\Omega/\mu m^2$ can be obtained if a heavily doped silicon layer is used in combination with a suitable metal system. This resistance is small enough that it can be neglected for most applications.

Any junction between dissimilar materials exhibits a contact potential. This rule applies to Ohmic contacts as well as to PN junctions and rectifying Schottky barriers. If all the contacts and junctions are held at the same temperature, then the sum of the contact potentials around any closed loop will equal zero. Contact potentials are, however, strong functions of temperature. If one of the junctions is held at a different temperature than the others, then its contact potential will shift and the sum of the contact potentials will no longer equal zero. This *thermoelectric effect* has significant implications for integrated circuit design.

Figure 1.17 shows a block of N-type silicon contacted on either side by aluminum. If one end of the block is heated, then a measurable voltage develops across the block due to the mismatch between the two contact potentials. This voltage drop, called the *Seebeck voltage,* is typically 0.1–1.0 mV/°C.[10] Many integrated circuits rely

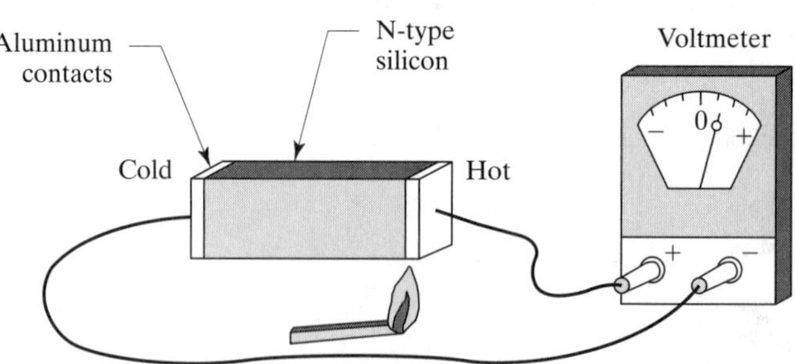

FIGURE 1.17 The thermoelectric effect produces a net measurable voltage if the two contacts are held at different temperatures.

[10] Lightly-doped silicon exhibits a higher Seebeck voltage; these values are taken from R.J. Widlar and M. Yamatake, "Dynamic Safe-Area Protection for Power Transistors Employs Peak-Temperature Limiting," *IEEE J. Solid-State Circuits*, SC-22, #1, 1987, p. 77–84.

upon voltages matching within a millivolt or two, so even small temperature differences are enough to cause such circuits to malfunction.

1.3 BIPOLAR JUNCTION TRANSISTORS

While diodes are useful devices, they cannot amplify signals, and almost all electronic circuits require amplification in one form or another. One device that can amplify signals is called a *bipolar junction transistor* (BJT).

Simplified structures of the two types of bipolar junction transistors are shown in Figure 1.18. Each transistor consists of three semiconductor regions called the *emitter, base,* and *collector.* The base is always sandwiched between the emitter and the collector. An NPN transistor consists of an N-type emitter, a P-type base, and an N-type collector. Similarly, a PNP transistor consists of a P-type emitter, an N-type base, and a P-type collector. In these simplified cross sections, each region of the transistor consists of a uniformly doped section of a rectangular bar of silicon. Modern bipolar transistors have somewhat different cross sections, but the principles of operation remain the same.

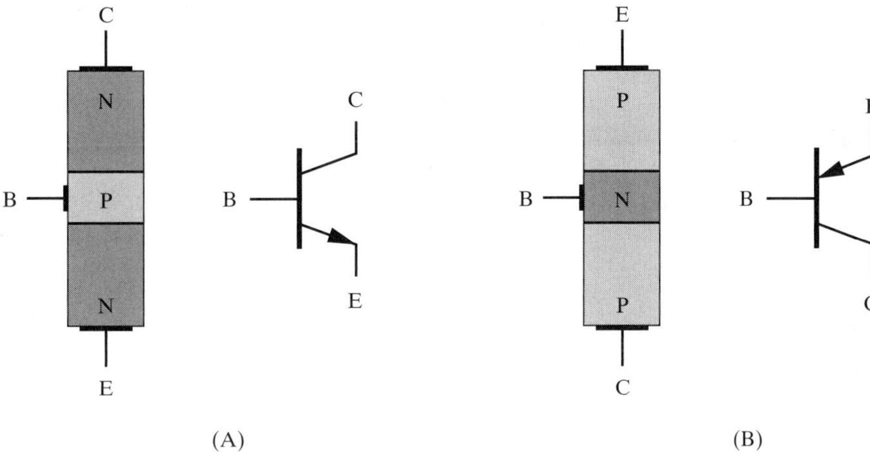

FIGURE 1.18 Structures and schematic symbols for the NPN transistor (A) and the PNP transistor (B).

Figure 1.18 also shows the symbols for the two types of transistors. The arrowhead placed on the emitter lead indicates the direction of conventional current flow through the forward-biased emitter-base junction. No arrow appears on the collector lead, even though a junction also exists between the collector and the base. In the simplified transistors of Figure 1.18, the emitter-base and collector-base junctions appear to be identical. One could apparently swap the collector and emitter leads without affecting the behavior of the device. In practice, the two junctions have different doping profiles and geometries and are not interchangeable.

A bipolar junction transistor can be viewed as two PN junctions connected back to back. The base region of the transistor is very thin (about 1–2 μm wide). When the two junctions are placed in such close proximity, carriers can diffuse from one junction to the other before they recombine. Conduction across one junction therefore affects the behavior of the other junction.

Figure 1.19A shows an NPN transistor with zero volts applied across the base-emitter junction and five volts applied across the base-collector junction. Neither junction is forward biased, so very little current flows through any of the three terminals of the transistor. A transistor with both junctions reverse biased (or zero biased) is said to be in *cutoff.* Figure 1.19B shows the same transistor with 10 micro-amps of

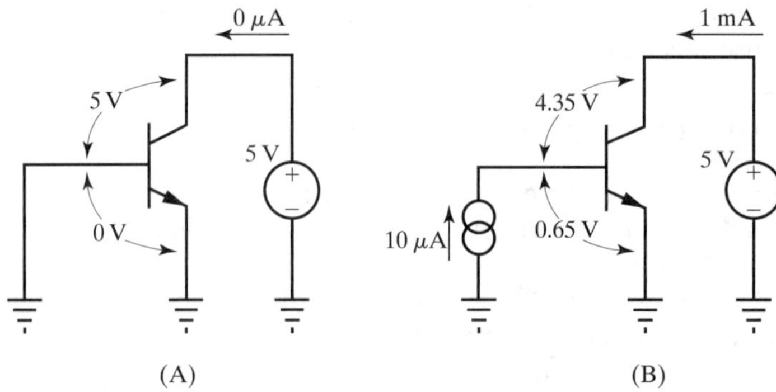

FIGURE 1.19 An NPN transistor operating in cutoff (A) and in the forward active region (B).

current injected into its base. This current forward biases the base-emitter junction to a potential of about 0.65 V. A collector current 100 times larger than the base current flows across the base-collector junction even though this junction remains reverse biased. This current is a consequence of the interaction between the forward-biased base-emitter junction and the reverse-biased base-collector junction. Whenever a transistor is biased in this manner, it is said to operate in the *forward active* region. If the emitter and collector terminals are interchanged so that the base-emitter junction becomes reverse-biased and the base-collector junction becomes forward-biased, the transistor is said to operate in the *reverse active* region. In practice, transistors are seldom operated in this manner.

Figure 1.20 helps explain why collector current flows across a reverse-biased junction. Carriers flow across the base-emitter junction as soon as it becomes forward biased. Most of the current flowing across this junction consists of electrons injected from the heavily doped emitter into the lightly doped base. Most of these electrons diffuse across the narrow base before they recombine. The base-collector junction is reverse biased, so minority carriers readily flow across it from base to collector. The electrons that cross the base-collector junction again become majority carriers flowing toward the collector terminal. The collector current consists of the electrons that successfully complete the journey from emitter to collector without recombining in the base.

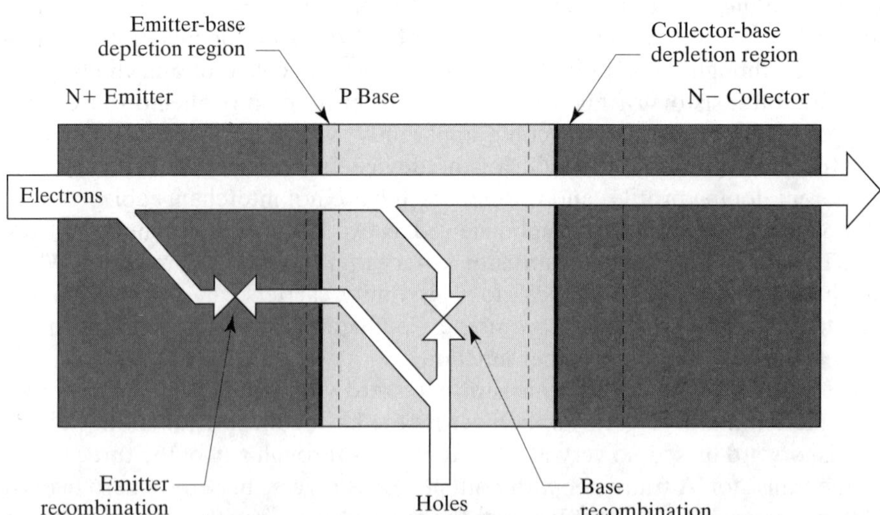

FIGURE 1.20 Current flow in an NPN transistor in the forward-active region.

Some of the electrons injected into the base do not reach the collector and instead recombine in the base region. Base recombination consumes holes that are replenished by a current flowing in from the base terminal. Some holes are also injected from the base into the emitter, where they rapidly recombine. These holes represent a second source of base terminal current. These recombination processes typically consume no more than 1% of the emitter current, so only a small base current is required to maintain the forward bias across the base-emitter junction.

1.3.1. Beta

The current amplification achieved by a transistor equals the ratio of its collector current to its base current. This ratio has been given various names, including *current gain* and *beta*. Likewise, different authors have used different symbols for it, including β and h_{FE}. A typical integrated NPN transistor exhibits a beta of about 150. Certain specialized devices may have betas exceeding 10,000. The beta of a transistor depends upon the two recombination processes illustrated in Figure 1.20.

Base recombination occurs primarily within the portion of the base between the two depletion regions, which is called the *neutral base* region. Three factors influence the base recombination rate: neutral base width, base doping, and the concentration of recombination centers. A thinner neutral base reduces the distance that the minority carriers must traverse and thus lessens the probability of recombination. Similarly, a more lightly doped base region minimizes the probability of recombination by reducing the majority carrier concentration. The *Gummel number* Q_B measures both of these effects. It is calculated by integrating the dopant concentration along a line traversing the neutral base region. In the case of uniform doping, the Gummel number equals the product of the base dopant concentration and the width of the neutral base. Beta is inversely proportional to the Gummel number.

The switching speed of transistors depends primarily on how quickly the excess minority carriers can be removed from the base, either through the base terminal or through recombination. Gold-doping is sometimes used to deliberately increase the number of recombination centers in bipolar junction transistors. The elevated recombination rate helps speed transistor switching, but it also reduces transistor beta. Few analog integrated circuits are built on gold-doped processes, because of their low betas.

Bipolar transistors typically use a lightly doped base and a heavily doped emitter. This combination helps ensure that almost all of the current injected across the base-emitter junction consists of carriers flowing from emitter to base and not *vice-versa*. Heavy doping enhances the recombination rate in the emitter, but this has little impact since so few carriers are injected into the emitter in the first place. The ratio of current injected into the emitter to that injected into the base is called the *emitter injection efficiency*.

Most NPN transistors use a thin, moderately doped base sandwiched between a thin, heavily doped emitter and a wide, lightly doped collector. The light collector doping allows a wide depletion region to form in the collector while simultaneously minimizing the intrusion of the depletion region into the base. This permits a high collector operating voltage without avalanching the collector-base junction. The asymmetric doping of emitter and collector helps explain why bipolar transistors do not operate well when these terminals are swapped. A typical integrated NPN transistor with a forward beta of 150 has a reverse beta of less than 5. This difference is primarily due to the drastic reduction in emitter injection efficiency caused by the substitution of a lightly doped collector for a heavily doped emitter.

Beta also depends upon collector current. Beta is reduced at low currents by leakage and by low levels of recombination in the depletion regions. At modest current levels, these effects become insignificant and the beta of the transistor climbs to a peak value determined by the mechanisms we have discussed. High collector currents cause beta to roll off due to an effect called *high-level injection.* When the minority carrier concentration approaches the majority carrier concentration in the base, extra majority carriers accumulate to maintain the balance of charges. The additional base majority carriers increase the apparent base doping, which in turn causes beta to decrease. Most transistors operate at moderate current levels to maximize beta, but power transistors must often operate in high-level injection because of size constraints.

The behavior of the PNP transistor is very similar to that of the NPN transistor. The beta of a PNP transistor is lower than that of an NPN of comparable dimensions and doping profiles, because the mobility of holes is lower than that of electrons. In many cases, the performance of the PNP is further degraded because of a conscious choice to optimize the NPN transistor at the expense of the PNP. For example, the material used to construct the base region of an NPN is often used to fabricate the emitter of a PNP. Since the resulting emitter is rather lightly doped, the emitter injection efficiency is low, and the onset of high-level injection occurs at moderate current levels. Despite their limitations, PNP transistors are very useful devices, and most bipolar processes support their construction.

1.3.2. I-V Characteristics

The performance of a bipolar transistor can be graphically depicted by drawing a family of curves that relate base current, collector current, and collector-emitter voltage. Figure 1.21 shows a typical set of curves for an integrated NPN transistor. The vertical axis measures collector current I_C, while the horizontal axis measures collector-to-emitter voltage V_{CE}. Multiple curves are superimposed upon the same graph, each representing a different base current I_B. This family of curves shows a number of interesting features of the bipolar junction transistor.

In the *saturation region,* the collector-emitter voltage remains so small that the collector-base junction is slightly forward-biased. The electric field that sweeps minority carriers across the collector-base junction still exists, so the transistor continues to conduct current. The collector-emitter voltage remains so low that Ohmic resistances in the transistor (particularly those in the lightly doped collector) become significant. The current supported in saturation is therefore less than that supported in the forward active region. The saturation region is of particular interest to integrated circuit designers because the forward biasing of the collector-base junction injects minority

FIGURE 1.21 Typical I-V plot of an NPN transistor.

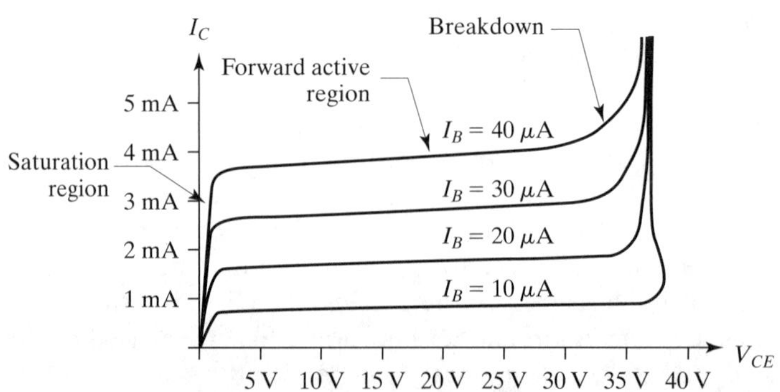

carriers into the neutral collector. Section 8.1.4 discusses in greater detail the effects of saturation upon integrated bipolar transistors.

The collector-emitter voltage in the forward active region is large enough to reverse bias the collector-base junction. Ohmic drops in the collector no longer significantly reduce the electric field across the collector-base junction, so the current flow through the transistor now depends solely upon beta. The slight upward tilt to the current curves results from the *Early effect.* As the reverse bias on the collector-base junction increases, the depletion region at this junction widens, and consequently the neutral base narrows. Since beta depends on neutral base width, it increases slightly as the collector-emitter voltage rises. The Early effect can be minimized by doping the collector more lightly than the base, so the depletion region extends primarily into the collector rather than into the base.

Beyond a certain collector-emitter voltage, the collector current increases rapidly. This *collector-to-base breakdown* limits the maximum operating voltage of the transistor. In the case of a typical integrated NPN transistor, this voltage equals some 30 V–40 V. The increased current flow results from either one of two effects, the first of which is avalanche breakdown. The collector-base junction will avalanche if it is sufficiently reverse-biased. A wide, lightly doped collector region can greatly increase the avalanche voltage rating, and discrete power transistors can achieve operating voltages of more than a thousand volts.

The second breakdown mechanism is *base punchthrough*. Punchthrough occurs when the collector-base depletion region reaches all the way through the base and merges with the base-emitter depletion region. Once this occurs, carriers can flow directly from emitter to collector, and current is limited only by the resistance of the neutral collector and emitter. The resulting rapid increase in collector current mimics the effects of avalanche breakdown.

Base punchthrough is often observed in high-gain transistors. For example, *super-beta* transistors use an extremely thin base region to obtain betas of 1000 or more. Base punchthrough limits the operating voltage of these devices to a couple of volts. Super-beta transistors also display a pronounced Early effect because of the encroachment of the collector-base depletion region into the extremely thin base. General-purpose transistors use wider base regions to reduce the Early effect, and their operating voltages are usually limited by avalanche instead of base punchthrough (Section 8.1.2).

1.4 MOS TRANSISTORS

The bipolar junction transistor amplifies a small change in input current to provide a large change in output current. The gain of a bipolar transistor is thus defined as the ratio of output to input current (beta). Another type of transistor, called a *field-effect transistor* (FET), transforms a change in input voltage into a change in output current. The gain of an FET is measured by its *transconductance*, defined as the ratio of change in output current to change in input voltage.

The field-effect transistor is so named because its input terminal (called its *gate*) influences the flow of current through the transistor by projecting an electric field across an insulating layer. Virtually no current flows through this insulator, so the gate current of a FET transistor is vanishingly small. The most common type of FET uses a thin silicon dioxide layer as an insulator beneath the gate electrode. This type of transistor is called a *metal-oxide-semiconductor* (MOS) transistor, or alternatively, a *metal-oxide-semiconductor field-effect transistor* (MOSFET). MOS transistors have replaced bipolars in many applications because they are smaller and can often operate using less power.

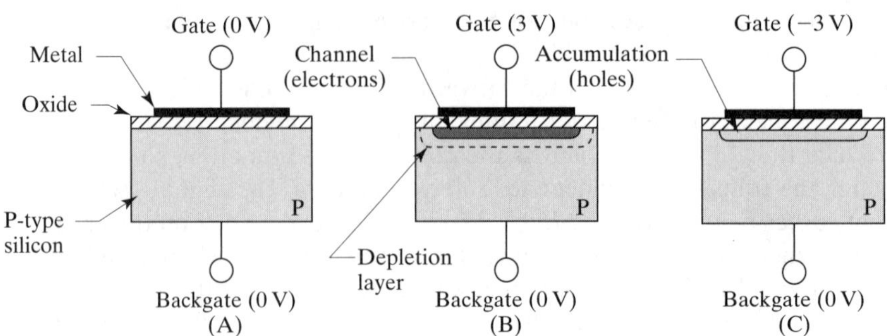

FIGURE 1.22 MOS capacitor: (A) unbiased ($V_{BG} = 0$ V), (B) inversion ($V_{BG} = 3$ V), (C) accumulation ($V_{BG} = -3$ V).

The MOS transistor can be better understood by first considering a simpler device called a *MOS capacitor*. This device consists of two electrodes, one of metal and one of extrinsic silicon, separated by a thin layer of silicon dioxide (Figure 1.22A). The metal electrode forms the *gate*, while the semiconductor slab forms the *backgate* or *body*. The insulating oxide layer between the two is called the *gate dielectric*. The illustrated device has a backgate consisting of lightly doped P-type silicon. The electrical behavior of this MOS capacitor can be demonstrated by grounding the backgate and biasing the gate to various voltages. The MOS capacitor of Figure 1.22A has a gate potential of 0 V. The difference in electron potential energies between the metal gate and the semiconductor backgate causes a small electric field to appear across the dielectric. In the illustrated device, this field biases the metal plate slightly positive with respect to the P-type silicon. This electric field attracts electrons from deep within the silicon up toward the surface, while it repels holes away from the surface. The field is weak, so the change in carrier concentrations is small and the overall effect upon the device characteristics is minimal.

Figure 1.22B shows what occurs when the gate of the MOS capacitor is biased positively with respect to the backgate. First, the majority carriers are repelled from the surface, forming a depletion region. As the bias is further increased, a point is reached where minority carriers are drawn up to the surface, where they form a thin layer of what appears to be silicon of the opposite doping polarity. The apparent reversal of doping polarity is called *inversion*, and the layer of silicon that inverts is called a *channel*. As the gate voltage increases still further, more electrons accumulate at the surface and the channel becomes more strongly inverted. The voltage at which the channel just begins to form is called the *threshold voltage* V_t. When the voltage difference between gate and backgate is less than the threshold voltage, no channel forms. When the voltage difference exceeds the threshold voltage, a channel appears.

Figure 1.22C shows what happens if the gate of the MOS capacitor is biased negatively with respect to the backgate. The electric field now reverses, drawing holes toward the surface and repelling electrons away from it. The surface layers of silicon appear to be more heavily doped, and the device is said to be in *accumulation*.

The behavior of the MOS capacitor can be utilized to form a true MOS transistor. Figure 1.23A shows the cross section of the resulting device. The gate, dielectric, and backgate remain as before. Two additional regions are formed by selectively doping the silicon on either side of the gate. One of these regions is called the *source* and the other is called the *drain*. Imagine that the source and backgate are both grounded and that a positive voltage is applied to the drain. As long as the gate-to-backgate voltage remains less than the threshold voltage, no channel forms. The PN junction formed between drain and backgate is reverse-biased, so very little current flows from drain to backgate. If the gate voltage exceeds the threshold voltage, a

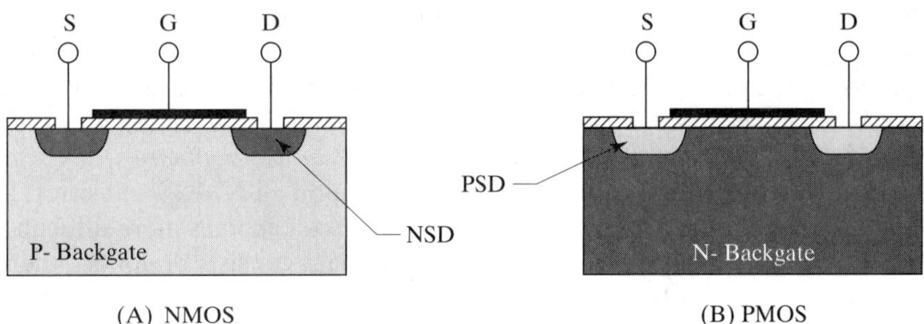

FIGURE 1.23 Cross sections of MOSFET transistors: NMOS (A) and PMOS (B). In these diagrams, S = Source, G = Gate, and D = Drain. The backgate connections, though present, are not illustrated.

channel forms beneath the gate dielectric. This channel acts like a thin film of N-type silicon shorting the source to the drain. A current consisting of electrons flows from the source across the channel to the drain. In summary, drain current will only flow if the gate-to-source voltage V_{GS} exceeds the threshold voltage V_t.

The source and drain of a MOS transistor are both N-type regions formed in the P-type backgate. In many cases, these two regions are identical and the terminals can be reversed without changing the behavior of the device. Such a device is said to be *symmetric*. In a symmetric MOS transistor, the labeling of source and drain becomes somewhat arbitrary. By definition, carriers flow out of the source and into the drain. The identity of the source and the drain therefore depends on the biasing of the device. Sometimes the bias applied across the transistor fluctuates and the two terminals swap roles. In such cases, the circuit designer must arbitrarily designate one terminal the drain and the other the source.

Asymmetric MOS transistors are designed with different source and drain dopings and geometries. There are several reasons why transistors may be made asymmetric, but the result is the same in every case. One terminal is optimized to function as the drain and the other as the source. If source and drain are swapped, then the performance of the device will suffer.

The transistor depicted in Figure 1.23A has an N-type channel and is therefore called an *N-channel MOS transistor*, or NMOS. *P-channel MOS* (PMOS) transistors also exist. Figure 1.23B shows a sample PMOS transistor consisting of a lightly doped N-type backgate with P-type source and drain regions. If the gate of this transistor is biased positive with respect to the backgate, then electrons are drawn to the surface and holes are repelled away from it. The surface of the silicon accumulates, and no channel forms. If the gate is biased negative with respect to the backgate, then holes are drawn to the surface, and a channel forms. The PMOS transistor thus has a negative threshold voltage. Engineers often ignore the sign of the threshold voltage, since it is normally positive for NMOS transistors and negative for PMOS transistors. An engineer might say, "The PMOS V_t has increased from 0.6 V to 0.7 V," when in actuality the PMOS V_t has shifted from −0.6 V to −0.7 V.

1.4.1. Threshold Voltage

The *threshold voltage* of a MOS transistor equals the gate-to-source bias required to just form a channel with the backgate of the transistor connected to the source. If the gate-to-source bias is less than the threshold voltage, then no channel forms. The threshold voltage exhibited by a given transistor depends on a number of factors, including backgate doping, dielectric thickness, gate material, and excess charge in the dielectric. Each of these effects will be briefly examined.

Backgate doping has a major effect on the threshold voltage. If the backgate is doped more heavily, then it becomes more difficult to invert. A stronger electric field is required to achieve inversion, and the threshold voltage increases. The backgate doping of an MOS transistor can be adjusted by adding dopant just beneath the surface of the gate dielectric to dope the channel region. This layer of doped silicon is formed by a process called ion implantation (Section 2.4.3), and the result is therefore called a *threshold adjust implant* (or V_t *adjust implant*).

Consider the effects of a V_t adjust implant upon an NMOS transistor. If the implant consists of acceptors, then the silicon surface becomes more difficult to invert and the threshold voltage increases. If the implant consists of donors, then the surface becomes easier to invert and the threshold decreases. If enough donors are implanted, the surface of the silicon can actually become counterdoped. In this case, a thin layer of N-type silicon forms a permanent channel at zero gate bias. The channel becomes more strongly inverted as the gate bias increases. As the gate bias is decreased, the channel becomes less strongly inverted, and at some point it vanishes. The threshold voltage of this NMOS transistor is actually negative. Such a transistor is called a *depletion-mode NMOS*, or simply a *depletion NMOS*. In contrast, an NMOS with a positive threshold voltage is called an *enhancement-mode NMOS*, or *enhancement NMOS*. The vast majority of commercially fabricated MOS transistors are enhancement-mode devices, but there are a few applications that require depletion-mode devices. A depletion-mode PMOS can also be constructed. Such a device will have a positive threshold voltage.

Depletion-mode devices should always be explicitly identified as such. One cannot rely on the sign of the threshold voltage to convey this information, because many engineers customarily ignore threshold polarities. Therefore, one should say "a depletion-mode PMOS with a threshold of 0.7 V," rather than "a PMOS with a threshold of 0.7 V." Many engineers would interpret the latter phrase as indicating an enhancement PMOS with a threshold of −0.7 V rather than a depletion PMOS with a threshold of +0.7 V. Explicitly referring to depletion-mode devices as such eliminates any possibility of confusion.

Special symbols are often used to distinguish between different types of MOS transistors. Figure 1.24 shows a representative collection of these symbols.[11] Symbols A and B are the standard symbols for NMOS and PMOS transistors, respectively. These symbols are not commonly used in the industry; instead, symbols C and D are preferred for NMOS and PMOS transistors, respectively. These symbols intentionally resemble NPN and PNP transistors. This convention helps highlight the essential

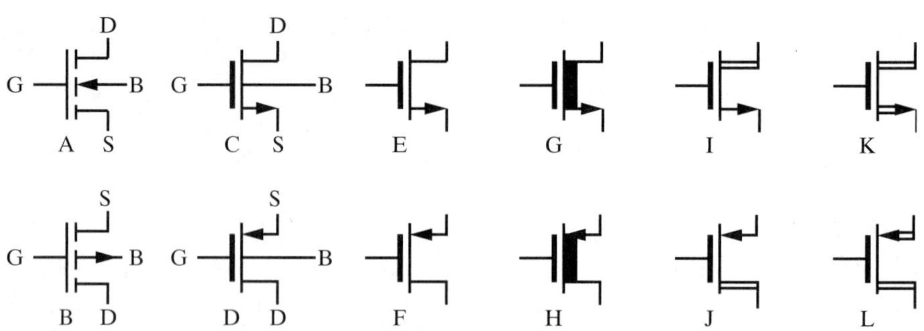

FIGURE 1.24 MOSFET symbols: A, B: standard symbols; C, D: industry symbols (four-terminal); E, F: industry symbols (three-terminal); G, H: depletion-mode devices; I, J: asymmetric high-voltage MOS symbols; K, L: symmetric high-voltage MOS symbols.

[11] Symbols A, B, E, F, G, and H are used by various authors; see A. B. Grebene, *Bipolar and MOS Analog Integrated Circuit Design* (New York: John Wiley and Sons, 1984), pp. 112–113; also P. R. Gray and R. G. Meyer, *Analysis and Design of Analog Integrated Circuits,* 3d ed. (New York: John Wiley and Sons, 1993), p. 60. The *J. Solid State Circuits* also uses three-terminal MOS symbols but differentiates PMOS devices by placing a bubble on their gate leads.

similarities between MOS and bipolar circuits. Symbols E and F are often employed when the backgates of the transistors connect to known potentials. Every MOS transistor has a backgate, so this terminal must always connect to something. Symbols E and F are potentially confusing, because the reader must infer the backgate connections. These symbols are nonetheless very popular because they make schematics much more legible. Symbols G and H are often used for depletion-mode devices, where the solid bar from drain to source represents the channel present at zero bias. Symbols I and J are sometimes employed for asymmetric transistors with high-voltage drains, and symbols K and L are used for symmetric transistors with high-voltage terminations for both source and drain. Versions of symbols G–L that include explicit backgate connections are also used. There are many other schematic symbols for MOS transistors; the ones shown in Figure 1.24 form only a representative sample.

Returning to the discussion of threshold voltage, the dielectric also plays an important role in determining the threshold voltage. A thicker dielectric weakens the electric field by separating the charges by a greater distance. Thus, thicker dielectrics increase the threshold voltage while thinner ones reduce it. In theory, the material of the dielectric also affects the electric field strength. In practice, most MOS transistors use pure silicon dioxide as the gate dielectric. This material can be grown in extremely thin films of exceptional purity and uniformity; no other material has comparable properties. Alternate dielectric materials therefore have very limited application.[12]

The gate electrode material also affects the threshold voltage of the transistor. As mentioned previously, an electric field appears across the gate dielectric when the gate and backgate are shorted together. This field is proportional to the contact potential that would exist if the gate and backgate materials touched. Most practical transistors use heavily doped polysilicon for the gate electrode. The threshold voltage of such a transistor can be varied to a limited degree by changing its gate doping.

A potentially troublesome source of threshold voltage variation comes from the presence of excess charges in the gate oxide or along the interface between the oxide and the silicon surface. These charges may consist of ionized impurity atoms, trapped carriers, or structural defects. The presence of trapped electric charge in the dielectric or along its interfaces alters the electric field and therefore the threshold voltage. If the amount of trapped charge varies with time, temperature, or applied bias, then the threshold voltage will also vary. This subject is discussed in greater detail in Section 4.2.2.

1.4.2. I-V Characteristics

The performance of an MOS transistor can be graphically illustrated by drawing a family of I-V curves similar to those used for bipolar transistors. Figure 1.25 shows a typical set of curves for an enhancement NMOS. The source and backgate were connected together to obtain these particular curves. The vertical axis measures drain current I_D, while the horizontal axis measures drain-to-source voltage V_{DS}. Each curve represents a specific gate-to-source voltage V_{GS}. The general character of the curves resembles that of the bipolar transistor shown in Figure 1.21, but the family of curves for an MOS transistor are obtained by stepping gate voltage, while those for a bipolar transistor are obtained by stepping base current.

---

[12] Devices have been fabricated using high-permittivity materials such as silicon nitride for the gate dielectric. Some authors use the term *insulated-gate field effect transistor* (IGFET) to refer to all MOS-like transistors, including those with nonoxide dielectrics.

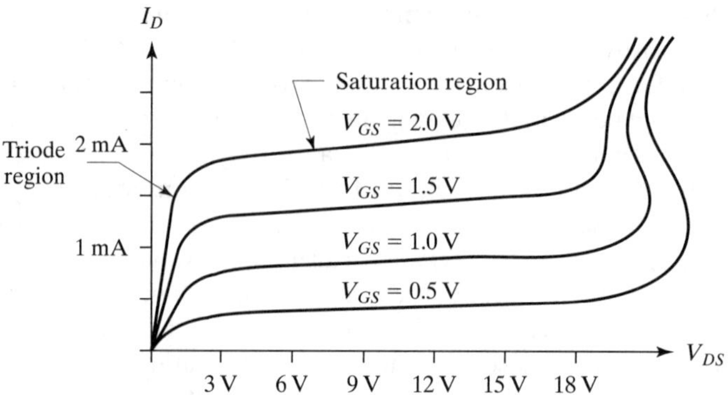

FIGURE 1.25 Typical I-V plot of an NMOS transistor.

At low drain-to-source voltages, the MOS channel behaves resistively, and the drain current increases linearly with voltage. This region of operation is called the *linear region, Ohmic region*, or *triode region.* This roughly corresponds to the saturation region of a bipolar transistor. At higher drain-to-source voltages, the rate of increase of the drain current diminishes. The drain current levels off to an approximately constant value when the drain-to-source voltage exceeds the difference between the gate-to-source voltage and the threshold voltage. This region is called the *saturation region,* and it roughly corresponds to the forward active region of a bipolar transistor. The term *saturation* thus has very different meanings for MOS and bipolar transistors.

The behavior of the MOS transistor in the linear region is easily explained. The channel acts as a film of doped silicon with a characteristic resistance that depends upon the carrier concentration. The current increases linearly with voltage, exactly as one would expect of a resistor. Higher gate voltages produce larger carrier concentrations and therefore lessen the resistance of the channel. PMOS transistors behave similarly to NMOS transistors, but, since holes have lower mobilities than electrons, the apparent resistance of the channel is considerably greater. The effective resistance of an MOS transistor operating in the triode region is symbolized $R_{DS(on)}$.

MOS transistors saturate because of a phenomenon called *pinch-off.* While the drain-to-source voltage remains small, the channel remains of a uniform thickness (Figure 1.26A). As the drain voltage rises, the drift current in the channel sweeps carriers into the drain and thins the drain end of the channel. Eventually, the drain end of the channel vanishes entirely, and the channel is said to have *pinched off* (Figure 1.26B). Carriers move down the channel propelled by the relatively weak

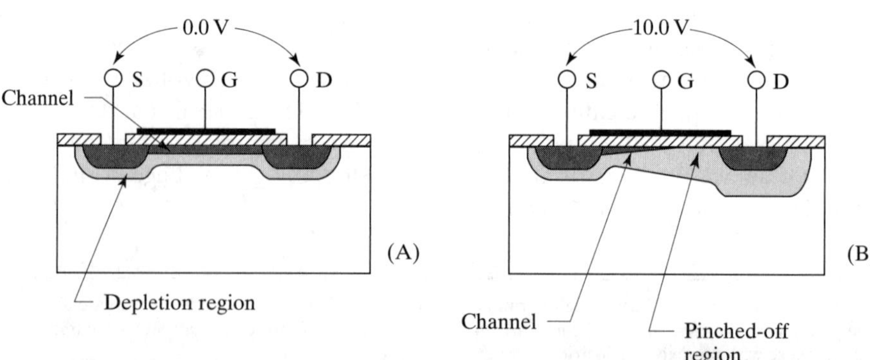

FIGURE 1.26 Behavior of a MOS transistor under bias: (A) $V_{DS} = 0$ V (triode region); (B) $V_{DS} = 10$ V (saturation region).

electric field along it. When they reach the edge of the pinched-off region, they are sucked across the depletion region by the strong electric field. The voltage drop across the channel does not increase as the drain voltage is increased; instead, the pinched-off region widens. Thus, the drain current reaches a limit and ceases to increase.

The drain current curves actually tilt slightly upward in the saturation region. This tilt is caused by *channel length modulation*, which is the MOS equivalent of the Early effect. Increases in drain voltage cause the pinched-off region to widen and the channel length to shorten. The shorter channel still has the same potential drop across it, so the electric field intensifies and the carriers move more rapidly. The drain current thus increases slightly with increasing drain-to-source voltage.

The I-V curves of Figure 1.25 were obtained with the backgate of the transistor connected to the source. If the backgate is biased independently of the source, then the apparent threshold voltage of the transistor will vary. If the source of an NMOS transistor is biased above its backgate, then its apparent threshold voltage increases. If the source of a PMOS transistor is biased below its backgate, then its threshold voltage decreases (becomes more negative). This *backgate effect*, or *body effect,* arises because the backgate-to-source voltage modulates the depletion region beneath the channel. This depletion region widens as the backgate-to-source differential increases, and it intrudes into the channel as well as into the backgate. A high backgate-to-source differential will thin the channel, which in turn raises the apparent threshold voltage. The intrusion of the depletion region into the channel becomes more significant as the backgate doping rises, and this in turn increases the magnitude of the body effect.

MOS transistors are normally considered majority carrier devices, which conduct only after a channel forms. This simplistic view does not explain the low levels of conduction that occur at gate-to-source voltages just below the threshold voltage. The formation of a channel is a gradual process. As the gate-to-source voltage increases, a depletion region first forms, then the gate attracts small numbers of minority carriers to the surface. The concentration of minority carriers rises as the voltage increases. When the gate-to-source voltage exceeds the threshold, the number of minority carriers becomes so large that the surface of the silicon inverts and a channel forms. Before this occurs, minority carriers can still move from the source to the drain by diffusion. This *subthreshold conduction* produces currents that are much smaller than those that would flow if a channel were present. However, they are still many orders of magnitude greater than junction leakages. Subthreshold conduction is typically significant only when the gate-to-source voltage is within about 0.3 V of the threshold voltage. This is sufficient to cause serious leakage problems in low-V_t devices. Some electrical circuits actually take advantage of the exponential voltage-to-current relationship of subthreshold conduction, but these circuits cannot operate at temperatures much in excess of 100°C because the junction leakages become so large that they overwhelm the tiny subthreshold currents.

As with bipolar transistors, MOS transistors can break down by either avalanche or punchthrough. If the voltage across the depletion region at the drain becomes so large that avalanche multiplication occurs, the drain current increases rapidly. Similarly, if the entire channel pinches off, then the source and drain will be shorted by the resulting depletion region and the transistor will punch through.

The operating voltage of an MOS transistor is often limited to a value considerably below the onset of avalanche or punchthrough breakdown by a long-term degradation mechanism called *hot carrier injection.* Carriers that traverse the pinched-off portion of the drain are accelerated by the strong electric field present here. When these hot carriers collide with atoms near the silicon surface, some of them are deflected up into the gate oxide, and a few of these become trapped.

Slowly, over a long period of operation, the concentration of these trapped carriers increases and the threshold voltage shifts. Hot hole injection occurs less readily than hot electron injection because the lower mobility of holes limits their velocity and therefore their ability to surmount the oxide interface. For this reason, NMOS transistors are frequently limited to lower operating voltages than PMOS transistors of similar construction. Various techniques have been devised to limit hot carrier injection (Section 12.1).

1.5 JFET TRANSISTORS

The MOS transistor represents only one type of field-effect transistor. Another is the *junction field-effect transistor* or JFET. This device uses the depletion regions surrounding reverse-biased junctions as a gate dielectric. Figure 1.27A shows a crosssection of an N-channel JFET. This device consists of a bar of lightly doped N-type silicon called the *body* into which two P-type diffusions have been driven from opposite sides. The thin region of N-type silicon remaining between the junctions forms the *channel* of the JFET. The two diffusions act as the *gate* and the *backgate*, and the opposite ends of the body form the *source* and the *drain.*

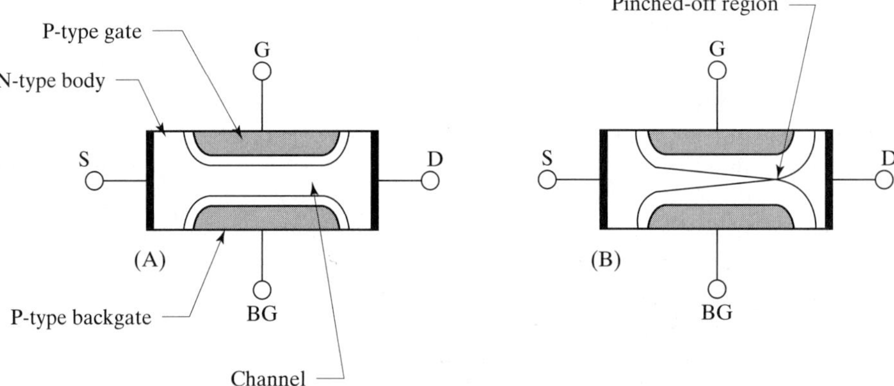

FIGURE 1.27 Cross sections of an N-channel JFET transistor operating in the linear region (A) and in saturation (B). In both diagrams, S = Source, D = Drain, G = Gate, and BG = Backgate.

Suppose that all four terminals of the N-channel JFET are grounded. Depletion regions form around the gate-body and backgate-body junctions. These depletion regions extend into the lightly doped channel, but they do not actually touch one another. A channel therefore exists from the drain to the source. If the drain voltage rises above the source voltage, then a current flows through the channel from drain to source. The magnitude of this current depends on the resistance of the channel, which in turn depends on its dimensions and doping. As long as the drain-to-source voltage remains small, it does not significantly alter the depletion regions bounding the channel. The resistance of the channel therefore remains constant and the drain-to-source voltage varies linearly with drain current. Under these conditions, the JFET is said to operate in its *linear region.* This region of operation corresponds to the linear (or triode) region of an MOS transistor. Since a channel forms at $V_{GS} = 0$, the JFET resembles a depletion-mode MOSFET rather than an enhancement-mode one.

The depletion regions at the drain end of the JFET widen as the drain voltage increases. The channel becomes increasingly constricted by the encroachment of the opposing depletion regions. Eventually, the depletion regions meet and pinch off the channel (Figure 1.27B). Drain current still flows through the transistor even though the channel has pinched off. This current originates at the source terminal and consists of majority carriers (electrons). These carriers move down the channel until they reach the pinched-off region. The large lateral electric field across this region draws the carriers across into the neutral drain.

Further increases in drain voltage have little effect once the channel has pinched off. The pinched-off region widens slightly, but the dimensions of the channel remain about the same. The resistance of the channel determines the magnitude of the drain current, so this also remains approximately constant. Under these conditions, the JFET is said to operate in *saturation.*

The gate and backgate electrodes also influence the current that flows through the channel. As magnitudes of the gate-body and backgate-body voltages increase, the reverse biases across the gate-body and backgate-body junctions slowly increase. The depletion regions that surround these junctions widen and the channel constricts. Less current can flow through the constricted channel, and the drain-to-source voltage required to pinch off the channel decreases. As the magnitudes of the gate and backgate voltages continue to increase, eventually the channel will pinch off even at $V_{DS} = 0$. Once this occurs, no current can flow through the transistor regardless of drain-to-source voltage, and the transistor is said to operate in *cutoff.* Strictly speaking, it is not the depletion region that blocks the carrier flow, but rather the electric field within the depletion region. In cutoff, the field is oriented so that the pinched-off portion of the N-JFET channel is negative with respect to the source, thus preventing electrons from flowing across the depletion region. By contrast, the pinched-off portion of a saturated N-JFET channel is positive with respect to the source, and electrons can therefore freely flow across the depletion region from channel to drain.

Figure 1.28 shows the I-V characteristics of an N-channel JFET whose gate and backgate electrodes have been connected to one another. Each curve represents a different value of the gate-to-source voltage V_{GS}. The drain currents are at their greatest when $V_{GS} = 0$, and they decrease as the magnitude of the gate voltage increases. Conduction ceases entirely when the gate voltage equals the *turnoff voltage* V_T. The turnoff voltage qualitatively corresponds to the threshold voltage of an MOS transistor. The comparison must not be taken too far, however, as the conduction equations of the two devices differ considerably.

The drain current curves of the N-JFET tilt slightly upward in saturation due to *channel length modulation.* This effect is analogous to that which occurs in MOS transistors. The pinched-off region of the JFET lengthens as the drain-to-source voltage increases. Any increase in the length of the pinched-off region produces a

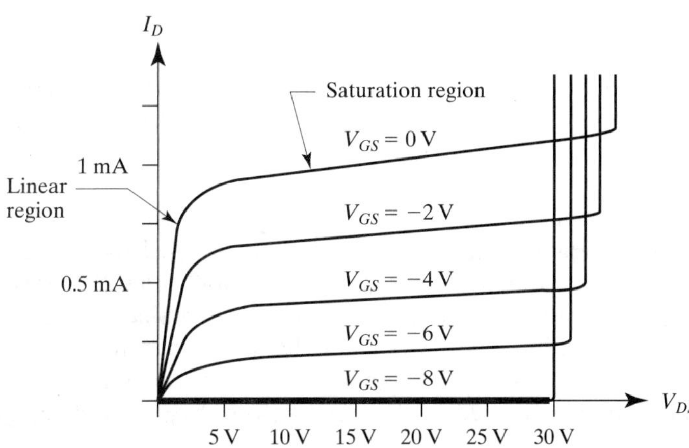

FIGURE 1.28 Typical I-V plot of an N-JFET transistor with $V_T = -8$ V.

corresponding decrease in the length of the channel. The effect of channel length modulation is usually quite small because the channel length greatly exceeds the length of the pinched-off region.

Either of two mechanisms can cause drain-to-source breakdown in the JFET. If the drain-to-source voltage is sufficient to deplete the device from drain contact to source contact, then punchthrough will occur. The dimensions of practical JFETs are such that they seldom, if ever, experience punchthrough before impact ionization begins within the depletion region. Once impact ionization begins, the device avalanches in much the same manner as a reverse-biased PN junction.

The source and drain terminals of a JFET can often be interchanged without affecting the performance of the device. The JFET structure of Figure 1.27A is an example of such a *symmetric* device. More complex JFET structures sometimes exhibit differences in source and drain geometries that render them *asymmetric*.

Most JFET structures short the gate and backgate terminals. Consider the device of Figure 1.27A. The channel is bounded on the left by the source, on the right by the drain, on the top by the gate, and on the bottom by the backgate. The drawing does not show what bounds the channel on the front or the rear. In most cases, these sides of the channel are also bounded by reverse-biased junctions that are extensions of the gate-body and backgate-body junctions. This arrangement necessitates shorting the gate and backgate.

Figure 1.29 shows the conventional schematic symbols for N-channel and P-channel JFET transistors. The arrowhead on the gate lead shows the orientation of the PN junction between the gate and the body of the device. The symbol does not explicitly identify the source and drain terminals, but most circuit designers orient the devices so that the drain of an N-JFET and the source of a P-JFET lie on top.

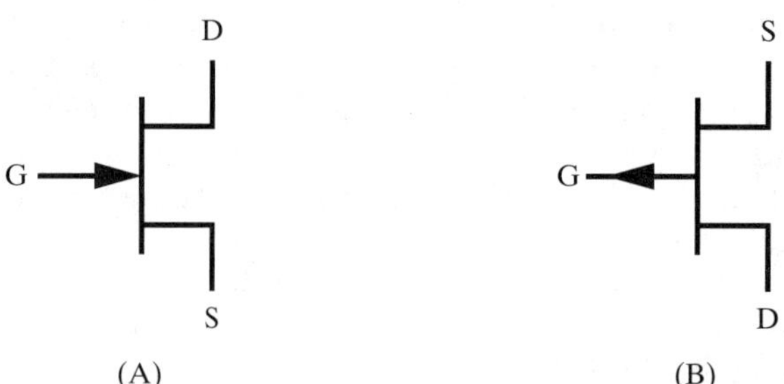

FIGURE 1.29 Symbols for an N-channel JFET (A) and a P-channel JFET (B).

1.6 SUMMARY

Device physics is a complex and ever-evolving science. Researchers constantly develop new devices and refine existing ones. Much of this ongoing research is highly theoretical and therefore lies beyond the scope of this text. The functionality of most semiconductor devices can be satisfactorily explained with relatively simple and intuitive concepts.

This chapter emphasizes the role of majority and minority carrier conduction across PN junctions. If a junction is reverse-biased, then the majority carriers on either side of it are repelled and a depletion region forms. If the junction is forward-biased, then majority carriers diffuse across and recombine to create a net current flow across the PN junction. The PN junction diode employs this phenomenon to rectify signals.

When two junctions are placed in close proximity, carriers emitted by one junction can be collected by the other before they can recombine. The bipolar junction transistor (BJT) consists of just such a pair of closely spaced junctions. The voltage across the base-emitter junction of the BJT controls the current flowing from collector to emitter. If the transistor is properly designed, then a small base current can control a much larger collector current. The BJT therefore serves as an amplifier capable of transforming weak signals into much stronger ones. Thus, for example, a BJT can amplify a weak signal picked up by a radio receiver into a signal strong enough to drive a loudspeaker.

The metal-oxide-semiconductor (MOS) transistor relies upon electrical fields projected across a dielectric to modulate the conductivity of a semiconductor material. A suitable voltage placed upon the gate of a MOS transistor produces an electric field that attracts minority carriers to form a conductive channel. The gate is insulated from the remainder of the transistor, so no gate current is required to maintain conduction. MOS circuitry can thus potentially operate at very low power levels.

The junction diode, the bipolar junction transistor, and the MOS transistor are the three most important semiconductor devices. Together with resistors and capacitors, they form the vast majority of the elements used in modern integrated circuits. The next chapter will examine how these devices are fabricated in a production environment.

1.7 EXERCISES

1.1. What are the relative proportions of aluminum, gallium, and arsenic atoms in intrinsic aluminum gallium arsenide?

1.2. A sample of pure silicon is doped with exactly 10^{16} atoms/cm^3 of boron and exactly 10^{16} atoms/cm^3 of phosphorus. Is the doped sample P-type or N-type?

1.3. The *instantaneous* velocity of carriers in silicon is almost unaffected by weak electric fields, yet the *average* velocity changes dramatically. Explain this observation in terms of drift and diffusion.

1.4. A layer of intrinsic silicon 1 μm thick is sandwiched between layers of P-type and N-type silicon, both heavily doped. Draw a diagram illustrating the depletion regions that form in the resulting structure.

1.5. A certain process incorporates two different N+ diffusions that can be combined with a P− diffusion to produce Zener diodes. One of the resulting diodes has a breakdown voltage of 7 V, while the other has a breakdown voltage of 10 V. What causes the difference in breakdown voltages?

1.6. When the collector and emitter leads of an integrated NPN transistor are swapped, the transistor continues to function, but exhibits a greatly reduced beta. There are several possible reasons for this behavior; explain at least one.

1.7. If a certain transistor has a beta of 60, and another transistor has a base twice as wide and half as heavily doped, then what is the approximate beta of the second transistor? What other electrical characteristics of the devices will vary, and how?

1.8. A certain MOS transistor has a threshold voltage of −1.5 V. If a small amount of boron is added to the channel region, the threshold voltage shifts to −0.6 V. Is the transistor PMOS or NMOS, and is it an enhancement or a depletion device?

1.9. If a depletion PMOS transistor has a threshold voltage of 0.5 V when constructed using a 200 Å oxide, will this threshold voltage increase or decrease if the oxide is thickened to 400 Å?

1.10. A certain NMOS transistor has a threshold voltage of 0.5 V; the gate-to-source voltage V_{GS} of the transistor is set to 2 V, and the drain-to-source voltage V_{DS} is set to 4 V. What is the relative effect of doubling the gate-to-source voltage versus doubling the drain-to-source voltage, and why?

1.11. A certain silicon PN junction diode exhibits a forward voltage drop of 620 mV when operated at a forward current of 25 μA at a temperature of 25°C. What is the approximate forward drop of this diode at −40°C? At 125°C?

1.12. Two JFET transistors differ only in the separation between their gate and backgate; in one transistor these two regions are twice as far apart as in the other transistor. In what ways do the electrical properties of the two transistors differ?

1.13. A voltage of 30 V is placed across a 1000 μm-long uniformly doped monocrystalline silicon resistor. Will this resistor obey Ohm's law? Why?

1.14. Based on an analogy with MOS transistors, suggest an appropriate symbol for a JFET having separate gate and backgate connections.

1.15. A reverse-biased PN junction exhibits a characteristic capacitance that diminishes as the voltage across the junction increases. Explain what causes this capacitance and its behavior with changing voltages.

2 *Semiconductor Fabrication*

Semiconductor devices have long been used in electronics. The first solid-state rectifiers were developed in the late nineteenth century. The galena crystal detector, invented in 1907, was widely used to construct crystal radio sets. By 1947, the physics of semiconductors was sufficiently understood to allow Bardeen and Brattain to construct the first bipolar junction transistor. In 1959, Kilby constructed the first integrated circuit, ushering in the era of modern semiconductor manufacture.

The impediments to manufacturing large quantities of reliable semiconductor devices were essentially technological, not scientific. The need for extraordinarily pure materials and precise dimensional control prevented early transistors and integrated circuits from reaching their full potential. The first devices were little more than laboratory curiosities. An entire new technology was required to mass produce them, and this technology is still rapidly evolving.

This chapter provides a brief overview of the process technologies currently used to manufacture integrated circuits. Chapter 3 then examines three representative process flows used for manufacturing specific types of analog integrated circuits.

2.1 SILICON MANUFACTURE

Integrated circuits are usually fabricated from *silicon,* a very common and widely distributed element. The mineral *quartz* consists entirely of silicon dioxide, also known as *silica.* Ordinary sand is chiefly composed of tiny grains of quartz and is therefore also mostly silica.

Despite the abundance of its compounds, elemental silicon does not occur naturally. The element can be artificially produced by heating silica and carbon in an electric furnace. The carbon unites with the oxygen contained in the silica, leaving more-or-less pure molten silicon. As this cools, numerous minute crystals form and grow together into a fine-grained gray solid. This form of silicon is said to be *polycrystalline* because it contains a multitude of crystals. Impurities and a disordered crystal structure make this *metallurgical-grade polysilicon* unsuited for semiconductor manufacture.

Metallurgical-grade silicon can be further refined to produce an extremely pure *semiconductor-grade* polysilicon. Purification begins with the conversion of the crude silicon into a volatile compound, usually trichlorosilane. After repeated distillation, the extremely pure trichlorosilane is reduced to elemental silicon by means of hydrogen gas. The final product is exceptionally pure, but still polycrystalline. Practical integrated circuits can only be fabricated from single-crystal material, so the next step consists of growing a suitable crystal.

2.1.1. Crystal Growth

The principles of crystal growing are both simple and familiar. Suppose a few crystals of sugar are added to a saturated solution that subsequently evaporates. The sugar crystals serve as seeds for the deposition of additional sugar molecules. Eventually the crystals grow to be very large. Crystal growth would occur even in the absence of a seed, but the product would consist of a welter of small intergrown crystals. The use of a seed allows the growth of larger, more perfect crystals by suppressing undesired nucleation sites.

In principle, silicon crystals can be grown in much the same manner as sugar crystals. In practice, no suitable solvent exists for silicon, and the crystals must be grown from the molten element at temperatures in excess of 1400°C. The resulting crystals are at least a meter in length and ten centimeters in diameter, and they must have a nearly perfect crystal structure if they are to be useful to the semiconductor industry. These requirements make the process technically challenging.

The usual method for growing semiconductor-grade silicon crystals is called the *Czochralski process.* This process, illustrated in Figure 2.1, uses a silica crucible charged with pieces of semi-grade polycrystalline silicon. An electric furnace raises the temperature of the crucible until all of the silicon melts. The temperature is then reduced slightly and a small seed crystal is lowered into the crucible. Controlled cooling of the melt causes layers of silicon atoms to deposit upon the seed crystal. The rod holding the seed slowly rises so that only the lower portion of the growing crystal remains in contact with the molten silicon. In this manner, a large silicon crystal can be pulled centimeter by centimeter from the melt. The shaft holding the crystal rotates slowly to ensure uniform growth. The high surface tension of molten silicon distorts the crystal into a cylindrical rod rather than the expected faceted prism.

The Czochralski process requires careful control to provide crystals of the desired purity and dimensions. Automated systems regulate the temperature of the

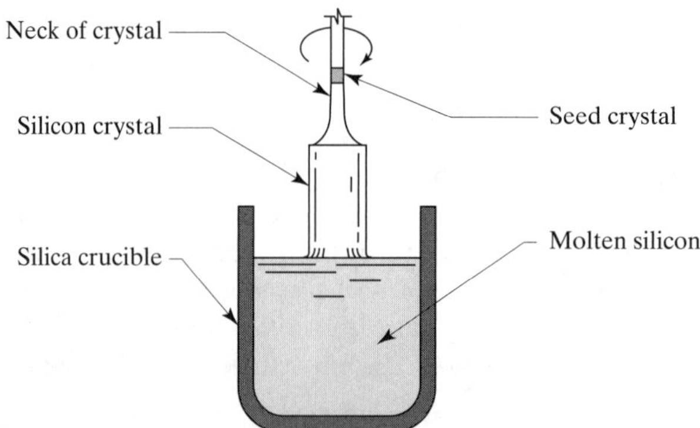

FIGURE 2.1 Czochralski process for growing silicon crystals.

melt and the rate of crystal growth. A small amount of doped polysilicon added to the melt sets the doping concentration in the crystal. In addition to the deliberately introduced impurities, oxygen from the silica crucible and carbon from the heating elements dissolve in the molten silicon and become incorporated into the growing crystal. These impurities subtly influence the electrical properties of the resulting silicon. Once the crystal has reached its final dimensions, it is lifted from the melt and is allowed to slowly cool to room temperature. The resulting cylinder of monocrystalline silicon is called an *ingot.*

Since integrated circuits are formed upon the surface of a silicon crystal and penetrate this surface to no great depth, the ingot is customarily sliced into numerous thin circular sections called *wafers.* Each wafer yields hundreds or even thousands of integrated circuits. The larger the wafer, the more integrated circuits it holds and the greater the resulting economies of scale. Most modern analog processes employ either 150 mm (6″) or 200 mm (8″) wafers. State-of-the-art digital processes now use 300 mm (12″) wafers. A typical ingot measures between 1 and 2 meters in length and can provide hundreds of wafers.

2.1.2. Wafer Manufacturing

The manufacture of wafers consists of a series of mechanical processes. The two tapered ends of the ingot are sliced off and discarded. The remainder is then ground into a cylinder, the diameter of which determines the size of the resulting wafers. No visible indication of crystal orientation remains after grinding. The crystal orientation is experimentally determined, and a flat stripe is ground along one side of the ingot. Each wafer cut from it will retain a facet, or *flat,* which unambiguously identifies its crystal orientation.

After grinding the flat, the manufacturer cuts the ingot into individual wafers, using a diamond-tipped saw. In the process, about one-third of the precious silicon crystal is reduced to worthless dust. The surfaces of the resulting wafers bear scratches and pockmarks caused by the sawing process. Since the tiny dimensions of integrated circuits require extremely smooth surfaces, one side of each wafer must be polished. This process uses a combination of mechanical and chemical polishing. The resulting mirror-bright surface displays the dark gray color and characteristic near-metallic luster of silicon.

2.1.3. The Crystal Structure of Silicon

Each wafer constitutes a slice from a single silicon crystal. The underlying crystalline structure determines how the wafer splits when broken. Most crystals tend to part along *cleavage planes* where the interatomic bonding is weakest. For example, a diamond crystal can be cleaved by sharply striking it with a metal wedge. A properly oriented blow will split the diamond into two pieces, each of which displays a perfectly flat cleavage surface. If the blow is not properly oriented, then the diamond shatters. Silicon wafers also show characteristic cleavage patterns that can be demonstrated using a scrap wafer, a pad of note paper, and a wooden pencil. Place the wafer on the notepad, and place the pad in your lap. Take a wooden pencil and press down in the center of the wafer, using the eraser. The wafer should split into either four or six regular wedge-shaped fragments, much like sections of a pie (Figure 2.2). The regularity of the cleavage pattern demonstrates that the wafer consists of monocrystalline silicon.

Figure 2.3 shows a small section of a silicon crystal drawn in three dimensions. Eighteen silicon atoms lie wholly or partially within the boundaries of an imaginary cube called a *unit cell.* Six of these occupy the centers of each of the six faces of the

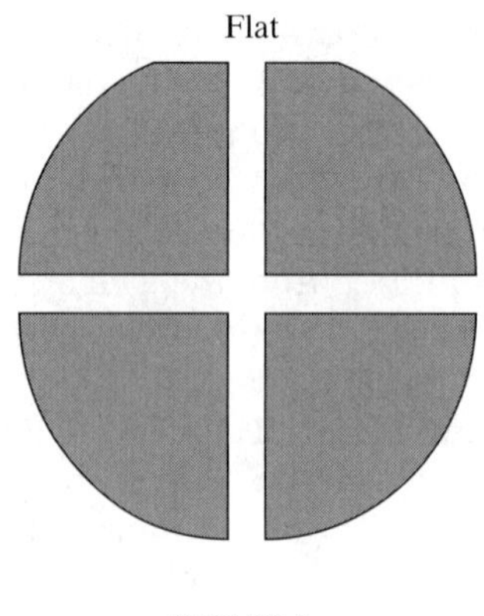

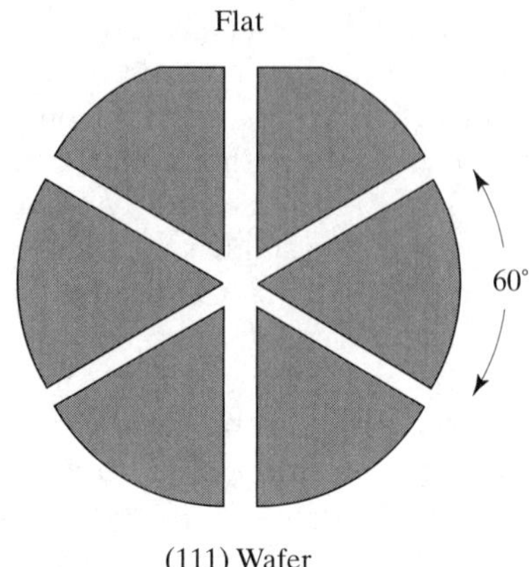

FIGURE 2.2 Typical fracture patterns for (100) and (111) silicon wafers. Some wafers possess a second, smaller flat that denotes crystal orientation and doping. These *minor flats* have not been illustrated.

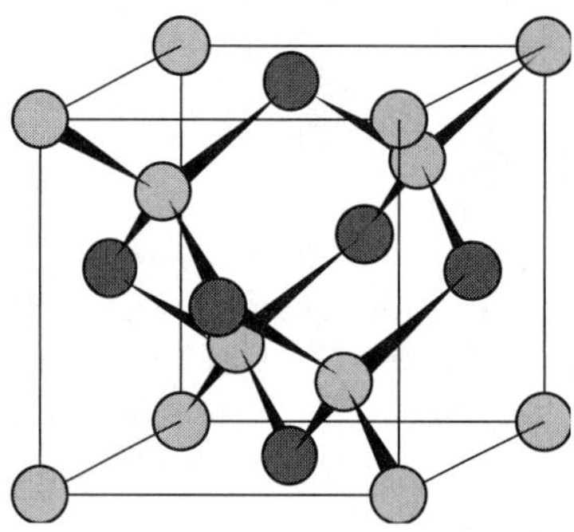

FIGURE 2.3 The diamond lattice unit cell displays a modified face-centered cubic structure. The face-centered atoms are shown in dark gray for emphasis.

cube. Eight more atoms occupy the eight vertices of the cube. Two unit cells placed side by side share four vertex atoms and a single face-centered atom. Additional unit cells can be placed on all sides to extend the crystal in all directions.

When the sawblade slices through a silicon ingot to form a wafer, the orientation of the resulting surface with respect to the unit cell determines many of the wafer's properties. A cut could, for example, slice across a face of the unit cell or diagonally through it. The pattern of atoms exposed by these two cuts differ, as do the electrical properties of devices formed into the respective surfaces. However, not all cuts made through a silicon crystal necessarily differ. Because the faces of a cube are indistinguishable from one another, a cut made across any face of the unit cell looks the same as cuts made across other faces. In other words, planes cut parallel to any face of a unit cube expose similar surfaces.

Because of the awkwardness of trying to describe various planes verbally, a trio of numbers called *Miller indices* are assigned to each possible plane passing through the crystal lattice. Figure 2.4 shows the two most important planar orientations. A plane parallel to a face of the cube is called a *(100) plane,* and a plane slicing diagonally through the unit cube to intersect three of its vertices is called a *(111) plane.* Silicon wafers are generally cut along either a (100) plane or a (111) plane. Although many other cuts exist, none of these have much commercial significance.

A trio of Miller indices enclosed in brackets denotes a direction perpendicular to the indicated crystal plane. For instance, a (100) plane has a [100] direction perpendicular to it and a (111) plane has a [111] direction perpendicular to it. Appendix B discusses how Miller indices are computed and explains the meaning of the different symbologies used to represent them.

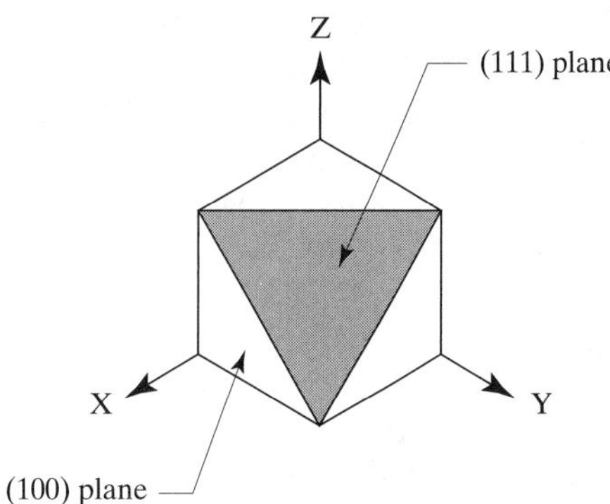

FIGURE 2.4 Identification of (100) and (111) planes of a cubic crystal.

2.2 PHOTOLITHOGRAPHY

The production of silicon wafers constitutes only the first step in the fabrication of integrated circuits. Many of the remaining steps deposit materials on the wafer or etch them away again. A variety of sophisticated deposition and etching techniques exist, but most of these are not *selective.* A nonselective, or *blanket,* process affects the entire surface of the wafer rather than just portions of it. The few processes that are selective are so slow or so expensive that they are useless for high-volume manufacturing. A technique called *photolithography* allows photographic reproduction of intricate patterns that can be used to selectively block depositions or etches. Integrated circuit fabrication makes extensive use of photolithography.

2.2.1. Photoresists

Photolithography begins with the application of a photosensitive emulsion called a *photoresist.* An image can be photographically transferred to the photoresist and a developer used to produce the desired masking pattern. The photoresist solution is usually *spun* onto the wafer. As shown in Figure 2.5, the wafer is mounted on a turntable spinning at several thousand revolutions per minute. A few drops of photoresist solution are allowed to fall onto the center of the spinning wafer, and centrifugal force spreads the liquid out across the surface. The photoresist solution adheres to the wafer and forms a uniform thin film. The excess solution flies off the edges of the spinning wafer. The film thins to its final thickness in a few seconds, the solvent rapidly evaporates, and a thin coating of photoresist remains on the wafer.

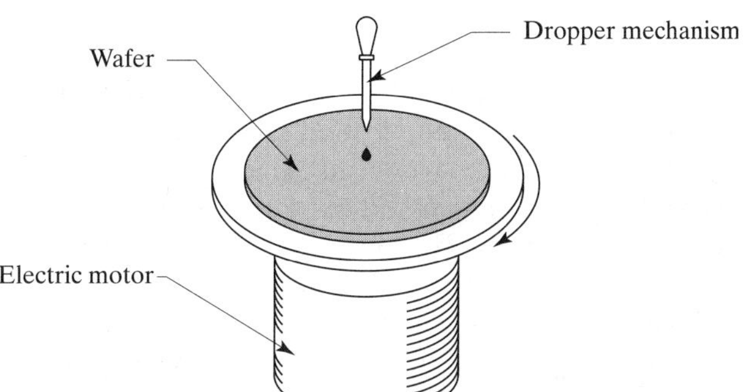

FIGURE 2.5 Application of photoresist solution to a wafer by spinning.

This coating is baked to remove the last traces of solvent and to harden the photoresist to allow handling. Coated wafers are sensitive to certain wavelengths of light, particularly ultraviolet (UV) light. They remain relatively insensitive to other wavelengths, including those of red, orange, and yellow light. Most photolithography rooms therefore have special yellow lighting systems.

The two basic types of photoresists are distinguished by what chemical reactions occur during exposure. A *negative resist* polymerizes under UV light. The unexposed negative resist remains soluble in certain solvents, while the polymerized photoresist becomes insoluble. When the wafer is flooded with solvent, unexposed areas dissolve and exposed areas remain coated. A *positive resist,* on the other hand, chemically decomposes under UV light. These resists are normally insoluble in the developing solvent, but the exposed portions of the resist are chemically altered so as to become soluble. When the wafer is flooded with solvent, the exposed areas wash away while the unexposed areas remain coated. Negative resists tend to swell during development, so process engineers generally prefer to use positive resists.

2.2.2. Photomasks and Reticles

Modern photolithography depends upon a type of projection printing conceptually similar to that used to enlarge photographic negatives. Figure 2.6 shows a simplified illustration of the exposure process. A system of lenses collimates a powerful UV light source, and a plate called a *photomask* blocks the path of the resulting light beam. The UV light passes through the transparent portions of the photomask and through additional lenses that focus an image on the wafer. The apparatus in Figure 2.6 is called an *aligner* since it must ensure that the image of the mask aligns precisely with existing patterns on the wafer.

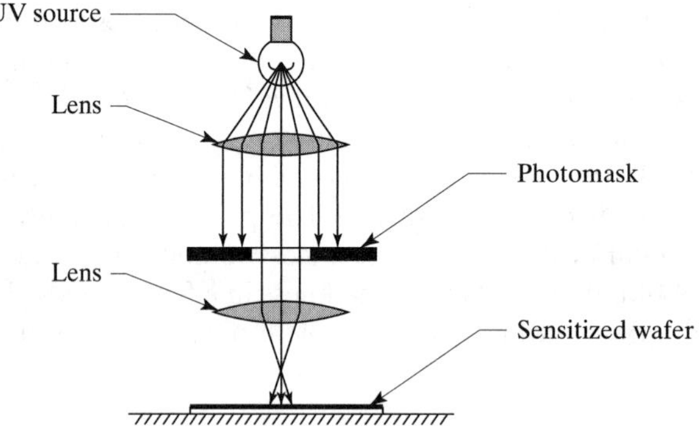

FIGURE 2.6 Simplified illustration of photomask exposure using an aligner.

The transparent plate used as the substrate of a photomask must be dimensionally stable or the pattern it projects will not align with those projected by previous masks. These plates most often consist of fused silica (often erroneously called quartz). After a thin layer of metal is applied to one surface of the plate, any one of various highly precise—but extremely slow and costly—methods are used to pattern the photomask. The image on the photomask is usually 5 or 10 times the size of the image projected onto the wafer. Photographic reduction shrinks the size of any defects or irregularities in the photomask and therefore improves the quality of the final image. This type of enlarged photomask is called either a 5X or a 10X *reticle*, depending on the degree of magnification employed.

A reticle can be used to directly pattern a wafer, but there are mechanical difficulties in doing so. The size of the photomask that an aligner can accept is limited by mechanical considerations, including the difficulty of constructing large lenses of the required accuracy. As a result, most commercial aligners accept a photomask about the same size as the wafer. A 5X reticle that could pattern an entire wafer in one shot would be five times the size of the wafer, and would therefore not fit in the aligner. Practical 5X or 10X reticles are constructed to expose only a small rectangular portion of the final wafer pattern. The reticle must be stepped across the wafer and exposures made at many different positions in order to replicate the pattern across the entire wafer. This process is called *stepping,* and an aligner designed to step a reticle is called a *stepper.* Steppers are slower than simple aligners and are therefore more costly. However, as wafers have grown larger, steppers have progressively replaced aligners.

There is a faster method of exposing wafers which can be used for integrated circuits that do not require extremely fine feature sizes. The reticle can be stepped, not onto a sensitized wafer, but instead onto another photomask. This photomask now bears a 1X image of the desired pattern. The resulting photomask, called a *stepped working plate,* can expose an entire wafer in one shot. Stepped working plates make photolithography faster and cheaper, but the results are not as precise as directly stepping the reticle onto the wafer. The small dimensions of modern integrated circuits have largely rendered stepped working plates obsolate.

Even the tiniest dust speck is so large that it will block the transfer of a portion of the image and ruin at least one integrated circuit. Special air filtration techniques and protective garments are routinely used in wafer fabrication, but some dust gets past all of these precautions. Photomasks are often equipped with *pellicles* on one or both sides to prevent dust from interfering with the exposure. Pellicles consist of thin transparent plastic films mounted on ring-shaped spacers that hold them slightly above the surface of the mask. Light passing through the plane of a pellicle is not in focus, so particles on the pellicle do not appear in the projected image. The pellicle also hermetically seals the surface of the mask and thereby protects it from dust.

2.2.3. Patterning

The exposed wafers are sprayed with a suitable developer, typically consisting of a mixture of organic solvents. The developer dissolves portions of the resist to uncover the surface of the wafer. A deposition or etch affects only these uncovered areas. Once the selective processing has been completed, the photoresist can be stripped away using a more aggressive mixture of solvents. Alternatively, the photoresist can be chemically destroyed by reactive ion etching in an oxygen ambient (Section 2.3.2). This procedure is called *ashing.*

Many important fabrication processes require masking layers that can withstand high temperatures. Since most practical photoresists are organic compounds, they are clearly unsuited to this task. Two common high-temperature masking materials are silicon dioxide and silicon nitride. These materials can be formed by the reaction of appropriate gases with the silicon surface. A photoresist can then be applied and patterned and an etching process used to open holes in the oxide or nitride film. Modern processing techniques make extensive use of oxide and nitride films for masking high-temperature depositions and diffusions.

2.3 OXIDE GROWTH AND REMOVAL

Silicon forms several oxides, the most important of which is *silicon dioxide* (SiO_2). This oxide possesses a number of desirable properties that together are so valuable that silicon has become the dominant semiconductor. Other semiconductors have better electrical properties, but only silicon forms a well-behaved oxide. Silicon dioxide can be grown on a silicon wafer by simply heating it in an oxidizing atmosphere. The resulting film is mechanically rugged and resists most common solvents, yet it readily dissolves in hydrofluoric acid. Oxide films are superb electrical insulators and are useful not only for insulating metal conductor patterns, but also for forming the dielectrics of capacitors and MOS transistors. Silicon dioxide is so important to silicon processing that it is universally known as *oxide*.

2.3.1. Oxide Growth and Deposition

If a silicon wafer is exposed to the air, atmospheric oxygen reacts with the silicon to form an oxide layer a few Angstroms thick. This *native oxide* is too thin for most applications, but much thicker layers of oxide can be grown by heating the silicon wafer in an oxidizing atmosphere. If pure dry oxygen is employed, then the resulting oxide film is called a *dry oxide*. Figure 2.7 shows a typical oxidation apparatus. The wafers are placed in a fused silica rack called a *wafer boat*. The wafer boat is slowly inserted into a fused silica tube wrapped in an electrical heating mantle. The temperature of the wafers gradually rises as the wafer boat moves into the middle of the heating zone. Oxygen gas blowing through the tube passes over the surface of each wafer. At elevated temperatures, oxygen molecules can actually diffuse through the oxide layer to reach the underlying silicon. There oxygen and silicon react, and the layer of oxide gradually grows thicker. The rate of oxygen diffusion slows as the oxide film thickens, so the growth rate decreases with time. As Table 2.1 indicates, high temperatures greatly accelerate oxide growth. Crystal orientation also affects oxidation rates, with (111) silicon oxidizing significantly faster than (100) silicon.[1] Once the oxide layer has reached the desired thickness (as gauged by time and temperature), the wafers are slowly withdrawn from the furnace.

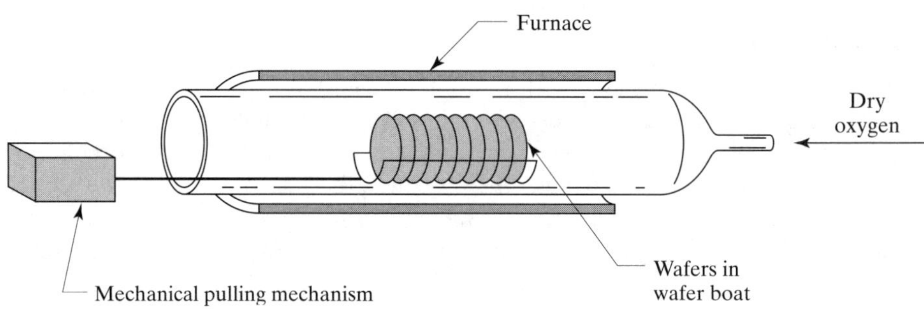

FIGURE 2.7 Simplified diagram of an oxidation furnace.

TABLE 2.1 Times required to grow 0.1 μm of oxide on (111) silicon.[2]

Ambient	800°C	900°C	1000°C	1100°C	1200°C
Dry O_2	30 hr	6 hr	1.7 hr	40 min	15 min
Wet O_2	1.7 hr	20 min	6 min		

[1] W. R. Runyan and K. E. Bean, *Semiconductor Integrated Circuit Processing Technology* (Reading, MA: Addison-Wesley, 1994), p. 84ff.

[2] Calculated from R. P. Donovan, "Oxidation," in R. M. Burger and R. P. Donovan, eds., *Fundamentals of Silicon Integrated Device Technology* (Englewood Cliffs, NJ: Prentice-Hall, 1967), pp. 41, 49.

Dry oxide grows very slowly, but it is of particularly high quality because relatively few defects exist at the oxide-silicon interface. These defects, or *surface states*, interfere with the proper operation of semiconductor devices, particularly MOS transistors. The density of surface states is measured by a parameter called the *surface state charge*, or Q_{SS}. Dry oxide films that are thermally grown on (100) silicon have especially low surface state charges and thus make ideal dielectrics for MOS transistors.

Wet oxides are formed in the same way as dry oxides, but steam is injected into the furnace tube to accelerate the oxidation. Water vapor moves rapidly through oxide films, but hydrogen atoms liberated by the decomposition of the water molecules produce imperfections that may degrade the oxide quality.[3] Wet oxidation is commonly used to grow a thick layer of *field oxide* where no active devices will be built. Dry oxidations conducted at higher-than-ambient pressures can also accelerate oxide growth rates.

Sometimes an oxide layer must be formed on a material other than silicon. For instance, oxide is frequently employed as an insulator between two layers of metallization. In such cases, some form of deposited oxide must be used rather than the grown oxides previously discussed. Deposited oxides can be produced by various reactions between gaseous silicon compounds and gaseous oxidizers. For example, silane gas and nitrous oxide react to form nitrogen gas, water vapor, and silicon dioxide. Alternatively, deposited oxides can be created by spinning on a solution of *tetraethoxysilane* (TEOS) and subsequently decomposing this by heat treatment. The resulting oxide layer is called a *spin-on glass* (SOG). Deposited oxides tend to possess low densities and large numbers of defect sites, so they are not suitable for use as gate dielectrics for MOS transistors. Deposited oxides are still acceptable for use as insulating layers between multiple conductor layers, or as protective overcoats.

Oxide films are brightly colored due to *thin-film interference*. When light passes through a transparent film, destructive interference between transmitted and reflected wavefronts causes certain wavelengths of light to be selectively absorbed. Different thicknesses of films absorb different colors of light. Thin-film interference causes the iridescent colors seen in soap bubbles and films of oil on water. The same effect produces the vivid colors visible in microphotographs of older generation integrated circuits. These colors are helpful in distinguishing various regions of an integrated circuit under a microscope or in a microphotograph. The approximate thickness of an oxide film can often be determined using a table of oxide colors.[4] Modern integrated circuits often exhibit little variation in color when seen under a microscope. This relatively drab appearance is due to the high degree of surface planarity required for fire-line photolithography, combined with a lack of silicon topography due to an absence of deep diffusions.

2.3.2. Oxide Removal

Figure 2.8 illustrates the procedure used to form a patterned oxide layer. The first step consists of growing a thin layer of oxide across the wafer. Next, photoresist is applied to the wafer by spinning. A subsequent oven bake drives off the final traces of solvent and hardens the photoresist for handling. After photolithographic exposure, the wafer is developed by spraying it with a solvent that dissolves the exposed

---

[3] Hydrogen incorporation due to wet oxidation conditions reduces the concentration of dangling bonds, but it increases the fixed oxide charge. The differences between wet and dry oxidation are therefore not as simplistic as the text may suggest.

[4] For a table, see W. A. Pliskin and E. E. Conrad, "Nondestructive Determination of Thickness and Refractive Index of Transparent Films," *IBM J. Research and Development*, Vol. 8, 1964, pp. 43–51.

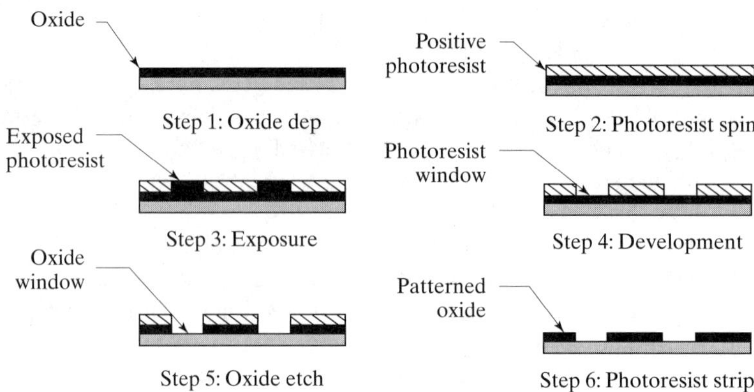

FIGURE 2.8 Steps in oxide growth and removal.

areas of photoresist to reveal the underlying oxide. The patterned photoresist serves as a masking material for an oxide etch. Having served its function, the photoresist is finally stripped away to leave the patterned oxide layer.

Oxide can be etched by either of two methods. *Wet etching* employs a liquid solution that dissolves the oxide, but not the photoresist or the underlying silicon. *Dry etching* uses a plasma or chemical vapor to perform the same function. Wet etches are simpler, but dry etches provide better linewidth control.

Most wet etches employ solutions of buffered hydrofluoric acid (HF). This highly corrosive substance readily dissolves silicon dioxide, but it does not attack either elemental silicon or organic photoresists. The etch process consists of immersing the wafers in a plastic tank containing the hydrofluoric acid solution for a specified length of time, followed by a thorough rinsing to remove all traces of the acid. Wet etches are *isotropic* because they proceed at the same rate laterally as well as vertically. The acid works its way under the edges of the photoresist to produce sloping sidewalls similar to those shown in Figure 2.9A. Since the etching must continue long enough to ensure that all openings have completely cleared, some degree of overetching inevitably occurs. The acid continues to erode the sidewalls as long as the wafer remains immersed. The extent of sidewall erosion varies depending upon etching conditions, oxide thickness, and other factors. Because of these variations, wet etching cannot provide the tight linewidth control required by modern semiconductor processes.

There are three classes of dry etching processes: reactive ion etching, plasma etching, and chemical vapor etching.[5] *Reactive ion etching* (RIE) will serve to illustrate

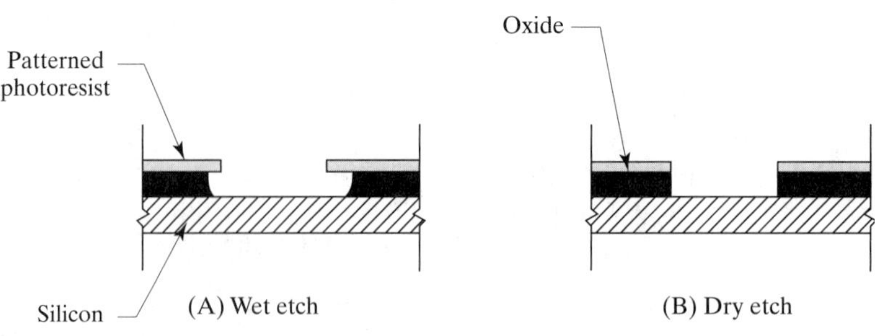

FIGURE 2.9 Comparison of isotropic wet etching (A) and anisotropic dry etching (B). Note the undercutting of the oxide caused by wet etching.

[5] Runyan, *et al.*, pp. 269–272.

the principles of dry etching. In a reactive ion etcher, a silent electrical discharge passed through a low-pressure gas mixture forms highly energetic molecular fragments called *reactive ions.* The etching apparatus projects these ions downward onto the wafer at high velocities. Because the ions impact the wafer at a relatively steep angle, etching proceeds vertically at a much greater rate than laterally. The *anisotropic* nature of reactive ion etching allows the formation of nearly vertical sidewalls such as those shown in Figure 2.9B. Figure 2.10 shows a simplified diagram of a reactive ion etching apparatus.

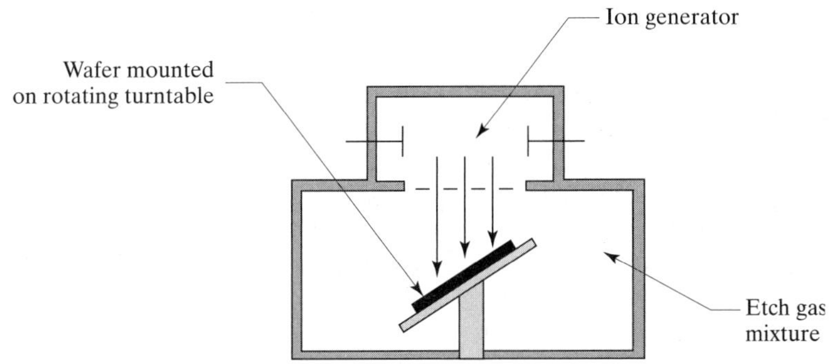

FIGURE 2.10 Simplified diagram of reactive ion etching apparatus.

The etch gas employed in the RIE system generally consists of a compound such as trichloroethane, perhaps mixed with an inert gas such as argon. The reactive ions formed from this mixture selectively attack silicon dioxide in preference to either photoresist or elemental silicon. Different mixtures of etch gases have been developed that allow anisotropic etching of silicon nitride, elemental silicon, and other materials.

Modern processes rely on dry etching to obtain tight control of submicron geometries that cannot be fabricated in any other way. The increased packing density and higher performance of these structures more than compensate for the complexity and cost of dry etching.

2.3.3. Other Effects of Oxide Growth and Removal

During a typical processing sequence, the wafer is repeatedly oxidized and etched to form successive masking layers. These multiple masked oxidations cause the silicon surface to become highly nonplanar. The resulting surface irregularities are of great concern because modern fine-line photolithography has a very narrow depth of field. If the surface irregularities are too large, then it becomes impossible to focus the image of the photomask onto the resist.

Consider the wafer in Figure 2.11. A planar silicon surface has been oxidized, patterned, and etched to form a series of oxide openings (Figure 2.11A). Subsequent thermal oxidation of the patterned wafer results in the cross section shown in Figure 2.11B. The opening that is left from the previous oxide removal initially oxidizes very rapidly, while the surfaces already coated with an oxide layer oxidize more slowly. The silicon surface erodes by about 45% of the oxide thickness grown.[6] The silicon under the previous oxide opening therefore recedes to a greater depth than the surrounding silicon surfaces. The thickness of oxide in the old opening will

---

[6] This value is the inverse of the *Pilling-Bedworth ratio,* which equals 2.2: G. E. Anner, *Planar Processing Primer* (New York: Van Nostrand Reinhold, 1990), p. 169.

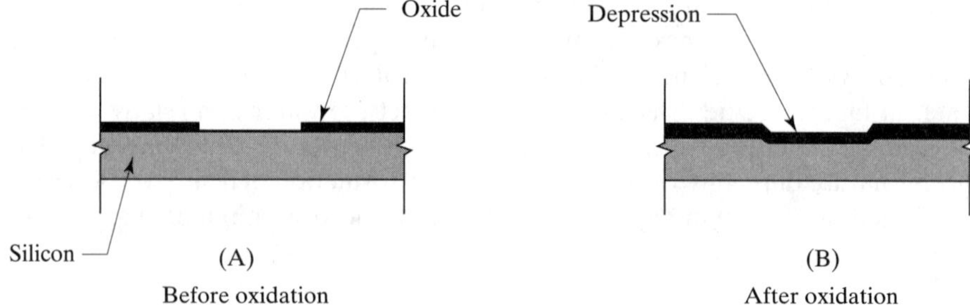

FIGURE 2.11 Effects of patterned oxidation on wafer topography.

always be less than that of the surrounding surfaces since these already have some oxide on them when growth begins. The differences in oxide thickness and in the depths of the silicon surfaces combine to produce a characteristic surface discontinuity called an *oxide step*.

The growth of a thermal oxide also affects the doping levels in the underlying silicon. If the dopant is more soluble in oxide than in silicon, during the course of the oxidation it will tend to migrate from the silicon into the oxide. The surface of the silicon thus becomes depleted of dopant. Boron is more soluble in oxide than in silicon, so it tends to segregate into the oxide. This effect is sometimes called *boron suckup.* Conversely, if the dopant dissolves more readily in silicon than in oxide, then the advancing oxide-silicon interface pushes the dopant ahead of it and causes a localized increase in doping levels near the surface. Phosphorus (like arsenic and antimony) segregates into the silicon, so it tends to accumulate at the surface as oxidation continues. This effect is sometimes called *phosphorus pileup* or *phosphorus plow.* The doping profiles of Figures 2.12A and 2.12B illustrate boron suckup and phosphorus plow, respectively. In both cases, the preoxidization doping profiles were constant and the varying dopant concentrations near the surface are solely due to segregation. The existence of these segregation mechanisms complicates the task of designing dopant profiles for integrated devices.

The doping of silicon also affects the rate of oxide growth. A concentrated N+ diffusion tends to accelerate the growth of oxide near it by a process called *dopant-enhanced oxidation.* This occurs because the donors interfere with the bonding of

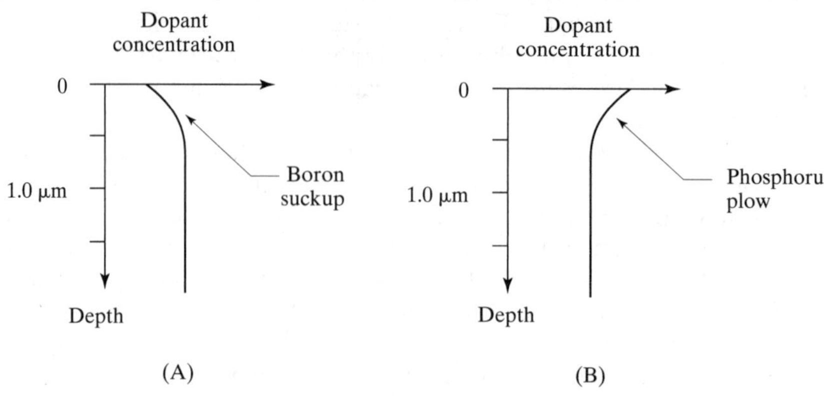

FIGURE 2.12 Oxide segregation mechanisms: (A) boron suckup and (B) phosphorus plow.[7]

[7] A. S. Grove, O. Leistiko, and C. T. Sah, "Redistribution of Acceptor and Donor Impurities During Thermal Oxidation of Silicon," *J. Appl. Phys.*, Vol. 35, #9, 1964, pp. 2695–2701.

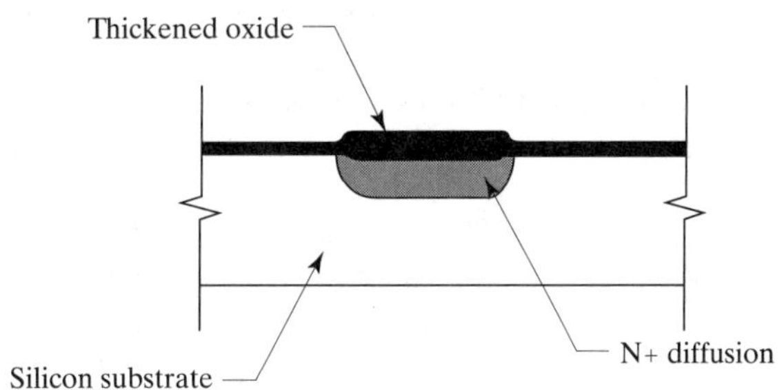

FIGURE 2.13 Effects of dopant-enhanced oxidation.

atoms at the oxide interface, causing dislocations and other lattice defects. These defects catalyze oxidation and thus accelerate the growth of the overlying oxide. This effect can become quite significant when a heavily doped N+ deposition occurs early in the process, before the long thermal drives and oxidations. Figure 2.13 shows a wafer in which a long thermal oxidation has been conducted after the deposition of an N+ region. The oxide over the N+ diffusion is actually thicker than the oxide over adjacent regions. Dopant-enhanced oxidation can be used to thicken the field oxide in order to reduce its capacitance per unit area. Thus, a capacitor formed over a deep-N+ diffusion will exhibit less parasitic capacitance between its bottom plate and the substrate than will a capacitor formed over lightly doped regions.

2.3.4. Local Oxidation of Silicon (LOCOS)

A technique called *local oxidation of silicon* (LOCOS) allows the selective growth of thick oxide layers.[8] The process begins with the growth of a thin *pad oxide* that protects the silicon surface from the mechanical stresses induced by subsequent processing (Figure 2.14). A nitride film is then deposited on top of the pad oxide. This nitride is patterned to expose the regions to be selectively oxidized. The nitride blocks the diffusion of oxygen and water molecules, so oxidation only occurs under the nitride windows. Some oxidants diffuse a short distance under the edges of the nitride, producing a characteristic curved transition region called a *bird's beak*.[9]

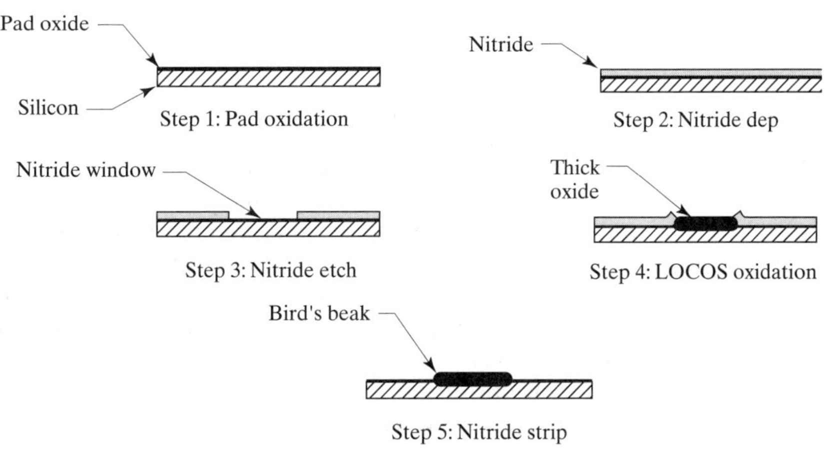

FIGURE 2.14 Local oxidation of silicon (LOCOS) process.

[8] "LOCOS: A New I.C. Technology," *Microelectronics and Reliability,* Vol. 10, 1971, pp. 471–472.

[9] E. Bassous, H. H. Yu, and V. Maniscalco, "Topology of Silicon Structures with Recessed SiO_2," *J. Electrochem. Soc.,* Vol. 123, #11, 1976, pp. 1729–1737.

Once oxidation is complete, the nitride layer is stripped away to reveal the patterned oxide.

CMOS and BiCMOS processes employ LOCOS to grow a thick *field oxide* over electrically inactive regions of the wafer. The areas not covered by field oxide are called *moat* regions because they form shallow depressions in the topography of the wafer. A very thin, high-quality gate oxide subsequently grown in the moat regions forms the gate dielectric of the MOS transistors.

A mechanism called the *Kooi effect* complicates the growth of gate oxide.[10] The water vapor typically used to accelerate LOCOS oxidation also attacks the surface of the nitride film to produce ammonia, some of which migrates beneath the pad oxide near the edges of the nitride window. There it reacts with the underlying silicon to form silicon nitride again (Figure 2.15). Since these nitride deposits lie beneath the pad oxide, they remain even after the LOCOS nitride is stripped. Removing the pad oxide prior to growing the gate oxide does not eliminate these deposits because this etch is selective to oxide, not to nitride. During gate oxidation, the nitride residues act as an unintentional LOCOS mask that retards oxide growth around the edges of the moat region. The gate oxide at these points may not be sufficiently thick to withstand the full operating voltage. The Kooi effect can be circumvented by first growing a thin oxide layer and then stripping it away. Because silicon nitride slowly oxidizes, this *dummy gate oxidation* removes the nitride residues and improves the integrity of the true gate oxide grown immediately afterward.

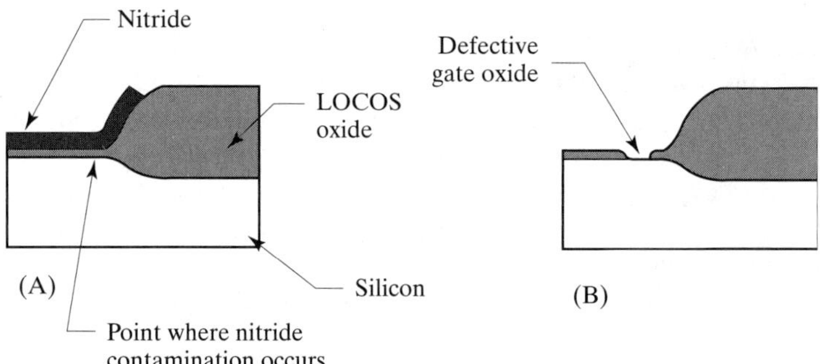

FIGURE 2.15 The Kooi effect is caused by nitride that grows under the bird's beak (A), preventing formation of gate oxide during subsequent oxidation (B).

2.4 DIFFUSION AND ION IMPLANTATION

Discrete diodes and transistors can be fabricated by forming junctions into a silicon ingot during crystal growth. Suppose that the silicon ingot begins as a P-type crystal. After a short period of growth, the molten silicon is counterdoped by the addition of a controlled amount of phosphorus. Continued crystal growth will now produce a PN junction embedded in the ingot. Successive counterdopings can produce multiple junctions in the crystal, allowing the fabrication of *grown-junction* transistors. Integrated circuits cannot be grown because there is no way to produce differently doped regions in different portions of the wafer. Even the manufacture of simple grown-junction transistors presents a challenge, because the thickness and planarity

---

[10] E. Kooi, J. G. van Lierop, and J. A. Appels, "Formation of Silicon Nitride at a Si-SiO_2 Interface during Local Oxidation of Silicon and during Heat-Treatment of Oxidized Silicon in NH_3 Gas," *J. Electrochem. Soc.*, Vol. 123, #7, 1976, pp. 1117–1120.

of grown junctions are difficult to control. Each counterdoping also raises the total dopant concentration. Some properties of silicon (such as minority carrier lifetime) depend upon the total concentration of doping atoms, not just upon the excess of one dopant species over the other. The repeated counterdopings therefore progressively degrade the electrical properties of the silicon.

Historically, the grown junction process was soon abandoned in favor of the much more versatile *planar process*. This process is used to fabricate virtually all modern integrated circuits as well as the vast majority of modern discrete devices. Figure 2.16 shows how a wafer of discrete diodes can be fabricated using planar processing. A uniformly doped silicon crystal is first sliced to form individual wafers. An oxide film grown on these wafers is photolithographically patterned and etched. A dopant source spun onto the patterned wafers touches the silicon only where the oxide has been previously removed. The wafers are then heated in a furnace to drive the dopant into the silicon, which forms shallow counterdoped regions. The finished wafer can be diced to form hundreds or thousands of individual diodes. The planar process does not require so many counterdopings of the silicon ingot, thereby allowing more precise control of junction depths and dopant distributions.

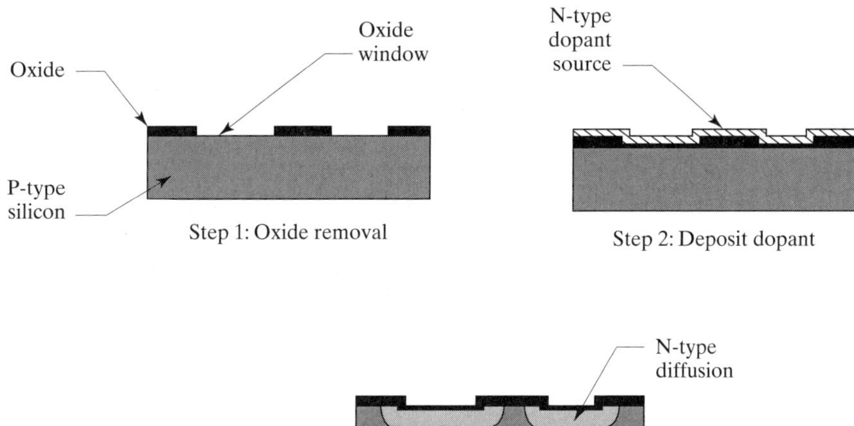

FIGURE 2.16 Formation of diffused PN-junction diodes, using the planar process.

2.4.1. Diffusion

Dopant atoms can move through the silicon lattice by thermal diffusion in much the same way as carriers move by diffusion (Section 1.1.3). The dopant atoms are heavier and more tightly bound to the crystal lattice, so temperatures of 800°C to 1250°C are required to obtain reasonable diffusion rates. Once the dopants have been driven to the desired junction depth, the wafer is cooled and the dopant atoms become immobilized within the lattice. A doped region formed in this manner is called a *diffusion*.

The usual process for creating a diffusion consists of two steps: an initial *deposition* (or *predeposition*) and a subsequent *drive* (or *drive-in*). Deposition consists of heating the wafer in contact with an external source of dopant atoms. Some of these diffuse from the source into the surface of the silicon wafer to form a shallow heavily doped region. The external dopant source is then removed and the wafer is heated to a higher temperature for a prolonged period of time. The dopants introduced during deposition are now driven down to form a much deeper and less concentrated diffusion. If a very heavily doped junction is required, then it is usually unnecessary to strip the dopant source from the wafer, and the deposition and subsequent drive can be conducted as a single operation.

Four dopants find widespread use in silicon processing: *boron, phosphorus, arsenic,* and *antimony.*[11] Only boron is an acceptor; the other three are all donors. Boron and phosphorus diffuse relatively rapidly, while arsenic and antimony diffuse much more slowly (Table 2.2). Arsenic and antimony are used where slow rates of diffusion are advantageous—for example, when very shallow junctions are desired. Even boron and phosphorus do not diffuse appreciably at temperatures below 800°C, necessitating the use of special high-temperature diffusion furnaces.

TABLE 2.2 Representative junction depths, in microns (10^{20} atoms/cm^3 source, 10^{16} atoms/cm^3 background, 15 min deposition, 1 hr drive).[12]

Dopant	950°C	1000°C	1100°C	1200°C
Boron	0.9	1.5	3.6	7.3
Phosphorus		0.5	1.6	4.6
Antimony			0.8	2.1
Arsenic			0.7	2.0

Figure 2.17 shows a simplified diagram of a typical apparatus for conducting a phosphorus diffusion. A long fused silica tube passes through an electric furnace that is constructed to produce a very stable heating zone in the middle of the tube. After the wafers are loaded into a wafer boat, they are slowly pushed into the furnace by means of a mechanical arrangement that controls the insertion rate. Dry oxygen is blown through a flask containing liquid phosphorus oxychloride ($POCl_3$, often pronounced "pockle"). A small amount of $POCl_3$ evaporates and is carried by the gas stream over the wafers. Phosphorus atoms released by the decomposition of the $POCl_3$ diffuse into the oxide film, forming a doped oxide that acts as a deposition source. When enough time has passed to deposit sufficient dopant in the silicon, the wafers are removed from the furnace and the doped oxide is stripped

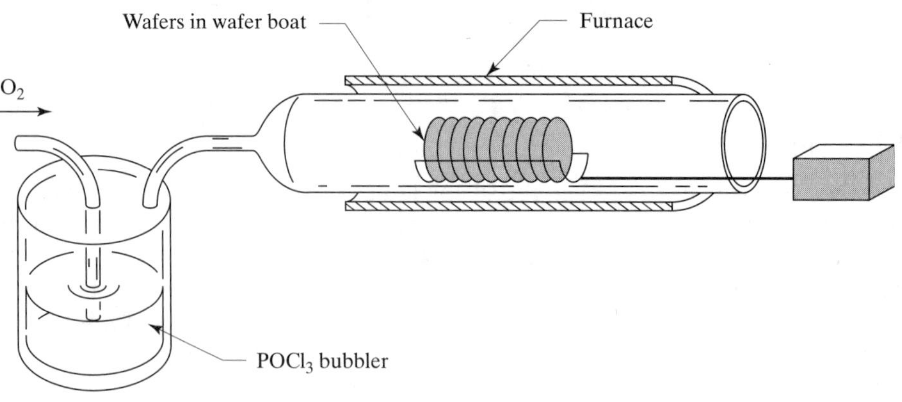

FIGURE 2.17 Simplified diagram of a phosphorus diffusion furnace using a $POCl_3$ source.

[11] These dopants were chosen because they readily ionize and because they are sufficiently soluble in silicon to form heavily doped diffusions. See F. A. Trumbore, "Solid Solubilities of Impurity Elements in Germanium and Silicon," *Bell Syst. Tech. J.,* Vol. 39, #1, 1960, pp. 205–233.

[12] Calculated using diffusivities from R. S. Muller and T. I. Kamins, *Device Electronics for Integrated Circuits,* 2d ed. (New York: John Wiley and Sons, 1986), p. 85.

away (a process called *deglazing*). The wafers are then reloaded into another furnace, where they are heated to drive the phosphorus down to form the desired diffusion. If a very concentrated phosphorus diffusion is desired, then the wafers need not be removed for deglazing prior to the drive. With suitable modifications to the dopant source, this apparatus can diffuse any of the four common dopants.

Many alternative deposition sources have been developed. A gaseous dopant such as diborane (for boron) or phosphine (for phosphorus) can be injected directly into the carrier gas stream. Thin disks of boron nitride placed between silicon wafers can serve as a solid deposition source for boron. In a high-temperature oxidizing atmosphere, a little boron trioxide outgases from these disks to the adjacent wafers. Various proprietary *spin-on glasses* are also sold as dopant sources. These are purchased in the form of a liquid. After the solution is spun onto a wafer, a brief bake drives out the solvent and leaves a doped oxide layer on the wafer. This so-called *glass* then serves as a dopant source for the subsequent diffusion.

None of these deposition schemes are particularly well controlled. Even with gaseous sources (which can be precisely metered), nonuniform gas flow around the wafer inevitably produces doping variations. For less-demanding processes such as standard bipolar, any of these schemes can give adequate results. Modern CMOS and BiCMOS processes require more accurate control of doping levels and junction depths than conventional deposition techniques can achieve. Ion implantation (Section 2.4.3) can provide the necessary accuracy at the expense of much more complex and costly apparatus. $POCl_3$ is still used to produce heavily doped N-type regions that cannot be economically fabricated by ion implantation.

2.4.2. Other Effects of Diffusion

The diffusion process suffers from a number of limitations. Diffusions can only be performed from the surface of the wafer, limiting the geometries that can be fabricated. Dopants diffuse unevenly, so the resulting diffusions do not have constant doping profiles. Subsequent high-temperature process steps continue the drive of previously deposited dopants, so junctions formed early in the process are driven substantially deeper during later processing. Dopants out-diffuse under the edges of the oxide windows, blurring the diffusion pattern. Diffusions interact with oxidizations due to segregation mechanisms, resulting in depletion or enhancement of surface doping levels. Diffusions even interact with one another since the presence of one doping species alters the diffusion rates of others. These and other complications make the diffusion process far more complex than it might at first appear.

Diffusion can produce only relatively shallow junctions. Practical drive times and temperatures limit junction depths to about 15 microns. Most diffusions will be much shallower. Since diffusions are typically patterned using an oxide mask, the cross section of a diffusion generally resembles that shown in Figure 2.18A. The

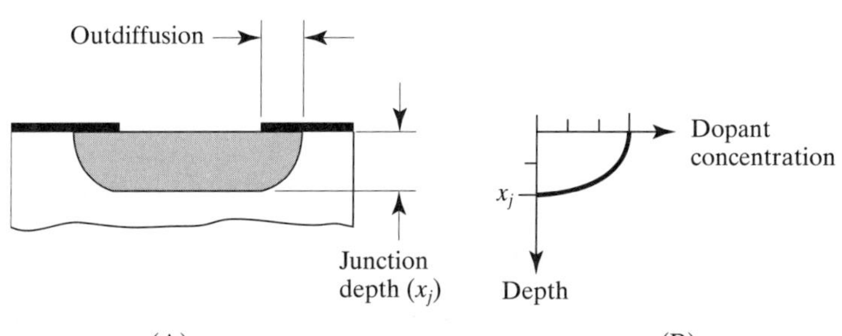

FIGURE 2.18 Cross section (A) and doping profile (B) of a typical planar diffusion.

dopant diffuses out in all directions at roughly the same rate. The junction moves laterally under the edges of the oxide window a distance equal to about 80% of the junction depth.[13] This lateral movement, known as *outdiffusion*, causes the final size of the diffused region to exceed the drawn dimensions of the oxide window. Outdiffusion is not visible under the microscope since the changes in oxide color caused by thin film interference correspond to the locations of oxide removals and not to the positions of the final junctions.

The doping level of a diffusion varies as a function of depth. Neglecting segregation mechanisms, dopant concentrations are highest at the surface and gradually lessen with depth. The resulting *doping profile* can be theoretically predicted and experimentally measured. Figure 2.18B shows the theoretical doping profile for a point in the center of the oxide window. This profile assumes that oxide segregation remains negligible, which is not always the case. Boron suckup may substantially reduce the surface doping of a P-type diffusion and can cause a lightly doped diffusion to invert to become N-type. Phosphorus pileup will not cause surface inversion, but it still affects surface doping levels.

As mentioned above, the rate of diffusion can be altered by the presence of other doping species. Consider an NPN transistor with a heavily doped phosphorus emitter diffused into a lightly doped boron base. The presence of high concentrations of donors within the emitter causes lattice strains that spawn defects. Some of these defects migrate to the surface, where they cause dopant-enhanced oxidation. Other defects migrate downward, where they accelerate the diffusion of boron in the underlying base region. This mechanism, called *emitter push*, results in a deeper base diffusion under the emitter than in surrounding regions (Figure 2.19A).[14] The presence of NBL beneath a diffusion may reduce the junction depth due to the intersection of the tail of the updiffusing NBL with the base diffusion. This effect is sometimes called *NBL push* in analogy with the better-known emitter push, even though the underlying mechanisms are quite different (Figure 2.19B). NBL push can interfere with the layout of accurate diffused resistors.

A similar mechanism accelerates the diffusion of dopants under an oxidization zone. The oxidation process spawns defects, some of which migrate downward to enhance the rate of diffusion of dopants beneath the growing oxide. This mechanism is

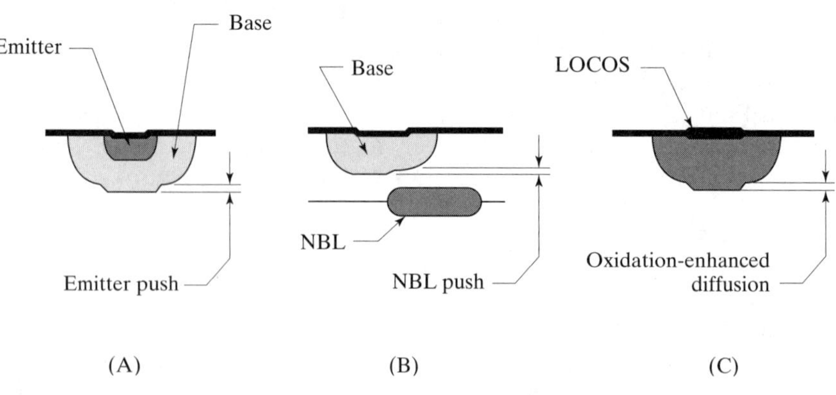

FIGURE 2.19 Mechanisms that can alter diffusion rates include emitter push (A), NBL push (B), and oxidation-enhanced diffusion (C).

[13] See D. P. Kennedy and R. R. O'Brien, "Analysis of the Impurity Atom Distribution Near the Diffusion Mask for a Planar p-n Junction," *IBM J. of Research and Development*, Vol. 9, 1965, pp. 179–186.

[14] A. F. W. Willoughby, "Interactions between Sequential Dopant Diffusions in Silicon—A Review," *J. Phys. D: Appl. Phys.*, Vol. 10, 1977, pp. 455–480.

called *oxidization-enhanced diffusion*.[15] It affects all dopants, and it can produce significantly deeper diffusions under a LOCOS field oxide than under adjacent moat regions (Figure 2.19C).

Even the most sophisticated computer programs cannot always predict actual doping profiles and junction depths because of the many interactions that occur. Process engineers must experiment carefully to find the proper recipe for manufacturing a given combination of devices on a wafer. The more complicated the process, the more complex these interactions become and the more difficult it is to find a suitable recipe. Since process design takes so much time and effort, most companies use only a few processes to manufacture all of their products. The difficulty of incorporating new process steps into an existing recipe also explains the reluctance of process engineers to modify their processes.

2.4.3. Ion Implantation

Due to the limitations of conventional diffusion techniques, modern processes make extensive use of *ion implantation*. An ion implanter is essentially a specialized particle accelerator used to accelerate dopant atoms so that they can penetrate the silicon crystal to a depth of several microns. Ion implantation does not require high temperatures, so a layer of patterned photoresist can serve as a mask against the implanted dopants. Implantation also allows better control of dopant concentrations and profiles than conventional deposition and diffusion. However, large implant doses require correspondingly long implant times. Ion implanters are also complex and costly devices. Many processes use a combination of diffusions and implantations to reduce overall costs.

Figure 2.20 shows a simplified diagram of an ion implanter. An ion source provides a stream of ionized dopant atoms that are accelerated by the electric field of a miniature linear accelerator. A magnetic analyzer selects the desired species of ion, and a pair of deflection plates scans the resulting ion beam across the wafer. A high vacuum must be maintained throughout the system, so the entire apparatus is enclosed in a steel housing.

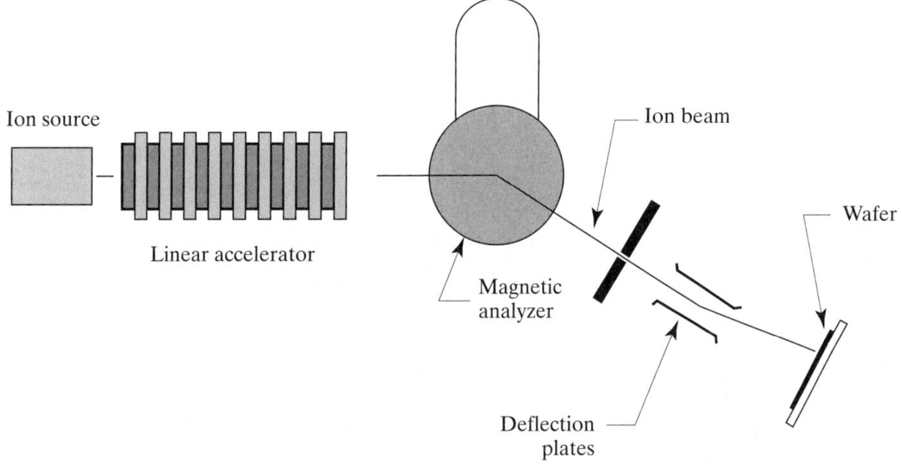

FIGURE 2.20 Simplified diagram of an ion implanter.[16]

---

[15] K. Taniguchi, K. Kurosawa, and M. Kashiwagi, "Oxidation Enhanced Diffusion of Boron and Phosphorus in (100) Silicon," *J. Electrochem. Soc.*, Vol. 127, #10, 1980, p. 2243–2248.

[16] The scheme shown is but one of several; see Anner, p. 313ff.

Once the ions enter the silicon lattice, they immediately begin to decelerate due to collisions with surrounding atoms. Each collision transfers momentum from a moving ion to a stationary atom. The ion beam spreads as it sheds energy, causing the implant to spread out (*straggle*) in a manner reminiscent of outdiffusion. Atoms are also knocked out of the lattice by the collisions, causing extensive lattice damage that must be repaired by *annealing* the wafer at moderate temperatures (800°C to 900°C) for a few minutes. The silicon atoms become mobile and the intact silicon crystal structure around the edges of the implant zone serves as a seed for crystal growth. Damage progressively anneals out from the sides of the implant zone toward the center. Dopants added by ion implantation will redistribute by thermal diffusion if the wafer is subsequently heated to a sufficiently high temperature. Therefore, a deep, lightly doped diffusion can be created by first implanting the required dopants and subsequently driving them down to the desired junction depth.

The dopant concentration provided by ion implantation is directly proportional to the *implant dose*, which equals the product of ion beam current and time. The dose can be precisely monitored and controlled, which allows for much better repeatability than is achievable by conventional deposition techniques. The doping profile is determined by the energy imparted to individual ions, a quantity called the *implant energy*. Low-energy implants are very shallow, while higher-energy implants actually place most of the dopant atoms beneath the surface of the silicon. Implant energies of several megavolts can be used to counterdope a region several microns beneath the surface so as to form a *buried layer*. By stacking a series of implants of progressively higher energies one on top of another and then annealing the resulting structure, one can form deep diffusions without prolonged heat cycles. These *chained implants* can be constructed with nearly vertical sidewalls.[17]

An ion implant can also be masked by a patterned deposited material such as polysilicon. This technique eliminates the alignment errors that inevitably occur when attempting to fabricate a structure by using multiple photolithographic exposures. The resulting *self-aligned* structures can therefore be constructed to very close dimensional tolerances. Figure 2.21 illustrates the creation of self-aligned MOS source/drain regions by ion implantation. A layer of polysilicon has been deposited and patterned on top of a thin gate oxide. The polysilicon not only forms the gate electrodes for MOS transistors but also simultaneously serves as an implant mask. The polysilicon blocks the implant from the region beneath the gate electrode,

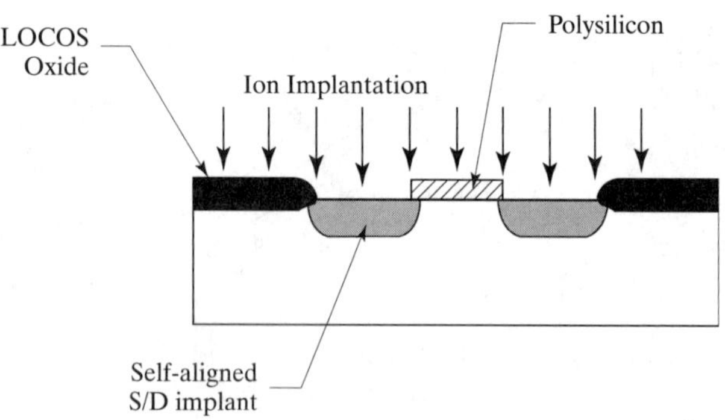

FIGURE 2.21 Self-aligned source and drain regions formed by ion implantation.

[17] R. K. Williams and M. E. Cornell, "The Emergence and Impact of DRAM-Fab Reuse in Analog and Power-Management Integrated Circuits," *Proc. of IEEE Bipolar/BIOMOS Circuits and Technologies Meeting*, 2002, pp. 45–52.

forming precisely aligned source and drain regions. The alignment of the source and drain with the gate is limited only by the small amount of straggle caused by the spreading of the ion beam. If self-aligned implants were not used, then photolithographic misalignments would occur between the gate and the source/drain diffusions, and the resulting overlap capacitances would substantially reduce the switching speed of the MOS transistors.

When the silicon lattice is viewed from certain angles, interstices between columns of silicon atoms, called *channels*, become visible. These disappear from view when the crystal is turned slightly. Channels are visible in both the (100) and the (111) silicon surfaces when these are viewed perpendicularly. If the ion beam were to impinge perpendicularly upon a (100) or a (111) surface, then ions could move deep into the crystal before scattering would commence. The final dopant distribution would depend critically upon the angle of incidence of the ion beam. To avoid this difficulty, most implanters project the ion beam onto the wafer at an angle of about 7°.

2.5 SILICON DEPOSITION AND ETCHING

Films of pure or doped silicon can be chemically grown on the surface of a wafer. The nature of the underlying surface determines whether the resulting film will be monocrystalline or polycrystalline. If the surface consists of exposed monocrystalline silicon, then this serves as a seed for crystal growth and the deposited film will also be monocrystalline. If the deposition is conducted on top of an oxide or nitride film, then no underlying crystalline lattice will exist to serve as a seed for crystal nucleation, and the deposited silicon will form a fine-grained aggregate of polycrystalline silicon (*poly*). Modern integrated circuits make extensive use of both monocrystalline and polycrystalline deposited silicon films.

A variety of etching processes have been developed for both monocrystalline and polycrystalline silicon. Of particular interest are the highly anisotropic trench etches that have recently become the technology of choice for isolating high-density integrated circuitry.

2.5.1. Epitaxy

The growth of a single-crystal semiconductor film upon a suitable crystalline substrate is known as *epitaxy*. The substrate normally consists of a crystal of the same material as the semiconductor that is to be deposited, but this need not always be the case. High-quality monocrystalline silicon films have been grown on wafers of synthetic sapphire or spinel, as these materials possess a crystal structure that is enough like silicon to allow crystal nucleation. The cost of synthetic sapphire or spinel wafers so greatly exceeds the cost of similar-sized silicon wafers that the vast majority of epitaxial depositions consist of silicon films grown on silicon substrates.

There are several different methods of growing epitaxial (*epi*) layers. One relatively crude method consists of pouring molten semiconductor material on top of the substrate, allowing it to crystallize for a short period of time, and then wiping the excess liquid off. The wafer surface can then be reground and polished to form an epitaxial layer. Obvious drawbacks to this *liquid-phase epitaxy* include the high cost of regrinding the wafer and the difficulty of producing a precisely controlled epi thickness.

Most modern epitaxial depositions use *low-pressure chemical vapor deposited* (LPCVD) epitaxy. Figure 2.22 shows a simplified diagram of an early type of LPCVD epi reactor. The wafers are mounted on an inductively heated carrier

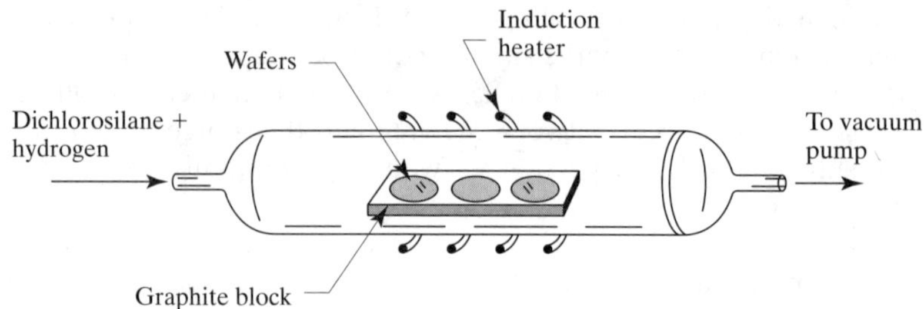

FIGURE 2.22 Simplified diagram of an epi reactor.[18]

block, and a mixture of dichlorosilane and hydrogen is passed over them. These gases react at the surface of the wafers to form a slow-growing layer of monocrystalline silicon. The rate of growth can be controlled by adjusting the temperature, pressure, and gas mixture used in the reactor. No polishing is required to render the epitaxial surface suitable for further processing, as vapor-phase epitaxy faithfully reproduces the topography of the underlying surface. The epitaxial film can also be doped *in situ* by adding small amounts of gaseous dopants such as phosphine or diborane to the gas stream.

There are several benefits of growing an epitaxial layer on the starting wafer. For one, the epi layer need not have the same doping polarity as the underlying wafer. For example, an N– epitaxial layer can be grown on a P– substrate—an arrangement commonly employed for standard bipolar processes. Another advantage of epitaxial silicon is that it is not contaminated by oxygen or carbon, as is Czochralski silicon. Multiple epitaxial layers can be grown in succession, and the resulting stack can be used to form transistors or other devices. The potential of epitaxy is limited chiefly by the slow rate of epi growth and by the expense and complexity of the required equipment, which are much greater than Figure 2.22 suggests.

Epitaxy also allows the formation of *buried layers*. An N+ buried layer constitutes a key step in most bipolar processes since it allows the construction of vertical NPN transistors with low collector resistances. Figure 2.23 depicts the growth of such an N-buried layer (NBL). Arsenic and antimony are the preferred dopants for forming an NBL because their slow diffusion rates minimize the outdiffusion of the buried layer during subsequent high-temperature processing. Antimony is

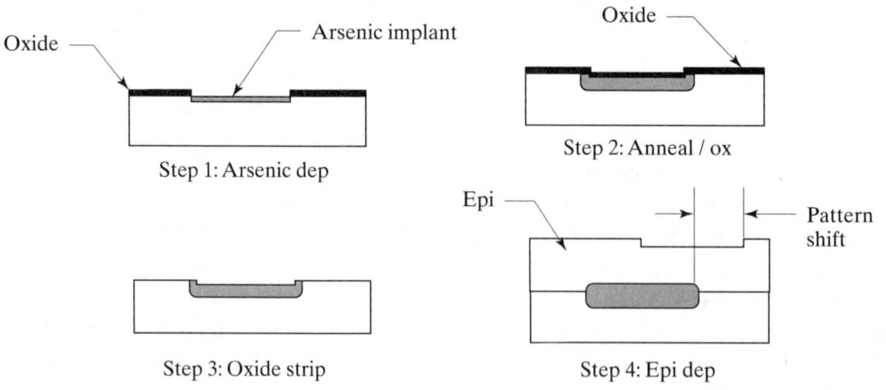

FIGURE 2.23 Formation of an N-buried layer (NBL), showing pattern shift.

[18] The horizontal tube reactor shown here has long been obsolete; for examples of more modern epi reactors, see C. W. Pearce, "Epitaxy," in S. M. Sze, ed., *VLSI Technology,* 2d ed. (New York: McGraw-Hill, 1988), pp. 61–65.

often chosen instead of arsenic because it exhibits less tendency to spread laterally during epitaxy (an effect called *lateral autodoping*).[19] Arsenic, on the other hand, has a higher solid solubility and can therefore produce more heavily doped buried layers. Buried layer fabrication begins with a lightly doped P-type wafer. This wafer is oxidized, and windows are patterned in the resulting oxide layer. Either arsenic or antimony is implanted through the windows, and the wafer is briefly annealed to eliminate the resulting implant damage. Thermal oxidation occurs during this anneal, and discontinuities form around the edges of the oxide windows. Next, all oxide is stripped from the wafer, and an N– epitaxial layer is deposited. The resulting structure consists of patterned N+ regions buried under an epitaxial layer.

As mentioned previously, oxidization during the anneal of the NBL causes slight silicon surface discontinuities to form around the edges of the oxide window. The epitaxial layer faithfully reproduces these discontinuities in the final surface of the wafer. Under a microscope, the resulting step forms a faintly visible outline called the *NBL shadow*. Subsequent photomasks are aligned to this discontinuity. An alternative alignment method uses infrared light to image the NBL doping through the overlying silicon, but this requires more complicated equipment.

The accretion of silicon atoms at the edge of the NBL shadow during epitaxy displaces it laterally, an effect called *pattern shift* (Figure 2.23).[20] The magnitude of the shift depends on many factors, including temperature, pressure, gas composition, substrate orientation, and tilt (See Section 7.2.4). When other layers are aligned to the NBL shadow, these must be offset to compensate for pattern shift.

2.5.2. Polysilicon Deposition

If silicon is deposited on an amorphous material, then no underlying lattice exists to align crystal growth. The resulting silicon film consists of an aggregate of small intergrown crystals. This *poly* film has a granular structure with a grain size dependent upon deposition conditions and subsequent heat treatment. The grain boundaries of the poly represent lattice defects, which can provide sneak paths for leakage currents. Therefore, PN junctions are not normally fabricated from poly. Polysilicon is often used to construct the gate electrodes of self-aligned MOS transistors because, unlike aluminum, it can withstand the high temperatures required to anneal the source/drain implants. In addition, the use of poly has led to better control of MOS threshold voltages due to the ability of phosphorus-doped polysilicon to immobilize ionic contaminants (Section 4.2.2). Suitably doped poly can be used to fabricate very narrow resistors that exhibit fewer parasitics than diffused devices. Heavily doped polysilicon can also be used as an additional metallization layer for signals that can tolerate the insertion of considerable resistance in the signal path.

A patterned poly layer is produced by first depositing polysilicon across the wafer with an apparatus similar to that employed for epitaxy (see Figure 2.22). The wafer is then coated with photoresist, patterned, and etched to selectively remove the polysilicon. Modern processes usually employ dry etching rather than wet etching because of the importance of precisely controlled gate dimensions.

---

[19] M. W. M. Graef, B. J. H. Leunissen, and H. H. C. de Moor, "Antimony, Arsenic, Phosphorus, and Boron Autodoping in Silicon Epitaxy," *J. Electrochem. Soc.*, Vol. 132, #8, 1985, pp. 1942–1954.

[20] M. R. Boydston, G. A. Gruber, and D. C. Gupta, "Effects of Processing Parameters on Shallow Surface Depressions During Silicon Epitaxial Deposition," in *Silicon Processing*, American Society for Testing and Materials STP 804, 1983, pp. 174–189.

2.5.3. Dielectric Isolation

The majority of integrated circuits use reverse-biased PN junctions to separate different electrical regions. Such integrated circuits are said to employ *junction isolation* (JI). This form of isolation is simple and inexpensive to fabricate, but it also has its limitations. The reverse-biased junctions that are supposed to isolate the various regions can be compromised by improper electrical biasing. Even if properly biased, junction isolation can still be defeated by hot carriers or ionizing radiation.

During the late 1960s and early 1970s, the military need for radiation-hardened integrated circuits led to the development of a new class of isolation systems, collectively known as *dielectric isolation* (DI). The first successful DI process was *silicon-on-sapphire* (SOS), which employs a sapphire substrate whose crystal structure is very similar to silicon. A layer of monocrystalline silicon is epitaxially deposited upon the sapphire. Various diffusions are made into this epitaxial layer with conventional processing techniques. Next, the individual active devices are isolated from one another by etching away the epi that separates them. This ensures that the devices are separated, not by reverse-biased junctions, but rather by an insulating dielectric—in this case, sapphire. Historically, the SOS process was never commercially viable because the sapphire substrates were too small and too expensive. Ways to use other insulators, such as silicon dioxide, were eventually developed. This led to the broadening of the term silicon-on-sapphire to *silicon-on-insulator* (SOI), a term that is essentially synonymous with dielectric isolation.

The immediate successor to silicon-on-sapphire was the silicon-DI process. Figure 2.24 illustrates the steps required to form a silicon-DI wafer. First, a lightly doped N-type wafer is etched to form cavities wherever isolation will eventually be required. Next, this wafer is oxidized, and a very thick layer of polysilicon is deposited on top of the oxide. The wafer is then turned over, so that the poly becomes the substrate. The monocrystalline N-type wafer is ground down until the poly protrudes through the surface. The remaining regions of monocrystalline silicon are now isolated from the poly substrate by the oxide layer (called a *buried oxide*, or BOX).[21] This process achieves dielectric isolation without the need for a sapphire substrate, but in its place is substituted a very challenging lap-and-polish process. If the wafer is not perfectly planar, or if the lap and polish is not perfectly uniform,

FIGURE 2.24 Steps in the fabrication of a silicon DI wafer.

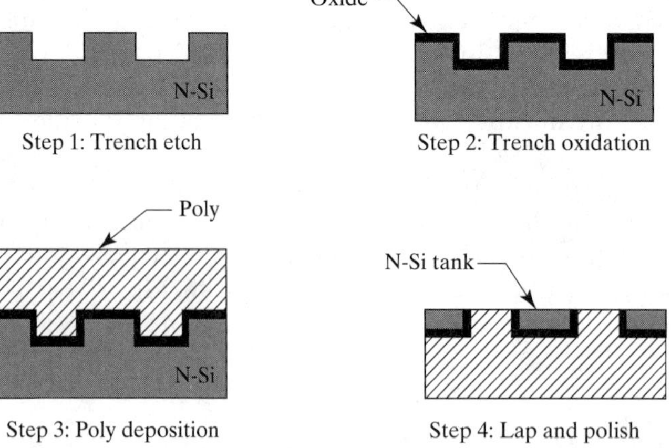

[21] D. J. Hamilton and W. G. Howard, *Basic Integrated Circuit Engineering* (New York: McGraw-Hill, 1975), pp. 88–89.

then some of the monocrystalline silicon regions will be exposed before others. This can lead to large manufacturing losses. Consequently, the silicon DI process is not economically competitive with junction isolation.

The subsequent history of DI technologies largely consists of finding cheaper ways to fabricate buried oxides. One clever solution uses high-energy ion implantation to form a layer of oxide beneath the surface of the silicon. The resulting SIMOX (separation by implanted oxygen) process produces a planar layer of oxide at a distance of 2–5 μm beneath the surface of the silicon. To complete the isolation system, insulating sidewalls must be formed around the active devices. The best means to this end is a process called *shallow trench isolation* (STI), which was independently developed to minimize isolation spacings in high-density digital processes. Shallow trench isolation (Figure 2.25) uses a highly anisotropic reactive ion etch to cut a nearly vertical groove in the surface of the silicon. The sidewalls of this groove are oxidized, and then polysilicon is deposited to fill up the remainder of the groove. The polysilicon is ground back to the surface of the surrounding wafer to ensure planarity.[22] Because only a small amount of excess polysilicon must be removed, the *chemical-mechanical polish* (CMP) process described in Section 2.6.4 can be used to accurately and economically perform this step. The polysilicon that fills the trenches has nearly the same coefficient of thermal expansion as silicon, so heating and cooling the die will not induce excessive mechanical stress. The highly planar surface that remains after STI processing is ideal for deposition of an advanced fine-line metallization system. The major problem with the SIMOX process is that it can only fabricate relatively thin layers of surface silicon. This in turn limits the allowable operating voltages of the resulting integrated circuits. If necessary, the surface layer of silicon can be thickened by epitaxy before the trench isolation is added. This also potentially allows the addition of a buried layer to the process.

Recently, a new generation of DI processes have been created based upon wafer bonding technology. *Wafer bonding* involves the fusion of two wafers to one another. One of the wafers is covered with a thin layer of oxide, while the other is not. The two wafers are stacked so that the layer of oxide lies between them, and then they

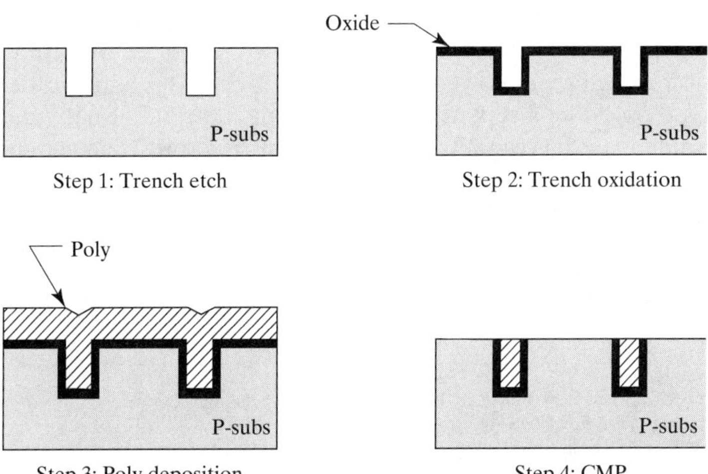

FIGURE 2.25 Steps in the fabrication of shallow trench isolation (STI).

[22] P. VanDerVoorn, D. Gan, and J. P. Krusius, "CMOS Shallow-Trench-Isolation to 50-nm Channel Widths," *IEEE Trans. Electron Devices*, Vol. 47, #6, 2000, pp. 1175–1182.

are heated until they fuse together. The temperatures required to induce fusion are far lower than those required to soften either silicon or oxide. Curiously, a small amount of moisture must be adsorbed on the surface of the oxide in order for wafer bonding to work. This probably means that wafer bonding is at heart a chemical process, although the exact details of its mechanism are not fully understood.[23] The wafer bonding process produces a thin layer of buried oxide sandwiched between two much thicker layers of silicon. One of these layers of silicon must be thinned so that trench isolation can be driven down to reach the BOX. Several methods of thinning have been proposed, including a lap-and-polish technique similar to that employed for silicon DI, a chemical etch-back technique that terminates on a heavily doped P-type buried layer, or, most ingeniously, a process called wafer cleaving.[24] *Wafer cleaving* involves the fabrication of a highly strained layer some distance beneath the surface of one of the wafers. This can be achieved by implanting hydrogen, or better, germanium, just prior to wafer bonding. Once the wafers have been bonded together, a thermal shock is used to induce the silicon to cleave along the strain layer, leaving a thin layer of monocrystalline silicon bonded on top of the buried oxide. Like SIMOX, wafer cleavage can only fabricate relatively thin layers of isolated silicon. However, it does open the possibility of stacking multiple layers of active silicon one above another to produce a true three-dimensional integrated circuit.[25]

Dielectric isolation is currently enjoying something of a renaissance due to the growing demand for high-voltage integrated circuits. SIMOX and wafer bonding appear to be the processes of choice for fabricating the buried oxide. Various elaborations of shallow trench isolation (some of which fabricate very deep, very narrow trenches) are favored for providing sidewall isolation. Given the prevalence of STI in advanced digital processes, there is some hope that DI processes can now compete on an even basis with the more established JI processes. Whether or not this will prove to be the case still remains to be seen.

2.6 METALLIZATION

The active components of an integrated circuit consist of diffusions, ion implantations, and epitaxial layers grown in or on a silicon substrate. When this processing is complete, the resulting components are connected to form the integrated circuit, using one or more layers of patterned wiring. This wiring consists of layers of metal and polysilicon separated by insulating material, usually deposited oxide. These same materials can also be used to construct passive components such as resistors and capacitors.

Figure 2.26 illustrates the formation of a typical *single-level-metal* (SLM) interconnection system. After the final implantations and diffusions, a layer of oxide is grown or deposited over the entire wafer, and selected areas are patterned and etched to create oxide windows exposing the silicon. These windows will form

---

[23] U. M. Gösele, M. Reiche and Q.–Y. Tong, "Wafer Bonding: An Overview," *4th Int. Conf. on Solid-State and Integrated Circuit Technology*, 1995, pp. 243–247.

[24] Hydrogen cleaving: B. Aspar, C. Lagahe, H. Moriceau, A. Soubie, E. Jalaguier, B. Biasse, A. Papon, A. Chabli, A. Claverie, J. Grisolia, G. Benassayag, T. Barge, F. Letertre, and B. Ghyselen, "Smart-Cut® Process: An Original Way to Obtain Thin Films by Ion Implantation," *Conf. on Ion Implantation Technology*, 2000, pp. 255–260. Germanium cleaving: M. I. Current, S. N. Farrens, M. Fuerfanger, S. Kang, H. R. Kirk, I. J. Malik, L. Feng and F. J. Henley, "Atomic-layer Cleaving with Si_XGe_Y Strain Layers for Fabrication of Si and Ge-Rich SOI Device Layers," *IEEE International SOI Conference*, 2001, pp. 11–12.

[25] L. Xue, C. C. Liu, H.-S. Kim, S. Kim and S. Tiwari, "Three-Dimensional Integration: Technology, Use, and Issues for Mixed-Signal Applications," *IEEE Trans. Electron Devices*, Vol. 50, #3, 2003, pp. 601–609.

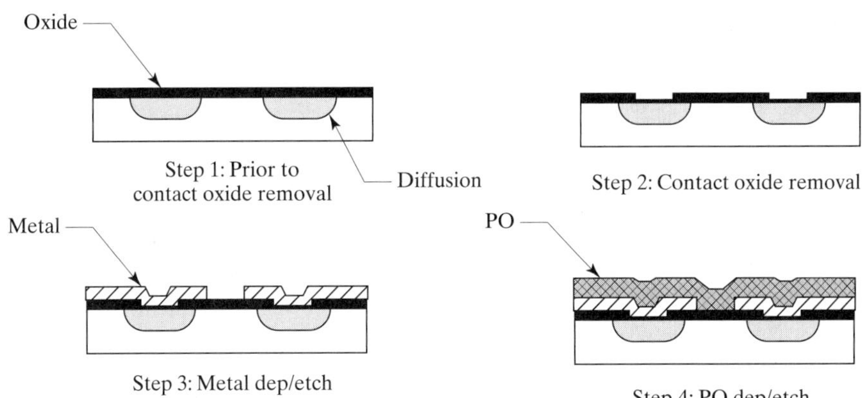

FIGURE 2.26 Formation of a single-level metal system.

contacts between the metallization and the underlying silicon. Once these contacts have been opened, a thin metallic film can be deposited and etched to form the interconnection pattern.

Exposed aluminum wiring is vulnerable to mechanical damage and chemical corrosion. An oxide or nitride film deposited over the completed wafer serves as a *protective overcoat* (PO). This layer acts as a conformal seal similar in principle to the plastic conformal coatings sometimes applied to printed circuit boards. Windows etched through the overcoat expose selected areas of the aluminum metallization so that bondwires can be attached to the integrated circuit.

The process illustrated in Figure 2.26 fabricates only a single aluminum layer. Additional layers of metallization can be sequentially deposited and patterned to form a multilevel metal system. Multiple metal layers increase the cost of the integrated circuit, but they allow denser packing of components and therefore reduce the overall die size. The savings in die area often compensate for the cost of the extra processing steps. Multiple metal layers also simplify interconnection and reduce layout time. Most modern analog processes incorporate three or four layers of metallization, while advanced digital processes may employ six or more layers of metal.

CMOS processes frequently employ low-resistivity polysilicon to form the gate electrodes of self-aligned MOS transistors. This material can serve as a free additional layer of interconnect. Even the lowest-sheet poly still has many times the resistance of aluminum, so the designer must take care to avoid routing high-current or high-speed signals in poly. Advanced processes may add a second and even a third layer of polysilicon. These additional layers are used to fabricate different types of MOS transistors, to form the plates of capacitors, and to construct polysilicon resistors. Each of these additional poly layers can be pressed into service as another layer of interconnect.

2.6.1. Deposition and Removal of Aluminum

Most metallization systems employ aluminum or aluminum alloys to form the primary interconnection layers. Aluminum conducts electricity almost as well as copper or silver, and it can be readily deposited in thin films that adhere to all of the materials used in semiconductor fabrication. A brief period of heating will cause the aluminum to alloy into the silicon to form low-resistance contacts.

Aluminum is usually deposited by *evaporation* by using an apparatus similar to the one shown in Figure 2.27. The wafers are mounted in a frame that holds their exposed surfaces toward a crucible containing a small amount of aluminum. When the

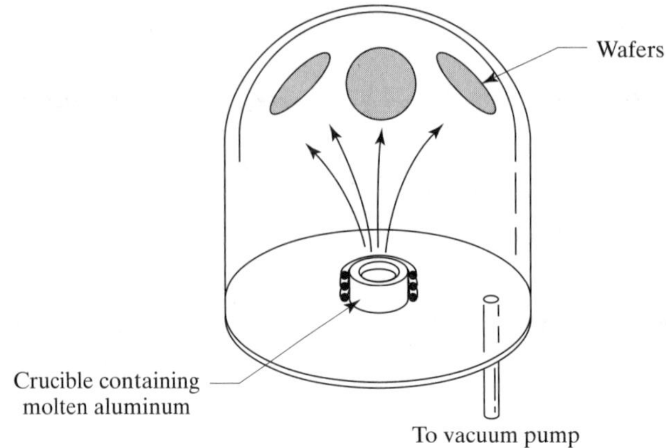

FIGURE 2.27 Simplified diagram of an aluminum evaporation apparatus.

crucible is heated, some of the aluminum evaporates and deposits on the wafer surfaces. A high vacuum must be maintained throughout the evaporation system to prevent oxidation of the aluminum vapor prior to its deposition upon the wafers. The illustrated evaporation system can only handle pure aluminum, but somewhat more complex systems can also evaporate selected aluminum alloys.

Aluminum and silicon alloy at moderate temperatures. A brief period of heating will form an extremely thin layer of aluminum-doped silicon beneath the contact openings. This process, called *sintering,* can achieve Ohmic contact to P-type silicon because aluminum acts as an acceptor. The aluminum-silicon alloy forms a shallow, heavily doped P-type diffusion that bridges between the metal and the P-type silicon. Less obviously, Ohmic contact also occurs when aluminum touches heavily doped N-type silicon. Junctions form beneath these contacts, but their depletion regions are so thin that carriers can surmount them by quantum tunneling. Rectification will occur if the donor concentration falls too low, so Ohmic contact cannot be established directly between aluminum and lightly doped N-type silicon. The addition of a shallow N+ diffusion will enable Ohmic contact to these regions.

Sintering causes a small amount of aluminum to dissolve in the underlying silicon. Some silicon simultaneously dissolves in the aluminum metal, eroding the silicon surface. Some diffusions are so thin that erosion can punch entirely through them, causing a failure mechanism called *contact spiking.* Historically, this was first observed in conjunction with the emitter diffusion of NPN transistors, so it is also called *emitter punchthrough*.[26] Contact spiking can be minimized by replacing pure aluminum metallization with an aluminum-silicon alloy. If the deposited aluminum is already saturated with silicon, then—at least in theory—it cannot dissolve any more. In practice, the silicon content of the alloy tends to separate during sintering to leave an unsaturated aluminum matrix. Careful control of sinter time and temperature will minimize this effect. Even so, the segregation of silicon forms small nodules that give the metal a rough, pebbly appearance under high magnification.

Another failure mechanism was encountered in high-density digital logic beginning in the early 1970s. As the dimensions of the integrated circuits were progressively reduced, the current density flowing through the metallization increased. Some devices eventually exhibited open-circuit metallization failures after many thousands of hours of operation at elevated temperatures. These failures were

---

[26] M.D. Giles, "Ion Implantation," in S.M. Sze, ed., *VLSI Technology,* 2[d] ed, (New York: McGraw-Hill, 1988), pp. 367–369.

eventually traced to a mechanism called *electromigration*[27] (Section 4.1.2). The addition of a fraction of a percent of copper to the aluminum alloy improves electromigration resistance by an order of magnitude. Most modern metal systems therefore employ either aluminum-copper-silicon or aluminum-copper alloys.

2.6.2. Refractory Barrier Metal

The feature sizes of integrated circuits have steadily shrunk, as ever-increasing numbers of components have been packed into approximately the same amount of silicon real estate. In order to obtain the necessary packing density, the sidewalls of contact and via openings have become increasingly steep. Evaporated aluminum does not deposit isotropically; the metal thins where it crosses oxide steps (Figure 2.28A). Any reduction in the cross-sectional area of a lead raises the current density and accelerates electromigration. A variety of techniques have been developed to improve step coverage on very steep sidewalls like those formed by reactive ion etching of thick oxide films.

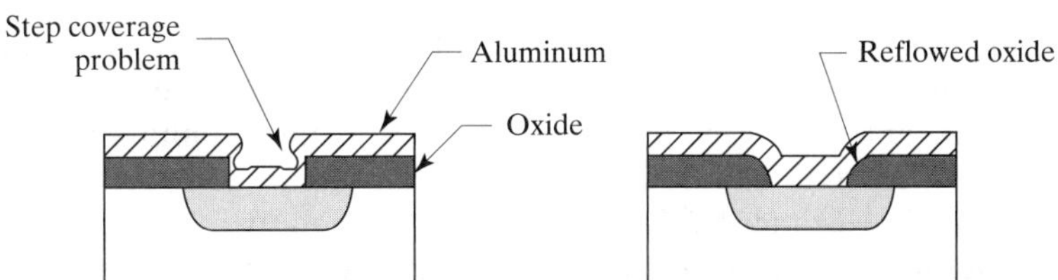

FIGURE 2.28 Step coverage of evaporated aluminum without reflow (A) and with reflow (B).

The step coverage of evaporated aluminum can be greatly increased by moderating the angle of the sidewalls. This can be achieved by heating the wafer until the oxide melts and slumps to form a sloped surface. This process is called *reflow* (Figure 2.28B). Pure oxide melts at too high a temperature to allow reflow, so phosphorus and boron are added to the oxide to reduce its melting point. The resulting doped oxide film is called either a *phosphosilicate glass* (PSG) or a *borophosphosilicate glass* (BPSG), depending on the choice of additives.

Reflow cannot be performed after aluminum has been deposited, because it cannot tolerate the temperatures required to soften PSG or BPSG. Therefore, while reflow can help improve the step coverage of first-level metal, it must be supplemented by other techniques in order to successfully fabricate a multilevel metal system. One option consists of using metals that deposit isotropically upon steeply sloped sidewalls, such as molybdenum, tungsten, or titanium. These *refractory barrier metals* possess extremely high melting points and are thus unsuited for evaporative deposition. A low-temperature process called *sputtering* can successfully deposit them. Figure 2.29 shows a simplified diagram of a sputtering apparatus. The wafers rest on a platform inside a chamber filled with low-pressure argon gas. Facing them is a plate of refractory barrier metal forming one of a pair of high-voltage electrodes. Argon atoms bombard the refractory metal plate. This knocks atoms loose that then deposit on the wafers to form a thin metallic film.

The sputtered refractory barrier metal film not only provides superior step coverage, but also virtually eliminates emitter punchthrough.[28] If step coverage were the only criterion for choosing a metal system, then aluminum could be entirely replaced

---

[27] J.R. Black, "Physics of Electromigration," *Proc. 12th Reliability Phys. Symp.*, 1974, p. 142.

[28] T. Hara, N. Ohtsuka, K. Sakiyama, and S. Saito, "Barrier Effect of W-Ti Interlayers in Al Ohmic Contact Systems," *IEEE Trans. Electron Devices*, Vol. ED-34, #3, 1987, pp. 593–597.

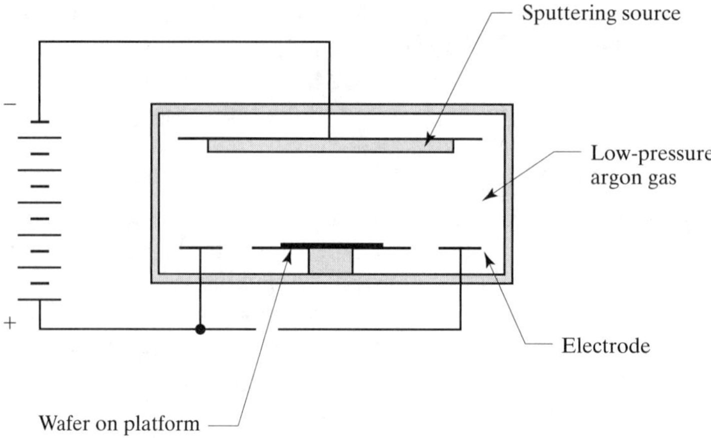

FIGURE 2.29 Simplified diagram of a sputtering apparatus.

by refractory barrier metal. Unfortunately, refractory metals have relatively high resistivities and cannot be deposited in thick films as easily as aluminum can. Most metal systems therefore employ a sandwich of both materials. A thin layer of refractory metal beneath the aluminum ensures adequate step coverage in the contacts where the aluminum metal drastically thins. Elsewhere, the aluminum reduces the electrical resistance of the metal leads. The relatively short sections of refractory barrier metal in the contacts do not contribute much resistance to the overall interconnection system. Refractory barrier metals are extremely resistant to electromigration, so the thinning of aluminum on the sidewalls of contacts and vias does not represent an electromigration risk.

As mentioned, refractory barrier metal virtually eliminates emitter punchthrough. The degree of alloying between silicon and refractory metals is negligible, and the aluminum cannot penetrate the barrier metal to contact the silicon. Most refractory barrier metal systems therefore employ aluminum-copper alloys rather than aluminum-copper-silicon, because aluminum-silicon alloying cannot occur.

Advanced metal systems require smaller metal linewidths, and these in turn require smaller vias. Smaller vias require steeper sidewalls, and these reduce aluminum step coverage. This problem becomes acute for vias less than about a micron across. A variety of alternative via fabrication techniques have been proposed. The most popular of these techniques involves the formation of chemical-vapor-deposited tungsten plugs in the vias. CVD tungsten is notable for its ability to deposit in narrow holes without voiding. Under the right conditions, the CVD tungsten can actually fill the via holes. Figure 2.30 shows the process used to fabricate CVD tungsten plug vias. First, a layer of refractory barrier metal is deposited to promote adhesion of the tungsten to the oxide and to protect the underlying aluminum. Next, tungsten is deposited over the refractory barrier metal. The tungsten completely fills the vias and then forms a more or less planar surface. This surface is then etched back to the oxide, leaving only the via holes filled with tungsten. A second layer of metallization can now be deposited on top of the via plugs.[29] This technique has several advantages. First, the thick tungsten plug ensures a low-resistance connection through the via. Second, since the metal that crosses the via is not thinned by the presence of a surface discontinuity, its current-carrying capability is not impaired by the presence

---

[29] P.E. Riley, T.E. Clark, E.F. Gleason, and M.M. Garver, "Implementation of Tungsten Metallization in Multilevel Interconnection Technologies," *IEEE Trans. Semiconductor Manufacturing*, Vol. 3, #4, 1990, pp. 150–157.

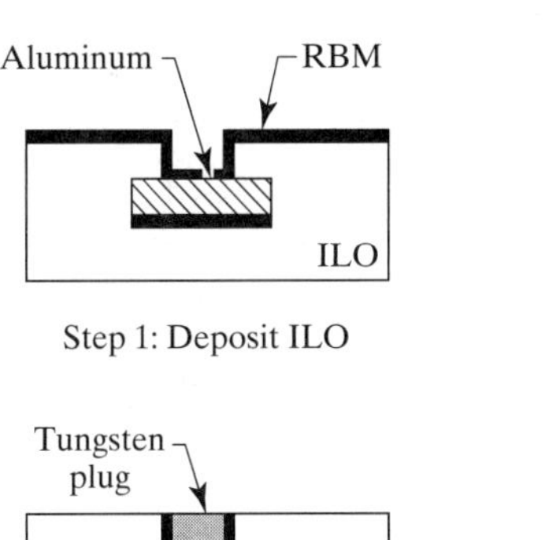

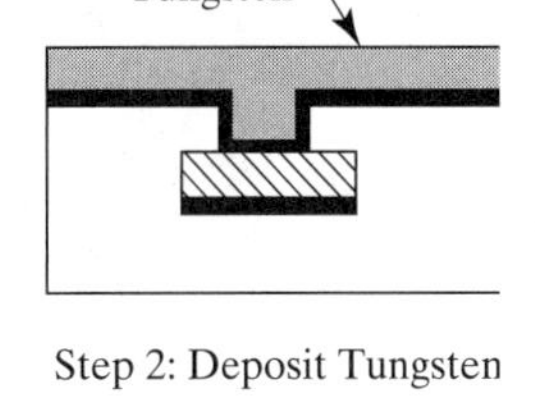

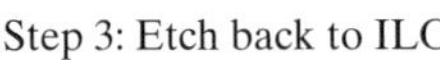

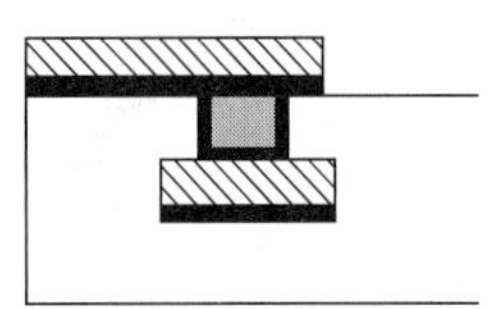

FIGURE 2.30 Steps in the formation of a tungsten-plug via system.

of vias. Third, the planar surface of the metal over the via allows the designer to stack vias one on top of another. Contacts can also be fabricated using tunsten plug technology.

2.6.3. Silicidation

Another modification of the standard metallization flow involves the addition of a silicide. Elemental silicon reacts with many metals, including platinum, palladium, titanium, cobalt, and nickel, to form compounds of definite composition. These *silicides* can form both low-resistance Ohmic contacts and, in the case of certain silicides, stable rectifying Schottky barriers. Thus, silicidation not only improves contact resistance—which can be a problem with barrier metal systems—but also allows the formation of Schottky diodes at no extra cost. Silicides have much lower resistivities than even the most heavily doped silicon, so they can also be used to reduce the resistance of selected silicon regions. Many MOS processes employ silicided poly (sometimes called *clad poly*) to form the gates of high-speed MOS transistors. Some of these processes also clad the source/drain regions of the transistors to reduce their resistance. Since most silicides are relatively refractory, their deposition does not preclude subsequent high-temperature processing. Silicided gates can thus be used to form self-aligned source/drain regions.

Figure 2.31 shows the steps required to deposit a platinum silicide layer in selected regions of the wafer. Immediately after the contacts are opened, a thin film of platinum metal is deposited across the entire wafer. The wafer is then heated to cause the portions of the platinum film in contact with silicon to react to form platinum silicide. The unreacted platinum can be removed using a mixture of acids called aqua regia. This procedure silicides both contact openings and any exposed polysilicon. If desired, an additional masking step can select which regions should receive silicide. Processes employing clad poly must incorporate a silicide block mask to fabricate polysilicon resistors. If this were not done, silicidation would turn all of the poly into a low-resistance material.

The silicidation reaction can occur only where silicon directly contacts the deposited metal, so the resulting silicide self-aligns to oxide openings or to the edges of polysilicon regions. Some authors refer to such a silicide film as a self-aligned silicide, or *salicide*.

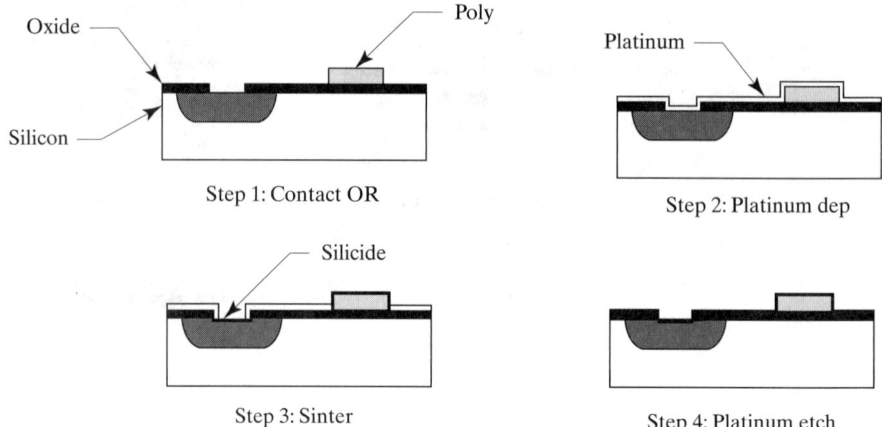

FIGURE 2.31 Silicidation process, showing both silicided contacts and silicided poly.

A typical silicided first-level metal system consists of a lowermost layer of platinum silicide, an intermediate layer of refractory barrier metal,[30] and a topmost layer of copper-doped aluminum. The resulting sandwich exhibits low electrical resistance, high electromigration immunity, stable contact resistance, and precisely controlled alloying depths. The three layers required to obtain all of these benefits are more costly than a simple aluminum alloy metallization, but the performance benefits are substantial.

A variety of silicides have been employed in metallization systems. Certain noble silicides, most notably those of platinum and palladium, are suitable for forming Schottky diodes. Unfortunately, these noble silicides decompose at relatively low temperatures, limiting the process options available after silicidation. Various refractory silicides can withstand much higher temperatures; but the Schottky barriers they produce exhibit very low forward voltages, and they are therefore useless as circuit components. Of the refractory silicides, titanium silicide has been favored because titanium reduces silicon dioxide and thereby ensures low-resistance contact to silicon even in the presence of thin layers of native oxide. Researchers have found that the resistivity of titanium silicide dramatically rises as the width of a silicided poly lead drops below about 1–2 μm. This effect has been attributed to a phase change that occurs during the annealing of the silicide. This phase change is accompanied by an increase in silicide grain size, and it is suppressed if the dimensions of the lead are inadequate to accomodate the larger grains. This results in the persistence of the higher resistivity phase. Certain other refractory silicides, notably those of nickel and cobalt, do not undergo annealing-induced phase changes and therefore do not exhibit linewidth-dependent resistivity increases.[31]

The resistance of silicided poly, while still much higher than that of metal, is sufficiently low to make it an attractive interconnection material. It cannot be used indiscriminately, but even limited amounts of poly interconnect can significantly decrease the size of digital logic cells. The use of silicided poly routing is especially valuable if only one or two layers of metal are available.

---

[30] The addition of a refractory barrier metal prevents the platinum silicide from reacting with the aluminum. This is not required for most refractory silicides; see Sze, pg. 409.

[31] T. Ohguro, S–I. Nakamura, M. Koike, T. Morimoto, A. Nishiyama, Y. Ushiku, T. Yoshitomi, M. Ono, M. Saito, and H. Iwai, "Analysis of Resistance Behavior in Ti- and Ni-Salicided Polysilicon Films," *IEEE. Trans. on Electron Devices*, Vol. 41, #12, 1994, pp. 2305–2317.

2.6.4. Interlevel Oxide, Interlevel Nitride, and Protective Overcoat

Figure 2.32 shows a cross section of a typical modern metallization system. The first layer of material above the silicon consists of thermally grown oxide. Upon this oxide lies a patterned polysilicon layer that will eventually form the gates of MOS transistors. On top of this poly lies a thin deposited oxide layer called a *multilevel oxide* (MLO) that serves to insulate the poly and to thicken the thermal oxide layer. Contact openings are etched through the MLO and thermal oxide to contact the silicon, and through the MLO to contact the poly. Following reflow, the contact openings are silicided to reduce contact resistance. Above the MLO lies the first layer of metal, consisting of a thin film of refractory barrier metal and a much thicker layer of copper-doped aluminum. Above the first metal layer lies another deposited oxide layer called an *interlevel oxide* (ILO), which insulates the first metal from the overlying second metal. Vias are etched through the ILO. On top of this lies the second layer of metal, again consisting of refractory barrier metal and copper-doped aluminum. The topmost and final layer consists of a compressive nitride film, which serves as a *protective overcoat* (PO). This metallization system has a total of six layers (one poly, two metals, MLO, ILO, and PO) and requires five masking steps (poly, contact, metal-1, via, metal-2, and PO). Some advanced processes may employ as many as three layers of polysilicon and seven layers of metal.

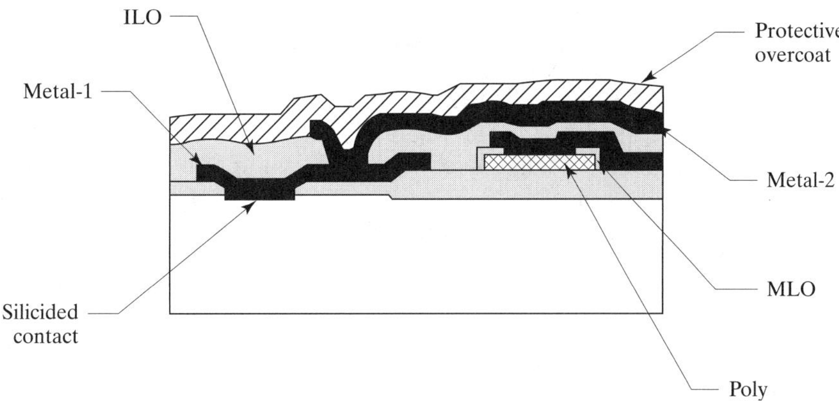

FIGURE 2.32 Cross section of a double-metal, single-poly metallization system.

Interlevel oxide layers are normally produced by low-temperature deposition—for example, by the reaction of silane and nitrous oxide or by the decomposition of tetraethoxysilane (TEOS). A relatively thick ILO helps minimize parasitic capacitances between layers of the conductor sandwich, but it can cause step coverage problems in via openings. As previously discussed, reflow is not possible once aluminum has been deposited, so a refractory barrier metal is often used to improve the step coverage of the second metal layer. Alternatively, tungsten-plug vias may be employed.

The deposition of patterned layers upon the wafer inevitably introduces variations in surface height. Each layer added to the conductor stack worsens the topography. Modern fine-line photolithography has a very narrow depth of field, so even tiny differences in surface height can render some features out of focus. The surface topography of the wafer can be minimized by special *planarization* techniques. The simplest of these involves the formation of an MLO or ILO layer through repeated applications of a spin-on glass (SOG). The SOG is deposited on the wafer in the form of a liquid film, the surface of which is pulled taut by surface tension. Recessed surface areas receive more than the average thickness of SOG, and elevated areas receive less. Each layer of SOG therefore reduces the nonplanarity to a certain degree. *Resist etch-back planarization* relies upon a similiar application of surface tension. A layer of resist is spun onto the wafer and baked. The resulting resist layer is thickest over recessed areas and thinnest over elevated areas. The resist is not patterned; instead, the wafers are simply loaded into a dry etch system adjusted to attack both resist and oxide. The surface of the resist gradually erodes away, progressively uncovering deeper and deeper layers of oxide. The highest areas of oxide are exposed longest to the etchant and therefore erode away the most.

Even the resist etch-back process does not produce the degree of planarity required for modern submicron processing. These processes instead use *chemical-mechanical polishing* (CMP). This technique uses an alkaline slurry of fine, soft abrasive particles. The alkaline solution attacks the oxide, softening it so that the abrasive particles can scrub it away. The CMP process selectively attacks the highest points on the oxide surface, while leaving recessed areas virtually untouched.[32] Although CMP is vastly superior to SOG and resist etchback, it is not perfect. Large areas devoid of underlying metal (or poly) produce low regions in the final surface. This effect, called *dishing*, becomes noticeable only over relatively large distances (ca. 1 mm). If this is objectionable, a pattern of dummy metal geometries (called *fill metal*) can be automatically added to the layout prior to mask generation.[33]

An excellent capacitor can be formed between two layers of metal or polysilicon. A thin insulating dielectric deposited between the plates completes the capacitor. The thinner this dielectric, the greater the resulting capacitance per unit area. One technique for forming a capacitor consists of depositing one polysilicon layer, oxidizing this to form a thin dielectric, and depositing a second polysilicon layer to complete the capacitor. Any region where the two poly layers overlap will form a capacitor consisting of two poly plates separated by the thin oxide dielectric. Oxide forms an ideal capacitor dielectric because it is a nearly perfect insulator, and very thin oxide films can be grown with little risk of pinholes or other defects. The capacitance achievable with oxide dielectrics is limited by the rupture voltage of the oxide; the thicker oxide layers required to withstand higher voltages have proportionately smaller capacitances per unit area.

One way to boost the capacitance per unit area for a given operating voltage consists of using a material with a higher dielectric constant. Silicon nitride, with a dielectric constant that is 2.3 times that of oxide, is a common choice for fabricating high capacitance-per-unit-area films. Unfortunately, nitride films are more prone to pinhole formation than are oxide films of equivalent thickness. Therefore, oxide and nitride films are sometimes combined to form a stacked dielectric with a dielectric constant between that of oxide and nitride. A typical oxide-nitride-oxide stacked dielectric can achieve a dielectric constant about twice that of oxide.

The protective overcoat consists of a thick deposited oxide or nitride film coating the entire integrated circuit. It insulates the uppermost metal layer from the outside world, so that (for example) a particle of conductive dust will not short two adjacent leads. The overcoat also helps to ruggedize the integrated circuit, a necessary precaution since the aluminum metallization is soft and deforms under pressure. The

---

[32] K.A. Perry, "Chemical Mechanical Polishing: The Impact of a New Technology on an Industry," *1998 Symp. on VLSI Technology Digest of Technical Papers*, 1998, pp. 2–5.

[33] B.E. Stine, D.S. Boning, J.E. Chung, L. Camilletti, F. Kruppa, E.R. Equi, W. Loh, S. Prasad, M. Muthukrishnan, D. Towery, M. Berman, and A. Kapoor, "The Physical and Electrical Effects of Metal-Fill Patterning Practices for Oxide Chemical-Mechanical Polishing Processes," *IEEE Tran. Electron Devices*, Vol. 45, #3, 1998, pp. 665–679.

protective overcoat also helps to block the entrance of contaminants. Both the aluminum metallization and the underlying silicon are vulnerable to certain types of contaminants that can penetrate the plastic encapsulation. A properly formulated protective overcoat can form a barrier to these contaminants. Heavily doped phosphosilicate glasses are sometimes used as protective overcoats, but most modern processes have switched to compressive nitride films, which offer superior mechanical hardness and chemical resistance.

2.6.5. Copper Metallization

Aluminum has been the metallization of choice for integrated circuits since the earliest days of the planar process. Aluminum has relatively little electrical resistance, it readily deposits in thin films, and it adheres well to oxide and silicon. However, by the late 1990s, the relentless advance of technology had stressed aluminum metallization to its limits. The resistance of aluminum, low as it is, becomes significant when leads are reduced to submicron dimensions or the operating currents are elevated to many amps. Electromigration also becomes a serious problem under these conditions. Copper offers significantly reduced resistance and much greater electromigration tolerance. Recently, metal systems that employ copper have begun to replace, or at least to supplement, conventional aluminum metallization.

The *dual damascene copper* process is specifically designed to fabricate narrow close-packed metal lines. This process uses conventional etching techniques to pattern an oxide layer, which then imposes its pattern upon the copper metallization. This approach has been chosen because the only available dry etching processes for copper require elevated temperatures that preclude the use of photoresist as a masking material.

The dual damascene process is best illustrated by the steps required to fabricate the second layer of metallization. First, an ILO is deposited across the wafer and is planarized by CMP. Next, a thin CVD nitride layer is deposited on top of the planarized ILO to act as an etch mask (Figure 2.33, Step 1). The nitride is then patterned using the via mask, and a second layer of ILO is deposited on top of the patterned nitride (Step 2). The wafer is now coated with photoresist and is patterned using the metal-2 mask. The exposed regions of oxide are removed using a selective dry etch. This etch terminates on exposed nitride as well as exposed metal-1 (Step 3). Next, a thin layer of electrically conductive titanium nitride is deposited, followed by a thin layer of copper that acts as the seed for subsequent electrolytic deposition of additional copper (Step 4). Finally, chemical-mechanical polishing is used to remove all metal above the ILO (Step 5). The resulting structure consists of two levels of copper, the upper level forming the second layer of metallization, and the lower level forming the underlying vias.[34] The term *dual damascene* refers to the simultaneous deposition of copper for leads and vias. This process is technically challenging, but appears to be commercially viable.

The *power copper* process was developed to reduce electrical resistance in high-current circuitry. Modern power integrated circuits require lead patterns that exhibit only a few milliohms of resistance. Conventional metallization is simply too thin to achieve such low resistances, regardless of how wide the leads are made, or how many layers of metallization are stacked one upon another. The power copper process begins with a fully processed wafer, including a protective overcoat. However,

---

[34] Y. Morand, M. Lerme, J. Palleau, J. Torres, F. Vinet, O. Demolliens, L. Ulmer, Y. Gobil, M. Fayolle, F. Romagna, and R. LeBihan, "Copper Integration in Self Aligned Dual Damascene Architecture," *1997 Symp. on VLSI Technology Digest of Technical Papers*, 1997, pp. 31–32.

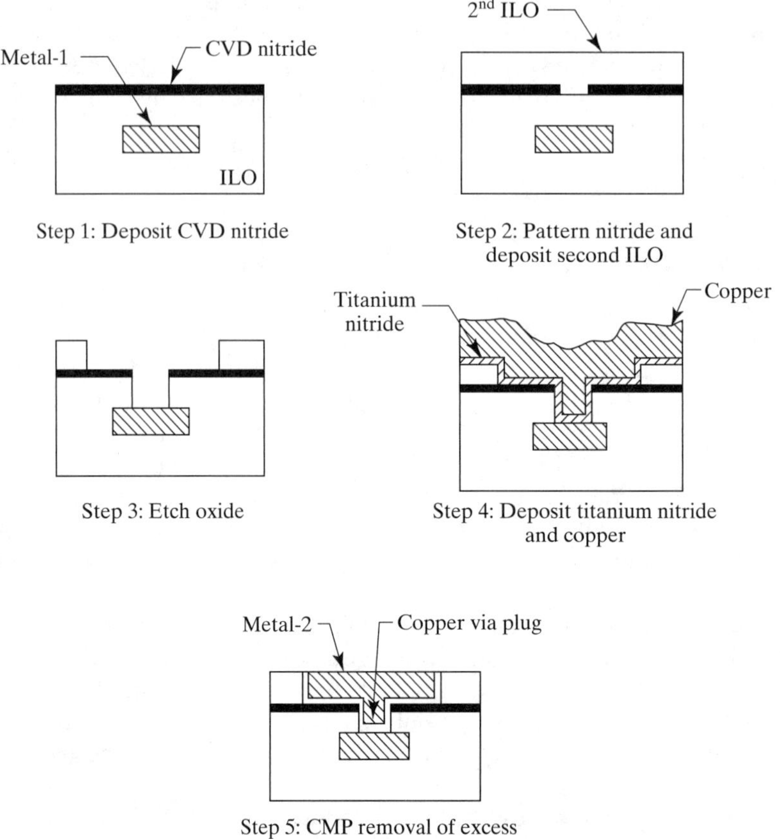

FIGURE 2.33 Steps involved in the formation of the second layer of a dual damascene copper metallization system.

the openings in this protective overcoat do not form bondpads, but rather vias to the power copper. A thin layer of copper is sputtered across the entire wafer to serve as a seed for the subsequent electrolytic deposition of some 25 μm of copper. A thin layer of nickel is plated on top of the copper, followed by a thin layer of palladium that serves to prevent oxidation and present a suitable surface for wire bonding. The wafer is then coated with photoresist and patterned. A simple wet etch forms the pattern of power copper leads (Figure 2.34). Both the lead width and the lead spacing are very coarse, but this matters little since the power copper only forms bondpads and massive high-current leads. The thick power copper cushions the underlying integrated circuit from the forces generated by wirebonding, so bondpads can be

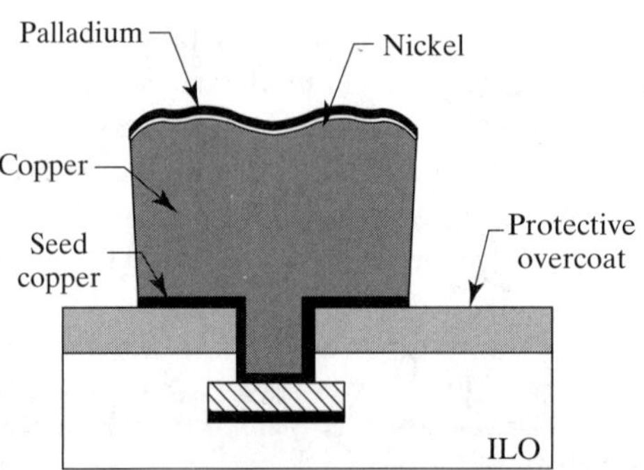

FIGURE 2.34 Cross section of a power copper metal lead and the underlying via through protective overcoat to aluminum metallization.

placed directly over active circuitry without fear of damage. This capability is known as *bond over active circuitry* (BOAC).[35] The power copper metallization is not itself protected by any overcoat, but the relatively large dimensions of the metal pattern give it a degree of immunity to corrosion.

2.7 ASSEMBLY

Wafer fabrication ends with the deposition of a protective overcoat, but there remain a number of manufacturing steps required to complete the integrated circuits. Since most of these steps require less-stringent cleanliness than wafer fabrication, they are usually performed in a separate facility called an *assembly/test site*.

Figure 2.35 shows a diagram of a typical finished wafer. Each of the small squares on the wafer represents a complete integrated circuit. This wafer contains approximately 300 integrated circuit dice arrayed in a rectangular pattern by the step-and-repeat process that created the stepped working plates. A few locations in the array are occupied by process control structures and test dice rather than actual integrated circuits.

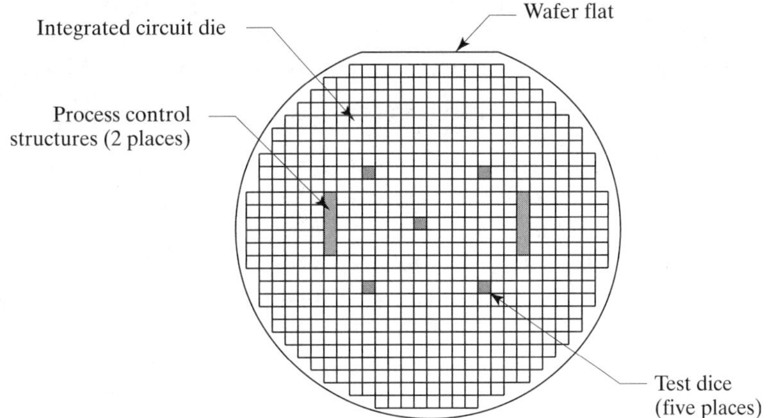

FIGURE 2.35 Pattern for a typical wafer created from a stepped working plate.

Process control structures consist of extensive arrays of transistors, resistors, capacitors, and diodes, as well as more specialized structures such as strings of contacts and vias. The wafer fab uses these structures to evaluate the success or failure of the manufacturing process. Automated testing equipment gathers data on every wafer, and any that fail to meet specifications are discarded. The data are also analyzed for statistical trends so that corrections can be implemented before the variances become large enough to cause yield losses. The process control structures are standardized, and the same ones are used for a wide range of products.

Test dice are used by design engineers to evaluate prototypes of an integrated circuit. Unlike process control structures, test dice are specific to a given product and in most cases are actually variations on the layout of the integrated circuit. A dedicated test metal mask allows probing of specific components and subcircuits that would be

---

35 T. Efland, D. Abbott, V. Arellano, M. Buschbom, W. Chang, C. Hoffart, L. Hutter, Q. Mai, I. Nishimura, S. Pendharkar, M. Pierce, C.C. Shen, C.M. Thee, H. Vanhorn, and C. Williams, "LeadFrameOnChip offers Integrated Power Bus and Bond over Active Circuit," *Proc. 2001 Int. Symp. Power Semiconductor Devices and ICs*, 2001, pp. 65–68.

difficult to access on the finished die. Sometimes a test contact or protective overcoat mask is also used, but in almost every case the test die shares the same diffusion masks as the integrated circuit. Test dice are normally created by adding a few more layers (e.g., test metal, test nitride) to the database containing the layout of the integrated circuit. These layers create a separate set of reticles that are used to expose a few selected spots on the stepped working plate. The wafer in Figure 2.35 contains only five test die locations. These locations become unnecessary when testing has been completed. Sometimes a new set of masks is created that replaces the test dice with product dice to gain an extra percent or two of yield. In other cases, the tiny increase in die yield cannot justify fabrication of new masks, so the test dice remain on production material.

Figure 2.35 depicts a wafer produced from a set of stepped working plates. Wafers created by *direct-step-on-wafer* (DSW)[36] processing rarely include any test dice because at least one test die must be included in every exposure. This would result in 20 or more test dice per wafer, which would consume a correspondingly large amount of area. If test dice are included in a DSW design, then the production mask set will almost certainly be modified to replace them with product dice to improve the die yield. The process control structures of a DSW wafer usually occupy the narrow strips between dice, called *scribe streets*, through which the saw blade passes to separate the individual dice. The process control structures are tested prior to saving, so their subsequent destruction is of no consequence.

As mentioned previously, all completed wafers are tested to determine whether the processing was performed correctly. If the wafers pass this test, then each die is individually tested to determine its functionality. The high-speed automated test equipment typically requires less than three seconds to test each die. The percentage of good dice depends on many factors, most notably the size of the die and the complexity of the process used to create it. Most products yield better than 80% and many yield in excess of 90%. High yields are obviously desirable because every discarded die represents lost profit. The equipment that tests the wafers also marks those that fail the test. Marking was formerly done by placing a drop of ink on each defective die, but modern systems eliminate the need for inking by remembering the location of the bad dice electronically.

Wafer-level testing, or *wafer probing*, requires contact to specific locations on the interconnection pattern of the integrated circuit. These locations are exposed through holes in the protective overcoat, allowing contact to be made with the help of an array of sharp metal needles, or *probes*. These probes are mounted on a board called a *probe card*. The automated test machine lowers the probe card until electrical continuity is established. The integrated circuit is tested, the card is lifted, and positioning servos move the wafer to align the next die underneath the probes.

Once the wafer has been completely tested, the individual dice are sawn apart using a diamond-tipped sawblade. Another automated system then selects the good dice from the scribed wafer for mounting and bonding. The rejected dice (including the remains of the process control structures and the test dice) are discarded.

2.7.1. Mount and Bond

Most manufacturers offer unmounted integrated circuit dice, but the sales of such *bare dice* are seldom large. Most customers simply do not have the equipment or

---

[36] The acronym *DSW* has also been used to stand for *direct slice write*, a process by which a scanned electron beam directly exposes the photoresist on a wafer. This process, more commonly called *direct write on wafer* (DWW), is strictly of academic interest because it is too slow to have any practical application in silicon processing. However, it is frequently used to fashion photomasks.

expertise needed to handle bare dice, let alone to package them. Packaging therefore falls in the province of integrated circuit manufacturing.

The first step in packaging an integrated circuit is mounting it on a *leadframe*. Figure 2.36 shows a somewhat simplified diagram of a leadframe for an eight-pin *dual-in-line package* (DIP), complete with a chip mounted on it. The leadframe itself consists of a rectangular *mount pad* that holds the die and a series of lead fingers that will eventually be trimmed to form the eight leads of the DIP. Leadframes usually come in strips, so several dice can be handled as a single assembly. They are either stamped out of thin sheets of metal, or they are etched using photographic techniques similar to those used to pattern printed circuit boards. The lead frame usually consists of copper or a copper alloy, often plated to resist corrosion and to promote solder adhesion. Copper is not an ideal material for leadframes because its coefficient of thermal expansion differs greatly from that of silicon. As the packaged part is heated and cooled, differential expansion of the die and the leadframe sets up mechanical stresses injurious to the performance of the die. Unfortunately, most of the materials that possess coefficients of expansion similar to silicon have inferior mechanical and electrical properties. Some of these materials are occasionally used for low-stress packaging of specialty parts; of these, a nickel-iron alloy called *Alloy-42* is probably the most commonly encountered (Section 7.2.9).

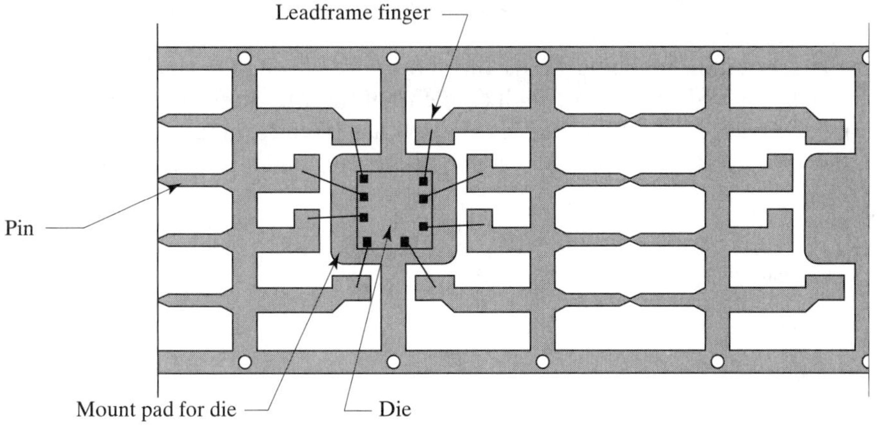

FIGURE 2.36 Simplified diagram of a leadframe for an 8-pin DIP.

The die is usually mounted to the leadframe with an epoxy resin. In some cases, the resin may be filled with silver powder to improve thermal conductivity. Epoxy is not entirely rigid, and this helps reduce the stresses produced by thermal expansion of the leadframe and die. Alternative methods exist that provide superior thermal union between the silicon and the leadframe, but at the cost of greater mechanical stress. For example, the backside of the die can be plated with a metal or metal alloy and soldered to the leadframe. Alternatively, a rectangle of gold foil called a *gold preform* can be attached to the leadframe; heating the die causes it to alloy with the gold preform to create a solid mechanical joint. Solder connections and gold preforms both allow excellent thermal contact between the die and the leadframe. Both also produce an electrical connection that can be used to connect the substrate of the die to a pin. Conductive epoxies improve thermal conductivity, but they cannot always be trusted to provide electrical connectivity.[37]

---

[37] R.L. Opila and J. D. Sinclair, "Electrical Reliability of Silver Filled Epoxies for Die Attach," *23rd International Reliability Physics Symp.*, 1985, pp. 164–172.

After the dice are mounted on leadframes, the next step is attaching bondwires to them. Bonding can only be performed in areas of the die where the metallization is exposed through openings in the protective overcoat; these locations are called *bondpads*. The probe card used for wafer probing makes contact to the bondpads for purposes of testing, but the probes may also make contact to a few pads that will not receive bondwires. Those pads reserved for testing purposes are usually called *testpads* or *probepads* to distinguish them from actual bondpads. Testpads are often made smaller than bondpads since probe needles can usually land in a smaller zone than can bondwires.

Bonding is performed by high-speed automated machines that use optical recognition to determine the locations of the bondpads. These machines typically employ 25 μm gold wire for bonding, although gold wires as small as 20 μm or as large as 50 μm are in common use. Aluminum wires up to 250 μm in diameter can also be employed, although these require rather different bonding machinery. Only one diameter and type of wire can be bonded at a time, so few dice use more. The most common arrangement consists of 25 μm gold bondwires on all pads. Multiple 25 μm wires bonded in parallel can carry higher currents or provide lower resistances without requiring a second bonding pass for larger diameter wire.

The most common technique for bonding gold wire is called *ball bonding*.[38] Since aluminum wire cannot be ball bonded, an alternate technique called *wedge bonding* has been developed for it. Figure 2.37 shows the essential steps of the ball-bonding process.

The bonding machine feeds the gold wire through a slender tube called a *capillary*. A hydrogen flame melts the end of the wire to form a small gold sphere, or *ball* (Figure 2.37, Step 1). Once a ball has been formed, the capillary presses down against the bondpad. The gold ball deforms under pressure, and the gold and

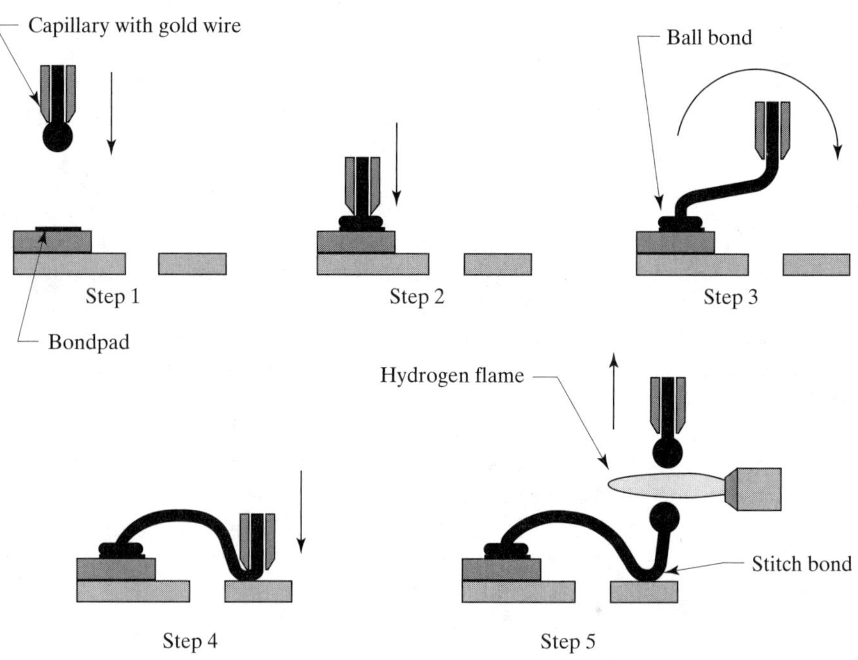

FIGURE 2.37 Steps in the ball-bonding process.

[38] B.G. Streetman, *Solid State Electronic Devices*, 2d ed. (Englewood Cliffs, NJ: Prentice-Hall, 1980), pp. 368–370.

aluminum fuse together to form a weld (Step 2). Next, the capillary lifts and moves to the vicinity of the lead finger (Step 3). The capillary again descends, smashing the gold wire against the lead finger. This causes the gold to alloy to the underlying metal to produce a weld (Step 4). Since no ball is present at this point, the resulting bond is called a *stitch bond* rather than a ball bond. Finally, the capillary lifts up from the lead finger and the hydrogen flame passes through the wire, causing it to fuse into two (Step 5). The bond is now complete, and another ball has been formed on the wire protruding from the capillary, allowing the process to be repeated. An automated bonding machine can perform these steps 10 times per second with great precision. The extreme speed and unerring accuracy of these machines produce huge economies of scale, and the entire bonding process costs no more than a penny or two.

Aluminum wire cannot be ball bonded because the hydrogen flame would ignite the fine aluminum wire. Instead, a small, wedge-shaped tool is used to supplement the capillary. When the capillary brings the wire into proximity with the bondpad, the tool smashes it against the pad to create a stitch bond. The process is repeated at the lead finger, and the tool is then held down against the lead finger while the capillary moves up. The tension created in the aluminum wire snaps it at its weakest point, immediately adjacent to the weld. The process can then be repeated as many times as necessary.

The ball-bonding process requires a square bondpad approximately two or three times as wide as the diameter of the wire. Thus, a 25 μm gold wire can be attached to a square bondpad about 50–75 μm across. Wedge bonding is more selective, and usually requires bondpads that are rectangular or trapezoidal in shape. These bondpads must lie in the same direction as the wedge tool. They are typically twice as wide and four times as long as the wire is thick. The exact rules for wedge bondpads can become quite complex, particularly for thicker aluminum wires.

Figure 2.36 shows a die mounted on a leadframe after the bonding process is complete. The bondwires connect various bondpads to their respective leads. Although the wires are quite small compared with the pins, each is still capable of carrying roughly an amp of current.

2.7.2. Packaging

The next step in the assembly process is injection molding. A mold is clamped around the leadframe, and heated plastic resin is forced into the mold from below. The plastic wells up around the die, lifting the wires away from it in gentle loops. Injection from the side or from the top usually smashes the wires against the integrated circuit and is therefore impractical. The plastic resin employed for integrated circuits cures rapidly at the temperatures used in molding and, once cured, it forms a rigid block of plastic.

When the molding process is complete, the leads are trimmed and formed to their final shapes. This is done in a mechanical press, using a pair of specially shaped dies that simultaneously trim away the links between the individual leads and bend them to the required shape. This step was in the past followed by solder dipping or solder plating, but current practice substitutes a noble metal plating (such as palladium) to eliminate lead from the packaging. The completed integrated circuits are now labeled with part numbers and other designation codes (which usually include a code identifying the date of manufacture and the lot number). The completed integrated circuits are tested again to ensure that they have not been damaged by the packaging process. Finally, the completed devices are packaged in tubes, trays, or reels for distribution to customers.

2.8 SUMMARY

Modern semiconductor processing takes advantage of the properties of silicon to manufacture inexpensive integrated circuits in huge volumes. Photolithography allows the reproduction of intricate patterns hundreds or thousands of times across each wafer, leading to enormous economies of scale.

Junctions can be formed by one of three means: epitaxial deposition, diffusion, or ion implantation. Low-pressure chemical-vapor-deposited (LPCVD) epitaxial layers can produce extremely high-quality silicon films with precisely controlled dopant concentrations. Diffusion of dopants from a surface source allows the formation of vast numbers of junctions using only a single photolithographic masking step. Ion implantation allows similar but more costly patterning of junctions with superior control of doping levels and distributions.

Many materials can also be deposited on the surface of the wafer. These include polycrystalline silicon (poly), oxide, nitride, and any of numerous metals and metal alloys. Typical semiconductor processes combine several diffusions into the bulk silicon with several depositions of materials onto the resulting wafer. The next chapter examines how the various techniques of semiconductor fabrication are combined to manufacture three of the most successful integrated circuit processes.

2.9 EXERCISES

2.1. When pressure is applied to the center of an unknown wafer, it splits into six segments. What can be definitely concluded from this observation? What may be reasonably conjectured?

2.2. Draw a diagram similar to that in Figure 2.4 illustrating the relationship between the (100) and (110) planes of a cubic crystal. (Refer to Appendix B, if necessary.)

2.3. Suppose a photomask consists of a single opaque rectangle on a clear background. A negative resist is used in combination with this mask to expose a sensitized wafer. Describe the pattern of photoresist left on the wafer after development.

2.4. Suppose a wafer is subjected to the following processing steps:

a. Uniform oxidation of the entire wafer surface.
b. Opening of an oxide window to the silicon surface.
c. An additional period of oxidation.
d. Opening of a smaller oxide window in the middle of the region patterned in step b.
e. An additional period of oxidation.

Draw a cross section of the resulting structure, showing the topography of both the silicon and the oxide surfaces. The drawing need not be made to scale.

2.5. Suppose a wafer is uniformly doped with equal concentrations of boron and phosphorus atoms. After a prolonged oxidation, will the surface of the silicon be N-type or P-type? Why?

2.6. Suppose that a wafer is uniformly doped with 10^{16} atoms/cm^3 of boron. This wafer is then subjected to the following processing steps:

a. Uniform oxidation of the entire wafer surface.
b. Opening of an oxide window to the silicon surface.
c. Deposition of boron and phosphorus, each at a source concentration of 10^{20} atoms/cm^3, using a 15-minute deposition and a one-hour drive at 1000°C.

Assuming that the two dopants do not interact with each other or with the oxide, draw a cross section of the resulting structure. Indicate the approximate depths of any junctions formed.

2.7. Phosphorus is diffused into a lightly doped wafer through an oxide window 5 μm square. If the resulting junction is 2 μm deep, then what is the width of the phosphorus diffusion at the surface?

2.8. Most ion implantation systems position the accelerator so that the ion beam impacts the wafer surface at a slight angle (often 7°). Explain the reason for this feature.

2.9. If the surface of the oxide layer covering a wafer is ground perfectly smooth, different regions of the wafer still exhibit different colors, but the NBL shadow vanishes. Explain these observations.

2.10. Draw a cross section of the following metallization system:

a. 1 μm-wide contacts through 5 kÅ oxide silicided with 2 kÅ platinum silicide.
b. First-level metal consisting of 2 kÅ RBM and 6 kÅ copper-doped aluminum.
c. 1 μm-wide vias 3 kÅ deep through highly planarized ILO.
d. Second-level metal consisting of 2 kÅ RBM and 10 kÅ copper-doped aluminum.
e. 10 kÅ protective overcoat.

Assume that the silicide surface is level with the surrounding silicon surface, and that the aluminum metal thins 50% on the sidewalls of the contacts and vias. The drawing should be made to scale.

2.11. Suppose that a die measures 60 by 80 mils, where one mil is one thousandth of an inch. Approximately how many of these dice could be fabricated on a 150 mm-diameter wafer? Assuming that 70% of these potential dice actually work, and that a finished wafer costs $250, compute the cost of each functional die.

2.12. Suggest an appropriate means of fabricating each of the following diffusions:

a. A shallow, heavily doped N-type source/drain diffusion.
b. A deep, lightly doped N-type well diffusion.
c. A deep, heavily doped N-type sinker diffusion.
d. A heavily doped arsenic buried layer.

2.13. Suggest a suitable silicide for each of the following processes:

a. A process that fabricates a minimum poly linewidth of 1 μm and which incorporates Schottky diodes.
b. A process that does not incorporate Schottky diodes and features a minimum poly linewidth of 2 μm.
c. A process that does not incorporate Schottky diodes and implements a minimum poly linewidth of 0.25 μm.

2.14. The silicon-on-sapphire process has an extremely nonplanar surface topography. Suggest a means of rectifying this difficulty that would enable the SOS process to utilize a modern fine-linewidth metallization system.

2.15. A proposed high-voltage dielectric isolation process requires a layer of lightly doped N-type silicon 25 μm thick to be placed atop a buried oxide 1 μm thick. As an additional complication, a patterned arsenic buried layer must be added, located just above the buried oxide. Suggest a possible means of fabricating this process.

3 *Representative Processes*

Semiconductor processing has evolved rapidly over the past 50 years. The earliest processes produced only discrete components, mainly switching diodes and bipolar transistors. The first practical integrated circuits appeared in 1960.[1] These consisted of a few dozen bipolar transistors and diffused resistors connected to form simple logic gates. By modern standards, these early integrated circuits were terribly slow and inefficient. Refinements were soon made, and by the mid-1960s bipolar integrated logic offered clear advantages over discrete logic. The first analog integrated circuits appeared at about the same time; these consisted of matched transistor arrays, operational amplifiers, and voltage references. The standard bipolar process that was created to support these products remains in use today.

Integrated bipolar logic was fast but power-hungry. MOS integrated circuits held out the promise of a low-power alternative, but the metal-gate MOS processes of the 1960s suffered from unpredictable threshold voltage shifts. This problem was eventually conquered through the development of polysilicon-gate MOS processes in the early 1970s. MOS logic soon replaced bipolar logic and created vast new markets for microprocessors and dynamic RAM chips. Analog CMOS circuits of this era touted greatly reduced operating currents but provided only mediocre performance, so standard bipolar remained the process of choice for high-performance analog integrated circuits.

By the mid-1980s, customers were demanding the integration of both digital and analog functions onto a single mixed-signal integrated circuit. A new generation of merged bipolar-CMOS (BiCMOS) processes were soon developed specifically for mixed-signal design. Although these processes are complex and costly, they offer a level of performance unachievable by other means. The world of analog integrated circuits has been dominated by these three processes: standard bipolar, polysilicon-gate CMOS, and analog BiCMOS. Although processing technology has evolved significantly since the 1980s, most modern processes are still derived from these three

---

[1] J. S. Kilby, "Invention of the Integrated Circuit," *IEEE Trans. on Electron Devices*, Vol. ED-23, #7, 1976, pp. 648–654.

archetypes. This chapter will analyze the implementation of a representative process of each type.

3.1 STANDARD BIPOLAR

Standard bipolar was the first analog integrated circuit process. Over the years, it has produced many classic devices, including the 741 operational amplifier, the 555 timer, and the 431 voltage reference. Even though these parts represent 30-year-old technology, they are still manufactured in vast quantities today.

Standard bipolar is seldom used for new designs. CMOS offers lower supply currents, BiCMOS provides superior analog performance, and various advanced bipolar processes yield faster switching speeds. But the knowledge gained through first developing and then refining standard bipolar will never become obsolete. The same devices reappear in every new process, along with many of the same parasitic mechanisms, design tradeoffs, and layout principles. This chapter will therefore begin with an overview of standard bipolar.

3.1.1. Essential Features

Standard bipolar was shaped by a conscious decision to optimize the NPN transistor at the expense of the PNP. This decision rested on the observation that the NPN transistor employs electron conduction while the PNP transistor relies on hole conduction. The lower mobility of holes reduces both the beta and the switching speed of PNP transistors. Given equivalent geometries and doping profiles, an NPN will outperform a PNP by more than two to one. Several additional processing steps are required to optimize both types of transistors simultaneously, so early processes optimized NPN transistors and avoided PNP transistors altogether. This decision met the requirements of bipolar logic, consisting as it does of NPN transistors, resistors, and diodes. When analog circuits were first constructed using standard bipolar, several types of PNP transistors were cobbled together from existing process steps. Although these transistors performed relatively poorly, they sufficed to design many useful circuits.

The standard bipolar process employs *junction isolation* (JI) to prevent unwanted currents from flowing between devices that are formed on the same substrate.[2] The components reside in a lightly doped N-type epitaxial layer deposited on top of a lightly doped P-type substrate (Figure 3.1). A deep-P+ *isolation diffusion* driven down to contact the underlying substrate provides isolation between components. Regions of N-epi separated from one another by isolation are called *tanks* or *tubs.* If

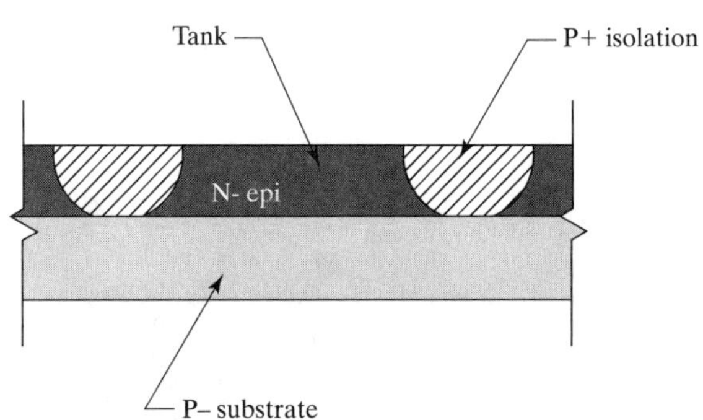

FIGURE 3.1 Cross section of the junction isolation system employed for standard bipolar.

[2] R. N. Noyce, U.S. Patent #2,981,877, 1961.

the isolation is biased to a potential equal to or below the lowest-voltage tank, then reverse-biased junctions surround every tank. The substrate forms the floor of these tanks, and isolation diffusions form their sidewalls.

Junction isolation has several significant drawbacks. The reverse-biased isolation junctions exhibit enough capacitance to slow the operation of many circuits. High temperatures can cause significant leakage currents, as can exposure to light or ionizing radiation. Unusual operating conditions can also forward bias the isolation junctions and inject minority carriers into the substrate. Despite these difficulties, junction-isolated processes can successfully fabricate most circuits.

3.1.2. Fabrication Sequence

The baseline standard bipolar fabrication sequence consists of eight masking operations. The significance of each step can be illustrated best by presenting the entire flow from starting material to finished wafer. Representative cross sections will be used to illustrate each step. When examining these cross sections (and all others in this text), keep in mind that the vertical scale of the drawings has been exaggerated by a factor of two to five for clarity. The lateral dimensions of a typical integrated device are so much greater than its vertical dimensions that a true-scale diagram would be virtually illegible. The substrate is also much thicker than depicted; the additional silicon serves to strengthen the wafer against warping and breakage.

Starting Material

Standard bipolar integrated circuits are fabricated on a lightly doped (111)-oriented P-type substrate. The wafers are usually cut several degrees off-axis to minimize distortion of the NBL shadow (*pattern distortion*).[3] The use of (111) silicon helps suppress a parasitic PMOS transistor inherent to the standard bipolar process. The N-epi forms the backgate of this parasitic, while a lead crossing the field oxide above the tank acts as its gate electrode. A base region within the tank forms the source, and the drain consists of either another base region or the P+ isolation (Figure 3.2). When the base diffusion acting as the source is biased to a high voltage relative to the metal lead, a P-type channel forms and allows current to flow from the base to the isolation. The threshold voltage of a MOS transistor formed under thick field oxide is called a *thick-field threshold.* The use of (111) silicon artificially elevates the PMOS thick-field threshold by introducing positive surface-state charges along the oxide-silicon interface.

N-Buried Layer

The first processing step grows a thin layer of oxide across the wafer. Photoresist spun onto this oxide is patterned using the N-buried layer (NBL) mask. After an

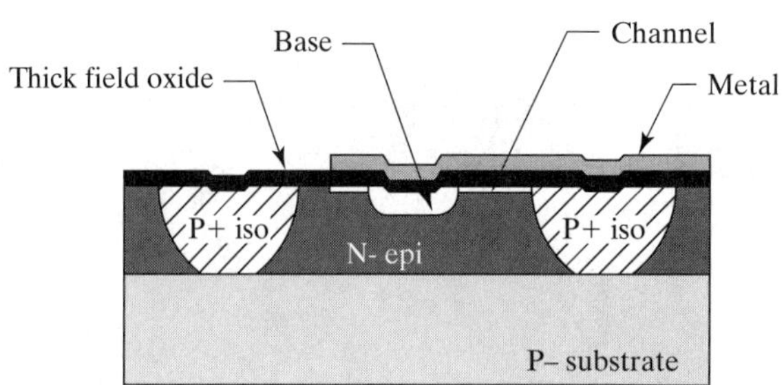

FIGURE 3.2 Parasitic PMOS formation in standard bipolar.

[3] W.R. Runyan and K.E. Bean, *Semiconductor Integrated Circuit Processing Technology* (Reading, MA: Addison-Wesley, 1994), p. 331.

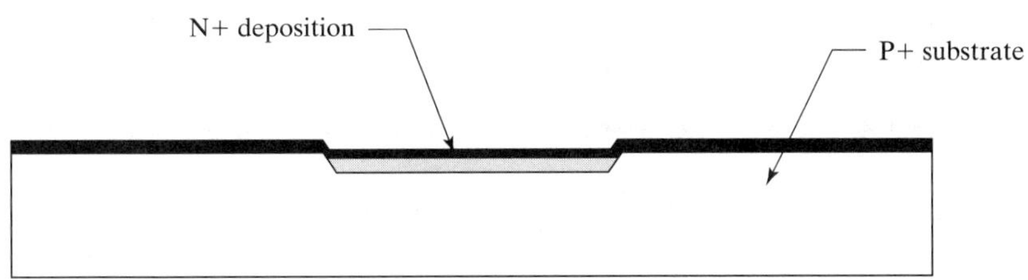

FIGURE 3.3 Wafer after anneal of NBL implant.

oxide etch opens windows to the silicon surface, ion implantation or thermal deposition transfers an N-type dopant into the wafer. The N-buried layer customarily consists of either arsenic or antimony because the low diffusivities of these elements limit up-diffusion during subsequent processing. The brief drive following the deposition serves two purposes: First, it anneals out lattice damage; and second, it grows a small amount of oxide that forms a slight discontinuity at the silicon surface (Figure 3.3). This discontinuity will later produce an NBL shadow to which other masks can align.

Epitaxial Growth

The oxide layer that remains on the wafer is stripped prior to the growth of some 10 to 25 μm of lightly doped N-type epi. Surface discontinuities propagate upward during epitaxial deposition at approximately a 45° angle. Upon completion of epitaxial growth, the NBL shadow will have shifted laterally a distance approximately equal to the thickness of the epi (Figure 3.4).

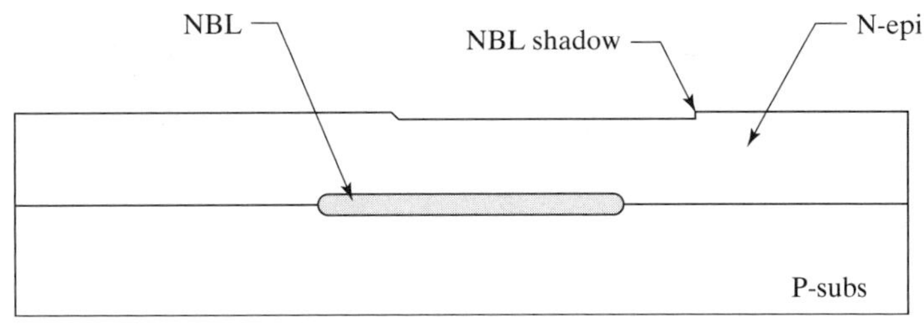

FIGURE 3.4 Wafer after epitaxial deposition. Note the pattern shift exhibited by the NBL shadow.

Isolation Diffusion

The wafer is again oxidized, coated with photoresist, and patterned using the *isolation* mask. This mask must be aligned to the NBL shadow by using a deliberate offset to correct for pattern shift. A heavy boron deposition followed by a high-temperature drive forces the isolation diffusion partway down through the epi layer. Oxidation also occurs during this drive, covering the isolation windows with a thin layer of thermal oxide. The drive stops before the isolation junction reaches the substrate, since later high-temperature processing steps (primarily the deep-N+ drive) will force the diffusion the rest of the way down. Figure 3.5 shows the wafer after this partial drive.

Deep-N+

A deep-N+ diffusion (sometimes called a *sinker*) allows low-resistance connection to the NBL. First, a photoresist is applied and patterned using the deep-N+ mask. A heavy phosphorus deposition followed by a high-temperature drive forms the deep-N+ sinkers. The drive not only causes the deep-N+ to diffuse down to meet the upward-diffusing NBL, but also completes the isolation drive. Sufficient time is

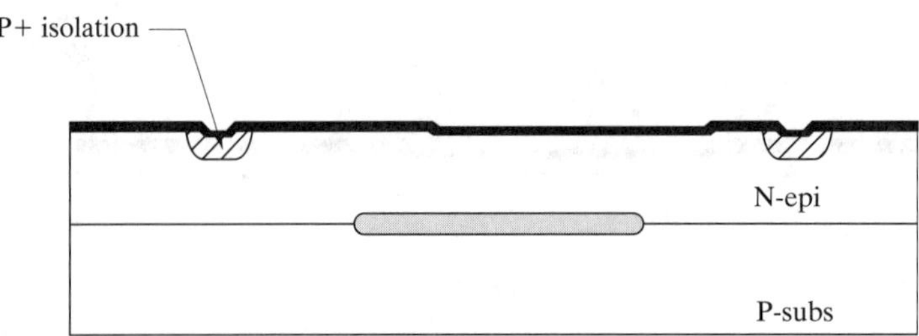

FIGURE 3.5 Wafer after isolation deposition and partial drive.

allowed to overdrive the junctions by about 25%. Without this overdrive, the bottom of the isolation and deep-N+ diffusions would be very lightly doped. The overdrive simultaneously reduces the vertical resistance through both the isolation and the deep-N+ sinkers. The deep-N+ drive also forms the thick field oxide.

Both deep-N+ and isolation diffusions approach their final junction depths during the deep-N+ drive (Figure 3.6). These junctions will diffuse slightly deeper during subsequent processing, but all of the later diffusions are fairly shallow compared to deep-N+ and isolation, and therefore the tank regions appear fully formed in Figure 3.6. The NBL regions are normally spaced some distance inside the isolation diffusion to increase the tank breakdown voltage. Otherwise, the N+/P+ junction formed by the intersection of NBL and isolation would avalanche at about 30 V.

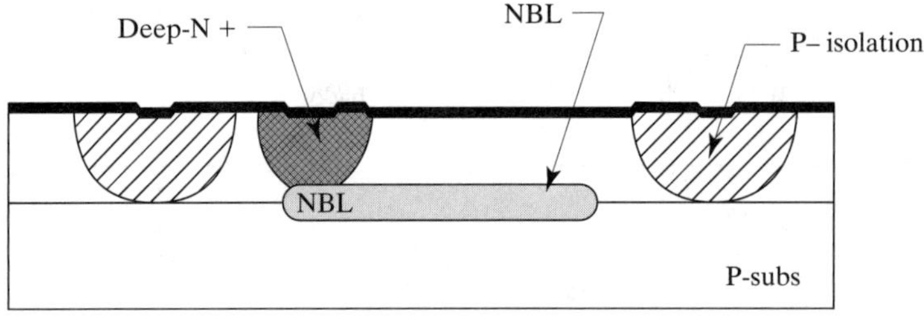

FIGURE 3.6 Wafer after isolation drive.

Base Implant

Next, photoresist spun onto the wafer is patterned using the base mask. An oxide etch clears windows through the field oxide to the silicon surface. A light boron implant conducted through these openings counterdopes the N-epi to form the base regions of the NPN transistors. Ion implantation allows precise control of base doping and thus minimizes process-derived beta variation. The subsequent drive anneals implant damage and sets the base junction depth. Oxide grown during this drive serves as a mask for the subsequent emitter deposition. Base is also implanted across the isolation regions to increase surface doping. This practice, called *base-over-isolation* (BOI), substantially increases the NMOS thick-field threshold without requiring the use of a separate channel stop. Figure 3.7 shows a cross section of the wafer following the base drive.

Emitter Diffusion

The wafer is again coated with photoresist and patterned using the emitter mask. A subsequent oxide etch exposes the silicon surface in regions where NPN emitters will form and in regions where Ohmic contact must be made to the N-epi or the deep-N+ diffusion. A very concentrated phosphorus deposition forms the emitter.

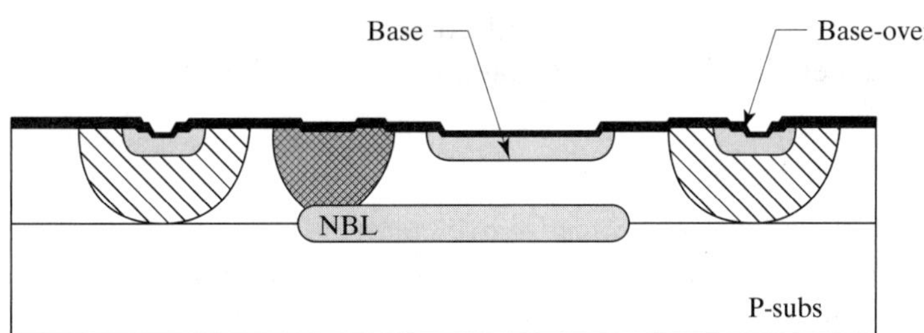

FIGURE 3.7 Wafer after base drive.

A $POCl_3$ source is often used, since precise control of emitter doping is unnecessary. A brief drive sets the final emitter junction depth and thereby determines the width of the active base region of the NPN transistors.

An oxide film grown over the emitter diffusion insulates it from subsequent metallization. Some processes employ dry oxidation for this step, but the short oxidation time results in a *thin emitter oxide* vulnerable to electrostatic discharges (Section 4.1.1). Alternatively, a wet oxidation can grow a *thick emitter oxide* that possesses a higher rupture voltage, or additional oxide may be added using a deposition process. Figure 3.8 shows a cross section of the wafer after the emitter drive.

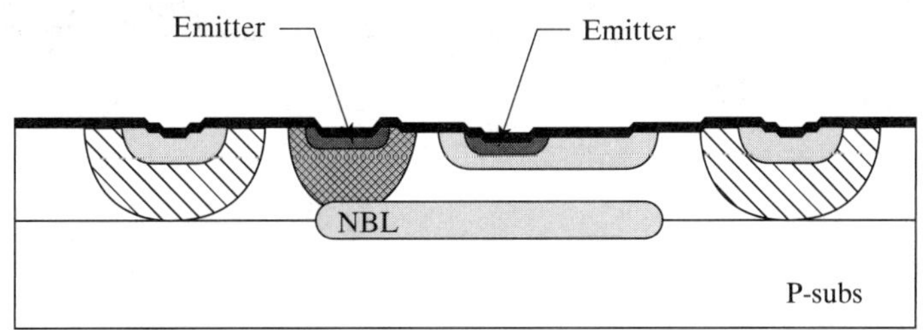

FIGURE 3.8 Wafer after emitter drive.

Many older processes also incorporate an *emitter pilot* step to provide a means of adjusting NPN beta. A dummy wafer that is inserted into the wafer lot before base implant and removed after emitter deposition is used to conduct an experimental emitter drive. By monitoring the performance of the base-emitter junction formed on the pilot wafer, the actual drive can be adjusted on a lot-by-lot basis to target the desired NPN beta.

Contact

All diffusions are now complete. The remaining steps form the metallization and apply the protective overcoat. The first step in this sequence forms contacts to selected diffusions. The wafer is again coated with photoresist, patterned using the contact mask, and etched to expose bare silicon. This process is sometimes called *contact OR,* in which OR stands for *oxide removal.*

Metallization

A layer of aluminum-copper-silicon alloy is evaporated or sputtered across the entire wafer. This metal system typically incorporates 2% silicon to suppress emitter punchthrough and 0.5% copper to improve electromigration resistance. Standard bipolar employs relatively thick metallization, typically at least 10 kÅ (1.0 μm) thick, to reduce interconnection resistance and decrease vulnerability to electromigration. The metallized wafer is patterned using the metal mask and etched to form the interconnection system.

Modern versions of the standard bipolar process usually incorporate more advanced metallization systems than the one described here, but the principles remain the same.

Protective Overcoat

Next, a thick layer of *protective overcoat* (PO) is deposited across the entire wafer. Compressive nitride protective overcoats provide excellent mechanical and chemical protection. Some processes use a *phosphosilicate-doped glass* (PSG) layer either beneath a compressive nitride or as a replacement for it. Since the deposition of the protective overcoat occurs at moderate temperatures, it also sinters the aluminum metallization.

Finally, a layer of photoresist is applied and patterned using the PO mask. A special etch opens windows through the protective overcoat to expose areas of metallization for bonding. This composes the final fabrication step; the wafer is now complete. Figure 3.9 shows a fully processed wafer. (The illustrated cross section does not include a bondpad opening.)

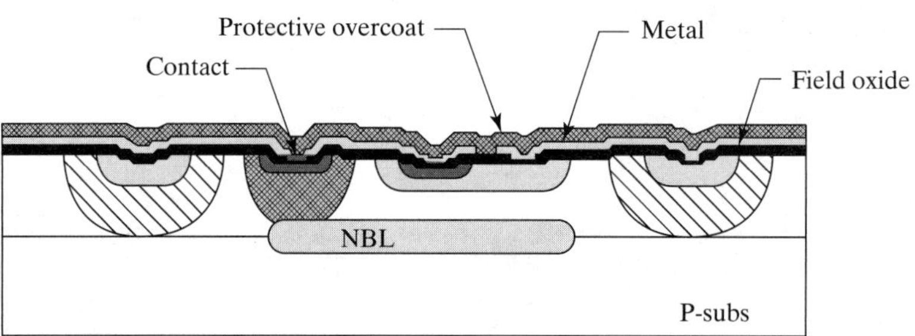

FIGURE 3.9 Completed standard bipolar wafer.

3.1.3. Available Devices

Standard bipolar was originally developed to provide bipolar NPN transistors and diffused resistors. A number of other devices can also be fabricated using the same process steps, including two types of PNP transistors, several types of resistors, and a capacitor.[4] These devices form a basic component set suitable for fabricating a wide variety of analog circuits. Section 3.1.4 will examine several additional devices that require extensions to the baseline process.

NOTE: The dimensions of standard bipolar devices are often specified in mils. A *mil* equals 0.001 inches. The relevant conversion factors are $1\,\text{mil} = 25.4\,\mu\text{m}$, and $1\,\text{mil}^2 \cong 645\,\mu\text{m}^2$. Another unit of measurement sometimes used to specify junction depths is the *sodium line,* which equals one-half of the wavelength of the sodium spectrum D-line ($1\ \text{line} = 0.295\,\mu\text{m}$).[5] Both the mil and the sodium line are gradually fading into obscurity, but these units are frequently encountered in older books and papers.

NPN Transistors

Figure 3.10 shows a representative layout and cross section of a minimum-area NPN transistor. The collector of the NPN consists of an N-epi tank, and the base and

---

[4] For a general overview of the standard bipolar process, see N. Doyle, "LIC Technology," *Microelectronics and Reliability,* Vol. 13, 1974, pp. 315–324.

[5] G.E. Anner, *Planar Processing Primer* (New York: Van Nostrand Reinhold, 1990), pp. 107–108.

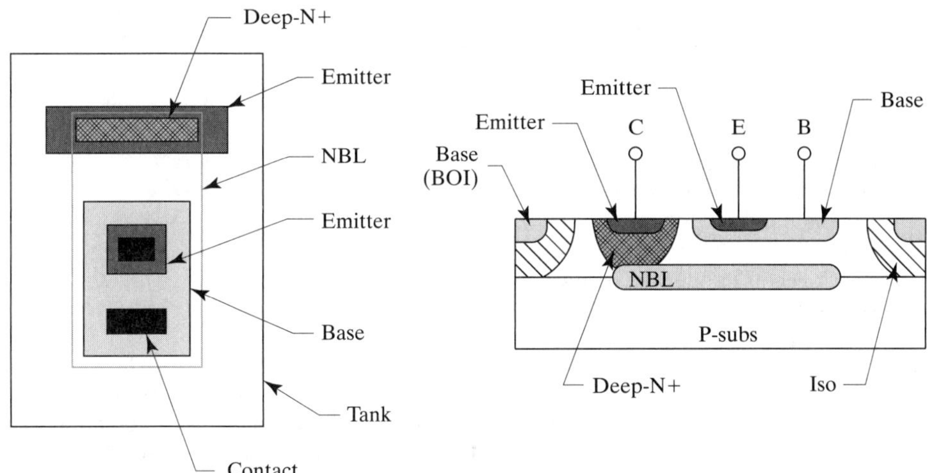

FIGURE 3.10 Layout and cross section of an NPN transistor with deep-N+ and NBL.[6]

emitter are fabricated by successive counterdopings. The carriers flow vertically from emitter to collector through the thin base region underneath the emitter diffusion. The difference between the base and emitter junction depths determines the effective base width. Since these dimensions are controlled entirely by diffusion processing, they are not subject to photolithographic misalignment, allowing a base width substantially smaller than the alignment tolerance. For example, a process with a 5 μm minimum feature size can easily fabricate a 2 μm base width.

The collector consists of lightly doped N-type epi lying on top of heavily doped NBL. The lightly doped epi allows the formation of a wide collector-base depletion region without excessive intrusion into the neutral base. This enables the transistor to support high operating voltages while simultaneously minimizing the Early effect (Section 8.2). NBL and deep-N+ create a low-resistance pathway to the portion of the epi layer beneath the transistor's active base. By this means, the collector resistance of a minimum NPN can be reduced to less than 100 Ω and the collector resistance of a power NPN can be reduced to less than 1 Ω.

The high concentration of donors in the NBL effectively halts the downward growth of the collector-base depletion region. The distance between the bottom of the base diffusion and the top of the NBL determines the maximum operating voltage of the NPN transistor. Thicker epi layers allow higher operating voltages, up to a limit set by the breakdown of the base diffusion sidewall (typically 50 to 80 V). The maximum operating voltage of a bipolar process is usually specified in terms of the avalanche voltage of the NPN, collector to emitter with base open (V_{CEO}). Depending on epi thickness and doping, this voltage can range from less than 10 V to more than 100 V.

The vertical NPN is the best device fabricated in the standard bipolar process. It consumes relatively little area and offers reasonably good performance. Circuit designers try to use as many of these transistors as possible. Table 3.1 lists typical device parameters for a minimum-emitter NPN transistor in a 40 V standard bipolar process.

The NPN transistor can also act as a diode, the characteristics of which depend upon the terminals chosen to form the anode and cathode. The least series resistance and fastest switching speeds occur when the base and collector form the anode

---

[6] In many standard bipolar processes, the isolation spacings are much wider than this illustration suggests; see Appendix C.1 for a brief discussion and some typical spacing rules.

TABLE 3.1 Typical vertical NPN device parameters.

Parameter	Nominal Value
Drawn emitter area	100 μm^2
Peak current gain (beta)	150
Early voltage	120 V
Collector resistance, in saturation	100 Ω
Collector current range for maximum beta	5 μA–2 mA
V_{EBO} (Emitter-base breakdown, collector open)	7 V
V_{CBO} (Collector-base breakdown, emitter open)	60 V
V_{CEO} (Collector-emitter breakdown, base open)	45 V

and the emitter forms the cathode. This configuration is sometimes called a *CB-shorted diode,* or a *diode-connected transistor.* Its only serious drawback consists of a low breakdown voltage, equal to the V_{EBO} of the transistor, or about 7 V. On the other hand, the relatively low V_{EBO} allows a suitably-connected transistor to serve as a useful Zener diode. The breakdown voltage of this structure varies somewhat due to doping variations and surface effects, so a tolerance of at least ±0.3 V should be allowed.

PNP Transistors

The standard bipolar process cannot fabricate an isolated vertical PNP because it lacks a P-type tank. A nonisolated vertical PNP transistor, called a *substrate PNP,* can be constructed using the substrate as a collector. The collector of this device always connects to the substrate potential of the die, which usually consists of either ground or the negative supply rail. Figure 3.11 shows a representative layout and cross section of this device.

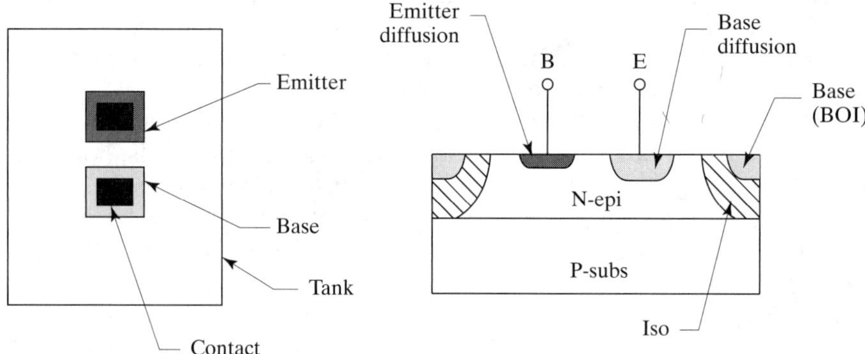

FIGURE 3.11 Layout and cross section of a substrate PNP transistor. The substrate forms the collector and is contacted through substrate contacts (not shown).

The base of the substrate PNP consists of an N-tank, and the emitter is fabricated from base diffusion. The collector current must exit through the substrate and the isolation. The collector contact does not have to reside next to the substrate PNP since all isolation regions interconnect electrically through the substrate. The resistance of the isolation and substrate are, however, substantial. Substrate contacts placed adjacent to the transistor help extract the collector current and thus minimize voltage drops in the substrate (*substrate debiasing*) that might otherwise impair circuit performance (Section 4.4.1).

The difference between the final epi thickness and the base junction depth determines the base width of a substrate PNP. As in the case of the vertical NPN, the base

Parameter	Lateral PNP	Substrate PNP
Drawn emitter area	100 μm^2	100 μm^2
Drawn base width	10 μm	N/A
Peak current gain (beta)	50	100
Early voltage	100 V	120 V
Typical operating current for maximum beta	5–100 μA	5–200 μA
V_{EBO} (Emitter-base breakdown, collector open)	60 V	60 V
V_{CBO} (Collector-base breakdown, emitter open)	60 V	60 V
V_{CEO} (Collector-emitter breakdown, base open)	45 V	45 V

TABLE 3.2 Typical PNP device parameters.

width is unaffected by photolithographic tolerances. NBL must be left out of the substrate transistor because its presence severely reduces beta. Deep-N+ therefore serves no useful function in a substrate PNP. An emitter diffusion placed beneath the collector contact ensures the surface doping concentration necessary to achieve Ohmic contact, while simultaneously thinning the oxide. The epi thicknesses and doping concentrations of the standard bipolar process are calculated to optimize the vertical NPN transistor, but the substrate PNP performs respectably (Table 3.2). The choice of names for the emitter and base diffusions is somewhat unfortunate, as the *emitter* of a substrate PNP consists of *base* diffusion.

The lack of an isolated collector limits the versatility of the substrate PNP. Another transistor, called a *lateral PNP*, trades off device performance for isolation. Figure 3.12 shows a representative layout and cross section of a minimum-geometry lateral PNP transistor. Both the collector and the emitter regions of the lateral PNP consist of base diffusions formed into an N-tank. As in the case of the substrate PNP,

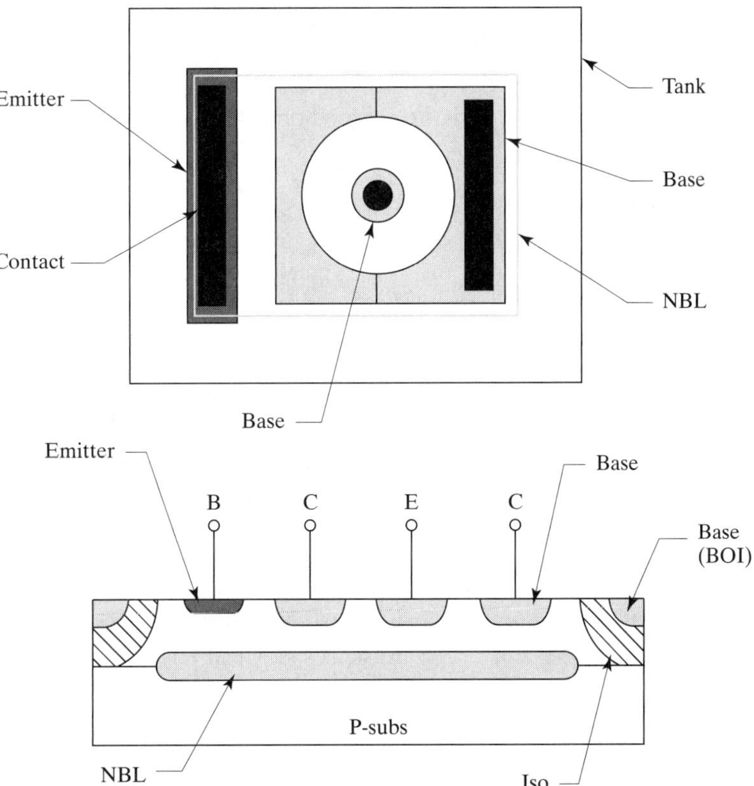

FIGURE 3.12 Layout and cross section of a lateral PNP transistor. The collector of the transistor appears twice on the cross section because it encircles the emitter.

this tank serves as the base of the transistor. Transistor action in the lateral PNP occurs laterally outward from the central emitter to the surrounding collector. The separation of the two base diffusions sets the base width of the transistor. The emitter and collector of the lateral PNP are said to *self-align* because a single masking operation forms both regions. The base width of the lateral PNP can be precisely controlled because photolithographic misalignment does not occur between self-aligned diffusions. The effective base width of the transistor is considerably less than the drawn base width because of outdiffusion. This consideration limits the drawn base width to a minimum of about twice the base junction depth. Narrow-base lateral PNP transistors exhibit low Early voltages and low-voltage punchthrough breakdown, so wider base widths are often employed.

Some percentage of the carriers injected by the lateral PNP's emitter will actually flow to the substrate rather than to the intended collector. This undesired conduction path forms a parasitic substrate PNP transistor. Unless this parasitic is somehow suppressed, most of the current injected by the emitter will find its way to the substrate, and the lateral PNP will exhibit very low apparent beta. For reasons that will be explained in Section 8.1.5, NBL largely blocks substrate injection and therefore boosts the lateral PNP beta.

Lateral PNP transistors have lower effective betas than their Gummel numbers would indicate. A large number of recombination centers reside at the oxide-silicon interface, especially in (111) silicon. The surface recombination rate thus far exceeds that in the bulk. Much of the current flow in the lateral PNP occurs near the surface and is therefore subject to these elevated recombination rates.[7] Despite these limitations, betas of 50 or more can be obtained. Lateral PNP transistors are also quite slow, due mainly to large parasitic junction capacitances associated with the base terminal.

Neither the lateral nor the substrate PNP transistor forms a true complement to the vertical NPN. Both are useful devices, but each has its drawbacks and limitations. Circuit designers tend to avoid routing active signal paths through PNP devices (especially laterals) because of their poor frequency response, but most analog circuits still contain PNP transistors in supporting roles. Table 3.2 lists typical device parameters for PNP transistors formed on a 40 V standard bipolar process.

Resistors

The baseline standard bipolar process does not include any diffusions intended specifically for fabricating resistors, but several types of resistors can be made using layers intended for other purposes. Typical examples include base, emitter, and pinch resistors, all three of which employ the relatively shallow base and emitter diffusions.

Each of the materials used to construct resistors possesses a characteristic *sheet resistance,* defined as the resistance measured across a square of the material contacted on opposite sides. Sheet resistance is customarily given in units of *Ohms per square* (Ω/□). It can be calculated from the thickness of the material and its doping concentration, but in the case of diffusions, nonuniform doping complicates these calculations (Section 5.2). In practice, sheet resistances are best determined by measuring sample resistors with known geometries constructed from the desired materials. Typical values for silicon diffusions range from 5 to 5000 Ω/□.

A *base resistor* consists of a strip of base diffusion isolated by an N-tank and connected so that it will reverse-bias the base-epi junction (Figure 3.13). Connecting the tank to the more positive end of the resistor will ensure isolation. Alternatively, the

---

[7] R. S. Muller and T. I. Kamins, *Device Electronics for Integrated Circuits,* 2d ed. (New York: John Wiley and Sons, 1986), pp. 366–368.

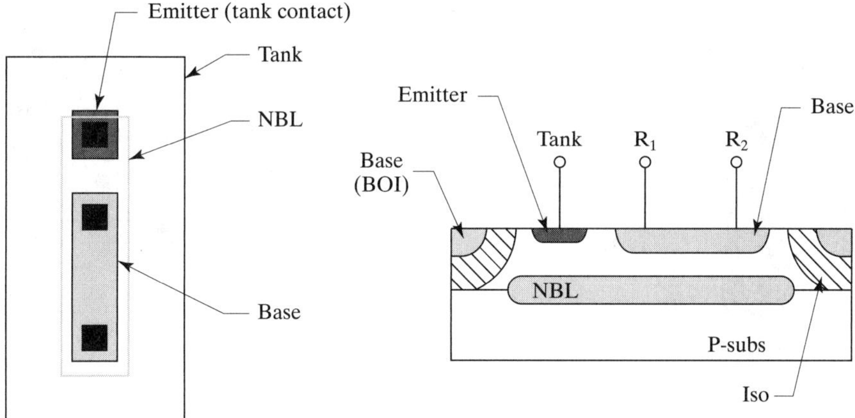

FIGURE 3.13 Layout and cross section of a base resistor.

tank can be connected to any point in the circuit biased to a higher voltage than the resistor. If a base resistor should forward-bias into its tank, a parasitic substrate PNP will inject current from the resistor into the substrate. NBL can help suppress this parasitic PNP should the base-epi junction momentarily forward-bias. Deep-N+ need not be added, because the tank terminal does not draw significant current. Most standard bipolar processes produce base resistors with sheet resistances of 150 to 250 Ω/□.

An *emitter resistor* consists of a strip of emitter diffusion isolated by a base diffusion enclosed within an N-tank (Figure 3.14). The base region is connected so that it will reverse-bias the emitter-base junction, while the tank is biased to reverse-bias the base-epi junction. The simplest way to achieve this goal consists of tying the base to the low-voltage end of the resistor and the tank to the high-voltage end. Various other connections are also feasible, so long as neither junction forward-biases. NBL is usually added to help suppress parasitic substrate PNP action. The emitter sheet resistance is relatively low (typically less than 10 Ω/□), and the breakdown of the emitter-base junction limits the differential voltage across the resistor to about 6 V.

A *pinch resistor* consists of a combination of base and emitter diffusions (Figure 3.15). The emitter forms a plate overlapping the middle of a thin strip of base.[8]

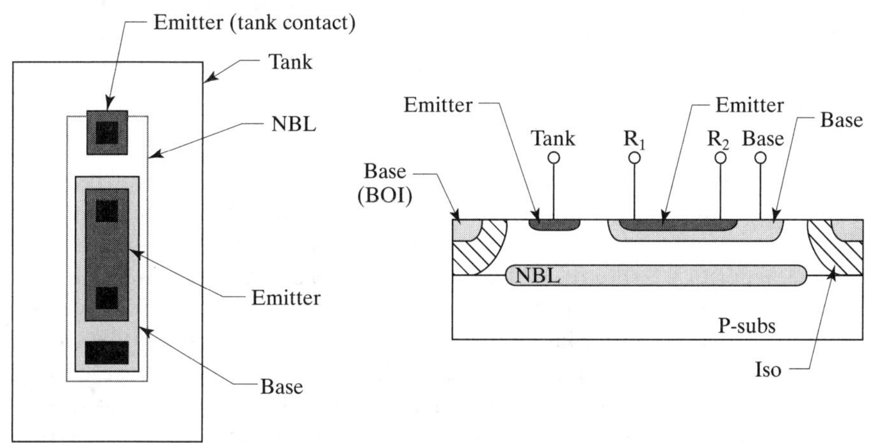

FIGURE 3.14 Layout and cross section of an emitter resistor. (Note the presence of bias terminals for both the tank and the base region.)

[8] R. P. O'Grady, "The 'Pinch' Resistor in Integrated Circuits," *Microelectronics and Reliability,* Vol. 7, 1968, pp. 233–236.

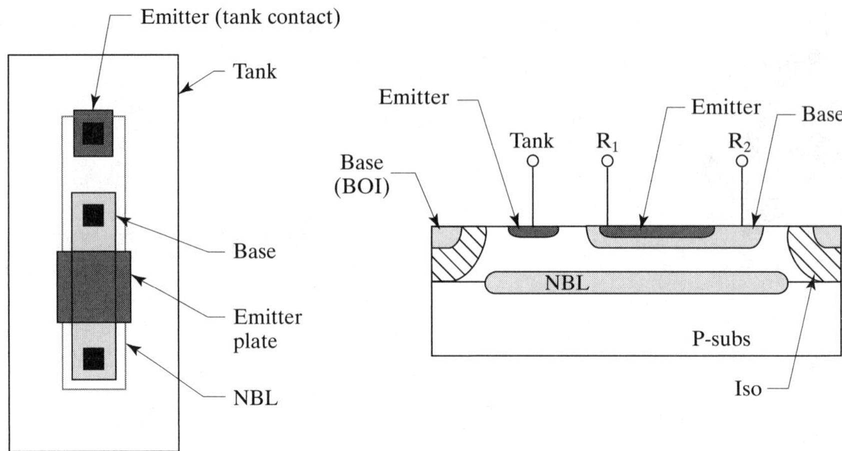

FIGURE 3.15 Layout and cross section of a base pinch resistor.

Contacts occupy the ends of the base strip, which project from under the emitter plate. The tank and emitter plate are both N-type and are therefore electrically united. A tank contact biases both to a voltage slightly more positive than the resistor in order to ensure isolation. The body of the resistor consists of the portion of the base diffusion beneath the emitter plate. This *pinched base* is thin and lightly doped, and therefore its resistance may exceed 5000 Ω/□. Emitter-base breakdown limits the differential voltage across the resistor to about 7 V. Pinch resistors are notoriously variable—much more so than either emitter or base resistors. Worst of all, these resistors exhibit severe voltage modulation. The intrusion of depletion regions into the neutral base tends to further pinch the resistor, causing it to act much like a JFET (Section 11.4). Pinch resistors find application in startup circuits and other noncritical roles, but their many drawbacks prohibit more widespread use. Table 3.3 compares the performance of emitter, base, and pinch resistors.

TABLE 3.3 Typical resistor device parameters.

Parameter	Emitter	Base	Pinch
Sheet resistance	5 Ω/□	150 Ω/□	3000 Ω/□
Minimum drawn width	8 μm	8 μm	8 μm
Breakdown voltage	7 V	50 V	7 V
Variability (15 μm width)	±20%	±20%	±50% or more

Capacitors

Standard bipolar was not intended to support capacitors. All of its oxide layers are so thick that they cannot be used to fabricate any but the smallest capacitors. However, the depletion region of a base-emitter junction exhibits a capacitance on the order of 0.8 fF/μm^2, which can be used to construct a so-called *junction capacitor* (Figure 3.16). This capacitor consists of a base diffusion overlapping an emitter diffusion, both placed in a common tank. The emitter diffusion shorts to the tank, and the base-tank capacitance therefore adds to the base-emitter capacitance. The emitter plate must be biased positively with respect to the base plate to maintain a reverse bias across the base-emitter junction, and the differential voltage across the capacitor must not exceed the emitter-base breakdown voltage (about 7 V). The resulting capacitance depends on bias and temperature and varies substantially (±50% or more). Junction capacitors are frequently used for

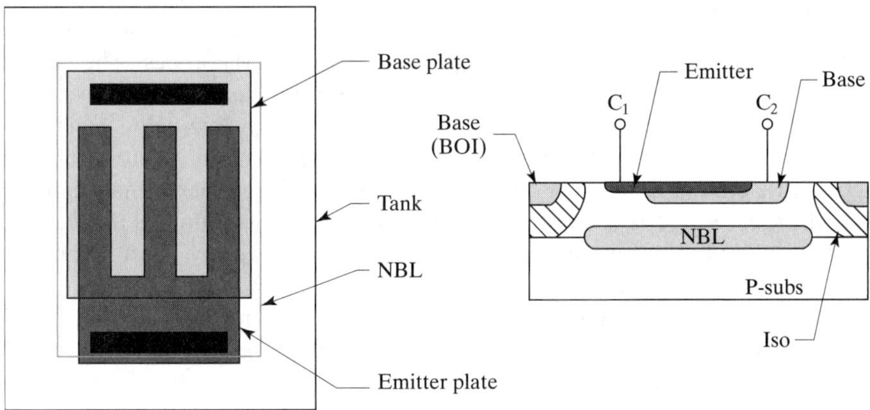

FIGURE 3.16 Layout and cross section of a junction capacitor.

compensating feedback loops, where their high capacitance per unit area makes up for their excessive variability.

3.1.4. Process Extensions

Standard bipolar has spawned a large number of process extensions, five of which are discussed in this section. They include up-down isolation, double-level metal, Schottky diodes, high-sheet resistors, and super-beta transistors. The benefits of each extension must be weighed against the increased cost and process complexity.

Up-Down Isolation

Standard bipolar employs deep-P+ junction isolation driven down through the epitaxial layer to the underlying substrate. Outdiffusion increases the width of the isolation by 20 μm or more, limiting how closely components can be packed together. One means of reducing outdiffusion uses a *P-buried layer* (PBL) to supplement the P+ isolation. The resulting *up-down isolation* consists of an isolation diffusion drawn coincident with the PBL. The isolation diffuses down from the surface, while the PBL diffuses up from the epi-substrate interface (Figure 3.17). Each only has to cross half the distance of a regular isolation diffusion, and outdiffusion is therefore cut approximately in half.

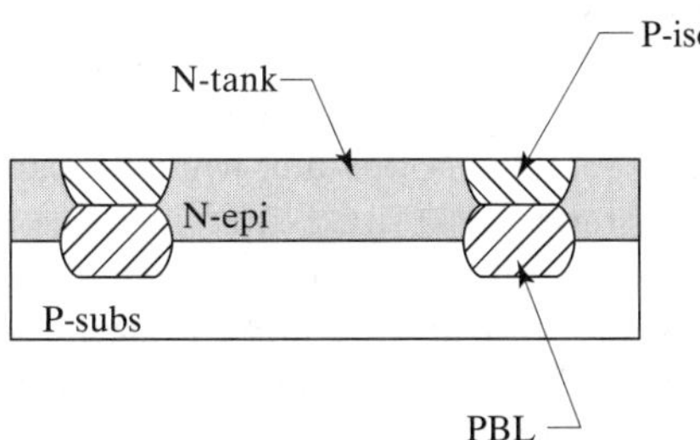

FIGURE 3.17 Cross section of a typical up-down isolation system.

Up-down isolation does have drawbacks. Lateral autodoping limits the PBL implant dose, so the final PBL becomes very lightly doped, and vertical resistance through the up-down isolation greatly exceeds that of conventional top-down isolation. The PBL also requires an additional masking step and diffusion. Up-down isolation saves 15 to 20% die area. Most modern versions of standard bipolar employ up-down isolation.

Double-Level Metal

Standard bipolar originated as a *single-level metal* (SLM) process. The lack of a second metal layer greatly complicated lead routing. Instead of crossing wires by means of jumpers, diffusions were employed to form low-value resistors, called *crossunders* or *tunnels* (Section 13.3.2). Many devices can be custom-tailored to incorporate crossunders at the cost of compromising device performance and increasing die area. Single-level routing requires a deep understanding of device and circuit operation and an intuitive sense of topological connectivity. Most layout designers require years to master these skills.

Double-level metal (DLM) can be added to a standard bipolar process at the cost of two extra masks: vias and metal-2. The thickness of the first metal layer is often reduced to simplify planarization. Double-level metal is a useful, if somewhat costly, option. Lead routing no longer requires the use of customized devices, allowing component standardization and a considerable reduction of layout time and effort. Since metallization consumes a great deal of area, double-level metal can also reduce die area by up to 30%. These benefits are so attractive that manufacturers now routinely employ double-level metal for all new designs. Very few new designs employ single-level metal, and the techniques associated with its use are gradually being forgotten.

Schottky Diodes

Standard bipolar originally used silicon-doped aluminum metallization. Modern processes usually employ a combination of silicidation and refractory barrier metallization to ensure adequate step coverage while maintaining low contact resistance. Along with its more obvious benefits, silicidation also offers the opportunity to fabricate reliable Schottky diodes. Although aluminum forms a rectifying Schottky barrier to lightly doped N-type silicon, the forward voltage of the resulting diodes varies unpredictably depending on annealing conditions. Certain silicides, most notably those of platinum and palladium, produce Schottky barriers with very stable and repeatable properties. The forward voltages of these Schottky diodes lie slightly below that of a moderately doped PN-junction diode, so they can serve as antisaturation clamps (Section 8.1.4).

Schottky diodes require contacts formed through thick-field oxide. A contact etch that just penetrates the field oxide will severely overetch the base and emitter contacts. Overetching can be prevented by performing two consecutive oxide removals, the first of which thins the field oxide over the Schottky contacts and the second of which creates the actual contact openings. The fabrication of Schottky diodes thus requires an additional masking step.

Figure 3.18 shows a typical Schottky diode layout. The anode consists of a rectifying contact to an N-epi tank while the cathode Ohmically contacts the same tank with the assistance of emitter diffusion. The addition of NBL and deep-N+ to the cathode greatly reduces the series resistance of the diode. The anode employs two concentric oxide removals, the larger of which thins the field oxide and the smaller of which forms the actual contact. This two-stage process not only eliminates overetching of base and emitter contacts but also improves the step coverage of metallization to the Schottky contact. This structure readily scales to provide diodes of any size. Electric field intensification at the exposed edges of the Schottky contact opening limits this simple structure's breakdown voltage and causes excessive reverse-bias leakage. Section 10.1.3 discusses methods of circumventing these problems.

High-Sheet Resistors

Accurate resistors in standard bipolar are usually fabricated from base diffusion. Because the sheet resistance of this material rarely exceeds 200 Ω/□, a typical die

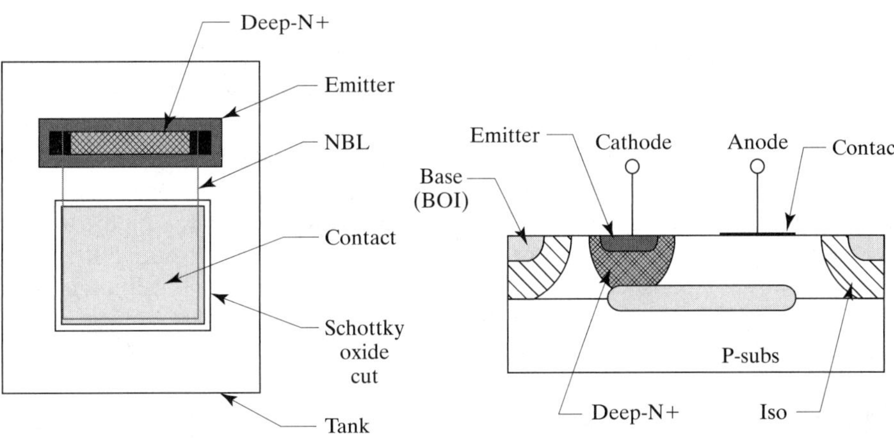

FIGURE 3.18 Layout and cross section of a Schottky diode.

can incorporate a total of only 200 to 500 kΩ of base resistance. Low-current circuits require more resistance than base diffusion can provide and more precision than pinch resistors can achieve. The only solution to this dilemma consists of adding a process extension to fabricate a precisely controlled high-sheet resistance (HSR) material.

The *high-sheet implant* consists of a shallow, lightly doped P-type implant. Depending on dose and junction depth, the sheet resistance of this implant can range from 1 to 10 kΩ/□. The larger sheet values suffer from surface depletion effects that cause resistance variations, so most processes employ HSR implants of 1 to 3 kΩ/□.

Figure 3.19 shows the layout and cross section of a typical HSR resistor. The body of the resistor consists of high-sheet implant, but small regions of base diffusion at either end of the resistor ensure Ohmic contact. The resistor occupies an N-tank and is isolated by the reverse-biased HSR-tank junction. The tank is often connected to the more positive end of the resistor, just as in the case of the base resistor. High-sheet resistors require one additional masking step and one dedicated implant. The cost of this process extension can usually be justified if the circuit includes more than 100 to 200 kΩ of resistance.

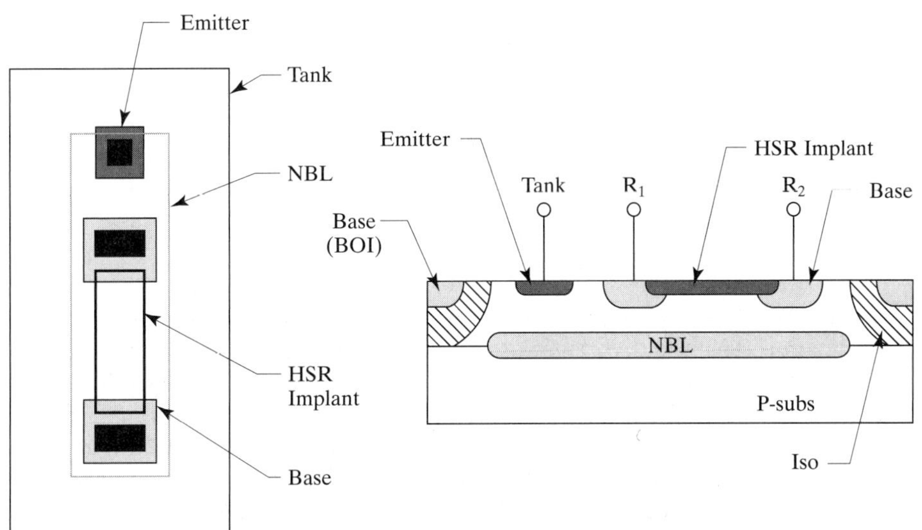

FIGURE 3.19 Layout and cross section of a high-sheet resistor.

Super-Beta Transistors

The standard bipolar NPN provides a reasonable compromise between high beta and adequate operating voltage. Beta can be greatly increased by narrowing the base width. If the process incorporates two separate emitter implants, then one can be optimized for beta and the other for operating voltage. The resulting *super-beta transistors* can achieve betas of 1000 to 3000 (Section 8.2.5). The use of an extremely thin and lightly doped base causes considerable depletion region intrusion into the neutral base and hence compromises not only operating voltage but also Early voltage. These highly specialized devices find application only in a limited range of circuits. For example, they have been used to fabricate bipolar operational amplifiers with extremely low input bias currents.

3.2 POLYSILICON-GATE CMOS

With the addition of two masking steps, standard bipolar can fabricate metal-gate PMOS transistors similar to those supported by early MOS processes (Figure 3.20). An N-tank serves as a backgate for the PMOS transistor; the backgate contact incorporates emitter diffusion to ensure Ohmic contact. None of the oxide layers in standard bipolar is sufficiently thin to serve as a gate oxide. This necessitates the addition of a special masking step that removes the field oxide from the channel region so that a short dry oxidation can form the thin gate oxide. Aluminum metal forms the gate electrode, while the source and drain consist of shallow P+ implants. Since standard bipolar does not include any suitable implant, another masking step patterns a special P-type source/drain (PSD) implant.

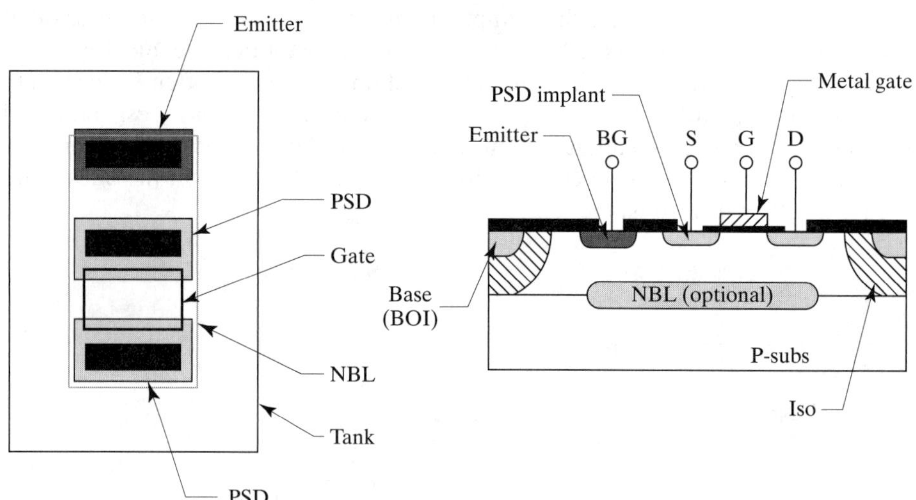

FIGURE 3.20 Layout and cross section of a metal-gate PMOS transistor constructed in standard bipolar.

Practical MOS transistors require threshold voltages that lie within relatively narrow limits. Threshold voltages of less than 0.5 V cause excessive leakage, while those of more than 1.5 V make it difficult to construct low-voltage circuitry. The threshold voltage of an unadjusted (or *natural*) PMOS usually lies between 2 and 4 V, necessitating a threshold adjust implant to shift it to the desired target of about 1 V. The V_t adjust implant is normally performed through the oxide opening patterned by the gate mask, just prior to the growth of the thin gate oxide. The threshold voltage must lie within ±0.5 V of target. The metal gate PMOS transistor has difficulty maintaining even this minimal degree of control. The excess surface state charges introduced by using (111) silicon constitute one source of threshold variation; mobile ion contamination (Section 4.2.2) represents another. Historically, these problems proved extremely difficult to overcome.

Metal-gate MOS transistors also suffer from excessive overlap capacitance. The gate electrode is patterned by a different mask than the source and drain diffusions are. The gate must therefore overlap the source and drain sufficiently to form a continuous channel, even in the presence of photolithographic misalignments. The overlap between the gate and source causes a gate-to-source capacitance C_{gs}, while the overlap between gate and drain causes a gate-to-drain capacitance C_{gd}. These parasitic capacitances slow the transistor because they must be charged and discharged during switching. The gate-to-drain capacitance is particularly deleterious because the voltage gain of the transistor multiplies its apparent value (a phenomenon called the *Miller effect*). These parasitic overlap capacitances must be minimized in order to allow the construction of high-speed logic circuitry.

A complementary NMOS transistor would greatly enhance the utility of the metal-gate PMOS process. Taken together, NMOS and PMOS transistors would allow the construction of versatile *complementary MOS* (CMOS) circuits. Unfortunately, standard bipolar cannot easily fabricate NMOS transistors, because they require a lightly doped P-type backgate that does not exist in this process. The NMOS threshold voltage on suitably doped (111) silicon is slightly negative, so a threshold-adjust implant is required to form an enhancement device. Yet another masking step is required to raise the thick-field threshold in the lightly doped P-type backgate around the NMOS transistors to prevent parasitic channel formation. The relatively poor performance of metal-gate CMOS cannot justify the cost of five additional masking steps, especially when a nine-mask polysilicon-gate CMOS process can fabricate vastly superior transistors. The next section examines the construction and performance of this alternate process.

3.2.1. Essential Features

The polysilicon-gate CMOS process is optimized to form complementary PMOS and NMOS transistors on a common substrate. It does not support the construction of bipolar transistors, and it offers only a limited range of passive components. Originally intended solely for manufacturing CMOS logic gates, with slight modifications this process can also fabricate a limited variety of analog circuits.

A key difference between polysilicon-gate CMOS and standard bipolar lies in the choice of substrate material. Standard bipolar employs (111) silicon to enhance the PMOS thick-field threshold by increasing the surface state density, while polysilicon-gate CMOS uses (100) silicon to reduce the surface state density in order to improve threshold voltage control. A second major innovation lies in the use of polysilicon rather than aluminum as the gate material. Polysilicon can safely withstand the high temperatures required to anneal the source/drain implants, so it can act as a mask to form self-aligned sources and drains. The effects of mobile ion contamination can also be minimized by doping the polysilicon gate with phosphorus. Poly gates thus offer not only faster switching speeds but also better control of threshold voltages.

The choice of threshold voltages forms one of the few differences between analog and digital CMOS processes. Most digital CMOS processes target threshold voltages between 0.8 V and 0.9 V with a variation of ±0.2 V.[9] Analog CMOS designers favor

[9] The information provided in the text applies to processes with operating voltages of 5 V or more. Lower-voltage processes require smaller threshold voltages and tighter control. For example, a 3 V process will typically target a threshold voltage of 0.6 ± 0.1 V.

maximizing headroom by targeting threshold voltages around 0.7 V. Since (100) silicon dictates the use of threshold-adjust implants in either case, the threshold voltages can often be retargeted by simply changing the implant dosage. Analog CMOS also rules out the use of blanket silicidation (Section 3.2.4). Neither of these requirements fundamentally modifies the polysilicon-gate CMOS process.

3.2.2. Fabrication Sequence

The baseline polysilicon-gate CMOS fabrication sequence consists of nine masking operations. The processing steps required to fabricate a finished wafer will be presented in the order in which they are performed. The cross sections used to illustrate this process employ a vertical exaggeration of between two and five, just as did the cross sections previously presented for standard bipolar.

Starting Material

CMOS integrated circuits are normally fabricated on a P-type (100) substrate doped with as much boron as possible in order to minimize substrate resistivity. This precaution helps provide a degree of immunity to *CMOS latchup* by minimizing substrate debiasing (Section 4.4.1). CMOS processes do not require NBL, so substrate doping is limited only by solid solubility.

Epitaxial Growth

The first step of the CMOS process consists of growing a lightly doped P-type epitaxial layer on the substrate. This epitaxial layer, typically some 5 to 10 μm thick, is considerably thinner than the one used for standard bipolar. NMOS transistors are formed directly into the epi layer, which serves as their backgate. Since this process needs no buried layers, epi-coated wafers can be stockpiled to serve as starting material for all types of products. Standard bipolar does not allow this economy of scale, since each product requires a uniquely patterned NBL.

In theory, CMOS processes do not require epitaxy since the MOS transistors can be grown directly into a P–substrate. Epitaxial processing increases costs, but it also improves latchup immunity by allowing the use of a P+ substrate. In addition, the electrical properties of the epitaxial layer are more precisely controllable than those of Czochralski silicon, resulting in better control of MOS transistor parameters.

N-Well Diffusion

After the wafer has been thermally oxidized, a layer of photoresist that has been spun onto it is patterned using the N-well mask. An oxide etch opens windows through which ion implantation deposits a controlled dose of phosphorus. A prolonged high-temperature drive creates a deep, lightly doped N-type region called an *N-well* (Figure 3.21). The N-well for a typical 20 V CMOS process has a junction depth of about 5 μm. Thermal oxidation during the well drive covers the exposed silicon with a thin layer of *pad oxide.*

FIGURE 3.21 Wafer after N-well drive.

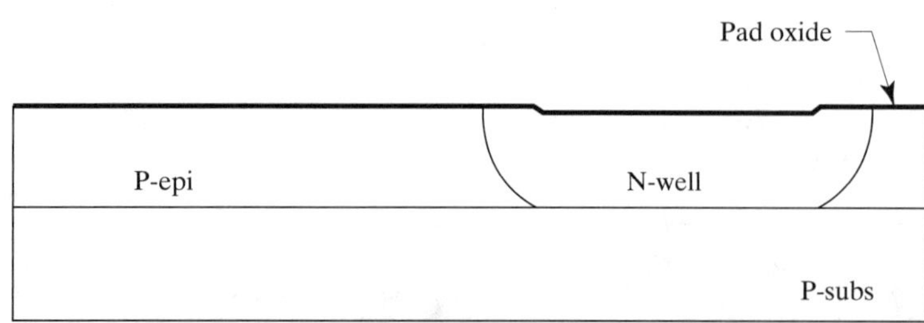

In an *N-well CMOS process,* such as that illustrated in Figure 3.21, NMOS transistors occupy the epi, and PMOS transistors reside in the well. The increased total dopant concentration caused by counterdoping the well slightly degrades the mobility of majority carriers within it. The N-well process therefore optimizes the performance of the NMOS transistor at the expense of the PMOS transistor. As a side effect, the N-well process also produces the grounded substrate favored by most circuit designers.

A *P-well CMOS process* uses an N+ substrate, an N− epitaxial layer, and a P-well. NMOS transistors are formed in the P-well and PMOS transistors in the epi. This process optimizes the PMOS transistor at the expense of the NMOS transistor, but the NMOS still outperforms its counterpart because electrons are more mobile than holes. A P-well process requires that the substrate connect to the highest-voltage supply instead of ground. Designs that employ multiple power supplies referenced to a common ground often have difficulty biasing an N-type substrate because of ambiguities in the sequencing of the supplies.

Both P-well and N-well CMOS processes exist. The N-well process offers a slightly better NMOS transistor, and it allows the use of a grounded substrate. N-well CMOS is also upwardly compatible with BiCMOS technology, as will become apparent later in this chapter. The N-well process has therefore been chosen to illustrate CMOS technology.

Inverse Moat

The CMOS process employs a thick-field oxide for much the same reasons as standard bipolar: It increases the thick-field threshold voltages, and it reduces parasitic capacitance between the metallization and the underlying silicon. Unlike standard bipolar, the baseline CMOS process employs LOCOS technology to selectively grow the field oxide, leaving only a thin pad oxide over the regions where active devices will be formed. The locally oxidized regions of the die are called *field regions,* while the areas protected from oxidation are called *moat regions.*

The LOCOS process uses a patterned nitride layer formed by first depositing nitride across the entire wafer, then patterning this nitride by using the inverse moat mask, and finally employing a selective etch to remove the nitride over the field regions (Figure 3.22). The photomask used for this step is called the *inverse moat* mask because it consists of a color reverse of the moat regions. In other words, the mask codes for areas where moat is absent, not where it is present.

The nitride layer used for LOCOS must lie on top of a thin oxide layer (called the *pad oxide*) because the conditions of nitride growth induce mechanical stresses that can cause dislocations in the silicon lattice. The pad oxide provides mechanical compliance that absorbs the strain and prevents it from damaging the silicon.

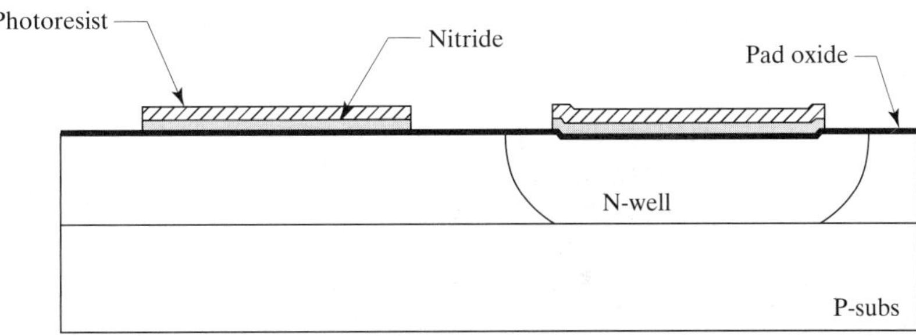

FIGURE 3.22 Wafer after nitride deposition and inverse moat pattern.

Channel Stop Implants

The CMOS process deliberately minimizes threshold voltages in order to produce practical MOS transistors. The LOCOS field oxide will raise the thick-field thresholds, but not by more than a few volts. Dopants are usually selectively implanted beneath the field regions to further raise the threshold voltages of the thick-field transistors. P-epi field regions receive a P-type *channel stop implant,* while N-well field regions receive an N-type channel stop implant. The formation of channel stops thus requires two successive ion implantations.

Various techniques have been developed to produce channel stops. The method presented here involves the use of a blanket boron implant followed by a patterned phosphorus implant. The boron implant uses the photoresist left from patterning the LOCOS nitride. This mask exposes the field regions where channel stops will be deposited, so all of these regions receive the blanket boron implant (Figure 3.23A). This step sets the thick-field threshold in the epi regions.

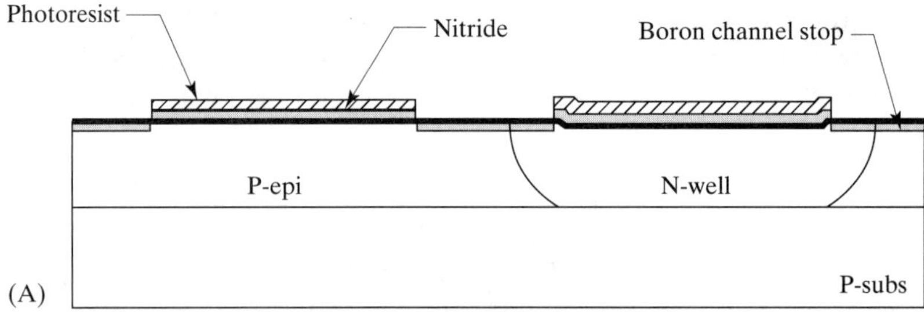

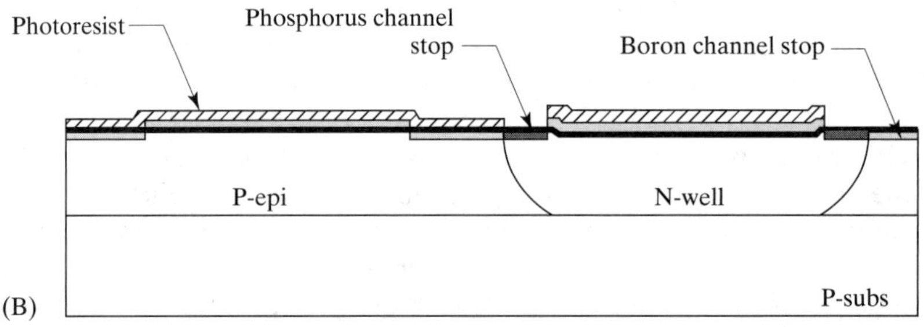

FIGURE 3.23 Wafer after blanket boron channel stop implant (A) and after selective phosphorus channel stop implant (B).

The wafer is again coated with photoresist immediately after the boron implant. The previous photoresist can remain in place, since the channel stop implant will not affect the moat regions that lie beneath it. The recoated wafer is patterned using the channel stop mask, exposing only N-well field regions. The subsequent phosphorus implant counterdopes the previous blanket boron implant and raises the NMOS thick-field threshold above the maximum operating voltage (Figure 3.23B). Following the phosphorus implant, all photoresist is stripped from the wafer in preparation for LOCOS oxidation.

LOCOS Processing and Dummy Gate Oxidation

Steam is often used to increase the rate of LOCOS oxidation; alternatively, the furnace pressure can be raised to 5 or 10 times atmospheric. After LOCOS oxidation, a suitable etchant strips away the remnants of the nitride block mask. Figure 3.24 shows a cross section of the resulting wafer. The curved transition region at the edges of the moat, called a *bird's beak*, results from oxidants diffusing under the edges of the nitride film.

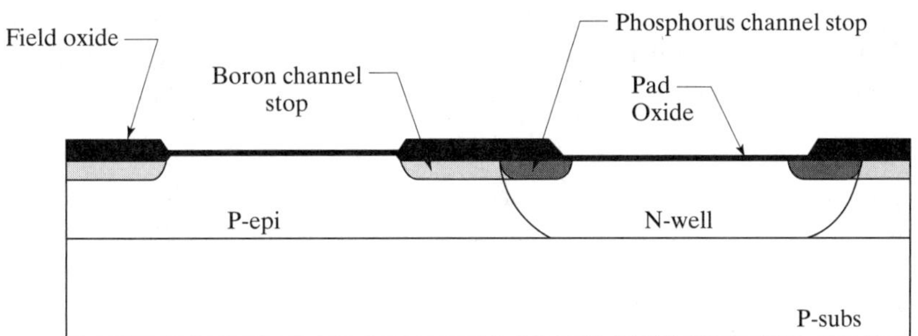

FIGURE 3.24 Wafer after LOCOS oxidation and nitride strip.

The *Kooi effect* (Section 2.3.4) causes nitride deposits to form underneath the pad oxide around the edges of the moat. These deposits can potentially cause gate oxide integrity failures, but they can be eliminated by a dummy gate oxidation. A brief etch strips away the thin pad oxide without substantially eroding the thick-field oxide. Next, a brief dry oxidation grows a thin layer of oxide called a *dummy gate oxide* (or *sacrificial gate oxide*) in the moat regions. Any nitride deposits that remain will gradually oxidize. All of the nitride will be consumed if the dummy gate oxidation continues for a sufficient length of time.

Threshold Adjust

The use of (100) silicon helps stabilize the threshold voltages of the MOS transistors, but the backgate dopings and gate electrode materials preclude the achievement of usable threshold voltages without threshold adjust implants. For example, the unadjusted PMOS threshold might range from −1.5 to −1.9 V, while the NMOS threshold might range from −0.2 to 0.2 V. One or two threshold adjust implants (also known as *V_t adjusts*) retarget the threshold voltages to the desired targets, usually 0.7 V for NMOS and −0.7 V for PMOS transistors.

Two methods exist for adjusting threshold voltages. The first method employs two separate implants, one to set the PMOS V_t and the other the NMOS V_t. The use of two implants allows independent optimization of both thresholds. Many processes do not require this degree of flexibility. These processes can use a single V_t adjust to simultaneously reduce the PMOS threshold and increase the NMOS threshold. If this implant is properly performed, a nominal threshold voltage of 0.7 to 0.9 V can be obtained for both types of MOS transistors. Figure 3.25 illustrates this approach.

After the wafer has been coated with photoresist, the V_t adjust mask is used to open windows over areas where MOS transistors will form. The boron V_t adjust implant penetrates the dummy gate oxide to dope the underlying silicon. After the V_t adjust implant, the dummy gate oxide is stripped away to reveal bare silicon in the moat regions.

The true gate oxidation employs dry oxygen to minimize excess charge incorporation due to surface states and fixed oxide charges. This oxidation must be very brief, because gate oxides are exceedingly thin. A 10 V MOS transistor typically requires a

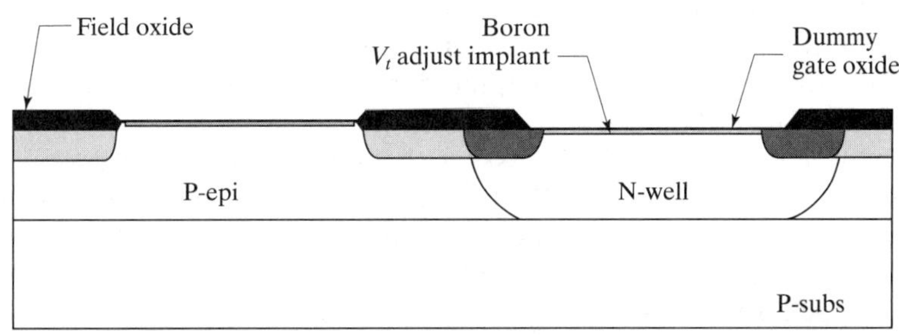

FIGURE 3.25 Wafer after V_t adjust implant.

300 Å (0.03 μm) gate oxide, while a 3 V transistor may employ an oxide less than 100 Å (0.01 μm) thick. This gate oxide will form the dielectric of the MOS transistors; it also covers the regions where source and drain implants will later occur.

Polysilicon Deposition and Patterning

The polysilicon layer used to form gate electrodes is heavily doped with phosphorus to reduce its resistivity to about 20 to 40 Ω/□. Although gate leads do not conduct DC current, switching transients do draw substantial AC current, and low-resistance gate polysilicon greatly improves the switching speeds of MOS circuitry. Phosphorus doping (rather than boron doping) produces threshold voltages compatible with a single-step V_t adjust. Phosphorus-doped gate polysilicon also minimizes threshold voltage variation due to mobile ions, allowing threshold voltage control of ±0.1 to 0.2 V. While it is possible to dope polysilicon during deposition, most processes first deposit intrinsic polysilicon and subsequently dope it by means of conventional deposition or implantation techniques.

The deposited polysilicon layer must now be patterned using the poly mask (Figure 3.26). Modern submicron processes can fabricate polysilicon gates less than 0.5 μm long, and any variation in gate length directly affects the transconductance of the resulting transistors. Thus, the patterning and etching of poly form the most critical photolithographic steps of a CMOS process. The simple process discussed here produces a minimum channel length of about 2 μm and therefore does not require as high a degree of precision as submicron processes, but polysilicon patterning still remains its most challenging photolithographic step.

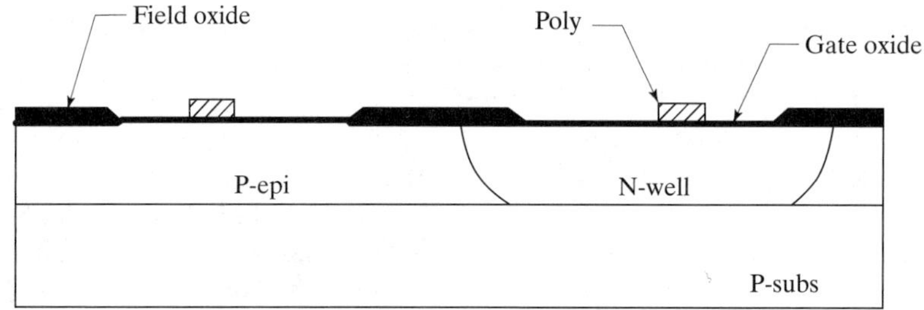

FIGURE 3.26 Wafer after polysilicon deposition and pattern. For simplicity, the channel stop and threshold adjust implants do not appear in this or subsequent cross sections.

Source/Drain Implants

The completed polysilicon gates now act as masks that self-align the source/drain implants for both the PMOS and NMOS transistors. These implants can be performed in either order. In the illustrated process, the N-type source/drain (NSD) implant occurs first, followed by the P-type source/drain (PSD) implant.

The NSD implant begins with the application of photoresist to the wafer, followed by patterning using the NSD mask. Shallow, heavily doped N-type regions are then formed by implanting arsenic through the exposed gate oxide. The polysilicon gate blocks this implant from the regions directly underneath the gate and therefore minimizes the gate/source and gate/drain overlap capacitances. Once the NSD implant has been completed, the photoresist residue is stripped from the wafer. The PSD implant begins with the application of a second photoresist layer patterned using the PSD mask. A shallow, heavily doped P-type region is formed by implanting boron through the exposed gate oxide. As with the NSD implant, the PSD implant self-aligns to the polysilicon, and the PMOS transistors also exhibit minimal overlap capacitance. Following the PSD implant, photoresist is again stripped from the wafer.

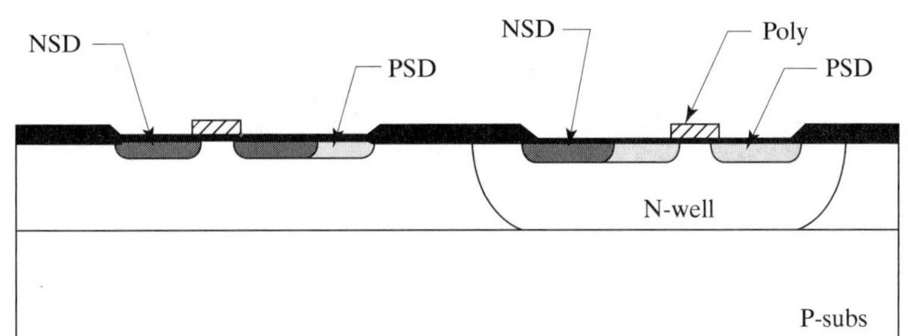

FIGURE 3.27 Cross section of the wafer after NSD and PSD implants and anneals. The backside contact implants abut the source implants to save space.

A brief anneal activates the implanted dopants and slightly thickens the oxide over the source and drain regions. This anneal is the final high-temperature step of the process, corresponding to the emitter drive of standard bipolar. Figure 3.27 shows a cross section of the wafer following the source/drain anneal.

Contacts

Despite further oxidation during the source/drain anneal, the oxide covering the moat regions remains thin and therefore vulnerable to oxide rupture. Most processes deposit a *multilevel oxide* (MLO) before contact patterning. The MLO thickens the oxide over the moat regions and at the same time coats and insulates the exposed polysilicon pattern. Metal leads can now run over moat regions and polysilicon gates without risk of oxide rupture.

After the wafer is again coated with photoresist, the contact regions are patterned using the contact mask. Ohmic contacts form to the heavily doped source and drain without difficulty, but the backgate regions are far too lightly doped to allow direct Ohmic contact. The addition of NSD and PSD implants in the vicinity of the backgate contacts overcomes this difficulty. Contacts opened over polysilicon allow contact to the gate electrodes.

Metallization

The shallow NSD and PSD diffusions are vulnerable to junction spiking. Most CMOS processes employ a combination of contact silicidation and refractory barrier metallization to ensure reliable contact to the source/drain regions. After formation of silicide in the contact openings, a thin film of refractory metal sputtered over the wafer precedes a much thicker layer of copper-doped aluminum. The metallized wafer is coated with photoresist and patterned, using the metal mask. A suitable etchant then removes unwanted metal to form the interconnection pattern. Most processes also include a second layer of metallization. In such a process, another layer of oxide deposited over the first metal pattern insulates it from the second metal pattern. This second deposited oxide is usually called an *interlevel oxide* (ILO). Some form of planarization minimizes the nonplanarities caused by the first metal pattern to ensure adequate second metal step coverage. Vias etched through the ILO connect to a second metal layer deposited and patterned in much the same way as the first. If the process includes additional metal layers, these are formed in the same manner as the second metal layer.

Protective Overcoat

A protective overcoat is now deposited over the final layer of metallization, both to provide mechanical protection and to prevent contamination of the die. The protective overcoat must resist penetration by mobile ions, so it normally consists of either a thick phosphosilicate glass (PSG), a compressive nitride layer, or both.

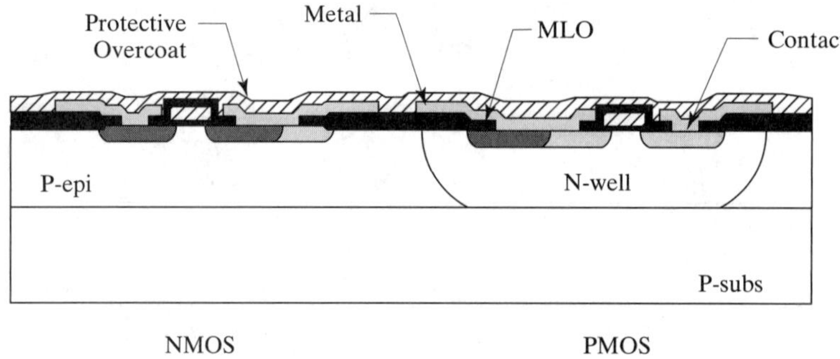

FIGURE 3.28 Cross section of the completed polysilicon-gate CMOS wafer.

After coating with photoresist, the wafer is patterned using the *protective overcoat* (PO) mask. A suitable etchant removes the overcoat over selected areas of metallization and allows attachment of bondwires to the integrated circuit. This composes the final fabrication step; the wafer is now complete. Figure 3.28 shows a cross section of the resulting wafer, with only a single level of metal for simplicity. No bondpad openings exist in the illustrated portion of the die. This cross section includes an NMOS transistor on the left and a PMOS transistor on the right.

3.2.3. Available Devices

Polysilicon-gate CMOS was originally developed to provide relatively low-voltage NMOS and PMOS transistors. The same process steps can fabricate several other components, including natural MOS transistors, a substrate PNP, several types of resistors, and a capacitor. Together these components allow the construction of a considerable variety of analog circuits. Section 3.2.4 examines process extensions that allow higher operating voltages and denser circuit packing.

NMOS Transistors

Figure 3.29 shows a representative layout and cross section of an NMOS transistor. The source and drain regions consist of NSD implants that self-align to the polysilicon gate. Since the backgate of the NMOS consists of the P-epi (and by extension the substrate), any substrate contact on the die will serve as a backgate terminal for the transistor. Many layouts actually include separate backgate contacts immediately adjacent to each NMOS transistor, even though these are not strictly necessary. The close proximity of these backgate contacts improves CMOS latchup immunity, and this arrangement ensures that an adequate number of substrate contacts are distributed throughout the layout. In cases where the source of the NMOS transistor connects to the substrate potential, a very compact layout can be achieved by

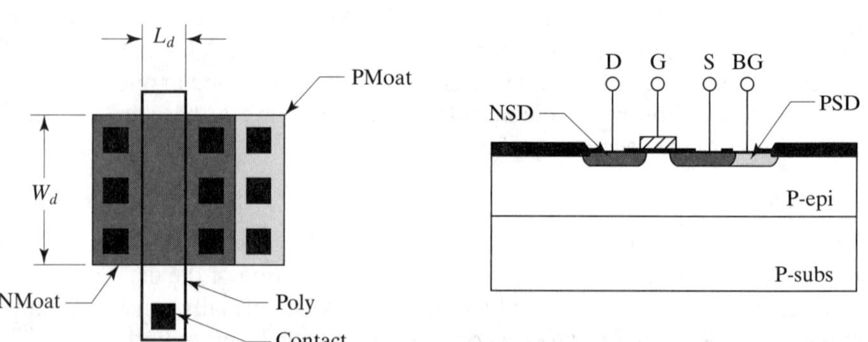

FIGURE 3.29 Layout and cross section of an NMOS transistor. The source and backgate of this transistor are shorted together using metal (not shown).

butting the PSD substrate contact against the NSD source. PSD and NSD cannot abut one another if they connect to different potentials because the resulting P+/N+ junction leaks excessively.

Figure 3.29 illustrates the common practice of coding CMOS transistors by using drawing layers called *NMoat* and *PMoat*. These layers do not correspond to individual masks but rather to combinations of several masks. A figure drawn on the NMoat layer simultaneously produces figures on both the NSD and moat masks. Similarly, a figure drawn on PMoat simultaneously generates figures on both the PSD and moat masks. The moat geometries are typically the same size as the NMoat and PMoat regions that generate them, while the NSD and PSD geometries are slightly oversized to ensure that the source/drain implant completely covers the moat opening. The use of PMoat and NMoat drawing layers simplifies the layout by reducing the number of figures required to draw a transistor.

The arrays of small square contacts shown in Figure 3.29 are characteristic of CMOS processes. The etch rate of oxide windows varies somewhat depending upon their size and shape, and these variations become particularly severe for very small openings. Many processes therefore allow only a single size of contact opening, usually consisting of a minimum-dimension square. Larger contacts must consist of arrays of minimum contacts rather than larger oxide openings.

In Figure 3.29, W_d and L_d denote the *drawn width* and *drawn length* of the NMOS transistor, respectively. The names given to these two dimensions may seem counterintuitive since the length of the gate is actually the width of the drawn poly silicon strip, but the channel length is customarily defined as the separation between the source and drain regions. The transconductance of a MOS transistor is approximately proportional to the ratio of the channel width divided by the channel length. Short channel lengths generate more transconductance per unit area, but analog circuits often employ longer channels to reduce channel length modulation.

Hot electron degradation limits the simple NMOS transistor of Figure 3.29 to relatively low operating voltages. The large electric field present across the pinched-off portion of a saturated MOS transistor's channel accelerates electrons to high velocities. A few of these hot electrons are injected into the overlying oxide, leading to a gradual degradation of transistor performance (Section 4.3.1).

Hot electron injection only occurs when the NMOS transistor operates in saturation with a relatively large drain-to-source bias. In the linear region, the drain-to-source voltage is too small to produce hot electrons, and in cutoff no conduction occurs. NMOS transistors used as switches experience hot electron injection only during switching transients. If the switching frequency remains fairly low, then the total quantity of hot electrons produced over the operating lifetime of the integrated circuit remains acceptably small. Two different operating voltages are often specified for NMOS transistors. Junction breakdown and punchthrough limit the *blocking voltage rating,* which applies to transistors used as switches and as components of low-frequency digital logic. The somewhat lower *operating voltage rating* determined by the onset of hot electron degradation applies to transistors that operate for an appreciable length of time in saturation (as do the majority of analog transistors).

The V_t adjust implant shifts the NMOS transistor threshold voltage from about zero volts to approximately 0.7V. The V_t adjust mask can block the implant from an NMOS transistor to form a *natural NMOS* possessing a relatively low threshold voltage. Natural NMOS transistors are used in certain analog circuits where the normal threshold is inconveniently large. Table 3.4 lists the device characteristics of NMOS and PMOS transistors constructed on a typical 2 μm, 10 V analog CMOS process.

TABLE 3.4 Typical polysilicon-gate CMOS device parameters.[10]

Parameter	NMOS	PMOS
Minimum channel length	2 μm	2 μm
Gate oxide thickness	400 Å	400 Å
Threshold voltage (adjusted)	0.7 V	−0.7 V
Threshold voltage (natural)	0 V	−1.4 V
Transconductance (W_d/L_d = 10/10)	50 μA/V^2	20 μA/V^2
Operating voltage	7 V	15 V

PMOS Transistors

Figure 3.30 shows a representative layout and cross section of a PMOS transistor. This device resides in an N-well that acts as its backgate. Any number of PMOS transistors can occupy the same well as long as their backgates all tie to the same potential. The relatively deep N-well outdiffuses substantially, and the layout dimensions associated with it become quite large. Merging PMOS transistors in the same well therefore saves substantial area. While the backgate of an NMOS transistor inherently connects to substrate, the backgate of a PMOS transistor can connect to any potential greater than or equal to its source. The N-well backgate thus provides an extra degree of freedom that analog designers frequently employ to enhance circuit performance.

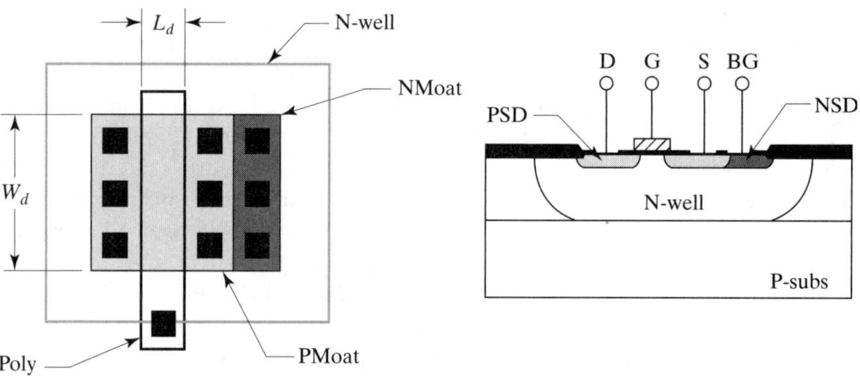

FIGURE 3.30 Layout and cross section of a PMOS transistor. The backgate and source of this transistor are shorted together using metal (not shown).

PMOS transistors are subject to *hot hole degradation,* but this causes fewer problems than hot electron degradation because holes are less mobile than electrons. A higher electric field and therefore a larger drain-to-source voltage is required to accelerate holes to velocities sufficient to inject charge into the oxide. Junction avalanche and punchthrough often limit PMOS transistors to voltages where hot hole degradation remains unimportant. Higher-voltage PMOS transistors will encounter hot-carrier problems similar to those of NMOS transistors.

A *natural PMOS* transistor can be fabricated by blocking the V_t adjust implant from the channel region of the device. Natural PMOS transistors possess inconveniently high threshold voltages, usually in excess of a volt. These transistors see only

---

[10] These parameters are roughly analogous to those of the 3 μm Advanced LinCMOS™ process described in R. K. Hester, L. Hutter, L. Le Toumelin, J. Lin, and Y. Tung, "Linear CMOS Technology," *TI Technical Journal,* Vol. 8, #1, 1991, pp. 29–41. (The trademark belongs to Texas Instruments.) The transconductance figures given in this text are those used in the Schichman-Hodges equation $I_d = \frac{1}{2}k\,(W/L)\,(V_{gs} - V_t)^2$.

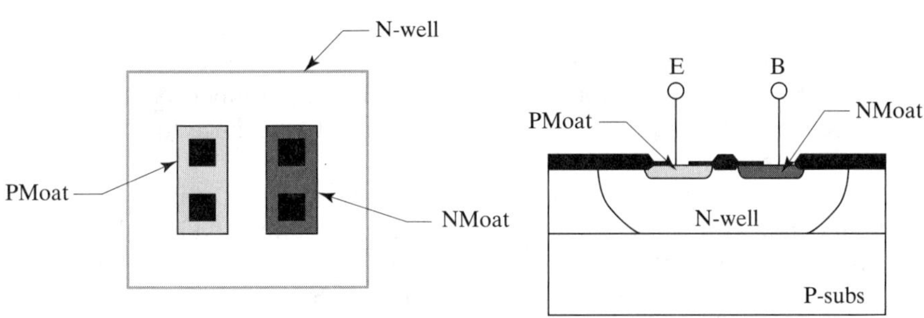

FIGURE 3.31 Layout and cross section of a substrate PNP transistor. The collector is connected by means of substrate contacts (not illustrated).

occasional use in analog circuit design. Table 3.4 lists typical device parameters for a 2 μm, 10 V PMOS transistor.

Substrate PNP Transistors

The only bipolar transistor available in an N-well process is a substrate PNP. Figure 3.31 shows a typical layout and cross section for this device. The emitter consists of a PSD implant formed in an N-well that acts as the base of the transistor. A slug of NSD provides Ohmic contact to the N-well. The collector of this device consists of the P+ substrate and P-epi surrounding the well.

Although CMOS processes do not deliberately optimize bipolar components, the substrate PNP can still provide reasonably good performance. Its beta may approach that of a standard bipolar device (50 to 100) on processes having relatively deep source/drain regions. Newer processes with very shallow source/drain implants may exhibit betas as low as 10–20.

Since this transistor injects current into the substrate, care must be taken to provide adequate substrate contact. The main component of collector resistance consists of the lightly doped P-epi layer interposed between the P+ substrate and the PSD diffusion beneath the substrate contacts. A large substrate contact area is necessary to prevent substrate debiasing. A typical CMOS integrated circuit incorporates enough substrate contact in the scribe street to accommodate 10 to 20 mA of substrate current. If higher substrate currents will occur, then unused areas of the die should be filled with substrate contacts.

Although a lateral PNP transistor can theoretically be constructed in an N-well process, the absence of NBL encourages substrate injection, and only a small fraction of the total emitter current reaches the intended collector. These transistors exhibit extremely low apparent gains, rendering them of limited use to the analog circuit designer.

Resistors

The most useful of the four resistors available in polysilicon-gate CMOS consists of doped polysilicon (Figure 3.32). Although gate polysilicon exhibits a sheet resistance of only 20 to 30 Ω/□, very narrow widths and spacings allow substantial resistance per unit area. A 2 μm process can produce polysilicon resistors as area-efficient as

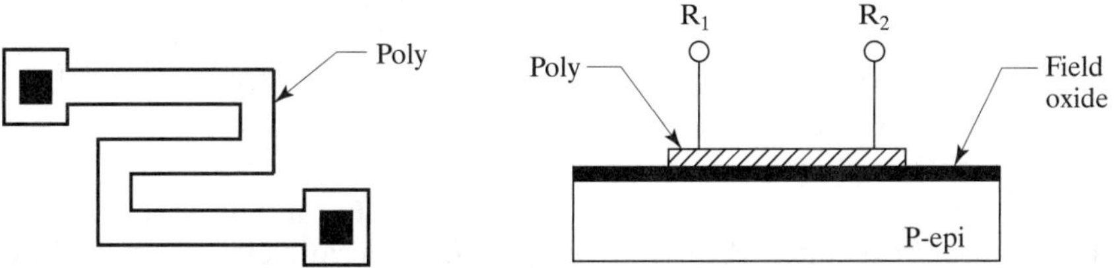

FIGURE 3.32 Layout and cross section of a poly resistor.

standard bipolar base resistors. Submicron processes can provide remarkable amounts of resistance from narrow polysilicon traces, but the tolerances and matching of such resistors leave much to be desired. In a clad-poly process, a silicide block mask becomes necessary to obtain sufficient sheet resistance to construct practical poly resistors.

The polysilicon resistor of Figure 3.32 consists of a strip of poly deposited on top of field oxide. Contacts at either end allow it to be connected into the circuit. Oxide completely isolates this resistor, enabling it to be biased in any manner desired. Poly resistors can withstand large voltages relative to the substrate (100 V or more) and can operate below substrate potential or above the positive supply voltage. The thick-field oxide also reduces parasitic capacitance between the resistor and the underlying substrate. Oxide isolation has one drawback: It isolates the resistor thermally as well as electrically. A poly resistor that dissipates sufficient power will experience permanent resistance variations due to self-induced annealing. Extreme power dissipation will melt or crack polysilicon long before diffused resistors of similar dimensions suffer damage. This behavior allows the construction of polysilicon fuses for wafer-level trimming, but it makes poly resistors undesirable for certain specialized applications, such as ESD protection.

Figure 3.33A shows the layout and cross section of an NSD resistor formed by contacting either end of a strip of NSD diffusion. NSD typically has a sheet resistance of 30 to 50 Ω/□. Avalanche breakdown of the relatively shallow NSD diffusion limits the operating voltage of this resistor, typically to no more than 10 to 15 V. A similar resistor can also be constructed from PSD (Figure 3.33B). This resistor consists of a strip of PSD diffusion contained in an N-well region. The well must be biased above the resistor to maintain isolation. The well is therefore connected either to the more positive end of the resistor or to a high-voltage node (for example, the positive supply). PSD resistors also suffer from limited sheet resistance and a relatively low breakdown voltage.

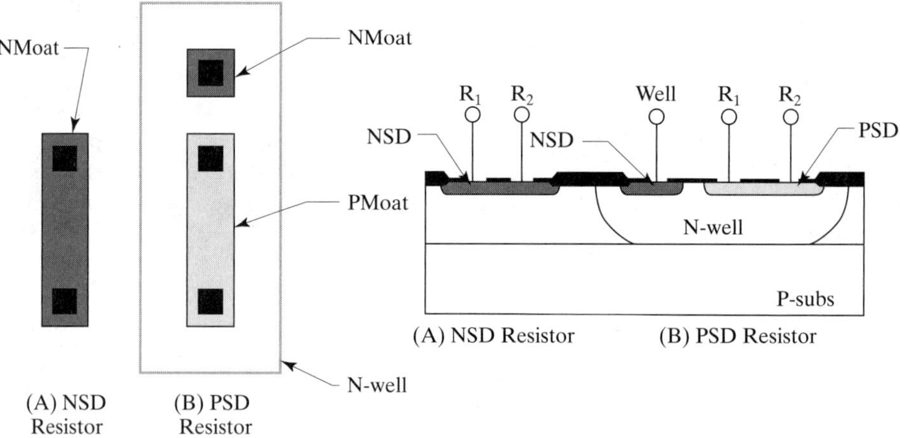

FIGURE 3.33 Layout and cross section of an NSD resistor (A) and a PSD resistor (B).

Another type of resistor consists of a strip of N-well with contacts at either end. NSD placed beneath these contacts prevents the formation of rectifying Schottky barriers. The large spacings required to allow for outdiffusion partially offset the high sheet resistance of the well (typically some 2 to 5 kΩ/□). Well resistors are notoriously variable. Slight differences in doping and outdiffusion, voltage modulation of the depletion regions, and surface effects can all cause significant variations in well resistance. Proper layout precautions can help minimize these sources

Parameter	Poly	PSD	NSD	Well
Sheet resistance	20 Ω/□	50 Ω/□	50 Ω/□	2000 Ω/□
Minimum drawn width	2 μm	3 μm	3 μm	5 μm
Breakdown voltage	>100 V	15 V	15 V	50 V
Variability (5 μm width)	±30%	±20%	±20%	±50% (10 μm)

TABLE 3.5 Typical resistor device parameters.

of variability, but most designers prefer to employ narrow polysilicon resistors. Table 3.5 summarizes the properties of the four types of resistors available in a typical CMOS process.

Capacitors

The gate oxide used to fabricate MOS transistors can also be employed to construct capacitors. One plate of the capacitor consists of doped polysilicon and the other of a diffusion, typically N-well. Figure 3.34 shows the layout and cross section of one type of MOS capacitor. The capacitance of this device typically ranges from 0.5 to 1.5 fF/μm^2, depending on oxide thickness. The tight control of gate oxide thickness characteristic of modern MOS processes results in typical tolerances of ±20% as long as the well electrode remains at least 1 V above the poly electrode. Failure to maintain adequate bias across the capacitor will cause a dramatic drop in capacitance (Section 6.2.2). The main drawbacks of these capacitors consist of excessive bottom-plate parasitic junction capacitance and series resistance, and of nonlinearity effects at certain voltages.

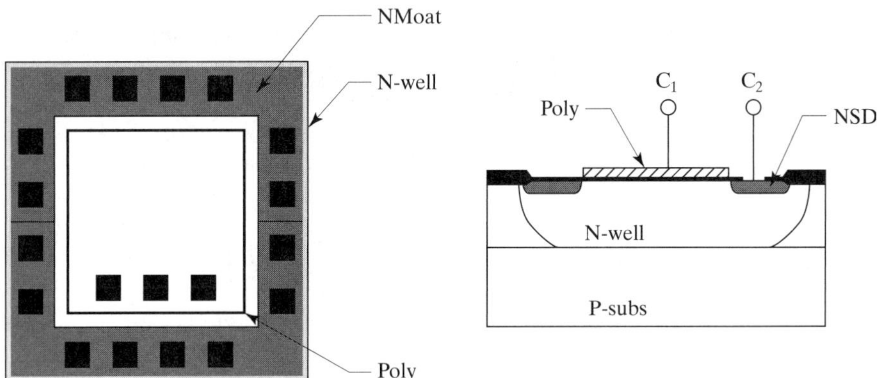

FIGURE 3.34 Layout and cross section of a PMOS capacitor.

3.2.4. Process Extensions

CMOS process extensions tend to focus on improving the PMOS and NMOS transistors rather than on constructing additional devices. One set of extensions seeks to provide higher operating voltages by suppressing hot carrier degradation. Another set of extensions focuses on reducing the size of the transistors to improve packing. Unlike standard bipolar, little emphasis is placed on providing a large number of specialized components, because most CMOS fabs build primarily digital products. The large expenditure of time and money required to construct additional devices cannot be justified by the relatively small volume of analog CMOS products. Analog BiCMOS presents an entirely different picture because this process caters specifically to analog design. Many of the features and process extensions of BiCMOS can be retrofitted into a CMOS process if sufficient economic incentive exists.

Double-Level Metal

The early CMOS processes used one metal and one poly layer (a combination sometimes called $1\frac{1}{2}$-*level metal*). Polysilicon can be used to jumper most signals, so routing is much easier than for single-level metal. Autorouting software still cannot efficiently route $1\frac{1}{2}$-level metallization, so most modern CMOS processes add at least a second layer of metal. (Many add more.) Analog CMOS layouts can benefit from using additional metal layers to increase packing density. Since planarization becomes more difficult as device sizes shrink, CMOS metallization tends to be substantially thinner than that used for standard bipolar. Higher-power CMOS circuits, such as output drivers, often benefit from combining multiple metal layers to form a thicker conductor.

Double-level metal adds two mask steps to the process: one for vias and one for metal-2. An *interlevel oxide* (ILO) deposited between the two metal layers provides insulation, and some form of planarization improves planarity for the second-level metal. Although these extra processing steps increase the cost of the wafer, most CMOS fabs routinely employ double-level metal for the majority of their products. Some processes use three, four, or even five metal layers to further reduce the area required for interconnection in high-density autorouted logic.

Shallow Trench Isolation

The polysilicon-gate CMOS process presented in the foregoing discussion uses LOCOS field oxidation. Lateral oxidant diffusion creates a gradual transition from the thin gate oxide to the thick field oxide. This transition region is often called a *bird's beak* because its cross-section vaguely resembles the head of a bird seen in profile. The thick field oxide forms the bird's neck, while the thin gate oxide forms its beak. The original process designers considered the bird's beak beneficial because it moderated what would otherwise have become a significant oxide step. As transistor dimensions shrank, the lateral dimensions of the bird's beak eventually became intolerably large. Modern CMOS processes have therefore had to turn to other isolation techniques, most notably *shallow trench isolation* (STI).

Shallow trench isolation can form a thick field oxide of arbitrary depth while leaving a planar surface (Section 2.5.3). The lateral dimensions of the trench equal only a fraction of its depth, and its sidewalls are nearly vertical. These features allow the fabrication of extremely small CMOS transistors. Modern digital processes combine STI with shallow buried oxides to achieve full dielectric isolation. These processes currently have little relevance to analog designers because of their extremely limited operating voltages and poor analog transistor characteristics.

The same techniques used to form shallow trench isolation for a digital process can be used to form somewhat deeper trench isolation for analog processes. Trench isolation provides many benefits. First, it allows the use of an arbitrarily thick oxide to achieve any desired thick field threshold. Second, it speeds the operation of bipolar transistors by eliminating the sidewall capacitance of the N-well/P-epi junction. Third, it greatly reduces lateral outdiffusion and therefore allows much tighter spacings between components. Fourth, trench isolation increases latchup immunity. Because the trench penetrates more deeply into the silicon than any grown oxide, the carriers must travel in downward arcs to reach adjacent devices. A longer carrier path implies weaker device interactions. Fifth, trench isolation produces a highly planarized surface compatible with advanced metal systems.

Although trench isolation has numerous advantages, it also encompasses certain disadvantages. As discussed in following sections, the abrupt transition between oxide and silicon makes trench isolation worthless as a field relief structure for extended-drain transistors. Secondly, etching and planarization usually impose

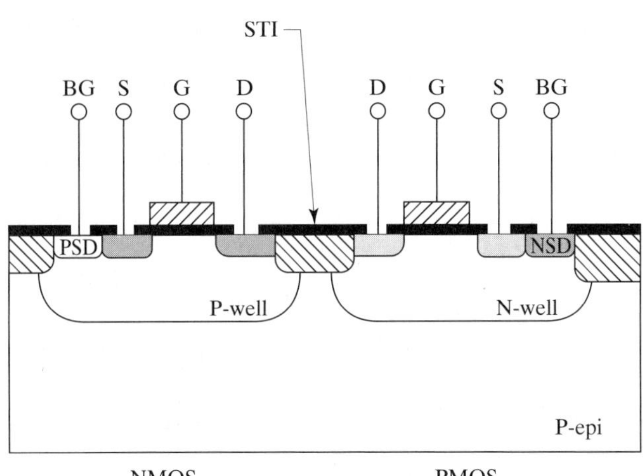

FIGURE 3.35 Cross section of NMOS and PMOS transistors using shallow trench isolation.

stringent layout restrictions. For example, most processes require that all trenches be of a single width and run either horizontally or vertically to ensure uniform etching. Corners often require special treatment, such as fillets or arcs. Deeper trench systems may not even allow two trenches to join because of planarization difficulties.

Figure 3.35 shows a cross section of a pair of transistors in a STI CMOS process. This process does not employ a moat mask, so NSD and PSD regions are defined by the intersection of their respective implants with the silicon surface. These junctions terminate on the trench sidewall, allowing the source/drain regions to self-align to both isolation and gate.

Silicidation

CMOS processes make extensive use of silicides. In addition to contacts, gate poly is often silicided to reduce its sheet resistance. Some processes even silicide source and drain diffusions to reduce their parasitic resistance. Processes with submicron feature sizes may use all of these forms of silicidation. Older processes with larger feature sizes cannot construct sufficiently fast transistors to justify siliciding their gates and source/drain regions, but they usually employ silicided contacts to prevent contact spiking.

A Schottky diode can be formed on an N-well CMOS process that employs platinum or palladium silicides. The Schottky anode consists of a silicided contact, while the cathode consists of N-well contacted by means of an NSD diffusion. The lack of a buried layer increases the internal resistance of this diode and prevents it from being used for high-current applications. The exposed edges of the silicide limit the reverse breakdown voltage, but techniques exist for suppressing edge breakdown without creating any additional process steps (Section 10.1.3). CMOS processes do not require a Schottky contact mask since a moat region can serve the same function. Unfortunately, most modern CMOS processes favor the use of refractory silicides, such as those of titanium or cobalt, over the noble silicides. Refractory silicides are not suitable for use in Schottky diodes because their low forward voltages cause excessive leakage.

Silicidation of the gate poly reduces its sheet resistance to about 2 Ω/□, which is too low for constructing most resistors. Poly resistors can still be formed in a silicided-gate process by adding a *silicide block mask*. This mask patterns the silicide layer, either by preventing metal deposition over resistors or by allowing its removal in these areas prior to sintering. The use of a silicide block mask complicates the silicidation process, but is necessary for most analog designs.

Some processes also silicide NSD and PSD diffusions to form so-called *clad moats*. These cannot be used as resistors since their sheet resistances approximate that of their silicide layer (typically about 2 Ω/□). A silicide block mask can prevent the silicidation of selected PSD and NSD regions to allow their use as resistors. A silicide block mask provides certain other benefits. Blocking silicidation of the emitter of a bipolar transistor often increases its beta (Section 8.3.1). Blocking silicidation from a portion of the source/drain diffusion of a MOS transistor often increases its ability to withstand ESD stresses.

Lightly Doped Drain (LDD) Transistors

If MOS transistors must operate at high drain-to-source voltages, then precautions become necessary to prevent hot carrier degradation. Typical 400 Å NMOS transistors with 3 μm channel lengths can indefinitely withstand 5 to 10 V, while 400 Å PMOS transistors of the same dimensions can withstand 15 to 20 V. Voltages beyond these limits require alternative structures.

The strong electric field across the pinched-off channel of a saturated MOS transistor causes hot carrier degradation. In a conventional transistor, the depletion region cannot intrude to any significant extent into the heavily doped drain. If the drain diffusion is more lightly doped, then the depletion region can extend into the drain as well as into the channel, and the electric field intensity will decrease. Such *lightly doped drain* (LDD) transistors can operate reliably at substantially higher drain-to-source voltages than can conventional *singly doped drain* (SDD) devices.

LDD transistors actually use two drain diffusions, one forming a lightly doped *drift region* near the edge of the gate, and the other forming a more heavily doped region beneath the contact. The heavily doped diffusion reduces the drain resistance of the structure and allows the transistor to retain most of the performance of a conventional SDD device. One process for forming LDD transistors uses an *oxide sidewall spacer* to self-align the two drain diffusions, enabling precise control of the width of the drift region.[11] Figure 3.36 illustrates the steps required to fabricate this structure. Immediately after patterning the polysilicon gate, a shallow implant self-aligned to the edges of the gate polysilicon deposits the lightly doped drain (N– S/D). The wafer is coated with a thick layer of isotropically deposited oxide. The oxide at the edges of the polysilicon gate is deeper than the oxide over adjacent regions of the wafer. An anisotropic dry etch removes most of the deposited oxide, but

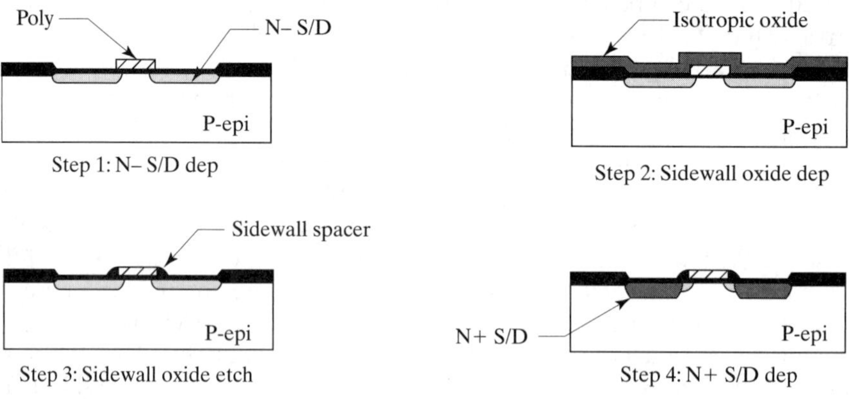

FIGURE 3.36 Steps required to fabricate an LDD NMOS transistor.

[11] R. H. Eklund, R. A. Haken, R. H. Havemann, and L. N. Hutter, "BiCMOS Process Technology," in A. R. Alvarez, ed., *BiCMOS Technology and Applications*, 2d ed. (Boston: Kluwer Academic, 1993), pp. 90–95.

some remains along the edges of the gates even after the planar surfaces have entirely cleared. The etch is halted so that filaments of oxide remain along either side of the gate; these form the desired oxide sidewall spacers. These spacers are approximately as wide as the polysilicon gate is thick. A second drain implant self-aligned to the edges of the oxide sidewall spacers forms the heavily doped portions of the LDD structure (N+ S/D). The width of the lightly doped drift region approximately equals the width of the sidewall spacer, typically about 0.5 μm.

Only the drain terminal of the NMOS transistor requires an LDD structure, but no simple way exists to block the formation of the sidewall spacer, so the source terminal of the NMOS also receives an LDD structure. The resulting transistor is *symmetric* in the sense that source and drain can be interchanged without affecting device performance. The PMOS transistor also receives the oxide sidewall spacers, but no lightly doped diffusion. For reasons discussed in Section 12.1.1, a channel forms underneath these sidewall spacers. The transistor, therefore, appears to have a slightly longer channel than its drawn dimensions suggest.

The LDD process described previously forms transistors that are suitable for drain-to-source voltages of 10 to 20 V. No additional masks are required if all transistors receive both source/drain implants (N− S/D and N+ S/D). Purchasing an additional mask to selectively block the N− S/D implant may provide some additional benefits. Short-channel transistors do not require LDD processing because they break down by punchthrough before hot carrier generation becomes significant. The presence of N− S/D in short-channel NMOS transistors therefore serves no useful purpose. The transistors can pack more densely if the drawn channel dimensions can be reduced without causing premature punchthrough. One way to achieve this goal consists of selectively blocking the N− S/D implant from the short-channel devices. By leaving out the N− S/D, the drawn channel length can be decreased by 0.5 to 1.0 μm without affecting the effective dimensions of the device. The purchase of separate N− S/D and N+ S/D masks can make a significant impact on high-density, low-voltage logic circuitry that does not require LDD transistors.

Extended-Drain, High-Voltage Transistors

Oxide sidewall spacers allow the construction of fairly conventional MOS transistors that can withstand drain-to-source voltages of 10 to 20 V. Higher voltages require a different approach to constructing the lightly doped drain region to protect against avalanche breakdown as well as hot carriers. Practical high-voltage MOS transistors can be constructed using only the existing masks of the standard N-well polysilicon-gate CMOS process. These *extended-drain* devices do not self-align, so they are inherently long-channel devices that exhibit substantial overlap capacitance. Even so, they suffice for constructing many high-voltage circuits.

Figure 3.37 shows a sample high-voltage, extended-drain NMOS. This transistor uses an N-well region as an extremely lightly doped drain. Since the well is both relatively deep and very lightly doped, it possesses a breakdown voltage in excess of 30 V. A plug of NSD provides Ohmic contact to the drain. The source of the transistor consists of an NSD region without the addition of N-well. Since source and drain employ different diffusions, this device is an *asymmetric* MOS transistor.

The gate oxide of a high-voltage MOS transistor presents something of a dilemma. The 15 V CMOS gate oxide is 300 to 400 Å thick and can safely withstand 20 to 25 V. Higher voltages will rupture the oxide and destroy the device. A separate gate oxidation can form a thicker oxide for the high-voltage devices, but increasing the oxide thickness decreases the device transconductance. The best solution consists of thickening the gate oxide only over the lightly doped drain where the highest field

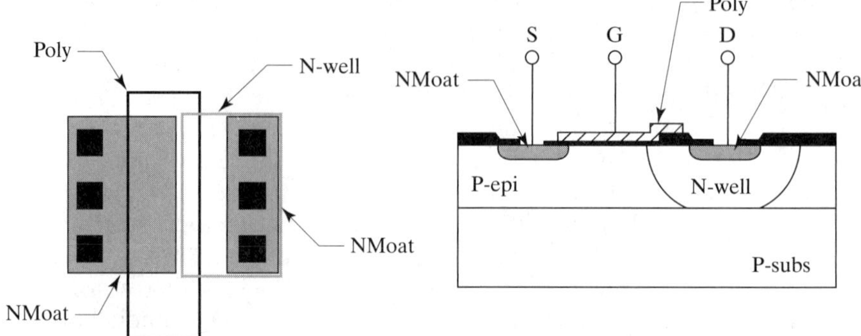

FIGURE 3.37 Layout and cross section of an extended-drain NMOS. The backgate is common to the substrate.

stresses occur (Figure 3.37). The LOCOS bird's beak offers a convenient means of fabricating this *field-relief structure*. No equivalent topologies exist in shallow trench isolation, making it difficult to fabricate high-voltage extended-drain transistors in STI processes. The thin gate oxide remains in place over the channel and ensures a high device transconductance.

High-voltage, extended-drain PMOS transistors use the P-type channel-stop as a lightly doped drain. These devices are discussed in Section 12.1.2.

3.3 ANALOG BiCMOS

The incessant demand for higher levels of integration has driven the evolution of ever more complex and costly processes. Not only must more devices fit into the same die area, but the performance of these devices must steadily improve to satisfy new applications. By the early 1980s, customers demanded *mixed-signal* integrated circuits incorporating both analog and digital subsystems on a common substrate. A typical mixed-signal integrated circuit contains 90 to 95% digital circuitry and 5 to 10% analog circuitry.[12] CMOS logic overwhelmingly outperforms bipolar logic in packing density and power requirements, so the first attempts to fashion mixed-signal circuits employed unmodified CMOS processes. Analog CMOS circuitry had been designed for decades, so few manufacturers envisioned any difficulty in building the last percentages of the mixed-signal system. But these manufacturers soon discovered that difficulty does not correspond to component count. Although the analog components compose only a few percent of the total devices, they often consume the majority of the design effort. The inferior performance of analog CMOS requires even more design resources to compensate for process inadequacies.

After a few years of failures and qualified successes, most manufacturers began to realize that the analog portions of a mixed-signal system require tailor-made components. The true benefits of using such components lie not only in improved performance, but also in faster cycle times and higher probabilities of success. The late 1980s saw the development of a new generation of processes specifically aimed at the construction of mixed-signal integrated circuits. These *analog BiCMOS* processes are usually based on CMOS process flows, but are augmented by the addition of bipolar transistors, high-sheet poly resistors, and other special devices.

---

[12] Reckoned by device count, not area. The relatively few analog devices on a mixed-mode integrated circuit tend to take up an inordinate amount of die area because analog circuitry cannot take full advantage of the reduced dimensions of modern transistors while retaining adequate performance.

3.3.1. Essential Features

Analog BiCMOS processes are characterized by their complexity. Most require at least 15 masks, and the more exotic variants use upward of 30 masks. The penalties of complexity include higher wafer costs, longer manufacturing times, and lower process yields. Set against these disadvantages are the benefits of higher-performance analog circuitry, reduced design effort, and faster design cycles. By the mid-1990s, the majority of new analog designs used some form of analog BiCMOS processing.

A typical BiCMOS process consists of a standard CMOS flow, with the addition of a minimum number of steps to construct adequate bipolar transistors. Deep-P+ isolation is not used because it requires an additional masking step. One alternative form of isolation that requires no additional masks uses the N-well to form the collector of the NPN transistor. The base and emitter then consist of successive counter-dopings of the well. The collector-epi junction serves to isolate this style of bipolar transistor, thus the name *collector-diffused-isolation* (CDI).[13] Figure 3.38 shows an NPN transistor constructed using CDI.

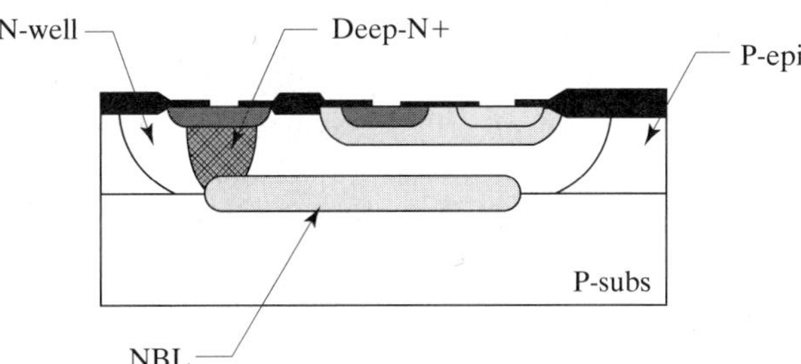

FIGURE 3.38 Details of collector-diffused-isolation (for an NPN transistor). Only the portions of the CDI system related to the collector are labeled.

A typical BiCMOS process follows the general outline of an N-well CMOS process, with the addition of three masking steps: NBL, deep-N+, and base. The reasons for the inclusion of each of these masking steps require some discussion.

NBL is the most reluctantly conceded mask, but also the most vital. Some CMOS fabs do not possess epi reactors and lack the experience necessary to implement buried layers. Predictably, attempts have been made to eliminate NBL, and equally predictably, the results have been unsatisfactory. NBL greatly reduces NPN collector resistance—one of the major parasitics of this device. NBL also provides higher NPN operating voltages on the thin epi favored by modern CMOS processing because it blocks vertical punchthrough breakdown. Furthermore, NBL suppresses parasitic substrate PNP action. Without it, lateral PNP transistors become almost impossible to construct. Any practical junction-isolated analog BiCMOS process will almost certainly include NBL.

Standard bipolar uses a deep-N+ sinker to further reduce the collector resistance of power NPN transistors. Although a power MOS transistor can substitute for a power NPN in many applications, deep-N+ remains necessary for creating low-resistance minority carrier injection guard rings. Furthermore, due to the graded nature of the well, CDI NPN transistors often exhibit excessive vertical resistance between the collector contact and the NBL. These transistors saturate prematurely, limiting low-voltage operation, complicating device modeling, and

---

[13] B. T. Murphy, S. M. Neville, and R. A. Pedersen, "Simplified Bipolar Technology and Its Application to Systems," *IEEE J. Solid-State Circuits,* Vol. SC-5, #1, 1970, pp. 7–14.

causing undesired substrate injection. Many BiCMOS processes include deep-N+, if only as a process extension.

The base diffusion sets the NPN transistor gain, breakdown voltage, and Early voltage. Some processes have attempted to construct NPN transistors by using layers scavenged from other devices, with mixed results. Attempts to construct NPN transistors by using the diffusions that normally form DMOS transistors have succeeded only in those processes that use a lightly doped DMOS backgate (Section 12.2.3). Many processes use a more heavily doped backgate to improve DMOS performance. The higher doping reduces the beta of NPN transistors fabricated from this backgate diffusion, often rendering them virtually useless. Extended-base transistors that use the P-epi as a base region have also been successfully constructed, but these devices have several drawbacks. They require more area than conventional CDI devices, and they often have low Early voltages because they lack a drift region (see Section 8.3.3). CMOS processes that employ multiple gate oxides usually include a shallow, moderately doped P-well for constructing the low-voltage thin-oxide transistors. This shallow P-well has proven an adequate replacement for a dedicated base implant in certain processes.

3.3.2. Fabrication Sequence

This section discusses an analog BiCMOS process based on N-well polysilicon-gate CMOS.[14] The N-well provides collector-diffused isolation; NBL, deep-N+, and base are added to create bipolar transistors. Double-level metal has been added to simplify interconnection. This process requires 15 masks, one of which is used twice for a total of 16 masking operations.

Starting Material

The substrate material chosen for analog BiCMOS consists of a P+ (100) substrate cut several degrees off the crystal axis to minimize pattern distortion. The use of NBL in conjunction with a P+ substrate dictates the insertion of an additional epitaxial deposition into the process. Without this step, the NBL would directly contact the substrate to form an N+/P+ junction with a very low breakdown voltage. A lightly doped P-epi layer some 20 μm thick therefore resides between substrate and NBL. Three factors determine the thickness of this first epi layer: the up-diffusion of the underlying substrate, the down-diffusion of the NBL, and the width of the depletion region required to withstand the maximum anticipated operating voltage (typically 30 to 50 V). The first epi deposits on an unpatterned wafer, so epi-coated wafers may be stockpiled for use as starting material. One could alternatively sacrifice the P+ substrate to eliminate the need for a first epi deposition, but the use of a lightly doped substrate makes the process very susceptible to latchup and substrate debiasing (Section 4.4.1).

N-Buried Layer

A brief thermal oxidation grows a thin layer of oxide across the wafer. This oxide is patterned using the N-buried layer (NBL) mask, and an oxide etch opens windows to the silicon surface. Ion implantation deposits an N-type dopant, either arsenic or antimony, in these windows. A brief drive conducted in an oxidizing ambient anneals out lattice damage and causes the formation of the surface discontinuity required for subsequent mask alignment.

---

[14] Eklund, *et al.*, pp. 120ff. Also see J. Erdeljac, B. Todd, L. Hutter, K. Wagensohner, and W. Bucksch, "A 2.0 Micron BiCMOS Process Including DMOS Transistors for Merged Linear ASIC Analog/Digital/Power Applications," *Proceedings 1992 Applied Power Elect. Conf.*, 1992, pp. 517–522.

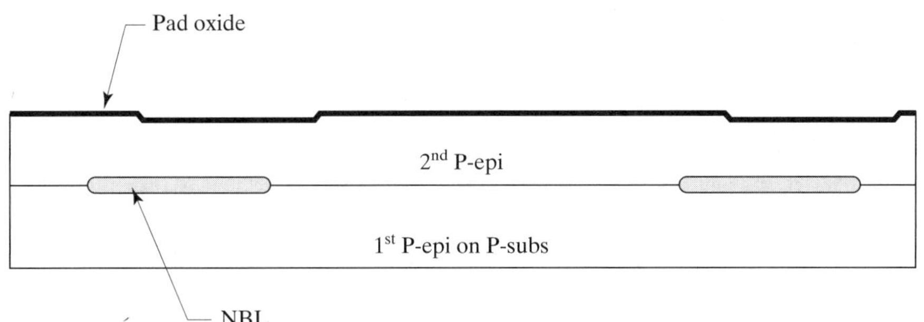

FIGURE 3.39 Wafer after second epitaxial deposition. The P+ substrate is not explicitly shown, but it lies beneath the first P-epi layer.

Epitaxial Growth

After the NBL anneal, the oxide is stripped and the wafers are returned to the epitaxial reactor for deposition of a second P− epi layer. Surface discontinuities propagate upward through the epi at approximately a 45° angle along a [100] axis determined by the tilt of the wafer. After epitaxial growth, the NBL shadow will have shifted laterally by about 10 μm. Figure 3.39 shows the wafer after the second epi deposition.

During the growth of the second epi, the reactant gases leach some of the NBL dopant from the surface of the wafer and redeposit it elsewhere, a process called *lateral autodoping.*[15] This mechanism can cause the formation of a thin layer of N-type silicon at the interface between the first and second epi layers, shorting adjacent wells. Autodoping can be limited by using antimony as the dopant or by conducting the epitaxy at reduced pressure. In either case, the BiCMOS NBL is liable to be somewhat more lightly doped than that used for standard bipolar.

N-Well Diffusion and Deep-N+

A thin layer of oxide is now grown and patterned using the N-well mask. Ion implantation deposits phosphorus, which is subsequently driven down to form the well diffusion. This drive stops before the well and NBL collide to permit the timely insertion of the deep-N+ deposition into the process flow. Additional oxide grown during the well drive allows patterning of the subsequent deep-N+ diffusion. After a heavy dose of phosphorus is implanted, the drive continues until the well and the deep-N+ both overlap the NBL by about 25% of their respective junction depths in order to minimize vertical resistance. Figure 3.40 shows a cross section of the wafer after the drive.

The N-well diffusion influences a number of device parameters for both PMOS and bipolar transistors. Compromises must be made between the two types of devices to the detriment of either or both. For example, short-channel PMOS transistors require a moderately doped well to suppress punchthrough, while bipolar transistors need a lightly doped well to form their collector drift region. The doping

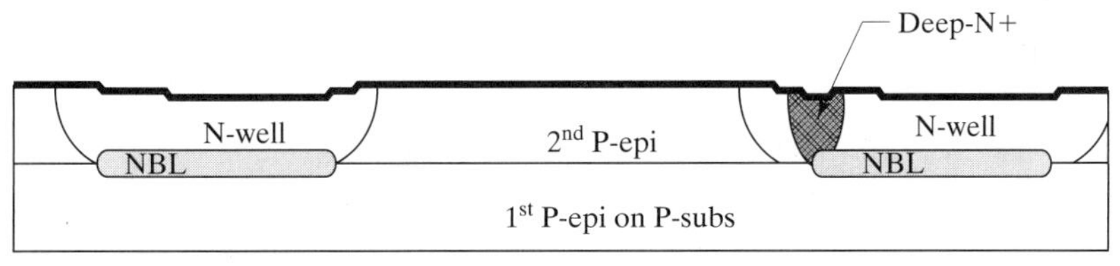

FIGURE 3.40 Wafer after N-well and deep-N+ deposition and drive.

[15] M. W. M. Graef, B. J. H. Leunissen, and H. H. C. de Moor, "Antimony, Arsenic, Phosphorus, and Boron Autodoping in Silicon Epitaxy," *J. Electrochem. Soc.,* Vol. 132, #8, 1985, pp. 1942–1954.

of the well therefore targets a compromise value that dictates a minimum channel length of 2 to 3 μm. If one desires to fabricate shorter channel lengths, then additional wells must be added to allow independent optimization of the MOS and bipolar components of the process.

Base Implant

A uniform, thin layer of pad oxide is grown after the remnants of the previous oxide patterns have been stripped from the wafer. The wafer is patterned using the base mask, and boron is implanted through the pad oxide to form P-type regions, which are subsequently annealed under an inert ambient. Later high-temperature processing steps complete the base drive and set the final base-junction depth. Figure 3.41 shows the wafer after the base anneal. Triple counterdoping degrades the beta of the CDI NPN by raising the total base dopant concentration and hence the recombination rate in the neutral base. This can be partially offset by using a relatively lightly doped base implant. The final base-sheet resistance therefore equals several times that of standard bipolar.

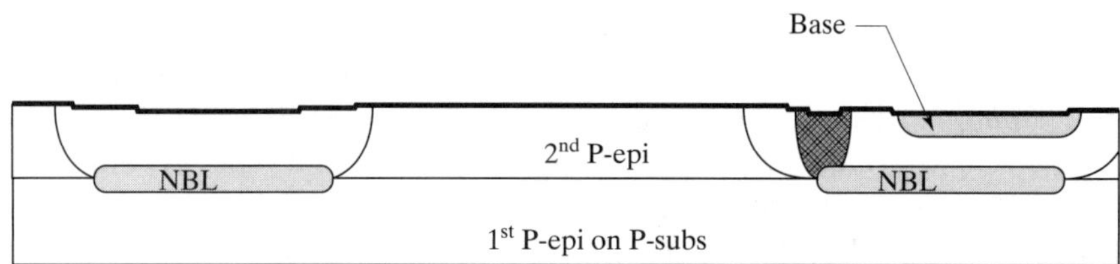

FIGURE 3.41 Wafer after base implant and anneal.

Inverse Moat

Analog BiCMOS uses the same LOCOS technology employed by polysilicon-gate CMOS. A thick layer of LPCVD nitride, patterned using the inverse moat mask, is etched to expose the regions where field oxide will eventually form (Figure 3.42). As in the case of polysilicon-gate CMOS, the MOS transistors occupy moat regions that are not covered by thick-field oxide. Moat regions also serve two additional purposes:

- Base regions are enclosed by moats to prevent oxidation-enhanced diffusion from overdriving the base.
- Schottky contacts are enclosed by moat to allow their etching to proceed simultaneously with base and emitter contacts.

The inverse moat mask consists of a color reverse of a combination of NMoat, PMoat, base, and Schottky contact drawing layers. Some processes generate the inverse moat mask automatically during pattern generation; other processes require the layout designer to code moat geometries over some or all of the aforementioned drawing layers.

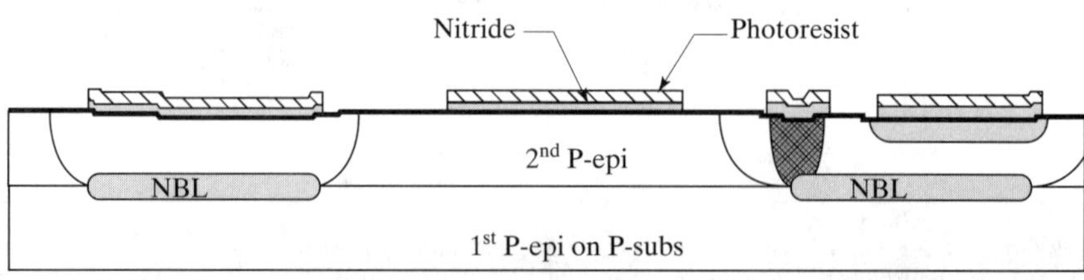

FIGURE 3.42 Wafer after nitride deposition and inverse moat etch.

Channel Stop Implants

Since analog BiCMOS uses (100) silicon, channel stop implants are required to raise the thick-field thresholds above the operating voltage. The channel stop strategy for analog BiCMOS parallels that for polysilicon-gate CMOS. A blanket boron channel stop adjusts the thick-field threshold over the P-epi, while a patterned phosphorus channel stop sets the thick-field threshold over well regions. The boron channel stop is implanted using the patterned photoresist left from the inverse moat masking operation. A second coat of photoresist is applied and patterned, using the channel stop mask. This mask exposes only the regions of N-well that will ultimately lie beneath the thick-field oxide. A phosphorus implant offsets the previously deposited boron and increases the surface concentration in these well regions. Figure 3.43 illustrates the wafer cross section after the completion of both channel stop implants and the subsequent photoresist removal.

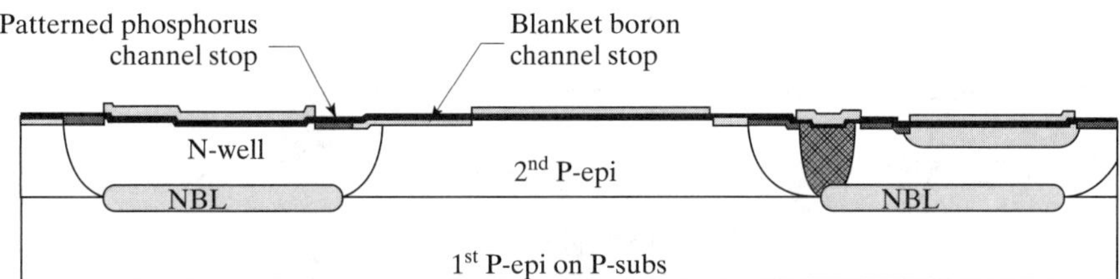

FIGURE 3.43 Wafer after channel stop implants.

LOCOS Processing and Dummy Gate Oxidation

The LOCOS oxidation employs either steam or elevated pressures to increase the rate of oxide growth. Afterward, the nitride layer and the underlying pad oxide are stripped away. A dummy gate oxidation removes any lingering nitride residue (Figure 3.44).

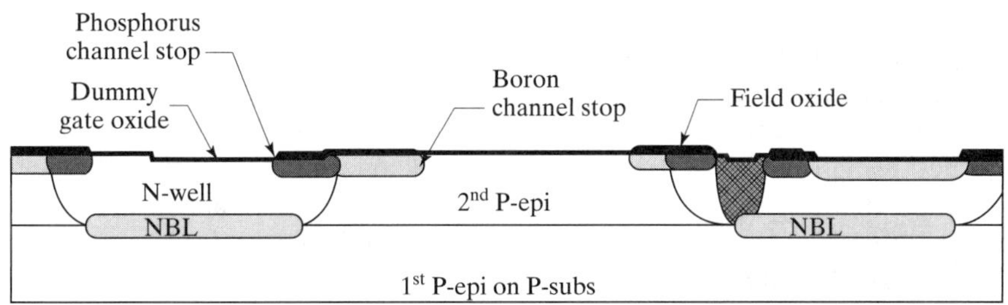

FIGURE 3.44 Wafer after LOCOS oxidation, nitride strip, and dummy gate oxidation.

Threshold Adjust

A single boron V_t adjust implant simultaneously raises the NMOS threshold and lowers the PMOS threshold. With suitable well and epi doping concentrations, both of these thresholds can be simultaneously adjusted to the desired target of 0.7 ± 0.2 V.

The threshold implant process consists of coating the wafer with photoresist, patterning the resist with the V_t adjust mask, and implanting the necessary dose of boron through the dummy gate oxide. The final gate dielectric consists of some 300 Å of high-quality dry oxide grown after the removal of the dummy gate oxide (Figure 3.44). Figure 3.45 shows a cross section of the resulting wafer.

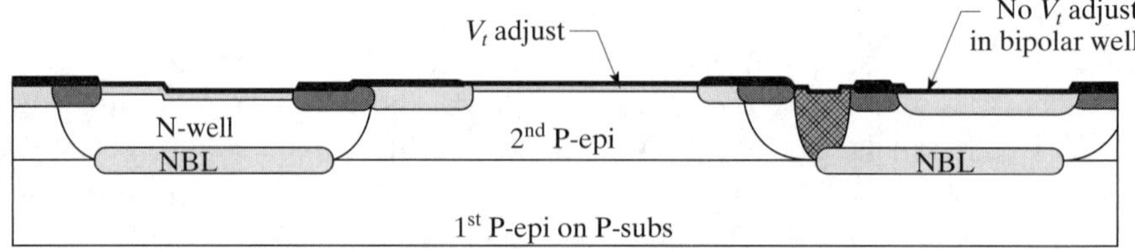

FIGURE 3.45 Wafer after V_t adjust implant.

Polysilicon Deposition and Pattern

The gates of the MOS transistors consist of heavily doped N-type polysilicon formed by depositing intrinsic polysilicon and subsequently doping it with a blanket phosphorus deposition. The patterning step uses the poly mask. Figure 3.46 shows the wafer after poly pattern.

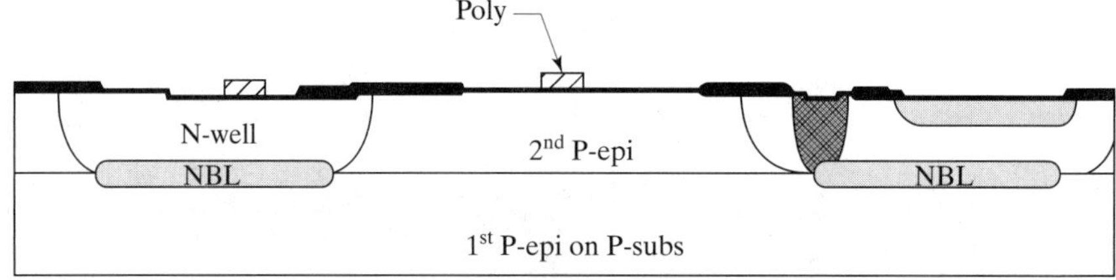

FIGURE 3.46 Wafer after polysilicon deposition and pattern. The channel stop and threshold adjust implants do not appear in this or subsequent cross sections.

Source/Drain Implants

Analog BiCMOS typically produces bipolar transistors with 10 to 20 V breakdown voltages. The MOS transistors should ideally be capable of withstanding similar voltages. The breakdown voltages of PSD and NSD can be raised to 15 to 20 V without difficulty. Punchthrough can be averted by increasing the minimum channel length to 2 to 3 μm. The addition of a lightly doped drain will suppress hot electron generation in the NMOS transistors.

Photoresist is spun onto the wafer and patterned using the N− S/D mask. A phosphorus implant forms lightly doped source and drain regions that self-align to the polysilicon gate. Isotropic deposition of oxide and subsequent anisotropic etching form sidewall spacers on either side of the gates. A second photoresist layer, patterned using the N+ S/D mask, defines a heavier and somewhat deeper N+ S/D implant that aligns to the edges of the oxide sidewall spacers. The lightly doped drain consists of that portion of the N− S/D implant that lies beneath the sidewall spacers. If all NMOS transistors receive LDD structures, then the N− S/D mask can be reused for the N+ S/D implant.

The PMOS transistor does not require a lightly doped drain, eliminating the need for a P− S/D implant. The P+ S/D implant occurs after the formation of the sidewall spacers, so the PMOS transistor channel length increases by twice the width of the sidewall spacer (Figure 3.47). This increase in width can be offset by reducing the drawn length of the PMOS gate proportionately.

Metallization and Protective Overcoat

The double-level metal flow requires five masks: contact, metal-1, via, metal-2, and protective overcoat. The contacts are silicided to control resistance and, coincidentally, to allow the formation of Schottky diodes. Refractory barrier metal ensures adequate step coverage across the nearly vertical sidewalls of contacts and vias. Copper-doped aluminum minimizes electromigration susceptibility, and a thick layer

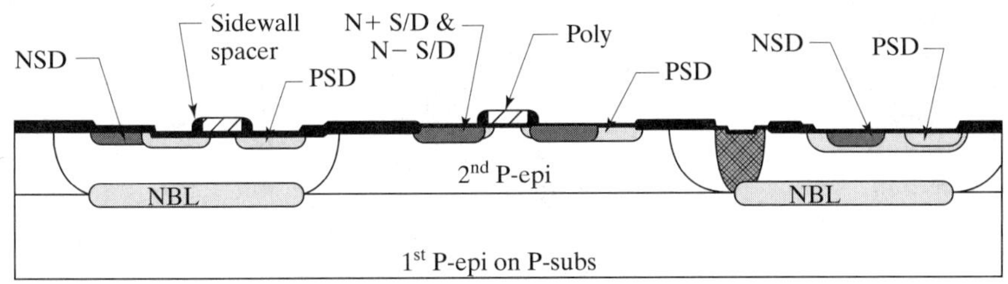

FIGURE 3.47 Wafer after N− S/D, N+ S/D, and P+ S/D implants.

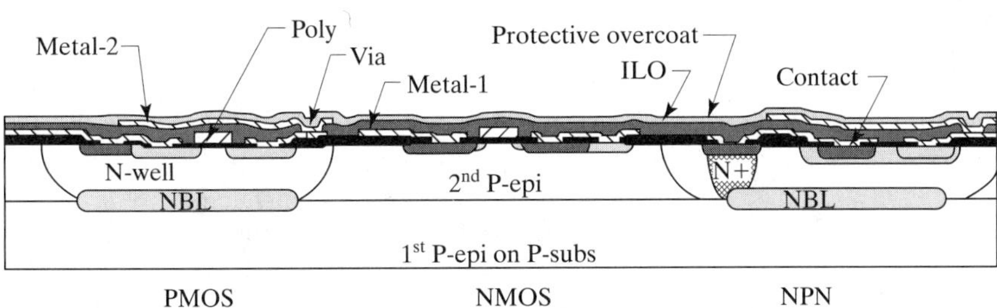

FIGURE 3.48 Completed wafer, showing metal-1 and metal-2 layers.

of compressive nitride protective overcoat provides a mechanical and chemical barrier between the metallization and the encapsulation. Figure 3.48 shows the cross section of a completed wafer, which includes NPN, NMOS, and PMOS transistors.

Process Comparison

Analog BiCMOS uses all of the same steps as polysilicon-gate CMOS. Three additional masks (NBL, deep-N+, and base) are inserted at opportune points in the process as shown by the shaded entries in Table 3.6. NBL deposition must occur before the growth of the second epi. Deep-N+ and base must occur early in the process because these deep diffusions require high temperatures and long drive times.

3.3.3. Available Devices

All of the devices available in polysilicon-gate CMOS also exist in analog BiCMOS. These devices include LDD NMOS and SDD PMOS transistors, four types of resistors (poly, PSD, NSD, and N-well), gate oxide capacitors, and Schottky diodes. Extended drain transistors can also be created without using any additional masks. Sections 3.2.3 and 3.2.4 discuss the preceding devices in detail. Additional analog BiCMOS components include a CDI NPN transistor, lateral and substrate PNP transistors, and base resistors.

NPN Transistors

Figure 3.49 shows the layout and cross section of a minimum-geometry NPN transistor. The collector of the NPN consists of an N-well region into which the base and emitter (consisting of NSD) are successively diffused. The inclusion of NBL beneath the active region of the transistor and the addition of a deep-N+ sinker help minimize collector resistance. NSD implanted on top of the sinker ensures Ohmic contact. In a similar manner, PSD allows contact to the lightly doped base. The general appearance of this transistor closely resembles that of the standard bipolar transistor in Figure 3.10. There are, however, several subtle differences.

Due to the use of the shallow NSD implant for the emitter, the gain of the transistor is reduced to roughly 80 (Section 8.3.1). Higher betas could be achieved, but only at the expense of degrading other device characteristics or adding additional process steps.

TABLE 3.6 Comparison of analog BiCMOS and polysilicon-gate CMOS processes.

Mask	Polysilicon-Gate CMOS	Analog BiCMOS
		First epi growth
1. NBL		NBL deposition/anneal
	Epi growth	Second epi growth
2. N-well	N-well deposition/drive	N-well deposition/drive
3. Deep-N+		Deep-N+ deposition/drive
	Pad oxidation	Pad oxidation
4. Base		Base implant/anneal
5. Inverse moat	Nitride deposition/pattern	Nitride deposition/pattern
	Blanket boron channel stop	Blanket boron channel stop
6. Channel stop	Patterned phosphorus channel stop	Patterned phosphorus channel stop
	LOCOS oxidation	LOCOS oxidation
	Nitride/pad oxide strip	Nitride/pad oxide strip
	Dummy gate oxidation	Dummy gate oxidation
7. V_t Adjust	Threshold adjust implant	Threshold adjust implant
	True gate oxidation	True gate oxidation
8. Poly-1	Polysilicon deposition	Polysilicon deposition
	Poly implant/anneal	Poly implant/anneal
9. NSD	N− S/D implant	N− S/D implant
	Sidewall spacer formation	Sidewall spacer formation
10. NSD (again)	N+ S/D implant	N+ S/D implant
11. PSD	P+ S/D implant	P+ S/D implant
	MLO deposition	MLO deposition
12. Contact	Contact OR	Contact OR
	Platinum sputter/sinter/etch	Platinum sputter/sinter/etch
13. Metal-1	1st metal deposition/etch	1st metal deposition/etch
	ILO deposition/planarization	ILO deposition/planarization
14. Via	Via etch	Via etch
15. Metal-2	2nd metal deposition/etch	2nd metal deposition/etch
16. PO	PO deposition/etch	PO deposition/etch

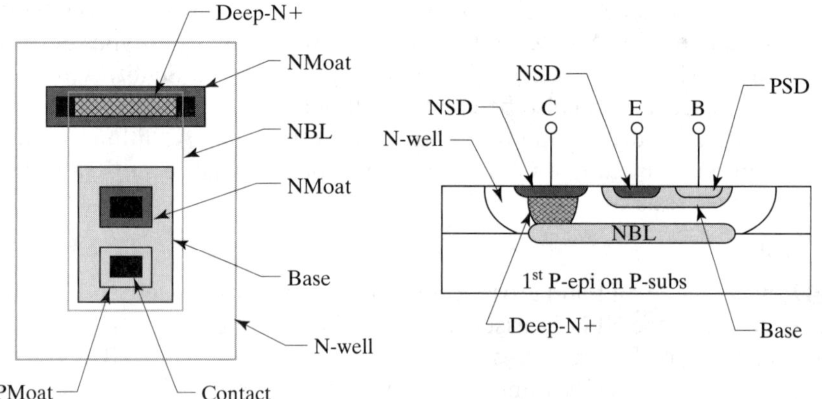

FIGURE 3.49 Layout and cross section of an NPN transistor with deep-N+ and NBL.

The graded well exhibits an extremely high collector resistance if deep-N+ is not placed beneath the collector contact. This resistance causes a soft transition from the saturation to the forward active region (Section 8.3.2). The transistor may also intrinsically saturate even when the terminal voltages seem to indicate that saturation cannot occur (Section 8.1.4). These problems can be prevented by adding a deep-N+ sinker. Transistors used as Zeners do not require deep-N+, even

Parameter	Standard Bipolar	Analog BiCMOS
Drawn emitter area	100 μm^2	16 μm^2
Peak current gain (beta)	150	80
Early voltage	120 V	120 V
Collector resistance, in saturation	100 Ω	200 Ω
Typical operating current range for maximum beta	5 μA–2 mA	1–200 μA
V_{EBO} (Emitter-base breakdown, collector open)	7 V	8 V
V_{CBO} (Collector-base breakdown, emitter open)	60 V	50 V
V_{CEO} (Collector-emitter breakdown, base open)	45 V	25 V

TABLE 3.7 NPN device parameters for standard bipolar and analog BiCMOS.

in analog BiCMOS, because the collectors of these devices do not conduct significant current.

The analog BiCMOS NPN transistor does not perform as well as the standard bipolar NPN transistor, but it still serves for most applications (Table 3.7). Analog BiCMOS also allows much smaller emitter areas than standard bipolar. This benefit does not translate into a proportional reduction in transistor area since many other spacings contribute to the size of the final device. Still, the minimum-geometry analog BiCMOS transistor requires only about 30% of the room of its standard bipolar counterpart.

PNP Transistors

Analog BiCMOS supports both substrate and lateral PNP transistors. Figure 3.50 shows the layout and cross section of a representative substrate PNP transistor. This device is constructed by implanting PSD into an N-well region. The PSD implant forms the emitter of the transistor, while N-well forms the base. A small NSD plug ensures Ohmic contact to the lightly doped N-well. The emitter of the substrate PNP cannot consist of base diffusion because this reaches so deeply into the well that it compromises the breakdown voltage of the transistor. The characteristics of the analog BiCMOS substrate PNP broadly resemble those of a substrate PNP formed in standard bipolar.

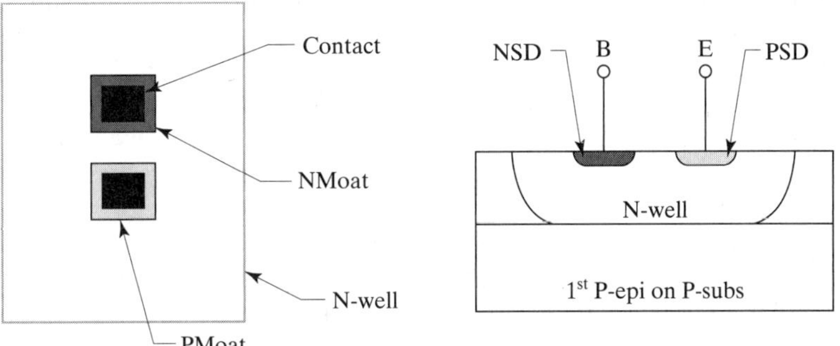

FIGURE 3.50 Layout and cross section of a typical substrate PNP transistor. The substrate acts as the collector.

Figure 3.51 shows a layout and cross section of a lateral PNP transistor constructed in analog BiCMOS. This device employs the base diffusion to form the emitter and collector, which both reside in an N-well region that forms the base. The addition of NBL to the transistor serves several purposes. First, it acts as a depletion stop, which allows the lateral PNP to withstand higher operating voltages without

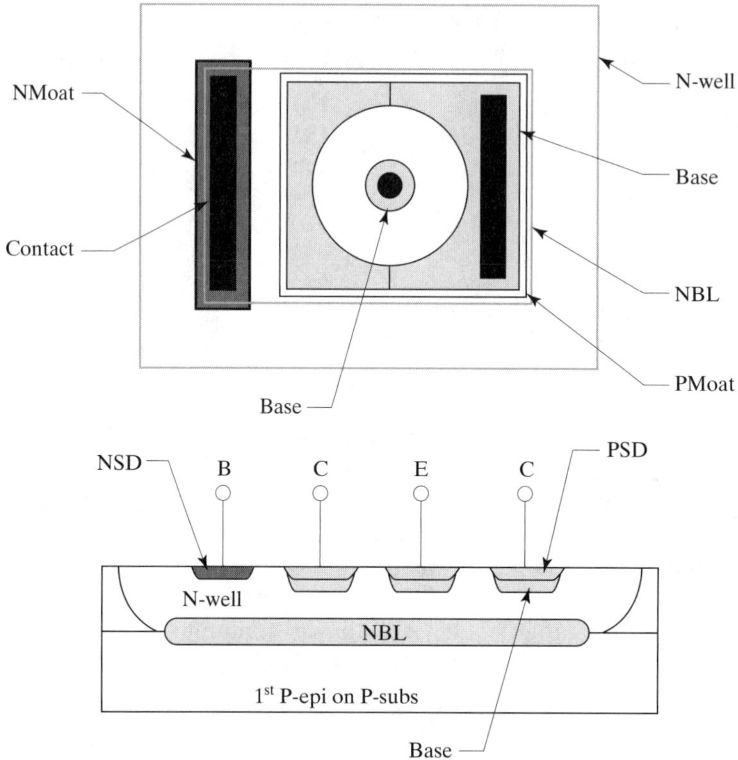

FIGURE 3.51 Layout and cross section of a lateral PNP transistor.

suffering punchthrough. Second, NBL blocks punchthrough breakdown and allows the base implant to be used in place of the shallower PSD implant. The deeper base implant enhances emitter sidewall injection and therefore raises the beta of the device. NBL also helps to minimize substrate injection. Without NBL, the lateral transistor would exhibit an apparent beta of less than 10; with NBL the beta can easily exceed 100.

Additional implants must surround the contacts since both the N-well and the base diffusion are too lightly doped to allow direct Ohmic contact. A rectangle of PSD encloses both the emitter and the collector, but this implant only penetrates the moat regions that form the actual collector and emitter regions of the device. The thick LOCOS oxide blocks the PSD implant from reaching the base of the transistor and prevents the P-type implant from shorting the emitter and collector. Similarly, an NSD plug allows Ohmic contact to the N-well. Deep-N+ is not normally required, although it may become necessary in larger transistors that conduct significant base current.

The minimum-geometry lateral PNP transistor can attain a peak beta well in excess of 100. Fine-line photolithography allows a narrower base width than standard bipolar and greatly reduces the area of a minimum emitter. As the emitter area decreases, the ratio of periphery to area increases. This enhances the desired lateral injection of carriers at the expense of undesired vertical injection. The graded nature of the well and the presence of the channel stop implant generate a doping gradient that forces carriers away from the surface, diminishing surface recombination losses. The use of (100) silicon instead of (111) silicon also reduces surface recombination. All of these effects together produce a transistor having a beta of up to 500.

The use of a lightly doped base implant for the emitter reduces emitter injection efficiency and causes a corresponding reduction in beta. Large PNP transistors

TABLE 3.8 Typical PNP device parameters for analog BiCMOS.

Parameter	Lateral PNP	Substrate PNP
Drawn emitter area	16 μm^2	16 μm^2
Drawn base width	5 μm	—
Peak current gain (beta)	120	100
Early voltage	80 V	100 V
Typical operating current for maximum beta	1–20 μA	1–50 μA
V_{EBO} (Emitter-base breakdown, collector open)	45 V	45 V
V_{CBO} (Collector-base breakdown, emitter open)	45 V	45 V
V_{CEO} (Collector-emitter breakdown, base open)	30 V	45 V

consist of arrays of many minimum emitters to achieve a high area-to-periphery ratio. Table 3.8 lists several important parameters for both types of PNP transistors available in analog BiCMOS.

Resistors

Analog BiCMOS base resistors consist of rectangles of base material occupying an N-well. Contact is made to either end of the resistor through a PSD plug. The well contact contains an NSD implant to increase the surface doping of the N-well. As with base resistors in standard bipolar, the addition of NBL serves not only to block possible minority carrier injection into the well during transients, but also to raise the operating voltage by preventing punchthrough between the resistor and the underlying substrate. Figure 3.52 shows the cross section and layout of a typical base resistor. The analog BiCMOS base diffusion is both thin and relatively lightly doped, and it typically exhibits a sheet resistance of 500 Ω/□. While the high sheet resistance provides compactness, it also complicates resistor layout because the diffusion becomes vulnerable to surface depletion effects (Section 5.3.3). Few designers actually employ base resistors; most prefer to pay the small additional cost for a process extension for high-sheet polysilicon resistors. Pinch resistors are also available, but they offer little or no advantage over minimum-width poly resistors.

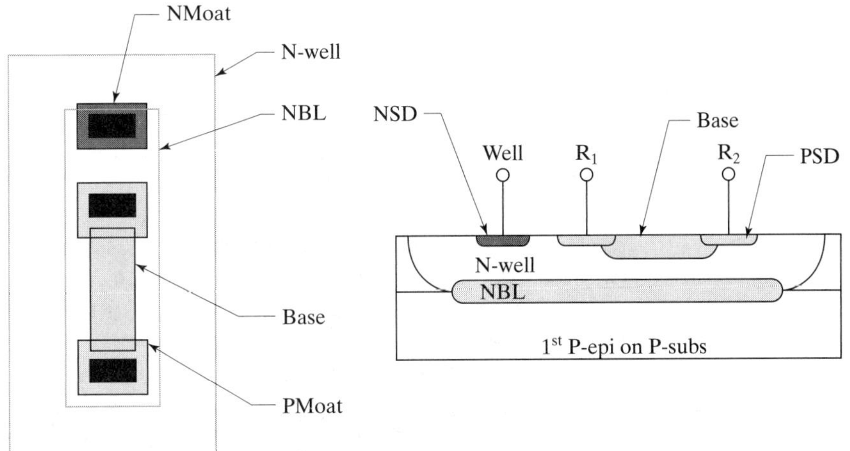

FIGURE 3.52 Layout and cross section of a base resistor.

3.3.4. Process Extensions

The BiCMOS technology that we have discussed builds upon the digital poly-gate CMOS processes of the early 1980s. Newer digital processes create smaller and faster transistors by adding innovations such as CMP planarization, CVD tungsten-plug vias, retrograde wells, high-angle lateral implants, shallow trench isolation, full dielectric isolation, and dual-polarity gate poly. Bipolar processes have evolved in a similar direction with the addition of such features as self-aligned polysilicon emitters, dielectric isolation, and tailored base regions consisting of silicon-germanium alloys.

Most of these innovations focus upon device size and speed to the exclusion of all other factors. The resulting processes are ill-suited for constructing anything other than digital and radio-frequency (RF) circuitry. Most analog designers continue to use older technologies that have long since become obsolete by digital standards. If past experience is any guide, the newer technologies will eventually diffuse into the analog realm. Indeed, several of these technologies have already done so. The remaining sections discuss two prominent examples of such technology transfers.

Advanced Metal Systems

The BiCMOS process described incorporates two levels of metallization, contact silicidation, and etched vias. This technology cannot reach pitches below about two microns.[16] Many processes, especially those that support dense CMOS logic, can greatly benefit from the addition of better metallization. Most of these benefits can be achieved without any modification to the silicon itself.

A typical advanced BiCMOS metal system fabricates either three or four layers of aluminum metallization. Each layer consists of perhaps 1 kÅ of refractory barrier metallization followed by 5 kÅ of copper-doped aluminum. Chemical-vapor-deposited (CVD) tungsten fills the contacts and vias. Chemical-mechanical polishing (CMP) planarization and fine-line photolithography enable a metal pitch of 0.6–1.0 μm. The CVD tungsten plugs allow contacts and vias to be stacked on top of each other, a practice called *nesting*. However, all contacts (and all vias) must be of a single uniform size. This requirement, combined with a widespread adoption of refractory silicides, precludes the fabrication of Schottky diodes in advanced-metallization processes.

Metal pitches below 0.6 μm offer few benefits to purely analog products. Most analog processes support operating voltages of at least 10–20 V, and the resulting isolation spacings preclude any further impact of pitch upon die size. On the other hand, some mixed-signal processes can make use of even the latest advancements in metallization, such as deep-submicron-pitch damascene copper.

Dielectric Isolation

Although dielectric isolation most often finds application in high-speed processes, it also provides compelling benefits for high-voltage components. The deep wells required to support voltages of 100 V or more require correspondingly large separations between components. These spacings can be greatly reduced by inserting trenches between components. Dielectric isolation also ensures that parasitic channels will not form between components, nor will minority carrier injection or substrate debiasing disturb circuit operation. A new generation of high-voltage analog processes uses wafer bonding and trench isolation to obtain all of these benefits at comparatively low cost.

Dielectric isolation can save varying amounts of space, depending upon how aggressively one wishes to deploy it. Figure 3.53 shows four separate cross sections of a portion of a vertical NPN transistor, each with a different variation on dielectric isolation. All four transistors consist of an NSD emitter diffused into a P-type base contained by an N-well collector. The four devices differ only in the means chosen to terminate the transistor against the trench. Figure 3.53A shows the most conservative

---

[16] The term *pitch* refers to the sum of metal lead width and lead-to-lead spacing; used without further qualification, it implies silicon dimensions rather than drawn dimensions.

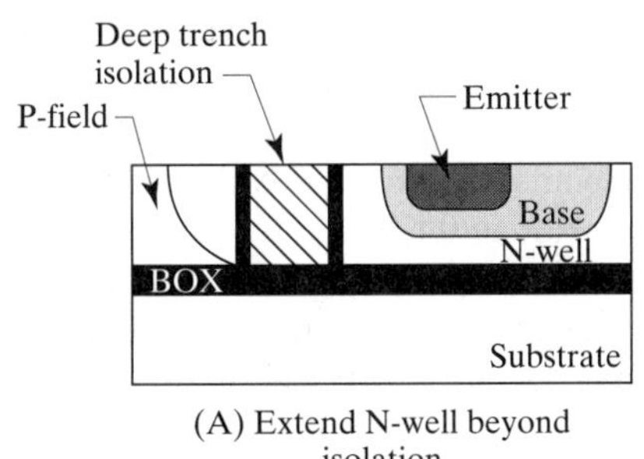

(A) Extend N-well beyond isolation

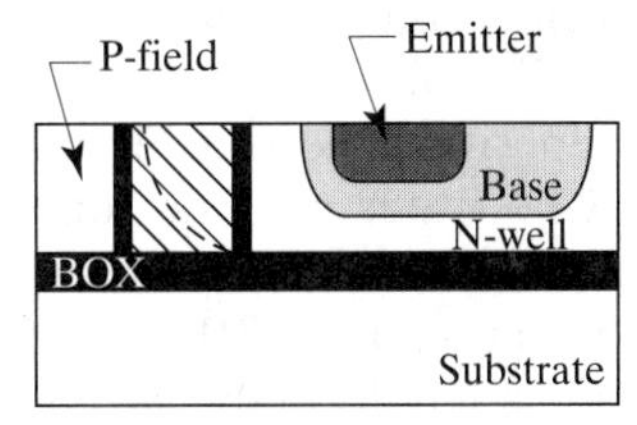

(B) N-well terminated inside trench

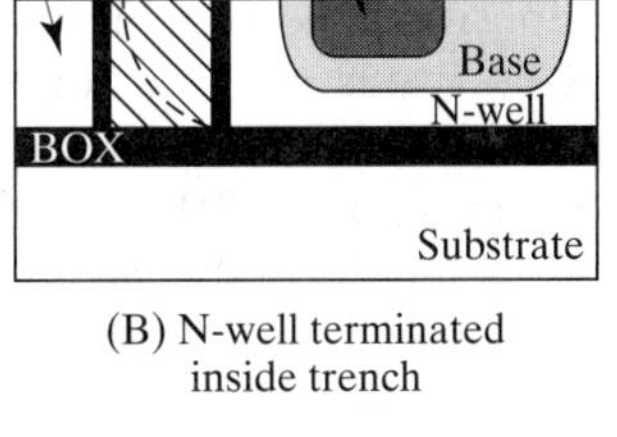

(C) Base and N-well terminated inside trench

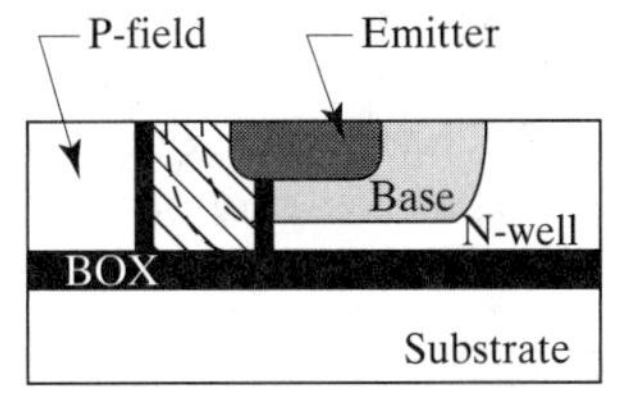

(D) Emitter, base, and N-well terminated inside trench

FIGURE 3.53 Four methods of terminating a vertical NPN transistor against an isolation trench.

termination scheme, in which all junctions are kept away from the isolation. In the case illustrated, the N-well has been extended beyond the trench, while the NBL stops short of the trench. This termination scheme ensures that breakdown cannot occur along the trench sidewalls. The fringe of N-well around the transistor continues to outdiffuse during the drive. This arrangement actually requires more space than conventional junction isolation.

Figure 3.53B shows a somewhat more aggressive style of termination in which junctions can be terminated within the trench, but not against it. In this particular case, the N-well that forms the collector terminates midway across the trench. Since the trench etch occurs before the N-well drive, this arrangement keeps the N-well dopant from diffusing outside of the trench. The NBL continues to stop slightly short of the trench because the buried layer diffuses laterally slightly during epitaxy; therefore, if the NBL were terminated in mid-trench, a small amount of arsenic or antimony might be left outside of the trench boundaries after etching. This termination scheme greatly improves device packing density while introducing only minimal risk.

Figure 3.53C shows an even more aggressive style of termination in which junctions are allowed to intersect the sidewalls of the trench. In this particular example, both the base and the N-well geometry are drawn to the middle of the trench. The collector-base junction therefore intersects the trench sidewall. This arrangement saves more space by eliminating the spacing that would otherwise exist between base and isolation. However, junctions that terminate against the isolation in this manner sometimes exhibit anomalously low breakdown or leakage.

Figure 3.53D shows the most aggressive style of termination, in which multiple junctions are allowed to intersect the same trench sidewall. In this transistor, both the collector-base junction and the emitter-base junction intersect the isolation sidewall. In addition to breakdown concerns, this style of device layout will exhibit reduced beta due to surface recombination along the sidewall.

Of the four termination schemes discussed, that of Figure 3.53B represents a sensible compromise between area savings and potential processing development risks. High-voltage components are inherently large, even after the space savings of dielectric isolation have been realized. Therefore, the more aggressive schemes of

Figures 3.53C–D introduce only marginal benefits to the overall design, while significantly complicating process development.

The fabrication of a dielectrically isolated analog BiCMOS wafer begins with the deposition and densification of approximately 1 μm of oxide upon the P+ substrate. A P− wafer is bonded on top of this oxide and cleaved to form a thin layer of active silicon over a 1 μm-thick buried oxide. After polishing, the wafer undergoes thermal oxidation. The NBL mask is used to selectively pattern a heavy antimony or arsenic implant to form the N-type buried layer. A brief heat treatment activates the dopant and anneals out lattice damage. Oxide is allowed to grow during the anneal process in order to ensure a surface discontinuity between NBL regions and the rest of the die. After annealing, the wafers are stripped of oxide and placed in the epi reactor for deposition of approximately 10 μm of lightly doped P-type silicon. The surface step at the periphery of the NBL propagates upwards during epitaxy, forming a characteristic discontinuity to which subsequent mask steps will align. Figure 3.54 shows a cross section of the wafer at this step of the process.

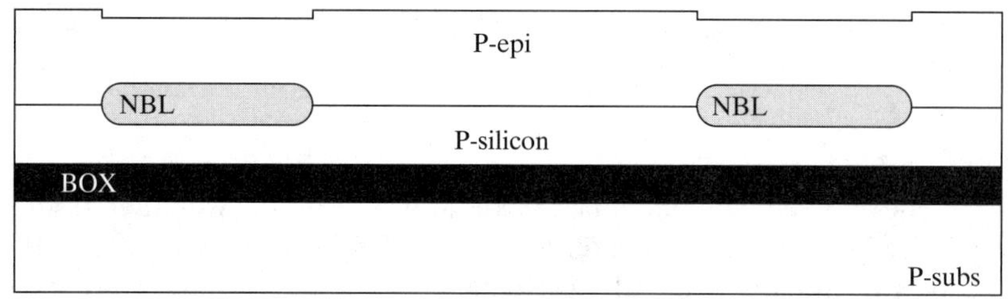

FIGURE 3.54 Cross section of DI BiCMOS wafer after NBL implant and epitaxial deposition.

Next, the wafers receive a brief oxidation followed by the deposition of a thick layer of CVD nitride. This nitride is patterned using the isolation mask, and a highly anisotropic plasma etch forms trenches beneath the nitride openings. After the remains of the photoresist have been stripped, a brief thermal oxidation forms a layer of insulating oxide along the sidewalls of the trenches. Additional CVD oxide helps thicken these oxide layers to enable them to withstand the full operating voltage. The trenches are then filled with polysilicon, and the excess is etched away to expose the nitride surface. After the nitride is stripped, the wafer is planarized by chemical-mechanical polishing. Figure 3.55 shows a cross section of the resulting wafer.

Processing then continues along the lines described in Section 3.3.1. Table 3.9 compares the first few steps of the DI BiCMOS process with the corresponding steps of its JI BiCMOS counterpart. The shaded entries are unique to the dielectrically isolated version of the process. The DI process requires one additional mask, which is used for fabricating the trench isolation.

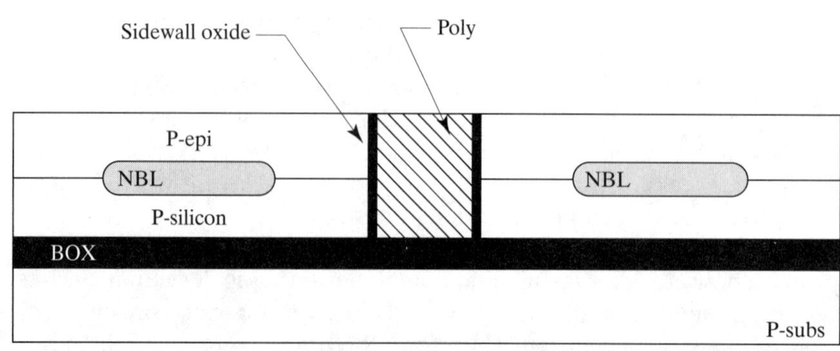

FIGURE 3.55 Cross section of DI BiCMOS wafer after trench fabrication.

TABLE 3.9 Comparison of JI BiCMOS and DI BiCMOS processes, up through deep-N+ deposition.

Mask	JI BiCMOS	DI BiCMOS
		Wafer bonding
1. NBL	First epi growth NBL deposition/anneal Second epi growth	NBL deposition/anneal Epi growth
1a. Isolation		Nitride deposition Trench pattern/etch Trench fill/planarization Pad oxidation
2. N-well	N-well deposition/drive	N-well deposition/drive
3. Deep-N+	Deep-N+ deposition/drive etc.	Deep-N+ deposition/drive etc.

As an example of the type of components that the DI BiCMOS process can fabricate, consider the fully-isolated vertical NPN transistor of Figure 3.56. This transistor occupies an isolated N-tank whose floor consists of a plate of NBL. The drawn N-well terminates along the centerline of the isolation trench, so this device uses the termination scheme depicted in Figure 3.53B. The remainder of the transistor appears very similar to the JI NPN transistor of Figure 3.49. In order to avoid etching and planarity problems at the four corners of the device, four relatively large-radius bends have been placed in the isolation trench.

The DI BiCMOS process offers a unique opportunity to fabricate a fully isolated vertical PNP transistor. Figure 3.57 shows the layout and cross section of this device.

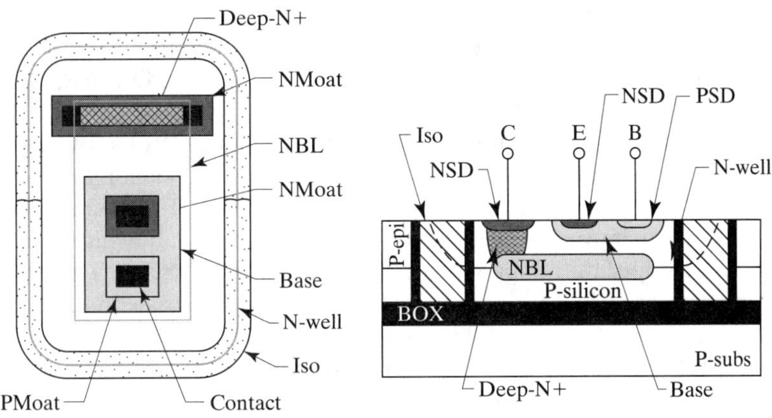

FIGURE 3.56 Layout and cross section of a DI BiCMOS vertical NPN transistor.

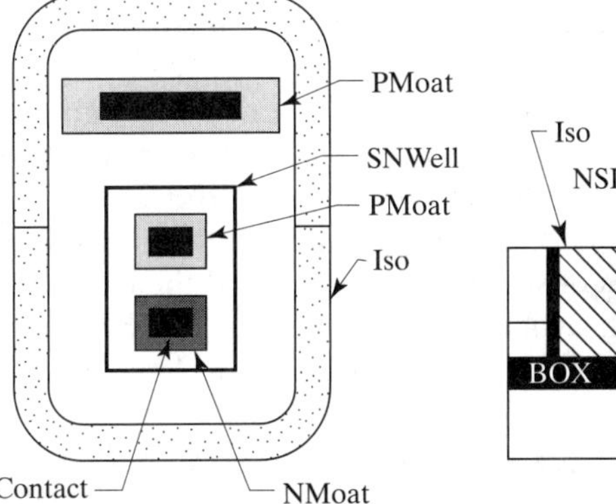

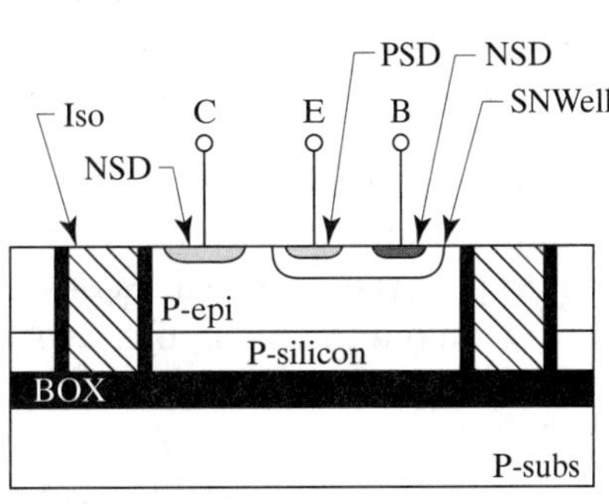

FIGURE 3.57 Layout and cross section of a DI BiCMOS vertical PNP transistor.

TABLE 3.10 Typical bipolar device parameters for a DI BiCMOS process.

Parameter	Vertical NPN	Vertical PNP
Drawn emitter area	16 μm^2	16 μm^2
Peak current gain (beta)	80	50
Early voltage	120 V	120 V
Collector resistance, in saturation	200 Ω	5000 Ω
V_{EBO} (Emitter-base breakdown, collector open)	8 V	8 V
V_{CBO} (Collector-base breakdown, emitter open)	50 V	40 V
V_{CEO} (Collector-emitter breakdown, base open)	25 V	25 V

The transistor uses the P-epi to form the collector, along with PSD to form a collector contact. The base region consists of a shallow N-well, here denoted as SNWELL. An NSD plug serves as the base contact, and a PSD geometry forms the emitter. This device represents the DI equivalent of the substrate PNP transistor of Figure 3.50. Table 3.10 summarizes the performance of these two devices.

3.4 SUMMARY

This chapter has examined three representative processes in detail: the standard bipolar process used to construct early analog integrated circuits, a polysilicon-gate CMOS process favored for digital logic, and an analog BiCMOS process, which merges the best of these two technologies onto a common substrate. Variations of these three processes fabricate the bulk of the low-cost, high-volume analog integrated circuits available today.

Standard bipolar is the oldest of the many available bipolar processes, most of which offer much faster switching speeds. The newer processes often replace junction isolation with partial or complete dielectric isolation to simultaneously reduce component area and minimize parasitic capacitance. Diffusions consist of much thinner layers, often created through innovative use of polysilicon as a doping source. Self-alignment and improved dimensional control allow the construction of far smaller geometries. The use of germanium-doped silicon for the base regions of bipolar transistors reduces their base transit times. These improvements have increased switching speeds by two orders of magnitude, allowing modern bipolar transistors to operate at speeds at or beyond fifty gigahertz. The highly specialized processes used to obtain such performance are fully as complex and costly as the most sophisticated of CMOS processes.

CMOS processes have undergone their share of evolutionary advances. Gate lengths have shrunk relentlessly as digital designers have sought to pack ever larger numbers of transistors onto limited amounts of silicon real estate. Once gate lengths fall below 1 μm, a variety of short-channel effects dictate the use of much more elaborate structures. The backgate doping levels must increase to combat premature punchthrough, and elaborate implantation strategies become necessary to confine the channel to the desired region immediately beneath the gate oxide. Cost and complexity have soared as processes push steadily deeper into the submicron regime seeking the elusive ultimate limits of device performance.

BiCMOS technologies seek to merge the best of both worlds. They are therefore heir to both the complexities of submicron CMOS and the elaborate bipolar structures required to obtain high switching speeds. Almost any bipolar or CMOS innovation can be merged into a BiCMOS process, and most have been.

3.5 EXERCISES

Refer to Appendix C for layout rules and process specifications.

3.1. Lay out the standard bipolar NPN transistor shown in Figure 3.10. Use a minimum-size square emitter. Allow room for all necessary metallization.

3.2. Draw a cross section of the standard bipolar NPN transistor shown in Exercise 3.1 to scale, assuming an epi thickness of 10 μm, NBL up-diffusion from the epi-substrate metallurgical junction of 3 μm, NBL down-diffusion from the epi-substrate metallurgical junction of 4 μm, an isolation junction depth of 12 μm, a deep-N+ depth of 9 μm, a base junction depth of 2 μm, and an emitter junction depth of 1 μm. Assume 80% outdiffusion where necessary. Oxide nonplanarities and deposited layers need not be shown.

3.3. Lay out the standard bipolar lateral PNP transistor shown in Figure 3.12. Use the minimum possible basewidth. Allow room for all necessary metallization, including a circular metal field plate connecting to the emitter contact and overhanging the inner edge of the collector region by 2 μm.

3.4. Lay out a 500 Ω standard bipolar base resistor following the example in Figure 3.13. Make the base resistor 8 μm wide, and widen the contacts as much as the width of the resistor allows.

3.5. Lay out a 25 kΩ standard bipolar base pinch resistor following the example in Figure 3.15. Assume that all of the resistance comes from the portion of the base beneath the emitter plate. Make the base strip 8 μm wide and extend the emitter plate beyond the sides of the base strip by at least 6 μm. NBL should overlap the base strip in the pinched region by at least 2 μm.

3.6. Lay out a fingered standard bipolar junction capacitor similar to the one shown in Figure 3.16. Make each of the three emitter fingers 50 μm long. The emitter plate should overlap the base by at least 6 μm; minimize all other dimensions.

3.7. Lay out the standard bipolar Schottky diode shown in Figure 3.18, assuming a contact opening that is 25 by 25 μm. Overlap metal over the Schottky contact layer (SCONT) by no less than 4 μm.

3.8. Lay out a 20 kΩ HSR resistor similar to that of Figure 3.19. Make the width of the HSR resistor 8 μm. The contacts should have the same width as the HSR resistor body. Assume that the base heads contribute negligible resistance, and compute the value of the resistor based only on the drawn length of the HSR segment between the base heads.

3.9. Lay out an NMOS transistor with a drawn width of 10 μm and a drawn length of 4 μm following the example in Figure 3.29. Allow room for all necessary metallization.

3.10. Draw a cross section of the NMOS shown in Exercise 3.9 to scale, assuming a well junction depth of 6 μm, PSD and NSD junction depths of 1 μm, a gate oxide thickness of 350 Å (0.035 μm), and a polysilicon thickness of 3 kÅ (0.3 μm). Ignore the V_t adjust and the channel stop implants. Assume 80% outdiffusion where necessary. Assume the silicon surface is planar, and ignore details of the metallization system.

3.11. Lay out a PMOS transistor with a drawn width of 7 μm and a drawn length of 15 μm, following the example in Figure 3.30. Assume NBL is not used. Include all necessary metallization.

3.12. Lay out a PSD resistor with a value of 200 Ω, following the example in Figure 3.33B. Make the resistor minimum width, and abut the well contact with one end of the resistor to save space.

3.13. Lay out a 3pF poly capacitor following the example in Figure 3.34. Include all necessary metallization. Contacts and vias can both reside on top of the poly plate.

3.14. Lay out the BiCMOS NPN transistor shown in Figure 3.49. The NBL should overlap the base region by at least 2 μm. Use minimum emitter dimensions and include all necessary metallization.

3.15. Draw a cross section of the NPN from Exercise 3.14 to proper scale, assuming the dimensions given in Exercise 3.10. Assume NBL diffuses upward by 3 μm and downward by 2 μm from the boundary between the first and second epi layers, which lies 7 μm below the silicon surface. In addition, assume a deep-N+ junction depth of 5 μm and a base junction depth of 1.5 μm. Ignore channel-stop implants and the effects of LOCOS field oxidation on surface planarity. Assume that the silicon surface is planar, and ignore the details of the metallization system.

3.16. Lay out the BiCMOS lateral PNP transistor shown in Figure 3.51. Assume a minimum basewidth. Unlike the device from Exercise 3.3, this transistor does not require a metal field plate over the base region. NBL should overlap the outer edge of the collector by at least 1.0 μm. Include all necessary metallization.

3.17. Lay out the resistor-transistor logic NOR gate shown in Figure 3.58A, using standard bipolar layout rules. Place Q_1 and Q_2 in the same tank, and use as small a plug of deep-N+ as possible to contact the collector of this tank. Assume the emitters of Q_1 and Q_2 are both minimum-size. Place R_1 in its own tank with the tank contact connected to V_{CC}. Provide at least one substrate contact. Surround this contact with base; this base region may touch, but not extend into, adjacent tanks. Label all inputs and outputs.

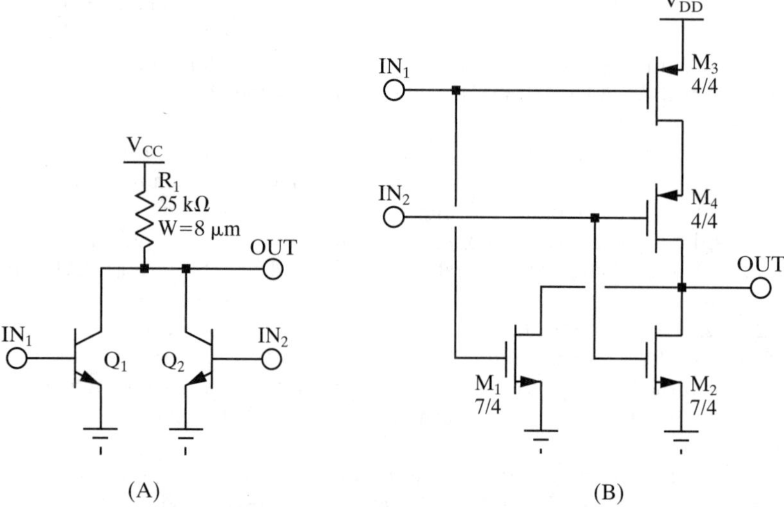

FIGURE 3.58 Circuits for Excercises 3.17 and 3.18.

3.18. Lay out the CMOS NOR gate shown in Figure 3.58B, using poly-gate CMOS layout rules. The W and L values for each transistor are shown on the schematic in the form of a fraction; 7/4 indicates a drawn width of 7 μm and a drawn length of 4 μm. Place all PMOS transistors in the same well, and connect this well to V_{DD}. Provide at least one substrate contact. Bring all inputs and outputs up to second-level metal, and label them appropriately.

3.19. Lay out the DI BiCMOS NPN transistor of Figure 3.56. Use minimum emitter dimensions and include all necessary metallization. Place as much NBL in the tank as possible. The layout rules for ISOL are as follows:

1.	ISOL width	6 μm exactly
2.	ISOL radius at corners	15 μm
3.	ISOL spacing to NBL	2 μm
4.	NWELL extends into ISOL	3 μm exactly
5.	ISOL spacing to DEEPN	6 μm
6.	ISOL spacing to BASE	4 μm
7.	ISOL spacing to PMOAT	2 μm
8.	ISOL spacing to NMOAT	2 μm

4 Failure Mechanisms

Integrated circuits are incredibly complex devices, and few of them are perfect. Most contain subtle weaknesses and flaws, which predispose them toward eventual failure. Such components can fail catastrophically and without warning after operating perfectly for many years. Engineers have traditionally relied on quality assurance programs to uncover hidden design flaws. Operation under stressful conditions can accelerate many failure mechanisms, but not every design flaw can be found by testing. The designer must therefore find and eliminate as many of these flaws as possible.

The layout of an integrated circuit contributes to many types of failures. If the designer knows about potential weaknesses, then safeguards can be built into the integrated circuit to protect it against failure. This chapter discusses a number of failure mechanisms that can be partially or entirely prevented by layout precautions.

4.1 ELECTRICAL OVERSTRESS

The term *electrical overstress* (EOS) refers to failures caused by the application of excessive voltages or currents to a component. Layout precautions can minimize the probability of four common types of EOS failures. *Electrostatic discharge* (ESD) is a form of electrical overstress caused by static electricity. The addition of special protective structures to vulnerable bondpads can minimize ESD failures. *Electromigration* is a slow wearout mechanism caused by excessive current densities; it can eventually cause open circuits or shorts between adjacent leads. Electromigration failures can be prevented by making leads wide enough to handle the maximum operating currents. *Dielectric breakdown* refers to the degradation and eventual failure of insulators subjected to excessive voltages or other forms of overstress. The *antenna effect* is a specific type of dielectric breakdown caused by charge accumulation on deposited conductors during etching or ion implantation. The problems posed by the antenna effect can be minimized by following specific design guidelines.

4.1.1. Electrostatic Discharge (ESD)

Almost any form of friction can generate static electricity. For example, if you shuffle across a carpet in dry weather and then touch a metal doorknob, a visible spark will leap from finger to doorknob. The human body acts as a capacitor, and the act of shuffling across a carpet charges this capacitance to a potential of 10,000 V or more. When a finger is brought near the doorknob, the sudden discharge creates a visible spark and a perceptible electrical shock. A discharge of less than 50 V will destroy the gate dielectric of a typical integrated MOS transistor. Voltages this low produce neither visible sparks nor perceptible electrical shock. Almost any human or mechanical activity can produce such low-level electrostatic discharges.

Proper handling precautions will minimize the risks of electrostatic discharge. ESD-sensitive components (including integrated circuits) should always be stored in static-shielded packaging. Grounded wrist straps and soldering irons can reduce potential opportunities for ESD discharges. Humidifiers, ionizers, and antistatic mats can minimize the buildup of static charges around workstations and machinery. These precautions reduce but do not eliminate ESD damage, so manufacturers routinely include special ESD protection structures onboard integrated circuits. These structures are designed to absorb and dissipate moderate levels of ESD energy without damage.

Special tests can measure the vulnerability of an integrated circuit to ESD. The three most common test configurations are called the human body model, the machine model, and the charged device model.[1] The *human body model* (HBM) employs the circuit shown in Figure 4.1A. When the switch is pressed, a 150pF capacitor charged to a specified voltage discharges through a 1.5 kΩ series resistor into the device under test (DUT). Ideally, each pair of pins would be independently tested for ESD susceptibility, but most testing regimens only specify a limited number of pin combinations to reduce test time. Each pair of pins is subjected to a series of positive and negative pulses; for example, three positive and three negative. After ESD stressing is complete, the part is tested to see if it still meets electrical specifications. Modern integrated circuits are routinely expected to survive 2 kV HBM. Specific pins on certain parts may be required to survive up to 25 kV HBM.

Figure 4.1B shows the circuit employed for the *machine model* (MM). A 200pF capacitor charged to a specified voltage discharges through a 0.5 μH series inductance into the DUT. As in the HBM test, each pin combination is subjected to a predetermined series of positive and negative pulses. With only a small inductance to limit the peak current, the machine model forms a much harsher test than the

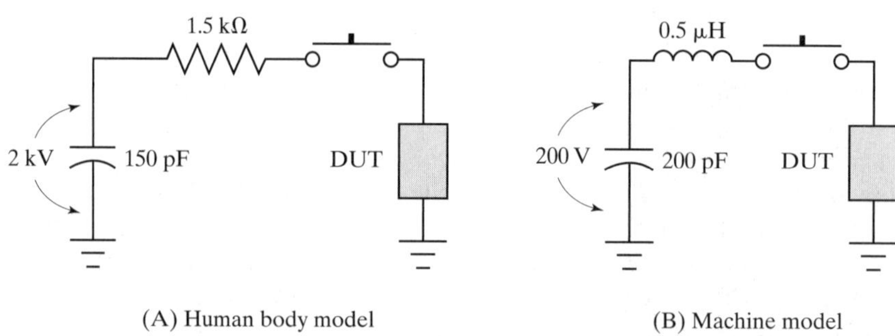

FIGURE 4.1 Representative ESD tests: 2 kV human body model (A) and 200 V machine model (B).

[1] Electrostatic Discharge Association specifications STM 5.1: *Electrostatic Discharge Sensitivity Testing—Human Body Model*, STM 5.2: *Electrostatic Discharge Sensitivity Testing—Machine Model*, and STM 5.3.1: *Electrostatic Discharge Sensitivity Testing—Charged Device Model.*

human body model. Few parts can survive more than 500 V under machine model testing.

A third ESD test called the *charged device model* (CDM) is gradually replacing the machine model. The charged device model places the integrated circuit package upside-down on a grounded metal plate and then charges the device to a specified voltage through a high-value resistor. A special probe then discharges one pin to a low-impedance ground. Researchers believe that this procedure more accurately models factory handling conditions than either the human body or the machine model. CDM testing produces very brief pulses of extremely high current. A typical testing regimen will specify 1 to 1.5 kV CDM testing.

Effects

Electrostatic discharge causes several different forms of electrical damage, including dielectric rupture, dielectric degradation, and avalanche-induced junction leakage.[2] In extreme cases, ESD discharges can even vaporize metallization or shatter the bulk silicon.

Less than 50 V will rupture the gate dielectric of a typical MOS transistor. The rupture occurs in nanoseconds, requires little or no sustained current flow, and is for all intents and purposes irreversible. The rupture typically shorts the gate and the backgate of the damaged transistor.[3] Capacitors that use thin insulating dielectrics are also vulnerable to this failure mechanism. An ESD discharge that strikes a pin connecting only to gates or capacitors will usually destroy these devices. If the pin also connects to diffusions, then these may avalanche before the gate oxide ruptures.

The integrity of a dielectric can be compromised by an ESD event that does not actually rupture it. The weakened dielectric can fail at any time, perhaps after hundreds or thousands of hours of flawless operation. Often the failure does not occur until the product has been delivered to the customer. Testing cannot screen out this type of delayed ESD failure; instead, vulnerable dielectrics must be protected against excessive voltages.

Although junctions are considerably more robust than dielectrics, they can still suffer ESD damage. An avalanching junction dumps a large amount of energy into a small volume of silicon. Extreme current densities can sweep metallization through contacts to short out underlying junctions. Excessive heating can also physically damage junctions by melting or shattering the silicon. These catastrophic forms of junction damage most often manifest themselves as short circuits. Avalanched junctions that do not fail outright usually exhibit increased leakage. Unlike overstressed dielectrics, damaged junctions will usually continue to operate without further degradation. Integrated circuits are often specified to have much larger leakages than are actually observed during testing to allow some margin for ESD-induced junction leakage. However, continued exposure to ESD events will often cause a junction to degrade beyond even these relaxed limits.

Preventative Measures

All vulnerable pins must have ESD protection structures connected to their bondpads. Some pins can resist ESD and therefore do not require additional protection. Examples include pins connected to substrate and to large diffusions, such as those found in large power transistors. These large junctions may be able to disperse and

---

2 Some of these are discussed in A. Amerasekera, W. van den Abeelen, L. van Roozendaal, M. Hannemann, and P. Schofield, "ESD Failure Modes: Characteristics, Mechanisms, and Process Influences," *IEEE Trans. Electron Devices*, Vol. 39, #2, 1992, pp. 430–436.

3 C. M. Osburn and D. W. Ormond, "Dielectric Breakdown in Silicon Dioxide Films on Silicon, II. Influence of Processing and Materials," *J. Electrochem. Soc.*, Vol. 119, #5, 1972, pp. 597–603.

absorb the ESD energy before it can damage other circuitry. Pins or devices that can withstand ESD events without the addition of ESD protection circuitry are said to be *self-protecting*.

Pins connecting to relatively small diffusions are vulnerable to ESD-induced junction damage. These junctions are simply not large enough to protect themselves. Certain junctions, most notably the base-emitter junctions of NPN transistors, are notoriously vulnerable to ESD damage. Avalanching the base-emitter junction of an NPN transistor permanently degrades its beta. A circuit designer can sometimes eliminate the vulnerable junctions by rearranging the circuit. Because ESD vulnerabilities are difficult to predict, cautious designers add protection devices to all pins that might be even remotely vulnerable.

Pins that connect only to gates of MOS transistors or to deposited capacitor electrodes are extremely vulnerable to ESD damage. Special input protection structures have been developed to protect dielectrics against HBM and MM events. The extremely high currents characteristic of CDM events require additional protection structures, called *CDM clamps*, to be placed near the vulnerable devices.

The thin emitter oxides employed in some standard bipolar processes are also susceptible to ESD-induced rupture. This vulnerability can be eliminated by ensuring that leads that connect to external bondpads do not cross any emitter region to which they do not connect. Alternatively, ESD structures similar to those used for protecting gates can protect the vulnerable circuits. Most modern versions of the standard bipolar process employ thick emitter oxides, which eliminate the need for these precautions.[4]

Considerable ingenuity is often required to formulate successful ESD structures for analog integrated circuits. A dozen or more protection circuits are often required to satisfy the large range of voltages and the many types of vulnerable devices found in analog circuits. The protection devices must also be evaluated to ensure that they do not interfere with the operation of the circuits they protect. Section 13.5 discusses several commonly employed ESD structures and explains how these can be modified to meet a variety of special requirements.

4.1.2. Electromigration

Electromigration is a slow wearout phenomenon caused by extremely high current densities. The impact of moving carriers with stationary metal atoms causes a gradual displacement of the metal. In aluminum, electromigration only becomes a concern when current densities approach $5 \cdot 10^5$ A/cm^2. Although this may seem a tremendous current density, a minimum-width lead in a submicron process can experience electromigration at currents of only a few milliamps.[5]

Effects

Despite its homogenous appearance, aluminum metal is a polycrystalline material. The individual crystals, or *grains,* normally abut one another. Electromigration causes metal atoms to gradually move away from the grain boundaries, forming voids between adjacent grains. This causes a decrease in the lead's effective cross-sectional area and raises the current density seen by the remainder of the lead. Additional voids form and gradually coalesce until they ultimately sever the lead. Metal displaced by voiding builds up small bumps called *hillocks*, or extrudes in sharp points called *dendrites*.

---

[4] Even thick emitter oxides can rupture under certain conditions; see "Dielectric Breakdown of Emitter Oxide," *Semiconductor Reliability News*, Vol. IV, #1, 1992, p. 1.

[5] Assuming a lead width of one micron and a thickness of 5000Å, a current of 2.5mA will produce a current density of $5 \cdot 10^5$ A/cm^2.

The addition of refractory barrier metal changes the observed modes of electromigration failure. Since the refractory metal is relatively resistive, most of the current initially flows through the aluminum. Once voiding finally severs the aluminum, the underlying refractory metal bridges the gap and continues to conduct current. Refractory metals are much less susceptible to the effects of electromigration, so the lead will not completely fail. Instead, the formation of voids in the aluminum causes the lead's resistance to gradually and erratically increase. More ominously, aluminum metal displaced by voiding sometimes forms dendrites that short adjacent leads together. The cross-sectional area of the aluminum portion of a lead therefore determines how much current it can safely conduct, regardless of the presence or absence of refractory barrier metal.

Refractory barrier metal is often used to prevent electromigration failures in contacts and vias because it ensures electrical continuity across steep sidewalls after the thin aluminum metallization at these points succumbs to electromigration-induced voiding. Lateral extrusion does not normally occur in contacts or vias since a contiguous sheet of metal covers the entire structure. Likewise, resistance changes caused by voiding are usually small compared with the inherent resistance of the contact or via structure. Recent studies have shown that the interfaces between barrier metal and aluminum in vias are especially susceptible to void formation. Similar problems have been observed in tungsten-plug vias.[6] The electromigration resistance of such nonhomogenous via systems may actually depend upon the direction of current flow through the vias. Experimental measurements are thus required to determine the true current-handling capabilities of vias and contacts.

Preventative Measures

The first line of defense against electromigration consists of process improvements. Aluminum metallization is now routinely doped with 0.5 to 4% copper to enhance electromigration resistance.[7] Copper accumulates at the grain boundaries, where it inhibits voiding by increasing the activation energy required to dislodge metal atoms from the lattice. Copper-doped aluminum exhibits 5 to 10 times the current handling capability of pure aluminum.[8] The electromigration resistance of leads can be further improved by using compressively stressed protective overcoats that confine the metal under pressure and inhibit void formation.[9]

Pure copper is much more resistant to electromigration than either pure aluminum or copper-doped aluminum. The exact lifetime improvement depends upon the nature of the interlevel dielectric and passivation as well as measurement conditions. Lifetimes of 40–100 times those of aluminum are typical.[10] Thick power copper metallization is thus virtually immune to electromigration. However, the submicron dimensions of damascene copper suggest that this metallization system will be constrained, at least to a certain degree, by electromigration limitations.

Processing techniques can minimize electromigration, but there remains some maximum current density that cannot be exceeded without risking eventual metallization

---

6 J. Tao, K. K. Young, N. W. Cheung, and C. Hu, "Comparison of Electromigration Reliability of Tungsten and Aluminum Vias Under DC and Time-Varying Current Stressing," *Proc. International Reliability Physics Symp.*, 1992, pp. 338–343.

7 I. Ames, F. M. d'Heurle, and R. E. Horstmann, "Reduction of Electromigration in Aluminum Films by Copper Doping," *IBM J. of Research and Development*, Vol. 14, #4, 1970, pp. 461–463.

8 S. S. Iyer and C. Y. Ting, "Electromigration Study of Al-Cu/Ti/Al-Cu System," *Proc. International Reliability Physics Symp.*, 1984, pp. 273–278. See also Lahri, *et al.*, p. 166.

9 J. R. Lloyd and P. M. Smith, "The Effect of Passivation on the Electromigration Lifetime of Al/Cu Thin Film Conductors," *J. Vacuum Science Technology A*, Vol. 1, #2, 1983, pp. 455–458.

10 T. C. Lee, M. Ruprecht, D. Tibel, T. D. Sullivan, and S. Wen, "Electromigration Study of Al and Cu Metallization Using WLR Isothermal Method," *Proc. International Reliability Symp.*, 2002, pp. 327–335.

failure. The design rules for each process thus define a maximum allowed current per unit width. Typical values are 2 mA/μm for leads that do not cross oxide steps and 1 mA/μm for those that do. These values depend on the thickness of the metallization and its composition, and on the anticipated operating temperature (Section 14.3.3). Consider a lead that must conduct 50 mA, following the electromigration limits specified above. If this lead routes across field oxide in order to avoid oxide steps, then it need be only 25 μm wide; otherwise, its width must increase to 50 μm. The lead cannot widen abruptly at the oxide steps, because the current only gradually flows out from a narrow lead into a wider one. The wider lead should extend beyond the oxide step in either direction for a distance at least twice its greatest width.

Most design rules also specify the maximum current allowed to flow through contacts and vias. A typical rule sets the current through a contact or via equal to the current that can flow through a lead of the same width. Following this principle, a 1 μm-wide via could conduct as much current as a 1 μm-wide lead, or, according to the numbers given above, 2.5 mA. Note that the currents specified here are only typical values and aren't necessarily representative of any given process.

Excessive current can also cause bondwires to overheat and fail. In practice, a typical 25 μm-diameter gold bondwire 1.25 mm in length can safely conduct about an amp, while a similar aluminum wire can conduct about 750 mA. These numbers depend upon both the diameter of the wire and its length, as considerable heat flows down the wire to the leadframe. If the anticipated currents exceed these limits, then the design will require larger-diameter bondwires or multiple bondwires placed in parallel (Section 14.3.3).

4.1.3. Dielectric Breakdown

Modern CMOS and BiCMOS processes use extraordinarily thin dielectric layers. The gate oxide of a typical 5 V CMOS transistor is only some 200 Å thick, and the gate oxide of an advanced 1.8 V CMOS transistor measures an incredibly thin 90 Å. Since the average length of a silicon-oxygen bond equals about 1.5 Å, a thickness of 90 Å represents only 60 atomic layers of oxide. Insulators as thin as these are exceedingly vulnerable to electrical overstress.

Effects

Dielectric breakdown involves a process called *tunneling* that allows carriers to travel short distances through seemingly insurmountable obstacles. A thin insulating layer represents just such an obstacle. The rate of tunneling diminishes exponentially with distance, and this relationship limits electron tunneling to distances of about 45 Å. Holes can only tunnel a fraction of this distance because of their larger effective mass.

Electrons can current directly across dielectrics less than about 45 Å thick (Figure 4.2A). This process, called *direct electron tunneling*, does not normally occur in gate oxides or capacitor dielectrics because they are too thick.

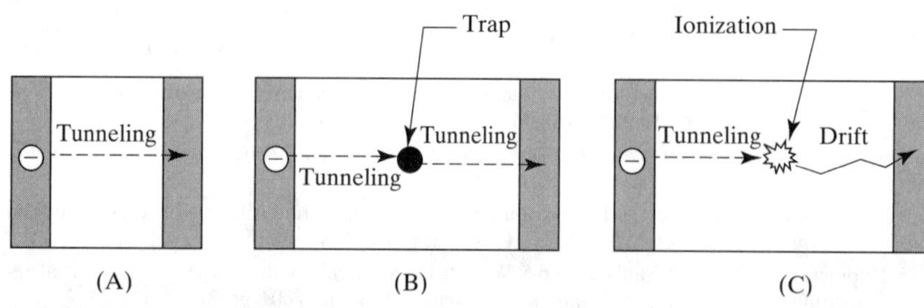

FIGURE 4.2 Tunneling mechanisms in gate oxides: direct electron tunneling (A), trap-assisted tunneling (B), and Fowler-Nordheim tunneling (C).

Electrons can tunnel across greater distances by means of *trap-assisted tunneling.* This process relies upon the presence of traps within the dielectric. These traps function as stepping stones for electrons attempting to tunnel from one side to the other. So long as the electrons can travel in a series of jumps of less than about 45 Å each, they can "percolate" across the trap-filled dielectric. High-quality dielectrics have so few traps that their average spacing greatly exceeds 45 Å. Trap-assisted tunneling can still occur from one side of the dielectric to traps near its center, and then on to the other side (Figure 4.2B). In high-quality dielectrics, this process cannot span distances of more than about 90 Å.[11] Poor-quality dielectrics have so many traps that carriers can jump from one trap to the next, allowing them to cross much greater distances. This process causes poor-quality dielectrics to leak, regardless of their thickness.

High-quality dielectrics thicker than about 90 Å can still experience a third variety of tunneling, called *Fowler-Nordheim tunneling*. This mechanism only occurs when an intense electric field exists within the dielectric. If an electron tunnels within such a field, it gains an energy proportional to the product of the field and the distance it travels. If the electron gains enough energy, it can travel freely through the rest of the dielectric (Figure 4.2C).

Fowler-Nordheim tunneling injects electrons into the dielectric, where the electric field accelerates them and transforms them into hot electrons. These fast-moving hot electrons slam into dielectric atoms, knocking loose valence electrons. This process generates holes, and the passage of these holes through the oxide produces traps. Therefore, dielectrics subjected to Fowler-Nordheim tunneling gradually deteriorate, and eventually begin to leak. The resulting current is called a *stress-induced leakage current* (SILC).[12]

Stress-induced leakage currents can lead to catastrophic failures if the dielectric remains under stress. Some of the electrons injected into the dielectric by trap-assisted tunneling escape their trap sites due to the energy they gain from the electric field. Subsequent collisions with dielectric atoms generate additional traps that in turn increase the leakage current. The weakest spots in the dielectric are disproportionately affected by this process, so as the magnitude of the leakage current increases, the area through which it flows diminishes. Eventually a tiny spot in the dielectric conducts so much current that it melts. Conductive material adjacent to the dielectric punches through this rupture point, producing an irreversible short-circuit failure.

Dielectrics subjected to large overstresses can rupture in nanoseconds. On the other hand, dielectrics subjected to borderline stresses may function for months or even years before failing. This delayed mode of failure is called *time-dependent dielectric breakdown* (TDDB). The vulnerability of dielectrics to TDDB depends critically upon their uniformity of thickness and composition.

Preventative Measures

All of the various forms of dielectric breakdown are due to excessive electrical stresses applied to gate oxides or other thin insulating layers. One preventative measure is obvious: the avoidance of excessive electrical stresses. Unfortunately, it is very difficult to determine precisely how much stress constitutes an excessive amount. The main problem is that neither the dielectric nor the electric field applied

---

[11] R. Bucksch, MSEE thesis, Friedrich-Alexander University, Erlangen-Nürnberg, 1977, pp. 15–16.

[12] L. Larcher, A. Paccagnella, and G. Ghidini, "A Model of the Stress Induced Leakage Current in Gate Oxides," *IEEE Trans. Electron Devices*, Vol. 48, #2, 2001, pp. 285–288. Also see P. M. Lenahan, J. J. Mele, J. P. Campbell, A. Y. Kang, R. K. Lowry, D. Woodbury, S. T. Liu, and R. Weimer, "Direct Experimental Evidence Linking Silicon Dangling Bond Defects to Oxide Leakage Currents," *Proc. International Reliability Physics Symp.*, 2001, pp. 150–155.

to it are uniform. The dielectric invariably has thinner and thicker regions, and the electric field may crowd towards certain points (such as at the sharp edges of conductors). The traps responsible for causing leakage are also not uniformly distributed. Reliable operation therefore requires a large safety margin. Thicker dielectrics are actually somewhat more fragile than thinner ones. The usual maximum stress allowed for a dry oxide of 300–500 Å thickness equals about 3.5–4 MV/cm, while the maximum stress allowed for thinner oxides equals about 4–4.5 MV/cm.

Various problems in the manufacturing process can reduce the *gate oxide integrity* (GOI) of a process. GOI problems are among the most difficult challenges facing a modern CMOS or BiCMOS wafer fab. Even with the most rigorous control measures, occasionally the process will produce a defective batch of material. This material will seem perfectly normal until its gate oxides are placed under large electrical stresses. Then one or two transistors out of the millions on the wafer will suddenly fail. It is very difficult to screen out GOI failures, and defective material often reaches customers. GOI problems have doubtless caused many of the sudden and unexpected failures of electronic appliances that are traditionally ascribed to power line transients.

A method has been developed that allows GOI defects to be detected before material is shipped to customers. This technique, called *overvoltage stress testing* (OVST), employs a precisely controlled overvoltage event to stress the gate oxide. The test voltage employed may be as much as twice the maximum specified operating voltage, but it is applied only once, and only for a very short time (perhaps 100 mS). If any of the devices on a chip fail due to the OVST, the unit is rejected. If several chips on a wafer fail OVST, then the entire wafer is rejected. If several wafers in a lot fail OVST, then the entire lot is rejected. OVST failures will alert the wafer fab to the existence of an otherwise-undetectable GOI problem. Once the wafer fab is aware of the situation, they can attempt to discover and correct the underlying problems.

Digital circuits are readily amenable to OVST, but analog circuits often shield vulnerable gate oxides from excessive applied voltages as part of their functionality. Therefore, special test modes must be incorporated into the design to allow the OVST to bypass protective circuitry. Since OVST is based upon a statistical methodology, only a fraction of all of the gate oxides need be tested.

Certain other factors can weaken a dielectric. Heavy metal atoms can interfere with the growth of an oxide layer, producing weak spots that reduce oxide integrity. Most processes use a Czochralski-grown substrate. Oxygen from the silica crucible dissolves in the Czochralski silicon. If the silicon is heated above 1000°C for a period of some hours, then the oxygen gathers in localized areas to form flecks of oxide called *oxygen precipitates*. These precipitates bind, or *getter*, heavy metal atoms, preventing them from interfering with surface oxidation. This process greatly improves gate oxide integrity.

Heavily doped N+ diffusions that occur early in the process can also act as getters. Deep-N+ and NBL both function in this manner. One can improve the gate oxide integrity of a device by coding regions of deep-N+ or NBL within a certain distance of the gate oxide. This gettering distance typically equals a hundred microns or so. Processes that do not benefit from oxygen precipitation, such as DI processes, may even add a deep-N+ step simply to improve gate oxide integrity. In such cases, layout rules may mandate the placement of blocks or strips of deep-N+ in proximity to MOS transistors.

The fact that deep-N+ getters heavy metal impurities suggests that oxides grown over deep-N+ will have reduced integrity. This effect has actually been observed.[13]

[13] Private communication, L. Hutter.

Although some processes allow the growth of capacitor oxides over deep-N+, the resulting devices have poorer oxide integrity than those grown over lightly doped silicon. Consequently, it is unwise to employ large areas of deep-N+ capacitor oxide without incorporating OVST into the design.

4.1.4. The Antenna Effect

Dry etching is known to deposit charges upon the surface of the wafer. Exposed conductors can collect an electrical charge that can damage thin gate dielectrics. This failure mechanism is called *process plasma-induced damage*, or, more colorfully, the *antenna effect*. The antenna effect generates stress-induced leakage currents that can lead to either immediate or delayed failure of the overstressed dielectrics.

Effects

The exact source of the electrical charges responsible for the antenna effect is a matter of some controversy. The plasma itself contains an equal number of positive and negative particles. However, various mechanisms can cause local fluctuations in charge densities within the plasma. Some of the proposed mechanisms include nonuniformities due to reactor design and AC plasma excitation, and an effect called *electron shading*, in which adjacent geometries block the isotropic electron flux to a greater degree than they block the anisotropic ion flux. Regardless of the precise mechanisms involved, experience has shown that both dry etching of conductor layers and the subsequent ashing of photoresist can cause plasma-induced damage.

The impact of the antenna effect must be evaluated for the etching and ashing of each conductor layer. Consider the case of polysilicon. During the initial stages of poly etching, the entire surface of the wafer is covered by an unbroken sheet of poly. Charge reaches this poly plate through all of the openings in the photoresist. Apparently, the fluctuations responsible for the antenna effect largely cancel one another out across the width of the wafer, for little damage occurs at this point. Partway through the etch process, the individual poly geometries separate from one another. Each geometry now picks up charge around its periphery, where the poly is exposed to the plasma. This charge is injected through the thin gate oxide. The vulnerability of a given geometry to the antenna effect therefore depends upon the ratio of its total perimeter to the active gate area beneath it. The larger this *peripheral antenna ratio*, the greater the risk of plasma-induced damage. Most processes define a maximum allowed peripheral antenna ratio for poly; a typical value is 100 μm^{-1}.

During the final stages of photoresist ashing, the entire surface of the poly pattern becomes exposed to the plasma. Each geometry now picks up charge across its entire surface and injects this charge through the thin gate oxide. The vulnerability of a given geometry to the antenna effect therefore depends upon the ratio of the total area to the active gate area beneath it. The larger this *areal antenna ratio*, the greater the risk of plasma-induced damage. Most processes define a maximum allowed areal antenna ratio for poly; a typical value is 500.

Each conductor layer is vulnerable to the antenna effect during etching and ashing, so each layer has its own peripheral and areal antenna ratios. Consider the case of metal-2. Near the end of the etch process, the individual metal-2 geometries become separated from one another. However, these geometries may be connected together through lower conductor layers. Therefore, the antenna effect cannot be evaluated on a geometry-by-geometry basis. Instead, one must define collections of electrically connected geometries called *nodes*. During the metal-2 etch, each node collects charge proportional to the metal-2 periphery exposed to the plasma and injects this charge through the active gate beneath poly geometries forming part of the node. Therefore, the metal-2 peripheral antenna ratio of a node equals the total metal-2 periphery of the node divided by the active gate area beneath the poly geometries of the node. Similarly, the evaluation of ashing damage depends upon the metal-2 areal antenna ratio, defined as the total metal-2 area of a node divided by the active gate area beneath the poly geometries of the node.[14]

A great deal of effort has been expended to understand the relationship between antenna ratios and gate dielectric damage, but much remains uncertain. Some researchers have uncovered evidence that PMOS gate oxides are considerably more sensitive to plasma-induced damage than NMOS gate oxides. Therefore, some processes define separate antenna ratios for each type of oxide. Other researchers have shown that oxide isolation greatly reduces plasma-induced damage, presumably by limiting the current that can flow through any given area of gate oxide.[15]

Preventative Measures[16]

Any node whose antenna ratio exceeds specifications must be reworked. The exact techniques employed depend upon which layer is involved. In the case of polysilicon, the ratio can be reduced by inserting metal jumpers. Consider the case shown in Figure 4.3A. This circuit contains a very long poly lead that crosses a minimum-size MOS transistor M_1. The antenna ratios of this poly geometry could clearly become very large. If, however, a short metal jumper is inserted in the poly lead next to the transistor, then the single poly geometry now becomes two separate geometries (Figure 4.3B). The geometry on the left (connecting to the gate of transistor M_1) has relatively small antenna ratios. The geometry on the right (connecting to the source/drain of transistor M_2) has zero antenna ratios because no gate oxide lies beneath it. Therefore, the addition of the metal jumper has eliminated any potential problem.

Metal layers are somewhat more difficult to evaluate because metal nodes can connect to diffusions that leak away the charge before it damages gate oxides. For processes that employ gate oxides thicker than about 400 Å, the source/drain junctions of the MOS transistors will typically avalanche before the gate oxides can be

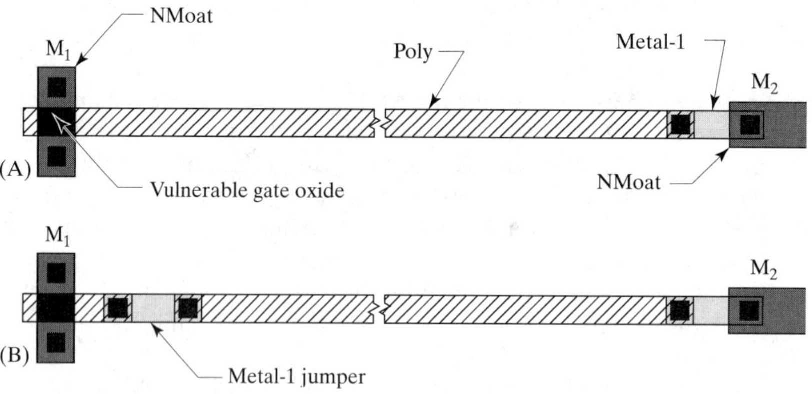

FIGURE 4.3 A layout susceptible to the antenna effect (A) can be made immune by the addition of a metal jumper (B).

---

14 C. T. Gabriel, "Gate Oxide Damage: A Brief History and a Look Ahead," *Proc. 6th International Symposium on Plasma Process-Induced Damage, 2001,* pp. 20–24. See also T. Watanabe and Y. Yoshida, "Dielectric Breakdown of Gate Insulator Due to Reactive Ion Etching," *Solid State Technology*, Vol. 27, #4, 1984, pp. 263–266.

15 A. C. Mocuta, T. B. Hook, A. I. Chou, T. Wagner, A. K. Stamper, M. Khare, and J. P. Gambino, "Plasma Charging Damage in SOI Technology," *Proc. 6th International Symposium on Plasma Process-Induced Damage*, 2001, pp. 104–107.

16 P. Simon, J-M. Luchies, and W. Maly, "Antenna Ratio Definition for VLSI Circuits," *Proc. 4th International Symposium on Plasma Process-Induced Damage*, 1999, pp. 16–20.

damaged. In such cases, any node that connects to a source/drain diffusion can generally be ignored when computing antenna ratios. If a metal node is found to have an excessive antenna ratio, the problem can be eliminated either by placing a jumper on a higher metal layer (as discussed previously in connection with poly), or by connecting a source/drain diffusion to the node. If the circuit does not include a transistor connected to the node, then a small structure called a *leaker* can be attached instead. Figure 4.4 shows examples of NSD/P-epi and PSD/N-well leakers. For thick-oxide processes, the NSD/P-epi leaker is preferred. This structure is essentially a diode whose anode is connected to the metal node and whose cathode is connected to the substrate. If the voltage on the node drops below the substrate potential, then the leaker will forward-bias and clamp the voltage. If the voltage on the node rises above substrate potential, then the NSD/P-epi junction will avalanche before the thick oxide is damaged.

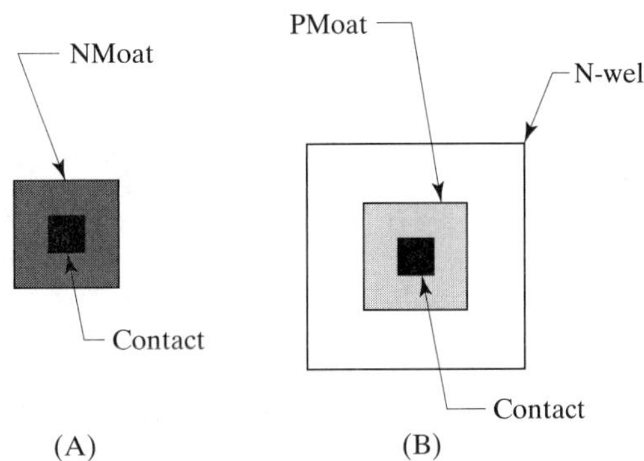

FIGURE 4.4 NSD/P-epi leaker (A) and PSD/N-well leaker (B).

Leakers for thin-oxide processes are somewhat more problematic. The avalanche voltage of an NSD/P-epi junction cannot be relied upon to protect a gate oxide much thinner than 400 Å. Experience has shown that nodes in thin-oxide processes can be protected by a combination of NSD/P-epi and PSD/N-well leakers. The NSD/P-epi leaker will forward-bias if the node drops below substrate potential. The PSD/N-well leaker will forward-bias if the node rises above the N-well potential, but the reverse-biased N-well/P-epi junction prevents currents from flowing through this structure during normal operation. During reactive ion etching, the light from the plasma reaction shines down on the wafer. This light encourages photogeneration within the depletion region of the N-well/P-epi junction, causing this junction to leak. This leakage helps carry away the charge injected onto the N-well. In order for the N-well/P-epi leaker to properly function, at least a portion of its periphery should remain uncovered by metal to a distance of at least 5–10 μm outside of drawn N-well. Whenever leakers are inserted, the circuit designer should be informed of their presence so that their effect upon circuit operation can be evaluated. In most cases, the leakers will not interfere with the circuit, but it is impossible to make blanket statements about what might or might not interfere with analog circuits.

4.2 CONTAMINATION

Plastic-encapsulated integrated circuits are vulnerable to certain types of contaminants. Assuming that a device has been properly manufactured, very low levels of contaminants will initially exist inside the plastic encapsulation. Plastic mold compounds have been carefully formulated to provide the highest possible degree of resistance to penetration by external contaminants, but no plastic is entirely impregnable. Contaminants seep in along the interface between the metal pins and the plastic, or they directly penetrate the plastic itself. Two major contamination issues faced by modern plastic-encapsulated dice are *dry corrosion* and *mobile ion contamination.*

4.2.1. Dry Corrosion

The aluminum metal system will corrode if exposed to ionic contaminants in the presence of moisture. Only trace amounts of water are necessary to initiate this so-called *dry corrosion.* Since moisture and ionic contaminants are both ubiquitous, integrated circuits must depend on their encapsulation to protect them. Early mold compounds had little moisture resistance. Newer compounds are more impermeable, but given enough time, moisture will eventually penetrate any plastic package.[17]

All modern integrated circuits are covered with a protective overcoat that acts as a secondary moisture barrier. Unfortunately, openings must be made through this overcoat to allow bondwires to be attached to the die. Fuse trim schemes often require additional openings in the protective overcoat. All of these openings represent potential pathways for contaminants to reach the die.

Effects

Water alone cannot corrode aluminum, but many ionic substances dissolve in water to form relatively corrosive solutions. Phosphosilicate glasses containing more than about 5% phosphorus represent a corrosion risk, as moisture can leach phosphorus from the glass to produce phosphoric acid.[18] This acid rapidly attacks and dissolves aluminum, causing open circuit failures. Many modern processes use nitride or oxynitride protective overcoats to ensure that moisture cannot reach the phosphosilicate glass that lies beneath. Alternatively, the phosphorus content of the glass can be reduced by using a combination of boron and phosphorus as dopants. Both of these elements reduce the softening point of a glass, so a *borophosphosilicate glass* (BPSG) will require less phosphorus to achieve the same softening point as a phosphosilicate glass.

Halogen ions in water solution can also corrode aluminum.[19] Common salt, or sodium chloride, provides an abundant source of chloride ions. Moisture seeping into the package of an integrated circuit can transport chloride ions to the surface of the die where they can begin to corrode the aluminum metal system.

The flame retardants added to the plastic encapsulation represent another potential source of ionic contaminants that can cause dry corrosion. Prior to about 2002, most mold compounds contained organobromine-based flame retardants. These substances begin decomposing at temperatures beyond 250°C, releasing bromide ions that can cause the same sort of corrosion problems more commonly associated with chloride ions.[20] Organobromine flame retardants are being replaced with other substances due to toxicity concerns. Ironically, some of the early substitutes relied upon phosphorus formulations that generated phosphoric acid in the presence of moisture. The consequent dry corrosion problems were reminescent of those previously encountered with phosphosilicate glass overcoats. The latest generation of so-called

---

17 J. E. Gunn and S. K. Malik, "Highly Accelerated Temperature and Humidity Stress Technique (HAST)," *Proc. International Reliability Symposium*, 1981, pp. 48–51.

18 W. M. Paulson and R. W. Kirk, "The Effects of Phosphorus-Doped Passivation Glass on the Corrosion of Aluminum," *Proc. International Reliability Physics Symposium*, 1974, pp. 172–179.

19 M. M. Ianuzzi, "Reliability and Failure Mechanisms of Non-hermetic Aluminum SIC's in an Environment Contaminated with Cl_2," *(sic), IEEE Trans. Comp. Hyb. Man. Tech.*, 6, 1983, pp. 191–201.

20 T. Raymond, "Avoiding Bond Pad Failure Mechanisms in Au-Al Systems," *Semiconductor International*, Sept. 1989, pp. 152–158.

green mold compounds seems to have conquered this problem, but, like all plastics, these are not completely impervious. Dry corrosion is, and will remain, a concern for all plastic-encapsulated integrated circuits.

Preventative Measures

Although contamination may seem completely beyond the control of the layout designer, several measures can be taken to minimize vulnerabilities in the protective overcoat. The designer should minimize the number and size of all PO openings. A production die should not include any openings that are not absolutely necessary for its manufacture. If the designer wishes to include additional testpads for evaluation, then these should occupy a special test mask. When the part is released to production, the test mask should be replaced by a production PO mask that seals the test pads under protective overcoat.

Metal should overlap bondpad openings on all sides by an amount sufficient to account for misalignment. The metal bondpads will then protect the underlying oxide from the entry of moisture and other contaminants. Openings made for polysilicon or metal fuses should be made as small as possible, and no circuitry of any sort except the fuse element itself should appear within the opening.

4.2.2. Mobile Ion Contamination

Many potential contaminants dissolve in silicon dioxide at elevated temperatures, but most lose their mobility at normal operating temperatures because they become bound into the oxide macromolecule. The alkali metals are exceptions to this rule and remain mobile in silicon dioxide even at room temperature.[21] Of these so-called *mobile ions*, sodium is by far the most common and the most troublesome.

Effects

Mobile ion contamination induces parametric shifts, most noticeably in MOS transistor threshold voltages. Figure 4.5A shows the gate oxide of an NMOS transistor contaminated by sodium during manufacture. The positively charged sodium ions are initially distributed throughout the oxide. An equal number of negatively charged ions (anions) are also introduced. Unlike the sodium ions, these anions remain rigidly locked into the oxide macromolecule.

Figure 4.5B shows the same gate dielectric after an extended period of operation under a positive gate bias. The positively charged gate electrode has repelled the

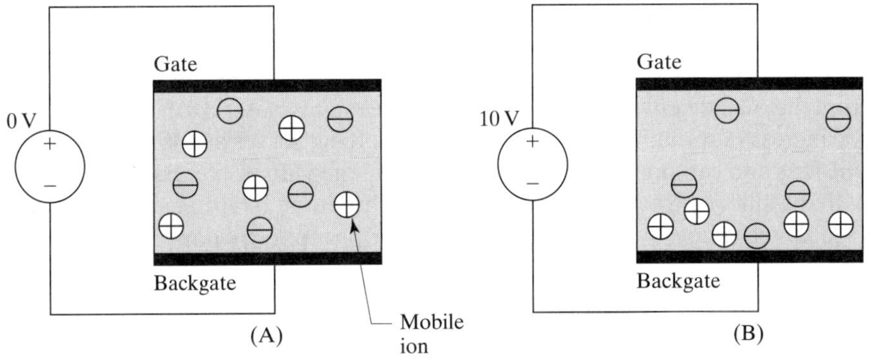

FIGURE 4.5 Behavior of mobile ions under bias: Ions that were randomly distributed through the oxide (A) shift in unison in response to a positive gate bias (B).

[21] Specifically, only lithium, sodium, and potassium (and, to a certain degree, hydrogen) qualify as mobile ions in silicon. The heavier alkali metals rubidium and cesium are far less mobile: B. E. Deal, "The Current Understanding of Charges in the Thermally Oxidized Silicon Structure," *J. Electrochem. Soc.*, Vol. 121, #6, 1974, pp. 198C–205C.

mobile sodium ions down toward the oxide-silicon interface. Since the negative ions do not move, the redistribution of the sodium ions results in a net separation of charges within the oxide.[22] The presence of positive charges near the channel of the NMOS transistor decreases its threshold voltage. The magnitude of the threshold voltage shift depends on sodium ion concentration, gate bias, temperature, and time. Many analog circuits require that threshold voltages match within a few millivolts. Even low concentrations of mobile ions can produce shifts of this magnitude.

Mobile ion contamination can produce long-term failures when the slow drift of threshold voltages eventually causes a circuit to exceed its parametric limits. If the faulty devices are removed from operation and are baked at 250°C for a few hours, the mobile ions redistribute and the threshold shifts vanish. This treatment is only temporary; the threshold drift returns as soon as electrical bias is restored. Although analog circuits are particularly susceptible to parametric shifts caused by mobile ions, even digital circuits will eventually fail if the threshold voltages shift too far. Early metal-gate CMOS logic was plagued by threshold voltage shifts caused by severe sodium contamination.

Preventative Measures

Some mobile ions inevitably become incorporated in an integrated circuit during manufacture. This source of contamination can be minimized by using purer chemicals and improved processing techniques. MOS processes typically take extraordinary steps to ensure process cleanliness, but these alone cannot entirely eliminate threshold voltage variations.

Manufacturers of metal gate CMOS attempted to stabilize threshold voltages by adding phosphorus to the gate oxide.[23,24] Phosphorus stabilization had the desired effect of immobilizing alkali metal contaminants, but it also introduced a new problem. The electrically charged phosphate groups shift slightly under strong electrical fields even though they are bound to the oxide macromolecule. Phosphosilicate glasses therefore exhibit the same problem that they were intended to cure! All is not lost, though, because this *dielectric polarization* is not as severe a problem as mobile ion contamination. The threshold shift caused by a given voltage bias remains relatively small—a few tens of millivolts. The threshold shifts caused by dielectric polarization are also much more predictable than those produced by mobile ions, so circuit designers can predict whether a given circuit configuration will be adversely affected or not.[25] The threshold voltage shifts were finally eliminated altogether by using phosphorus-doped polysilicon gates rather than phosphorus-doped gate oxides. Phosphorus-doped polysilicon immobilizes alkali metals in much the same way as phosphorus stabilization without the added complication of dielectric polarization.

Moisture seeping into the integrated circuit's package can transport sodium in from the outside environment. Improved packaging materials can slow, but not stop, the ingress of sodium ions. The protective overcoat serves as a further barrier to mobile ions and can prevent them from reaching the vulnerable oxide layers in contact with the silicon. Protective overcoats typically consist of either silicon nitride, which is relatively impermeable to mobile ions, or phosphorus-doped glasses, which can

---

[22] N. E. Lycoudes and C. C. Childers, "Semiconductor Instability Failure Mechanisms Review," *IEEE Trans. of Reliability*, Vol. R-29, #3, 1980, pp. 237–249.

[23] M. Kuhn and D. J. Silversmith, "Ionic Contamination and Transport of Mobile Ions in MOS Structures," *J. Electrochem. Soc.*, Vol. 118, 1971, pp. 966–970.

[24] S. R. Hofstein, "Stabilization of MOS Devices," *Solid-State Electronics*, Vol. 10, 1967, pp. 657–670.

[25] E. H. Snow and B. E. Deal, "Polarization Phenomena and Other Properties of Phosphosilicate Glass Films on Silicon," *J. Electrochem. Soc.*, Vol. 113, #3, 1966, pp. 263–269.

immobilize them. The protective overcoat therefore serves as a final line of defense against impurities entering the die from outside.

Any opening through the protective overcoat represents a potential route for mobile ion contamination to enter the die. The metallization normally seals the bondpad openings, but scars left by probe needles can puncture the metal and expose the interlevel oxide (ILO) beneath. A minimum number of probe pads should be used, and these should not be placed near sensitive analog circuitry. Fuse openings through the protective overcoat also represent vulnerable points that should be kept away from sensitive analog circuitry.

The scribe street surrounding the die typically consists of bare silicon because other materials either fracture or clog the saw blade. Contaminants can seep laterally into exposed oxide layers abutting the scribe street. Special structures, called *scribe seals*, placed around the periphery of the die can slow the ingress of contaminants. Figure 4.6A illustrates a typical scribe seal for a single-level-metal CMOS process. The first component of this scribe seal consists of a narrow contact strip surrounding the active area of the die. This contact must be a continuous ring uninterrupted by any gaps in order for it to block the lateral movement of mobile ions through the field oxide. A P-type diffusion placed underneath this contact allows it to double as a substrate contact. This arrangement is very convenient, as the metal plate forming part of the scribe seal also carries the substrate lead around the periphery of the die. The scribe seal also provides a guaranteed minimum area of substrate contacts.

The scribe seal also contains a second contamination barrier formed by flapping the protective overcoat into the scribe street directly on top of the exposed silicon. Any mobile ions attempting to penetrate the scribe seal must first surmount this flap-down and next pass the continuous contact ring before reaching the active regions of the die. Most processes prohibit direct contact between nitride and silicon because compressive stresses in the nitride spawn defects in the silicon lattice. The flap-down of protective overcoat over the scribe street is permitted because the scribe street does not contain any active circuitry that could be damaged by defects. Nitride should still not touch exposed silicon inside the active area of the die because defects spawned by the damaged silicon can propagate for some distance and may affect adjacent components.

Figure 4.6B shows a scribe seal for a double-level-metal CMOS process. This seal includes a third barrier consisting of a continuous via ring placed just inside the contact ring. This via ring helps prevent contaminants from entering the ILO between the two metal layers. In a triple-level-metal process, a second via ring would be added to protect the second ILO layer between metal-2 and metal-3.

The scribe seals shown in Figure 4.6 can protect almost any die, but the substrate contacts may require different diffusions, depending on the process flow. For example,

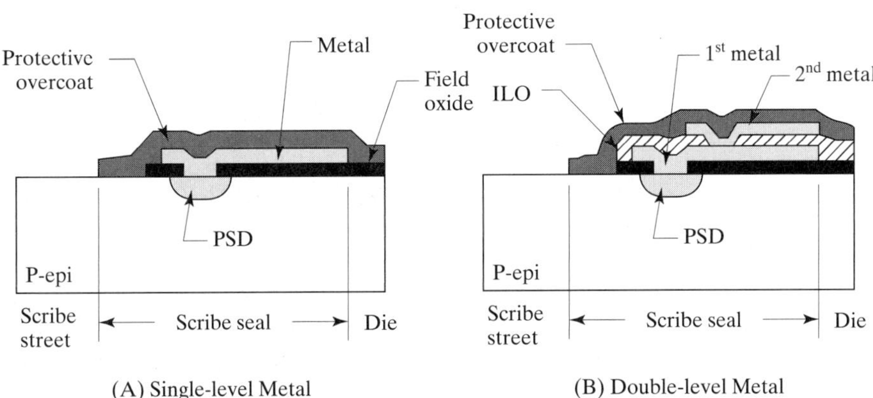

FIGURE 4.6 Scribe seals for single- and double-level-metal variants of a CMOS or BiCMOS process. Depending on the manufacturer, various diffusions may also be placed over the scribe street.

standard bipolar would substitute a combination of P-isolation and base for the PSD rings of Figure 4.6. The P-isolation would probably extend to the edge of the active die because most designers surround the die with a ring of substrate contacts placed underneath the grounded metallization. The functionality of the scribe seals remains the same regardless of the exact diffusions used.

Dielectric-isolated processes usually employ one or more isolation rings as part of their scribe seals. These rings are not actually intended to block the entry of mobile ions. Instead, they help prevent delamination along the interface between the buried oxide (BOX) and the superficial silicon. This interface is weaker than one might expect, because wafer bonding does not produce a perfect molecular union between the layers. Stresses generated during sawing and assembly can cause delamination, which typically begins at an edge and proceeds inwards. Isolation seal rings can help stop this delamination process before it intrudes into the active area of the die. Because sharp corners concentrate stress, large-radius fillets are usually applied to the four corners of the isolation seal rings.

4.3 SURFACE EFFECTS

Surface regions of high electric field intensity can inject hot carriers into the overlying oxide. Surface electric fields can also induce the formation of parasitic channels. Both of these mechanisms are referred to as *surface effects* because they occur at the interface between the silicon and the overlying oxide.

4.3.1. Hot Carrier Injection

Under ordinary circumstances, virtually no carriers possess the 5 eV or so required to surmount the oxide-silicon interface. Oxide therefore acts as a nearly perfect insulator. However, if an intense electric field occurs close to the surface of the silicon, some of the hot carriers generated by this field have sufficient energy to enter the oxide. This mechanism, called *hot carrier injection* (HCI), can cause serious reliability problems in MOS transistors.

Effects

When a MOS transistor operates in saturation, most of its drain-to-source voltage appears across the narrow pinched-off portion of its channel. This region consequently experiences an intense lateral electric field. While it is true that the width of the pinched off region increases at higher voltages, this effect is insufficient to compensate for the increased drain-to-source voltage. The electric field therefore increases. This field accelerates carriers crossing the pinched-off region, producing hot carriers that can potentially be injected into the oxide.

Figure 4.7 shows the behavior of an NMOS transistor under a large drain-to-source voltage. Electrons flow down the channel from source to drain. Reaching the pinched-off region, they encounter an intense electric field that accelerates them to high velocities. The resulting hot electrons slam into lattice atoms and ricochet off at all angles. A tiny percentage of these ricochets travel upwards and punch through the oxide-silicon interface. These carriers generate a tiny gate current, typically no more than a few picoamps. This seemingly inconsequential current has major long-term consequences for device reliability.

Figure 4.7 also shows a flow of holes from the pinched-off region to the substrate. These holes are generated when hot electrons slam into lattice atoms and knock free their valence electrons. Each such impact ionization event generates an electron–hole pair. The electrons generated in this manner simply add to the swarm of

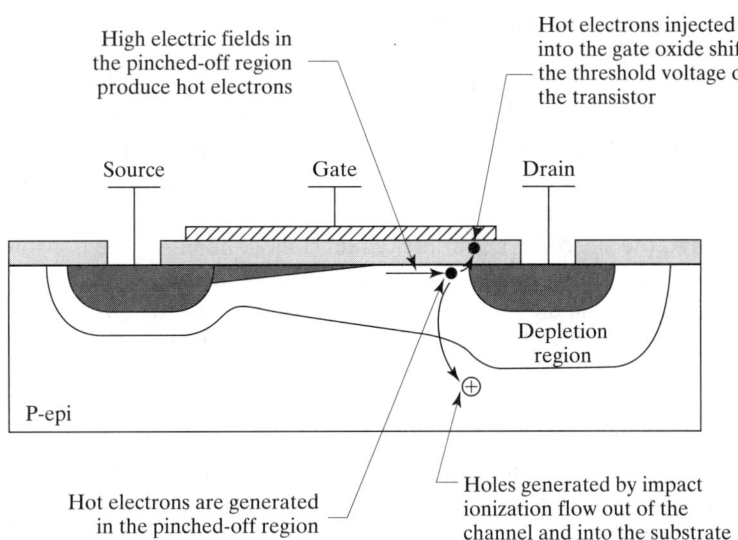

FIGURE 4.7 Simplified diagram showing the mechanism responsible for hot electron injection in an NMOS transistor.

electrons heading for the drain. The holes, on the other hand, move in the opposite direction: towards the source, and towards the substrate. Those holes that reach the substrate contribute to a current flowing out of the backgate terminal of the device. Since less energy is required to generate an electron–hole pair than to surmount the oxide-silicon interface, the backgate current is several orders of magnitude larger than the gate current. This backgate current is often used as an indicator of hot-carrier injection because it is much easier to measure than the tiny gate current.[26]

The rate of hot carrier injection within a transistor depends upon its biasing. Most obviously, hot carrier injection is a function of drain-to-source voltage. Below some critical value, virtually no injection occurs. Above this voltage, hot carrier injection increases more or less exponentially. Less obviously, hot carrier injection also depends upon gate-to-source voltage. Higher gate-to-source voltages generate higher drain currents, but they also broaden the pinched-off region and thus reduce the peak electric field. For most transistors, hot carrier injection reaches a maximum when the gate-to-source voltage equals approximately 40% of the drain-to-source voltage.[27]

Hot carrier injection also depends upon the polarity of the transistor. Because holes have lower mobilities than electrons, they are more difficult to accelerate to high velocities. Therefore, PMOS transistors are much less subject to hot-carrier injection than are NMOS transistors. A PMOS limited by hot carrier injection can operate at roughly twice the drain-to-source voltage of an NMOS with similar dimensions and dopings.

Device doping plays a major role in determining the critical voltage at which hot carrier injection begins. Lighter backgate doping widens the pinched-off region of the channel and thus diminishes the electric field. Therefore, more lightly doped backgates can support higher operating voltages.

Hot carrier injection causes a gradual decrease in threshold voltages. This shift reduces the magnitude of an enhancement NMOS threshold voltage, while it increases the magnitude of an enhancement PMOS threshold voltage. This behavior suggests that hot carrier injection causes an accumulation of positive charge within

---

[26] S. Tam, P.-K. Ko, C. Hu, and R. S. Muller, "Correlation between Substrate and Gate Currents in MOSFET's," *IEEE Trans. on Electron Devices*, Vol. ED-29, #11, 1982, pp. 1740–1744.

[27] C. C.-H. Hsu, D.-S. Wen, M. R. Wordeman, Y. Taur, and T. H. Ning, "A Comprehensive Study of Hot Carrier Instability in P- and N-Type Poly-SI Gated MOSFET's," *IEEE Trans. on Electron Devices*, Vol. 41, #5, 1994, pp. 675–680.

the oxide. The exact mechanism by which this positive charge forms was only understood relatively recently. The key factor turns out to be the annealing process used to minimize surface state charges in MOS transistors. Without this anneal, a substantial positive charge exists along the oxide-silicon interface. This *fixed oxide charge* is caused by silicon atoms at the oxide interface that lack their normal complement of four bonds. These so-called *dangling bonds* represent defects that can become positively charged. During the anneal, hydrogen atoms diffuse through the oxide and bond to the silicon atoms at the defect sites, eliminating the dangling bonds. Hot carriers injected into the oxide can break the relatively weak silicon-hydrogen bonds, regenerating the dangling bonds and the corresponding positive fixed oxide charge.[28]

The parametric shifts caused by hot carrier injection can be partially reversed by baking the unbiased units at temperatures of 200 to 250°C for several hours. These temperatures enhance the diffusion of hydrogen atoms and allow them to reattach to the dangling bonds. Temperatures of roughly 400°C will completely reverse the parametric shift, but plastic packaging cannot endure this treatment. As in the case of mobile ions, these apparent cures are only temporary. As soon as bias is restored, hot carrier generation resumes and the threshold voltages begin to drift again.

Preventative Measures

Threshold voltage shifts can be minimized by redesigning the affected devices, by choosing the conditions under which they operate, or by resizing them. All three of these approaches are frequently used in analog design.

Most approaches to device redesign focus upon reducing the electric field within the pinched-off region of the drain. Widening the pinched-off region proportionately reduces the electric field. The most obvious way to widen the pinched-off region consists of reducing the backgate doping. Altering the doping of the entire backgate region would entail undesirable consequences, including a substantial increase in minimum channel lengths, increased channel length modulation, and a reduced safe operating area (Section 12.2.1). Much better results can be achieved by reducing the backgate doping only in the immediate vicinity of the drain electrode. By this means, one can constrain the pinched-off region within a relatively well-defined volume that depletes out at low drain-to-source voltages. The RESURF transistor (Section 12.2.3) illustrates this approach.

Alternatively, one can reduce the doping of the drain periphery. This allows the depletion region to extend further into the drain rather than into the channel. This technique reduces the peak electric field without altering the backgate doping. The resulting device can retain a short channel and does not suffer any increase in channel length modulation or any reduction in safe operating area. The lighter drain doping does, however, increase the drain resistance in the linear region of operation. The lightly doped drain (LDD) transistor (Section 3.2.4) illustrates this approach.

Another device redesign focuses upon reducing the impact of hot carriers upon dangling bonds. Annealing the wafer in deuterium rather than hydrogen reduces the rate of hot carrier degradation by a factor of ten or more. Deuterium, an isotope of hydrogen, has an atomic weight roughly twice that of normal hydrogen. This increase in mass causes deuterium to strongly resist desorption by hot carrier excitation through a rather esoteric mechanism called the *giant isotope effect*.[29] Consequently, the threshold voltage of a deuterium-annealed device shifts much more gradually than that of a hydrogen-annealed device.

---

[28] K. Hess, I. C. Kizilyalli, and J. W. Lyding, "Giant Isotope Effect in Hot Electron Degradation of Metal Oxide Silicon Devices," *IEEE Trans. on Electron Devices*, Vol. 45, #2, 1998, pp. 406–416.

[29] ibid.

Hot carrier injection can also be minimized by proper selection of operating conditions. The total threshold shift depends upon the drain-to-source voltage applied across the transistor and the total charge flowing through the transistor. No injection occurs when the transistor operates in cutoff, where the drain current equals zero; or in the linear mode, where the drain-to-source voltage remains extremely small. Digital circuits operate almost exclusively in one of these two modes. Analog circuits, by contrast, contain MOS transistors that continuously operate in saturation. Furthermore, digital circuits are much less sensitive to small variations in threshold voltage than are analog circuits. Taken together, these factors mean that a given transistor can operate at higher drain-to-source voltages in digital applications than in analog applications. Many processes therefore specify two separate drain-to-source voltage ratings for MOS transistors: one for cases where they are used in digital circuits, and the other for cases where they operate continuously in saturation, as sometimes occurs in analog circuits.

Long channel devices also gain some measure of protection against hot carrier injection. Hot carriers are still produced, but only in the vicinity of the drain. The rest of the channel remains unaffected, minimizing the overall impact of hot carriers upon transistor parameters. A few extra volts of operating margin can often be obtained by increasing the channel length by a few microns.

4.3.2. Zener Walkout

Although hot carrier injection is most often associated with MOS transistors, the same process also occurs in Zener diodes and bipolar transistors. The observed modes of failure differ, but the underlying mechanisms are much the same.

Effects

Avalanching junctions also produce large numbers of hot carriers. Avalanche occurs near the surface in most diffused junctions because the dopant concentrations are largest there (Figure 4.8A). Some of the hot carriers produced by the avalanche process are injected into the overlying oxide. These carriers break silicon-hydrogen bonds, causing the regeneration of positive fixed oxide charges. According to the classical model of Zener walkout, this charge electrostatically induces a gradual widening in the depletion region at the surface (Figure 4.8B). The avalanche voltage slowly increases during operation, a phenomenon called *Zener walkout.*[30] NPN base-emitter junctions are particularly susceptible to Zener walkout, and they usually exhibit shifts

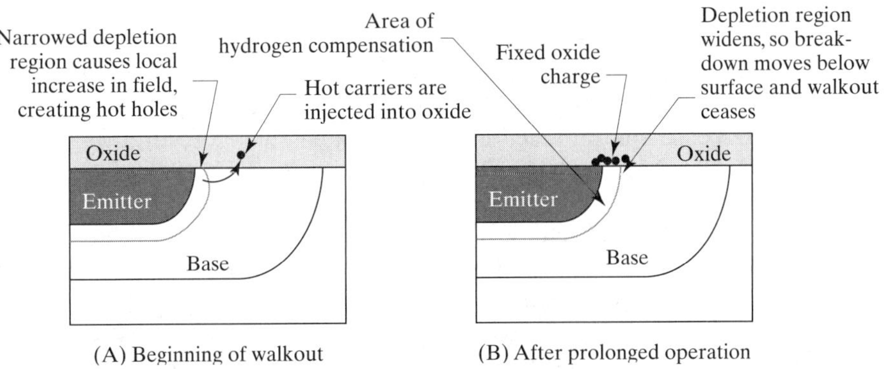

FIGURE 4.8 Simplified diagrams showing the classical model of Zener walkout: (A) initial condition of junction, in which hot carrier production occurs near the surface; (B) condition of junction after extended period of operation.

[30] R. W. Gurtler, "Avalanche Drift Instability in Planar Passivated p-n Junctions," *IEEE Trans. on Electron Devices*, Vol. ED-15, #12, 1968, pp. 980–986. See also G. Blasquez, G. Barbottin, and V. Boisson, "A Review of Passivation-Related Instabilities in Modern Silicon Devices," in G. Barbottin and A. Vapaille, eds., *Instabilities in Silicon Devices, Volume 2: Silicon Passivation and Related Instabilities*, (Amsterdam: North-Holland, 1989), pp. 459–460.

of several tenths of a volt.[31] Much larger shifts have been reported for certain processes. A 200–250°C unbiased bake will partially (but not completely) reverse Zener walkout. In some cases, a partial reversal may occur if the device is left unbiased for an extended period of time at room temperature. These apparent cures are only temporary, as walkout resumes when the diode is biased into avalanche.

Recent research has focused upon the role of the hydrogen atoms that are liberated when hot carriers regenerate the fixed oxide charge. Some of these desorbed hydrogen atoms migrate down into the silicon. P-type silicon contains acceptors that have valence electron deficiencies; these acceptors are in many ways similar to the dangling bonds at the oxide interface. Migrating hydrogen atoms can weakly bond to acceptors, and, once this occurs, the acceptors no longer generate mobile holes. This process is called *hydrogen compensation*. The desorption of hydrogen at the oxide interface can lead to a reduction in the doping concentration within the P-type silicon through hydrogen compensation. This in turn widens the depletion region.[32]

The hydrogen desorption model explains many observations that the classical model cannot. For example, some diodes that are operated continuously in breakdown first exhibit a rapid walkout, then a stabilization of the breakdown voltage for a long period of time, and finally a very gradual reversal of the walkout (*Zener walkback*). According to the hydrogen desorption model, walkback occurs when the supply of hydrogen from the interface is exhausted. The gradual diffusion of hydrogen through the silicon diminishes hydrogen compensation near the junction, causing the depletion region to narrow. The same mechanism probably also accounts for the reduction in walkout observed on processes employing refractory barrier metal and silicides.[33] Titanium and its silicide both have a strong affinity for hydrogen and therefore tend to immobilize, or *getter*, hydrogen liberated in their vicinity. This gettering mechanism could well reduce the supply of hydrogen that reaches the underlying silicon.

Preventative Measures

Hot carrier injection only happens when avalanche occurs in close proximity to the oxide-silicon interface. Ordinary base-emitter Zeners avalanche at the surface, so they commonly exhibit several hundred millivolts of Zener walkout. Alternative structures have been designed in which avalanche is confined to subsurface layers. These *buried Zeners* do not exhibit Zener walkout. Section 10.1.2 presents examples of several styles of buried Zeners, most of which require additional processing steps to fabricate.

Field plates have been suggested as a means of stabilizing surface Zeners. A *field plate* is a conductor biased to generate an electric field across the underlying oxide so as to control depletion or accumulation in the silicon. In effect, the field plate forms the gate electrode of a MOS capacitor. For example, if a field plate is placed over a reverse-biased base-emitter junction, and the plate is biased positive with respect to the base diffusion, then it will tend to deplete the surface of the base. Providing that the voltage differential between the field plate and the base diffusion exceeds some fraction of the base thick field threshold, the base diffusion beneath the plate will deplete. Smaller voltage differentials still widen the depletion region, if only slightly. This effect should therefore drive avalanche breakdown beneath the surface. The

---

[31] W. Bucksch, "Quality and Reliability in Linear Bipolar Design," *TI Technical Journal*, Nov. 1987, pp. 61–69. See also R. W. Gurtler, *et. al.*

[32] P. K. Gopi, G. P. Li, G. J. Sonek, J. Dunkley, D. Hannaman, J. Patterson, and S. Willard, "New Degradation Mechanism Associated with Hydrogen in Bipolar Transistors under Hot-Carrier Stress," *Appl. Phys. Lett.*, Vol. 63, #9, 1993, pp. 1237–1239. See also C.-T. Sah, J. Y-C. Sun, and J. J-T. Tzou, "Deactivation of the Boron Acceptor in Silicon by Hydrogen," *Appl. Phys. Lett.*, Vol. 43, #2, 1983, pp. 204–206.

[33] W. Bucksch, unpublished manuscript, 1988.

usual approach to constructing this field plate involves connecting it to the emitter terminal of the Zener so as to take advantage of the breakdown voltage to bias the field plate. Unfortunately, the emitter-base breakdown voltage is far too low to have the desired effect. Experimental results generally confirm the ineffectiveness of field plates, although some designers claim to see small benefits. In all probability, these benefits arise, not from the electric field generated by the field plate, but rather from hydrogen gettering caused by refractory barrier metal.

Given that surface devices are vulnerable to stray fields generated by charge spreading, the addition of an emitter field plate to a surface Zener seems to be a wise precaution. However, in the absence of experimental evidence, one should not assume that the field plates reduce the magnitude of Zener walkout. If used, the field plate on a base-emitter Zener should connect to the emitter terminal. The field plate should consist of the lowest layer of metallization, and it should extend beyond the edges of the drawn junction by several microns to allow for outdiffusion and fringing fields. If necessary, the base contact can be moved back a few microns to allow room for the emitter field plate (Figure 4.9).

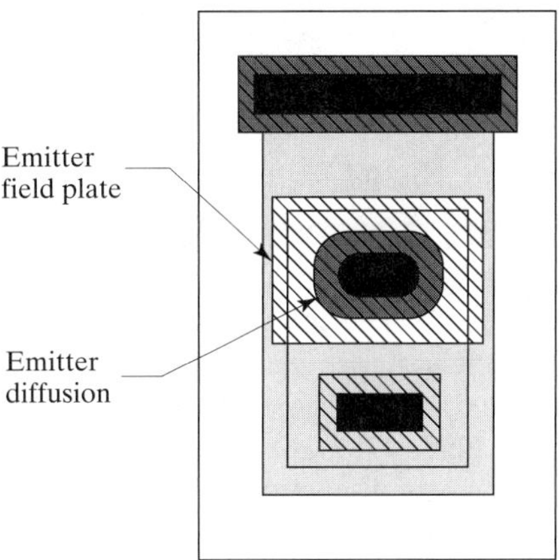

FIGURE 4.9 Emitter field plate applied to a base-emitter Zener.

4.3.3. Avalanche-Induced Beta Degradation

Avalanching the base-emitter junction of a bipolar transistor can significantly reduce its beta. Not all transistors are equally vulnerable. For example, the standard-bipolar vertical NPN transistor is more susceptable than its lateral PNP counterpart. Also, polysilicon-emitter transistors are more vulnerable than diffused-emitter transistors.

Effects

Avalanche-induced beta degradation causes a reduction in beta at low collector currents, but it does not significantly affect beta at moderate-to-high collector currents. The degradation occurs in a matter of seconds if the transistor actually experiences emitter-base avalanche. The same effects occur at somewhat lower biases, only much more slowly. An unbiased bake at 200–250°C can partially restore the lost beta. This apparent fix only lasts until the device again experiences large emitter-base reverse biases.

Avalanche-induced beta degradation is caused by the same mechanisms responsible for Zener walkout. Consider the case of a diffused-emitter NPN transistor. The doping concentrations are highest near the surface, so the depletion region is thinnest here. This in turn ensures that the highest electric field occurs adjacent to the oxide-silicon interface. The application of a large reverse bias across the emitter-base junction generates hot electrons. These hot electrons are injected into the oxide-silicon interface, where they desorb hydrogen and regenerate dangling bonds. These dangling bonds act as recombination centers and consequently increase recombination within the emitter-base depletion region. This in turn causes a degradation in low-current beta.[34]

Poly-emitter transistors (Section 8.3.5) behave somewhat differently than diffused-emitter transistors. In a poly-emitter transistor, the emitter diffusion is extremely thin. Immediately above this diffusion lies an interface between the monocrystalline silicon and the polycrystalline emitter contact. Dangling bonds exist along this interface, and these are passivated by hydrogen in the same manner as those along the oxide-silicon interface. The entire area of the emitter-base depletion region is affected. By contrast, only a small portion of the emitter-base depletion region of a diffused-emitter transistor lies in close proximity to the surface. This difference explains why the poly-emitter transistor exhibits a heightened sensitivity to avalanche-induced beta degradation.

Another form of beta variation has been observed in some poly-emitter transistors. In these devices, the medium-current beta increases slightly after operation at large collector currents. This effect appears to be caused by hydrogen atoms that migrate into the polysilicon-emitter interface. The hydrogen atoms presumably tie off dangling bonds and thus reduce recombination in the emitter-base depletion region. Not all poly-emitter devices exhibit this effect.[35]

Preventative Measures

Avalanche-induced beta degradation occurs because of intense fields at the surface of the base-emitter junction. One could theoretically design devices in which the intense fields are confined to subsurface regions. In practice, this cannot be done without either compromising the beta of the transistor or adding additional process steps.

The trap sites responsible for beta degradation in the poly-emitter transistor can be passivated by dopant atoms. Thus, the higher the doping within the emitter polysilicon, the less susceptable the transistor becomes to avalanche-induced beta degradation. Arsenic also appears somewhat more effective than phosphorus in passivating dangling bonds. Unfortunately, the effect of increased doping does not appear to be sufficient to eliminate avalanche-induced beta degradation.

In practice, avalanche-induced beta degradation can be avoided by reducing the reverse-biased emitter-base voltage rating of the device. As a general rule, diffused-emitter transistors should not operate beyond 75% of their base-emitter avalanche voltage, or V_{EBO}. Poly-emitter transistors are more susceptible, and should generally not operate beyond 50% of their V_{EBO}. These voltage ratings should not be exceeded for even short periods of time.

Base-emitter junctions connected to pins are particularly susceptible to beta degradation caused by ESD strikes. Special ESD protection clamps similar to those used to protect gate oxides from CDM events may be required. Alternatively, circuits can sometimes be redesigned to avoid connection of base-emitter junctions to pins.

4.3.4. Negative Bias Temperature Instability

Surface effects also cause another form of long-term variation that primarily affects PMOS transistors. This mechanism causes a gradual shift in the threshold voltage

---

[34] B. A. McDonald, "Avalanche Degradation of h_{FE}," *IEEE. Trans. on Electron Devices*, Vol. ED-17, #10, 1970, pp. 871–878.

[35] A. J. Melia, "Current Gain Degradation Induced by Emitter-Base Avalanche Breakdown in Silicon Planar Transistors," *Microelectronics and Reliability*, Vol. 15, #6, 1976, pp. 619–623.

when the gate is biased negatively with respect to the source and backgate. Higher temperatures accelerate the process, so it has been dubbed *negative bias temperature instability* (NBTI).

Effects

Negative bias temperature instability occurs when the gate of a PMOS transistor is biased negatively with respect to the silicon. This biasing condition occurs whenever the transistor conducts, so it is by no means an unusual situation. NBTI causes the threshold voltage to shift to a more negative value, or in other words, the absolute value of the threshold voltage increases. The rate at which the threshold voltage drifts depends upon both the magnitude of the gate-to-source voltage and the temperature at which the device is operated. As with most surface effects, NBTI can be at least partially reversed by baking the unbiased units at temperatures of 250°C for a few hours.

The mechanism that causes negative bias temperature instability is not entirely understood. Given that the threshold voltage becomes more negative, a positive fixed oxide charge is undoubtedly involved. Current theories suggest that this charge is generated by holes that are drawn up to the oxide-silicon interface, where they somehow react to form the positive fixed oxide charge. Research has shown that this reaction only occurs if the gate oxide has been exposed to the atmosphere between the time of its growth and the deposition of the overlying polysilicon. Since water vapor is strongly adsorbed onto oxide, some researchers believe that the reaction that generates the positive charges must involve water.[36]

Another form of bias temperature instability has also been observed in certain processes. This bias temperature instability occurs when the gate of the PMOS is biased positive with respect to the source and backgate. This effect is therefore called *positive bias temperature instability* (PTBI). Although not fully understood, PTBI appears to involve the generation of positive fixed oxide charges at the interface between the gate poly and the oxide. These charges cause depletion effects within the poly-gate electrode. Only processes with relatively lightly doped gate poly are vulnerable to PTBI, as heavier gate doping prevents any significant depletion into the gate electrode.

Preventative Measures

Negative bias temperature instability poses a significant challenge to analog circuit designers. Many analog circuits rely upon precise matching of MOS transistor threshold voltages, and any mechanism that causes threshold voltages to shift imperils this matching. The designer can always attempt to operate critical transistors under similar biasing conditions, but this precaution can only go so far towards minimizing the impact of threshold variations.

Negative bias temperature stability could be eliminated if the atmosphere could be kept away from the gate oxide until the gate polysilicon was deposited upon it. Unfortunately, the traditional design of processing equipment renders this impossible. The machinery used to deposit polysilicon cannot grow oxide, nor can the machinery used to grow oxide deposit polysilicon. The best practical solution lies in minimizing the exposure of the gate oxide to atmospheric moisture through improved handling techniques.

Positive bias temperature instability rarely causes much trouble. The biasing conditions responsible for this form of bias temperature instability can usually be

---

[36] C. E. Blat, E. H. Nicollian, and E. H. Poindexter, "Mechanism of Negative-Bias-Temperature Instability," *J. Appl. Phys.*, Vol. 69, #3, 1991, pp. 1712–1720.

avoided by simple circuit modifications. Even if this were not the case, the effects of PTBI are usually far smaller than those of NTBI because the gate is usually much more heavily doped than the backgate.

4.3.5. Parasitic Channels and Charge Spreading

Any conductor placed above the silicon surface can potentially induce a *parasitic channel.* If the conductor bridges two diffusions, then a leakage current can flow through the channel from one diffusion to the other. Most parasitic channels are relatively long and cannot conduct much current, but even small currents can cause parametric shifts in low-power analog circuitry. Channels can sometimes form even in the absence of a conductor due to a mechanism called *charge spreading.* The addition of channel stops or field plates can suppress parasitic channel formation and so protect vulnerable circuitry.

Effects

Both PMOS and NMOS parasitic channels exist. A PMOS parasitic channel can form across any lightly doped N-type region, such as an N-tank in a standard bipolar process or an N-well in a CMOS or BiCMOS process. An NMOS parasitic channel can form across any lightly doped P-type region, such as the P-epi of a CMOS or BiCMOS process, or the lightly doped P-type isolation of standard bipolar processes. Both of these types of parasitic channels can cause a great deal of trouble.

PMOS parasitic channels can form underneath leads crossing lightly doped N-type regions. Consider a metal lead that crosses an N-tank containing a base diffusion (Figure 4.10A). The lead acts as the gate of a PMOS transistor and the N-tank as its backgate. The base region forms the source of the transistor and the isolation serves as its drain. A channel will form if the voltage difference between the lead and the base region exceeds the threshold voltage of the parasitic MOS transistor.[37] Since the thick-field oxide serves as the gate dielectric of this transistor, its threshold voltage is called the *PMOS thick-field threshold.* If the process has a 40 V PMOS thick-field threshold, then the base must be biased at least 40 V above the lead in order for a channel to form beneath the lead. A similar condition applies to any other potential parasitic PMOS: The P-type region serving as the source must rise above the conductor acting as the gate by a voltage in excess of the PMOS thick-field threshold.

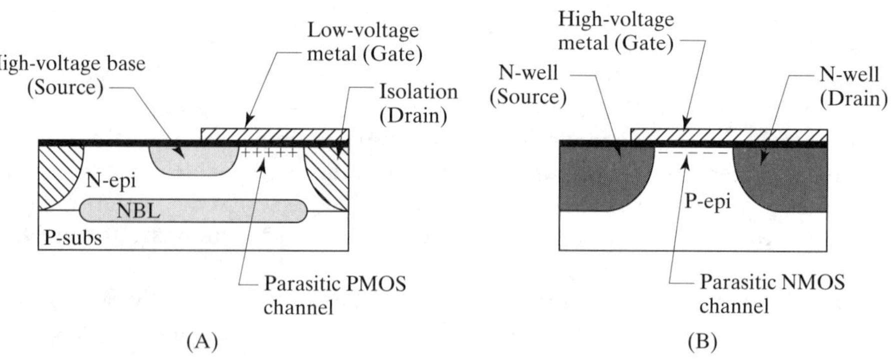

FIGURE 4.10 Parasitic PMOS in a standard bipolar process (A) and parasitic NMOS in an N-well CMOS process (B).

[37] "Bipolar Field Inversion." *Semiconductor Reliability News*, Vol. 3, #1, 1991, p. 7.

NMOS parasitic channels can form underneath leads crossing lightly doped P-type regions. Figure 4.10B shows a parasitic NMOS channel forming on an N-well CMOS die. This channel forms beneath the lead crossing the lightly doped P-epi. The lead acts as the gate and the P-epi as the backgate. Two adjacent wells serve as the source and the drain. A channel will form if the voltage difference between the gate and the source exceeds the NMOS thick-field threshold. In this case, the voltage on the lead must exceed the voltage on the N-well acting as the source by an amount equal to or greater than the NMOS thick-field threshold. Similar conditions apply to any other potential parasitic NMOS: The conductor serving as the gate must rise above the N-type region serving as the source by a voltage equal to or greater than the NMOS thick-field threshold.

The thick-field threshold voltages of a process depend on a number of factors, including conductor material, oxide thickness, substrate crystal orientation, doping levels, channel dimensions, and processing conditions. Most processes quote only one value for the thick-field threshold, this being a minimum value obtained from a worst-case combination of conductors and diffusions. Other processes have undergone more extensive characterization to determine separate thick-field voltages for each combination of conductor and diffusion.

Designers sometimes invoke the body effect (Section 1.4.2) as justification for approaching or even exceeding the thick-field threshold. The body effect increases the apparent threshold voltage of the transistor when the backgate-source junction is reverse-biased. For example, the backgate of the parasitic PMOS in Figure 4.8A is probably biased to a higher voltage than the base. Unfortunately, backgate biasing cannot be relied on for any significant aid. The body effect is most significant in heavily doped backgates, whereas the backgate of a parasitic MOS is usually rather lightly doped. Furthermore, the threshold shift produced by the body effect varies as the square root of the backgate-to-source bias, so even a large backgate bias may not buy more than a few volts of margin.

Engineers once believed that channels could only form beneath conductors, but experience has shown otherwise. Channels can form whenever a suitable source and drain exist, even if no conductor exists to act as a gate. The mechanism underlying the formation of such channels is called *charge spreading*, and although some details still remain unclear, the basic principles are well understood.[38, 39] The oxide and nitride films covering an integrated circuit are nearly perfect insulators. Electric current cannot flow through an insulator, but static electrical charges can accumulate on the surface of an insulator or along the interface between two dissimilar insulators. These static charges are not entirely immobile and can slowly shift or spread under the influence of electrical fields. In integrated circuits, the interface between the protective overcoat and the plastic encapsulation is susceptible to this phenomenon. If a nitride protective overcoat is used, then the oxide-nitride interface is also vulnerable, although to a lesser degree. The rate of movement of such charges depends on temperature and on the presence of contaminants. Higher temperatures greatly accelerate charge spreading, as does the presence of even trace amounts of moisture.[40]

Charge spreading requires the presence of static electrical charges at the insulating interface. Experience has shown that these charges do exist and that they consist primarily of electrons, but the mechanisms that generate them are not fully

---

[38] D. G. Edwards, "Testing for MOS IC Failure Modes," *IEEE Trans. Rel.*, R-31, 1982, pp. 9–17.

[39] Lycoudes, *et al.*, p. 240ff.

[40] E. S. Schlegel, G. L. Schnable, R. F. Schwarz, and J. P. Spratt, "Behavior of Surface Ions on Semiconductor Devices," *IEEE Trans. on Electron Devices*, Vol. ED-15, #12, 1968, pp. 973–980.

understood. Hot carrier injection certainly contributes to charge spreading, but integrated circuits that do not produce hot carriers still exhibit charge spreading. Ionic contaminants may contribute to certain cases of charge spreading, particularly where higher voltages are involved.[41] A variety of other hypothetical mechanisms have been postulated. In practice, the source of the static charge is less important than its consequences.

Figure 4.11A shows a cross section of a standard bipolar die susceptible to charge spreading. The base region inside the tank is biased above the PMOS thick-field threshold, and therefore acts as the source of a parasitic PMOS transistor. The tank containing this base region is also, of necessity, biased above the PMOS thick-field threshold. Electrons present in the overlying insulating layers will tend to migrate toward the positively charged tank. Eventually, enough electrons may accumulate over the tank to induce a channel (Figure 4.11B). In effect, the static charge generated by charge spreading behaves as the gate electrode of an MOS transistor.

Standard bipolar appears to be more susceptible to charge spreading than does CMOS, probably because of less stringent process cleanliness. CMOS processes must minimize ionic contamination to maintain threshold voltage control; no such requirement exists for standard bipolar. The presence of mobile ions in the field oxide appears to amplify the effects of charge spreading, probably due to a dipole separation mechanism in which the positively charged mobile ions are attracted to the surface of the oxide, leaving behind negative charges in close proximity to the silicon surface (Figure 4.5).

Charge spreading produces parasitic PMOS transistors because it involves the accumulation of negative charges. The sources of these parasitic transistors consist of any P-regions that operate at voltages exceeding the PMOS thick-field threshold. The most vulnerable devices contain large, high-voltage P-regions operating at low currents—for example, matched high-voltage HSR resistors. Failures tend to occur after long periods of high-temperature operation under bias. Moisture increases the mobility of surface charges, so environmental tests designed to detect moisture sensitivity often uncover charge spreading problems. The resulting parametric shifts resemble those produced by hot carrier injection, in that they can be partially or completely reversed by baking the unbiased units at 200 to 250°C for several hours. The high temperature causes the accumulated static charges to disperse and restores an equilibrium between mobile ions and their fixed countercharges. This treatment does not constitute a permanent cure, because the parametric drifts resume as soon as bias is restored.

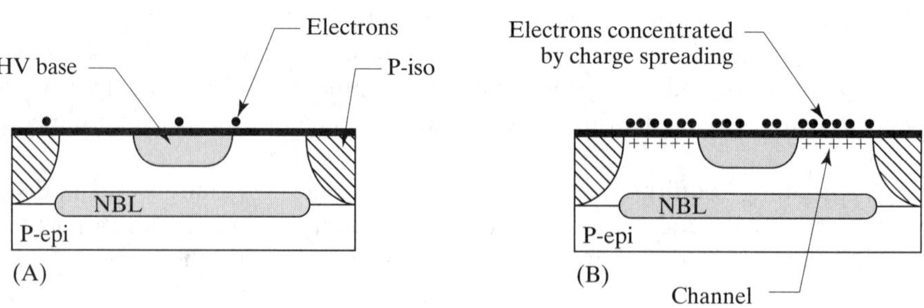

FIGURE 4.11 Cross section of a standard bipolar structure susceptible to charge spreading: (A) before and (B) after an extended period of operation under bias.

[41] H. J. Bruggers, R. T. H. Rongen, C. P. Meeuwsen, and A. W. Ludikhuize, "Reliability Problems due to Ionic Conductivity of IC Encapsulation Materials under High Voltage Conditions," *Proc. 11th Int. Symp. Power Semiconductor Devices and IC's*, 1999, pp. 197–200.

Preventative Measures (Standard Bipolar)

NMOS channel formation can be suppressed in standard bipolar by coding base over all isolation regions. This *base-over-isolation* (BOI) requires no additional die area because the spacings required by the isolation are much larger than those required by the base. The BOI can therefore coincide with the isolation, or even slightly overlap it. Not all standard bipolar processes employ base-over-isolation; some already have a sufficiently heavily doped isolation diffusion to suppress channel formation.

Standard bipolar devices are susceptible to the formation of PMOS channels through charge spreading. Any tank that contains a P-type diffusion biased above the PMOS thick-field threshold requires protection in the form of field plates, channel stops, or a combination of both. Conservative designers usually derate the thick-field threshold of standard bipolar to account for this process's known propensity for charge spreading. For example, a designer might field plate and channel stop high-voltage P-regions operating above 30 V, even though the process has a rated PMOS thick-field threshold of 40 V.

Figure 4.12 shows an example of a high-voltage HSR resistor vulnerable to parasitic channel formation. The tank containing the resistor connects to the positive supply (VCC) to ensure isolation. A lead must route across the tank to connect to some adjacent low-voltage circuitry. A PMOS channel will form beneath this lead as soon as the voltage difference between the resistor and the lead rises above the PMOS thick-field threshold.

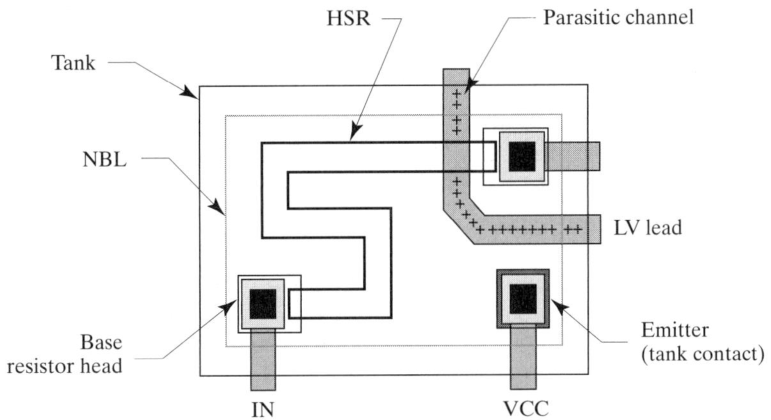

FIGURE 4.12 Example of a layout susceptible to parasitic PMOS channel formation.

CMOS processes use channel stop implants to raise the thick-field thresholds. Standard bipolar does not include channel stop implants, but an emitter can be coded over selected regions of the N-tank to serve the same purpose. Figure 4.13A shows how emitter diffusions can disrupt the parasitic channels formed beneath a low-voltage lead. Each of the two minimum-width emitter strips disrupts a channel that would otherwise conduct current from the resistor to the isolation.

The emitter bars in Figure 4.13A extend slightly beyond the leads in either direction. These extensions will sever the channel even if the metal and the emitter misalign. The electric field also fringes out to either side of the lead. These fringing fields rarely extend laterally more than two or three times the oxide thickness, so the overlap of the emitter bar over the lead should equal the maximum photolithographic misalignment plus twice the oxide thickness. Assuming a two-level misalignment of 1 μm and a 10 kÅ thick-field oxide, the emitter bar should extend about 3 to 5 μm

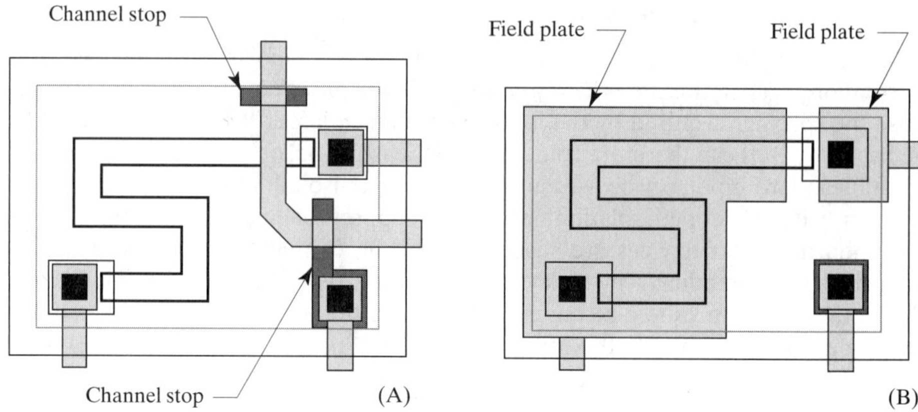

FIGURE 4.13 Two methods for preventing parasitic PMOS channels: (A) channel stops prevent channel formation beneath leads but do not stop charge spreading, and (B) field plates provide relatively complete coverage, except possibly in the gap between the plates.

beyond the lead on either side. These emitter bars are often called *channel stops,*[42] but they should not be confused with the blanket *channel stop implants* used in CMOS and BiCMOS processes. Channel stops are sometimes called *guard rings,* although this term is more properly applied to minority carrier guard rings (Section 4.4.2). Channel stops are also called *channel stoppers.*

The channel stops in Figure 4.13A cannot, by themselves, prevent charge spreading. Even if a channel stop entirely encircles a serpentine resistor like that in Figure 4.13, parasitic channels can still form between its turns. If additional channel stops are placed between the turns, parasitic effects could still alter the effective width of the resistor by inverting the silicon along its edges. Some other technique must be used to supplement channel stops, especially for high-voltage diffused resistors.

Field plating can provide comprehensive protection against both parasitic channel formation and charge spreading. A field plate consists of a conductive electrode placed above a vulnerable diffusion and biased to inhibit channel formation.[43] Figure 4.13B shows an HSR resistor with field plates added. The low-voltage lead has been rerouted, and a large plate of metal has been placed over the body of the resistor and connected to its positive terminal. The metal lead connecting to the negative end of the resistor has also been enlarged to protect the head of the resistor protruding beyond the main field plate. Both of these field plates must overlap the resistor enough to allow for outdiffusion, misalignment, and fringing fields. Assuming a two-level misalignment of 1 μm, a maximum outdiffusion of 2 μm, and a maximum fringing distance of 2 μm, the total overlap must equal 5 μm. Since the field plate consists only of metal, it can extend to fill the required area without enlarging either the resistor or its tank.

A field plate operates by providing an intentional gate for at least a portion of the MOS channel. This gate is biased to prevent the gate-to-source voltage of the parasitic transistor from exceeding the thick-field threshold. The presence of the conductive plate prevents the accumulation of static charges and thus suppresses charge spreading. Field plates also prevent modulation of carrier concentrations in the underlying silicon by acting as electrostatic shields. They therefore provide excellent protection against all types of electrostatic interactions, including conductivity modulation and noise coupling from overlying leads. The use of field plates in preference to channel stops is encouraged where possible.

Most field plates contain gaps in which channels can still form. In the resistor of Figure 4.13B, a gap remains between the two field plates covering the resistor. Two

---

[42] J. Trogolo and S. Sutton, "Surface Effects and MOS Parasitics," unpublished report, 1988, p. 13ff.

[43] Trogolo, *et al.,* p. 13ff.

methods exist for blocking these gaps. One method consists of flaring, or *flanging*, the ends of the field plate to elongate the channel as much as possible (Figure 4.14A). The close proximity of the parallel field plates induces a lateral electric field that sweeps static charges out of this region. The longer the potential channel, the greater the margin of safety provided by the flanges. The second method bridges the gaps between the field plates with short channel stops (Figure 4.14B). The emitter strips used for this purpose must overlap the field plates sufficiently to account for misalignment. This technique combines the strengths of a field plate with those of a channel stop to provide ironclad protection at all points.

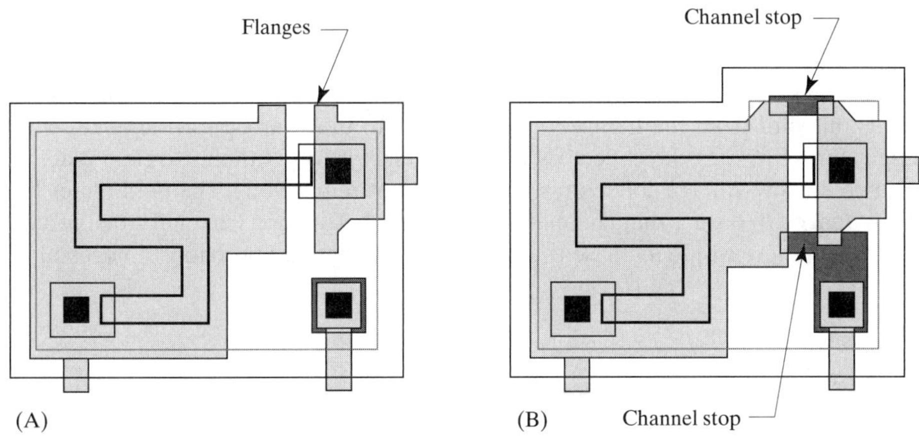

FIGURE 4.14 Improved field plating schemes: (A) flanged field plates and (B) combination of field plates and channel stops.

The resistors in Figures 4.13 and 4.14 illustrate another important principle of field plating: The field plate biased to the highest potential should cover as much of the resistor as possible. If the low-voltage field plate were extended, it would provide less protection to the high-voltage end of the resistor. If the voltage difference between the tank and the field plate exceeds the thick-field threshold, then the field plate will actually induce channel formation. The field plate should extend from the high-voltage terminal of a vulnerable resistor to encompass as much of the resistor body as possible. The low-voltage terminal of the resistor should have just enough field plating to cover and protect the contact head. Matched resistors may require a slightly different field-plating strategy (Section 7.2.12).

Figure 4.15 shows an interesting situation that sometimes occurs when laying out resistors. The two terminals of this device connect to a high potential and a low potential, respectively. The high-voltage end of the resistor needs protection against

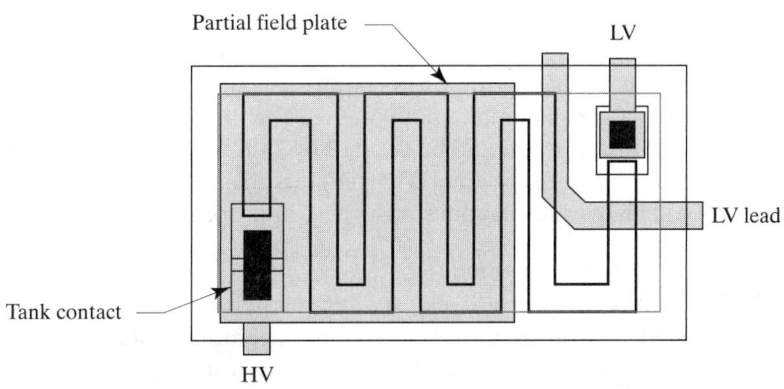

FIGURE 4.15 Example of partial field plating.

charge spreading, but the low-voltage end does not. Since the voltage drops linearly along the resistor, the field plate has been terminated partway down its length. Partial field plates should extend well beyond the point where the voltage drops below the thick-field threshold. In the case of Figure 4.15, as much of the resistor as possible has been field plated, even though much of it apparently serves no useful function. The large safety margin obtained by this means costs nothing and helps ensure that the device will work even under worst-case conditions.

Figure 4.16 shows another example of selective field plating involving a multiple-collector lateral PNP transistor. The emitter, base, and one collector operate at voltages in excess of the thick-field threshold, while the remaining collector operates at a low voltage. The emitter field plate extends out from the emitter across the exposed surface of the base to a point just beyond the inner edge of the collector. The field plate need only overlap the collector by an amount equal to the maximum misalignment minus outdiffusion; certainly no more than 2 to 3 μm. A second field plate extends outward from the high-voltage collector to block any parasitic channel that might form between the collectors, or from collector to isolation. No field plate surrounds the low-voltage collector, since it does not require one. The field plates have been flanged to ensure that channels cannot form in the gaps. Channel stops could be added, but these would increase the size of the tank and are probably unnecessary.

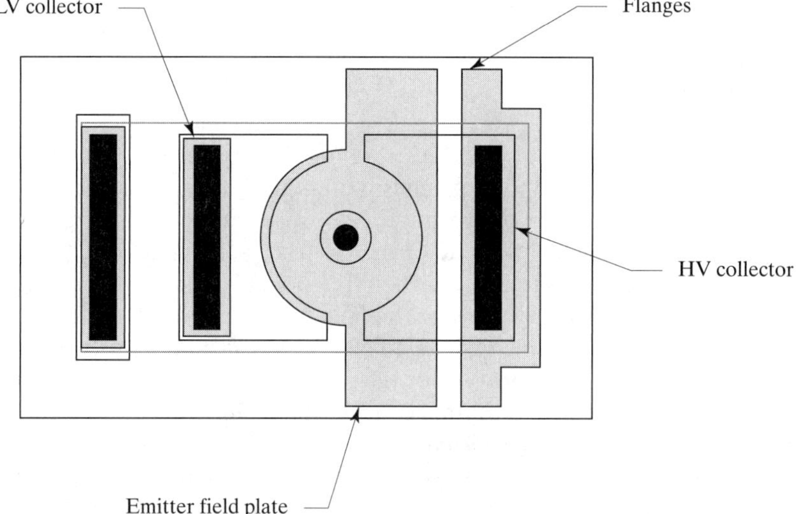

FIGURE 4.16 Field-plated, split-collector, lateral PNP with one low-voltage collector and one high-voltage collector. Flanging suppresses parasitic formation in the gaps between field plates.

To summarize, any P-type region biased in excess of the thick-field threshold acts as the source of a parasitic PMOS transistor. Field plates and channel stops ensure that no parasitic channels form from a high-voltage P-type region to any adjacent P-type diffusion. Field plating protects most of the device, while channel stops or flanges protect gaps left in the field plating. The official thick-field threshold of standard bipolar processes should be derated by 25% to provide an additional margin of safety against charge spreading. The chapters that follow describe additional examples of field plates and channel stops where appropriate.

Preventative Measures (CMOS and BiCMOS)

CMOS and BiCMOS processes usually incorporate channel stop implants to raise the thick-field threshold above the nominal operating voltage. The voltage rating of an N-well CMOS process is usually defined by NSD/P-epi breakdown, PSD/N-well breakdown, or gate oxide rupture, but some structures can withstand much higher voltages. The high N-well/P-substrate breakdown allows PMOS transistors to operate at elevated backgate voltages. These transistors will function normally, as long as the drain-to-source voltage does not exceed the PSD/N-well breakdown voltage or the N-well punchthrough voltage. Similarly, an extended-drain NMOS using N-well as a lightly doped drain can withstand the full N-well/P-substrate breakdown voltage applied to its drain terminal.

The lightly doped N-well inverts in much the same manner as the lightly doped N-epi tanks of standard bipolar. Any N-well region containing a P-type diffusion biased above the thick-field threshold becomes vulnerable. As before, PMOS parasitic channels can be suppressed by field plates, channel stops, or a combination of both. The flanged field plate (Figure 4.14A) is especially attractive because the tighter CMOS layout rules allow a narrower gap between the flanges. The stronger lateral electric field makes close-spaced flanges particularly effective at preventing the accumulation of static charges. Flanged fieldplates are the method of choice for suppressing parasitic PMOS channels in high-voltage CMOS and BiCMOS structures.

Charge spreading is less prevalent in CMOS processes than in bipolar ones, probably because of improved process cleanliness. Many CMOS designers consequently take a rather cavalier approach to field plates and channel stops. Such indiscretion is unwise considering the greatly reduced operating currents characteristic of modern CMOS designs. These processes are indeed vulnerable to charge spreading, as recent experience has shown.[44] One special case does exist where charge spreading can be safely ignored. Most CMOS processes list a lower thick-field threshold for poly than for metal because the oxide layer beneath the poly consists only of thick-field oxide and MLO, while that beneath the metal contains an added layer of deposited ILO. Static charges can only accumulate at the interface between two dissimilar materials, so the lower thick-field thresholds associated with thinner oxides do not have any significance for charge spreading. Charge spreading becomes significant only at voltages beyond the thick-field threshold of the uppermost conductor layer.

Poly leads can induce parasitic channels if they run across an N-well containing a P-diffusion biased above the poly thick-field threshold. Figure 4.17A shows a typical

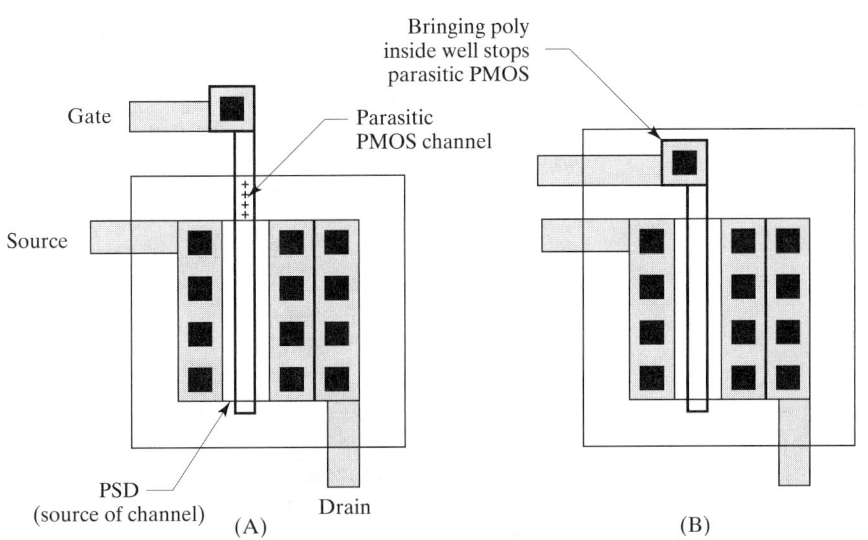

FIGURE 4.17 The parasitic PMOS channel beneath a poly lead (A) can be eliminated by pulling poly inside the well (B).

[44] S. Merchant, private communication, 2003.

example of a vulnerable structure consisting of the poly-gate lead from a high-voltage PMOS transistor extending across the well and into the surrounding isolation. The well forms the backgate of the parasitic PMOS, the poly acts as the gate, the sources are the PSD regions of the PMOS transistor, and the drain is the P-epi isolation. As long as the backgate potential does not exceed the metal thick-field threshold, the channel can be interrupted by stopping the polysilicon lead short of the drawn edge of the well (Figure 4.17B). The channel can form only beneath the polysilicon lead, so a complete channel cannot form if the lead does not bridge the gap between source and drain. Charge spreading is unlikely to occur so long as the voltages involved do not exceed the highest thick-field threshold of the process. The minimum spacing between the poly and the drawn edge of the N-well should equal the photolithographic misalignment allowance, plus an extra 3 to 5 μm to account for fringing fields and outdiffusion.

The lightly doped P-type epi can also invert if a high-voltage lead runs across it (Figure 4.10B). The source and drain of this parasitic NMOS consist of two adjacent N-wells; the high-voltage lead acts as the gate, and the P-epi acts as the backgate. A parasitic channel will form if the voltage differential between the lead and an adjacent well exceeds the NMOS thick-field threshold. A channel stop can be inserted by running a thin bar or ring of PSD material down the center of the P-epi beneath the high-voltage lead. The PSD should extend beyond either edge of the lead by an amount sufficient to account for misalignment, plus an additional 2 to 3 μm to allow for fringing effects. In many cases, the N-well to N-well spacing can accommodate a minimum-width PSD channel stop with little or no increase in the spacing between adjacent wells. A thin ring of PSD material can then encircle each well (Figure 4.18). This ring not only stops any possible leakage caused by charge spreading but also allows complete freedom to route the leads in any pattern desired. If the PSD rings are drawn when the wells are placed, or if they are automatically produced during mask generation, then the designer can subsequently ignore NMOS channel formation. These PSD rings correspond to the base-over-isolation (BOI) scheme used for the same purpose in standard bipolar designs.

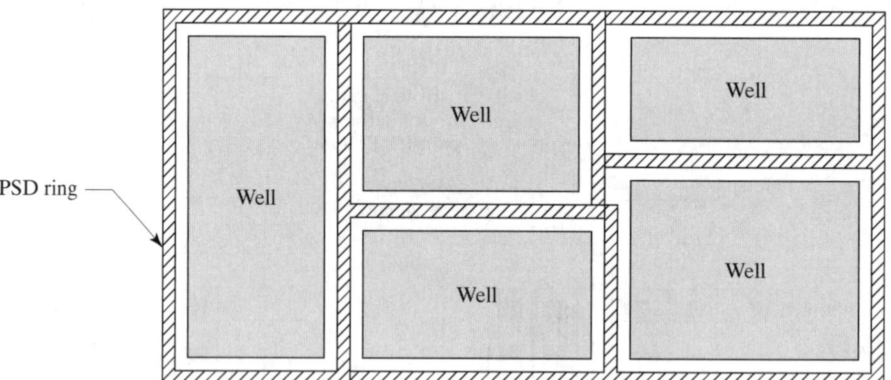

FIGURE 4.18 Sample layout showing the use of PSD rings to prevent NMOS channels.

4.4 PARASITICS

All integrated circuits contain electrical elements not required for their operation. These include reverse-biased isolation junctions, and resistances and capacitances between various diffusions and depositions. The circuit does not benefit from the presence of these *parasitic components,* but they can sometimes adversely affect its operation.

Parasitics are responsible for a number of different types of electrical failures. For example, capacitive coupling can inject noise into sensitive circuitry. The type of parasitics that will be discussed in this section concern the forward biasing of junctions that normally remain reverse-biased. When these junctions forward bias, current begins to flow between circuit nodes that normally remain isolated from one another. If these currents are small and the circuit is relatively insensitive to their presence, then these leakages may produce only subtle parametric shifts. Larger currents can catastrophically disrupt the operation of the circuit. The malfunctioning circuit may actually *latch up,* causing it to continue malfunctioning even after the removal of the triggering event. Latchup can cause physical destruction of an integrated circuit due to excessive power dissipation and consequent overheating. Even if the circuit does not self-destruct, normal operation can only be restored by interrupting the power.

Two important parasitic mechanisms involve currents flowing through the substrate. *Substrate debiasing* occurs when parasitic currents induce voltage drops in a resistive substrate. If these voltage drops become large enough, they can forward-bias one of the isolation junctions. The forward-biased junction then injects current into other circuit nodes, causing potentially catastrophic malfunctions. *Minority carrier injection* occurs when a forward-biased junction injects minority carriers into the isolation, a tank, or a well. Some of these carriers diffuse several hundred microns before recombining and can easily cross reverse-biased junctions that block majority carrier flow.

Dielectrically isolated processes are subject to a mechanism analogous to substrate debiasing whenever multiple devices are integrated into a common isolation region. Currents flowing through the lightly doped superficial silicon can debias it. This situation frequently occurs on high-voltage oxide-isolated processes where the low-voltage components are integrated together in the P-field silicon. Because the superficial silicon is both thin and lightly doped, debiasing within this layer can easily become a serious problem.

4.4.1. Substrate Debiasing

Substrate debiasing becomes a problem when currents flowing through the substrate generate voltage drops of a few tenths of a volt or more. This substrate current consists of majority carriers that cannot surmount reverse-biased isolation junctions, but sufficient debiasing may cause one or more isolation junctions to forward-bias and inject minority carriers into active circuitry.

Figure 4.19 shows a typical example of substrate debiasing in a standard bipolar process. Substrate PNP transistor Q_1 injects its collector current I_C directly into the substrate. This current then flows laterally to substrate contact SC_1. Because

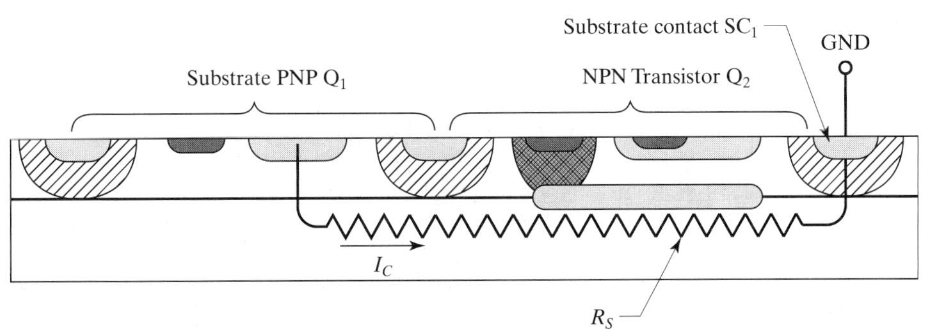

FIGURE 4.19 Cross section of a standard bipolar die showing potential substrate debiasing caused by substrate resistance R_s.

of the presence of substrate resistance R_s, the substrate voltage immediately under NPN transistor Q_2 rises. Only a few hundred millivolts of substrate debiasing are necessary to forward-bias the collector-substrate junction of a saturated common-emitter NPN.

Effects

The voltage required to forward-bias a PN junction depends on both current density and temperature. Table 4.1 lists typical forward-bias voltages for the collector–substrate junction of a minimum-area NPN transistor constructed in a standard bipolar process. This table is useful for estimating susceptibility to substrate debiasing. For example, a circuit using 100 μA minimum currents can probably tolerate 1 μA of leakage. If it must operate at 125°C, then Table 4.1 indicates that substrate debiasing must not exceed 0.35 V. If the same circuit has to operate at 150°C, then it can tolerate no more than 0.30 V of debiasing.

TABLE 4.1 Forward voltages for a typical collector-substrate junction of a minimum NPN transistor in standard bipolar, as a function of temperature and current.[45]

Current	25°C	85°C	125°C	150°C
10 nA	0.43 V	0.29 V	0.19 V	0.13 V
100 nA	0.49 V	0.36 V	0.27 V	0.22 V
1 μA	0.55 V	0.43 V	0.35 V	0.30 V
10 μA	0.61 V	0.50 V	0.43 V	0.39 V
100 μA	0.67 V	0.57 V	0.51 V	0.47 V

Figure 4.20 depicts the cross section of a standard bipolar wafer containing a single substrate current injector and a single substrate contact. R_1 models the lateral resistance through the substrate, while R_2 models the vertical resistance beneath the substrate contact. The total resistance of the substrate R_s equals the sum of the lateral and vertical components: $R_s = R_1 + R_2$.

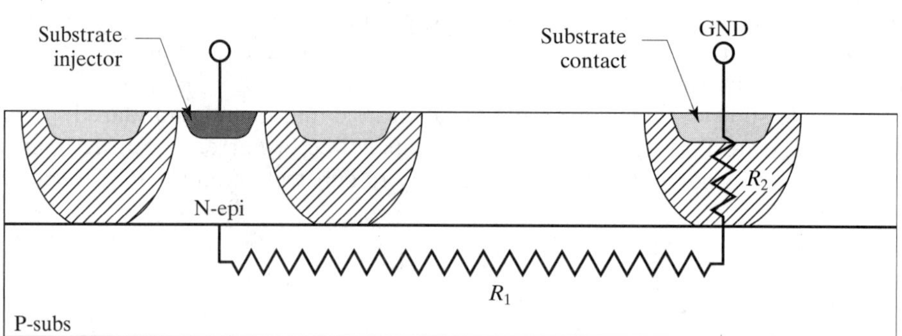

FIGURE 4.20 Simplified model of substrate debiasing in a standard bipolar process.

The relative magnitudes of R_1 and R_2 depend upon the process. Standard bipolar uses a lightly doped substrate and a heavily doped isolation diffusion, so $R_1 \gg R_2$. The value of R_1 depends on various geometric factors, including the cross-sectional area of both the injector and the substrate contact, as well as the distance between

[45] Based on $V_{BE}(150°C, 1\ \mu A) = 0.3$ V.

them. Typical values of R_1 range from hundreds to thousands of Ohms,[46] while R_2 rarely exceeds 10 Ω.[47] The presence of a network of isolation diffusions criss-crossing the die complicates the computation of substrate resistance because the low sheet resistance of the isolation (usually about 10 Ω/□) allows each substrate contact to extract current from a large area of isolation. This effect complicates the computations to such an extent that only empirical measurement or sophisticated computer simulation can yield accurate results.

Two points should be kept in mind when using lightly doped substrates. First, substrate resistance always increases with separation. A substrate contact placed adjacent to the injector will extract some of the current before it ever reaches the substrate. Contacts placed farther away require the current either to flow through long stretches of isolation or to pass through the highly resistive substrate. Second, a substrate contact to a heavily doped isolation diffusion draws current not only from the substrate immediately beneath it but also from adjoining stretches of isolation. This effectively magnifies the area of substrate contacts, allowing even a minimum-size contact to have an effective area of many hundreds of square microns. Because of this effect, a scattering of minimum-size substrate contacts throughout the die will have a much lower effective resistance than a single large contact.

CMOS and BiCMOS processes usually employ a heavily doped substrate and a lightly doped epi, so $R_1 \ll R_2$. The value of R_1 is usually so small that it can safely be ignored. The value of R_2 depends on the thickness of the epi layer and its resistivity. A typical value is about 600 kΩ/μm^2. This value can be used to compute the area of substrate contacts required for a CMOS or BiCMOS design, as explained in the following section.

Even on a heavily doped substrate, contacts placed immediately adjacent to a substrate injector will exhibit less resistance than ones placed far away. This *proximity effect* falls off rapidly with distance, and substrate contacts placed hundreds of microns away are no more effective than those placed on the opposite side of the die. The proximity effect occurs because carriers can flow directly to the adjacent contact, rather than having to flow down to the substrate, across, and up to a distant contact. Contacts placed immediately adjacent to a substrate injector can also help prevent localized debiasing of the highly resistive isolation, protecting adjacent tanks from injection from the isolation sidewalls.

Preventative Measures

Integrated circuits should inject as little current into the substrate as possible, as this not only minimizes substrate debiasing but also helps limit noise and crosstalk caused by modulation of the substrate potential. The collector current of substrate PNP transistors flows directly into the substrate, so these devices should be used sparingly, and no single device should conduct more than a milliamp or two. Lateral PNP and vertical NPN transistors can inject large substrate currents when

---

[46] An estimate of the lateral resistance R_1 can be obtained by examining the spreading resistance $R_{sp} = \rho/d$, where ρ is resistivity and d is the diameter of the points of contact. Spreading resistance assumes a semi-infinite slab of uniformly doped material and a probe separation much wider than the probe diameter; these conditions are only approximately met by typical substrate contacts. Assuming that the cross-sectional areas of injector and substrate contact are 25 μm^2 each, this yields d = 28.7 μm. A substrate with a resistivity of 10 Ω · cm would have a spreading resistance of 3.48 kΩ. More accurate results can be obtained by applying various correction factors: G. A. Gruber and R. F. Pfeifer, "The Evaluation of Thin Silicon Layers by Spreading Resistance Measurements," *National Bureau of Standards Special Publication 400–10*, Spreading Resistance Symposium, NBS, Gaithersburg, Maryland: June 1974.

[47] The vertical resistance through single-diffused isolation can be approximated by dividing the diffusion into multiple layers of constant doping. Computations for a diffusion with a surface doping of 10^{20} cm^{-3}, a minimum dopant concentration of 10^{17} cm^{-3}, and a depth of 5 μm yield a resistance of about 4 Ω/mil^2.

they saturate, but techniques have been developed to minimize this problem (Sections 8.1.4–5). The exact requirements for substrate contacts depend on the nature of the substrate and isolation:

Heavily doped substrates. The contacts in the scribe seal can usually extract 5 to 10mA without undue debiasing. If higher substrate currents are anticipated, then the total area of contacts required can be computed using the following formula:[48]

$$A_c = 10\frac{\rho t_{epi} I_s}{V_d} \quad [4.1]$$

This formula assumes a uniform lightly doped isolation, such as the P-epi of N-well CMOS and BiCMOS processes. A_c represents the required total area of substrate contacts in μm^2, ρ is the resistivity of the epi in $\Omega \cdot$cm, t_{epi} is the epi thickness in microns, I_s is the maximum substrate current in milliamps, and V_d is the maximum allowable debiasing in volts (from Table 4.1). The thickness of the epi is reduced by up-diffusion of dopants from the underlying substrate and from the presence of a heavily doped (if thin) contacting diffusion, such as PSD. Consider a die with a P-epi resistivity of 10 $\Omega \cdot$cm and an effective epi thickness of 7 μm. If the substrate must conduct 20 mA without more than 0.3 V of debiasing, then 47,000 μm^2 of substrate contacts are required. Subtracting the area of substrate contacts in the scribe seal yields the required area of additional contacts. These can be inserted wherever space exists in the layout. As a precaution against localized debiasing, substrate contacts should ring any device injecting more than 1 mA.

Substrate contacts should always include whatever P-type diffusions are available. The addition of a relatively deep diffusion, such as a base implant or a P-well, can greatly reduce vertical resistance. The P-type channel stop implant can also serve in this capacity. In order to implant a channel stop beneath a PMoat region, one must explictly code a geometry on the channel stop mask. If this is not feasible, then the designer can replace large PMoat geometries with dense arrays of small PMoat plugs. Either approach can significantly reduce substrate contact resistance without requiring any additional area.

Lightly doped substrates with heavily doped isolation. No simple formula exists for computing the area of substrate contacts required to protect a lightly doped substrate from debiasing. A scattering of substrate contacts across the die will, when combined with the scribe seal, handle at least 5 to 10 mA. Any device that injects 100 μA or more should have substrate contacts located nearby, and any device that injects 1 mA or more should be ringed with as much substrate contact as possible. Sensitive low-current circuitry should reside at least 250 μm away from any substantial source of substrate injection, since debiasing on lightly doped substrates tends to localize around the point of injection. Once the layout has been completed, additional substrate contacts should be scattered throughout the layout wherever room exists. A large number of small substrate contacts scattered throughout the layout will prove more effective than a few large contacts. Even with all of these precautions, designs that inject more than 10 mA into the substrate may experience debiasing. Apart from adding more substrate contacts or moving sensitive circuits away from substrate injectors, the only remedies for such problems are the addition of a heavily doped substrate or the use of backside contacting.

---

[48] This formula is derived from the fundamental equation $R = \rho t/A$. It neglects fringing effects, which tend to reduce the effective resistance of small substrate contacts. The formula provides a first-order approximation of the worst-case substrate resistance.

Lightly doped substrates with lightly doped isolation. A few processes use a lightly doped substrate in combination with a very resistive isolation. This situation can arise when a BiCMOS design is constructed on a lightly doped substrate to save costs. Such designs cannot rely on the scribe seal to extract more than a few milliamps of substrate current. Large numbers of substrate contacts scattered across the die will help extract substrate current, but some degree of localized substrate debiasing is almost inevitable. Sensitive circuits should be located far away from major sources of substrate injection. Since substrate modulation can inject substantial noise into high-impedance circuitry, consider placing wells under resistors and capacitors to isolate them from substrate noise coupling. Sensitive MOS circuitry may also employ NBL to isolate NMOS transistors from the substrate (Section 11.2.2). In some cases, it may be possible to add strips of heavily doped material to the isolation without increasing the well-to-well spacings (Figure 4.18). This strategy effectively converts the design into one that uses a lightly doped substrate in conjunction with a heavily doped isolation. This stratagem substantially reduces the number and area of substrate contacts required to extract large substrate currents. Backside contact can also provide a large reduction in substrate resistance, but it is difficult to obtain Ohmic contact to a lightly doped substrate unless a backside diffusion is performed to increase the surface doping concentration.

Dielectrically isolated substrates. Processes that employ dielectric isolation can experience problems analogous to substrate debiasing. The thin superficial silicon layer is very prone to debiasing unless buried layers are added to reduce its sheet resistance. Most processes avoid adding a P-buried layer beneath the P-field, leaving this region vulnerable to debiasing. This vulnerability would not be of much concern if every device resided in its own tank, but designers naturally wish to merge devices to save space. Debiasing problems can be avoided by following a few common-sense rules. Any device that can potentially inject more than a few microamps of current into the P-field should reside in its own tank. Examples of devices that are best isolated by themselves include bipolar transistors, Schottky diodes, devices connected to pins, and power MOS transistors. Sensitive circuitry should be isolated from the rest of the P-field to minimize noise coupling. Large numbers of backgate contacts should be scattered through the P-field among merged devices, just as if one were dealing with a lightly-doped P-epi on top of a lightly doped substrate. One simple way to ensure a sufficient number of backgate contacts is to place a backgate contact beside every NMOS transistor. While this practice arguably provides more backgate contacts than necessary, it avoids the need to go back and add substrate contacts after a circuit is laid out.

4.4.2. Minority-Carrier Injection

Junction isolation relies on reverse-biased junctions to block unwanted current flow. The electric fields set up by depletion regions repel majority carriers, but they cannot block the flow of minority carriers. If any isolation junction forward-biases, it will inject minority carriers into the isolation. Many of these carriers recombine, but some eventually find their way to the depletion regions isolating other devices.

Effects

Figure 4.21 shows a cross section of a standard bipolar circuit. Suppose that the collector of NPN transistor Q_1 connects to a pin of the integrated circuit, and that the external circuitry experiences occasional transient disturbances that pull current out of this pin. If transistor Q_1 is off, then these transients pull its tank below ground, forward-biasing the collector-substrate junction of Q_1 and injecting minority carriers (electrons) into the substrate. Most of these carriers recombine, but some diffuse across to other tanks, such as T_1.

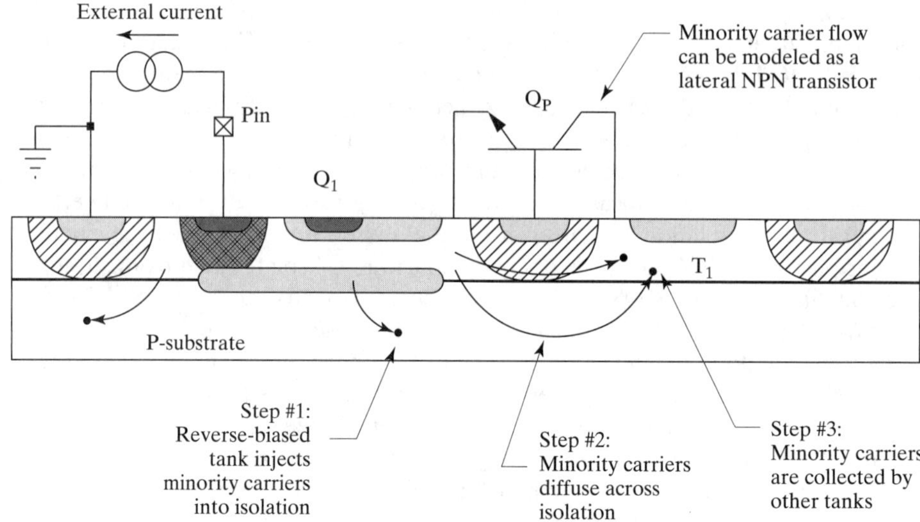

FIGURE 4.21 Example of minority-carrier injection into the substrate of a standard bipolar process. Lateral NPN transistor Q_P models the transit of minority-carriers across the isolation.

The transit of minority carriers across the isolation is analogous to the flow of minority carriers through a bipolar transistor. The tank pulled below ground acts as the emitter of lateral NPN transistor Q_P. The isolation and substrate act as the base of this transistor, and any other reverse-biased tank acts as a collector. Each reverse-biased tank forms a separate parasitic transistor corresponding to Q_P. The betas of these parasitic lateral NPN transistors are very low because most of the minority carriers recombine in transit. The parasitic bipolar between two adjacent tanks might have a beta of 10, but the beta between two widely separated tanks might not even reach 0.001. Even such low gains can cause circuit malfunctions. Suppose that a forward-biased tank injects a minority current of 10 mA into the substrate. If the parasitic associated with another tank has a beta of 0.01, then this tank will collect 100 μA of current—easily enough to disrupt the operation of a typical analog circuit.

Substrate contacts cannot, by themselves, stop minority-carrier injection, since minority carriers travel by diffusion and not by drift. Minority carriers are collected only by reverse-biased junctions. However, substrate contacts still provide majority carriers to feed recombination. Since most minority carriers recombine in the isolation, substrate contacts remain necessary to prevent substrate debiasing.

In some cases, minority-carrier injection can cause a circuit to latch up. Early CMOS processes suffered from a form of this malady that has since come to be called *CMOS latchup*.[49] Figure 4.22A shows the cross section of a portion of a CMOS die consisting of an NMOS transistor M_1 and a PMOS transistor M_2. In addition to these two desired MOS transistors, this layout contains two parasitic bipolar transistors. Lateral NPN transistor Q_N's emitter is the source of M_1, its base is the isolation, and its collector is the N-well of M_2. Lateral PNP transistor Q_P's emitter is the source of M_2, its base is the N-well, and its collector is the isolation. Figure 4.22B shows the two parasitic bipolar transistors drawn in a more familiar fashion. In this schematic, R_1 represents the well resistance of M_2, and R_2 represents the substrate resistance. These two resistors normally ensure that both bipolar transistors remain off. As long as this remains the case, neither parasitic conducts any current and the integrated circuit works as intended. When a transient disturbance turns on either transistor, the current flowing through this device will turn on the

[49] R. R. Troutman, "Recent Developments in CMOS Latchup," *IEDM Tech. Dig.*, 1984, pp. 296–299.

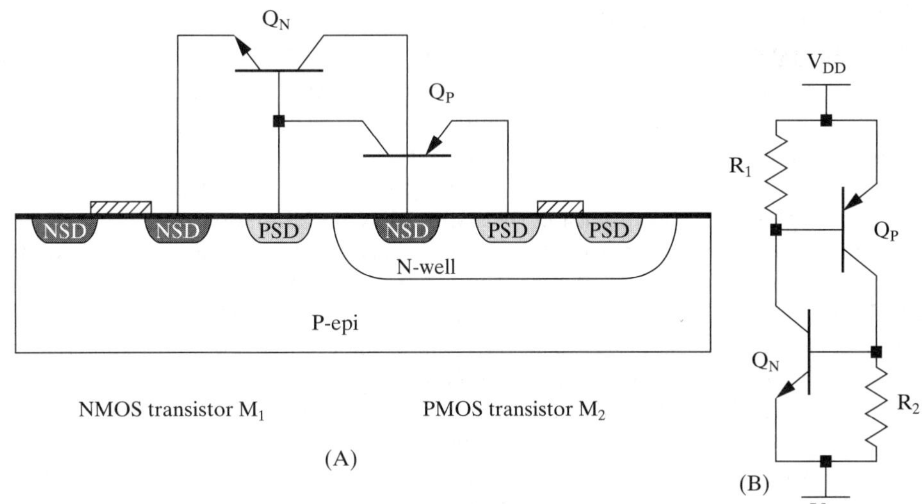

FIGURE 4.22 (A) Cross section of a CMOS die showing diffusions that form the two parasitic bipolar transistors Q_P and Q_N; (B) equivalent schematic showing these transistors along with well resistance R_1 and substrate resistance R_2.

other parasitic as well. Each transistor then supplies the other's base current. This process becomes self-sustaining if the product of the betas of transistors Q_N and Q_P exceeds unity. Once this happens, the circuit is said to have *latched up*, and it will remain in this state until power is removed. The integrated circuit can actually conduct so much current that it overheats and self-destructs. Even if this does not occur, latchup causes circuit malfunctions and excessive supply current consumption.

CMOS latchup can be triggered in one of two ways. If the source of NMOS transistor M_1 is pulled below ground, it will inject minority carriers (electrons) into the substrate, turning on parasitic transistor Q_N. This transistor will then turn on Q_P. Alternatively, the source of PMOS transistor M_2 may be pulled above the well. It will then inject minority carriers (holes) into the well and will turn on parasitic transistor Q_P. This transistor then turns on Q_N. Some authors explain CMOS latchup by comparing the four regions involved (PMOS source, PMOS backgate, NMOS backgate, NMOS source) to the four-layer PNPN device called a *silicon-controlled rectifier* (SCR). Electrically, an SCR is equivalent to the coupled bipolar transistors of Figure 4.22B. Therefore the SCR model of CMOS latchup is essentially the same as the bipolar model presented here.

The obvious way to stop CMOS latchup consists of reducing the beta of either or both parasitic transistors. If the product of these betas is less than unity, then latchup cannot occur. This is usually achieved by increasing layout spacings, which in turn increases the width of the neutral base regions of the parasitic lateral transistors. Alternatively, the amount of dopant present in the neutral base region of one (or both) parasitic transistors may be increased. Both of these approaches increase the Gummel number of one or both transistors and reduce the beta product.

Although many CMOS processes claim immunity to latchup, these claims are true only in a somewhat narrow sense. The PNPN structure inherent in such a process lacks sufficient gain to establish regenerative feedback, but minority-carrier injection still occurs. The collected carriers can still cause circuit malfunctions, and if positive feedback exists in the circuit, these malfunctions can still cause a form of latchup. The significance of this observation is frequently underestimated. Any integrated circuit that experiences unanticipated minority-carrier injection can potentially latch up. Even if it does not actually do so, it is still likely to malfunction. Not only do electrons injected into the substrate pose a potential threat, but so do holes unintentionally injected into wells or tanks.

Preventative Measures (Substrate Injection)

Fundamentally, there are four ways to defeat minority-carrier injection: (1) eliminate the forward-biased junctions that cause the problem, (2) increase the spacing between components, (3) increase doping concentrations, and (4) provide alternate collectors to remove unwanted minority carriers. All of these techniques provide some benefit, and in combination they can correct almost any minority-carrier injection problem.

The simplest solution, at least in theory, consists of eliminating the forward-biased junctions that inject minority carriers. This goal is often very difficult to achieve. In a standard bipolar process, tanks must not go below substrate by more than about 0.3 V or they will inject minority carriers into the substrate. In an N-well CMOS process, no well and no NSD region residing in the epi may go below substrate potential by more than about 0.3 V. If the voltage on a pin slews rapidly, parasitic inductance can cause transients that pull the pin above supply or below ground. The faster the node slews, the smaller the parasitic inductance required to cause such transients. Modern switching speeds have become so fast that the inductance of pin and bondwire alone often cause objectionable transients. Substrate injection has become difficult, if not impossible, to eliminate.

Minority-carrier injection into the substrate will cause fewer problems if potential injectors are separated from sensitive circuitry. In most designs, only a few devices connect to pins. With a little forethought, these devices can be placed far away from sensitive circuitry. In many cases, the layout will naturally favor this sort of separation. For example, power transistors inject minority carriers during transients. Since these transistors form part of the output circuitry, they will typically be placed far away from sensitive input circuitry to minimize electrical and thermal feedback. This same arrangement also minimizes the circuit's vulnerability to minority-carrier injection.

Additional dopant added to the isolation regions of the die will reduce the gain of the parasitic lateral bipolars. CMOS and BiCMOS processes often employ P+ substrates for just this reason. All other factors being equal, a process incorporating a heavily doped substrate will provide greater immunity to electrical upsets than one that uses a lightly doped substrate. However, a heavily doped substrate cannot, by itself, prevent minority carriers from moving laterally through isolation regions separating adjacent tanks or wells. In order to obtain the full benefits of the heavily doped substrate, the process must use a heavily doped isolation, or the designer must add suitable guard rings.

The isolation doping can also be increased by adding a deep-P+ diffusion. Most CMOS processes do not include any suitable diffusion. Some BiCMOS processes include one for constructing certain components (such as DMOS transistors)—usually as part of a process extension. Standard bipolar processes sometimes offer a deep-P+ process extension for constructing high-current lateral PNP transistors. If a suitable diffusion exists, it can be placed in the isolation regions of the die to help increase the isolation doping. This technique can help offset the very light doping of the PBL portion of an up-down isolation system, and can minimize lateral conduction of minority carriers between adjacent tanks or wells.

Minority carriers are collected in disproportionate numbers by reverse-biased junctions near the point of injection. Not only do carriers have less distance to travel to reach a nearby junction, but the nearer junctions also block the flow of carriers to more distant ones. Designers can take advantage of this *shadow effect* to erect deliberate barriers to the flow of minority carriers by placing reverse-biased junctions between the point of injection and vulnerable diffusions. A reverse-biased junction used in this manner is called a *minority-carrier-collecting guard ring.* These guard rings can be further subdivided into two categories, depending upon which polarity of carriers they collect. An *electron-collecting guard ring* (ECGR) collects minority electrons from a P-type region, while a *hole-collecting guard ring* (HCGR) collects minority holes from an N-type region. Figure 4.23 shows a typical layout for an electron-collecting guard ring in a standard bipolar process. Tanks T_1 and T_2 connect to pins that may experience voltage transients. These are surrounded by a third tank, T_3, that collects a significant fraction of the minority electrons injected by T_1 and T_2. Tanks T_1 and T_2 connect to pins, so it is quite natural to place them along one side of the die or even in a corner (as shown in Figure 4.23). This not only minimizes the length of interconnecting leads, but also eliminates the need for guard rings along two edges.

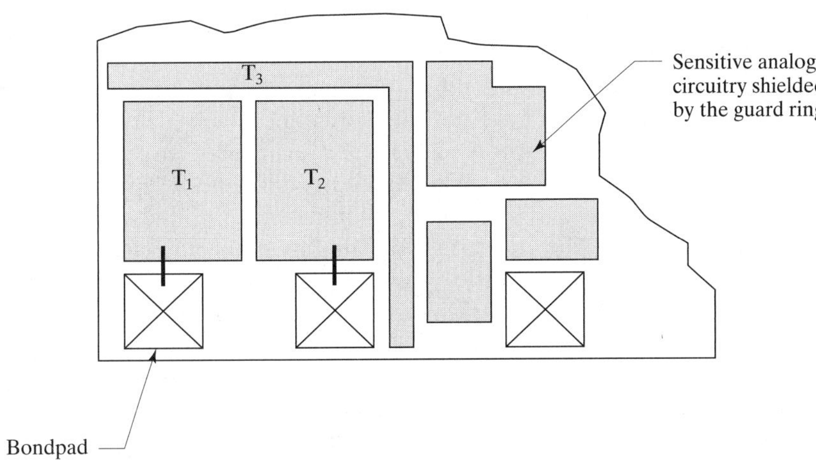

FIGURE 4.23 Sample electron-collecting guard ring (T_3) implemented in a standard bipolar process.

The key to designing efficient minority-carrier-collecting guard rings consists of making them deep, wide, and low-resistance. The deeper the guard ring, the larger the fraction of passing minority carriers it can collect. N-well makes a more effective electron-collecting guard ring than NSD, and an epi tank makes a better electron-collecting guard ring than an emitter diffusion. If the guard ring can be connected to produce a large reverse bias, then the depletion region surrounding it will widen and the collection surface will be forced even deeper into the silicon. Thus, electron-collecting guard rings connected to the positive supply become more effective than those connected to substrate potential. Since diffusing minority carriers move randomly, some will actually be collected by the bottom of the guard ring's depletion region. A wider guard ring will therefore collect more minority carriers than a narrow one will. Also, narrow diffusions do not penetrate as deeply as wide diffusions because dopants become diluted by lateral dispersion. A diffusion two or three times wider than minimum will contain enough dopant to obtain the maximum possible junction depth. Low resistance also helps improve the effectiveness of a guard ring, especially if it cannot be strongly reverse-biased. Collected carriers can forward-bias a high-resistance guard ring and cause it to re-inject minority carriers. The lower the vertical resistance of the guard ring, the larger the current it can collect before it saturates and re-injects.

Figure 4.24 shows cross sections of two minority-carrier guard rings designed to collect electrons injected into the substrate. Figure 4.24A shows the standard electron-collecting guard ring for standard bipolar.[50] This guard ring includes all four available N-type materials: N-epi, deep-N+, NBL, and emitter. The NBL helps obtain the

---

[50] W. Davis. *Layout Considerations*, unpublished manuscript, 1981, p. 53.

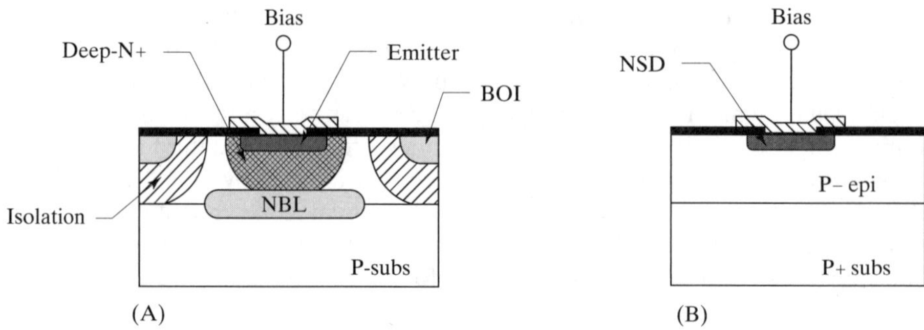

FIGURE 4.24 Cross sections of two representative electron-collecting guard rings: (A) standard bipolar[51] and (B) CMOS.

maximum possible junction depth, while deep-N+ and emitter reduce the vertical resistance. The wider this structure, the more effectively it will collect minority carriers. Most designers compromise between efficiency and area by making the deep-N+ strip no more than twice minimum width and by spacing the other layers accordingly. If possible, this guard ring should connect to the highest supply voltage available on the die. The guard ring will still function connected to substrate potential, but it may saturate unless all parts of the guard ring connect to the substrate terminal by a direct metal run. This is probably not practical in a single-level metal process since gaps must be left in the metallization to allow leads to pass through. Single-level-metal electron-collecting guard rings should connect to the positive supply and should contain as few gaps as possible.

BiCMOS layouts can produce electron-collecting guard rings similar to those in Figure 4.24A, although in this case N-well replaces the N-epi tank. Electron-collecting guard rings are considerably more difficult to construct in CMOS-only processes. The N-well becomes extremely resistive in the absence of deep-N+ and NBL, and most CMOS devices operate at relatively low voltages. Figure 4.24B shows an alternate CMOS minority-carrier guard ring. NSD has a relatively low resistance, but it is too shallow to capture more than a small percentage of the electrons in the substrate. The wider the NSD strip, the more effective the guard ring will be. A width of at least 8 to 10 μm is recommended, although narrower guard rings do provide some benefit. The NSD guard ring should, if possible, connect to a supply pin. The reverse bias across the NSD-epi junction drives the depletion region deeper into the epi and increases the apparent depth of the guard ring. In low-voltage processes, a strongly reverse-biased NSD guard ring will often generate secondary carriers due to impact ionization. This problem can be minimized by connecting the low-voltage NSD guard ring to ground instead of to a power supply.

Guard rings of the type shown in Figure 4.24A can reduce substrate injection by a factor of 10 to 100, provided that they are used in conjunction with a heavily doped substrate. The P–/P+ interface between the lightly doped epi and the heavily doped substrate repels minority carriers (Section 8.1.5), constraining them to remain within the relatively thin epi layer. This greatly improves the collection efficiency of the guard ring.[52] A simple modification to the electron-collecting guard ring of Figure 4.24A can further increase its attenuation. Instead of connecting the guard ring directly to a supply voltage, it is connected back to the substrate so that a majority-carrier current flows through the substrate beneath the guard ring (Figure 4.25). This

[51] C. Jones, "Bipolar Parasitics," unpublished report, 1988, p. 43.

[52] L. S. White, G. R. M. Rao, P. Linder, and M. Zivitz, "Improvement in MOS VLSI Device Characteristics Built on Epitaxial Silicon," in *Silicon Processing,* American Society for Testing and Materials STP 804, 1983, pp. 190–205.

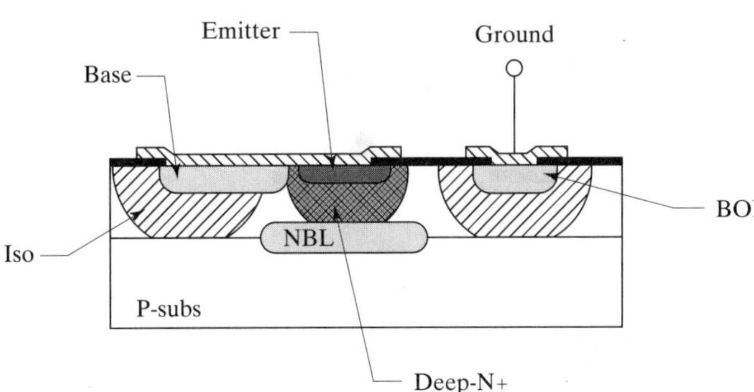

FIGURE 4.25 Cross section of an improved electron-collecting guard ring for collecting electrons injected into the substrate.

type of guard ring is intended to protect against minority carriers originating on only one side of the ring—in this case, the right side. The deep-N+ sinker in the center of the guard ring collects most of these minority carriers. The resulting current flows out of the sinker and into the attached metal plate. The base/iso diffusion at the left side of the structure re-injects this current into the substrate in the form of majority carriers. Since the nearest substrate contact lies on the other side of the structure, the majority carriers flow back underneath the guard ring. This current locally debiases the substrate and creates an electric field that opposes minority-carrier flow. The minority carriers are forced upward and toward the guard ring, where they are ultimately collected, or they are held in the substrate until they recombine. The inventor[53] claims an attenuation factor in excess of 1 million for this structure. While this degree of attenuation may not be achieved in every process, this guard ring will provide more attenuation than those shown in Figure 4.24, especially at higher currents.

The modified guard ring in Figure 4.25 suffers from several drawbacks. It provides enhanced attenuation of minority carriers flowing in only one direction—in this case, from right to left. Substrate contacts cannot be placed near the guard ring on the side facing the injected carriers. This structure also relies upon deliberate debiasing of the substrate, which could potentially forward-bias adjacent junctions. The principle behind this style of guard ring also applies to the design of ordinary guard rings of the sort shown in Figure 4.24. In all cases, it is better to place the electron-collecting guard ring adjacent to the injector, and to place substrate contacts inside of the guard ring. The majority-carrier substrate current that flows under the guard ring generates an electric field opposing the flow of minority carriers, thereby enhancing the performance of the guard ring.

Minority-carrier guard rings can also prevent holes injected into a tank from reaching the substrate and debiasing it. This situation can occur whenever a bipolar transistor saturates, regardless of whether this transistor is an NPN or a PNP (Section 8.1.4–5). Power transistors can easily inject tens or even hundreds of milliamps into the substrate. A heavily doped layer such as NBL can reduce minority-carrier injection from a P-type region through an N-well or N-epi tank to substrate. NBL also helps to minimize tank or well resistance and therefore makes it more difficult to develop the debiasing required to trigger CMOS latchup. CMOS processes generally do not incorporate NBL due to the cost of the extra masking step and to manufacturing difficulties associated with the fabrication of buried layers. Standard bipolar and analog BiCMOS processes frequently use NBL to reduce NPN collector resistance. If NBL exists, it should be added to all tanks or wells that can tolerate its presence, in order to minimize substrate injection and to improve latchup immunity.

[53] F. Van Zanten, U.S. Patent # 4,466,011, 1984.

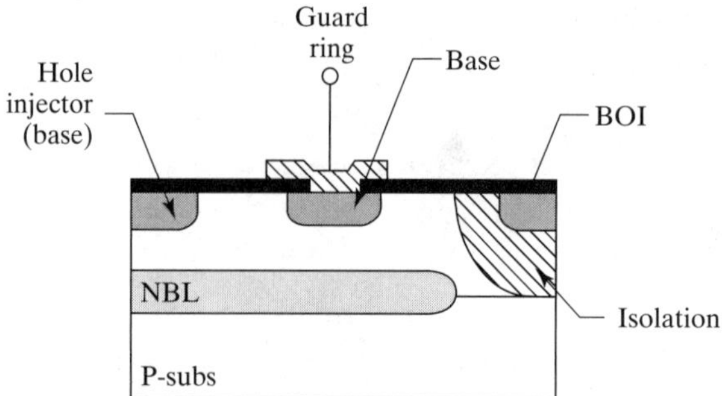

FIGURE 4.26 Cross section of a hole-collecting guard ring for a standard bipolar process.

Figure 4.26 shows a hole-collecting guard ring constructed in a standard bipolar process. The NBL placed in the bottom of the tank forms a vital part of this guard ring. An electric field appears at the interface between the NBL and the N-epi due to the outdiffusion of majority electrons from the heavily doped NBL into the lightly doped N-epi. The resulting built-in potential biases the NBL positive with respect to the N-epi. Positively charged holes are therefore repelled upwards from the interface. Even if a hole succeeds in entering the heavily doped NBL, it will probably recombine there. The guard ring of Figure 4.26 also includes a reverse-biased base ring placed around the point of hole injection. Holes that attempt to move laterally to the isolation must pass this ring. Most are collected before they can successfully traverse the narrow gap between the reverse-biased base region and the NBL beneath it. This guard ring will become more effective if the base is connected to a low potential to drive the depletion region beneath it deeper into the N-epi.

The ability of an N+/N− interface to repel holes suggests another possibility for constructing a guard ring. A deep, heavily doped diffusion such as deep-N+ can be used in conjunction with NBL to block the flow of minority holes without actually collecting them. The holes will eventually recombine in the N-epi. The electrons required to sustain the recombination process will then flow in from the tank contact. This structure is called a *hole-blocking guard ring* (HBGR). Figure 4.27 shows an example of a hole-blocking guard ring constructed in a standard bipolar process. Circuit designers often favor blocking guard rings over collecting guard rings because they do not waste power by discarding current to ground.

A hole-blocking guard ring will only function effectively if it presents a barrier to hole flow at every point. A potential weakness exists where the NBL plate and the deep-N+ ring unite. The bottom of the deep-N+ diffusion and the edges of the NBL are both relatively lightly doped. In order to maximize the doping in this region, one

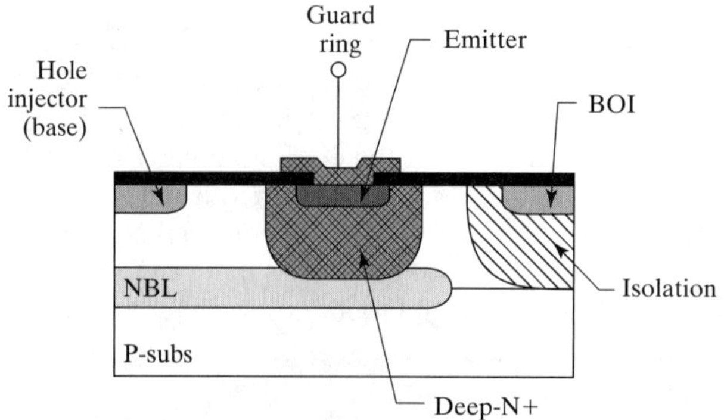

FIGURE 4.27 Cross section of a hole-blocking guard ring for a standard bipolar process.

should always extend the NBL to the outer edge of the deep-N+ ring. Some designers even extend the NBL slightly beyond the deep-N+, but this degree of conservatism is usually unnecessary.

Multiple guard rings can be combined to provide even more protection. For example, two hole-collecting guard rings can be placed concentric to one another. This configuration is often used when the hole current is so large that power dissipation becomes a concern. The inner guard ring connects to the tank contact, while the outer ring connects to ground. The inner ring will collect most of the hole current, but it will debias and become ineffectual if the current grows too large. When this happens, the outer ring will collect the remaining holes. The inner ring operates at zero bias and therefore does not dissipate much power. The outer ring operates at a large reverse bias, but it only collects a small fraction of the carriers. Together, the two guard rings combine the benefits of low power dissipation and high collection efficiency.

One can also place a hole-collecting guard ring like that of Figure 4.26 inside a hole-blocking guard ring like that of Figure 4.27. This combination offers an alternative means of preventing debiasing from rendering a zero-biased guard ring ineffectual. If the guard ring debiases, then the holes that it fails to collect will be blocked by the outer hole-blocking guard ring. Even more elaborate schemes have been proposed, but most of these are not worth the space they consume.

Effective hole collection guard rings cannot be constructed in a pure CMOS process due to the absence of NBL. Without NBL, holes can flow freely down to the substrate as well as laterally to the isolation. Attempts to collect the holes by using PSD rings are mere futile gestures. The vertical path is so much shorter than the lateral path that most of the holes flow down to the substrate before they can reach a PSD ring. This remains true even if the PSD ring is placed immediately adjacent to the injector. Not only are the geometries of the process working against the guard ring, but the graded nature of the well actually generates a weak built-in potential that urges the holes to flow downwards rather than laterally. CMOS processes must therefore rely upon low-resistance substrate contacts to extract any hole current injected into the substrate.

BiCMOS processes can construct hole guard rings similar to those in Figures 4.26 and 4.27, but these rings may or may not prove effective. In order for the NBL to block vertical hole flow, a significant built-in potential must exist at the NBL/N-well interface. This built-in potential largely depends upon the doping concentration on the lighter side of the interface, or in other words, in the N-well. High-voltage processes that use deep, lightly doped wells will generate sufficient built-in potential to block hole flow, while low-voltage processes that use shallow, heavily doped wells will not. Section 13.2 discusses this problem in more detail and presents several types of hole guard rings tailored for use in BiCMOS processes.

Many advanced CMOS and BiCMOS processes include features specifically designed to minimize CMOS latchup. One approach uses a high-energy implant to deposit additional dopant at the bottom of wells, effectively creating a buried layer without the cost and complexity of epitaxy. The resulting structure is called a *retrograde well*. Low-voltage shallow-well processes have made extensive use of retrograde wells to reduce resistance, contain vertical depletion, and provide an N+/N− interface. Even the most energetic implants can only reach a couple of microns into the silicon, so high-voltage processes must continue to use conventional buried layer technologies.

Trench isolation greatly improves latchup immunity. The insertion of a trench lengthens the path that minority carriers must take, and therefore reduces the gain of the parasitic bipolar transistor. Trenches are especially valuable when used in combination with an N+/N− interface that hinders vertical carrier flow. Retrograde well implants conducted in combination with shallow trench isolation can produce dramatic improvements in latchup immunity.[54] Taking this concept to its logical conclusion, full dielectric isolation utterly eliminates latchup—but only if one places the isolation trenches in the right places!

Preventative Measures (Cross-Injection)

Circuit upsets caused by minority-carrier injection into a tank or well can be eliminated by placing each potential emitter of minority carriers in its own tank or well. As a rule, any PMOS transistor whose source connects to an external pin should occupy its own well. Similarly, any base resistor, HSR resistor, or lateral PNP collector connecting to a pin is best placed in its own tank. The small amount of extra space required to construct separate tanks or wells will be amply repaid by the elimination of even one parasitic. If, on the other hand, several devices connect to a common pin, then these can all occupy a common tank or well.

The hole-collection rings discussed previously have been designed to minimize injection of holes into the substrate. Another type of minority-carrier guard ring can prevent holes injected by one device from interfering with the operation of other devices in the same tank or well, a problem called *cross-injection.* Consider the case where two lateral PNP transistors occupy a common tank. If either transistor saturates, some fraction of the carriers it emits will be collected by the adjacent transistor. The resulting increase in collector current may disturb the operation of the circuit, particularly if the devices were intended to match one another. Cross-injection can be prevented by placing each transistor in its own tank, but this wastes area because of the large spacings associated with the isolation diffusion. A more compact solution employs a type of minority carrier guard ring called a *P-bar* (Figure 4.28).[55]

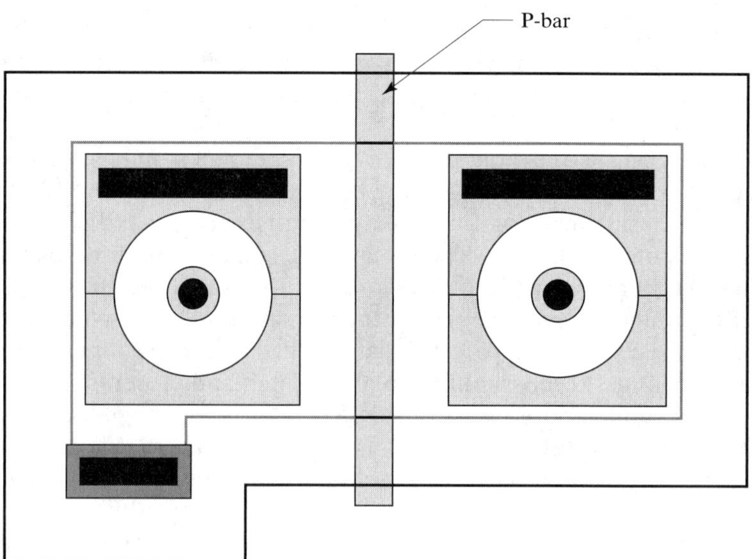

FIGURE 4.28 Example of a P-bar used to prevent cross-injection between two lateral PNPs.

A P-bar consists of a minimum-width strip of base diffusion placed between the two transistors.[56] Each end of the P-bar extends out into the isolation far enough to guarantee electrical contact. This arrangement ensures that the P-bar electrically

---

[54] W. Morris, "Latchup in CMOS," *Proc. 41st International Reliability Physics Symp.*, 2003, pp. 76–84.

[55] Davis, p. 27; Jones, p. 10.

[56] Jones, p. 10.

connects to the isolation without requiring contacts. Now suppose that the lateral PNP on the left side of the P-bar saturates and begins to inject holes into the tank. In order for these holes to reach the lateral PNP transistor on the right, they must first pass underneath the P-bar. The base diffusion forming the bar reaches fairly deeply into the epi and leaves little room for carriers to pass underneath it. Most of the holes traveling from left to right will be collected by the P-bar and shunted to ground. This structure thus acts as a specialized type of hole-collecting guard ring. The presence of NBL beneath the P-bar ensures a low-impedance path for base current passing from the right-hand transistor to the tank contact at the lower left corner of the tank. The tank contact on the left side of this structure therefore suffices for both transistors.

Although the collection efficiency of P-bars can only be determined through empirical measurement, several observations are in order. As the tank bias increases, the depletion region surrounding the bar deepens and progressively pinches off the N-epi underneath it. Devices operating at high tank-to-substrate potentials therefore obtain a higher degree of isolation from a P-bar than devices operating near substrate potential. A wider P-bar also increases collection efficiency, in part because the pinched portion of the tank becomes wider and in part because the wider base region diffuses deeper into the epi. Even a minimum-width P-bar provides a high degree of isolation against minority-carrier cross-injection due to the up-diffusion of the underlying NBL and the formation of a depletion region beneath the P-bar.

The P-bar has many applications. Bipolar circuits often contain current mirrors composed of lateral PNP transistors with a common base connection. These transistors often occupy a common tank, but if one transistor saturates, then the currents provided by the adjacent transistors increase. P-bars placed between the saturating transistor and the adjacent devices will prevent this effect without unduly enlarging the tank. Another common application consists of an NPN driving either a lateral or a substrate PNP transistor, in which the collector of the NPN connects to the base of the PNP. Minority-carrier conduction from the PNP to the NPN can initiate positive-feedback latchup by triggering the SCR inherent in this structure. A P-bar placed between the transistors may suppress the latchup, although this is not guaranteed unless the collection efficiency of the bar exceeds the reciprocal of the beta product of the two transistors.

P-bars also find use in CMOS processes, where they typically consist of PMoat. This type of P-bar exhibits a lower collection efficiency than its bipolar counterpart due to the shallowness of the PMoat diffusion and the absence of NBL. The lack of a buried layer greatly increases the well resistance beneath the P-bar, so prudence dictates the inclusion of well contacts on both sides of the bar. This structure can help increase a circuit's latchup immunity without requiring separate wells. If one PMOS transistor in the tank has a source or drain connecting to an outside terminal, then a transient can potentially forward-bias this PMoat into the well. The resulting minority-carrier injection can disturb adjacent transistors and can even lead to latchup. The strategic placement of a few minimum-width PMoat P-bars can provide considerable protection against this sort of cross-injection without consuming as much area as separate wells require.

Another type of minority-carrier guard ring called an *N-bar* can also protect against minority-carrier cross-injection. An N-bar consists of a strip of deep-N+ placed between two devices occupying a common tank (Figure 4.29).[57] The N-bar typically serves as a tank contact for the devices around it since the spacings surrounding the deep-N+ are large enough to allow room for both emitter diffusion and a contact. The doping gradient surrounding the N-bar repels holes, and most of the carriers that overcome this gradient recombine inside the deep-N+ before they

---

[57] Davis, p. 31.

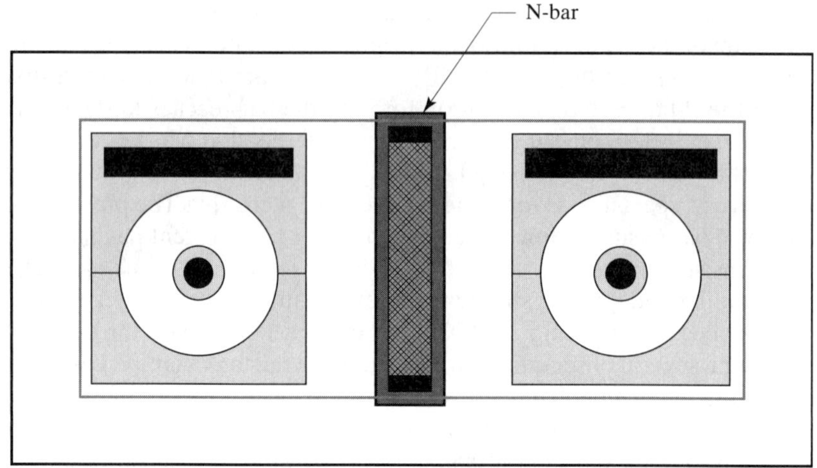

FIGURE 4.29 Example of an N-bar used to simultaneously provide tank contact and to minimize cross-injection between two lateral PNP transistors.

pass through it. Unfortunately, the N-bar generally stops short of the P-isolation on either side of the tank to avoid forming an N+/P+ junction that would break down at a relatively low voltage. These gaps allow minority carriers to bypass the N-bar, so an N-bar usually exhibits a lower efficiency than a P-bar. Still, the combination of a highly efficient collector contact with a moderately effective minority-carrier guard ring sometimes finds applications in high-current lateral PNP current mirrors and similar circuits.

4.4.3. Substrate Influence

The substrate of a dielectrically isolated process is electrically insulated from the superficial silicon by the buried oxide (BOX) sandwiched between them. However, this does not mean that the bias applied to the substrate cannot affect devices constructed in the superficial silicon. The sandwich of substrate, BOX, and superficial silicon can be viewed as a MOS transistor. The substrate acts as the gate of this transistor, the BOX as its dielectric, and the superficial silicon as its backgate. The voltage difference between the substrate and the superficial silicon generates an electric field that can deplete or enhance the bottom of the superficial silicon. This effect is called *substrate influence*.

Effects

The substrate of a dielectrically isolated die should be connected to the lowest-voltage pin, which, for a single-supply circuit, is usually the ground pin. Substrate influence causes no trouble so long as this connection is present. If the connection is absent, or if it somehow fails during operation, then an electrostatic charge will accumulate on the substrate. The magnitude of this charge fluctuates erratically over time. Consequently, the thickness of the depletion region above the BOX also fluctuates with time. This depletion region may alter breakdown voltages or cause unexpected parametric variations. In particular, supply current variations are likely to occur. The precise mechanisms behind these parametric variations are often obscure, but their cause is easily verified. If they vanish when the substrate connection is re-established, then they are caused by substrate influence.

Preventative Measures

Parametric variations due to substrate influence can be eliminated by establishing a reliable connection to the substrate. Most dielectrically isolated processes provide no means of gaining access to the substrate from the top of the die. Therefore, contact must be made from the leadframe, through what is called a *backside contact*. Three requirements must be met in order for a backside contact to function properly. First, the layer of oxide that grows on the back of the wafer during processing must be removed. The backgrind operation that reduces the thickness of the wafer prior to assembly will meet this requirement. Second, the die must be attached to the leadframe with a conductive material. Choices include gold eutectic, solder, and silver-filled conductive epoxy. Most processes opt to use conductive epoxy since this is the cheapest and lowest-stress approach. Third, an electrical connection must be made between the die mount pad and the lowest-voltage pin. There are two approaches to achieving this connection: downbonds and fused leadframes.

A *downbond* consists of a bondwire that connects from a lead finger to the mount pad. This connection gets its name from the fact that the mount pad is usually depressed below the level of the lead fingers. Downbonds have a number of drawbacks. They require considerable space on the mount pad to allow room for the capillary to form the bond. This often requires mounting the die off-center or using an oversized leadframe. Downbonds are also vulnerable to shearing caused by delamination between the plastic mold compound and the flat, featureless metal surface of the mount pad. This type of delamination may detach the downbond without causing any other noticeable consequences. In a worst-case scenario, this can lead to intermittent contact during temperature cycling. Downbonds to the substrate of a dielectrically isolated die are also difficult to test. One possible approach is to use a pair of downbonds to connect the lowest-voltage pin to the die (Figure 4.30A). The first down-bond connects the pin to the mount pad, and the second downbond connects the mount pad to the die. The absence of either downbond breaks the continuity between die and pin. Unfortunately, this approach will not reliably detect intermittent downbonds.

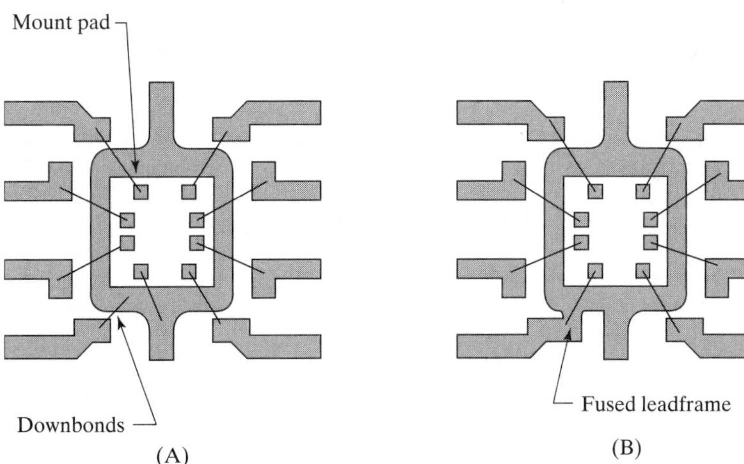

FIGURE 4.30 Backside contact can be established using a downbond (A) or a fused leadframe (B).

Another approach to backside contact uses a *fused leadframe* in which the appropriate pin is joined to the mount pad (Figure 4.30B). A fused leadframe ensures positive electrical contact between pin and mount pad. Unfortunately, analog integrated circuits rarely follow any specific pin ordering. Different products will probably require that different pins be fused to the mount pad. Therefore, fused leadframes must often be custom-designed for the products that require them.

TABLE 4.2 Summary of failure mechanisms.

Failure Mechanism	Symptoms	Corrective Actions*
Electrostatic discharge (ESD)	Gate oxide ruptures either immediately or after delay, junctions shorted or leaky.	***Add ESD protection devices, do not route leads over thin emitter oxide.***
Electromigration	Open or short circuits after long-term operation, usually at high temperature.	Use copper-doped aluminum, use refractory barrier metal, ***use adquate lead widths, use adequate bondwires***.
Dielectric breakdown	Dielectrics rupture either immediately upon application of voltage or after a delay.	***Include provision for OVST, include gettering structures, avoid using grown oxides on top ofdeep-N+***.
Antenna effect	Small gate oxides connected to large conductors suffer delayed failure.	***Reduce ratio of conductor area to gate oxide area, add diodes***.
Dry corrosion	Open circuit failures, moisture accelerates failure.	Use nitride PO, ***minimize PO openings***.
Mobile ions	Threshold shifts under high-temp biased operation, relaxes after unbiased bake.	Use phosphosilicate glass, use polysilicon gate MOS, ***minimize PO openings, use adequate scribe seals***.
Hot carrier injection	Threshold shifts under high-temp biased operation, relaxes after unbiased bake.	***Limit drain-source voltages, use LDD structures, use long-channel devices***.
Zener walkout	Breakdown voltage drifts, relaxes after unbiased bake.	***Use buried Zener*** (if available).
Avalanche-induced beta degradation	Beta of bipolar transistor decreases after emitter-base junction reverse-biased.	***Avoid excessive emitter-base reverse bias***.
Negative-bias temperature instability	PMOS threshold voltage shifts during operation.	Avoid damp oxides, ***bias matched PMOS transistors at same drain-source bias***.
Parasitic channels & Charge spreading	Leakage currents at high voltage. If they appear after high-temp biased operation and relax after high-temp bake, charge spreading is responsible.	Use (111) silicon, add channel stop implants, ***add base-over-iso, use channel stops, use field plates***.
Substrate debiasing	Latchup, parametric shifts that occur under specific biasing conditions.	***Maximize substrate contact, place contacts near injectors***.
Minority-carrier injection into the substrate	Latchup, parametric shifts that occur under specific biasing conditions.	Use P+ substrate, ***maximize substrate contact, separate sensitive circuitry, add NBL to shared wells, use deep-P+ in isolation, add guard rings***.
Minority carrier cross-injection	Latchup, mismatches between merged devices.	***Use P-bar or N-bar, place devices in separate tanks or wells***.
Substrate influence	Parametric drifts on dielectrically isolated device.	***Ensure verifiable connection to substrate, use conductive epoxy die attach***.

* Possible solutions listed in ***bold italics*** are under the control of circuit and layout designers; the remaining solutions can only be implemented by process engineers.

4.5 SUMMARY

This chapter discusses a number of common failure mechanisms of integrated circuits. Table 4.2 summarizes these mechanisms, along with typical symptoms and suggested corrective actions. Even a cursory glance at the table reveals the interdisciplinary nature of the subject. Some mechanisms are primarily electrical, while others depend upon chemical or electrochemical processes. Some of these failure mechanisms require a knowledge of device physics to counteract them, while others require knowledge of processing and packaging technology. Only by amassing a working knowledge of many fields can one hope to design integrated circuits that will function reliably over a lifetime of use.

4.6 EXERCISES

Refer to Appendix C for layout rules and process specifications.

4.1. A certain copper-doped aluminum alloy can safely operate at current densities of $5 \cdot 10^5$ A/cm^2. If the metallization thickness equals 8 kÅ, but thins by 50% when passing over oxide steps, then how much current can a 10 μm-wide lead carry across an oxide step?

4.2. Propose a scribe seal structure for a single-level-metal standard bipolar process. Draw a cross section of this structure and explain the purpose of each of its components.

4.3. Lay out a 15 kΩ, 8 μm-wide HSR resistor. Field plate the resistor as well as possible, including flanges where necessary. The field plate should overhang HSR by at least 6 μm and base by at least 8 μm.

4.4. Modify the layout from Exercise 4.3 to include channel stops constructed from emitter diffusion. Assume that the channel stops must overlap the field plates by 4 μm.

4.5. Construct a minimum-size, standard-bipolar, lateral PNP, using a circular emitter geometry. Fully field-plate both the emitter and the collector, leaving space for base metallization. Assume the emitter field plate must overlap the collector by 2 μm and the collector field plate must overhang the collector by 8 μm.

4.6. Compute the area of substrate contacts necessary to extract 25 mA from a die that uses an 8 μm-thick, 10 Ω · cm, P-type epi layer on top of a 0.01 Ω · cm P-type substrate. Assume a maximum allowed debiasing of 0.3 V.

4.7. Lay out a standard-bipolar NPN transistor with a 20 μm by 40 μm emitter. Arrange the transistor to minimize the distance between the emitter and collector contacts. The transistor should include deep-N+ in the collector to reduce collector resistance. Place an electron-collecting guard ring around this transistor, following the cross section shown in Figure 4.24.

4.8. Lay out a 2000/5 PMOS transistor. Divide the transistor into a sufficient number of fingers to obtain a roughly square aspect ratio. Construct a hole-collecting guard ring similar to that in Figure 4.26 that encircles the PMOS transistor.

4.9. Lay out an example of a P-bar separating two minimum-size standard bipolar lateral PNP transistors. The P-bar should extend at least 4 μm into the isolation to ensure electrical continuity.

4.10. Several failed devices have been de-encapsulated (*decapped*) for microscopic examination. Suggest at least one failure mechanism consistent with each of the following observations:

a. A metal trace from a bond pad has melted open.
b. A greenish deposit covers the bond pads.
c. The gate oxide of a minimum-size NMOS has ruptured at one point, shorting the poly gate to the underlying epi.
d. A thin, dark filament appears across the base of a large NPN transistor. The transistor's base-collector junction appears shorted.

4.11. A new high-voltage, low-current bipolar operational amplifier has just completed burn-in testing. Sample units were operated under bias at 150°C for 1000 hours. Parametric testing reveals that the input offset voltages of the amplifier have drifted several millivolts during testing, and the supply currents have increased by 20%. What failure mechanisms might be responsible for these symptoms, and how can the designer determine what to fix?

4.12. Add a hole-blocking guard ring to the PMOS transistor of Exercise 4.8, placing it outside of the hole-collecting guard ring.

4.13. A diffused-emitter BiCMOS process has a guaranteed minimum emitter-base breakdown voltage of 8.0 V. A circuit designer proposes to operate an NPN transistor with the emitter-base junction transiently reverse biased to no more than 7 V. What potential reliability risks does this pose?

4.14. A low-voltage CMOS operational amplifier with a PMOS input stage shows a gradual shift in input offset voltage during operation. What failure mechanisms might account for this drift?

5 Resistors

Resistors provide specific and controlled amounts of electrical resistance. They are useful in a variety of applications, ranging from current limiting to voltage division. Analog circuits usually include many resistors, so it is fortunate that they are relatively easy to integrate. Although the tolerance of any particular integrated resistor is relatively poor ($\pm 20\%$), the tracking between matched pairs of integrated resistors is excellent ($\pm 0.1\%$). Laser-trimmed thin-film resistors can achieve tolerances of better than $\pm 0.1\%$, but only at the cost of additional processing steps.

Most processes offer a choice of several different resistor materials. Some are better suited to fabricating high-value resistors and others to fabricating low-value ones. The precision and temperature variation of different materials also varies widely. Circuit designers usually select appropriate materials for each resistor and mark their schematics accordingly. Sometimes different symbols identify the various types of resistors; sometimes the type of each resistor is printed beside it. The choice of resistor materials can have tremendous impact on circuit performance, so substitutions should not be made without careful consideration of the consequences.

5.1 RESISTIVITY AND SHEET RESISTANCE

A current flowing through a conductor causes a voltage drop to appear across it. This relationship is quantified by *Ohm's law*,

$$V = IR \tag{5.1}$$

where V is the voltage drop across the conductor, I is the current flowing through it, and R is a constant of proportionality called the *resistance* of the conductor. Resistance arises because carriers flowing through the conductor collide with the atoms that constitute it. These collisions cause the carriers to lose energy, which in turn represents a reduction in electrical potential or voltage.

TABLE 5.1 Selected prefixes of the International System of Units (SI).

Name of Prefix	Value	SI Symbol	SPICE Symbol
atto-	10^{-18}	a	
femto-	10^{-15}	f	F
pico-	10^{-12}	p	P
nano-	10^{-9}	n	N
micro-	10^{-6}	μ	U
milli-	10^{-3}	m	M
kilo-	10^{3}	k	K
mega-	10^{6}	M	MEG
giga-	10^{9}	G	
tera-	10^{12}	T	

The International System of Units[1] (SI) defines the *Ohm* (Ω) as the standard unit of resistance. As with other SI quantities, a system of prefixes allows this unit to be scaled upward or downward. Table 5.1 lists most of the prefixes used by engineers, along with their official SI abbreviations and those used by the simulation program SPICE.[2]

The value of a resistor can be computed, given its dimensions and composition. Each material possesses a characteristic *resistivity,* usually measured in $\Omega \cdot$cm. Resistivity is the inverse of *conductivity,* so if one of these two properties is known, then the other can be determined. (A resistivity of 10 $\Omega \cdot$cm implies a conductivity of $0.1(\Omega \cdot \text{cm})^{-1}$, and *vice versa.*) Conductors have very low resistivities, while doped semiconductors have moderate resistivities (Table 5.2). The resistivity of a true insulator such as silicon dioxide is virtually infinite.

Figure 5.1 shows a simple resistor constructed from a rectangular slab of a homogeneous material having resistivity ρ. This resistor is contacted at either end by perfectly conductive plates. If the slab of resistive material has length *L,* width *W,*

TABLE 5.2 Resistivities of selected homogeneous materials.[3]

Material	Resistivity $\Omega \cdot$ cm (25°C)
Copper, bulk	$1.7 \cdot 10^{-6}$
Gold, bulk	$2.4 \cdot 10^{-6}$
Aluminum, thin film	$2.7 \cdot 10^{-6}$
Aluminum (2% silicon)	$3.8 \cdot 10^{-6}$
Platinum silicide	$3.0 \cdot 10^{-5}$
Silicon, N-type ($N_d = 10^{18}$ cm^{-3})	0.25
Silicon, N-type ($N_d = 10^{15}$ cm^{-3})	48
Silicon, intrinsic	$2.5 \cdot 10^{5}$
Silicon dioxide	$\sim 10^{14}$

1 The International System of Units, or *Système Internationale (SI)*, is more commonly called the metric system.

2 The acronym SPICE stands for *Simulation Program with Integrated Circuit Emphasis*, the most familiar and widely used circuit simulator. Developed by Larry Nagel and others under the supervision of D.O. Pederson at the University of California at Berkeley, SPICE was first released in 1972.

3 Resistivities vary substantially depending on the conditions of preparation; for example, bulk resistivities of pure materials are considerably smaller than the thin-film values. Values for Cu, Au, Al, PtSi: W. R. Runyan and K. R. Bean, *Semiconductor Integrated Circuit Processing Technology* (Reading, PA: Addison-Wesley, 1994), pp. 535, 546, 548. Doped silicon: W. R. Thurber, R. L. Mattis and Y. M. Liu; *National Bureau of Standards Special Publication 400-64:* 1981, p. 42. Intrinsic silicon: B. G. Streetman, *Solid State Electronic Devices,* 2d ed. (Englewood Cliffs, NJ: Prentice-Hall, 1980), p. 443. SiO_2: Runyan, *et al.,* p. 63.

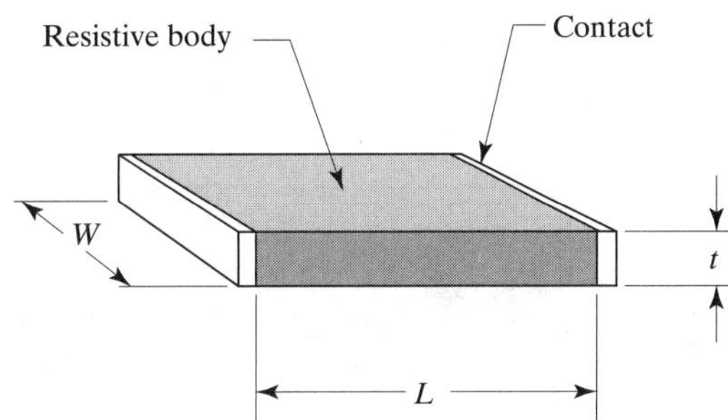

FIGURE 5.1 Layout of a simple resistor consisting of a rectangular slab of resistance material contacted by perfectly conductive terminations.

and thickness *t*, then its resistance *R* equals

$$R = \rho \frac{L}{Wt} \quad [5.2]$$

Integrated resistors consist of diffusions or depositions that can usually be modeled as films of constant thickness. It is therefore customary to combine resistivity and thickness into a single term called the *sheet resistance* R_s. In the case of a homogeneous material, $R_s = \rho / t$. The formula for resistance can now be rewritten as follows:

$$R = R_s\left(\frac{L}{W}\right) \quad [5.3]$$

Resistors are often specified in terms of their (L/W) ratio, which, although technically dimensionless, is usually assigned fictitious units of *squares* (□). A resistor with equal length and width contains one square; a resistor with a length twice its width consists of two one-square resistors in series, or two squares, and so forth. The sheet resistance R_s is usually given in units of Ohms per square (Ω/□). The value of a resistor can be found by multiplying the number of squares it contains by its sheet resistance. For example, a resistor containing 10 squares of 150 Ω/□ material will have a value of 1.5 kΩ.

Although the sheet resistance of a homogeneous film can be easily computed, many integrated resistors consist of inhomogeneous diffusions. No simple formula exists for determining the sheet resistance of such a diffusion. One can determine the sheet resistance of an ideal Gaussian diffusion by using Irwin's graphs,[4] but real diffusions do not necessarily follow these idealized profiles. In practice, sheet resistances of diffusions are usually determined by empirical measurement instead of by computation.

5.2 RESISTOR LAYOUT

Figure 5.2 shows the layout of the simplest possible thin-film resistor, consisting of a simple rectangle of resistance material with contacts at either end. The low resistance of a contact effectively shorts out the material underneath it. Almost all of the current exits the contact along its inner edge, facing the main body of the resistor. The *drawn length* of the resistor L_d therefore equals the distance measured from the inner edge of one contact to the inner edge of the other. Similarly, the width of the strip of resistance material is called its *drawn width* W_d. The drawn length and width

4 J. C. Irvin, "Resistivity of Bulk Silicon and Diffused Layers in Silicon," *Bell Syst. Tech. J.*, Vol. 41, #2, 1962, pp. 387–410.

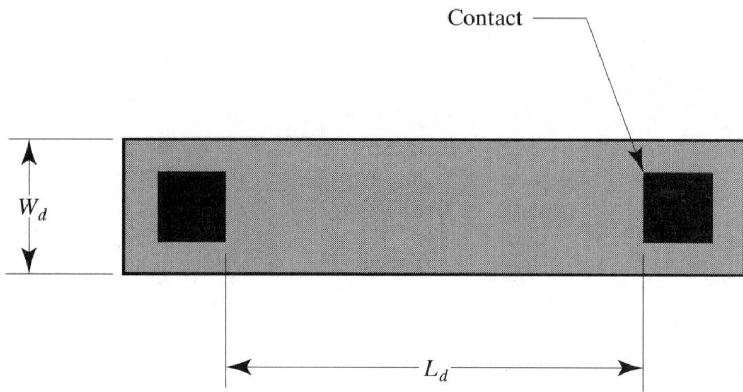

FIGURE 5.2 Layout of a simple strip resistor.

can be used to determine the approximate value of the resistor with the aid of Equation 5.3. However, there are several factors at work in an integrated resistor that do not occur in the simple resistor in Figure 5.1. Photolithography and etching can cause oxide openings to grow or shrink slightly. Outdiffusion can widen a resistor and thus reduce its resistance. Nonuniform current flow near the contacts can increase its resistance. An examination of each of these terms will show which must be taken into account in order to accurately predict the value of a resistor.

The most significant corrections to the resistor equation are those associated with width rather than length, because most resistors are much narrower than they are long. The resistor equation can therefore be rewritten as follows:

$$R = R_s\left[\frac{L_d}{W_d + W_b}\right] \tag{5.4}$$

The *width bias* W_b models the difference between the drawn width and the effective width. Outdiffusion adds about 20% of the junction depth to the drawn width of a diffused resistor.[5] For example, a base resistor with a junction depth of 1.25 μm will have a width bias of about 0.25 μm due to outdiffusion. This causes about a 5% error in the value of a 5 μm-wide base resistor—enough to be worth attempting to correct. The width bias for a given diffusion can be determined experimentally by measuring a set of resistors of varying widths. Layout rules sometimes include tables of width biases. If available, these should be used when calculating resistor values.

Equation 5.4 implicitly assumes uniform current flow through the resistor. The layout in Figure 5.2 violates this assumption because the contacts do not extend entirely across the ends of the resistor. The current must crowd inward as it approaches the contacts, making the actual value of the resistor slightly larger than that estimated by width and length. The effect of lateral nonuniform current flow can be computed using the formula[6]

$$\Delta R = \frac{R_s}{\pi}\left[\frac{1}{k}\ln\left(\frac{k+1}{k-1}\right) + \ln\left(\frac{k^2-1}{k^2}\right)\right] \tag{5.5}$$

where $k = W_e/(W_e - W_c)$, with W_e being the effective width of the resistor and W_c the width of the contact. The effective width of the resistor equals the sum of the

[5] A. B. Glaser and G. E. Subak-Sharpe, *Integrated Circuit Engineering* (Reading, MA: Addison-Wesley, 1977), p. 127. See also P. R. Gray and R. G. Meyer, *Analysis and Design of Analog Integrated Circuits,* 3d ed. (New York: John Wiley and Sons, 1993), p. 139; D. J. Hamilton and W. G. Howard, *Basic Integrated Circuit Engineering* (New York: McGraw-Hill, 1975), p. 150.

[6] C. Y. Ting and C. Y. Chen, "A Study of the Contacts of a Diffused Resistor," *Solid State Elect.,* Vol. 14, 1971, p. 433–438. This formula is strictly valid only for $W_c \gg W_e - W_c$.

drawn width W_d and the width bias W_b. The quantity ΔR represents the increase in resistance caused by nonuniform current flow at both ends of the resistor. For example, a resistor 5 μm wide containing 3 μm-wide contacts at either end will be 0.05 squares longer than Equation 5.3 predicts. Since most resistors are at least 10 squares long, this factor usually causes less than 1% error and is therefore inconsequential.

Current also flows nonuniformly in the vertical dimension as it enters and exits the resistor contacts. The current must bend upward to exit through the surface of the resistor, and it crowds toward the inside edges of the contacts as it does so. This current crowding produces a slight increase in overall resistance. This crowding effect is usually considered part of the *contact resistance* between the resistor and its metallization; see Section 5.3.4.

In summary, the width bias is usually important, and the effects of nonuniform current are usually not. Instead of applying corrections for nonuniform current flow, the designer should make the resistors long enough to avoid its influence. If the resistors are at least five squares long, then the effects of nonuniform current flow will probably total less than 5% and can usually be neglected.

Large resistors are often folded, producing so-called *serpentine* or *meander* resistors (Figure 5.3). These resistors usually employ rectangular turns (Figure 5.3A) rather than circular turns (Figure 5.3B). Rectangular turns not only are easier to draw, but also allow the spacing between the turns of the resistor to be easily adjusted. A circular end segment can be split into two arcs to allow the insertion of an additional resistor segment, but this often requires redrawing the resistor.

Current does not flow uniformly around the bends in a serpentine resistor. Each square corner adds approximately 0.56 squares.[7] Neglecting process biases and end effects, the value of the resistor of Figure 5.3A is

$$R = R_s\left(\frac{2A + B}{W} + 1.12\right) \qquad [5.6]$$

The contribution of a corner square is usually rounded to $\frac{1}{2}$ square—thus the often-quoted rule "a corner counts half a square." The slight errors implicit in this assumption rarely have any practical significance.

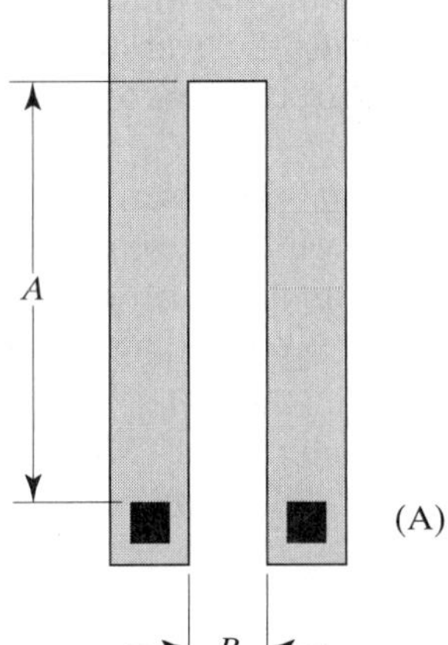

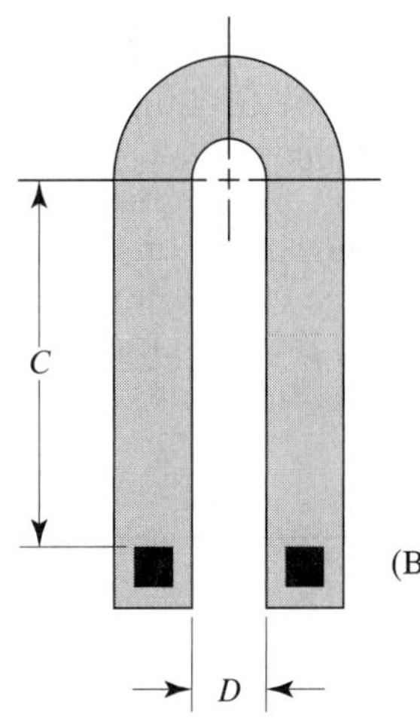

FIGURE 5.3 Layout of serpentine resistors with (A) rectangular turns and (B) circular turns. In the case of circular turns, the spacing, *D*, is assumed to equal the width, *W*, of the resistor.

[7] Glaser, *et al.*, p. 118. Grebene cites a value of 0.53: Grebene, p. 140. Reinhard cites one of 0.65: D. K. Reinhard, *Introduction to Integrated Circuit Engineering* (Boston: Houghton Mifflin, 1987), p. 191.

The 180° circular end segment of Figure 5.3B adds 2.96 squares to the resistor.[8] Neglecting process biases and end effects, the resistance of this structure equals

$$R = R_s\left(\frac{2C}{W} + 2.96\right) \qquad [5.7]$$

The contribution of a circular end is usually rounded to 3 squares.

Sometimes a resistor becomes so narrow that contacts cannot reside inside it without violating design rules. This problem is usually overcome by enlarging the ends of the resistor to form heads around the contacts. The resulting structure is called a *dogbone* or *dumbbell resistor* because of its characteristic shape (Figure 5.4). The *drawn length* L_d of a dogbone resistor is measured from contact to contact, and the *drawn width* W_d is measured across the body of the resistor. The approximate value of the resistor can then be computed using Equation 5.4. The effects of bends in the resistor can be handled in the same manner as for the strip resistor.

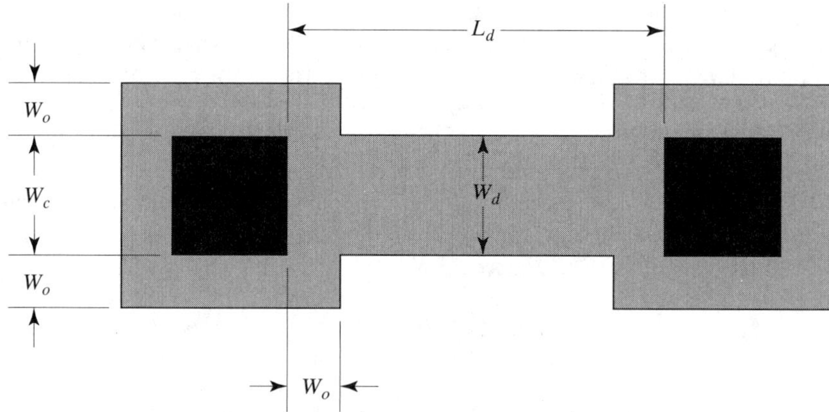

FIGURE 5.4 The dimensions of a dogbone resistor required for computing its value.

The effects of lateral nonuniform current flow differ for dogbone and strip resistors. In a strip resistor, such as that in Figure 5.2, the current crowds inward toward the contact, and the effective value of the resistor increases. In a dogbone resistor, the current spreads out as it enters the head, and the effective value of the resistor decreases. This effect can be minimized by making the width of the contact W_c equal to the width of the resistor W_d and by minimizing the overlap W_o of the resistor head over the contact. Table 5.3 lists published correction factors ΔR for two dogbone heads (one at either end of the resistor).

The correction factor ΔR is usually less than 0.3 squares because most strip resistors use an overlap of less than half the drawn width. The contact resistance (Section 5.3.4) incorporates the effects of nonuniform current flow in the vertical plane. These corrections are negligible if the resistor is more than about five squares long.

TABLE 5.3 Correction factors ΔR for dogbone resistors.[9]

W_o	W_c	ΔR
W_d	W_d	−0.7□
$\frac{1}{2}W_d$	W_d	−0.3□

[8] Glaser, *et al.*, p. 118. Grebene's value appears to be in error: Grebene, p. 140.

[9] Reinhard, p. 192.

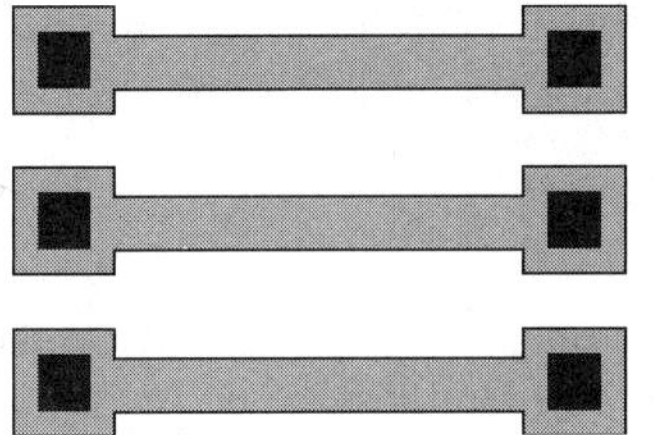
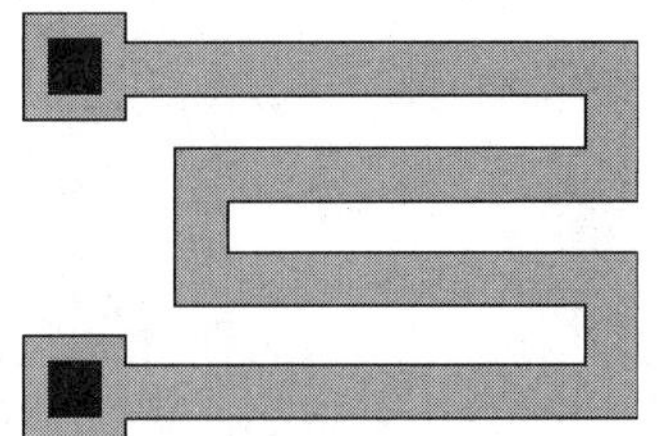

FIGURE 5.5 Sample dogbone and serpentine resistors, showing the poorer packing density caused by the presence of enlarged heads.

Dogbone resistors do not pack as densely as strip or serpentine resistors, although proper arrangement of the segments can minimize this problem (Figure 5.5). Many designers consider dogbone resistors to be more accurate than strip or serpentine resistors of the same width. True, the errors caused by nonuniform current flow are smaller for dogbones than for either strip or serpentine resistors, but this does not actually improve the accuracy of the resistor. Matched resistors should always be laid out in identical sections, and as long as each section has the same layout, it does not matter whether it uses dogbone heads or not.

5.3 RESISTOR VARIABILITY

The value of a resistor depends on numerous factors, including process variability, temperature, nonlinearity, and contact resistance. Other, less-significant factors that primarily affect resistor matching include orientation, stress and temperature gradients, thermoelectric effects, nonuniform etch rates, NBL push, voltage modulation, charge spreading, hydrogen compensation, and PSG polarization. These less-significant factors will be discussed in Chapter 7.

5.3.1. Process Variation

The value of a resistor depends upon its sheet resistance and its dimensions. Sheet resistances vary because of fluctuations in film thickness, doping concentration, doping profiles, and annealing conditions. The dimensions of a resistor vary because of photolithographic inaccuracies and nonuniform etch rates.

Modern processes maintain sheet resistances within ±20% or ±25%,[10] except for pinch resistors. Base pinch resistors are formed from the region corresponding to the neutral base of an NPN transistor. Base pinch sheet resistance strongly correlates with NPN beta because both are functions of the doping and thickness of the neutral base region. A high NPN beta indicates a high base pinch sheet and *vice versa.* Most processes specify a 3:1 variation (±50%) in beta, and base pinch resistors vary by a similar amount.

Linewidth control is a measure of dimensional variation introduced by photolithography and processing. It depends only weakly on width for feature sizes of a micron or more. In other words, if a 5 μm feature can be held to a tolerance of ±1 μm, then so can a 25 μm feature. Consequently, linewidth control measured as a percentage of feature size improves with increasing feature size.[11] Most processes can maintain linewidth control to within about ±20% of their minimum feature size. For example, a standard bipolar process with a minimum feature size of 5 μm will probably have a linewidth control of about ±1 μm.

---

[10] The variability figures used in this section represent the 4.5-sigma limits of a Gaussian distribution.

[11] This rule is based upon the reluctance of manufacturers to invest in equipment which exceeds the requirements that are absolutely necessary to fabricate a process. It breaks down in cases where the cheapest available equipment exceeds the minimum requirements of the process. This situation is becoming quite common for higher-voltage processes.

If the sheet resistance variation and the linewidth control are known, then the actual tolerance for a resistor of effective width W_e can be computed using the equation[12]

$$\delta R + \frac{C_L}{W_e} + \delta R_s \qquad [5.8]$$

where δR is the tolerance of the resistor, C_L is the linewidth control of the applicable layer, and δR_s is the variability of the sheet resistance. The equation assumes that the resistor is long enough to ignore length variations. Suppose that a resistor with an effective width of 2 μm is drawn on a layer with a linewidth control of ±0.25 μm, using a material with a sheet resistance variation of ±25%. Equation 5.8 predicts that the resulting resistor will vary by ±38%. If the effective width were increased to 10 μm, then the variability would be reduced to ±28%.

This information can be summarized as a set of design guidelines:

- Where tolerance does not matter, use minimum-width resistors and expect variations of about ±50%. Diffused resistors should not be made narrower than about 150% of their junction depth, or the amount of dopant may be inadequate to achieve the targeted depth.
- Where moderately precise tolerances are required, use resistors two or three times as wide as the minimum feature size and expect variations of ±35%.
- Where highly precise tolerances are needed, use resistors approximately five times as wide as the minimum feature size and expect variations of ±30%.

These rules assume ±25% sheet variation and a linewidth control of ±20% of the minimum feature size. Base pinch resistors are exceptions to the above rules since their values depend primarily on their sheet variation, and increased widths offer little benefit. The width of a base pinch resistor should equal at least 150% of minimum to ensure that enough dopant diffuses down to form a pinched region. Resistor matching follows somewhat different rules than resistor tolerance (see Section 7.3.)

These rules also require modification when dealing with deep diffusions such as N-well. The precision of deeply diffused resistors depends as much on outdiffusion as it does on linewidth control. In this case, the width required for a given accuracy should be computed using the junction depth or the minimum feature size, whichever is larger. Thus, an 8 μm-deep N-well resistor should be made at least 16 μm wide to obtain moderate accuracy.

5.3.2. Temperature Variation

Resistivity depends on temperature in a complex, nonlinear manner. Like any nonlinear function, this one can be expanded into a polynomial series. Unless considerable accuracy is required or the temperature varies widely, only the first two terms of the series are significant:

$$R(T) = R(T_0)[1 + 10^{-6}\,TC(T - T_0)] \qquad [5.9]$$

Here, $R(T)$ is the resistance at the desired temperature T, $R(T_0)$ is the resistance at some other temperature T_0, and TC is the *temperature coefficient of resistivity* (TCR) in parts per million per degree Celsius (ppm/°C). A factor of 10^{-6} has been inserted into Equation 5.9 to balance the units involved. Table 5.4 lists typical linear temperature coefficients for several common integrated resistance materials. Most

---

[12] Much of the variability in resistor sheets and linewidth control is due to unpublicized process biases; these add linearly rather than in quadrature. Glaser, *et al.* correctly use a linear sum but do not offer justification: Glaser, *et al.*, p. 121. C_L and δR_S are usually quoted as 4.5-sigma values, in which case δR is also a 4.5-sigma value.

Material	TCR, ppm/°C
Aluminum, bulk	+3800
Copper, bulk	+4000
Gold, bulk	+3700
160 Ω/□ Base diffusion	+1500
7 Ω/□ Emitter diffusion	+600
5 kΩ/□ Base pinch diffusion	+2500
2 kΩ/□ HSR implant (P-type)	+3000
500 Ω/□ Polysilicon (4 kÅ N-type)	−1000
25 Ω/□ Polysilicon (4 kÅ N-type)	+1000
10 kΩ/□ N-well	+6000

TABLE 5.4 Typical temperature coefficients of resistivity for selected materials at 25°C.[13]

of these values are reasonably accurate between 0 and 50°C. The temperature coefficients of polysilicon are exceptions; they depend on annealing conditions and can vary significantly from the values listed in Table 5.4.

In cases where greater accuracy is required, the first three terms of the polynomial series can be retained, giving the equation

$$R(T) = R_0(T)[1 + 10^{-6}TC_1(T - T_o) + 10^{-6}TC_2(T - T_0)^2] \quad [5.10]$$

where TC_1 is the linear temperature coefficient of resistivity, in ppm/°C, and TC_2 is the quadratic temperature coefficient of resistivity, in ppm/°C^2. The quadratic temperature coefficient is typically much smaller than the linear temperature coefficient, but it can still have a dramatic effect upon resistance across large variations in temperature.

The temperature coefficients of resistivity are usually computed by statistical curvefitting. This process optimizes the values of the coefficients for a specific equation and a specific range of temperatures. Trouble may occur if the coefficients are used in the wrong equation, or if the applicable temperature range is exceeded. Particular care must be taken when using Equation 5.10 if $TC_2 > 0.001 \cdot TC_1$, as the resistance computed by the equation will rapidly become inaccurate beyond the temperature limits for which the coefficients were computed.

Matched resistors must consist of the same material to ensure that temperature variations do not upset their matching. In addition, the TCRs of different diffusions are sometimes used by circuit designers to temperature-compensate circuits. For these reasons, different resistance materials cannot be arbitrarily substituted for one another.

5.3.3. Nonlinearity

Ideal resistors exhibit a linear relationship between voltage and current. Practical resistors always exhibit some degree of nonlinearity; in other words, their resistance varies with applied voltage. Nonlinearity, or *voltage modulation,* arises from several sources, including self-heating, high-field velocity saturation, and depletion region encroachment.

---

[13] Temperature coefficients depend strongly on the conditions of fabrication and measurement, so the values given here are only approximations. Values for Al, Cu, and Au determined by linear interpolation of data from G. W. C. Kay and T. H. Laby, *Tables of Physical and Chemical Constants,* 15th ed. (Essex, England: Longman Scientific and Technical, 1986), pp. 117–118. Base, emitter, pinch, and HSR diffusions: Gray, *et al.,* p. 139. Base and HSR diffusions: Grebene, pp. 138, 153. Poly: W. A. Lane and G. T. Wrixon, "The Design of Thin-Film Polysilicon Resistors for Analog IC Applications," *IEEE Trans. on Electron Devices,* Vol. 36, No. 4, 1989, pp. 738–744. See also discussion and curves for various diffusions in Hamilton, *et al.,* pp. 277–279; and P. Norton and J. Brandt, "Temperature Coefficient of Resistance for p- and n-type Silicon," *Solid State Electronics,* Vol. 21, 1978, pp. 969–974.

A resistor dissipates power equal to the product of the voltage across it and the current through it. This dissipation causes internal heating. Plastic packages conduct heat rather poorly, so even small amounts of internal heating can cause substantial temperature increases inside the package. Most resistors have relatively large temperature coefficients, and even modest increases in temperature can cause significant resistance variations. Suppose 10 mA flows through a 1 kΩ HSR resistor, which consequently dissipates 100 mW. Assuming a thermal impedance of 80°C/W and a temperature coefficient of 3000 ppm/°C, this dissipation results in an 8°C temperature rise and a 2.4% increase in resistance.

Poly resistors are especially vulnerable to self-heating because they reside on top of the thick-field oxide, which acts as a thermal insulator between the poly resistors and the silicon substrate. Very little heat can escape upward through the protective overcoat and encapsulation, so the temperature rise ΔT between the resistor and the silicon substrate, in degrees Celsius, equals

$$\Delta T = 71\frac{V^2 t_{ox}}{R_s L} \qquad [5.11]$$

where R_s is the sheet resistance of the poly in Ω/□, t_{ox} is the thickness of the field oxide in Angstroms (Å), L is the length of the resistor in microns, and V is the voltage applied across the resistor. Equation 5.11 can also be applied to deposited thin-film resistors. The temperature-induced nonlinearity equals the product of the temperature rise and the TCR of the resistor. This nonlinearity can usually be neglected for resistors that experience less than 1°C of self-heating.

The rate of carrier drift in weak electric fields is proportional to the electric field intensity. As the field increases, the drift velocity of the carriers eventually becomes diffusion-limited, and the resistance begins to increase. The critical electric field intensity where this nonlinearity begins equals approximately 0.2 V/μm for electrons and 0.6 V/μm for holes.[14] The electric field should be kept well below these critical intensities to minimize nonlinearity. Assuming a safety factor of two, the minimum resistor length L_{min} equals

$$L_{min} = (6.7\ \mu m/V) \cdot V_{max} \quad \text{for N-type silicon,} \qquad [5.12A]$$

$$L_{min} = (3.3\ \mu m/V) \cdot V_{max} \quad \text{for P-type silicon,} \qquad [5.12B]$$

where V_{max} is the maximum voltage applied across the resistor. Resistors shorter than L_{min} can be constructed, but their resistance will increase at high voltages.

Poly resistors may also exhibit nonlinearities if they are so short that appreciable voltage drops can appear across individual poly grains. Under these conditions, the resistance of the barrier regions between grains becomes a function of the voltage drop across the resistor. These nonlinearities can be neglected if the resistor length is at least a thousand times the diameter of an individual grain.[15] Since most polysilicon films have grain sizes of approximately 0.5 to 1 μm, this means that precision poly resistors should be made at least 50 to 100 μm long to avoid nonlinearity effects. The length of poly resistors should also satisfy Equations 5.12A and 5.12B.

Additional voltage nonlinearities occur in lightly doped diffused resistors, especially base pinch resistors, due to modulation of the depletion regions around

---

[14] Muller, *et al.* give curves showing high-field mobility saturation from which these values can be obtained by examination: R. S. Muller and T. I. Kamins, *Device Electronics for Integrated Circuits,* 2d ed. (New York: John Wiley and Sons, 1986), p. 36.

[15] Lane, *et al.,* p. 741.

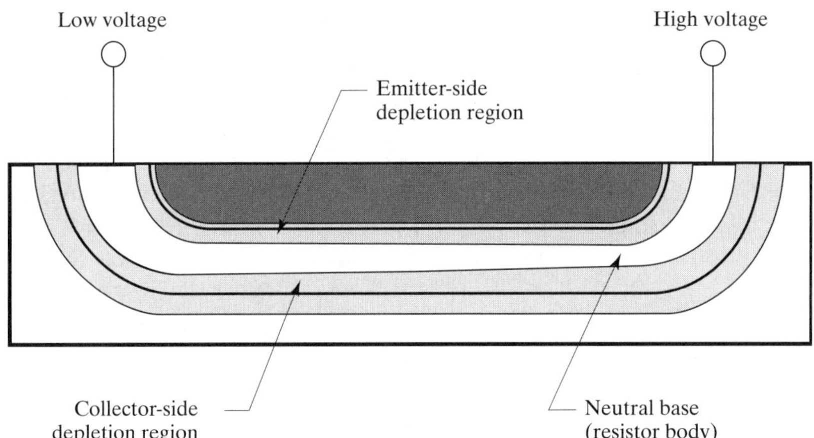

FIGURE 5.6 Cross section of a base pinch resistor showing the intrusion of the depletion regions into the neutral base. Notice that the high-voltage end of the resistor narrows slightly.

reverse-biased junctions. Figure 5.6 shows a cross section of a base pinch resistor. The depletion regions widen toward the high-voltage end of the resistor because the reverse biases are largest there. In other words, the resistor is pinched more severely at the high-voltage end than at the low-voltage end. This pinching effect becomes more pronounced as the voltage across the resistor increases. Voltage nonlinearities of up to 1%/V have been reported for HSR,[16] and similar (or larger) values can be expected for base pinch resistors. This form of voltage nonlinearity is often modeled by treating the resistor as a JFET with a large pinchoff voltage.

Heavily doped resistors do not experience this pinching effect because the depletion regions extend only a minute distance into them. Emitter resistors, for example, show practically no voltage modulation.

Depletion regions also cause an increase in resistance when significant tank bias is applied. As the voltage difference between the resistor and the tank increases, the depletion regions widen and the resistance increases. This effect, called *tank modulation,* is analogous to backgate modulation in a FET. A 160 Ω/□ base resistor experiences a tank modulation of about 0.1%/V.[17] If the voltage between the base resistor and its tank were to increase by 10V, then the resistance would increase by 1%. The tank modulation for a 2 kΩ/□ high-sheet resistor is about 1%/V, and base pinch resistors have tank modulations of at least 1%/V. Matched resistors either require carefully controlled tank biasing or relatively low-sheet materials.

Another form of voltage modulation occurs when leads cross a lightly doped resistor. The electric fields generated by the leads cause carriers to redistribute in the body of the resistor in much the same way that the field generated by an MOS gate redistributes carriers in the backgate. This *conductivity modulation* can cause several percent variation in the value of a 2 kΩ/□ HSR resistor. HSR resistors are especially susceptible to this effect because they are both very thin and very lightly doped. Accurate HSR resistors should be field plated to minimize variations caused by conductivity modulation. Split field plates (Section 7.2.12) are recommended because they reduce nonlinearities caused by conductivity modulation from the field plates themselves. Poly resistors show less conductivity modulation than diffused resistors because they are usually much more heavily doped. Poly resistors with sheet resistances of 1 kΩ/□ or less generally do not suffer from conductivity modulation.

---

16 W. Bucksch, "Quality and Reliability in Linear Bipolar Design," *TI Tech. J.,* Vol. 4, #6, 1987, pp. 61–69. See also Grebene, p. 153.

17 Value for 200 Ω/□ base: Hamilton, *et al.,* p. 155.

5.3.4. Contact Resistance

Each resistor contains at least two contacts, and each of these adds some resistance to the overall structure. This resistance results from the presence of a potential barrier between the resistance material and the metallization. Although carriers can tunnel through this barrier, they lose some energy in the process. This energy loss varies with current and is therefore best described as a specific resistance of the contact interface, measured in $\Omega \cdot \mu m^2$. This *contact resistance* depends on the nature of the materials in contact and on processing conditions. Contact resistance can vary greatly from lot to lot, and designers should assume that it varies from essentially zero to the specified maximum unless the design rules state otherwise. The resistance R_c added by a single contact having width W_c and length L_c equals[18]

$$R_c = \frac{\sqrt{R_s \rho_c}}{W_c} \coth\left(L_c \sqrt{R_c / \rho_c}\right) \qquad [5.13]$$

where R_s is the sheet resistance of the resistor material, ρ_c is the specific contact resistance, and coth() represents the hyperbolic cotangent function. Equation 5.13 also accounts for current crowding in the vertical dimension (Section 5.2). Table 5.5 lists typical observed specific contact resistances for a number of contact systems.

TABLE 5.5 Typical contact resistances for various contact systems.[19]

Contact System	Contact Resistance, $\Omega \cdot \mu m^2$
Al-Cu-Si to 160 Ω/□ base	750
Al-Cu-Si to 5 Ω/□ emitter	40
Al-Cu/Ti-W/PtSi to 160 Ω/□ base	1250
Al-Cu/Al-Cu (Via)	5
Al-Cu/Ti-W/Al-Cu (Via)	5

Because contact resistances are strongly process-dependent, these values are only indicative of what any given fab can actually guarantee.

The aluminum-copper-silicon (Al-Cu-Si) metal system employed by older processes exhibits significant contact resistance, especially for lightly doped materials such as 160 Ω/□ base diffusion. The use of refractory barrier metal significantly increases contact resistance variability due to sintering problems. The addition of a silicide layer beneath the barrier metal (Al-Cu/Ti-W/PtSi) eliminates this problem and gives contact resistances only slightly larger than those of aluminum-copper-silicon. Modern CMOS and BiCMOS processes use silicided contacts in combination with heavily doped PSD and NSD diffusions to achieve contact resistances similar to those of emitter ($40\ \Omega \cdot \mu m^2$ or better).

Consider a 1 kΩ base resistor constructed with 8 × 8 μm contacts in a standard bipolar process using an Al-Cu-Si metal system. Equation 5.13 indicates that each contact adds 43 Ω. The resistor therefore gains 86 Ω, or about 9% of its value. Standard bipolar base resistors that total less than 10 squares per segment may require oversized contacts to avoid excessive variability due to contact resistance. In such cases, it is usually better to place several longer resistor segments in parallel rather than to use dogbone-style resistor heads to accomodate enlarged contacts.

---

[18] H. Murrmann and D. Widmann, "Current Crowding on Metal Contacts to Planar Devices," *IEEE Trans. on Electron Devices* ED-16, #12, 1969, pp. 1022–1024.

[19] Murrmann, *et al.* give a value of 650 $\Omega \cdot \mu m^2$ for 150 Ω/□ base; D'Andrea, *et al.* give a value of 1000 $\Omega \cdot \mu m^2$ for 180 Ω/□ base and 900 $\Omega \cdot \mu m^2$ for 140 Ω/□ base: G. D'Andrea and H. Murrmann, "Correction Terms for Contacts to Diffused Resistors," *IEEE Trans. on Electron Devices*, ED-17, 1970, pp. 484–485. Emitter: Murrmann, *et al.* Schottky contact value: by extrapolation from base value using data in Bucksch. Via values are the author's estimate. For a theoretical analysis, see Runyan, *et al.*, pp. 522ff.

5.4 RESISTOR PARASITICS

No practical resistor is ever completely isolated from its environment. Capacitive and inductive coupling inevitably occur at higher frequencies, and some types of resistors also experience junction leakage. Circuit designers can model these interactions by replacing each integrated resistor with a subcircuit containing several ideal components. One of these components is an ideal resistor, while the remainder are *parasitic components* that represent the undesired but unavoidable interactions of the resistor with the rest of the die. These *subcircuit models* are also of interest to layout designers because they illustrate the limitations of various types of resistors.

Figure 5.7 shows the cross section of a typical polysilicon resistor layout. The resistor is surrounded on all sides by oxide, an excellent insulator that exhibits virtually no leakage. The oxide also acts as a capacitive dielectric that couples the resistor to adjoining components. Most poly resistors are laid out on field oxide having a capacitance of about 0.05 fF/μm^2 (Section 6.1). Ignoring fringing effects, a resistor that is 5 μm wide and contains 100 squares has a total substrate capacitance of about 125 fF. This capacitance is distributed uniformly along the resistor, so it cannot be accurately modeled by a single capacitor. This *distributed capacitance* can be approximated using the π-section circuit of Figure 5.8A in which C_1 and C_2 are ideal capacitors, each representing half of the distributed capacitance. If a single π-section does not adequately model the distributed capacitance, then multiple π-sections can be used. Figure 5.8B shows a model incorporating two π-sections: R_1 and R_2 are ideal resistors with each being equal to half of the total resistance, and C_1, C_2, and C_3 are ideal capacitors; C_1 and C_3 each equal one-fourth of the distributed capacitance, and C_2 equals one-half of it.[20]

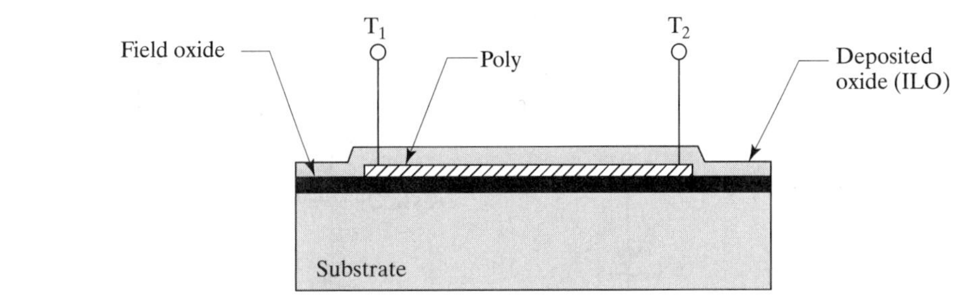

FIGURE 5.7 Cross section of a polysilicon resistor.

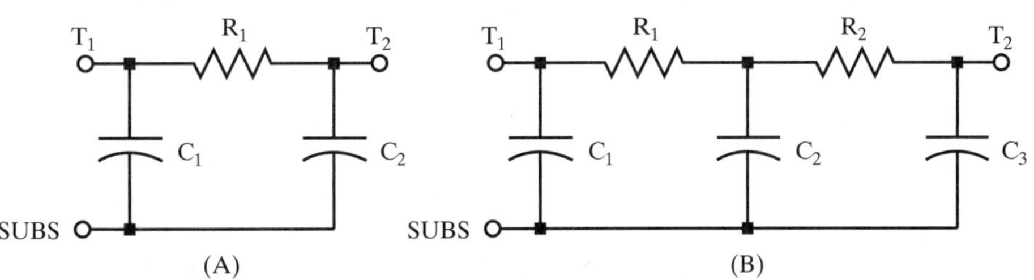

FIGURE 5.8 Subcircuit models for polysilicon resistors that approximate distributed substrate capacitance, using π-sections: (A) single π-section and (B) dual π-section.

Leads routed across a polysilicon resistor introduce additional parasitic capacitances. The capacitance of the interlevel oxide (ILO) typically equals that of the field oxide, or about 0.5 fF/μm^2. Thus, a 3 μm lead crossing a 5 μm-wide resistor at right angles typically produces about 7.5 fF of coupling capacitance. This is an extremely small capacitance, but even so it can couple noise into high-impedance

[20] Y. Tsividis, *Mixed Analog-Digital VLSI Devices and Technology* (New York: McGraw-Hill, 1996), p. 166.

circuitry. Noisy signals should not be routed over polysilicon resistors in delicate analog circuits such as voltage references or low-noise amplifiers. Section 7.2.8 gives additional reasons to avoid routing metal over matched resistors.

Figure 5.9 shows a simplified cross section of a diffused resistor. One or more reverse-biased junctions isolate this resistor from the remainder of the die. Tank contacts are required to maintain the necessary reverse bias across these junctions.

The parasitics of this diffused resistor consist chiefly of reverse-biased junctions: one between the resistor and the tank, and a second between the tank and the substrate. These junctions form distributed structures that are frequently modeled using π-sections. Figure 5.10 shows two single π-section subcircuit models for the resistor of Figure 5.9.[21]

The subcircuit of Figure 5.10A includes an ideal resistor and three diodes. D_1 and D_2 each model half of the total area of the resistor-tank junction, while D_3 models the full area of the tank-substrate junction. This subcircuit remains reasonably accurate as long as the tank resistance is relatively small compared with the resistance of R_1. In the case of a more resistive tank (for example, a PSD resistor in N-well), the subcircuit of Figure 5.10B is preferable. This model includes a resistor, R_2, that models the effects of tank resistance, and diodes, D_3 and D_4, that model the distributed nature of the tank-substrate junction. Both of these subcircuits can incorporate additional π-sections to enhance their accuracy at higher frequencies.

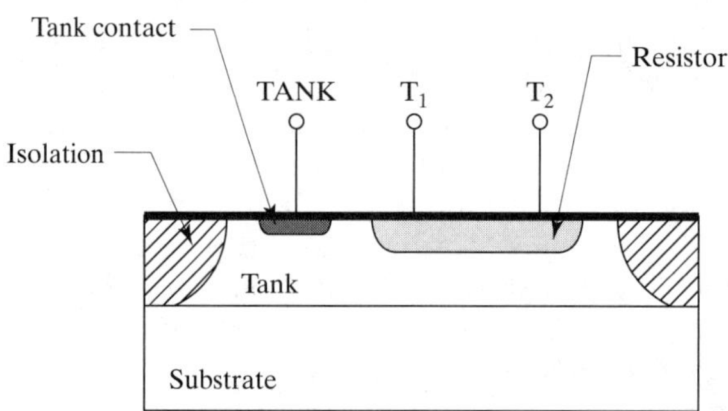

FIGURE 5.9 Cross section of a typical diffused resistor.

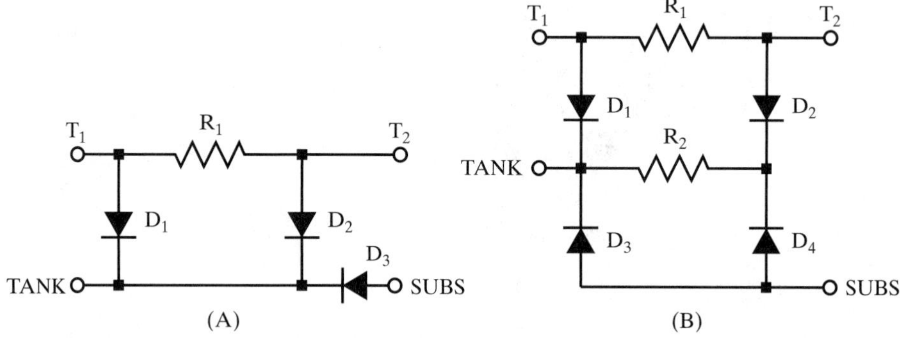

FIGURE 5.10 Subcircuit models for diffused resistors: (A) neglecting tank resistance and (B) including tank resistance.

The reverse-biased diodes associated with a diffused resistor cause several undesirable effects. If the resistor-tank junction ever becomes forward biased, it will inject minority carriers into the tank. This can trigger latchup (Section 4.4.2). Even if latchup does not occur, large currents may flow through the tank contact. The proper biasing of resistor tanks requires considerable thought. The three tank biasing

[21] For a similar discussion see Hamilton, *et al.,* pp. 160–182.

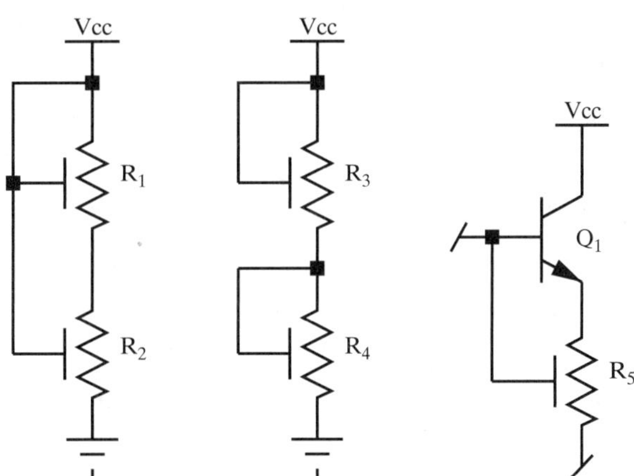

FIGURE 5.11 Several methods of biasing the tank connections of P-type diffused resistors.

schemes shown in Figure 5.11 illustrate several common arrangements for connecting the tanks of P-type resistors.

The resistor divider formed by R_1 and R_2 is biased as simply as possible: Both tanks connect to a power supply that is more positive than either resistor. This reverse-biases the tank-substrate junctions, but each resistor sees a different relative tank bias and experiences a different amount of tank modulation. The resulting resistance variations cause the divider ratio to vary with supply voltage. The resistor divider formed by R_3 and R_4 illustrates a tank connection that is better suited for matched resistor dividers. Each tank connects to the positive end of its respective resistor. If the two resistors are equal in value, then the relative tank biases seen by each are also equal, and they will experience the same tank modulation. This technique can be extended to encompass resistors of unequal values by dividing them into multiple sections occupying separate tanks.

R_5 illustrates yet another method of biasing a resistor tank. Assuming that transistor Q_1 always conducts, the tank of R_5 will be biased one base-emitter drop above the positive terminal of the resistor. This connection illustrates one of the many possible configurations in which the tank connects neither to the resistor nor to a supply. Like all too many of these configurations, this one has the potential to forward-bias the resistor-tank junction. Suppose that the transistor is conducting and that R_5 is biased several volts above ground. If the base of Q_1 is suddenly pulled low, the resistor-tank junction will momentarily forward-bias. Connections that do not tie back to the positive end of the resistor nor to a supply should be carefully analyzed for sneak conduction paths similar to this one.

The reverse-biased junctions that isolate a diffused resistor can also suffer avalanche breakdown. This is of particular concern for emitter resistors and base pinch resistors since these usually avalanche at about 7 V. Such resistors should not operate beyond about 2/3 of the applicable breakdown voltage, or about 4.5 V. If the bias across a resistor may exceed this voltage, then it should be constructed in multiple segments placed in separate tanks. The depletion regions associated with reverse-biased junctions also have considerable capacitance. This capacitance depends upon junction doping and reverse bias and is typically 1 to 5 fF/μm^2. This value is substantially larger than the 0.5 fF/μm^2 typically associated with poly resistors, so the high-frequency performance of diffused resistors cannot match that of poly resistors.

Most designers favor poly resistors because the absence of junctions makes parasitics less of a concern. However, the sheet resistance and temperature coefficient of polysilicon are not always as desirable as those of a particular type of diffused resistor. For example, N-well resistors are often more compact than heavily doped poly resistors. Thus, diffused resistors still see occasional use even in processes that offer polysilicon resistors.

5.5 COMPARISON OF AVAILABLE RESISTORS

Most processes offer several types of resistors optimized for different applications. This section compares the various types of resistors presented in Chapter 3, and also presents several additional resistors that are useful for special applications.

5.5.1. Base Resistors

Base resistors are available in standard bipolar (Figure 3.13) and also in analog BiCMOS (Figure 3.52). Base sheets in standard bipolar typically range from 100 to 200 Ω/□, allowing Ohmic contact to be made directly to the resistor. Processes that do not silicide the contacts often have rather high contact resistances. The base sheet in BiCMOS and advanced bipolar processes is usually somewhat higher than standard bipolar (300 to 600 Ω/□), so reliable Ohmic contact requires the addition of a more heavily doped diffusion beneath the contacts. The resistance of the resulting composite structure can be computed using formulas similar to those used for HSR resistors (Section 5.5.4).

The base diffusion is best suited for laying out resistors in the range of 50 Ω to 10 kΩ. Larger resistors are usually constructed from HSR, smaller ones from emitter. Base sheet control is relatively precise, and base resistors are doped heavily enough to minimize tank modulation. These considerations often outweigh the area savings possible with HSR resistors, especially for precisely matched ones.

Field plates should be employed for high-voltage resistors to prevent charge spreading (Section 4.3.5). Leads can be routed across base resistors without causing significant conductivity modulation. Care should still be taken in routing noisy signals across base resistors because the oxide over the resistor is thinner than the field oxide, and capacitive coupling may inject noise into the circuitry attached to the resistor.

Base resistors must be placed in a suitable tank, consisting of either an N-epi region in standard bipolar or an N-well in analog BiCMOS. This tank should contain as much NBL as possible to reduce tank resistance. NBL can also help minimize noise coupling between resistors occupying a common tank by providing a low-impedance path from the tank to a clean supply node. NBL not only helps minimize debiasing, but it also acts as a barrier to minority carriers and consequently enhances latchup immunity. In the absence of NBL, the base diffusion also pushes somewhat deeper and therefore becomes more resistive. The NBL geometry should overlap the resistors sufficiently to ensure that all portions of the resistor experience the same amount of NBL push. An NBL overlap of 5 to 8 μm will usually suffice.

Base resistors are often merged with other devices in a common tank. In order to prevent possible latchup problems, these devices should not inject minority carriers into the shared tank. For example, a resistor placed in the same tank as a lateral PNP can collect minority carriers emitted by the transistor if it saturates. Resistors can safely be merged with other resistors and with NPN transistors that do not saturate. If NPN transistors are included in a merged tank, a plug of deep-N+ should be used to contact the NBL to prevent tank debiasing and noise coupling (Section 13.1). If only resistors are present in a tank, then the deep-N+ plug can be omitted to save space.

Base resistors are probably the best general-purpose resistors available in standard bipolar. When offered, poly resistors are usually preferred over base resistors because they have smaller parasitic capacitances and because they have no tank junctions that might forward-bias.

5.5.2. Emitter Resistors

Emitter resistors are available in standard bipolar (Figure 3.14) and in some varieties of analog BiCMOS. Their sheet resistances typically range from 2 to 10 Ω/□, so there is no difficulty in directly contacting the emitter diffusion. Because of its relatively low sheet, emitter is only suitable for relatively small resistors (0.5 to 100 Ω). Larger resistors are usually constructed from base or HSR. Voltage modulation and conductivity modulation are both negligible in emitter resistors.

Capacitive coupling between an emitter resistor and an overlying lead can become significant. Processes that employ thin emitter oxides may exhibit oxide capacitances of up to 0.5 fF/μm^2, but processes that employ thick emitter oxide will have only a small fraction of this capacitance. Thin emitter oxides are also vulnerable to rupture during ESD events. Leads connecting to outside pins should not route across thin-oxide emitter diffusions unless they connect to them (Section 4.1.1). Thick emitter oxides are less vulnerable to rupture, so leads connecting to outside pins can safely cross them.

Emitter resistors must reside in a suitable tank. The usual arrangement consists of an emitter resistor enclosed within a base diffusion, which is in turn enclosed in a tank or an N-well region (Figure 3.14). When this configuration is employed, the base diffusion must connect to an equal or lower voltage than the emitter resistor. In turn, the tank enclosing the base diffusion must connect to a voltage equal to or higher than the base. This can be achieved by connecting the base diffusion to the low-voltage end of the resistor and connecting the tank to the high-voltage end. The emitter resistor must not be biased more than about two-thirds of the avalanche voltage, or about 4 V in a standard bipolar process. This limitation can be circumvented by dividing the resistor into several sections, each contained in its own base region.

Emitter resistors need not be enclosed in a base region to isolate them from the surrounding tank because the N-epi sheet resistance is much larger than the emitter sheet resistance. Even though the emitter resistor may electrically connect to the tank, the difference in resistivities ensures that most of the current flows through the resistor and not through the tank. Considerable space can be saved by eliminating the base diffusion (Figure 5.12). This layout is particularly suited for emitter resistors employed as tunnels. A *tunnel,* or *crossunder,* is a low-value resistor used to jump leads on a die with only a single level of metal (Section 13.3.3). Tunnels should consume minimal area and achieve a very low resistance, something that the layout of Figure 5.12 does admirably.

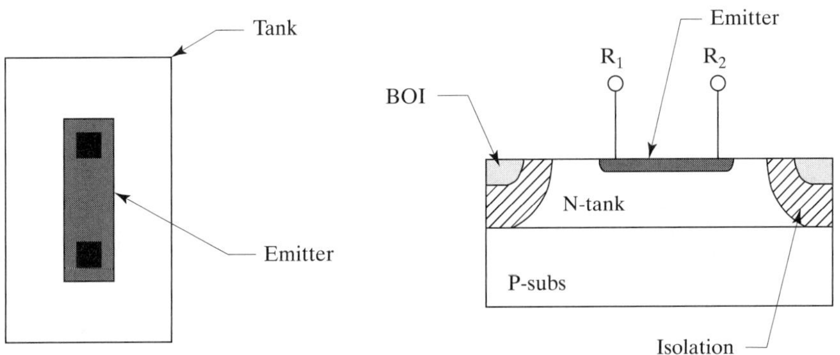

FIGURE 5.12 Layout and cross section of an alternative style of emitter resistor that eliminates the enclosing base diffusion to save spacc. (Compare with Figure 3.14.)

In analog BiCMOS, emitter resistors can be placed directly into the P-epi. This tankless style of emitter resistor can also be constructed in standard bipolar by coding base entirely across the tank that contains an emitter resistor. Unfortunately, this structure has a breakdown voltage of about 7 V that decreases to about 5 V if the tank is omitted. Emitter resistors formed directly into isolation also tend to leak. An emitter resistor can safely reside in isolation when it serves as a tunnel for a lead operating at substrate potential (Section 13.3.3). Leakage is obviously of no concern in this case, but care must still be taken to ensure that debiasing does not interfere with proper circuit operation.

Emitter resistors are frequently employed in standard bipolar to ballast power transistors and to serve as current sense resistors. They are also used extensively as tunnels in single-level-metal processes. Emitter resistors are uncommon in BiCMOS layouts since low-sheet polysilicon resistors are more compact and less susceptible to parasitics.

5.5.3. Base Pinch Resistors

Base pinch resistors can be constructed in standard bipolar (Figure 3.15) and in analog BiCMOS. Their effective sheet resistance is typically 2 to 10 kΩ/□, allowing very compact layouts. Pinch resistors are typically used to construct high-value resistors that cannot be economically implemented using other diffusions. The sheet resistance of base pinch resistors is poorly controlled, and a process variation of ±50% is typical. NBL should always be placed beneath a base pinch resistor because NBL push helps increase its sheet resistance.

The tank modulation of base pinch resistors is on the order of 1%/V or greater. Base pinch resistors can only be matched if they are of identical dimensions and if they see the same relative tank biasing. The biasing scheme of resistors R_3 and R_4 in Figure 5.11 must be used instead of the scheme of resistors R_1 and R_2. Failure to ensure equal tank biases may cause mismatches of up to ±20%. Even with perfectly matched layouts, errors of ±5% may result from the inherent lack of base pinch sheet control.[22] Superior matching can be achieved using even the narrowest high-sheet resistors, so if this implant is available, consider using it to replace base pinch resistors.

Like emitter resistors, base pinch resistors are limited to about 4 V by the breakdown voltage of the base-emitter junction. Pinch resistors can withstand larger differential voltages if they are constructed in multiple segments. Since the emitter pinch plate connects to the tank, each segment of the resistor must occupy a separate tank. This is very wasteful of area, and constitutes yet another reason for considering alternative resistors such as epi-pinch resistors or narrow HSR resistors.

In summary, the pinch resistor is a marginal device used primarily to construct compact high-value resistors. These see frequent use in noncritical roles, as (for example) base turn-off resistors for transistors. Occasionally, designers will employ base pinch resistors to compensate for beta variation in NPN transistors because base pinch sheet resistance correlates with NPN beta. For most other applications, HSR resistors are superior to pinch resistors.

5.5.4. High-Sheet Resistors

High-sheet-resistance (HSR) implants are available as extensions for most standard bipolar processes. These implants exhibit sheet resistances of 1 to 10 kΩ/□, depending upon implant dose, junction depth, and subsequent annealing conditions. The temperature coefficient of HSR resistors can be minimized by performing an incomplete anneal.[23] HSR resistors consist of light, shallow boron implants into an N-epi tank (Figure 3.19). Ohmic contact cannot be made directly to the HSR implant

---

22 Grebene cites a value of ±6%, and Gray, *et al.* one of ±5%: Grebene, p. 147; Gray, *et al.*, p. 139.

23 J. L. Stone and J. C. Plunkett, "Recent Advances in Ion Implantation—State of the Art Review," *Solid State Technol.*, Vol. 9, #6, 1976, pp. 35–44.

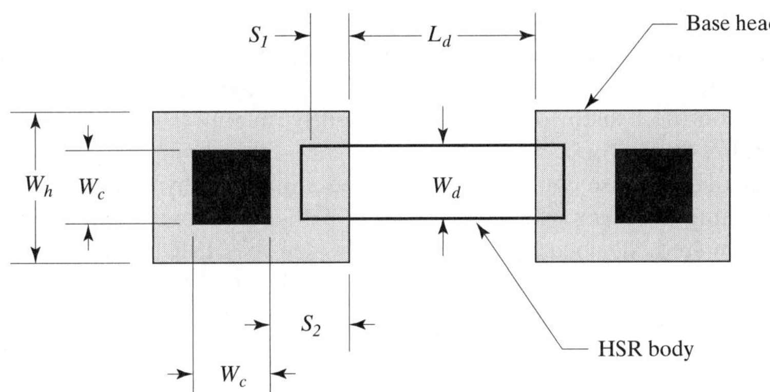

FIGURE 5.13 Layout of an HSR resistor, showing relevant dimensions.

because it is both too lightly doped and too shallow, so the heads of the resistor consist of base diffusions. Figure 5.13 shows the layout and dimensions of a sample HSR resistor.

The drawn length L_d of an HSR resistor is customarily measured between the inner edges of the two base heads. The value of the resistor equals

$$R = R_s\left[\frac{L_d - L_b}{W_d + W_b}\right] + 2R_h \qquad [5.14]$$

where R_s is the sheet resistance of the HSR implant, W_b is the width bias, and L_b is the length bias. The width bias applies to the HSR implant forming the resistor body, while the length bias applies to the separation between the base heads. The resistance of each base head, R_h, is computed separately since it does not depend on the HSR sheet resistance. The junction depth of the HSR implant generally does not exceed 0.5 μm, so the major contributors to the width bias are usually photolithography and etching. The length-bias term accounts for the outdiffusion of the base heads and is about 20% of the junction depth of the base diffusion. This term can usually be ignored since it is so small. The resistance of the base heads is difficult to predict because of nonuniform current flow. The following equation forms a useful approximation:

$$R_h = kR_{sb}\frac{S_2 + W_{bb}/2}{W_b + W_{bb}} \qquad [5.15]$$

Here, R_{sb} is the base sheet resistance and W_{bb} is the width bias computed for base resistors. The arbitrary constant k accounts for nonuniform current flow in the resistor and typically equals about 0.7. This constant approaches one as the head is elongated, as is sometimes done to accommodate lead routing.

High-voltage HSR resistors are notoriously prone to charge spreading. The sheet resistance of HSR is far greater than that of base, so the currents flowing through are smaller and the effects of leakage become magnified. All HSR resistors operating at voltages approaching or exceeding the thick-field threshold require careful field plating. The thin, lightly doped HSR implant is also vulnerable to voltage modulation effects. For example, a merged HSR resistor divider connected to a 20 V supply will experience severe mismatches. This same divider functions properly if the resistors are segmented and placed in separate tanks so that each segment experiences the same tank modulation. As this example suggests, the effect of tank connections is much more critical for HSR resistors than for deeper and more heavily doped base resistors. A merger that works well with base resistors may cause unacceptable errors if the resistors are converted to HSR.

Conductivity modulation also affects HSR resistors. Leads in single-level-metal designs must often route over HSR resistors for topological reasons. Conductivity modulation will occur if the voltage on such a lead differs significantly from the voltage on the resistor. This effect frequently magnifies noise coupling between leads and resistors. Since base diffusions experience much less conductivity modulation than HSR, noise coupling can often be minimized by elongating the base head and running the leads over base diffusion rather than over HSR implant (Figure 5.14). Elongated base heads are also useful when the HSR resistor is too short to accommodate all of the leads that must cross it. The resistance of an elongated base head can be approximated using Equation 5.15, where the constant k is set equal to one.

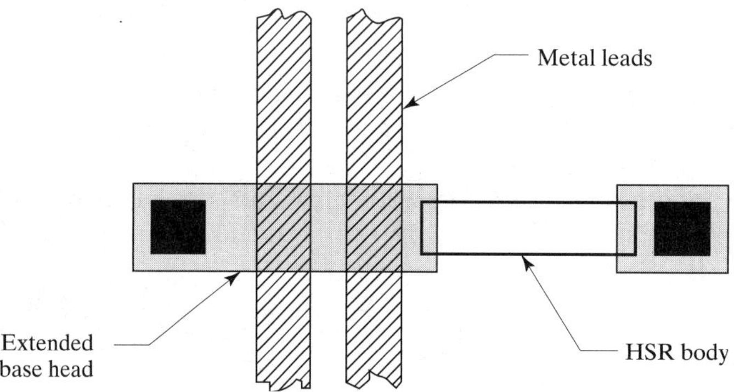

FIGURE 5.14 Example of an HSR resistor with an extended head.

The shallowness of the HSR implant limits the avalanche voltage of HSR resistors to a small fraction of the planar breakdown. Most HSR resistors can only withstand 20 to 30 V. By comparison, base resistors can typically withstand 50 to 60 V, despite their heavier doping. The HSR breakdown can be increased a few volts by filleting all of the corners in order to reduce the electric field intensity at these points. Much higher operating voltages can be achieved by dividing the HSR resistor into multiple segments, placing each segment in a separate tank, and connecting each tank to the positive end of the enclosed segment. This arrangement not only enables the resistor to withstand operating voltages up to the tank-substrate breakdown but also reduces voltage modulation.

As with base resistors, HSR resistors should always have NBL placed underneath them. This not only helps prevent tank debiasing but also acts as a barrier to minority-carrier injection if a resistor momentarily forward-biases into the tank. The sheet resistance of the shallow HSR implant is not materially altered by the presence or absence of NBL, but HSR is more affected by NBL shadow than is base. HSR resistors can be merged with other devices in a common tank, provided that none of the merged devices injects minority carriers into the shared tank.

HSR implants come in a range of sheet resistances. The smaller sheets (such as 1 k$\Omega/\square$) are of little value because they cannot save enough die area to justify their cost. The larger sheets (such as 5 k$\Omega/\square$) are extremely vulnerable to charge spreading and conductivity modulation. The optimal sheet resistance lies in the neighborhood of 2 k$\Omega/\square$, which allows substantial area savings without having to worry too much about field plates or lead routing. If higher sheet resistances must be employed, then all of the resistors should be field plated to prevent surface charges from modulating them. These field plates also prevent the inadvertent routing of leads across the HSR resistors. For these higher sheet values, consider sectioning larger resistors and applying separate field plates to each section to lessen the conductivity modulation caused by the plates. Split-field plates (Section 7.2.12) may also prove useful for precisely matched resistors. Voltage modulation and tank modulation worsen as the sheet resistance increases, so tank connections must be carefully scrutinized.

HSR resistors are useful for packing large amounts of resistance into a limited die area. They are less variable than base pinch resistors, and they have a much larger sheet resistance than base. Most standard bipolar designs operating at supply currents of less than a milliamp make extensive use of HSR. CMOS and BiCMOS processes rarely implement HSR resistors because doped polysilicon offers a superior alternative. Although the poly sheet resistances commonly offered by these processes are lower than HSR sheet resistances (500 Ω/□ versus 2 kΩ/□), the narrower poly widths and spacings usually enable the construction of more compact layouts. Polysilicon does not experience tank modulation, and conductivity modulation is generally negligible for sheet resistances of less than 1 kΩ/□.

5.5.5. Epi Pinch Resistors

An *epi pinch resistor* resembles a base pinch resistor in that it consists of a resistive region pinched between two junctions to increase its sheet resistance. In this case, the resistive layer consists of N-epi pinched between the underlying substrate and an overlying base diffusion (Figure 5.15). The substrate dopant diffuses upward during the isolation drive to produce an effective sheet resistance of 5 to 10 kΩ/□.

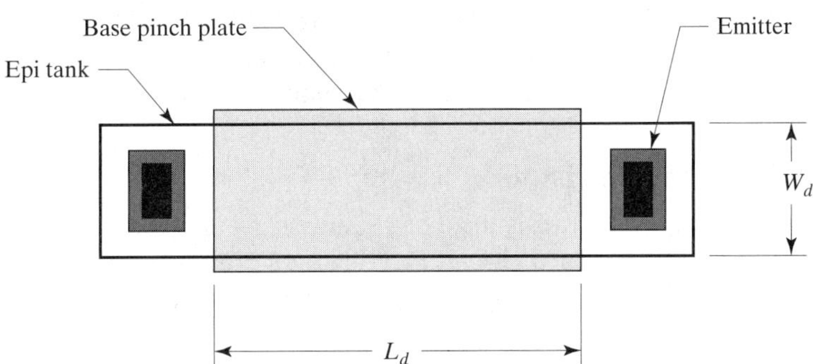

FIGURE 5.15 Layout of an epi pinch resistor (epi-FET).

As one would expect of a highly resistive material, the pinched epi suffers from severe voltage modulation effects. At higher voltages, the positive end of the resistor can deplete entirely through. Once the resistor pinches off, current flow ceases to increase and becomes largely independent of voltage bias. The *pinchoff voltage* at which this occurs depends upon epi thickness, epi doping, and base junction depth. For a typical 40 V standard bipolar process, the epi pinchoff voltage lies between 20 and 40 V.

The epi pinch resistor is an example of a class of devices that pinch a lightly doped semiconductor region between closely spaced reverse-biased junctions. This device is actually a JFET (Section 1.5). The epi pinch resistor is also called an *epi-FET.* (See Section 11.2.)

The voltage modulation of epi pinch devices is so severe that they are rarely used as ordinary resistors. Instead, they find almost exclusive application in startup circuits, where they provide a trickle of current. The voltage modulation and eventual pinchoff are desirable in this application because they limit the increase of the startup current with voltage and thus reduce the current consumption of the part at higher voltages. Epi pinch resistors are not laid out with any particular precision because of their extremely large sheet variability (at least ±50%, not counting voltage modulation). The value of the resistor is usually computed using Equation 5.3, using the drawn width and length shown in Figure 5.15. The width bias is negative because the out-diffusion of the isolation regions encroaches upon the pinched resistance layer.

Epi pinch resistors are often laid out in a serpentine pattern. Many of these layouts use the circular turns illustrated in Figure 5.3(b). This practice is sometimes incorrectly described as an attempt to prevent premature avalanche breakdown of the sharp corners, although fillets have little effect on the breakdown voltages of deep diffusions. Actually, the rounding due to outdiffusion is so drastic that it transforms a rectangular turn into something resembling a circular turn. Circular turns are less affected by this rounding, and their values are therefore easier to compute.

Pinch resistors can also be constructed in a BiCMOS process by using base to pinch N-well. The layout appears identical to that in Figure 5.15, but narrower drawn widths are allowed because the width bias is now positive. The width of a well pinch resistor should still equal at least 150% of the nominal well junction depth, or the amount of dopant may not suffice to invert the P-epi beneath the base plate. The overlap of the base plate over the N-well must also be increased to account for the outdiffusion of the N-well. In the absence of specific layout rules, this overlap should be at least 150% of the epi thickness.

A structure analogous to an epi pinch resistor can also be constructed in a dielectrically isolated process. This resistor consists of a tank containing a strip of lightly doped N-type silicon, overlaid with a P-type implant, such as a shallow P-well. The lightly doped N-type silicon is pinched between the P-type implant and the buried oxide. This structure is very sensitive to substrate influence and variations in superficial silicon thickness. Epi and well pinch resistors are not used in large numbers. They provide a convenient means of obtaining small currents for startup circuits, but their inherent variability and nonlinearity usually preclude their application elsewhere.

5.5.6. Metal Resistors

Although the sheet resistance of aluminum metallization is small, it is by no means insignificant. Standard bipolar metallization is about 10 to 15 kÅ thick and exhibits a sheet resistance of 20 to 30 mΩ/□. The smaller feature sizes of CMOS processes usually dictate metal thicknesses of considerably less than 10 kÅ, with correspondingly larger sheet resistances.

Metal resistors typically have values between 50 mΩ and 5 Ω. Resistors in this range are used for constructing current sense circuits and for ballasting large power bipolar transistors. A metal resistor can be laid out either as a straight run or in a serpentine pattern. The resistor should reside over field oxide to prevent oxide steps from causing variations in the metal sheet resistance. In a multiple-level-metal process, resistors can be constructed using any metal layer. Leads on a higher metal layer can route over a resistor placed on a lower metal layer, but leads on lower levels of metal or poly should not route beneath a resistor placed on a higher level of metal, because they may introduce nonplanarities in the resistor.

Accurate voltage sensing across metal resistors requires special techniques. The leads of the resistor are only extensions of the resistance layer, so any excess lead length causes significant errors in the sense voltage. Therefore, two sets of leads are employed: one pair to conduct current through the resistor and the other pair to sense the voltage developed across it. Voltage drops in the current leads do not alter the voltage difference across the sense leads, and no significant voltage drops occur in the sense leads because little or no current flows through them. The use of separate current-carrying and voltage-sensing leads is called a *Kelvin connection* (Section 14.3.2).

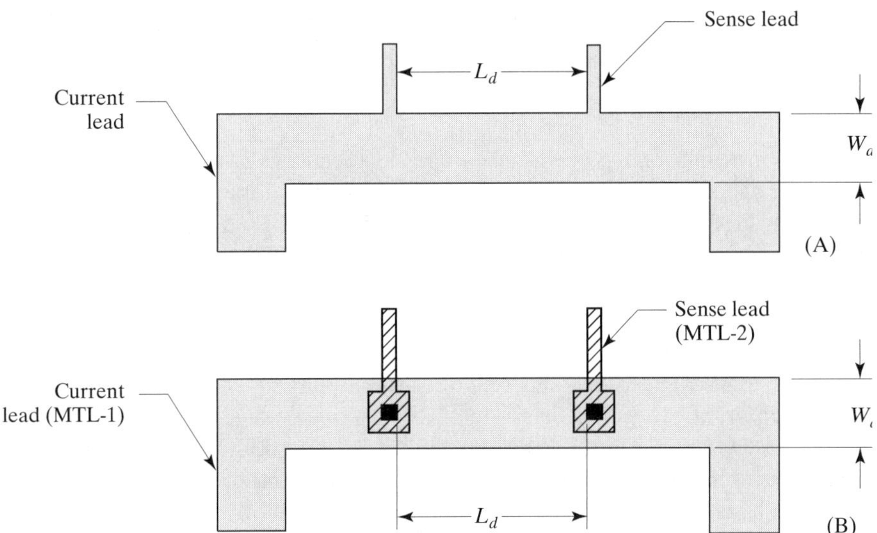

FIGURE 5.16 Two styles of Kelvin-connected metal sense resistors: (A) single-level-metal layout and (B) double-level-metal layout.

Figure 5.16A shows one means of providing Kelvin connections for a metal resistor. The resistor is implemented using a single level of metal, and the sense points are simply taps connecting into the side of the resistor. These sense leads should be as narrow as possible to ensure that the length of the resistor is determined by the spacing between the leads and not by their width. The body of the resistor should extend beyond either sense lead by a distance at least equal to the resistor width, and preferably by at least twice this width. These extensions promote uniform current flow in the vicinity of the sense leads and ensure accurate voltage sensing.[24]

Figure 5.16B shows an alternative layout that uses a second layer of metal to tap the center of the resistor. The sense leads should occupy the upper metal layer to preserve the planarity of the resistor. This layout is less sensitive to variations in current flow near the sense points. If the resistor is laid out to extend past both sense points for a sufficient distance, then the layouts of Figure 5.16A and 5.16B will have similar accuracies.

The values of metal resistors depend primarily upon the thickness of the metal layer, and only secondarily upon its composition. Providing that the resistor is sufficiently wide to ignore width bias, the sheet control of the resistor will approximately equal the variation in metal thickness, or about ±20% for most processes. This value compares very favorably with those for other types of resistors, but a potential pitfall exists. Many processes do not routinely monitor metal sheet resistance. This raises the possibility that the process might drift out of control without anyone recognizing it. Such problems can be minimized by promptly communicating any unexpected shift in metal sheet resistance to the appropriate process control personnel.

Many designers favor metal resistors for current sense circuits. Not only can they handle large currents with minimal voltage drops, but their temperature coefficient can prove quite useful. A simple bipolar transistor circuit called a ΔV_{BE} generator (Section 9.2.4) can create a small voltage with a temperature coefficient of approximately 3300 ppm/°C. This temperature coefficient largely compensates the TCR of aluminum.

---

[24] In practice, one must often settle for less-optimal layouts; see B. Murari, "Power Integrated Circuits: Problems, Tradeoffs, and Solutions," *IEEE J. Solid-State Circuits,* Vol. SC-13, #6, 1978, pp. 307–319.

This happy coincidence allows the design of very simple and surprisingly accurate current limit circuits.

5.5.7. Poly Resistors

Polysilicon resistors are available in CMOS (Figure 3.32) and BiCMOS processes. The poly used for constructing MOS gates is heavily doped to improve conductivity and has a sheet resistance of about 25 to 50 Ω/□. Lightly doped polysilicon can have sheet resistances of hundreds or even thousands of Ohms per square. Intrinsic or lightly doped polysilicon can be doped with NSD and PSD implants to provide additional choices of sheet resistance. The NMoat and PMoat coding layers cannot be used to dope polysilicon resistors because these layers generate moat geometries as well as implant regions. Instead, one must use coding layers that produce geometries only on the NSD and PSD masks.

The resistivity of polysilicon depends not only on doping but also on grain structure. The boundaries between grains interfere with the orderly flow of carriers and raise the resistivity of the material. Small-grained poly films therefore exhibit higher resistivities than large-grained films. These differences are most pronounced in lightly doped poly, which may exhibit a resistivity that is several orders of magnitude greater than similarly doped monocrystalline silicon.

The heterogeneous nature of polysilicon also affects its temperature coefficient of resistivity. Lightly doped poly exhibits a strongly negative TCR, while heavily doped poly has a positive TCR. For example, 4 kÅ 500 Ω/□ poly has a TCR of about −1000 ppm/°C, while 4 kÅ 70 Ω/□ poly has a TCR of about 500 ppm/°C. A suitably doped poly film will exhibit a TCR of zero. For 4 kÅ poly, this point lies in the neighborhood of 200 Ω/□.[25] Although this allows the construction of poly resistors having very small linear temperature coefficients, these devices still retain a significant quadratic temperature coefficient (TC_2). Process variations will also affect the temperature coefficient of the poly. In practice, it is rarely possible to hold temperature coefficients to tolerances of better than ±250 ppm/°C. Even so, this represents a significant advance over diffused resistors with temperature coefficients of several thousand ppm/°C.

Figure 5.17 shows a poly high-sheet resistor composed of lightly doped N-type polysilicon. NSD implants coded around either end of the resistor produce low-resistivity heads for the contacts. If the resistors must match accurately, these implants should overlap the poly so that the entire width of the resistor receives the dopant. Otherwise, the NSD implantation need only overlap the contacts sufficiently to ensure electrical continuity.

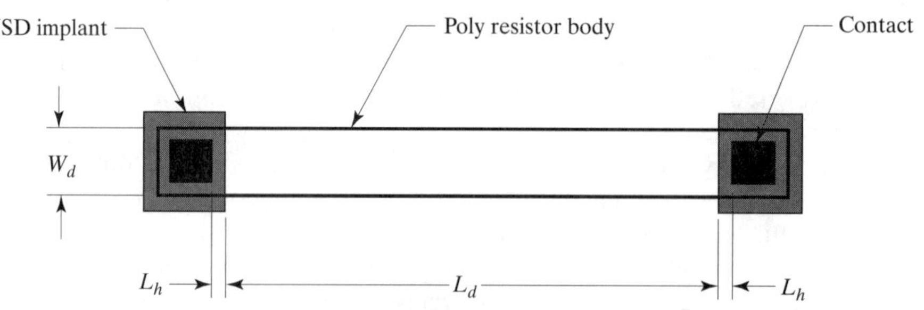

FIGURE 5.17 High-sheet poly resistor with implanted heads.

[25] Values taken from Lane, *et al.*, p. 740.

The total resistance of a high-sheet poly resistor can be computed by dividing it into sections and calculating the sheet resistance of each section:

$$R = R_s\left[\frac{L_d - 2L_b}{W_d + W_b}\right] + 2R_h\left[\frac{L_h + 2L_b}{W_d + W_b}\right] \qquad [5.16]$$

Here, R_s is the sheet resistance of the poly used to construct the resistor body, R_h is the sheet resistance of the poly used to construct the heads, and L_h is the overlap of the implant over the contact. Unlike HSR resistors, poly resistors do not exhibit nonuniform current flow at the boundaries between the body of the resistor and its heads, because these have the same width. The effects of nonuniform current flow at the contacts can be analyzed using the same techniques used for diffused resistors (Equations 5.5 and 5.13).

The width bias W_b models the oversizing or undersizing that occurs during photolithography and poly etching. This bias can equal as much as a micron, so it can have a significant effect upon narrow poly resistors. By the same token, narrow resistors will exhibit large process variations due to linewidth control. Most processes can control poly dimensions to within about 10% of the minimum width. For example, a process that can fabricate a 0.8 μm gate will probably have a poly linewidth control of about 0.08 μm. Although this is a remarkable degree of precision, narrow poly resistors still experience considerable variability. The effects of linewidth variation can be minimized by increasing the resistor width to several times minimum. Extremely narrow poly resistors may also experience increased resistance variability due to the growth of individual grains across the entire width of the resistor (an effect sometimes called *bamboo poly*). This effect rarely occurs in resistors more than 2 μm wide.

The length bias L_b represents the intrusion of the NSD dopant into the body of the resistor. This term has a much smaller impact on resistance than the width bias because the resistor is typically much longer than it is wide. Therefore, the length bias is often ignored.

Poly resistors should reside on top of field oxide. This not only reduces the parasitic capacitance between the resistor and the substrate but also ensures that oxide steps do not cause unexpected resistance variations. If the parasitic capacitance of the field oxide is still too large for a given application, consider using a second poly layer (if available) since the interlevel oxide will further reduce the parasitic capacitance. In some BiCMOS processes, deep-N+ can be coded underneath a resistor to thicken the field oxide by dopant-enhanced oxidation.[26] If this technique is employed, make sure to extend the drawn boundaries of the deep-N+ several microns beyond the resistor on all sides to ensure that it will reside on planar oxide.

Poly resistors do not tolerate transient overloading as well as monocrystalline silicon does. The oxide surrounding a poly resistor does not conduct heat well, and excessive power dissipation can cause localized overheating (Section 5.3.3). At temperatures beyond 250°C, the annealing process resumes. This can produce irreversible changes in resistance due to the movement of grain boundaries or the activation of incompletely incorporated dopants. In extreme cases, polysilicon resistors may experience a phenomenon similar to what occurs in Zener zaps.[27]

Some processes require additional mask layers in order to produce poly resistors. For example, some processes silicide all of the gate poly, rather than only the portions beneath the contacts. In such a *clad-poly* process, the silicide lowers the poly

---

[26] Not all processes allow poly resistors on top of deep-N+ because of planarization concerns.

[27] D. M. Petković and N. D. Stojadinović, "Polycrystalline Silicon Thin-film Resistors with Irreversible Resistance Transition," *Microelectronics Journal*, Vol. 23, #1, 1992, pp. 51–58.

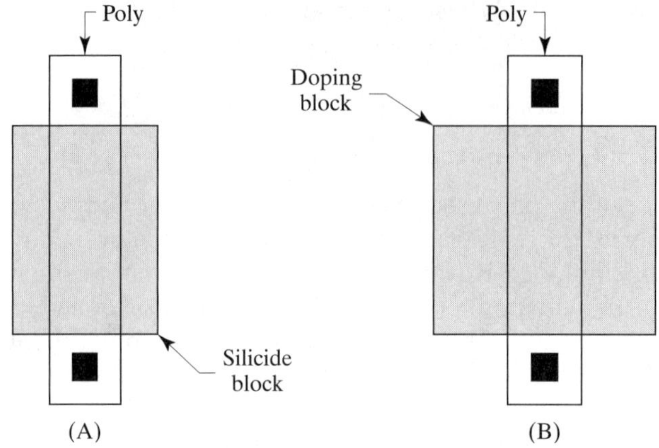

FIGURE 5.18 Layouts of (A) a silicide-blocked poly resistor and (B) a gate-doping blocked poly resistor.

sheet resistance to about 2 Ω/□. Silicided poly cannot provide enough resistance for most applications, so either one must use other types of resistors, or a special mask must be added to the process to remove silicide from the body of the resistor. This *silicide block mask* must be coded across the body of the resistor (Figure 5.18A). The ends of the resistor extend beyond the silicide block mask to ensure low-resistance contact. The illustrated resistor is uniformly doped using NSD, but other doping arrangements are possible. The sheet resistance of a silicide-blocked resistor can be computed using Equation 5.16, but the sheet resistance of the head material, R_h, now becomes so low that the second term in the equation is often neglected. Similarly, the length bias is usually negligible. The equation for the resistance of a silicide-blocked resistor therefore becomes

$$R = R_S\left[\frac{L_d}{W_d + W_b}\right] \quad [5.17]$$

Many processes use a blanket phosphorus gate doping that reduces the poly sheet resistance to about 20 Ω/□. This sheet resistance is too low for many applications, so some of these processes include a *gate doping block mask* (Figure 5.18B). This mask blocks the blanket phosphorus implant from the body of the resistor. The heads protrude from beneath the block mask, allowing them to receive the heavy phosphorus implant. The body of the resistor must be doped N-type in order to establish Ohmic contact to the N-type heads. This is usually achieved by inserting a light blanket phosphorus implant just prior to the patterning of the gate doping block mask. Because the dopant tends to spread significantly during annealing, the block mask geometry must overlap the sides of the resistor by 3–5 μm. The resistance of the gate doping-blocked resistor can be computed using Equation 5.16. The length bias L_b for a gate-doping blocked resistor can easily amount to several microns.

Certain dopants, including phosphorus and arsenic, tend to diffuse through polysilicon faster than through monocrystalline silicon. These dopants preferentially diffuse along grain boundaries, rather than through the body of the silicon grains. This phenomenon is responsible for the large required overlap of the gate block mask over the body of the poly resistor, as well as for the large value of the length bias. Grain-boundary diffusion can be suppressed by implanting the poly with oxygen before annealing. This technique holds the potential for significantly reducing both length bias and gate block mask overlap of poly.[28]

[28] R. Saito, Y. Sawahata, and N. Momma, "A Novel Scaled-Down Oxygen-Implanted Polysilicon Resistor for Future Static RAM's," *IEEE Trans. on Electron Devices*, Vol. 35, #3, 1988, pp. 298–301.

Poly forms the best resistors available on most processes. Even if the poly sheet is only half or a third of that of a diffusion, the narrower poly pitch will probably result in a smaller layout. Poly resistors do not experience tank modulation, and conductivity modulation is generally a minor concern as long as the sheet resistance does not exceed 1 kΩ/□.

5.5.8. NSD and PSD Resistors

Diffused resistors can be constructed using the NSD and PSD implants of a CMOS or BiCMOS process (Figure 3.33). These resistors usually exhibit sheet resistances of 20 to 50 Ω/□, and they are almost immune to voltage modulation. NSD and PSD are shallow diffusions that avalanche at relatively low voltages due to sidewall curvature. NSD resistors residing in the P-epi are limited by the NSD/epi breakdown voltage, but PSD resistors can operate at relatively high voltages when segmented and placed in separate wells.

Some processes silicide the moat regions to reduce their resistance. The use of such a *clad moat* technology prevents the formation of useful NSD and PSD resistors without using a silicide block mask. If a silicide block is used, the heads of the resistor must remain silicided to ensure proper contact. The value of such a resistor can be computed from Equation 5.17.

NSD and PSD resistors are not often used because most CMOS and BiCMOS processes offer poly resistors with equal or greater sheet resistances. NSD and PSD resistors see occasional use in ESD devices because their parasitic junction diodes can serve as voltage clamps. Diffused resistors can also withstand larger power transients than poly resistors because the thermal conductivity of silicon far exceeds that of the field oxide. Again, this consideration primarily applies to resistors used in transient suppressors and ESD structures. High-sheet poly resistors are preferable to NSD and PSD resistors for most other applications.

5.5.9. N-Well Resistors

Sometimes a large resistor must be fabricated in a CMOS process that lacks high-sheet poly. A high-value resistor can be created using a stretch of N-well contacted at either end by NMoat regions (Figure 5.19A). By itself, the N-well exhibits a sheet resistance of as much as 10 kΩ/□. Even higher sheet resistances can be produced by pinching the well with PSD (Figure 5.19B). The PSD implant forms a reasonably effective pinch plate despite its shallow junction depth because most of the conduction

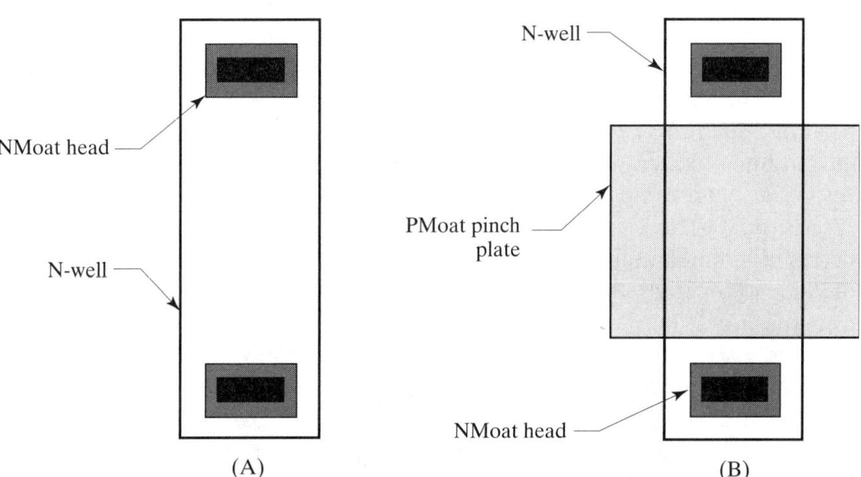

FIGURE 5.19 (A) N-well resistor and (B) N-well resistor with PSD pinch plate (field plates not shown).

normally occurs in the uppermost—and most heavily doped—portions of the well. In analog BiCMOS, base (or shallow P-well) can be substituted for PSD to produce still higher sheet resistances. The resulting devices closely resemble epi-pinch resistors, but they usually possess much higher initial resistances (20 kΩ/□ or more) and lower pinchoff voltages (20 V or less). A base pinch plate can be made still more effective by coding base implant alone rather than in conjunction with moat. Oxidation-enhanced diffusion forces the base junction deeper into the silicon and produces an even thinner pinched channel. This structure can produce pinchoff voltages of 10 V or less, depending on the depths and dopings of N-well and base.

N-well resistors without pinch plates are used for many of the same applications as base pinch resistors. They exhibit similar process variabilities and tank modulation effects, but N-well resistors usually have even larger temperature coefficients than base pinch resistors (6000 ppm/°C versus 2500 ppm/°C). N-well resistors without pinch plates may suffer from surface depletion and inversion unless they are properly field plated. The field plate should extend beyond the drawn outline of the N-well far enough to account for outdiffusion; in practice this means that the field plate must overlap the drawn N-well by a distance slightly greater than the junction depth.

Pinched N-well resistors are much less susceptible to conductivity modulation and charge spreading, because the pinching material acts as a field plate and protects the underlying resistor from surface effects. The heads of the resistor protruding from beneath the pinch plate remain vulnerable. The metal leads contacting each head of the resistor should extend to cover the exposed portions of the N-well.

When laying out N-well resistors, remember that the well does not achieve its full junction depth unless the drawn geometry is at least twice as wide as the well is deep. N-well resistors with widths less than this will exhibit progressively higher sheet resistances. Base-pinched N-well resistors are especially vulnerable to this effect because the pinched well region is so thin. In extreme cases, the base may succeed in punching entirely through a narrow N-well resistor, causing it to appear as an open circuit.

5.5.10. Thin-Film Resistors

Integrated resistors are usually fabricated from materials optimized for other uses, so they do not perform as well as discrete resistors fashioned from materials selected specifically for use in resistors. These specialized materials can also be deposited as thin films on the surface of an integrated circuit. The resulting *thin-film resistors* easily outperform diffused resistors and poly resistors. Thin-film resistors can achieve temperature coefficients of less than 100 ppm/°C with nearly perfect linearity, and they can be adjusted to tolerances of better than ±0.1% by laser trimming. These resistors are desirable components for high-performance analog applications, but the specialized process technologies they require are costly and are not always available.

Thin-film resistors can be made from a wide variety of materials, the most common of which are *nichrome* (a nickel-chromium alloy) and *sichrome* (a silicon-chromium mixture). Resistors made from these materials have varying sheet resistances and temperature coefficients, but all utilize essentially the same layout (Figure 5.20). These resistors do not require contacts, since the thin-film material is deposited immediately prior to the uppermost layer of metallization. Any top-level metal lead touching the resistor body will contact the resistor. Top-metal leads cannot cross thin-film resistors without shorting to them, but lower levels of metal are not so constrained. These resistors should be laid out on field oxide because any steps in the underlying surface may cause significant variability.

The value of a thin-film resistor can be computed using Equation 5.3. The length of the resistor is measured between the inner edges of the two metal plates forming

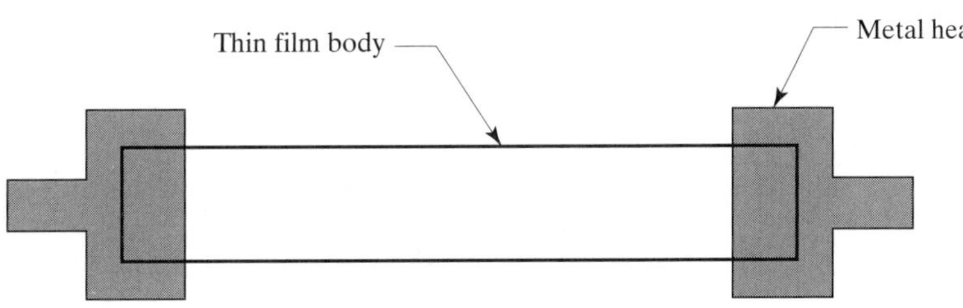

FIGURE 5.20 Layout of a thin-film resistor.

its heads. These should overhang the resistor body sufficiently to ensure that misalignment will not leave any portion of the resistor head uncovered. No corrections are required for nonuniform current flow at the ends of the resistor. Serpentine resistors use the correction factors previously discussed.

Some processes use an alternative style of thin-film resistor that requires a special contact mask. The layout of these resistors resembles the layout of a poly resistor (Figure 3.32), except that a layer of sichrome or nichrome replaces the poly. This style of resistor is preferred when the selectivity of the metal etch process is insufficient to ensure the integrity of the thin-film resistor beneath areas where metal is etched away. This problem frequently arises when dry etching a metal system, as dry etches often exhibit relatively poor selectivity.

Thin-film resistors are preferable to all others, so designs that employ them will not require most other types of resistors. Diffused resistors do, however, have better power handling capabilities. Thin-film resistors subjected to overheating often exhibit permanent changes in sheet resistance due to annealing of their crystalline structure. Extreme overheating can even cause open-circuit failures. Diffused resistors have larger volumes and are in intimate contact with the silicon die, which serves to provide additional heat-sinking. They are therefore preferred to thin-film resistors for applications subject to severe transient overloads—for example, ESD protection circuitry.

5.6 ADJUSTING RESISTOR VALUES

One often finds small variable resistors called *trimmers* scattered around printed circuit boards. Manual or robotic adjustment of these resistors allows a circuit to be adjusted after assembly to achieve tighter tolerances than can be economically obtained using high-precision components. Analog integrated circuits also make considerable use of adjustable resistors to counter process variations and to provide for design uncertainties. The relatively large amounts of process variability inherent in high-volume wafer fabrication can prevent high-precision circuitry from meeting specification. In such cases, each integrated circuit must be adjusted after it is manufactured by *trimming* the value of onc or more components. Trimming is usually performed during wafer-level testing, and it usually requires specialized test equipment and procedures.

Circuit design is fraught with uncertainties, so experienced designers often allow room for adjustments after the initial design is evaluated. A redesign of a circuit that affects only component values is called a *tweak*. Tweaks adjust the mean of a distribution, whereas trimming minimizes its variance. All tweaked units receive the same adjustment, whereas each trimmed unit receives individualized adjustment.

5.6.1. Tweaking Resistors

Integrated circuits are normally laid out with little regard for future modification. This approach is justifiable, since most changes affect the entire mask set and most designs require several passes to meet specifications. Resistor tweaks represent an exception to this rule because they need not affect the entire mask set. If the resistors are properly designed, their values can be tweaked by altering only one mask. Tweakable resistors can drastically reduce the time required to obtain fully parametric devices. Several wafers can be held out of the first wafer lot just prior to the step required to tweak the resistors. After sample wafers have been evaluated, a new mask can be made incorporating corrected resistor values. The time required to complete the processing of the tweaked wafers is usually much less than the time required to fabricate a new wafer lot.

There are four commonly used methods for tweaking resistors: sliding contacts, sliding heads, trombone slides, and metal options. Each technique involves a different mask, and no one method can adjust all types of resistors.

Sliding Contacts

Sliding contacts are the simplest type of resistor tweak.[29] Figure 5.21 shows two examples of resistors that incorporate sliding contacts.

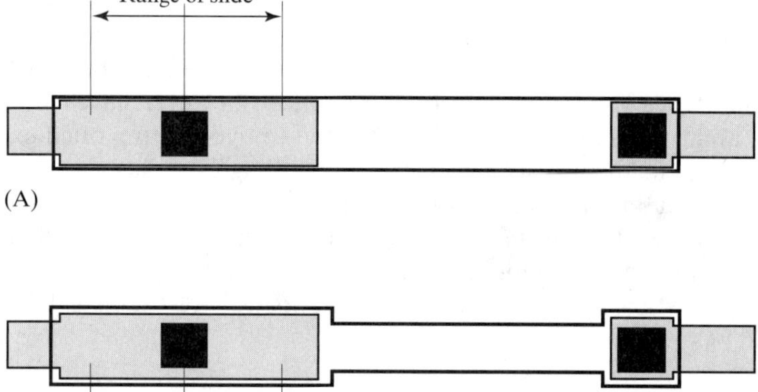

FIGURE 5.21 Two styles of sliding contacts: (A) without heads and (B) with heads.

Sliding contacts are easiest to implement for resistors without heads (Figure 5.21A). The body of the resistor is extended so that one contact can slide inward or outward. The metal plate for this contact is also elongated so that it covers the contact no matter where the latter is located. This precaution prevents having to purchase a new metal mask to perform the tweak. The sliding contact should initially reside at a point midway between the limits of travel (Figure 5.21A).

Sliding contacts are somewhat more difficult to construct if the resistor uses enlarged heads. A sliding contact can still be made as long as the material of the head has the same resistivity as the material of the body (Figure 5.21B). The head must be enlarged to accommodate the sliding contact, and this complicates the computation of its resistance. An approximate value can be obtained by assuming that the resistor consists of two sections in series: one narrow and one wide. The total resistance equals the sum of the resistances of each segment. Nonuniform current flow at the juncture of the two segments introduces a small error, but this can be ignored

[29] Grebene, p. 156.

because the resistor will probably be tweaked anyway. The sliding contact removes or adds a certain length to the wider segment. Movement of the contact does not significantly affect the nonuniformities of current flow, either at the joint where the head meets the body of the resistor or at the contacts.

Sliding contacts work well for resistors whose body and heads consist of the same material, as do standard bipolar base and emitter resistors. Sliding contacts are much less useful if the head is constructed of a low-sheet material, as is the case for most resistors with sheet resistances of more than about 200 Ω/□. The sliding contact can only make trifling changes in such a resistor, as it can only move within the confines of the low-sheet material used to form the resistor heads. High-sheet resistors are usually tweaked using *sliding heads.*

Sliding Heads

Figure 5.22 illustrates the layout of a resistor incorporating a sliding head at its left end. The body of this resistor consists of a high-sheet material such as HSR implant or lightly doped poly. The heads consist of a lower sheet material used to ensure Ohmic contact. The resistance can be reduced by extending the sliding head further into the resistor body. If sufficient room is provided to pull the head back, then the total resistance can also be increased by sliding the head toward its contact.

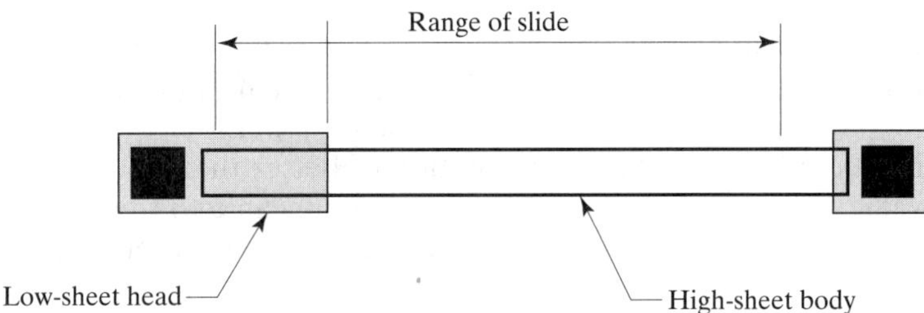

FIGURE 5.22 Layout of a resistor with a sliding head.[30]

The sliding-head resistor can be modeled as a series combination of two separate resistors, one representing the body and the other representing the heads. Although nonuniform current flow causes slight errors in the calculation, these can be ignored because the resistor will probably require adjustment anyway.

Sliding heads are commonly used for adjusting HSR implant and high-sheet polysilicon resistors. The same concept works for pinch resistors, except that the pinch plate slides forward or backward instead of the heads. Poly resistors that use a silicide block mask can also be adjusted by sliding the silicide block toward or away from the contact.

Trombone Slides

Serpentine resistors can be adjusted by sliding turns inward or outward, a technique colorfully referred to as a *trombone slide.* This adjustment alters the total length of the resistor without changing the number of turns it contains (Figure 5.23). Room left adjacent to the resistor allows for its extension. If the resistor occupies a tank or is enclosed in an implant, then these geometries should also enclose the region into which the resistor can slide.

Metal Options

Another method that enables limited tweaking of resistors involves dividing the resistor into multiple sections. The majority of these sections are connected in series to

---

[30] *Ibid.*

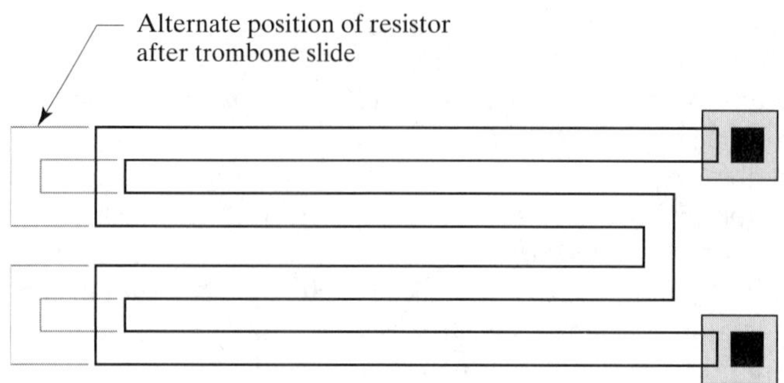

FIGURE 5.23 Resistor adjustment by means of a trombone slide.

form the final resistor, but several sections are left unconnected as spares. These spares include contacts, each covered by just enough metal to meet the minimum design rules. The resistance can be adjusted by adding or removing segments, using a new metal mask. The flexibility of this tweaking scheme is limited by the number of possible combinations of the segments. Segments can be joined in parallel as well as in series, but some potential combinations are undesirable because they do not properly balance the thermoelectric potentials generated by the contacts (Section 7.2.11).

5.6.2. Trimming Resistors

Resistors can be trimmed using fuses, Zener zaps, or laser trims. Fuses and Zener zaps act as programmable switches, allowing a network of resistors to be reconfigured in much the same way as a metal option. Laser trimming can provide incremental adjustment with a resolution of better than $\pm 0.1\%$ of the initial resistor value. However, this technique is only applicable to thin-film resistors, and it requires the assembly/test site to purchase automated laser trimmers.

Fuses

A *fuse* is simply a short section of minimum-width metal or polysilicon connected between two bondpads. It is programmed, or *blown,* by passing a large current between the bondpads, causing the fuse material to vaporize. After programming, the fuse becomes an open circuit.

Figure 5.24A shows a typical example of a metal fuse, which consists of nothing more than a constricted segment in a wide metal lead. A small opening in the

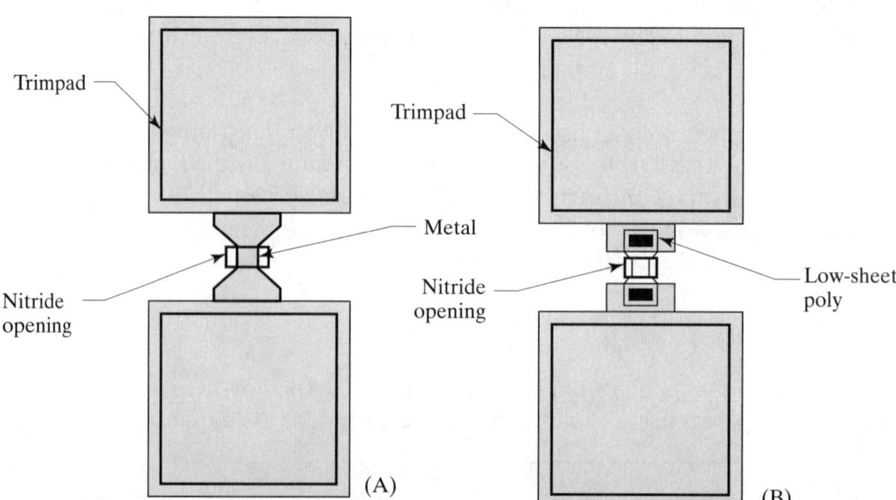

FIGURE 5.24 Layouts of (A) a typical metal fuse and (B) a polysilicon fuse.

protective overcoat above the fuse allows the vaporized metal to escape during programming. This opening represents a potential pathway for contaminants to enter the die, but if it is omitted, the process of blowing the fuse can crack the protective overcoat. These cracks allow the metal vapor to escape, but they sometimes extend into adjacent circuitry. Most manufacturers agree that the inclusion of an opening in the protective overcoat is a small price to pay to prevent fuse regrowth and overcoat cracking. These openings should be made as small as possible to minimize the ingress of contaminants. No metal (besides the fuse itself) should intrude into the opening, or even approach within a few microns of it, to prevent the possibility of leakage caused by ejected material bridging the gap from the fuse to the intruding metal.

Figure 5.24B shows a typical polysilicon fuse. The polysilicon is laid out as a small resistor connecting two bondpads. As with the metal fuse, a small opening allows the vaporized silicon to escape without rupturing the protective overcoat. This opening is sometimes unnecessary if the polysilicon layer is sufficiently thin. In such cases, insufficient material is vaporized to rupture the protective overcoat, and instead a small bubble forms about the fuse when it is blown. Because the formation of this bubble stresses the surrounding material, no metal should cross over the fuse, or even approach within a few microns of it. The polysilicon fuse must be heavily doped to obtain a sufficiently low resistance to allow programming. The fuse must have a total resistance of no more than 200 Ω to ensure reliable programming with a 10 V pulse.

The pads used to program fuses are not normally bonded, so they need only be large enough to allow the probe needles to reliably land on them. These pads are sometimes called *trimpads* to distinguish them from true bondpads. Despite their reduced dimensions, trimpads still require substantial die area. Some processes allow active circuitry to reside underneath trim pads, in which case very little area is wasted. Trim pads must be placed so that the probe needles can reach them; this usually requires that they reside around the periphery of the die. Reducing the number of trimpads saves die area and reduces the cost of the integrated circuit. The number of trimpads required to blow the fuses can be minimized if several fuses are connected in series or in parallel.

Aluminum fuses are easy to program. Aluminum melts at a relatively low temperature (660°C) but boils at a much higher temperature (2470°C).[31] By the time vaporization begins, most of the metal is already molten and is readily ejected. A pulse of several hundred milliamps for a period of a few milliseconds will reliably blow the fuse. The ejected aluminum tends to splatter onto the probe needles and sometimes fouls them so badly that shorts occur. The probe card may require occasional cleaning to alleviate this problem. Metal fuses incorporating refractory barrier metal can also be programmed reliably; the aluminum is ejected first, but the thin refractory metal soon follows.

Polysilicon fuses are more difficult to program. Silicon melts at a much higher temperature than aluminum (1410°C versus 660°C), and it is brittle and prone to fracturing from the stresses of rapid heating. The polysilicon may crack before it begins to melt unless the programming current pulse has an extremely fast rise time (<25 nS). Such fractures cause the current flow to stop and prevent proper programming. Mechanical stresses can cause cracked fuses to reform at any time. This type of reliability failure can be prevented by ensuring that the programming pulse has a sufficiently fast rise time to blow the fuse before cracks appear.

Fuses that incorporate openings in the protective overcoat are not suited for programming after encapsulation, as the plastic seals the opening and prevents the

---

[31] Melting and boiling point data: R. C. Weast, ed., *CRC Handbook of Chemistry and Physics,* 62d ed. (Boca Raton, FL: CRC Press, 1981), pp. B-2–B-48. Values are rounded to the nearest 10 degrees.

ejection of conductive material. Fuses without openings in the protective overcoat can be programmed after encapsulation, but the need for large programming currents makes it difficult to design compact programming circuitry. Typically, these designs use large FET switches to steer the currents that blow the fuses. Since the programming currents can amount to several hundred milliamps, these FETs can be as large as the trimpads used to program fuses during wafer probing.

The programming process can cause high voltages to appear on circuitry connected to the fuse. A large current flows through the fuse in the instant before it blows, and the sudden interruption of this current can cause voltage transients due to parasitic inductances. These transients can avalanche junctions or damage thin gate oxides. The magnitude of the transients can be reduced by minimizing the loop area of the programming circuit. Especially difficult cases may benefit from the addition of integrated Zener clamps. Circuit designers can help minimize the impact of fuse programming transients by placing the fuses on the least-vulnerable end of a resistor. For example, the resistor normally trimmed in a Brokaw bandgap[32] connects between the emitter of an NPN transistor and ground. The trims should be placed on the grounded end of the resistor so that transients must pass through the rest of the resistor before reaching the base-emitter junction. This same connection also saves a trimpad because the lowest fuse connects to ground.

Several fuses in combination can provide additional trim resolution. The resistor segments should be *binary weighted* to ensure that the achievable values of resistance are uniformly spaced across the trim range. This not only provides the maximum possible trim range for a given precision, but also simplifies the design of the test program by allowing the use of a binary search algorithm. Binary weighting can be implemented in either of two ways. If the voltage across a resistor requires trimming, then the resistors should be connected in series and weighted $R_{lsb}:2R_{lsb}:4R_{lsb}:8R_{lsb}\ldots$, where R_{lsb} is the value of the *least significant bit* (LSB) of the trim network.[33] Figure 5.25A shows a 3-bit binary-weighted voltage trim scheme. If the current through a resistor requires trimming, then the resistors should be connected in parallel and weighted in the ratio $R_{msb}:R_{msb}/2:R_{msb}/4:R_{msb}/8\ldots$, where R_{msb} is the value of the *most significant bit* (MSB) resistor (Figure 5.25B).

A precision trim scheme often requires a very small LSB resistance. Minimum-width resistors of less than about 1 kΩ may overheat during trimming, causing resistance variations and, in extreme cases, outright failures. Smaller resistors can be implemented as parallel combinations of two or more resistors, each having a relatively large resistance, but this technique becomes impractical for resistors much smaller than 200 Ω. A technique called *differential trimming* can implement arbitrarily small LSB resistances. This scheme uses two resistors per trim bit rather than one. The two resistors are connected in parallel while the fuse remains intact. Blowing the fuse disconnects one of the resistors and leaves the other to conduct the current alone. The effective LSB resistance equals the difference between the two resistor values. Figure 5.26 shows an example of differential trim applied to the LSB of a two-bit, series-connected trim scheme. If R_A has a value of 1 kΩ and R_B has a value of 250 Ω, then the parallel combination of these resistors equals 200 Ω. The difference between R_B acting alone and the two resistors acting in parallel therefore equals 50 Ω.

Differential trimming requires two trimpads per bit, while standard trimming requires only one pad per bit. The additional trimpads required for differential trim

---

[32] A. P. Brokaw, "A Simple Three-Terminal IC Bandgap Reference," *IEEE J. Solid-State Circuits,* SC-9, 6, 1974, pp. 388–393.

[33] See Grebene, pp. 156–158.

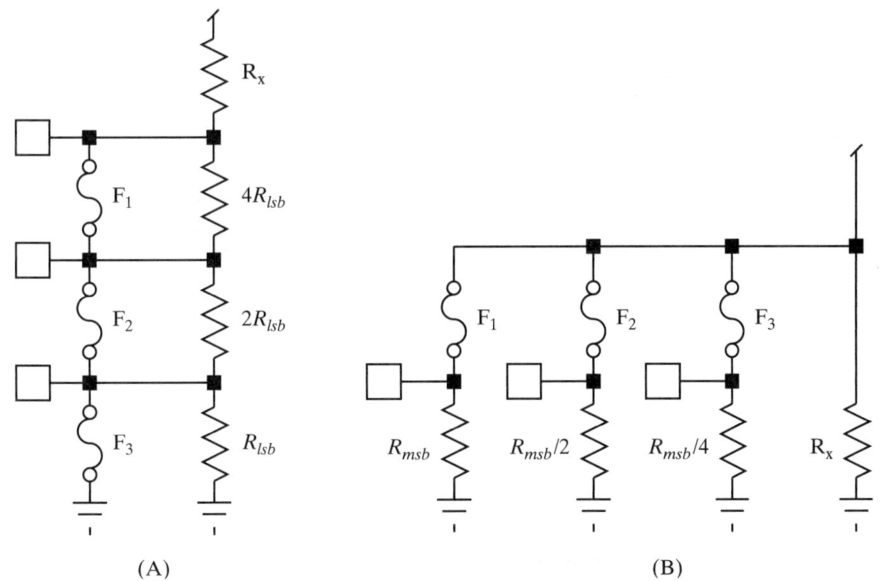

FIGURE 5.25 Two different binary-weighted resistor trim schemes using fuses: (A) series-connected and (B) parallel-connected. Both cases assume that the ground pad is used to program the fuses.

make this technique uneconomic for all but the smallest resistors. Resistors smaller than about 500 Ω usually benefit from differential trimming, while larger resistors are better implemented as standard trims.

Fuse trim schemes require trim pads around the periphery of the die, but precision resistors normally reside in the interior to minimize mechanical stresses. Long leads are required to connect fuses at the edge of the die to resistors in the middle. These leads not only waste die area but can also pick up noise from other circuitry. If CMOS transistors are available, then these can be used as switches to reconfigure the resistor network. These transistors can, in turn, be controlled by fuses placed around the periphery of the die. Since the fuse leads no longer connect directly to the trim resistors, they are less susceptible to noise, and their length and routing become less critical. Care must be taken to ensure that the CMOS transistors have small on-resistances compared with the trim resistors. The gate drive voltage for these transistors must be derived from a well-regulated supply, since this voltage will modulate their on-resistance. The design of remotely programmed trim networks is beyond the scope of this text, but the foregoing comments should convey the general concept.

Some designers have attempted to use remotely programmed poly fuses to implement *look-ahead trimming.* A voltage sufficient to switch the transistor but inadequate to blow the fuse can be used to test the results of programming the fuse before committing to it. Unless the circuitry is specifically designed to switch at low voltages, the look-ahead process may overstress the poly fuse, causing it to increase in value to the point where it can no longer be reliably programmed.[34]

Zener Zaps

Zener diodes short-circuit when severely overloaded, and this phenomenon forms the basis of the trim device called a *Zener zap.* The Zener diodes connect across segments of the resistor network in the same manner as the fuses shown in Figures 5.25 and 5.26. The Zeners must be oriented so that they remain reverse-biased during normal operation, and the voltage placed across each Zener must not exceed about two-thirds the base-emitter breakdown voltage. The Zeners appear as open circuits until they are

[34] D. W. Greve, "Programming Mechanism of Polysilicon Resistor Fuses," *IEEE Trans. on Electron Devices,* Vol. ED-29, #4, 1982, pp. 719–724.

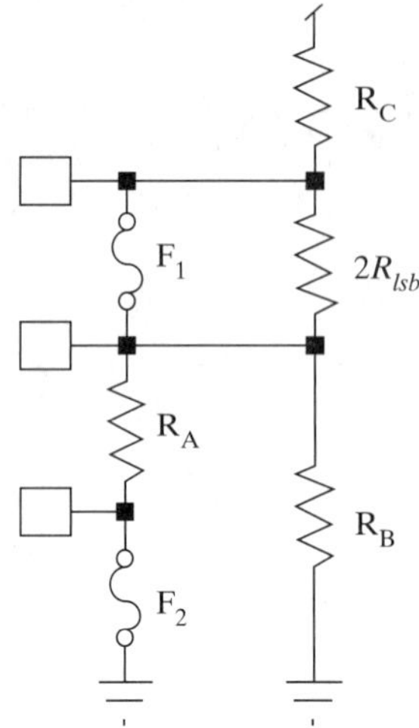

FIGURE 5.26 Differential trim scheme applied to the LSB fuse, F_2, of a series-connected binary-weighted trim network.

programmed, after which they appear as shorts. The act of programming a Zener is called *zapping,* so these Zeners are called *zap Zeners*, or *Zener zaps*.

Figure 5.27A shows the layout of a Zener zap constructed in a standard bipolar process. The device has the same basic structure as a small NPN transistor. The collector and emitter terminals of the NPN together form the cathode of the Zener, and the base terminal serves as the anode. Because the device is used as a Zener, little or no current flows through its collector contact. Deep-N+ is therefore unnecessary and can be omitted to save space. The emitter and base contacts should be placed as close as layout rules allow to facilitate the zapping process. The emitter should extend as close to the base contact as possible to minimize the series resistance of the Zener. Although the illustrated Zener zap structure uses a separate collector contact, an alternate layout involves stretching the emitter so that it extends beyond the base diffusion and thus shorts to the tank. This layout theoretically consumes less space, but since the tanks can extend underneath the adjacent trimpads, the two layouts actually consume similar amounts of area.

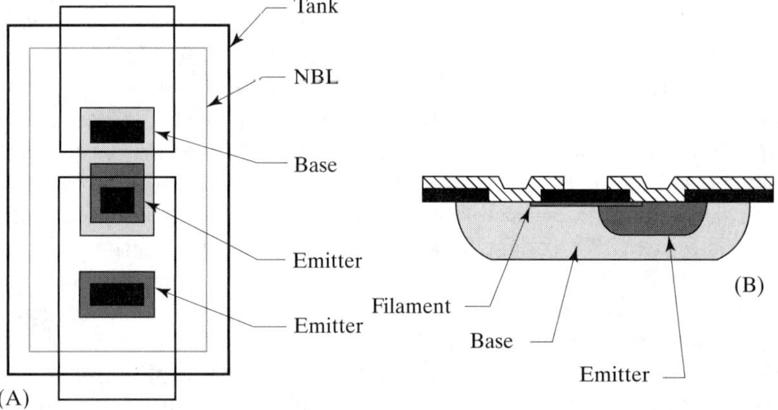

FIGURE 5.27 Zener zap constructed using base and emitter diffusions from a standard bipolar process: (A) layout of unprogrammed Zener and (B) cross section of programmed Zener.

Programming involves forcing a large reverse current through the diode to avalanche its base-emitter junction.[35] A programming current of about 250 mA will

[35] G. Erdi, "A Precision Trim Technique for Monolithic Analog Circuits," *IEEE J. Solid-State Circuits,* SC-10, 1975, pp. 412–416.

result in a total drop of as much as 10 to 20 V across the Zener, much of this due to internal series resistance. The resulting power dissipation in a very limited volume of silicon results in extreme localized heating. The aluminum metal contacting the silicon melts, and a molten filament of aluminum-silicon alloy flows underneath the oxide and bridges the gap between the contacts (Figure 5.27B). Once this filament forms, the resistance of the Zener zap drops to a few Ohms.

Zener zaps require the same arrangements of pads as fuses, so the networks illustrated in Figures 5.25 and 5.26 are applicable to Zeners as well as to fuses. Because larger voltages are generally required to program Zeners than are required to program fuses, care must be taken to ensure that the circuitry connected to the trim network can withstand a momentary overvoltage condition. As long as a few kilohms of resistance lie between the zaps and the remainder of the circuit, programming will generally cause no harm. If necessary, diodes or Zeners can be used to clamp the voltage seen by delicate circuitry.

Zener zapping involves the formation of an aluminum-silicon alloy, and this in turn presumes that aluminum is directly touching the silicon. The presence of refractory barrier metal or silicide between the aluminum metal and the silicon interferes with the zapping process. Experimental fabrication of Zener zaps on a process using refractory barrier metal demonstrated that the Zeners could be zapped, but only with difficulty.[36] The programming current had to be nearly doubled, and some wafer lots of material resisted zapping. The performance of refractory-barrier-metal Zener zaps is consistent with recently discovered evidence that refractory barrier metal forms an inhomogenous filament structure.[37] Zener zaps should not be implemented on processes that use refractory barrier metal or silicides.

Unlike fuses, Zener zaps do not require openings in the protective overcoat. This not only eliminates a potential pathway for contaminants to enter the die but also raises the possibility of trimming packaged units. While post-package trimming is certainly possible, in practice it has rarely been implemented because of the large number of pins (or alternatively, the large power devices) required for zapping.

Very short emitter resistors can be zapped by the same mechanism as Zeners. Attempts have been made to regulate the zapping process to provide infinite adjustability. Since the molten filament moves at a finite speed, it is theoretically possible to halt the programming process before the gap between the contacts is fully bridged.[38] In practice, the filament moves so quickly and so erratically that it is difficult to control, and this scheme cannot be recommended for production use.

EPROM Trims

Many CMOS and BiCMOS processes include provision for nonvolatile memory in the form of *electrically programmable read-only memory* (EPROM) and *electrically erasable programmable read-only memory* (EEPROM). These programmable devices behave somewhat like fuses and Zener zaps in that they can be made either conductive or nonconductive by application of an electrical signal. However, the currents required to program EPROM and EEPROM are tiny compared with those required to program fuses and Zener zaps. Therefore, EPROM and EEPROM are ideally suited for applications that require post-package trimming.

---

[36] F. W. Trafton, private communication.

[37] A. J. Walker, K. Y. Le, J. Shearer, and M. Mahajani, "Analysis of Tungsten and Titanium Migration During ESD Contact Burnout," *IEEE Trans. on Electron Devices*, Vol. 50, #7, 2003, pp. 1617–1622.

[38] R. L. Vyne, W. F. Davis, and D. M. Susak, "A Monolithic P-channel JFET Quad Op Amp with In-Package Trim and Enhanced Gain-Bandwidth Product," *IEEE J. Solid-State Circuits*, Vol. SC-22, #6, 1987, pp. 1130–1138. Also see R. L. Vyne, "Method for resistor trimming by metal migration," US Patent 4 606 781, Aug. 1986.

EPROM elements can be programmed by application of an electrical signal, but they can only be erased by exposure to heat or ultraviolet (UV) light. EPROM used for trimming typically includes no provision for erasure. Such *one-time programmable* (OTP) EPROM can avoid the otherwise burdensome requirements for UV-transparent protective overcoats and packaging. The OTP EPROM is erased during manufacture by the high-temperature sinter that completes the fabrication of the metallization. During programming, selected EPROM cells are programmed to render them conductive. The EPROM cells take the place of fuses in a remotely programmed trim scheme. Section 11.3.1 presents the details of EPROM construction and briefly discusses the associated programming circuitry.

EEPROM elements are similar to EPROM, except that they can be erased by an electrical signal as well as by heat or UV light. EEPROM elements are sometimes used to construct trim networks that can be repeatedly programmed to different states. Such networks are occasionally necessary in complicated analog systems where multiple sets of trims interact with one another. Section 11.3.2 presents the details of EEPROM construction and programming.

In general, if a process supports a form of EPROM memory that requires no additional masking steps, then trimming should be conducted using EPROM. If a process does not include EPROM, then it will probably use fuses. Zener zaps are generally found only on older standard bipolar processes that do not use silicides or refractory barrier metals.

Laser Trims

Another method of trimming uses a laser to alter the resistance of a thin conductive film. These films commonly consist of *nichrome* (a nickel-chromium alloy) or *sichrome* (a silicon-chromium mixture). The laser beam causes localized heating, which alters the grain structure or the chemical composition of the material to drastically increase its resistance. In the case of sichrome, the programming process appears to involve the segregation of chromium into narrow filaments isolated by more resistive material. Although the protective overcoat remains intact, a small bubble forms beneath it about the area struck by the laser.[39] Each shot of the laser affects a circular zone about 3 to 10 μm in diameter. By moving the laser incrementally while performing a series of shots, a continuous line of high-resistance material can be formed. It is possible to virtually sever a resistor by this process, allowing

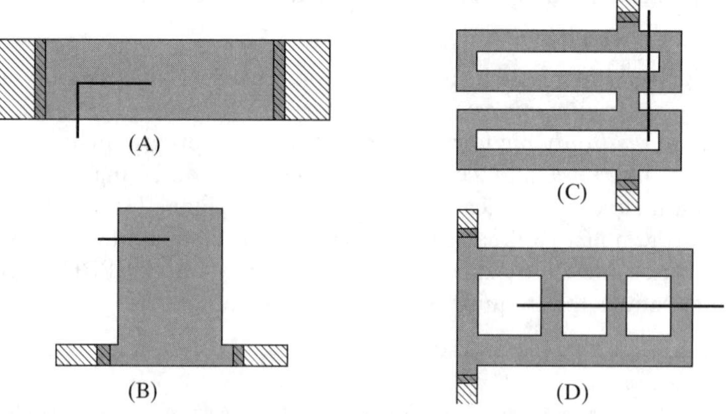

FIGURE 5.28 Four different schemes for laser-trimming thin-film resistors: (A) notched bar, (B) tophat, (C) looped layout, and (D) ladder layout.[40] The heavy black lines show the path of the laser beam through the resistor.

---

[39] E. Coyne, "Laser Interaction with SiCr Thin Film Resistors: 'The Bubble Theory," *Proc. 41ST International Reliability Physics Symp.*, 2003, pp. 553–558.

[40] After Glaser, *et al.*, p. 358.

discrete adjustments to a network of resistor segments. Alternatively, the value of the resistor can be continuously monitored and the trim halted once the desired resistance has been achieved. Continuous trimming allows finer resolution, but it also alters the temperature coefficient of the resistor, because current continues to flow through portions of the material that have been altered by heat from the laser beam. The change in temperature coefficient is proportional to the increase in resistance caused by trimming, but it rarely exceeds 100 ppm/°C. Discrete trimming entirely avoids this problem because current only flows through links that have not been altered by exposure to the laser beam.

Figure 5.28A shows one common style of continuously trimmed, thin-film resistor. The laser beam first moves laterally across the resistor. When the resistance has increased to about 90% of the target value, the laser beam begins to move down the length of the resistor. This longitudinal cut causes a more gradual increase in resistance and therefore allows finer trim resolution. This trim technique can produce tolerances of better than ±0.1%, but it requires resistors at least 30 to 50 μm wide. Figure 5.28B shows another style of layout frequently employed for continuous trimming. Discrete trims usually employ loop or ladder networks similar to those of Figures 5.28C and D. The value of the resistor will depend on how many of its segments have been severed. The segments of a discrete network can be made as narrow as the thin-film material can be patterned, but they must lie some 10 μm apart in order to allow the laser to sever only one link at a time.

Lasers are also used to sever metal and poly links. Unlike electrical fuses, these laser-ablated links are normally covered by a thin layer of oxide or oxynitride. The laser penetrates this overcoat and heats the fuse to the point of vaporization. The resulting pressure shatters the overcoat and allows the fuse material to explosively erupt outward. The extreme temperatures and pressures generated by laser ablation allow a very clean cut with minimal splattering. The networks used for laser trimming are similar to those shown in Figure 5.25, but they require no trimpads. Differential trimming is unnecessary and there is no danger of damaging sensitive circuitry. The width of the laser-ablated link is quite critical. Excessively narrow links cannot be successfully ablated because insufficient material exists to produce the pressure necessary to rupture (or bubble) the overcoat. Very wide links require multiple laser shots and tend to splatter. Typical laser-ablated links are about 1 μm wide by 15 μm long. The links must reside far enough from adjacent circuitry to prevent the laser from interfering with other devices. A spacing of 10 to 15 μm usually provides adequate clearance for the trimming process.

5.7 SUMMARY

Resistors are the most common type of passive component in analog integrated circuits, and many different types are available. For standard bipolar products, lower values of resistance are obtainable using base and emitter resistors, while higher values usually require HSR implant resistors. Extremely low-value resistors can be fabricated using aluminum metallization, while pinched structures can provide high resistances as long as their nonlinearity is not a concern. CMOS and BiCMOS processes typically offer doped polysilicon resistors that are superior to diffused resistors. Metal resistors are still used for very low-value resistors, and well resistors are sometimes used to obtain large values of resistance when high-sheet poly is unavailable. Thin-film resistors offer superior temperature stability at a higher manufacturing cost.

The value of most resistors can be determined by means of a simple computation involving width and length. A width correction factor may be needed to account for process size adjusts and outdiffusion. Each rectangular 90° turn inserted into a serpentine resistor adds approximately a half square. Each circular 180° turn adds approximately three squares. The corrections for nonuniform current flow near the contacts can usually be ignored because these rarely amount to more than a half square.

Provision can be made for tweaking resistors by means of sliding contacts, sliding heads, trombone slides, and metal option jumpers. Trimming can be performed by means of fuses, Zener zaps, nonvolatile memory, or laser adjustment. Laser trimming of thin-film resistors can produce extremely precise and stable resistors, while Zener zaps and nonvolatile memory potentially allow post-package trimming for canceling package shifts.

5.8 EXERCISES

Refer to Appendix C for layout rules and process specifications.

5.1. What is the sheet resistance of a 7 kÅ aluminum film if its resistivity equals 2.8 μΩ · cm?

5.2. Suppose a standard bipolar base resistor has a value of 2 kΩ, a sheet resistance of 160 Ω/□, a uniform width of 8 μm, and two 5 × 5 μm contacts. Compute the change in resistance caused by each of the following:
- a. A width bias of 0.4 μm.
- b. Nonuniform current flow.
- c. Contact resistance, assuming Al-Cu-Si metallization.

5.3. Lay out a 20 kΩ minimum-width serpentine standard bipolar base resistor. Place the two contacts as close to one another as possible. Assume a width bias of 0.4 μm and ignore all other sources of error except the contributions of turns. The resistor tank should have a roughly square aspect ratio.

5.4. Using the guidelines of Section 5.3, recommend a width for each of the following types of resistors, assuming a moderate degree of accuracy is required:
- a. Standard bipolar base.
- b. Standard bipolar HSR.
- c. Analog BiCMOS base.
- d. Analog BiCMOS poly-2.

5.5. If a 2 kΩ/□ HSR resistor has a value of 34.4 kΩ at 25°C, calculate its value at 125°C, given a TCR of +2200 ppm/°C. What percentage change does this represent?

5.6. Suppose that a 200 Ω/□ poly resistor must support a voltage differential of 5 V. How short could this resistor be made before voltage nonlinearities become a concern? Consider both self-heating and granularity.

5.7. Lay out a 30 kΩ HSR resistor with a width of 8 μm. Account for the effects of width bias, base head resistance, and contact resistance. Assume a width bias of 0.2 μm, a length bias of 0.15 μm, and a base width bias of 0.4 μm. The metallization system is Al-Cu/Ti-W/PtSi. Compute the change in resistance caused by each of the factors considered when designing the resistor.

5.8. Lay out a serpentine high-sheet poly-2 resistor with a width of 4 μm and a value of 150 kΩ. Assume the body of the resistor has a sheet resistance of 600 Ω/□ and the NSD-doped heads have a sheet resistance of 50 Ω/□. To allow for misalignment, the overlap of NSD over contact must equal at least 2 μm. Account for turns, but ignore all other correction factors.

5.9. Design a 5 kΩ standard bipolar base resistor with a sliding contact allowing a ±10% adjustment of its value. Use a drawn resistor width of 8 μm and a width bias of 0.4 μm. Account for turns, but ignore all other correction factors.

5.10. Design a 25 kΩ standard bipolar HSR resistor incorporating a sliding head allowing a ±20% adjustment of its value. Assume that the resistor width equals 8 μm and that the HSR width bias equals 0.2 μm. Account for turns, but ignore all other correction factors.

5.11. Design a PSD-doped poly-2 resistor with a value of 50 kΩ and a width of 4 μm. Allow an adjustment of ±25% in value by means of a trombone slide. Account for turns, but ignore all other correction factors.

5.12. Design a polysilicon fuse. Use a strip of minimum-width poly-1 as the fuse, and place a 5 × 5 μm opening in the protective overcoat over the fuse. Contact either end of the fuse with an array of at least four contacts. Keep metal a minimum of 2 μm away from the opening in the protective overcoat. Assume the trimpads require 75 × 75 μm openings in the protective overcoat.

5.13. Lay out a series-connected binary-weighted resistor network consisting of four resistors, the smallest of which equals 1 kΩ. Assume all of these resistors consist of PSD-doped poly-2 with a width of 4 μm. Make all resistors from one or more 1 kΩ segments to ensure precise matching. Assume a width bias of 0.2 μm and ignore all other correction factors. Connect an array of four polysilicon fuses (designed in Exercise 5.12) to the resistor array to complete the trim network.

5.14. Lay out a standard bipolar Zener zap. Assume that the emitter contact has a width of 8 μm and minimize all other dimensions. Assume the trimpads require 75 × 75 μm openings in the protective overcoat. The tank and NBL geometries can reside beneath the trimpads.

5.15. For the differential trim network of Figure 5.26, determine the values of resistors R_A and R_B required to produce a 20 Ω change in resistance when fuse F_2 is blown. Minimize the total resistance $R_A + R_B$ as far as is consistent with good design practice.

5.16. Suppose a certain resistor has a linear TCR of 700 ppm/°C and a quadratic TCR of 60 ppm/°C^2. If this resistor has a value of 4700 Ω at 100°C, what will its resistance be at 125°C?

6 Capacitors and Inductors

Capacitors are a class of passive elements useful for coupling AC signals and for constructing timing and phase shift networks. They are relatively bulky devices that store energy in electrostatic fields. The microscopic dimensions of integrated circuits preclude the fabrication of more than a few hundred picofarads of capacitance. Even this tiny amount suffices for certain crucial applications, particularly for compensating feedback loops. Most analog integrated circuits contain at least one capacitor.

Inductors are another class of passive elements that store energy in the form of electromagnetic fields. Inductors are extremely bulky devices, so only a few tens of nanohenries of inductance can be integrated on a practical integrated circuit. These tiny inductors become useful only at frequencies beyond 100MHz. Certain radio-frequency (RF) circuits have made use of integrated inductors, but few analog integrated circuits will contain them.

6.1 CAPACITANCE

If charge is added to a conductor, this charge generates an electric field. The appearance of the electric field implies a change in the electrical potential of the conductor. This relationship can be quantified by the equation

$$Q = CV \tag{6.1}$$

where Q is the electrical charge placed upon the conductor, V is the voltage differential generated by the addition of the charge, and C is a constant of proportionality called the *capacitance*. A *capacitor* is an electrical device that is designed to introduce a known and precise amount of capacitance.

The International System of Units (SI) defines the *Farad* (F) as the standard unit of capacitance. A Farad is an extremely large amount of capacitance. Most discrete circuits employ capacitors ranging from a few picofarads to a few thousand microfarads.[1] No more than a few hundred picofarads can be economically integrated, so larger capacitors must reside off-chip. Most systems use a number of discrete capacitors in conjunction with each analog integrated circuit.

All of the capacitors used in integrated circuits are *parallel-plate capacitors,* which consist of two conductive plates called *electrodes* attached to either side of a slab of insulating material called the *dielectric* (Figure 6.1). In the simple parallel-plate capacitor, the two electrodes are assumed to have the same dimensions and to reside directly opposite one another.

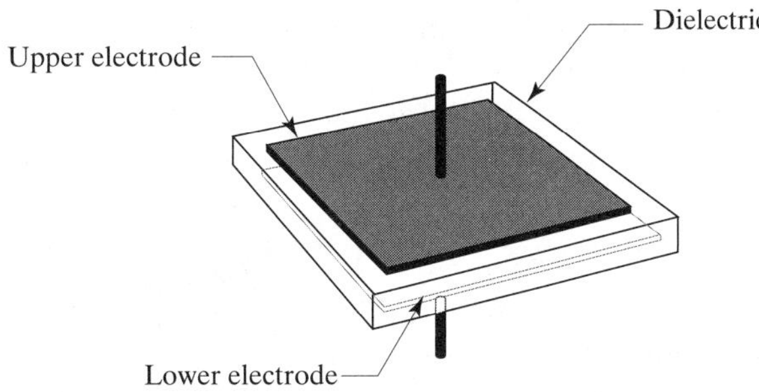

FIGURE 6.1 Construction of a simple parallel-plate capacitor.

The value of the simple parallel-plate capacitor can be computed with the approximate equation

$$C \cong 0.0885\frac{A\varepsilon_r}{t} \qquad [6.2]$$

where C is the capacitance in picofarads (pF), A is the area of either electrode in square microns (μm^2), t is the thickness of the dielectric in Angstroms (Å), and ϵ_r is a dimensionless constant called the *relative permittivity* or the *dielectric constant.* ϵ_r depends on the nature of the dielectric. Table 6.1 lists the relative permittivities of

TABLE 6.1 Relative permittivities and dielectric strengths of selected materials.[2]

Material		Relative Permittivity (Vacuum = 1)	Dielectric Strength (MV/cm)
Silicon		11.8	30
Silicon dioxide (SiO_2)	Dry oxide	3.9	11
	Plasma	4.9	3–6
	TEOS	4.0	10
Silicon nitride (Si_3N_4)	LPCVD	6–7	10
	Plasma	6–9	5

[1] The correct abbreviations for picofarads and microfarads are pF and μF. Historically, a number of other abbreviations were used, including μμF for picofarads and mF and mFd for microfarads. The use of such non-standard abbreviations should be avoided. At the same time, it is inadvisable to use the abbreviation mF for millifarads because of the historical meaning attached to this term.

[2] Values for SiO_2, Si_3N_4 are from A. C. Adams, "Dielectric and Polysilicon Film Deposition," in S. M. Sze, ed., *VLSI Technology,* 2d ed. (New York: McGraw-Hill, 1988), pp. 259, 263. Critical field for silicon taken from D. J. Hamilton and W. G. Howard, *Basic Integrated Circuit Engineering* (New York: McGraw-Hill, 1975), p. 135. See also W. R. Runyan and K. E. Bean, *Semiconductor Integrated Circuit Processing Technology* (Reading MA: Addison-Wesley, 1994), pp. 67–68 for a discussion of the distribution of breakdown voltages.

several materials that are frequently used in integrated circuits. The entries for oxide and nitride list a range of values because the properties of these materials depend on deposition conditions.

Consider a capacitor with a plate area of 0.1 mm^2 constructed using a 200 Å (0.02 μm) dry oxide film. If the dielectric has a relative permittivity of about four (as is usually the case) then the capacitance will equal about 180 pF. This example helps explain why it is so difficult to integrate capacitors of more than a few hundred picofarads.

Reducing the thickness of the dielectric increases its capacitance, but it also increases the electric field imposed across it. Sufficiently intense electric fields can cause dielectric breakdown (Section 4.1.3). To prevent catastrophic failure, the electric field across the dielectric must never exceed a critical value called the *dielectric strength.* Table 6.1 lists the dielectric strengths of various materials in megavolts per centimeter (MV/cm). The maximum voltage V_{max} that a parallel plate capacitor can withstand equals

$$V_{max} = 0.01tE_{crit} \tag{6.3}$$

where t is the dielectric thickness in Angstroms (Å) and E_{crit} is the dielectric strength in MV/cm. According to this formula, the maximum voltage that a 200 Å dry oxide can withstand equals about 20 V. Long-term reliability requires this value to be derated by at least 50%, so a 200 Å oxide is usually rated for 10 V operation. Oxide films grown on polysilicon may require further derating because of the presence of microscopic irregularities at the polysilicon/oxide interface. The presence of these *asperities* causes localized intensification of the electric field and reduces the dielectric strength of the oxide.[3] A 200 Å dry oxide grown on polysilicon might therefore be rated for no more than 5 V.

When the thickness of the dielectric has been reduced as far as the operating voltage allows, then only a high-permittivity dielectric can further increase the capacitance per unit area. Certain materials, such as barium strontium titanate, have relative permittivities of several thousand.[4] Although these materials can be deposited on an integrated circuit, the costs involved render them economical for only a few applications. Designers must instead turn to other, more commonly available, materials. Silicon nitride is often chosen because it has a permittivity roughly twice that of oxide. Unfortunately, thin nitride films are prone to the formation of *pinholes*—small areas of inadequate thickness that compromise the dielectric strength of the film. Some processes sandwich a nitride layer between two oxide layers to obtain a composite dielectric less susceptible to pinhole formation.[5] The effective relative permittivity ϵ_{eff} of an oxide-nitride composite dielectric can be computed from the formula

$$\varepsilon_{eff} = \frac{t_{ox} + t_{nit}}{\left(\frac{t_{ox}}{\varepsilon_{ox}}\right) + \left(\frac{t_{nit}}{\varepsilon_{nit}}\right)}, \tag{6.4}$$

[3] N. Klein and O. Nevanlinna, "Lowering of the Breakdown Voltage of Silicon Dioxide by Asperities and at Spherical Electrodes," *Solid-State Electronics,* Vol. 26, #9, 1983, pp. 883–892.

[4] The actual capacitances per unit area achievable with high-permittivity dielectrics are not as large as relative permittivities suggest because the dielectric strength of a material decreases as its permittivity increases. See J. McPherson, J. Kim, A. Shanware, H. Mogul, and J. Rodriguez, "Proposed Universal Relationship between Dielectric Breakdown and Dielectric Constant," *Proc. International Electron Devices Meeting,* 2002, pp. 633–636.

[5] The exact mechanism by which oxidation improves dielectric integrity is uncertain, but may involve charge trapping; see K. K. Young, C. Hu, and W. G. Oldham, "Charge Transport and Trapping Characteristics in Thin Nitride-Oxide Stacked Films," *IEEE Electron Device Letters,* Vol. 9, #11, 1988, pp. 616–618.

where t_{ox} and t_{nit} are the thicknesses of oxide and nitride, and ϵ_{ox} and ϵ_{nit} are their relative permittivities. For example, if 200 Å of nitride with a relative permittivity of 7.5 is sandwiched between two 50 Å oxide films with relative permittivities of 3.9, then the composite has an effective relative permittivity of 5.7. The resulting film has the dielectric strength of 300 Å dry oxide, yet it has 50% more capacitance per unit area.

Capacitors that use oxide or oxide-nitride dielectrics are identified by a bewildering array of different names. An *oxide capacitor* employs silicon dioxide as its dielectric. This oxide is usually grown on a lower electrode consisting of either a silicon diffusion or a polysilicon deposition. The upper plate usually consists of metal or doped polysilicon. An *ONO capacitor* resembles an oxide capacitor, except that it employs an oxide-nitride-oxide composite dielectric to obtain a higher capacitance per unit area. *Poly-poly capacitors* employ two polysilicon electrodes in combination with either an oxide or an ONO dielectric. *MOS capacitors* consist of a thin layer of grown oxide formed on a silicon diffusion that serves as one of the electrodes. The other electrode consists of either metal or doped polysilicon. If gate oxide is used to form a MOS capacitor, the resulting structure is often called a *gate oxide capacitor.* Despite their many names, all of these structures are variations upon a common theme: that of the *thin-film capacitor.*

The value of a thin-film capacitor may vary due to voltage modulation effects within its electrodes, but its maximum possible capacitance depends solely on the dielectric. This *dielectric capacitance* can be computed using Equation 6.2. If the two electrodes are of different sizes, then the common area of the two plates is used in the equation. For example, in the hypothetical thin-film capacitor of Figure 6.2, only the cross-hatched area where the two electrodes overlap contributes to the capacitance; therefore, the effective area of the electrodes equals 300 μm^2.

Equation 6.2 slightly underestimates the true capacitance because the electric field is not entirely confined to the region between the electrodes. The field actually flares out around the edges—an effect called *fringing* (Figure 6.3). The fringing field increases the apparent width of the capacitor plates by an amount proportional to the thickness of the dielectric. This effect is usually ignored when constructing thin-film

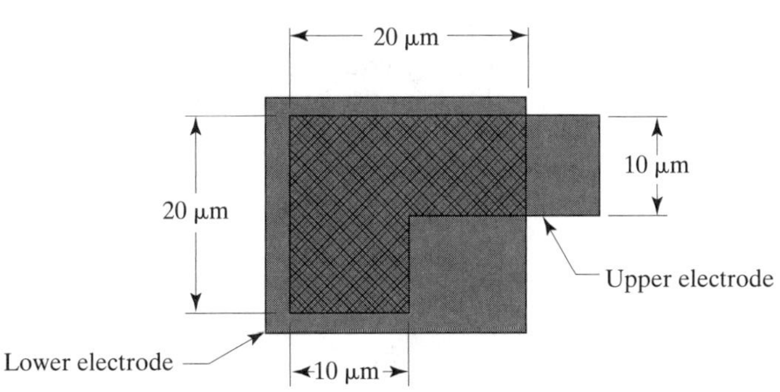

FIGURE 6.2 Hypothetical example of a thin-film capacitor. The crosshatched region where the two plates intersect forms the effective area of the capacitor plates, or in this case 300 μm^2.

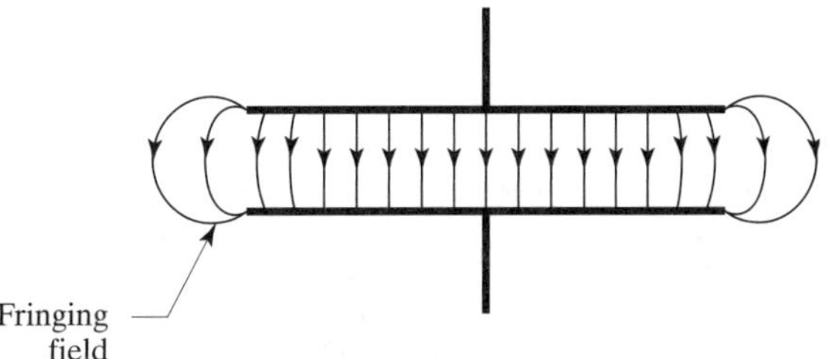

FIGURE 6.3 Illustration of the fringing field surrounding a parallel-plate capacitor embedded in a dielectric of constant permittivity.

capacitors because the thickness of the dielectric is much less than the dimensions of the electrodes. For example, consider a capacitor using a 500 Å dielectric. Assuming that the capacitor has circular plates 25 μm in diameter, the error caused by fringing fields equals about 0.7%.[6] For larger capacitors, or those that employ thinner dielectrics, the effects of fringing fields are even smaller.

Another type of integrated capacitor is the *junction capacitor,* which uses the depletion region surrounding a reverse-biased junction as a dielectric. The permittivity and dielectric strength of silicon are about three times those of oxide, so junction capacitors can obtain high capacitances per unit area. The benefits of compact size are offset by extreme voltage nonlinearity caused by variations in depletion region width with applied bias. The *zero-bias capacitance* C_{j0} serves as a measure of the value of the capacitor. As the reverse bias across the junction increases, its depletion region widens and its capacitance decreases.

The zero-bias capacitance can be computed using Equation 6.2, given the thickness of the depletion region. In the case of an abrupt junction between a heavily doped region and a uniform lightly doped one, the zero-bias depletion region width W_0 (in Angstroms) equals[7]

$$W_0 \cong 3.10^{11} \sqrt{1/N} \tag{6.5}$$

where N is the concentration of dopant atoms per cubic centimeter on the lighter side of the junction. In practice, this equation only applies to shallow, heavily doped diffusions formed in a lightly doped silicon layer (for example, NSD in P-epi). Deeper junctions exhibit considerable grading of the doping profile, and Equation 6.5 ceases to even approximate reality. No simple closed-form equation exists for computing the depletion region widths of diffused junctions, but Lawrence and Warner[8] have published junction capacitance curves for diffusions into a constant background concentration.

The area of junction capacitors is also relatively difficult to compute. Figure 6.4 shows the three-dimensional profile of a typical planar diffusion. As the dopant is driven down, it outdiffuses in all directions to form curved sidewalls. These sidewalls intersect each other to form rounded (*filleted*) corners.

The area of a diffused junction consists of three components: the area of the bottom surface, which approximately equals the area of the oxide window shown in gray in Figure 6.4; the area of the sidewalls, which is proportional to the perimeter of the drawn

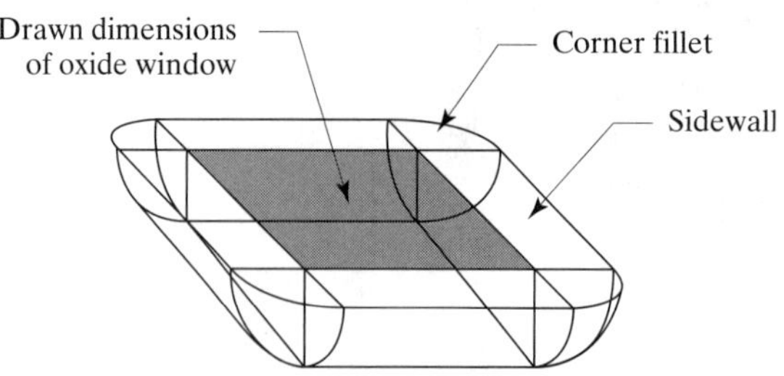

FIGURE 6.4 Three-dimensional view of a diffused junction, showing the sidewalls and filleted corners produced by outdiffusion beyond the drawn dimensions of the oxide window.

---

6 Computed from a formula in C. H. Séquin, "Fringe Field Corrections for Capacitors on Thin Dielectric Layers," *Solid State Electronics,* Vol. 14, 1971, pp. 417–420.

7 This equation assumes that the built-in potential is 0.7 V, which is a reasonable approximation for light-to-moderate doping levels.

8 H. Lawrence and R. M. Warner, Jr. "Diffused Junction Depletion Layer Capacitance Calculations," *Bell System Tech. J.*, Vol. 34, 1955, pp. 105–128.

geometry; and the area of the corner fillets. Approximating the sidewalls as cylindrical segments and neglecting the corner fillets, the total area of the junction equals

$$A_{total} = A_d + \frac{\pi}{2} x_j P_d \qquad [6.6]$$

where A_d and P_d are the drawn area and perimeter of the oxide window, and x_j is the junction depth of the diffusion.[9]

While one can use the Lawrence-Warner curves to predict the capacitance of a diffused junction, the results are approximate and the process is tedious. A simpler method of determining junction capacitances uses the empirical equation

$$C_{total} = C_a A_d + C_p P_d \qquad [6.7]$$

where the *areal capacitance* C_a represents the capacitance per unit area, and the *peripheral capacitance* C_p represents the capacitance per unit periphery. The values of these two constants are determined by measuring two or more junction capacitors with very different perimeter-to-area ratios. This technique is considerably more accurate than *a-priori* computations because the experimentally determined constants take into account the vast majority of nonidealities.

Junction capacitors are a staple of standard bipolar design because this process does not produce thin oxides suitable for use as capacitor dielectrics. The base-emitter junction usually provides the highest capacitance per unit area. Experimental measurements on a 40 V standard bipolar process with a 160 Ω/□ 2 μm-deep base gave $C_a = 0.82\ \text{fF}/\mu\text{m}^2$ and $C_p = 2.8\ \text{fF}/\mu\text{m}$.[10]

Junction capacitors customarily employ one of two competing styles of layout. The *plate capacitor* (Figure 6.5A) maximizes junction area, while the *comb capacitor* (Figure 6.5B) maximizes junction periphery. The comb capacitor will have more capacitance per unit area than the plate capacitor if the spacing between the fingers S_f is sufficiently small. Quantitatively, the comb layout is superior whenever the following inequality is met:[11]

$$S_f < \frac{2C_p}{C_a} \qquad [6.8]$$

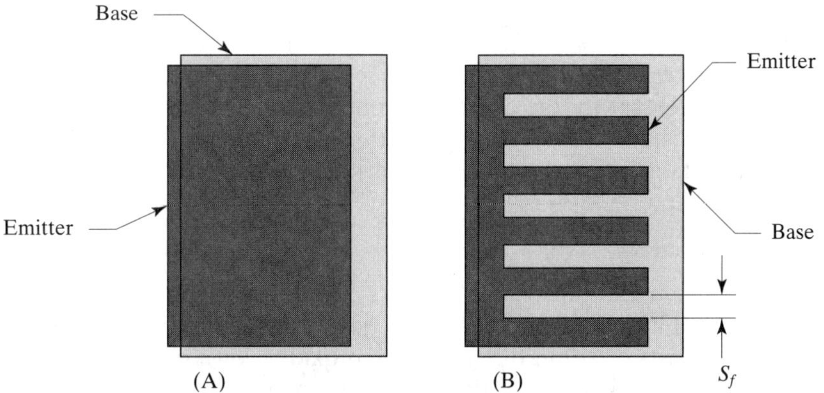

FIGURE 6.5 Two different styles of diffusion capacitor: (A) plate and (B) comb. The tank, NBL, contact, and metal layers are omitted for clarity.

9 This formula and the technique associated with its use are discussed at some length in Hamilton, *et al.*, pp. 129–135.

10 F. W. Trafton and R. A. Hastings, "A Study of Emitter-Base Junction Capacitance," unpublished paper, 1989.

11 *Ibid.*

In practice, most versions of standard bipolar can obtain slightly more capacitance from the comb capacitor. Comb capacitors are therefore a familiar sight on older analog layouts. They rarely appear on CMOS and BiCMOS designs because thin-film capacitors can provide equal or larger capacitances per unit area with fewer parasitics than junction capacitors.

A number of different symbols have been used to represent capacitors on schematics. Figure 6.6A shows the standard symbol for a generic capacitor. On schematics, this symbol is usually supplemented by annotations indicating the type of capacitor, its value, and the identity of each of its plates. Figure 6.6B shows a symbol originally used for tubular foil capacitors. The curved electrode represented the outside layer of foil, which was usually grounded to form an electrostatic shield for the rest of the capacitor. This symbol is frequently used to represent integrated capacitors because the two plates are easily distinguished from one another. The curved plate usually (but not always) represents the bottom electrode of an integrated capacitor.[12] The symbol of Figure 6.6C represents a junction capacitor. The arrowhead indicates the P-type electrode (*anode*), while the unadorned plate indicates the N-type electrode (*cathode*).

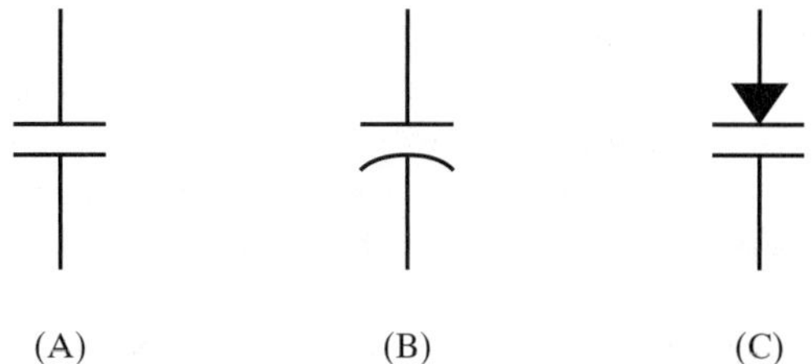

FIGURE 6.6 Typical schematic symbols for capacitors.

Junction capacitors are sometimes represented as PN junction diodes with values given in terms of junction areas. This practice is understandable from a simulation point of view, since the junction capacitor is modeled as a PN diode. On the other hand, the layouts used for diodes and junction capacitors differ because the capacitor always remains reverse-biased, while a true PN diode operates under forward as well as reverse bias. Many designers therefore use the symbol of Figure 6.6C to distinguish junction capacitors from diodes.

6.1.1. Capacitor Variability

Integrated capacitors display considerable variability, due mostly to process variation and voltage modulation. There are several lesser sources of variability that only become important for the construction of accurately matched capacitors. These include electrostatic fields and fringing effects, nonuniform etch rates and gradients in doping, film thickness, temperature, and stress. An analysis of these lesser effects appears in Chapter 7.

Process Variation

Both thin-film and junction capacitors experience significant process variations, the causes of which are unique to each type of capacitor. In MOS capacitors, the dielectric consists of a thin film of silicon dioxide grown on monocrystalline silicon. The thickness of this film rarely exceeds 500 Å, and in low voltage processes it may be less than 100 Å thick. Since the silicon-oxygen bond is approximately 1.5 Å

[12] The curved plate of a capacitor is sometimes used to denote the upper electrode of an integrated capacitor because this plate is outermost and thus corresponds to the outside foil of a tubular capacitor. Because of the potential for confusion, the plates should always be explicitly labeled.

long,[13] these films consist of no more than a few hundred atomic monolayers. Much research and development has been directed toward achieving precise control of thin oxide dielectrics. Modern CMOS processes routinely control gate oxide capacitance to within ±20%, and some processes can maintain ±10%.[14]

Dielectrics deposited or grown on polysilicon or metal electrodes are less well controlled than gate oxide. The permittivity of the dielectric film depends not only on thickness, but also on composition, which can vary substantially depending on the conditions of growth or deposition. ONO dielectrics are particularly variable because they are formed by a three-step process consisting of initial oxide growth, followed by nitride deposition and subsequent surface oxidization. Each of these steps introduces its own uncertainties, so ONO capacitors typically vary by at least ±20% over process.

Junction capacitors are usually constructed from base and emitter diffusions. The emitter-base depletion region width depends on numerous factors, including the average base doping concentration, the base doping profile, and the emitter junction depth. These factors produce at least ±20% variation in a plate capacitor. A comb layout varies even more than a plate layout for several reasons. First, the peripheral capacitance depends more strongly on emitter junction depth than does the areal capacitance. Second, the peripheral capacitance is more susceptible to surface effects such as oxide charge modulation and boron suckup. Third, the intersecting tails of adjacent emitter diffusions modulate the base doping between the fingers and, consequently, vary the peripheral capacitance. Comb capacitors generally vary by at least ±30% over process. These tolerances do not include the effects of voltage modulation and temperature variation.

Voltage Modulation and Temperature Variation

Ideally, the value of a capacitor should not depend on the bias placed across it. Junction capacitors do not even approximate this ideal because the reverse bias placed across the junction modulates the width of its depletion region. Similar effects are observed in many thin-film capacitors because one (or both) electrodes consist of doped silicon subject to depletion effects. MOS capacitors are particularly vulnerable to depletion modulation because their lower electrode is lightly doped and therefore easily depleted, but even poly-poly capacitors with relatively heavily doped electrodes exhibit small voltage nonlinearities due to depletion of the polysilicon. These effects completely vanish if both electrodes consist of metal or silicide.

Figure 6.7 shows the general character of the voltage variation exhibited by a junction capacitor. The capacitance gradually decreases from the zero-bias value C_{j0}

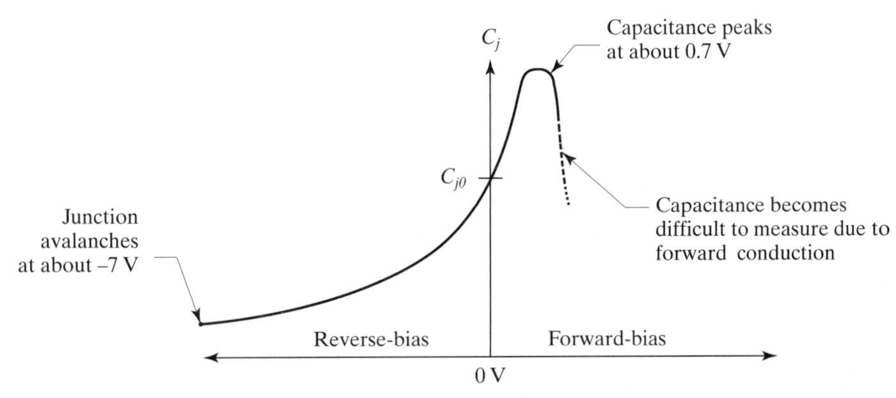

FIGURE 6.7 General behavior of base-emitter junction capacitance under bias. The minimum capacitance just prior to avalanche equals about 40 to 50% of the zero-bias value C_{j0}.

[13] R. C. Weast, ed., *Handbook of Chemistry and Physics,* 62d ed. (Boca Raton, FL: CRC Press, 1981), p. F-178.

[14] The variability figures cited in this section represent the 4.5-sigma limits of a Gaussian distribution.

as the reverse-bias increases, because the depletion region gradually widens. Eventually the electric field across the depletion region becomes so intense that the junction avalanches, which occurs in standard bipolar base-emitter junctions at about −7 V. The capacitance of a forward-biased junction actually increases, because the depletion region narrows as the external bias begins to counteract its built-in potential. As the forward-bias approaches the built-in potential, the depletion region collapses and the junction capacitance drops away rapidly.[15] The forward-bias enhancement of junction capacitance is not particularly useful because forward-biased diodes conduct current. Even a forward-bias of only 0.3 V will cause noticeable conduction at higher temperatures, so most designers completely avoid forward-biasing junction capacitors.

Junction capacitors are often used as compensation capacitors because their voltage variability rarely impairs their usefulness in this role. The compensation capacitor is sized so that its absolute minimum value still stabilizes the circuit. This requires that the nominal zero-bias capacitance equal about three times the minimum value needed to stabilize the circuit. Given such a large safety factor, the designer need not worry too much about the exact size or shape of the capacitor.

MOS capacitors may also exhibit strong voltage modulation effects. Figure 6.8 shows the capacitance curve for an NMOS transistor configured as a MOS capacitor. Majority carriers drawn up from the bulk silicon accumulate beneath the gate oxide when the gate is biased negative with respect to the backgate. The capacitance of the device in accumulation is determined solely by the gate dielectric, as given by Equation 6.2. When the gate is biased positively, majority carriers are repelled from the surface and a depletion region begins to form. As the bias increases, this depletion region widens and its capacitance decreases. Once the gate bias equals the threshold voltage, sufficient minority carriers will have been drawn up from the bulk to invert the surface. After inversion occurs, larger forward biases only increase the concentration of minority carriers and do not affect the width of the depletion region. Therefore, the capacitance levels off at a new,

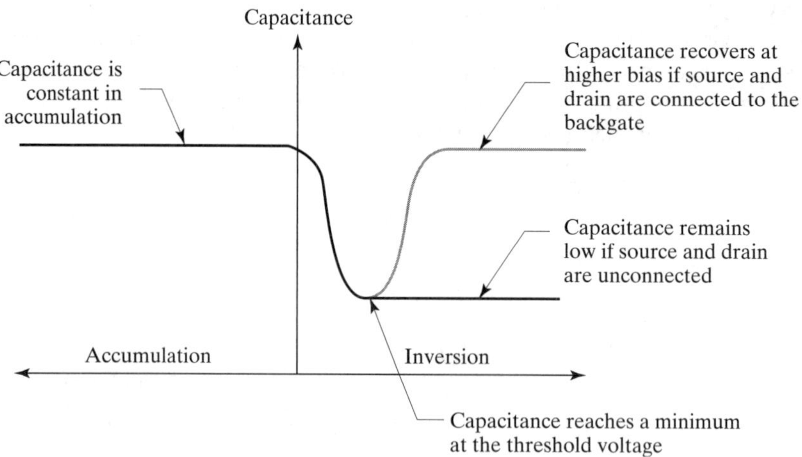

FIGURE 6.8 General behavior of a MOS transistor employed as a capacitor. Different curves are obtained depending on the connection of the source and the drain.

[15] The usual depletion-capacitance formulas indicate that capacitance asymptotically approaches infinity at the built-in potential. This is incorrect; actually, capacitance peaks at the built-in potential at a large but finite value; see B. R. Chawla and H. K. Gummel, "Transition Region Capacitance of Diffused p-n Junctions," *IEEE Trans. on Electron Devices*, ED-18, 1971, pp. 178–195.

lower value.[16] This minimum capacitance C_{min} may equal less than 20% of the gate oxide capacitance.[17]

The preceding analysis applies as long as the source and drain diffusions are absent or unconnected. If these diffusions exist and are connected to the backgate, then the behavior of the MOS transistor capacitance is somewhat more complicated. Once strong inversion occurs, a conducting channel shorts the source and drain terminals. This channel then becomes the lower plate of the capacitor, and the capacitance rises to again equal the gate oxide capacitance (Figure 6.8).

MOS transistors used as capacitors are generally biased outside the capacitance dip centered on the threshold voltage. The source and drain diffusions are unnecessary if the device operates in accumulation, but they must be present and electrically connected to the backgate if it is to achieve full capacitance in inversion. An NMOS transistor operates in accumulation if the gate is biased negative to the backgate and in inversion if the gate is biased positive to the backgate. Similarly, a PMOS transistor operates in accumulation if the gate is biased positive to the backgate and in inversion if the gate is biased negative to the backgate. One can usually determine by examination whether or not source and drain diffusions are needed for a given MOS capacitor. If in doubt, the diffusions can always be included because they cause no harm even if the device operates in accumulation.

Junction and MOS capacitors both have one electrode formed of lightly doped silicon that is prone to depletion modulation, so these devices exhibit considerable voltage variation. Most other types of capacitors use highly conductive electrodes and exhibit much less voltage modulation. Heavily doped silicon electrodes still exhibit a small amount of voltage modulation, usually no more than 50 to 100 ppm/V.[18] This small voltage modulation becomes significant only in precisely matched capacitors operating at different voltages, as is the case in charge-redistribution DACs. Capacitors with electrodes formed from metal or, in the case of the lower plate, silicided poly typically have voltage modulations of less than 5 ppm/V.[19]

6.1.2. Capacitor Parasitics

All integrated capacitors have significant parasitics. The desired capacitance results from the electrostatic interaction between two large-area electrodes. These same electrodes also electrostatically couple to the rest of the integrated circuit, producing unwanted parasitics. The parasitic capacitances associated with one plate usually outweigh those associated with the other, so the orientation of the capacitor becomes quite important.

Figure 6.9A shows a simple subcircuit model of the parasitics associated with a poly-poly capacitor. This model also applies to other types of capacitors whose electrodes are both deposited layers. Ideal capacitor C_1 represents the desired capacitance

---

16 Actually, the capacitance may recover somewhat if the measurements are performed at very low frequencies. This phenomenon is caused by modulation of generation and recombination within the inversion region due to the electric field projected by the gate electrode. This effect is of no practical significance to the layout designer.

17 It is rarely necessary to evaluate C_{min}, because MOS capacitors are normally operated to retain the full gate oxide capacitance. Most device physics texts discuss the evaluation of C_{min}; for example, see B. G. Streetman, *Solid State Electronic Devices,* 2d ed. (Englewood Cliffs, NJ: Prentice-Hall, 1980), p. 296 ff.

18 A linear coefficient of -148.3 ppm/V and a quadratic coefficient of -9.1 ppm/V^2 were reported by R. H. Eklund, R. A. Haken, R. H. Havemann, and L. N. Hutter, "BiCMOS Process Technology," in A. R. Alvarez, ed., *BiCMOS Technology and Applications*, 2d ed. (Boston: Kluwer Academic, 1993), p. 123.

19 A linear coefficient of 1.74 ppm/V and a quadratic coefficient of -0.4 ppm/V^2 were reported by Eklund, *et al.,* p. 123.

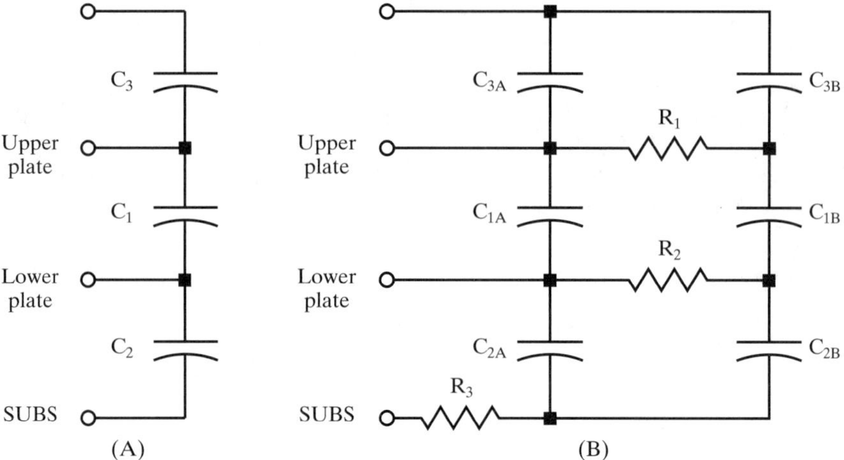

FIGURE 6.9 Subcircuit models for poly-poly capacitors: (A) a simple model without series resistance, and (B) a model incorporating series resistance using π-sections.

of the structure. C_2 represents the parasitic capacitance between the lower electrode and the substrate. The value of this parasitic can be calculated using the area of the lower plate and the thickness of the field oxide. Capacitor C_3 represents the parasitic capacitance associated with the upper plate. This capacitance is usually much smaller than C_2, and it becomes significant only if another conductor overlies the capacitor. No leads should route across a capacitor unless they connect to it, not only because they add unwanted capacitance but also because of the potential for noise coupling. C_3 can become significant when a metal shield is placed over the capacitor to help improve matching. (See Section 7.2.12.) In this case, the capacitance C_3 coupling the upper plate of the capacitor to the shield can be computed using the area of the upper electrode and the thickness of the interlevel oxide (ILO).

The series resistance of the polysilicon electrodes may become significant at higher frequencies. The subcircuit model of Figure 6.9B incorporates this series resistance by dividing all of the capacitors into single π-sections. Resistor R_1 models the series resistance of the upper plate, while R_2 models that of the lower plate. Capacitors C_1, C_2, and C_3 are divided into equal sections C_{1A}/C_{1B}, C_{2A}/C_{2B}, and C_{3A}/C_{3B}. Resistor R_3 represents the finite resistance of the substrate, which is often as large as—or larger than—the series resistance of either plate.

Capacitors using diffused electrodes are modeled by replacing the parasitic capacitors with diodes. These diodes remain reverse-biased under normal operating conditions, but leakage currents still flow through them, and these may become significant at higher temperatures. Much larger currents will flow if the diodes even momentarily forward-bias. Since forward conduction involves significant minority carrier flow, inadvertent forward-biasing of junction capacitors can cause latchup.

Figure 6.10A shows a subcircuit model for a base-emitter junction capacitor. The emitter plate is usually connected to the tank, placing the base-emitter and base-collector junctions in parallel. Although the base-collector junction does not add much capacitance, every bit helps. Diodes D_1 and D_2 model the paralleled base-emitter and base-collector junctions. Diodes D_3 and D_4 model the collector-substrate junction. Resistor R_1 models the distributed resistance of the base plate, which is greatly increased by the pinching action of the emitter plate. The resistance of the emitter plate is negligible because the emitter sheet resistance is so much smaller than that of the pinched base. Resistor R_2 models the resistance of the substrate. This model does not include parasitic capacitances coupled to the emitter plate, but these can be modeled as ideal capacitances if desired.

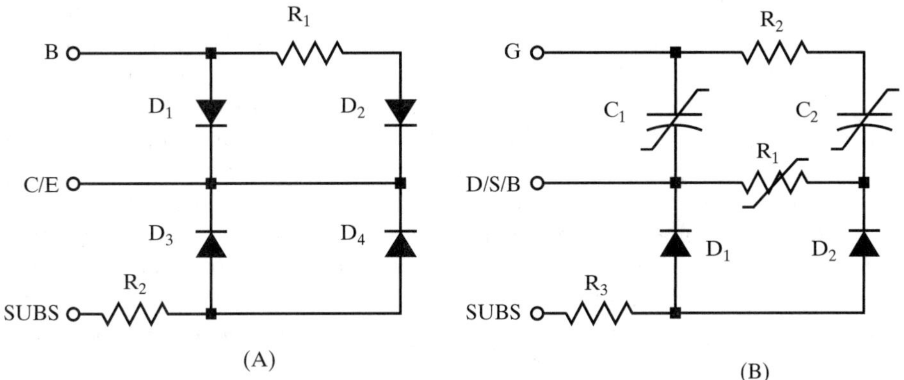

FIGURE 6.10 Subcircuit models for (A) a junction capacitor where C/E denotes the collector/emitter electrode and B denotes the base electrode; and (B) a MOS or gate oxide capacitor where G denotes the deposited gate electrode and D/S/B denotes the drain/source-backgate electrode.

Figure 6.10B shows a subcircuit model for a MOS capacitor. The capacitance of the structure is modeled using two voltage-dependent capacitors C_1 and C_2, each representing half of the capacitance of the MOS structure. R_1 models the distributed resistance of the lower plate of the capacitor, which, in accumulation, consists of well resistance and, in inversion, consists of channel resistance. In either case, this resistance depends strongly upon voltage. R_2 models the resistance of the top-plate electrode, which is often negligible in comparison with R_1. Diodes D_1 and D_2 represent the junction isolating the N-well from the P-epi. Each of these diodes has an area equal to half of the well-epi junction area. Resistor R_3 models the resistance of the substrate contact system that extracts current from the epi side of the well-epi junction. The diagonal slashes through R_1, C_1, and C_2 indicate that these components are voltage dependent.

6.1.3. Comparison of Available Capacitors

Most processes offer only a limited selection of capacitors. Standard bipolar offers base-emitter junction capacitors, and (with the addition of one extra masking step) thin oxide capacitors. CMOS processes are usually limited to MOS capacitors, but some of these processes offer extensions to build poly-poly capacitors. BiCMOS processes offer greater flexibility, as they can typically fabricate both MOS and base-NSD junction capacitors, and they may also provide poly-poly capacitors. This section analyzes the strengths and weaknesses of each of these types of capacitors.

Base-Emitter Junction Capacitors

Base-emitter junction capacitors exist in both standard bipolar (Figure 3.16) and analog BiCMOS processes. They provide excellent capacitance per unit area at zero bias (typically 0.8 fF/μm^2), but this capacitance falls away with increasing reverse bias. A reverse bias of only −1 V causes the capacitance to drop to 75% of its zero-bias value. The capacitance loss slows as the bias increases, so a reverse bias of −5 V reduces the capacitance by 50%. Most circuits apply several volts of reverse bias to the capacitor, so the effective capacitance per unit area of base-emitter junction capacitors equals approximately 0.5 fF/μm^2. The extreme variability of junction capacitors limits them to noncritical roles such as noise filters and compensation capacitors for closed-loop feedback circuits.

The layout designer must decide whether to use a plate layout (Figure 6.5A) or a comb layout (Figure 6.5B). If the areal and peripheral capacitances are known, then Equation 6.8 will indicate which layout requires the least space. If the values of the areal and peripheral capacitances are not known, then the plate layout should be favored over the comb layout because the value of the latter depends strongly on its peripheral capacitance, which is difficult to compute on an *a-priori* basis.

The series resistance of a junction capacitor is greatly increased by the pinching action of the emitter plate. The best structure for minimizing this resistance consists of a series of minimum-width emitter fingers interdigitated with base contacts. This configuration minimizes the effects of base pinch resistance at the cost of increased area. The comb structure provides a reasonably low series resistance because the unpinched base regions between the emitter fingers are much less resistive than the pinched base regions underneath them. The emitter fingers can run either horizontally or vertically; the orientation that gives the shortest fingers also gives the minimum series resistance. (The layout in Figure 6.5B follows this rule, but the one in Figure 3.16 does not.) The plate layout (Figure 6.5A) has by far the highest series resistance.

Junction capacitors are normally laid out inside a tank. Contact must be made to this tank to ensure that the base-collector junction remains reverse-biased. This contact also places the base-collector junction in parallel with the base-emitter junction and thus slightly increases the total capacitance. The tank contact can be made by simply extending the emitter plate beyond the base plate (Figure 3.16). The isolation junction acts as a parasitic capacitance tied to the cathode (emitter) plate, as represented by diodes D_1 and D_2 in Figure 6.10B. The addition of NBL to the tank substantially increases this capacitance. NBL serves no useful function in a junction capacitor, so it is often omitted to minimize the bottom-plate parasitic capacitance. If the anode (base) plate is connected to substrate potential, then this parasitic junction capacitor appears in parallel with the desired capacitance. In this case, NBL may be added to maximize the available capacitance per unit area.

If the anode connects to substrate potential, then the base diffusion can extend out into the isolation to save area (Figure 6.11). The base contact can occupy any portion of the base diffusion, even those portions over isolation. In most processes, the emitter cannot reside over isolation and must instead occupy a tank following the appropriate layout rules. The capacitor of Figure 6.11 consists of two sets of fingers branching from a common tank/emitter contact in the middle; this arrangement helps minimize finger lengths and parasitic resistances.

A base-NSD junction capacitor can be constructed in the P-epi of a BiCMOS process. The P-epi is so lightly doped that it does not significantly affect the breakdown of the base-NSD junction. The well diffusion can be omitted as long as the

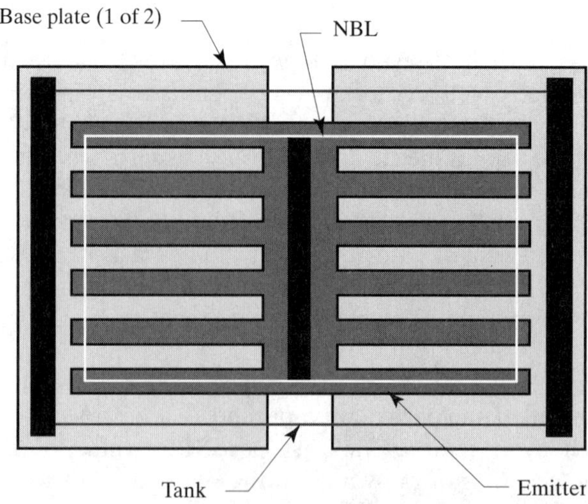

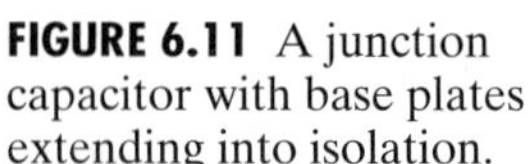
FIGURE 6.11 A junction capacitor with base plates extending into isolation.

anode plate of the junction capacitor connects to substrate potential. The capacitance of the structure increases slightly if N-well is coded beneath it due to the added capacitance of the base/N-well junction. This increase in capacitance rarely exceeds 20% of the total, making it debatable whether the N-well is worth the space it consumes.

While more capacitance can be obtained from a junction held under a slight forward-bias, it is very difficult to prevent forward conduction at higher temperatures. Forward-biasing a junction capacitor will cause substantial current flow, although this is not necessarily destructive to the device. Certain circuit configurations actually use forward conduction to clamp the voltage across the capacitor, but the vast majority of junction capacitors are intended to remain reverse-biased at all times.

Junction capacitors have relatively small breakdown voltages. The standard bipolar base-emitter junction typically avalanches at about 6.8 V, so the reverse bias across such a junction capacitor should not exceed 4–5 V. Avalanche breakdown should be avoided because it increases junction leakage by generating surface trap sites that promote recombination in the depletion regions.

The value of a junction capacitor can be increased slightly by placing a metal plate over the emitter to produce a capacitor, using the emitter oxide as its dielectric. If the process employs a thin emitter oxide, then a substantial amount of capacitance can be obtained in this manner, but yields may be reduced by the tendency of thin emitter oxides to form pinholes. Thick emitter oxide does not have this vulnerability, but it provides so little capacitance per unit area that the metal plate produces no appreciable benefit. Most modern processes use thick emitter oxides and thus do not significantly benefit from the addition of a metal plate to junction capacitors.

The die symbolization is often placed over junction capacitors because these devices contain some of the largest areas of unmetallized silicon in the layout. This practice is not harmful, although it does preclude the placement of a metal plate over the capacitor. The symbolization must meet all applicable design rules since it occupies an electrically active region of the die.

Processes that do not experience excessive emitter-isolation leakage can use junction capacitors formed by placing emitter diffusion over isolation regions. Since isolation outdiffuses much farther than emitter, an emitter geometry can be placed directly over an existing isolation region. This technique can produce 100 to 500 pF of junction capacitance with little or no increase in die area. The emitter plate of the capacitor has a relatively low resistance, so the entire capacitor need not be metallized to ensure proper operation. This is an important consideration for single-level-metal designs because leads must cross the emitter-in-iso capacitor to connect adjacent components to one another.

MOS Capacitors

A MOS transistor can be pressed into service as a capacitor, but its lightly doped backgate increases its parasitic resistance. Better results are obtainable using a thin oxide dielectric formed on a heavily doped diffusion. MOS capacitors are sometimes constructed in standard bipolar, using the emitter diffusion as the lower electrode. Unless the process forms an exceptionally thin emitter oxide, an additional masking step is required to produce a suitable dielectric oxide.

MOS transistors are ill-suited for use as capacitors, but on CMOS processes they are often the only choice. A MOS transistor used as a capacitor should be biased to avoid the capacitance dip near the threshold voltage (Figure 6.8). This places the device in either of two favorable biasing modes: *accumulation* or *strong inversion.* Accumulation requires biasing the gate of an NMOS negative to its backgate, or biasing the gate of a PMOS positive to its backgate. A constant bias of at least one volt will ensure that the transistor operates within a relatively linear portion of its capacitance curve, limiting voltage variation to about ±10%. Source and drain electrodes serve no function and can be eliminated as long as the device operates solely in accumulation. Figure 3.34 shows a MOS capacitor of this sort.

An NMOS transistor enters strong inversion when its gate is biased positive to its backgate by the sum of its threshold voltage plus 1 V. A PMOS transistor operates in strong inversion when its gate is biased negative to its backgate; again, the bias should exceed the threshold voltage by at least a volt. A MOS capacitor operating in inversion requires source/drain electrodes to contact the channel. These electrodes normally connect to the backgate terminal. The layout of an inversion-mode capacitor is identical to that of a regular MOS transistor (Figures 3.29, 3.30).

A MOS transistor operated as a capacitor has substantial series resistance, most of which is associated with the lower electrode (resistor R_1 in Figure 6.10B). This resistance can be minimized by using a fairly short channel length, ideally 25 μm or less. If the source and drain diffusions are omitted, then the backgate contact can run entirely around the gate (as illustrated in Figure 3.34).

Figure 6.12 shows one style of MOS capacitor compatible with standard bipolar processing. The capacitor dielectric consists of a thin oxide formed by an etch and regrowth process controlled by a special masking step. The lower electrode of the capacitor consists of an emitter diffusion enclosed in a tank. The upper electrode is formed from first-level metal. The sheet resistance of the emitter diffusion is so low that voltage modulation and series resistance can both be neglected.

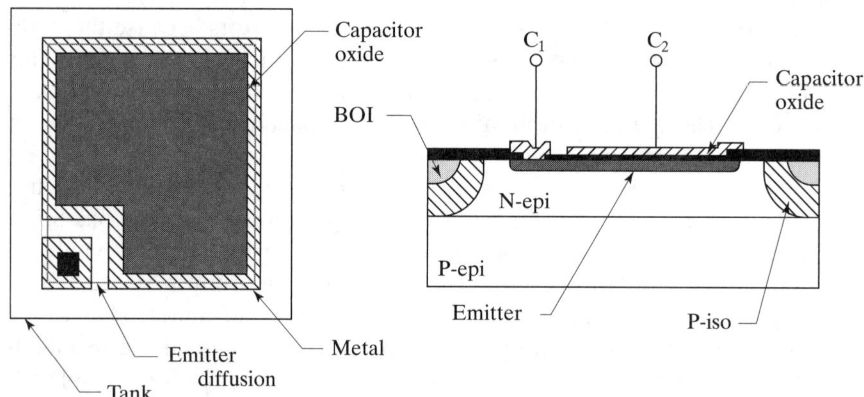

FIGURE 6.12 Layout and cross section of a MOS capacitor constructed in a standard bipolar process using a capacitor oxide mask.

The emitter plate of a MOS capacitor can be formed directly into the standard bipolar isolation, but the resulting N+/P+ junction has considerable parasitic capacitance, and it often exhibits excessive leakage. These difficulties can be circumvented by connecting the emitter plate to the same potential as the isolation. If the emitter plate must connect to a different potential, then it should be enclosed in a tank. This tank does not require the addition of NBL, which can be omitted to further reduce the parasitic capacitance between the emitter plate and the isolation.

Alternatively, the lower (emitter) plate of the MOS capacitor can reside inside a base region connected to the upper electrode. This configuration places the base-emitter junction capacitance in parallel with the thin oxide capacitance of the MOS capacitor to obtain a very high capacitance per unit area—often more than 1.5 fF/μm^2. This type of structure is called a *sandwich capacitor* or a *stacked capacitor.* Like junction capacitors, sandwich capacitors exhibit extreme variability and low breakdown voltages. They are principally used for large compensation and supply line bypass capacitors.

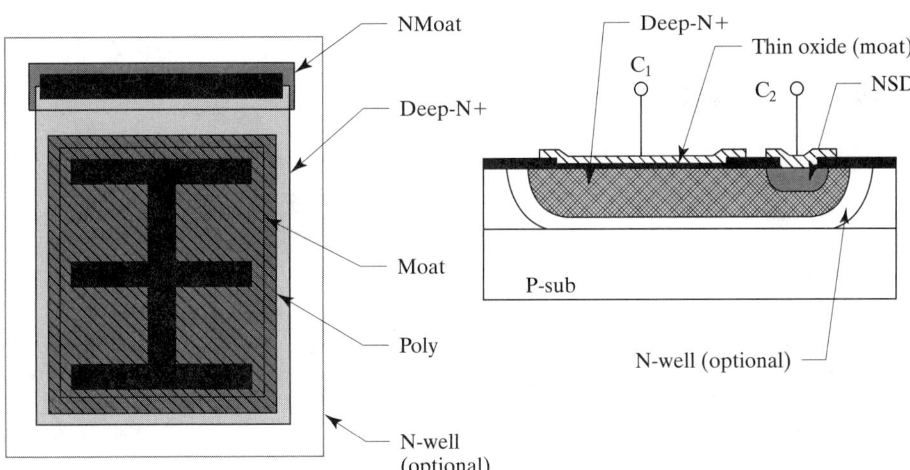

FIGURE 6.13 Layout and cross section of a deep-N+ MOS capacitor constructed in an analog BiCMOS process.

MOS capacitors can also be formed in a BiCMOS process. Since the NSD implant follows gate oxide growth and poly deposition, the lower plate must consist of some other diffusion, usually deep-N+ (Figure 6.13). Deep-N+ has a higher sheet resistance than NSD (typically 100 Ω/□), so the lower plate parasitic resistance can be substantial. The heavily concentrated N-type doping often thickens the gate oxide by 10 to 30% through dopant-enhanced oxidation, resulting in higher working voltages but lower capacitance per unit area. Unfortunately, oxides grown above deep-N+ often exhibit reduced oxide integrity due to defects generated by the heavy doping. The deep-N+ is often placed inside an N-well to reduce parasitic capacitance to the substrate. The N-well can be omitted if the larger parasitic capacitance and lower breakdown voltage of the deep-N+/P-epi junction are tolerable.

Regardless of how a MOS capacitor is constructed, its two electrodes are never wholly interchangeable. The lower plate always consists of a diffusion with substantial parasitic junction capacitance. This junction capacitance can be eliminated only by connecting the lower plate of the capacitor to substrate potential. The upper plate of the MOS capacitor consists of a deposited electrode that has relatively little parasitic capacitance. A circuit designer usually attempts to connect a MOS capacitor so that the lower plate connects to the driven node (the one with the lower impedance). Swapping the two electrodes of a MOS capacitor may load a high-impedance node with unwanted parasitics and can potentially cause circuit malfunctions.

Poly-Poly Capacitors

Both junction and MOS capacitors use diffusions as their lower electrodes. The junction isolating the diffused electrode exhibits substantial parasitic capacitance and restricts the voltages that can be applied to the capacitor. These limitations can be circumvented by using a deposited material such as polysilicon for both electrodes. Many CMOS and BiCMOS processes already incorporate multiple polysilicon layers, so poly-poly capacitors do not necessarily require any additional masking steps. For example, many processes blanket-dope the gate poly and add a second poly for constructing high-sheet resistors. The gate poly can serve as the lower electrode of a poly-poly capacitor, while the resistor poly (doped with a suitable implant) can form the upper electrode. The upper electrode can be doped using either the NSD or the PSD implant (Figure 6.14). The implant that yields the lowest sheet resistance will produce the best capacitor, since heavier doping not

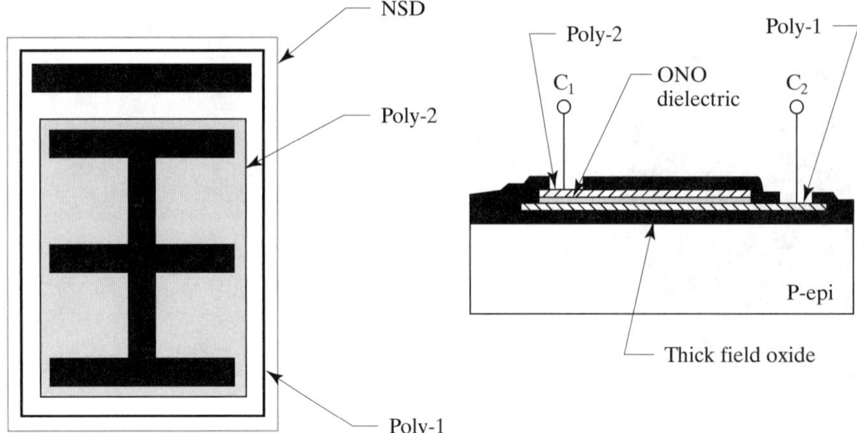

FIGURE 6.14 Layout and cross section of a poly-poly ONO capacitor. The entire capacitor has been enclosed in NSD because the gate poly is also N-type and the additional dopant only further reduces its sheet resistance.

only reduces series resistance, but also minimizes voltage modulation due to poly depletion.

Poly-poly capacitors always require at least one additional process step. Even if both of the electrodes consist of existing depositions, the capacitor dielectric is unique to this structure and consequently requires a process extension. The simplest way to form this dielectric is to eliminate the interlevel oxide (ILO) deposition that normally separates the two polysilicon layers, and in its place substitute a thin oxide grown on the lower polysilicon electrode. Using this technique, a capacitor forms wherever the two poly layers lie on top of one another. This is not a serious limitation as long as the second polysilicon layer is not used for interconnection.

Oxide has a relatively low permittivity. A higher permittivity, and therefore a higher capacitance per unit area, can be obtained with a stacked oxide-nitride-oxide (ONO) dielectric. The first step in forming the ONO dielectric consists of thermal oxidation of the lower polysilicon electrode. A chemical-vapor-deposited (CVD) nitride layer grown on top of the polysilicon is superficially oxidized to form the final composite dielectric. Unwanted regions of nitride can be removed by using a suitable etch after patterning the second poly.

The capacitors in Figures 6.13 and 6.14 use fingered contacts for the upper electrode. Alternatively, a sparse array of contacts can be speckled across the upper plate of the capacitor. A dense array of contacts unnecessarily slows down design software without providing any significant benefit. Some processes also allow the use of a single, large contact opening. All three of these styles of contacts provide a relatively low series resistance over the entire poly-2 plate. The lower electrode of the capacitor in Figure 6.14 is contacted along only one edge. Its series resistance can be reduced by ringing the entire structure with contacts. An even lower series resistance can be obtained by breaking the poly-2 plate into strips and interdigitating these strips with poly-1 contacts.

Poly-poly capacitors normally reside over field oxide. Oxide steps should not intersect the structure, because they can cause surface irregularities in the lower capacitor electrode. Not only can these irregularities cause localized thinning of the dielectric, but they also concentrate the electric field. Both of these effects compromise the dielectric integrity of the capacitor.

Although oxide steps should not intersect a poly-poly capacitor, these capacitors are sometimes enclosed entirely within a deep-N+ diffusion. The heavy phosphorus doping accelerates LOCOS oxidation and produces a thicker field oxide that reduces the parasitic capacitance between the lower plate of the capacitor and the substrate. If the deep-N+ region connects to a quiet low-impedance node, such as analog ground, it will shield the lower capacitor electrode from substrate noise. A similar shield constructed from N-well is useful for situations where deep-N+ is not available, or where the layout rules do not allow capacitors to reside on top of deep-N+ due to planarization difficulties.

If there is a choice of dielectrics for use with poly-poly capacitors, several additional points should be considered. Composite dielectrics experience hysteresis effects at high frequencies (10 MHz or above) due to the incomplete redistribution of static charges along oxide-nitride interfaces.[20] If the value of the capacitor must remain constant regardless of frequency, then pure oxide dielectrics are preferable to ONO composite dielectrics. Oxide dielectrics typically have lower capacitance per unit area, but this is not always undesirable. Larger plate areas improve matching, so low-capacitance dielectrics are useful for improving the matching of small capacitors.

The voltage modulation of poly-poly capacitors is relatively small, as long as both electrodes are heavily doped. An unsilicided poly-poly capacitor typically exhibits a voltage modulation of 150 ppm/V. The temperature coefficient of a poly-poly capacitor also depends on voltage modulation effects and is typically less than 250 ppm/°C.[21] Both of these values will increase if either or both electrodes are lightly doped.

The two plates of a poly-poly capacitor are not wholly interchangeable. The upper electrode usually has less parasitic capacitance than the lower electrode. ONO capacitors may also exhibit asymmetric breakdown characteristics. The dielectric rupture process is triggered by field-aided emission of electrons from the surface of the negative electrode. The lower electrode of the ONO capacitor contains asperities produced by thermal oxidation, while the upper electrode lies on a smooth deposited oxide. The ONO capacitor therefore ruptures at a higher voltage if the lower electrode is biased positively with respect to the upper electrode. Proper orientation may be critical for certain applications, as an improperly oriented ONO capacitor may have its breakdown voltage reduced by 50% or more.

Stack Capacitors

The voltage coefficient of a thin-film capacitor can be greatly reduced by using highly conductive materials for both plates. The deposited dielectric usually consists of CVD oxide, either alone or in conjunction with CVD nitride. Grown oxides exhibit poor dielectric integrity unless they are densified by heat treatment. Aluminum cannot withstand the temperatures required for densification, so thin-dielectric capacitors must use some other material for the lower plate. Silicided poly is often chosen for this role. Capacitors that employ metal upper plates and silicided lower plates can achieve voltage coefficients of as little as 2 ppm/V.[22]

Capacitors can also be created by using the interlevel oxide (ILO) as a dielectric. This oxide is on the order of 5–10 kÅ thick, so densification is not required in order to obtain adequate voltage ratings. In fact, the ILO will typically withstand higher voltages than any other available dielectric. Of course, the high voltage rating also implies a low capacitance per unit area. Nearly 30,000 μm^2 of 10 kÅ ILO are required to construct a 1 pF capacitor.

---

[20] J. W. Fattaruso, M. de Wit, G. Warwar, K.-S. Tan, and R.K. Hester, "The Effect of Dielectric Relaxation on Charge-Redistribution A/D Converters," *IEEE J. Solid-State Circuits*, Vol. 25, #6, 1990, pp. 1550–1561.

[21] J. L. McCreary, "Matching Properties, and Voltage and Temperature Dependence of MOS Capacitors," *IEEE J. Solid-State Circuits*, Vol. SC-16, #6, 1981, pp. 608–616.

[22] C. Kaya, H. Tigelaar, J. Paterson, M. de Wit, J. Fattaruso, D. Hester, S. Kiriakai, K.-S. Tan, and F. Tsay, "Polycide/Metal Capacitors for High-Precision A/D Converters," *Proc. International Electron Devices Meeting*, 1988, pp. 782–785.

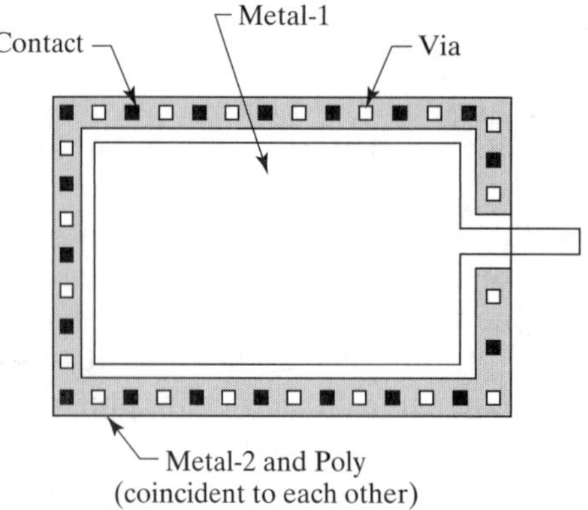

FIGURE 6.15 Metal-metal-poly stack capacitor.

The problem of low capacitance per unit area can be partially overcome by interleaving multiple metal layers together to form a *stack capacitor*. Figure 6.15 shows an example of a metal-metal-poly stack. This capacitor contains two sections connected in parallel. The lower section consists of ILO sandwiched between poly and metal-1. The upper section consists of ILO sandwiched between metal-1 and metal-2. Assuming that MLO and ILO have similar thicknesses, this structure will have twice the capacitance per unit area of a simple metal-metal capacitor. If the process supports additional layers of metallization, then these can be added to further increase the capacitance of the stack.

The plates of a stack capacitor often exhibit radically different parasitics. In the stack of Figure 6.15, the metal-1 electrode exhibits almost no parasitic capacitance because it lies sandwiched between the other two electrodes. On the other hand, the poly/metal-2 electrode has a relatively large parasitic capacitance because the poly plate couples to the substrate through the multilevel oxide (MLO). One must take special care to connect a stack capacitor in the proper orientation to ensure that parasitics do not interfere with proper circuit operation.

Another type of stack capacitor combines a poly-poly capacitor with a gate oxide capacitor (Figure 6.16). This structure places two thin dielectrics in parallel to produce an extremely high capacitance per unit area. As with the metal-metal-poly capacitor, the two electrodes have radically different parasitics. The poly plate, sandwiched between metal and silicon, exhibits relatively little parasitic capacitance. The

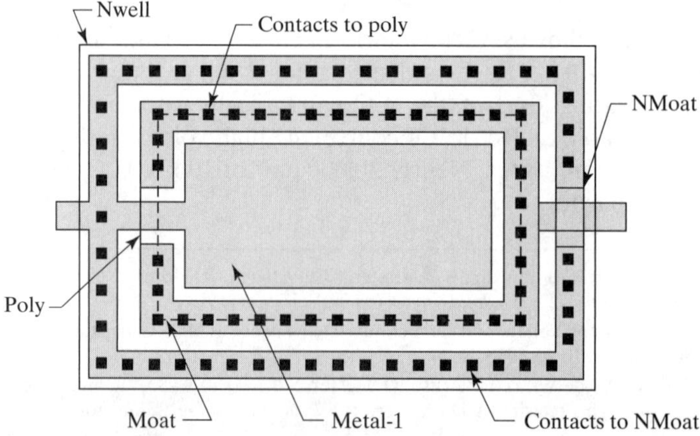

FIGURE 6.16 Metal-poly-silicon stack capacitor.

N-well/poly plate sees the large parasitic capacitance generated by the N-well/substrate junction.

Lateral Flux Capacitors

All of the capacitors so far discussed generate electric fields oriented perpendicularly to the surface of the die. Some capacitors generate fields oriented parallel to the surface; these are called *lateral flux capacitors*. Figure 6.17A shows a cross section of a lateral flux capacitor constructed from an array of alternating conductor strips. All of the strips marked "A" connect together to form the anode, and all the strips marked "C" connect together to form the cathode. Each anode strip is surrounded by cathode strips. Similarly, each cathode strip is surrounded by anode strips. Vertical electric fields link each strip to those above and below it, while lateral electric fields link each strip to those on either side. The total capacitance of this structure therefore equals the sum of a vertical capacitance and a horizontal capacitance. The lateral capacitance becomes significant when the sum of conductor thickness and dielectric thickness becomes of the same order as the sum of conductor width and conductor spacing. For metal-metal capacitors, both conductor and dielectric thicknesses are typically on the order of 0.5 μm. This implies that metal-metal lateral flux capacitors are useful only for metal systems with submicron pitches.

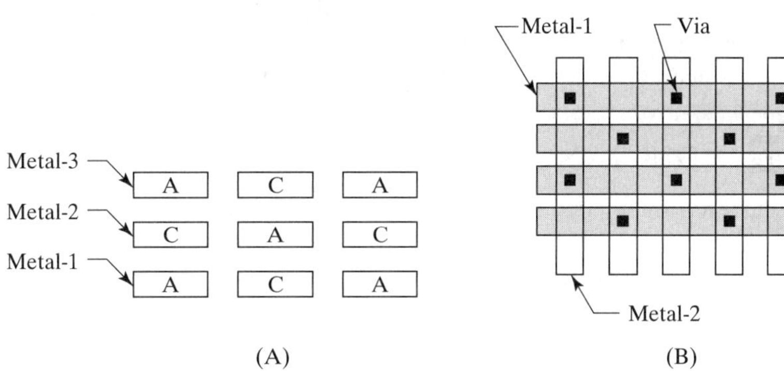

FIGURE 6.17 Cross section of interdigitated strip lateral flux capacitor (A) and layout of woven lateral flux capacitor (B).

A variety of alternative styles of lateral flux capacitors have been proposed.[23] Most of these structures are recognizable as variations on the basic interdigitated strip lateral flux capacitor of Figure 6.17A. The woven lateral flux capacitor of Figure 6.17B simply rotates the two sets of strips until they lie perpendicular to one another. Vias can then be added to interconnect the portions of the electrode that lie on different conductor levels. This structure foregoes some of the vertical capacitance generated by interdigitated strips in favor of reduced plate resistance and inductance. Another alternative structure uses fractal geometries to help increase electrode periphery without a proportional increase in resistance and inductance.

The lateral spacings between conductors ensure that lateral flux capacitors have relatively thick dielectrics. One can therefore obtain more capacitance per unit area by using a horizontal-parallel-plate capacitor with a thin dielectric. Most designs therefore favor the standard types of capacitor layouts. Lateral flux capacitors come into their own for high-frequency design, where they offer a simple way to create fully-isolated capacitors without requiring special processing.

---

[23] H. Samavati, A. Hajimiri, A. R. Shahani, G. N. Nasserbakht, and T. H. Lee, "Fractal Capacitors," *IEEE J. Solid-State Circuits*, Vol. 33, #12, 1998, pp. 2035–2041. See also R. Aparicio and A. Hajimiri, "Capacity Limits and Matching Properties of Lateral Flux Integrated Capacitors," *IEEE 2001 Custom Integrated Circuits Conf.*, pp. 365–368.

High-Permittivity Capacitors

Large capacitors can be fabricated in a relatively small area if they employ a high-permittivity dielectric. The relative permittitivies of silicon dioxide and silicon nitride are about 3.9 and 7, respectively. Discrete capacitors have long used other materials to obtain higher relative permittivies. Tantalum capacitors, for example, use tantalum pentoxide for their dielectric. This material has a relative permittivity of 22, and it exhibits a low incidence of pinhole defects even in very thin layers.[24] A thin-film tantalum capacitor can be constructed by first depositing a layer of tantalum and then partially oxidizing it to form the dielectric. The deposition of metal on top of the tantalum pentoxide completes the capacitor. This structure offers much higher permittivity than silicon dioxide and silicon nitride, but it requires special processing that most wafer fabs are not equipped to perform.

Discrete ceramic capacitors use materials such as barium strontium titanate to achieve relative permittivities of several hundred or more. Most of these high-permittivity ceramics are difficult to deposit in thin high-quality films. Their permittivities tend to vary greatly depending upon the conditions of deposition and annealing. Also, many of these materials exhibit profound temperature variabilities. They may also exhibit capacitance shifts over time (*aging*) and severe hysteresis effects (*soakage*). Thus, while these materials theoretically allow the integration of nanofarads of capacitance, the resulting devices are useless for precision circuitry. Most integrated capacitors used in analog circuits will continue to be constructed out of silicon dioxide and silicon nitride for the foreseeable future.

6.2 INDUCTANCE

A current flowing through a conductor generates a magnetic field around it. Energy flows into or out of this magnetic field as the magnitude of the current changes. These energy flows produce voltage drops along the conductor. The relationship between current and voltage can be quantified by the equation

$$V = L\frac{\Delta I}{\Delta t} \qquad [6.9]$$

where $\Delta I/\Delta t$ is the time rate of change of the current, V is the voltage developed as a consequence of the changing current, and L is a constant of proportionality called the *inductance*. An *inductor* is a circuit element designed to provide a known and precise amount of inductance.

The International System of Units (SI) defines the *Henry* (H) as the standard unit of inductance. A Henry represents a relatively large amount of inductance. Typical integrated inductors have values of a few tens of nanohenries. Such small inductors are virtually useless at frequencies below 100 MHz, so traditional analog integrated circuits do not use integrated inductors. Some *radio-frequency* (RF) integrated circuits operate at frequencies of a gigahertz or more. Some of these RF integrated circuits do employ integrated inductors as components of tuned circuits, despite their relatively poor performance compared to discrete devices.

The simplest inductor consists of a circular loop of wire (Figure 6.18A). The inductance of this loop equals

$$L = \mu_r\mu_0 r\left[\ln\left(\frac{8r}{a}\right) - 2\right] \qquad [6.10]$$

[24] Value for ε_r from A. B. Glaser and G. E. Subak-Sharpe, *Integrated Circuit Engineering* (Reading, MA: Addison Wesley, 1977), p. 355.

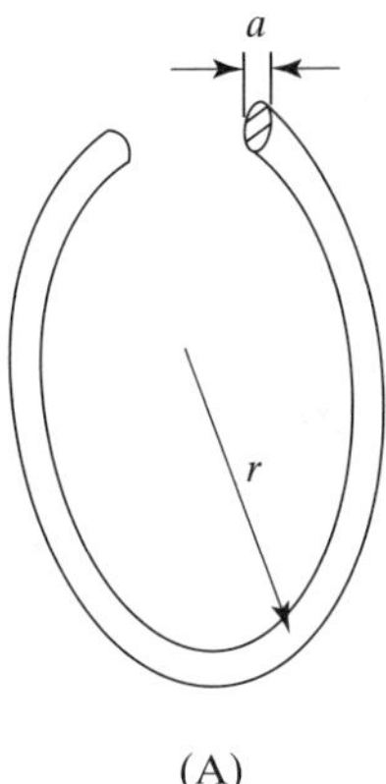

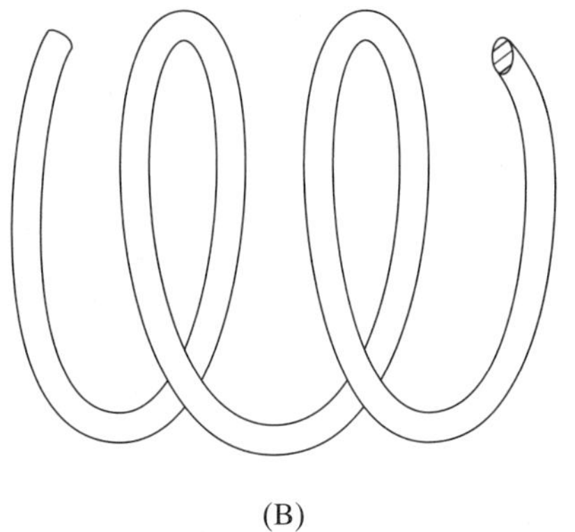

FIGURE 6.18 Geometries of inductors: (A) circular loop, (B) solenoid.

where r is the radius of the loop and a is the radius of the wire.[25] The constant μ_0, called the *permeability of free space*, equals 1.26 μH/m. The constant μ_r, called the *relative permeability*, varies depending upon the material surrounding the inductor. Most materials have relative permeabilities that lie very close to unity. The only exceptions to this rule are *ferromagnetic* materials such as iron and nickel, which have large relative permeabilities. None of the materials used in ordinary integrated circuits are ferromagnetic, so the equations presented in the remainder of this chapter assume $\mu_r = 1$.

The case of a circular loop inductor is of interest because it provides a rough estimate of the inductances contributed by bondwires. A typical bondwire measures some 25 μm in diameter by 1 mm in length. The inductance of a 1 mm-diameter loop of 25 μm-diameter wire equals about 5.6 nH. This computation correctly suggests that bondwires contribute several nanohenries of inductance each.

A circular loop doesn't provide much inductance. Discrete inductors get around this difficulty by stacking multiple loops of wire on top of one another to form a helical structure called a *solenoid* (Figure 6.18B). Each loop of wire in the solenoid is called a *turn*. The magnetic field generated by each turn also passes through all the others; they are said to be *magnetically coupled*. This coupling effectively multiplies the inductance of one turn by the total number of turns. Therefore, the inductance of the entire solenoid is proportional to the square of the number of turns. This square-law relationship makes it relatively easy to obtain large inductances with the use of solenoids.

Solenoids are very difficult to integrate, so alternative structures called *planar inductors* have been developed. A planar inductor consists of a spiral lead with a jumper to its innermost end. Figure 6.19 shows planar inductors having circular, octagonal,

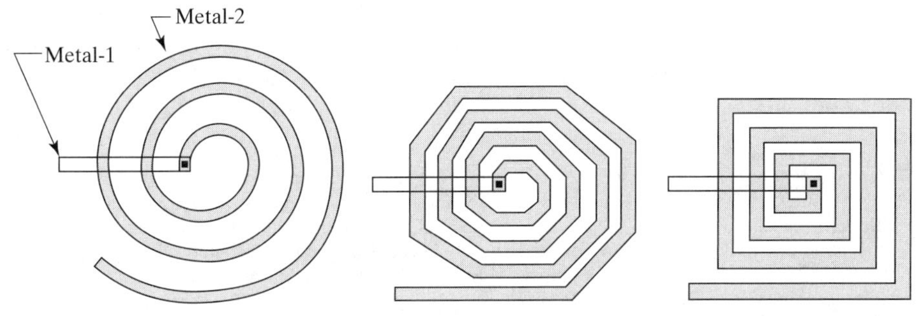

FIGURE 6.19 Planar inductors: (A) circular, (B) octagonal, and (C) square.

[25] F. W. Grover, *Inductance Calculations: Working Formulas and Tables* (New York: Van Nostrand, 1947).

and square turns, respectively. Circular planar inductors are difficult to digitize, so most designers prefer the square and octagonal layouts.

The multiple turns of a planar inductor are less effective than the multiple turns of a solenoid. There are two reasons for this shortfall. First, not all of the turns of a planar inductor are of the same diameter. The inner turns are smaller and therefore exhibit less inductance. Second, not all of the magnetic field generated by the large outer turns passes through the small inner turns (and vice versa). Therefore, the multiplication of inductance due to magnetic coupling is somewhat diminished.

The value of a square or octagonal planar inductor can be estimated using the formula

$$L = \frac{K_1 \mu_0 N^2 (d_o + d_i)}{2\left(1 + K_2 \dfrac{d_o - d_i}{d_o + d_i}\right)} \quad [6.12]$$

where N is the number of turns in the inductor, d_o is its outside diameter, d_i is its inside diameter, and K_1 and K_2 are constants.[26] For a square planar inductor, $K_1 = 2.34$ and $K_2 = 2.75$. For an octagonal planar inductor, $K_1 = 2.25$ and $K_2 = 3.55$. The inside diameter d_i can be calculated from the formula

$$d_i = d_o - 2Np \quad [6.13]$$

where p, the pitch of the metallization, equals the sum of the width of a turn and the spacing between adjacent turns. Consider a square planar inductor 300 μm on a side composed of 10 turns, each 9 μm wide and spaced 1 μm apart. The inner diameter of this inductor equals 100 μm. The value of this inductor will equal about 18.6 nH. This example gives some idea of the amount of inductance that can be integrated in a practical structure.

In addition to simple inductors such as those of Figure 6.19, one can also construct multiple inductors that are magnetically coupled to one another. Such an arrangement is called a *transformer*. Integrated transformers are rarely used, so the details of their construction will be omitted.[27]

6.2.1. Inductor Parasitics

Integrated inductors are plagued by a host of parasitics. As high-frequency equipment, such as cellphones, has proliferated, designers have only reluctantly begun to use integrated inductors. A brief overview of inductor parasitics will shed some light on the difficulties they have encountered.

Eddy current losses are perhaps the most troublesome inductor parasitics. The fluctuating magnetic fields generated by an inductor set up circulating currents within nearby conductors. These *eddy currents* leach energy from the magnetic field. The overall effect upon the circuit resembles the loss caused by inserting series resistance. The magnitude of eddy losses in an integrated inductor depends upon the resistivity of the underlying silicon. Resisitivities of less than about 10 $\Omega \cdot$cm exhibit objectionable

---

[26] S. S. Mohan, M. del Mar Hershenson, S. P. Boyd, and T. H. Lee, "Simple Accurate Expressions for Planar Spiral Inductances," *IEEE J. Solid-State Circuits*, Vol. 34, #10, 1999, pp. 1419–1424. Another formula for square inductors can be found in S. Jenei, B. K. J. C. Nauwelaers, and S. Decoutere, "Physics-Based Closed-Form Inductance Expression for Compact Modeling of Integrated Spiral Inductors," *IEEE J. Solid-State Circuits*, Vol. 37, #1, 2002, pp. 77–80.

[27] For those interested in integrated transformers, see J. R. Long, "Monolithic Transformers for Silicon RF IC Design," *IEEE J. Solid-State Circuits*, Vol. 35, #9, 2000, pp. 1368–1382.

losses. The simplest way to avoid eddy losses is to construct the inductor over lightly doped silicon. Since the magnetic field penetrates many microns into the silicon, the epi and the underlying substrate must both be lightly doped.

Figure 6.20 shows a simple lumped-element model of an integrated inductor constructed on lightly doped silicon. This model splits the parasitic capacitance between the inductor metallization and the underlying silicon into two equal halves, C_1 and C_2, that form a single π-section with the desired inductance L. Resistor R_S represents the series resistance of the inductor metallization. Resistors R_1 and R_2 represent the substrate resistance from beneath the coil to whatever substrate contacts are present. These substrate resistances will be quite large due to the use of a lightly doped substrate.

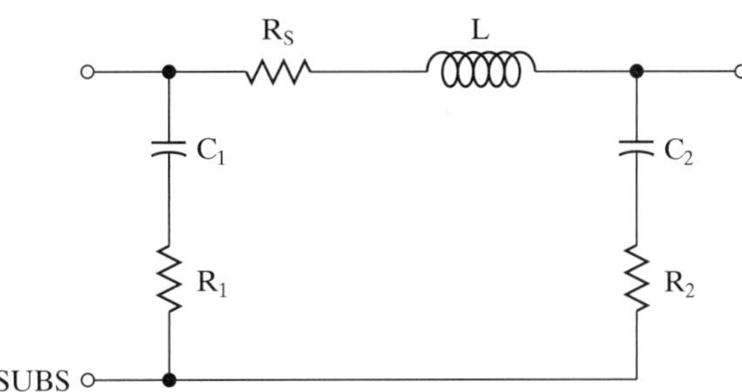

FIGURE 6.20 Subcircuit model for an integrated inductor constructed on lightly doped silicon.

The apparent series resistance R_S of the inductor increases at high frequencies due to *current crowding*. A mechanism called the *skin effect* is partially responsible for current crowding. The electromagnetic fields generated by a flowing current can penetrate a conductor only to a certain depth. At low frequencies, this depth of penetration is so great that the current flows uniformly through all portions of the conductor. The depth of penetration diminishes at higher frequencies, forcing the current to flow only in the outer portions of the conductor adjacent to its surface. Eddy currents generated within the metallization of the inductor also contribute to current crowding. The effects of current crowding become significant beyond a critical frequency f_{crit}, which equals

$$f_{crit} \approx \frac{pR_s}{2\,\mu_0 w^2} \qquad [6.14]$$

The sheet resistance of 5 kÅ aluminum metallization equals about 50 m$\Omega/\Box$. If the inductor uses a pitch of 10 μm and a width of 9 μm, then $f_{crit} = 2.4$ GHz. Wider leads will have lower critical frequencies. Whenever possible, one should set the width of inductor metallization so that f_{crit} lies above the frequency of operation.[28]

For an inductor whose inner diameter equals approximately a third of its outer diameter, the effective series resistance R_S equals

$$R_S = R_{DC}\left[1 + \frac{1}{10}\left(\frac{f}{f_{crit}}\right)^2\right] \qquad [6.15]$$

Circuit designers often use a quantity called the *quality factor*, or Q, to quantify the effect of inductor parasitics. Q is defined as the ratio of the maximum amount of

[28] Equations 6.14 and 6.15 are both taken from W. B. Kuhn and N. M. Ibrahim, "Analysis of Current Crowding Effects in Multiturn Spiral Inductors," *IEEE Trans. Microwave Theory and Techniques*, Vol. 49, #1, 2001, pp. 31–38.

energy stored in a system to the amount of energy lost during one cycle of operation. If the substrate is sufficiently resistive, then the parasitic capacitances and substrate eddy currents can be neglected and the quality factor of a planar inductor becomes

$$Q = \frac{2\pi f L}{R_S} \tag{6.16}$$

In most cases, parasitic capacitances and eddy currents cannot actually be neglected; therefore, Equation 6.16 represents something of an oversimplification. However, examination of this equation reveals several important generalities concerning Q. First, the smaller the parasitics, the higher the value of Q becomes. An ideal inductor would have infinite Q, but all practical inductors have finite Q. Secondly, the value of Q depends upon frequency. In most practical inductors, Q increases with frequency until a peak is reached, and then it rolls off again due to current crowding and parasitic capacitance. The peak quality factors of integrated inductors range from a low of about one to a high of about forty. By comparison, air-core discrete inductors can easily obtain quality factors of a hundred or more.

6.2.2. Inductor Construction

Eddy losses pose difficult problems for the designers of integrated inductors. Most processes favor heavily doped substrates to reduce debiasing and improve latchup immunity, but these substrates greatly reduce Q. A variety of ideas have been tried to minimize eddy losses in heavily doped substrates. One approach suspends the inductor in air by etching away the silicon beneath it to form a deep cavity.[29] This type of micromachining is not compatible with conventional integrated circuit processing, nor with standard plastic encapsulation.

The capacitive parasitics associated with an integrated inductor can be minimized by raising the inductor above the surface of the silicon. This goal can be achieved by the use of thicker interlevel dielectrics, or by simply using the uppermost level of metal to construct the body of the inductor. The jumper to the innermost turn of the inductor should always be placed beneath the body of the inductor, as it is so short that it contributes only a small fraction of the total capacitance.

Thicker metallization will also improve the quality factor by reducing the series resistance. One way to obtain thicker metallization is to strap several layers of metal together with vias. This approach clearly conflicts with the goal of minimizing capacitive parasitics by moving the inductor to the highest possible level of metal. However, in a process that includes four or five levels of metal, one can usually benefit by combining the topmost two or three metal levels together to form the inductor.

Some RF processes include a relatively thick layer of copper metallization deposited on top of the protective overcoat. This metallization simultaneously reduces both parasitic capacitance and resistance. The copper used for this purpose resembles the power copper metallization described in Section 2.6.5, but it is only a couple of microns thick.

Many investigators have attempted to find ways to increase on-chip inductances, but to little avail. The obvious approach involves application of a high-permeability thin-film material over the inductor, but no known ferromagnetic material has acceptably low core losses beyond 100 MHz. An approach that has had some success involves stacking planar inductors on different metal layers. Unfortunately, the thin dielectric between the metal layers generates interwinding capacitances that degrade Q.

---

[29] J. Y.-C. Chang, A. A. Abidi, and M. Gaitan, "Large Suspended Inductors on Silicon and Their Use in a 2-μm CMOS RF Amplifier," *Electron Device Letters*, Vol. 14, #5, 1993, pp. 246–248.

Currently, the best integrated inductors provide values of 10–100 nH with quality factors of about 40 at frequencies of up to several gigahertz. A great deal of research is currently aimed at improving these figures, but it is doubtful if any significant strides can be made without resorting to exotic processing technologies.

Guidelines for Integrating Inductors[30]

Most layout designers will never have to create an integrated inductor. Only those who work in the relatively specialized field of RF design will likely encounter them. Even these designers will probably never have to compute the value of an inductor by hand, as computer programs are usually employed to optimize the relevant geometric parameters. The following general guidelines apply to the layout of integrated inductors regardless of their exact geometry:

1. Use the highest-resistivity substrate available.
   Substrate resistivities of less than about 10 $\Omega \cdot$cm will generate significant eddy losses that will reduce the quality factor. If a choice of substrate resistivities exists, choose the highest resistivity possible.

2. Place inductors on the highest possible metal layers.
   The body of the inductor should reside on the highest possible layers of metallization. The jumper to the innermost turn should always be placed below the body of the inductor, never above it. These precautions help minimize parasitic capacitances.

3. Consider strapping two or three metal layers together.
   If the process offers three or four layers of metal, consider using two or three of these layers in parallel to construct the body of the inductor. By strapping these layers together with vias, one can effectively reduce the sheet resistance of the metallization and thus increase the Q of the inductor. Avoid using first-level metal to form any portion of the body of the inductor, as it resides in close proximity to the silicon.

4. Keep all unconnected metal well away from inductors.
   As a general rule, unconnected metal should be kept away from the inductor by a distance of at least five times the width of the inductor metallization. This rule helps minimize eddy current losses caused by intrusion of conductive materials within the magnetic field generated by the inductor.

5. Avoid excessively wide or narrow metallization.
   For inductors operating at 1–3 GHz, a width of approximately 10–15 μm is optimal. Narrower leads have excessive resistances, and wider leads are vulnerable to current crowding. Equation 6.14 can provide some guidance as to the optimal width for inductor metallization.

6. Use the narrowest possible spacing between turns.
   Narrower spacings enhance magnetic coupling between the turns of the inductor, producing larger inductor values and higher quality factors.

7. Do not fill the entire inductor with turns.
   The magnetic field generated by the spiral of metallization becomes most intense at the center of the inductor. If turns of the inductor occupy this region, they will generate eddy current losses. The inside diameter of the inductor should equal at least five times the width of the inductor metallization. For larger inductors, the inner diameter should equal at least a third of the outer diameter.

---

[30] Rules 2–7 are based on J. R. Long and M. A. Copeland, "The Modeling, Characterization and Design of Monolithic Inductors for Silicon RF IC's," *IEEE J. Solid-State Circuits*, Vol. 32, #3, 1997, pp. 357–369.

8. Do not place metal plates over or under inductors.
   A metal plate placed over or under an inductor offers the opportunity for eddy currents to circulate. The larger the area of the plate, the greater these eddy currents become. Ideally, metal plating should never be placed over or under an inductor. If this is not possible (as is the case with fill metal), then the plate should be broken up either by dividing it into an array of smaller geometries or by cutting a pattern of slots into it.[31]

9. Do not place junctions beneath an inductor.
   The presence of a junction in close proximity to an inductor can produce unwanted device interactions. The high-frequency AC signals coupled into the junction from the inductor may be rectified, resulting in parasitic losses or the injection of unexpected currents into the diffusions. The same rules that apply to unconnected metal also apply to junctions.

10. Keep inductor leads short and direct.
    The leads of an inductor contribute parasitics of their own, so their length and area should be minimized. The leads should also occupy the highest possible metal layer to minimize their parasitic capacitance relative to the substrate.

6.3 SUMMARY

Capacitance is not as readily integrated as resistance. Barring the use of exotic materials such as titanates, only a few hundred picofarads of capacitance can economically be integrated on a single die. Even this relatively small amount of capacitance suffices for many applications, including timers, capacitive dividers, and active filters.

Integrated capacitors generally fall into two categories: those using thin insulating films as dielectrics (*thin-film capacitors*) and those using reverse-biased junctions as dielectrics (*junction capacitors*). Properly constructed thin-film capacitors have fewer parasitics than junction capacitors, but they usually require additional processing steps. Most standard bipolar processes offer a base-emitter junction capacitor as part of the baseline process and a MOS capacitor as a process extension. CMOS processes always offer a MOS capacitor since a MOS transistor can be pressed into service in this role. This device exhibits an undesirable dip in capacitance near the threshold voltage and thus requires careful biasing. Many CMOS and BiCMOS processes also offer a poly-poly capacitor, using a thin oxide or ONO stack dielectric specifically tailored for this application. Although the fabrication of this capacitor increases the process complexity and cost, its superior performance can often justify its inclusion.

The absolute accuracy of capacitors is relatively poor. Variations in doping and junction depth can cause junction capacitors to vary by up to $\pm30\%$. Dimensional variations cause thin-film capacitors typically to vary by at least $\pm10\%$. Capacitors are difficult to trim because the usual trim structures add excessive amounts of parasitic capacitance. If necessary, laser link trimming (Section 5.6.2) can be applied to most types of capacitors. The vast majority of circuits either trim resistors or current sources to compensate for capacitor variability, or they use topologies sensitive only to the matching of capacitors rather than to their absolute values.

Inductors are even more difficult to integrate. Only about a hundred nanohenries can be integrated, and the resulting devices exhibit poor quality factors, especially if

---

[31] C. P. Yue and S. S. Wong, "On-chip Spiral Inductors with Patterned Ground Shields for Si-Based RF ICs," *IEEE Journal of Solid-State Circuits*, Vol. 33, #5, 1998, pp. 743–752.

they are constructed above a low-resistivity substrate. Despite these disadvantages, integrated planar inductors in the form of spirals of metal have found application in RF integrated circuits.

6.4 EXERCISES

Refer to Appendix C for layout rules and process specifications.

6.1. Assume that a thermal oxide film with a relative permittivity of 3.9 can safely withstand a field of $5 \cdot 10^5$ V/cm. How thick must the oxide be made to withstand an operating voltage of 15 V? What is the capacitance of the resulting film in fF/μm^2?

6.2. What is the relative permittivity of a composite dielectric consisting of 60Å of dry oxide, 220 Å of plasma nitride, and a further 50 Å of dry oxide, assuming an oxide permittivity of 3.9 and a nitride permittivity of 6.8? What operating voltage can this dielectric withstand? What percentage improvement in capacitance will the composite dielectric provide over a pure oxide capacitor having the same working voltage?

6.3. Determine the approximate zero-bias junction capacitance of a device consisting of a square NSD region diffused into P-epi, given that the oxide window through which the NSD is implanted measures 10 × 20 μm, the NSD junction depth equals 0.9 μm, and the P-epi doping concentration equals $5 \cdot 10^{16}$ atoms/cm^3. Include the effects of sidewall capacitance.

6.4. A junction capacitor with a drawn area of 5800 μm^2 and a drawn periphery of 300 μm has a zero-bias junction capacitance of 6.45 pF. A second junction capacitor with a drawn area of 3000 μm^2 and a drawn periphery of 670 μm has a zero-bias junction capacitance of 4.92 pF. What are the areal and peripheral capacitances for this type of junction capacitor? What must the spacing between fingers equal in order for a comb capacitor of this type to obtain a higher capacitance per unit area than a plate capacitor?

6.5. Lay out a junction capacitor with a nominal zero bias capacitance of 10 pF, using the standard bipolar layout rules of Appendix C. Assume $C_a = 0.82$ fF/μm^2 and $C_p = 2.8$ fF/μm. Justify your choice of layout style (comb or plate).

6.6. Lay out a 5 pF standard bipolar thin oxide capacitor. A special process extension will be used to produce a 450 Å oxide with a relative permittivity of 3.9. The layout rules for the thin oxide layer TOX are as follows:

1. TOX width 10 μm
2. EMIT overlap of TOX 4 μm
3. METAL overlap TOX 4 μm

6.7. Lay out a poly-poly capacitor having a minimum guaranteed capacitance of 20 pF. Use a sparse array of contacts spaced 10 μm apart to contact the poly-2 plate. Dope the poly-2 plate by using NSD. Contact the poly-1 plate on at least three sides, and include all necessary metallization.

6.8. Construct a 5 pF MOS capacitor from a PMOS transistor laid out according to the baseline CMOS process of Appendix C. Assume that the capacitor will operate in inversion, and use a channel length of no more than 20 μm to minimize bottom-plate resistance. Use a sparse array of contacts spaced 10 μm apart to contact the poly plate.

6.9. An integrated circuit uses two 5 kÅ thick layers of aluminum separated by a 10 kÅ interlevel oxide. The layout rules for this metal are as follows:

1. MET1 width 1 μm
2. MET1 spacing to MET1 0.8 μm

Would a lateral flux capacitor offer significant advantage over a horizontal plate capacitor? Why or why not?

6.10. An RF process uses a 15 kÅ top-level aluminum metallization to fabricate inductors. The layout rules for this metal are as follows:

1. TMET width 1 μm
2. TMET spacing 0.8 μm

Compute the dimensions of a square planar inductor that provides 30 nH at 1.5 GHz. Lay out this inductor.

7 Matching of Resistors and Capacitors

Most integrated resistors and capacitors have tolerances of ±20 to 30%. These tolerances are much poorer than those of comparable discrete devices, but this does not prevent integrated circuits from achieving a high degree of precision matching. All of the devices in an integrated circuit occupy the same piece of silicon, so they all experience similar manufacturing conditions. If one component's value increases by 10%, then all similar components experience similar increases. The ratio between two similar components on the same integrated circuit can be controlled to better than ±1%, and, in many cases, to better than ±0.1%. Devices specifically constructed to obtain a known, constant ratio are called *matched devices.*

Analog integrated circuits usually depend on matching to obtain much of their precision and performance. Any number of mechanisms can interfere with matching. Most of these mechanisms are known, and layout designers have devised ways to minimize their impact. This chapter covers the design of matched resistors and capacitors. Much of this information also applies to the matching of other components, such as bipolar transistors (Section 9.2), diodes (Section 10.3), and MOS transistors (Section 12.3).

7.1 MEASURING MISMATCH

The *mismatch* between two components is usually expressed as a deviation of the measured device ratio from the intended device ratio. Suppose a designer lays out a pair of matched 10 kΩ resistors. After fabrication, one pair of these resistors is found to equal 12.47 kΩ and 12.34 kΩ. The ratio between these resistors equals 1.0105, or approximately 1% more than the intended ratio of 1.0000. This pair of resistors therefore exhibits a mismatch of approximately 1%.

The concept of mismatch also applies to devices having ratios other than 1:1. The mismatch between any two devices equals the difference between the ratio of their measured values and the ratio of their intended values, divided by the ratio of their intended values. The final step, division by the ratio of intended values, normalizes the result so that it becomes independent of ratio. If the intended values are X_1 and

X_2 and the measured values are x_1 and x_2, then the mismatch δ equals

$$\delta = \frac{(x_2/x_1) - (X_2/X_1)}{(X_2/X_1)} = \frac{X_1 x_2}{X_2 x_1} - 1 \qquad [7.1]$$

Equation 7.1 computes the mismatch of one specific pair of devices. The same measurements performed on a second pair of devices will yield a different mismatch. Measurements of a large number of device pairs will produce a random distribution of mismatches. An analysis of the mismatch distribution of a small sample of devices allows the designer to determine the percentages of units that are likely to fail specifications. In order for this analysis to yield valid results, the sample must fairly represent the capabilities of the process. Ideally, the sample should consist of 50 to 100 devices drawn from random locations on at least 10 wafers that are drawn randomly from at least three wafer lots. In practice, one must often rely on a sample drawn from a single wafer lot. When selecting this sample, consider the following guidelines:

- The sample should include no fewer than 20 devices, and preferably at least 30 devices.
- The sample should include devices drawn from at least three wafers. Each wafer should contribute approximately the same number of devices to the sample.
- The wafers should be selected from various positions in the wafer lot. A three-wafer sample should include one wafer from the front of the lot, one from the middle, and one from the back.
- The sample devices should be selected from random locations on each wafer.
- If possible, the sample should include wafers from more than one wafer lot.
- Wafers that have been misprocessed or reworked do not fairly represent the process and should not be used for characterization.
- If at all possible, the sample units should be packaged using the same lead frames and encapsulation as production material.

Once a sample has been selected and all of the sample units have been measured, the resulting data must be statistically analyzed. The theory of such analyses is beyond the scope of this text, but the simplified procedure given next provides a good example of the techniques and terminology involved.

Suppose the sample includes N units, the mismatches of which are $\delta_1, \delta_2, \delta_3 \ldots \delta_N$, as computed by Equation 7.1. Mismatches can be either negative or positive quantities. The signs of these mismatches are significant and must be retained in order for the following computations to have meaning. Based on the computed mismatches, one can derive an average mismatch m_δ. This average, or *mean,* consists of the sum of all the mismatches divided by the number of sample units N, or

$$m_\delta = \frac{1}{N} \sum_{i=1}^{N} \delta_i \qquad [7.2]$$

where the sigma function $\Sigma()$ represents the sum of all of the individual terms. Once the mean has been computed, one can determine the *standard deviation of the mismatches* s_δ:

$$s_\delta = \sqrt{\frac{1}{N-1} \sum_{i=1}^{N} (\delta_i - m_\delta)^2} \qquad [7.3]$$

The mean m_δ is a measure of the *systematic mismatch,* or *bias,* between the matched devices. Systematic mismatches are caused by mechanisms that influence all of the samples in the same manner. Consider the case of a pair of matched 2 kΩ and 4 kΩ base resistors laid out as single strips of base diffusion with contacts at either end. Both of these resistors have the same contact resistance, which, for purposes of argument, will be assumed to equal 100 Ω. This contact resistance causes a 5% increase in the 2 kΩ resistor, but only a 2.5% increase in the 4 kΩ resistor. Every pair of resistors shows the same imbalance, so the contact resistance represents a systematic mismatch. This mismatch could be eliminated by dividing both resistors into segments of 2 kΩ. The 2 kΩ resistor would contain one segment with 100 Ω of contact resistance, which produces a 5% increase in its value. The 4 kΩ resistor would contain two segments with a total of 200 Ω of contact resistance, which produces a 5% increase in value. Since both segmented resistors experience the same percentage increase in value, they do not exhibit any systematic mismatch.

The standard deviation quantifies *random mismatch* caused by statistical fluctuations in processing conditions or material properties. These fluctuations are an unavoidable part of semiconductor manufacture, but their magnitude can sometimes be reduced if one can identify their underlying causes. Figure 7.1 shows a histogram of the mismatches of 30 units. The exact appearance of the histogram will change as additional units are added, but the values of the mean and the standard deviation fluctuate relatively little once the distribution contains 20 or 30 units.

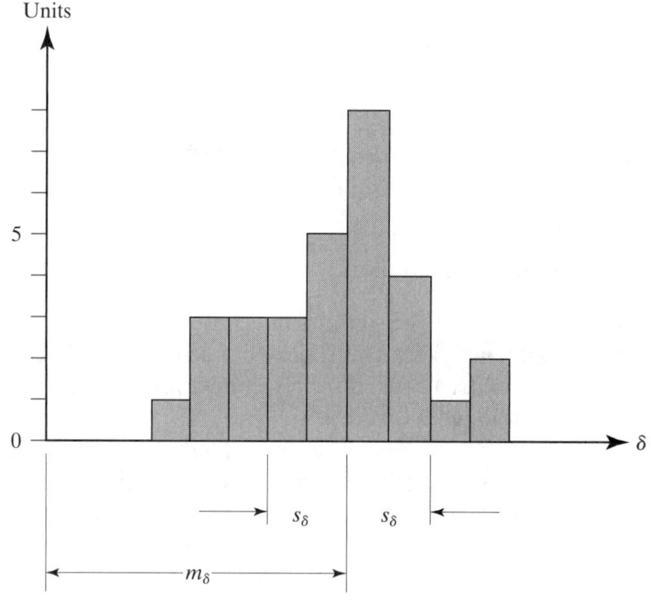

FIGURE 7.1 Histogram of the mismatch, δ, of 30 units, showing mean, m_δ, and standard deviation, s_δ.

Once the mean and standard deviation have been computed using Equations 7.2 and 7.3, these can be used to predict worst-case mismatches, using one of several indices. The *three-sigma mismatch* equals the sum of the absolute value of the mean plus three times the standard deviation. The *six-sigma mismatch* equals the sum of the absolute value of the mean plus six times the standard deviation. Suppose a pair of resistors shows a mean mismatch of −0.3% with a standard deviation of 0.1%. The three-sigma mismatch of these resistors equals 0.6%, while their six-sigma mismatch equals 0.9%. Less than 1% of all units should have mismatches greater than the three-sigma value, and virtually no units should have mismatches larger than the six-sigma value. The mismatch figures given in this text are all 4.5-sigma values, which are commonly used in the industry to quantify process capabilities.

7.2 CAUSES OF MISMATCH

Random mismatches stem from microscopic fluctuations in dimensions, dopings, oxide thicknesses, and other parameters that influence component values. Although these statistical fluctuations cannot be entirely eliminated, their impact can be minimized through proper selection of component values and device dimensions. Systematic mismatches stem from process biases, contact resistances, nonuniform current flow, diffusion interactions, mechanical stresses, temperature gradients, and a host of other causes. A major goal of designing matched components consists of rendering them insensitive to various sources of systematic error. The following sections discuss the major known causes of mismatch and techniques for combating them.

7.2.1. Random Variation

All components exhibit microscopic irregularities in their dimensions and composition. These irregularities fall into two general categories: those that occur only along the edges of a device, and those that occur throughout a device. The former are called *peripheral variations* because they scale with device periphery, while the latter are called *areal variations* because they scale with device area. The matching of most integrated devices depends primarily upon areal variations.

Statistical theory suggests that areal variations can be modeled by an equation of the form

$$s = m\sqrt{\frac{k}{2A}} \qquad [7.4]$$

where m and s are the mean and standard deviation of some parameter of a device having active area A. The constant of proportionality k is called the *matching coefficient*. The magnitude of this coefficient depends upon the sources of mismatch. Different types of components will have different matching coefficients, and the differences between them are often dramatic. Components that appear similar, but which are fabricated in different processes, will also have different matching coefficients. In order to provide quantitatively correct results, one must know the matching coefficients for the specific type of device and the specific process in question. However, much information of value can be derived from a study of the equations, even if one does not know the exact value of the matching coefficients.

The standard deviation of the mismatch s_δ between two components equals

$$s_\delta = \sqrt{\left(\frac{s_1}{m_1}\right)^2 + \left(\frac{s_2}{m_2}\right)^2} \qquad [7.5]$$

where m_1 and m_2 are the means of the parameter of interest for each device, and s_1 and s_2 are the standard deviations of this parameter. Equations 7.4 and 7.5 form the basis for the computation of random mismatches in all types of integrated components. This section will cover random mismatch in capacitors and resistors. Section 9.2.1 covers bipolar transistors, Section 10.3 covers diodes, and Section 12.3 covers MOS transistors.

Capacitors[1]

Consider the case of a parallel-plate capacitor. The parameter of interest is obviously capacitance. If we neglect systematic errors, then the mean of the capacitance equals the value of the capacitor, C. Furthermore, the device area A is proportional to its capacitance. Taking both of these factors into consideration, Equation 7.4 reduces to

$$s = \sqrt{\frac{k_C C}{2}} \qquad [7.6]$$

where k_c is the *mismatch coefficient of capacitance*, which has units of capacitance. This equation can be used to derive the mismatch between any two capacitors. The simplest case consists of two matched capacitors, each of value C. Substituting Equation 7.6 into Equation 7.5 gives the standard deviation of the mismatch between the capacitors:

$$s_\delta = \sqrt{\frac{k_C}{C}} \qquad [7.7]$$

This equation illustrates the fundamental relationship that exists between capacitor size and capacitor matching. In order to reduce the random mismatch by a factor of N, the capacitance must increase by a factor of N^2. Attempts to reduce random mismatch to arbitrarily small values quickly run into a point of diminishing returns. Extremely accurate matched capacitors can only be obtained by trimming.

Equation 7.7 covers only the case of two equal matched capacitors. A more general relationship for the mismatch between two arbitrary capacitors C_1 and C_2 equals

$$s_\delta = \sqrt{\frac{k_C(C_1 + C_2)}{2C_1C_2}} \qquad [7.8]$$

Examination of this equation reveals that the smaller of two matched capacitors contributes the majority of the mismatch. This has ominous implications for large capacitance ratios. If the smaller of the two capacitors is made large enough to ensure reasonable matching, then the larger of the two capacitors will consume excessive area. Some designers attempt to circumvent this problem by constructing the smaller capacitor from several devices connected in series. This solution would work quite nicely were it not for bottom-plate parasitics. These parasitics make it very difficult to construct a capacitance of definite and controlled value by series connection. If at all possible, one should avoid circuits that require matching large ratios of capacitance.

Resistors[2]

Consider the case of a simple rectangular resistor. The area A of this resistor equals its length L times its width W. Similarly, the value of this resistor R equals its sheet resistance R_S times its length divided by its width. Combining these two

---

1 J. B. Shyu, G. C. Temes, and F. Krummenacher, "Random Error Effects in Matched MOS Capacitors and Current Sources," *IEEE J. Solid-State Circuits*, Vol. SC-19, #6, 1984, pp. 948–956. See also J. L. McCreary, "Matching Properties, and Voltage and Temperature Dependence of MOS Capacitors," *IEEE J. Solid-State Circuits*, Vol. SC-16, #6, 1981, pp. 608–616.

2 W. A. Lane and G. T. Wrixon, "The Design of Thin-Film Polysilicon Resistors for Analog IC Applications," *IEEE Trans. on Electron Devices*, Vol. 36, #4, 1989, pp. 738–744.

relationships yields

$$A = \frac{R}{R_S} W^2 \qquad [7.9]$$

Substituting this quantity into Equation 7.4 gives

$$s = \frac{1}{W}\sqrt{\frac{k_R R}{2}} \qquad [7.10]$$

where k_r is the *mismatch coefficient of resistance*, which has units of $\Omega \cdot \mu m^2$. This equation can be used to determine the mismatch between any two resistors. The simplest case consists of two matched resistors having equal widths and equal resistances R. Substituting Equation 7.10 into Equation 7.5 gives the standard deviation of mismatch between the resistors:

$$s_\delta = \frac{1}{W}\sqrt{\frac{k_R}{R}} \qquad [7.11]$$

This equation illustrates the two fundamental relationships that govern resistor matching. First, random mismatch scales inversely with the square root of resistance. This is the same relationship as that which governs capacitor matching. In both cases, one must quadruple the value of the components to halve their mismatch. Second, random mismatch scales inversely with resistor width. Doubling the width of a pair of matched resistors while holding their values constant will halve the random offset. This leads to the conclusion that large-value matched resistors can be made much narrower than small-value matched resistors.

Equation 7.11 only covers the case of two matched resistors of equal values and equal widths. A more general relationship between two resistors of equal width gives

$$s_\delta = \frac{1}{W}\sqrt{\frac{k_R(R_1 + R_2)}{2R_1R_2}} \qquad [7.12]$$

This relationship is quite similar to the one derived for capacitors, in that the mismatch between two resistors of unequal sizes primarily depends upon the smaller of the two resistors. This situation often arises in voltage dividers with large attenuation ratios. For example, a voltage divider consisting of a 100 kΩ resistor in series with a 10 kΩ resistor will produce a 10:1 attenuation ratio, but this ratio will vary considerably because of the presence of the small 10 kΩ resistor. The mismatch can be reduced by increasing the width of both resistors, but this requires a large amount of area. A better approach consists of constructing the smaller resistor out of a number of parallel segments of equal value. If R_1 and R_2 are two resistors, R_1 consisting of a single segment of width W, and R_2 consisting of N_s segments each of width W and value N_sR_2, then

$$s_\delta = \frac{1}{W}\sqrt{k_R\left(\frac{1}{R_1} + \frac{1}{N_s^2 R_2}\right)} \qquad [7.13]$$

In the case of the 10:1 attenuator just discussed, the overall mismatch can be approximately halved by simply constructing R_2 of two parallel 20 kΩ segments rather than a single 10 kΩ segment. This solution requires far less area than the alternative approach of doubling the widths of both R_1 and R_2.

The value of the mismatch coefficient k_R depends upon the nature of the resistors under consideration. For poly resistors, researchers have shown that

$$k_R = \eta R_S d_g^2 \qquad [7.14]$$

where R_S is the sheet resistance of the poly, d_g represents the mean diameter of the poly grains, and η is a dimensionless constant that typically equals two.[3] Equation 7.14 holds only so long as the resistor is considerably wider than the mean grain diameter. If this is not the case, then matching becomes erratic. Since most poly contains grains that are considerably less than a micron in width, poly resistors that are more than a micron wide will normally follow the equations discussed in this section.

7.2.2. Process Biases

The dimensions of geometries fabricated in silicon never exactly match those in the layout database because the geometries shrink or expand during photolithography, etching, diffusion, and implantation. The difference between the drawn width of a geometry and its actual measured width constitutes the *process bias*. Process biases can introduce major systematic mismatches in poorly designed components. Consider the case of two matched poly resistors having widths of 2 μm and 4 μm, respectively. Suppose that poly etching introduces a process bias of 0.1 μm. The ratio of the actual widths equals (2 + 0.1)/(4 + 0.1), or 0.512. This represents a systematic mismatch of no less than 2.4%! Since most processing steps have biases of at least 0.1 μm, the layout designer must ensure that all matched devices are insensitive to process biases. In the case of resistors, process biases can be virtually eliminated by simply making both resistors the same width.

Process biases can also affect the length of a resistor. The length of most resistors is determined by the placement of their contacts. Suppose that these contacts have a process bias of 0.2 μm. If one matched resistor is 20 μm long and the other is 40 μm long, then the mismatch due to this bias equals (20 + 0.2)/(40 + 0.2), or 0.503. This represents a systematic mismatch of about 0.5%. The simplest way to avoid this bias consists of dividing both matched resistors into segments of the same size. If the resistors of the previous example were laid out in 20 μm segments, then the ratio of the resistors would equal (20 + 0.2)/[2 · (20 + 0.2)], or exactly 0.5. The same stratagem has already been shown to eliminate systematic mismatches due to contact resistances and nonlinear current flow at the ends of the resistors. Section 7.2.10 explains how to divide matched resistors into arrays of optimally sized segments.

Capacitors also experience systematic mismatches caused by process biases. Suppose a pair of poly-poly capacitors, one measuring 10 × 10 μm and the other 10 × 20 μm, both experience a poly etch bias of 0.1 μm. The actual area of the 10 × 10 μm capacitor equals $(10.1)^2$, or 102.1 μm^2, while the actual area of the 10 × 20 μm capacitor equals (10.1 · 20.1), or 203.01 μm^2. The ratio of these two areas equals 0.5029, which represents a systematic mismatch of 0.6%.

In theory, matched capacitors become insensitive to process biases when their area-to-periphery ratios equal one another. In the case of two capacitors of the same value, this can be achieved by using the same geometry for both capacitors. Identical matched capacitors are usually laid out as squares. The problem becomes somewhat

[3] R. Thewes, R. Brederlow, C. Dahl, U. Kollmer, C. G. Linnenbank, B. Holzapfl, J. Becker, J. Kissing, S. Kessel, and W. Weber, "Explanation and Quantitative Model for the Matching Behavior of Poly-Silicon Resistors," *Proc. International Electron Devices Meeting*, 1998, pp. 771–774.

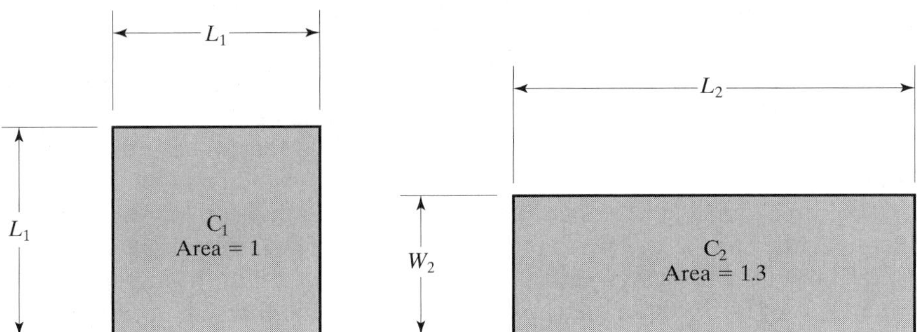

FIGURE 7.2 Matching capacitors by using identical area-to-periphery ratios.

more difficult if the capacitors have values that are not in simple ratio. Although the smaller capacitor should still be laid out as a square, the larger capacitor must be laid out as a rectangle. Suppose the smaller capacitor, C_1, has dimensions L_1 by L_1 (Figure 7.2). The larger capacitor, C_2, should have a length L_2 and a width W_2 equal, respectively, to[4]

$$L_2 = \frac{C_2}{C_1}\left(1 + \sqrt{1 - \frac{C_1}{C_2}}\right) \qquad [7.15A]$$

and

$$W_2 = \frac{C_2}{C_1}\left(1 - \sqrt{1 - \frac{C_1}{C_2}}\right) \qquad [7.15B]$$

Although Equations 7.15A and 7.15B theoretically eliminate systematic mismatches due to process bias, things do not work out so nicely in practice. Process biases are not constant quantities; they actually depend on the dimensions of the geometries in question. The process biases experienced by a rectangular capacitor do not precisely equal those experienced by a square capacitor. This problem becomes increasingly severe for larger ratios. Rectangular capacitors also increase the contribution of peripheral fluctuations to random mismatches. In practice, capacitors with ratios of more than 1.5:1 should not be constructed using Equations 7.15A and 7.15B. In such cases, the designer should instead resort to using arrays of matched capacitor segments, or *unit capacitors*.[5]

Not all systematic mismatches are due to process biases. Other mechanisms that produce systematic mismatches include pattern shift, etch variations, proximity effects, hydrogenation, diffusion interactions, mechanical stresses, thermal gradients, thermoelectrics, voltage modulation, charge spreading, and dielectric polarization. Later sections will discuss each of these topics.

7.2.3. Interconnection Parasitics

The leads that connect components into a circuit can introduce systematic mismatches. Ideally, leads should contribute negligible resistance and capacitance to the circuit, but real leads exhibit sufficient nonidealities to disturb the matching of both precision resistors and capacitors. Proper attention to layout can reduce or eliminate the impact of these nonidealities.

---

4 Y. Tsividis, *Mixed Analog-Digital VLSI Devices and Technology* (New York: McGraw-Hill, 1996), pp. 220–223.

5 M. J. McNutt, S. LeMarquis, and J. L. Dunkley, "Systematic Capacitance Matching Errors and Corrective Layout Procedures," *IEEE J. Solid-State Circuits*, Vol. 29, #5, 1994, pp. 611–616.

When constructing precisely matched resistors, one must consider lead resistance, particularly that of jumpers between segments in a resistor array. Aluminum metallization typically displays a sheet resistance of 50–80 mΩ/□. Jumpers between distant segments might each contain perhaps 100 squares of metal, which represent 5–8 Ω of resistance. Vias can contribute 2–5 Ω each. Jumpers can thus exhibit resistances of as much as 20 Ω.

The impact of lead resistance is obviously greatest for low-value resistors. Consider the case of a resistor array whose segments each contain 20 squares of 50 Ω/□ poly. Each segment has a resistance of 1 kΩ. A 10 Ω jumper would contribute 1% of the value of such a segment. On the other hand, if the segments were constructed of 500 Ω/□ poly, then the same jumper would contribute only 0.1% of the value of a segment. Generally speaking, one should pay close attention to jumper resistances whenever dealing with segments of less than 500 Ω. When working with particularly precise resistor arrays, jumper resistance may become a factor even if the individual segments each measure as much as 1 kΩ.

One can obviously minimize the impact of lead resistance by increasing the resistance of the individual segments. However, this frequently consumes excessive amounts of space. One alternative approach seeks to minimize jumper resistance by reducing jumper lengths where possible and by placing multiple vias where one would normally employ but one. Another alternative seeks to match jumper resistances. In many cases, vias contribute the majority of jumper resistance, so a high degree of matching can be achieved by simply inserting pairs of vias into every jumper, regardless of whether they are actually required or not (Figure 7.3). This approach does not always work as well as one might expect because via resistances can vary considerably. Of course, one can always insert two or more vias at every point where a single via would have otherwise been considered sufficient.

Precisely matched capacitors can easily suffer from systematic mismatches caused by the parasitics introduced by the leads. For example, a metal lead running over 10 kÅ MLO has a capacitance of about 0.035 fF/μm^2. A lead 1 μm wide by 200 μm long would thus exhibit about 7 fF of capacitance if one neglects the contribution of fringing capacitance. This represents 0.7% of the value of a 1 pF unit capacitor.

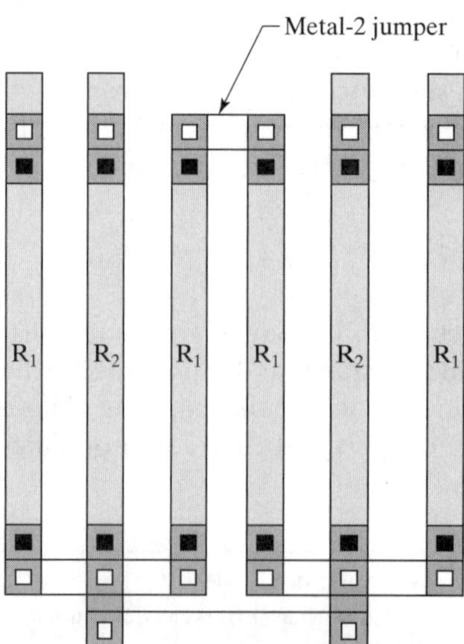

FIGURE 7.3 Insertion of vias into every jumper helps match jumper resistances and thus improves overall resistor matching.

Lead capacitance issues can be minimized by increasing the size of the individual capacitors, but this frequently proves impractical, either because of space considerations, or because a circuit requires specific values of capacitance. In such cases, one should evaluate the impact of lead capacitance. If the effects are objectionable, then the lengths of the individual leads should be adjusted so that their ratios match the ratios of the corresponding capacitor segments. The effective length of a lead can be increased either by inserting jogs (Figure 7.4A) or by adding dead-end branches (Figure 7.4B). In either case, all portions of the interconnecting leads should be of equal width to ensure that fringing capacitance contributions match. If leads include segments on more than one layer of metal, then matching requires that every lead has the same ratios of metal on the various layers. Similarly, if a portion of one lead is covered by another layer of metal, then each lead must include a similar percentage of metal covered in a similar fashion.

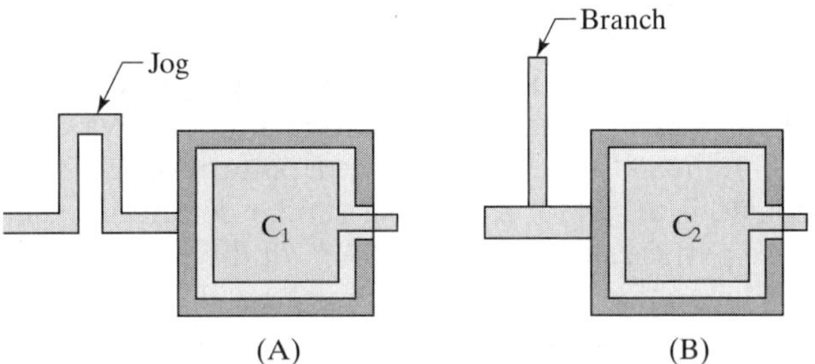

FIGURE 7.4 Methods of increasing the length of a capacitor lead: (A) inserting jogs and (B) inserting dead-end branches.

7.2.4. Pattern Shift

As discussed in Section 2.5.1, surface discontinuities left from the thermal annealing of the N-buried layer (NBL) propagate up through the monocrystalline silicon layer deposited during vapor-phase epitaxy. The resulting surface discontinuities become faintly visible under an optical microscope, particularly under lateral illumination. This image, called the *NBL shadow*, serves as a registration marker for the alignment of subsequent diffusions.

Process engineers have long been aware that surface discontinuities present in the substrate are not always faithfully reproduced in the final silicon surface. These discontinuities are frequently displaced laterally during epitaxial growth (Figure 7.5A). This effect is called *pattern shift*. Sometimes the various edges of the discontinuity shift by different amounts, causing *pattern distortion* (Figure 7.5B). Occasionally, the

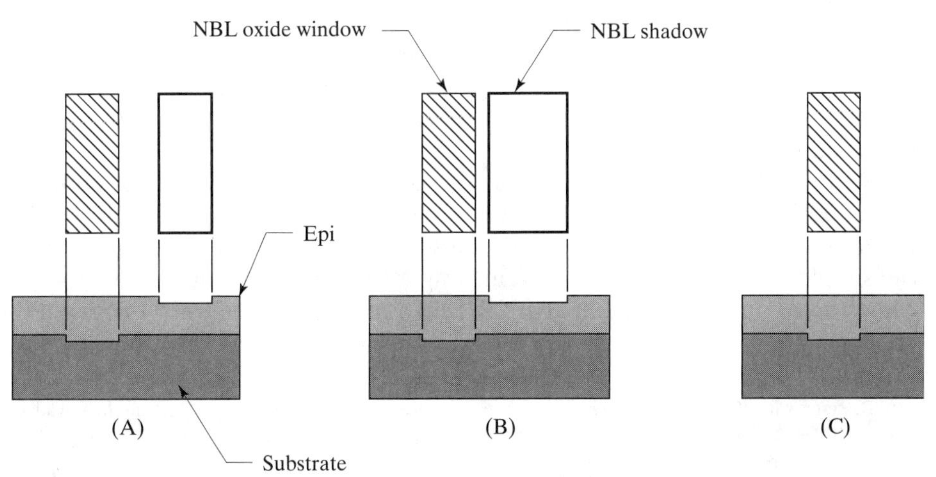

FIGURE 7.5 Effects of epitaxy on surface discontinuities: (A) pattern shift, (B) pattern distortion, and (C) pattern washout.

surface discontinuities completely vanish during the course of epitaxy, causing *pattern washout* (Figure 7.5C).

Pattern shift, distortion, and washout are all manifestations of the same underlying phenomenon. During vapor-phase epitaxy, reactant molecules adsorb on the silicon surface and move laterally until they find suitable locations where they can incorporate into the growing lattice. The exposed microsteps, formed by the intersection of the crystal lattice with the surface, encourage crystal growth in a specific direction and cause the surface topography to shift as epitaxy continues. Wafers that are (111)-oriented experience relatively severe pattern shift and distortion, which can be minimized by tilting the plane of the wafer by approximately 4° around a <110> axis.[6] Wafers that are (100)-oriented experience significant pattern distortion, but no pattern shift. The use of slightly tilted (100)-oriented wafers minimizes pattern distortion at the price of introducing pattern shift.

The magnitude of pattern shift depends upon the mobility of adsorbed reactants and the crystal orientation. Higher pressures, faster growth rates, and the presence of chlorine as a substituent in the reactant all favor increased pattern shift, while higher temperatures tend to reduce it. LPCVD deposition of dichlorosilane on (111) silicon tilted 4° will induce a pattern shift of 50 to 150% of the thickness of the epi along a <211> axis, while similar conditions using silicon tetrachloride will induce pattern shifts of 100 to 200% of the epi thickness.[7]

Due to the many variables involved, the direction and magnitude of pattern shift in a specific process can be determined only through experimental observation. A high-quality optical microscope with at least 100X magnification and a reflective illuminator are required. The shadow is usually clearest in the vicinity of minimum-geometry NPN transistors; it appears as a faint dark line not associated with any corresponding change in the oxide color. Once the shadow has been identified, the NBL shift can be estimated by comparison to features of known dimensions, such as contacts or narrow resistors. Metal lines are not recommended as a reference for comparisons because their process biases are often quite large. The deposition of interlevel oxide (ILO) obscures the NBL shadow, and planarization renders it completely invisible. Wafers used for measuring NBL shift must therefore be removed from the process prior to ILO deposition and planarization.

Pattern shift becomes a potential concern whenever matched devices are laid out in a process that employs a patterned buried layer, such as the NBL of standard bipolar and analog BiCMOS. If the NBL shadow intersects a component, it may interfere with diffusion and implantation and may cause subtle shifts in component value. Not all components are necessarily affected by pattern shift. Capacitors generally do not incorporate NBL, and polysilicon resistors usually reside over field oxide. On the other hand, diffused resistors are usually enclosed in tanks or wells containing NBL. HSR resistors are especially susceptible because the extremely thin high-sheet implant is less tolerant of slight surface discontinuities than are deeper diffusions.

Figure 7.6 depicts four matched base resistors laid out in a common tank containing NBL. This process exhibits a pattern shift to the right, so the NBL shadow intersects the leftmost resistor. There are several ways to avoid this intersection. The simplest approach consists of removing NBL from beneath the components. While this certainly eliminates the NBL shadow, it also needlessly increases the

---

[6] W. R. Runyan and K. E. Bean, *Semiconductor Integrated Circuit Processing Technology* (Reading, MA: Addison-Wesley, 1994), p. 331.

[7] For further discussion, see Runyan, *et al.,* pp. 331–333.

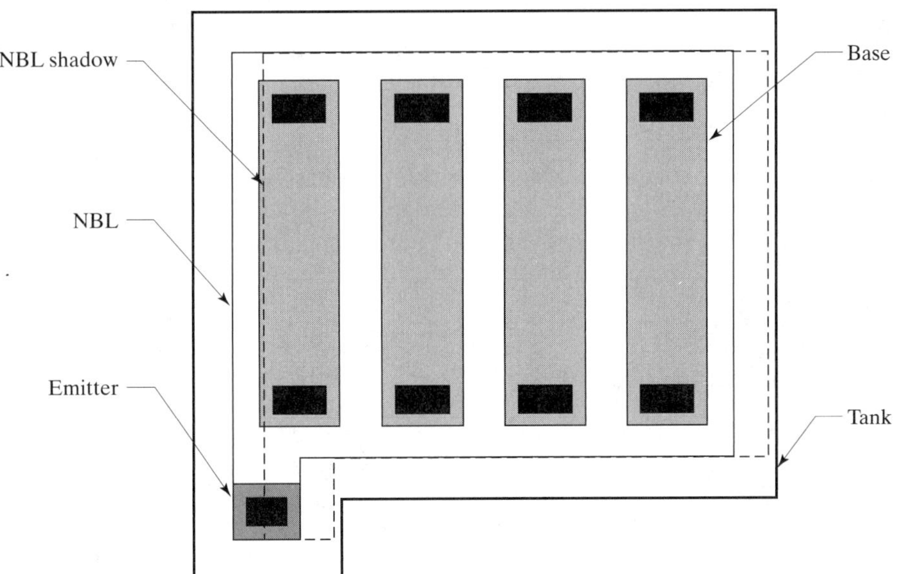

FIGURE 7.6 Layout showing an intersection of the NBL shadow with the leftmost base resistor.

tank resistance and potentially leaves the circuit vulnerable to latchup. A better approach relies on knowledge of the direction of pattern shift. If the NBL shadow shifts to the right, then the overlap of the left edge of the NBL over the components can be increased so that the NBL shadow cannot fall across them, even with worst-case pattern shifts and misalignments. This usually requires that the NBL overlap the components by at least 120% of the nominal pattern shift. If no experimental data on the direction of the pattern shift exists, then the NBL must overlap all sides of the devices vulnerable to encroachment by the NBL shadow. In the layout in Figure 7.6, only the left and right sides of the resistors are vulnerable, since the tops and bottoms of the resistors play no role in determining their values. If no information on the magnitude of the pattern shift exists, then the NBL should overlap the components by at least 150% of the nominal epi thickness.

7.2.5. Etch Rate Variations

Poly resistors are created by etching a doped polysilicon film. The etching rate depends, at least to some extent, on the geometry of the poly openings. Larger openings grant more access to the etchant and thus clear more quickly than small openings. Consequently, sidewall erosion occurs to a greater degree around the edges of a large opening than around the edges of a small one. This effect causes widely separated poly geometries to have smaller widths than closely packed geometries do. Consider the case of three polysilicon resistors with no other nearby regions of poly (Figure 7.7). The edges of the resistors facing outward form the

FIGURE 7.7 Variations in etch rate in an array of supposedly matched resistors. The exposed outer edges of the resistors experience overetching relative to the protected inner edges.

sidewalls of a vast opening that etches quickly and clears early. The edges of the resistors facing inward form sidewalls of narrow slits that etch more slowly and clear later. The middle resistor lacks outward-facing edges, so it has a slightly larger final width than either of the other resistors.

Although small, these variations in etch rate can produce serious systematic mismatches. Suppose that a 10 kÅ polysilicon film etches 90% anisotropically, resulting in an undercut of 0.1 μm. The difference between the undercutting of the outward-facing and inward-facing edges will equal only a small fraction of the total undercut, perhaps 0.02 μm. As small as this value seems, it still represents 0.5% of the width of a 4 μm resistor.

When a number of polysilicon strips are arrayed side by side, only the strips on the ends of the array experience etch rate variations. *Dummy resistors* (or *etch guards*) are often added to either end of an array of matched resistors to ensure uniform etching (Figure 7.8).[8] The dummy resistors may be constructed in one of two ways. *Unconnected dummies* are simply strips of polysilicon placed on either side of the array (Figure 7.8A). The spacing between the dummy segments and the adjacent resistors must match the spacing between the resistors of the array. The width of poly geometries has little effect on their etch rates, so the dummies can be made much narrower than the resistors they protect. This scheme has the slight disadvantage that the dummy segments remain electrically unconnected. Since the oxide isolating these segments is an exceptionally good insulator, a static electrical charge can accumulate on the dummy segments. This charge might influence the behavior of adjacent resistors. Any possibility of electrostatic modulation can be eliminated by connecting the dummies to ground or to some other suitable low-impedance node (Figure 7.8B). This precaution is normally unnecessary.

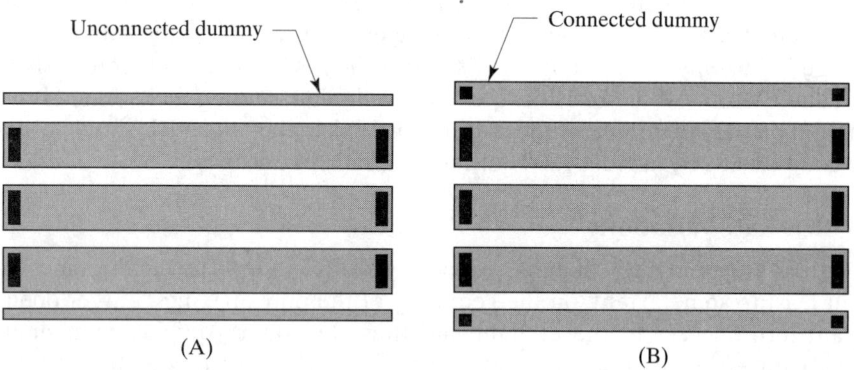

FIGURE 7.8 Examples of (A) unconnected dummy resistors and (B) connected dummy resistors.

A less desirable style of dummy resistor consists of a continuous ring of poly looped around the resistor array. Dry etching employs intense electromagnetic fields to generate and direct the reactive ions. These fields can interact with the loop of polysilicon to produce circulating currents that could affect etch rates during the final minutes of etching. Instead of an unbroken loop, consider using separate dummy segments like those in Figure 7.8. If a loop must be employed, then a gap should be left at some point to interrupt circulating currents.

Poly-poly capacitors experience the same etch rate variations as poly resistors. When matching arrays of capacitors, additional dummy capacitors should be placed

---

[8] Y. Tsividis, pp. 229–231. Tsividis's structure includes dummies, but is not connected to neutralize thermoelectrics (Section 7.2.11).

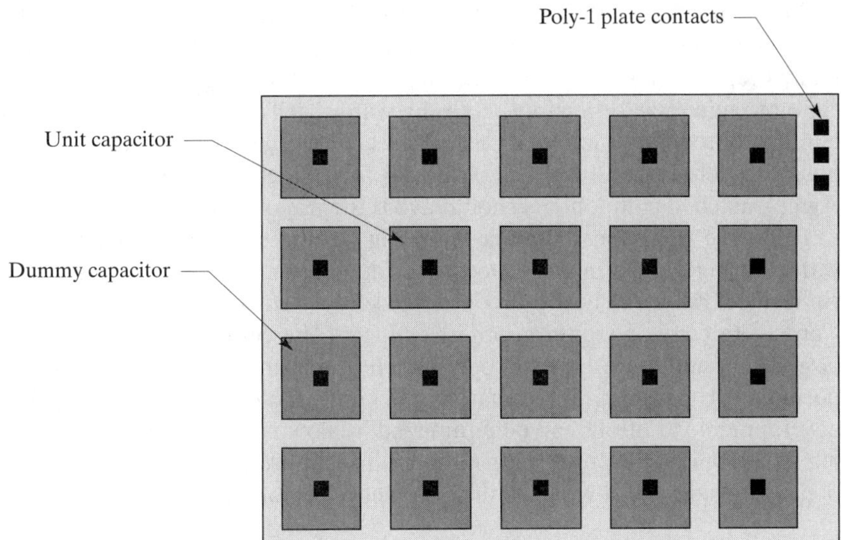

FIGURE 7.9 Matched capacitor array employing grounded dummies. The metal-2 electrostatic shield covering this array has been omitted for clarity.

around the edges of the array.[9] Figure 7.9 shows an array of six matched poly-poly capacitors with grounded dummies. Individual strips are again employed instead of a continuous ring. Notice that the poly-2 (shown in dark gray) has been drawn so that the spacing from dummy poly-2 to capacitor poly-2 matches the spacing between capacitor poly-2 regions. The dummies should always be electrically connected so that they can shield the matched capacitors from stray electrostatic fields (Section 7.2.12).

Designers often use duplicates of the matched capacitors as dummy devices, presumably to provide better matching. The size of the dummies actually has little or no effect upon etch rates. As long as the array is covered by a metal plate to block fringing fields (Section 7.2.12), full-size dummy capacitors are unnecessary.

7.2.6. Photolithographic Effects

Photolithographic processing can introduce systematic mismatches in several different ways. Optical interference and sidewall reflections can occur during exposure, and etch rate variations can occur during development. These mechanisms cause linewidth variations that are of particular concern to those constructing narrow geometries such as resistors.

Light waves constructively and destructively interfere with one another as they pass through narrow slits, such as those of a diffraction grating. These effects are responsible for the iridescent play of colors from the surface of a compact disc, or from the surface of an integrated circuit containing regular structures such as power MOS transistors. Similar effects can occur during photolithographic exposure. Far-UV light sources are used to generate shorter wavelengths than possessed by visible light, so only the narrowest of features exhibit significant interference effects. In practice, devices with dimensions of 1 μm or greater in all directions are largely immune to interference-induced mismatches. Matched components should not employ submicron dimensions unless absolutely necessary, so interference rarely constitutes much of a problem.

---

[9] Y. Tsividis, p. 228.

Reflections from the sidewalls of openings can also occur during photolithographic exposure. Photomasks have employed an *antireflective coating* (ARC) for many years. Similar coatings are often spun onto the wafer before photoresist application to minimize reflections from the wafer itself. These precautions have largely eliminated reflection-induced mismatches, but very narrow geometries continue to pose problems because of the potential for reflections from adjacent structures. Again, matched components should avoid the use of submicron geometries.

Etch rate variations may occur during photoresist development. Spinning the wafer during development exacerbates development rate variations near the periphery of the wafer because centrifugal forces cause the developer to flow outwards. The edges of resist geometries facing the center of the wafer receive the largest amount of fresh developer, and thus etch the fastest. Researchers have reported a 0.4% systematic mismatch in 0.4 μm-wide diffused resistors.[10] Photoresist development variations can be minimized by slowing the rate of wafer rotation during development, but the best solution is to employ dummies on all matched components, regardless of whether they are deposited or diffused.

7.2.7. Diffusion Interactions

The dopants that form a diffusion do not all reside within the boundaries of its junction. Consider a P-type diffusion made into an N-type epi. The concentration of acceptors in the middle of the diffusion greatly exceeds the concentration of donors, so this region of silicon is P-type. The acceptor concentration decreases as one moves outward, but the donor concentration remains constant. At the *metallurgical junction,* the acceptor concentration exactly equals the donor concentration. Beyond the junction, the acceptor concentration drops below the donor concentration and the silicon becomes N-type. The acceptor concentration falls to negligible levels some distance outside the junction. The portion of the dopant that lies outside the metallurgical junction is called the *tail* of the diffusion.

The tails of two adjacent diffusions will intersect one another. Assuming that both diffusions are of the same polarity, their tails add and each diffusion reinforces the other. Both diffusions will have slightly lower sheet resistances and slightly greater widths than they would have had in isolation from one another. Exactly the opposite situation occurs if the diffusions are of opposite polarities. The two intersecting tails counterdope one another, causing both diffusions to exhibit slightly higher sheet resistances and slightly narrower widths.

The effects of diffusion interactions on matching resemble those discussed previously for variations in poly etch rate. The resistors occupying the ends of the array will have slightly different values than the resistors occupying the middle of the array. This source of systematic mismatch can be eliminated by adding dummy resistors to either end of the array. These diffused dummy resistors must have exactly the same width as the other resistors to ensure that their dopant profiles match. The dummy resistors should be connected to prevent the formation of floating diffusions that might exacerbate latchup sensitivity (Figure 7.10).

Layouts, even those of ordinary unmatched resistors, can frequently be designed to eliminate diffusion interactions without significantly increasing die area. For example, Figure 7.11A shows a serpentine resistor that has been rather carelessly laid out. Not only are the spacings between the turns inconsistent, but the base heads also lie immediately adjacent to the resistor body. The layout of Figure 7.11B minimizes

[10] S. Hausser, S. Majoni, H. Schligtenhorst, and G. Kolwe, "Mismatch in Diffusion Resistors Caused by Photolithography," *IEEE Trans. on Semiconductor Manufacturing*, Vol. 16, #2, 2003, pp. 181–186.

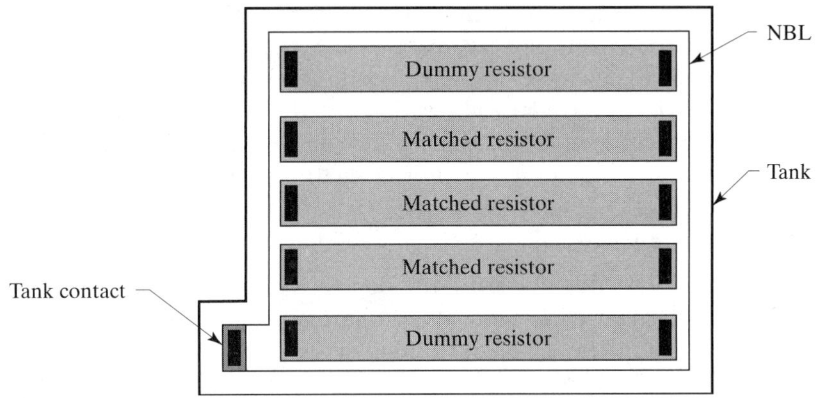

FIGURE 7.10 Matched array of diffused base resistors including grounded dummies.

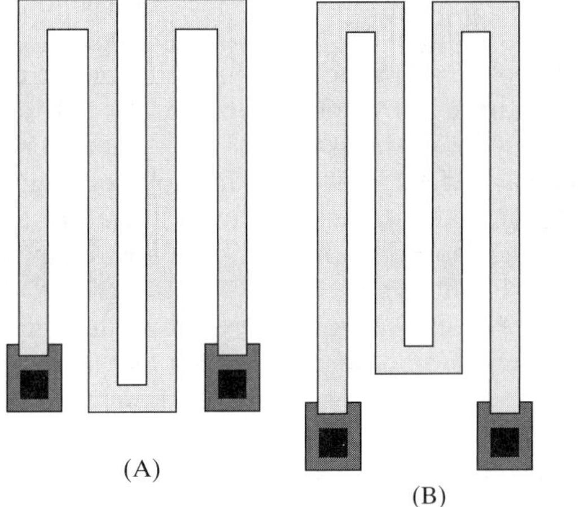

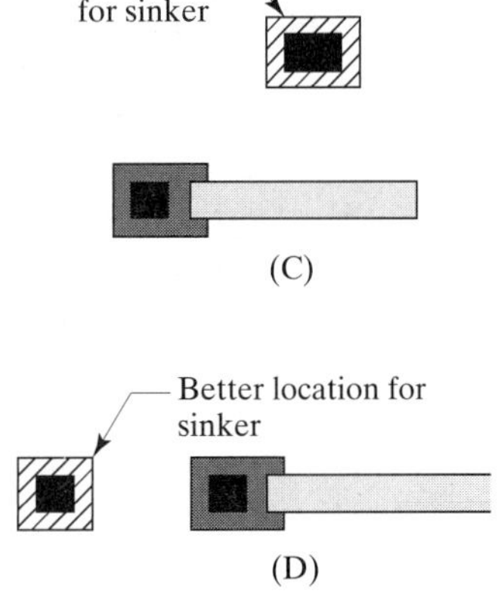

FIGURE 7.11 Examples of additional opportunities for reducing diffusion interactions: (A) poor serpentine resistor layout and (B) improved layout; (C) poor placement of deep-N+ sinker and (D) improved placement.

these potential sources of diffusion interactions by extending the heads of the resistor slightly outside the array. This modification requires little or no additional area because the tighter packing in the serpentine helps compensate for the room consumed by the extended base heads.

The layout of Figure 7.11C illustrates another type of diffusion interaction. An HSR resistor has been merged into a large tank containing a deep-N+ sinker. The deep-N+ diffusion is located immediately adjacent to the resistor body. Deep-N+ outdiffuses further than most diffusions and so has more opportunity to produce undesired interactions. The heavy phosphorus doping of the deep-N+ region can spawn lattice defects that can enhance diffusion rates in adjacent areas by a mechanism similar to emitter push (Section 2.4.2). Severe diffusion interactions may occur when these two mechanisms reinforce each other. Figure 7.11D shows a more prudent layout that places the deep-N+ sinker behind one head of the resistor, where it will have little or no effect on the resistance of the device.

Recently, another mechanism that produces effects resembling diffusion interactions has been reported.[11] This mechanism involves scattering of ions from the edges of the photoresist during high-energy ion implantations, such as those required to

[11] T. B. Hook, J. Brown, P. Cottrell, E. Adler, D. Hoyniak, J. Johnson, and R. Mann, "Lateral Ion Implant Straggle and Mask Proximity Effect," *IEEE Trans. on Electron Devices*, Vol. 50, #9, 2003, pp. 1946–1951.

fabricate retrograde wells. Ions may scatter more than 1 μm from the edge of the photoresist. These ions possess relatively low energies, and they therefore dope the surface layers of silicon adjacent to the photoresist. Shallow diffusions placed near the drawn edges of a retrograde well may experience anomalous doping variations due to this mechanism. These mismatches can be avoided by placing matching diffusions at least 2–3 μm inside the drawn edges of retrograde wells.

7.2.8. Hydrogenation

Hydrogen is introduced during the deposition and etching of the metal system. Hydrogen seeps into the oxide layers, where it acts as a mobile ion. When hydrogen atoms diffuse to the edge of silicon regions, they can influence device operation by either of two mechanisms. First, hydrogen atoms can tie up dangling bonds and thereby eliminate surface states. Second, hydrogen can actually diffuse into the silicon itself, where it forms weak molecular complexes with boron atoms. These complexes do not ionize at room temperature, so the affected boron atoms no longer act as acceptors. This mechanism is called *hydrogen compensation.*

Hydrogen can affect the value of poly resistors both through elimination of dangling bonds at grain interfaces and through hydrogen compensation. The latter mechanism occurs only in P-type resistors, and even there it plays just a minor role due to the dominance of grain boundary effects in determining poly resistivity.[12]

Hydrogen cannot diffuse through metal. Furthermore, some materials used in metal systems, such as titanium, strongly adsorb hydrogen. Therefore, variations in hydrogenation often develop between metallized and exposed areas of the die. In particular, the presence of metal plates or leads above poly resistors can significantly affect their values. Systematic mismatches between metallized and unmetallized resistors in excess of 1% have been observed. Hydrogenation is currently thought to be the primary cause of these mismatches, although metallization-induced stresses undoubtedly also play a role. One clue that suggests that hydrogenation plays a predominant role comes from the observation that resistors several microns from the edges of metal plates exhibit systematic mismatches similar to, but of lesser magnitude than, those experienced by resistors actually under the plates.[13] This observation can be explained by the fact that the titanium-tungsten refractory barrier metal employed by the metallization system in question strongly adsorbs hydrogen. This probably sets up a hydrogenation gradient that extends for a distance of several microns beyond the edges of the metal plate. Resistors that fall within this dehydrogenated zone experience systematic mismatches that are inversely proportional to the distance from the metal plate that produced the dehydrogenation.

The interconnection of segments in a resistor array can produce significant metallization-induced mismatches. The resistor array of Figure 7.12A shows a popular style of interconnection in which the jumpers between the segments have been "folded in" over the resistors to save space. The presence of differing amounts of metal coverage over individual resistor segments can lead to metallization-induced mismatches. The resistor array of Figure 7.12B illustrates an alternative style of interconnection in which jumpers are "folded out" to minimize the amount of metal overlapping the active area of the resistors. Experience shows that the folded-out style of interconnection generally produces better matching, even when precautions are taken to match the metal coverage of individual segments of a folded-in array by

---

[12] M. Rydberg and U. Smith, "Long-Term Stability and Electrical Properties of Compensation Doped Poly-Si IC-Resistors," *IEEE Trans. on Electron Devices*, Vol. 47, #2, 2000, pp. 417–426.

[13] D. Briggs, private communication, 2003.

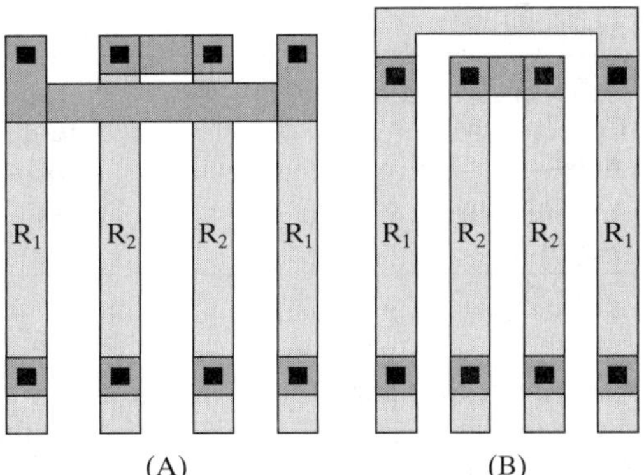

FIGURE 7.12 Comparison of resistor array interconnection strategies: (A) folded-in jumpers and (B) folded-out jumpers.

adding dummy metal traces. Where folded-out arrays require vias, these may introduce mismatches of their own (Section 7.2.3).

Another approach that can virtually eliminate metallization-induced mismatches employs a metal-1 plate that covers as much of the array as possible, leaving only the contact heads projecting from beneath either side of the plate. Metal-2 jumpers can be folded in over the metal-1 shield without inducing hydrogenation mismatches. The metal plate also minimizes mechanical stress concerns by introducing a certain degree of mechanical compliance that relieves strains set up by the upper level of metallization. Some researchers have found metal-1 jumpers can be folded in beneath a metal-2 shield, presumably because the metal-2 shield blocks the hydrogenation of resistors beneath it.[14] This observation may not apply to all processes. In any event, the shield should always be connected to a quiet node of the circuit so that it can serve as an electrostatic shield (Section 7.2.12).

Given the choice, one should consider using phosphorus resistors rather than boron-doped resistors. Phosphorus is not subject to hydrogen compensation. Furthermore, phosphorus selectively accumulates at grain boundaries, where it tends to reduce the concentration of dangling bonds.[15] However, before choosing phosphorus-doped poly as the resistance material of choice, one should first examine the random mismatch between similar resistors of every available type to determine whether the phosphorus-doped poly exhibits more random mismatch than other available materials. This situation has arisen in certain processes, presumably due to differences in grain size or incomplete dopant activation.

7.2.9. Mechanical Stress and Package Shift

Silicon is piezoresistive in that it exhibits changes in resistivity under stress. Variations in stress will thus generate mismatches between precision resistors. Capacitors remain largely unaffected because mechanical stress has little effect upon the dimensions and permittivity of common dielectrics. Although well-matched capacitors consequently exhibit less systematic mismatch than well-matched resistors, not all circuits can rely upon matched capacitors. Layout techniques have been developed for minimizing the stress sensitivity of resistors. The severity of the problem varies depending upon the method of packaging.

Metal cans and hermetically sealed ceramic packages generate the lowest stresses. Although the metal header and the silicon die have vastly different coefficients

---

14 J. Smith, private communication, 2003.

15 Rydberg, et al.

TABLE 7.1 Coefficients of thermal expansion (CTE) for several materials used in packaging integrated circuits.[16]

Material	CTE ppm/°C
Epoxy encapsulation (typical)	24
Copper alloys	16–18
Alloy-42	4.5
Molybdenum	2.5
Silicon	2.5

of thermal expansion (Table 7.1), the epoxy used for die attachment absorbs the resulting mechanical strain. Precision integrated circuits are often packaged in metal cans or ceramic packages despite the cost and inconvenience, as there is simply no other way to obtain such a low-stress environment.

The vast majority of integrated circuits are packaged in plastic, which offers a combination of low cost and mechanical ruggedness. Unfortunately, plastic encapsulation places severe stresses upon the die. As Table 7.1 indicates, the coefficient of thermal expansion of plastic encapsulants is approximately 10 times that of silicon. The epoxy resin is injected into the mold at high temperatures (typically 175°C). Chemical changes occur within the heated resin, causing it to rapidly solidify. As the encapsulated device cools, the difference between the coefficients of thermal expansion of silicon and epoxy generates residual stresses that remain permanently frozen into the packaged device. Measurements of electrical parameters before and after packaging reveal differences called *package shifts* that are proportional to the level of residual stresses.

The stresses that generate package shifts fall into two general categories. The first of these consists of stresses that affect the die as a whole, and which vary only gradually across the surface of the die. Section 7.2.10 discusses these stresses and the means by which their effects can be mitigated. The second category of stresses involve those which are highly localized and whose effects are largely random in character. These stresses arise because of the addition of fillers to plastic encapsulants. The majority of a typical encapsulant formulation consists of small particles of silica, many of which are angular fragments ranging from 15 to 150 μm in diameter. If a filler particle happens to appear directly above a matched component, then, as the plastic contracts after encapsulation, the filler particle is driven into the surface of the die. The resulting *filler-induced stresses* are localized in regions some tens of microns in diameter, but they are of extremely large magnitude, as is evidenced by the fact that they can shear entirely through narrow upper level metal leads.[17] Depending upon the distribution of filler particles, which is essentially random, individual units may exhibit drastically different package shifts. Mismatches produced by filler-induced stresses exhibit means and standard deviations of similar magnitudes, whereas those produced by stresses upon the die as a whole exhibit means that are much larger than their standard deviations. Filler-induced stresses have been observed to induce mismatches with standard deviations as high as 2%.[18]

---

[16] Values for epoxy, molybdenum, and silicon: R. E. Thomas, "Stress-Induced Deformation of Aluminum Metallization in Plastic Molded Semiconductor Devices," *IEEE Trans. on Components, Hybrids, and Manufacturing Technology,* Vol. CHMT-8, #4, 1985, pp. 427–434. Values for Alloy-42 and copper from "Leadframe Materials" *Semiconductor Reliability News,* Vol. 8, #9, September 1996, p. 5.

[17] P. Yalamanchili and V. Baltazar, "Filler Induced Metal Crush Failure Mechanism in Plastic Encapsulated Devices," *Proc. 37th Annual International Reliability Physics Symposium*, 1999, pp. 341–346.

[18] R. Pendse and D. Jennings, "Parametric Shifts in Devices: Role of Packaging Variables and Some Novel Solutions," *Proc. 40th Electronic Components Technology Conf.*, 1990, pp. 322–326.

Postpackage trimming has often been touted as a solution for package shifts in general, and filler-induced stresses in particular. Unfortunately, the benefits of postpackage trimming are often overstated. The initial variation exhibited by packaged units can indeed be almost entirely eliminated, but the magnitude of package shifts varies dramatically with temperature. Package shifts may almost vanish if a unit is heated from 25°C to 125°C, and they can easily double if a unit is cooled from 25°C to −40°C. Clearly, the impact of package shifts upon temperature drift is as much a concern as their impact upon initial mismatch. Some designers have speculated that multiple-temperature testing could be used to calibrate out both the initial mismatch and the temperature drift of individual units, but this is not the case. Package shifts gradually increase when devices are operated at high temperatures for long periods of time. This effect is one of a number of instabilities collectively called *long-term drift*. It is believed to arise from gradual chemical changes that occur in the heated plastic that cause it to gradually contract. Whatever the exact nature of these changes, they act to increase package shift with time. The usual test of long-term drift involves a 1000-hour bake at 125°C. Little data on long-term drift has been published, but one researcher reported an approximate 30% increase in mean drain-current mismatch in MOS transistors, and there seems no reason to expect package shifts in resistors to behave any differently.[19] Although postpackage trimming undeniably has some value, it is by no means a panacea for package shift.

Another approach that significantly reduces filler-induced stresses involves the application of a special overcoat that provides mechanical compliance. Filler particles press into this overcoat, but it elastically deforms and thus absorbs the majority of the resulting stresses. In order to be effective, the overcoat must be of the same order of thickness as the filler particle diameter, or at least 10–30 μm. Excellent results have been obtained with dropper-applied silicones and patterned polyimide films. One group of researchers reported a threefold decrease in random package shift after the application of a 10 μm-thick polyimide coating.[20] Thick power copper metallization can also be used as an effective mechanical compliance layer.[21] No form of overcoat can provide much reduction in large-scale mechanical stresses that affect the entire die, but proper layout precautions can greatly minimize the impact of such large-scale stresses (Section 7.2.10). On the other hand, layout measures are of little value in combating the random variations produced by filler-induced stresses.

Power packaging requires an intimate thermal union between the die and its leadframe (or header) to minimize heat buildup. The die attach for a power package consists of either silver-filled epoxy, solder, or gold eutectic. Silver-filled epoxies do not provide quite as good a thermal and electrical union as either of the other alternatives, but epoxy mounting produces significantly lower residual stresses. Solder mounting is commonly employed for large metal tab or can packages. Since solder does not adhere to silicon, the back side of the die must be plated with a sputtered or evaporated metal film. Gold eutectic bonding uses a thin strip of gold foil, called a *gold preform*, placed between the die and the header. When heated, the gold preform alloys to both materials. Neither solder nor gold alloy has much mechanical compliance, so the thermal mismatch between the copper header and the silicon die generates extreme mechanical stresses.

---

19 H. Ali, "Stress-Induced Parametric Shift in Plastic Packaged Devices," *IEEE Trans. on Components, Packaging and Manufacturing Technology—Part B*, Vol. 20, #4, 1997, pp. 458–462.

20 P. Yalamanchili, et al.

21 B. Abesingha, G. A. Rincón-Mora, and D. Briggs, "Voltage Shift in Plastic-Packaged Bandgap References," *IEEE Trans. on Circuits and Systems—II: Analog and Digital Signal Processing*, Vol. 49, #10, 2002, pp. 681–685.

Stresses generated by solder or gold eutectic mounting can be minimized by constructing the leadframe or header out of a material with a coefficient of thermal expansion similar to that of silicon, such as molybdenum or alloy-42 (a nickel-iron alloy containing 42% nickel). Molybdenum headers are very expensive, while alloy-42 is brittle and exhibits poor thermal and electrical conductivity. Most power products continue to use copper leadframes or headers despite the package shifts they generate.

7.2.10. Stress Gradients

Mechanical stresses that affect the whole of the die give rise to regular patterns of stress. These patterns, and the effects they produce upon electrical parameters, can be minimized by proper layout precautions. This section will first examine the impact of stress upon silicon in more detail, and will then consider methods of canceling the resulting systematic variations.

Piezoresistivity

The piezoresistivity of (100)-oriented silicon varies with orientation and doping.[22] An N-type (100) silicon wafer exhibits maximum piezoresistivity along <100> axes and minimum piezoresistivity along <110> axes. N-type diffused and implanted resistors therefore exhibit minimum stress sensitivity if they lie along <110> axes. One of the <110> axes of the wafer lies parallel to the major wafer flat, while the other <110> axis lies perpendicular to it (Figure 7.13). Since dice are laid out in rows and columns relative to the wafer flat, the X- and Y-axes of the layout correspond to the desired <110> directions. The stress sensitivity of N-type monocrystalline resistors can therefore be minimized by laying them out either horizontally or vertically.

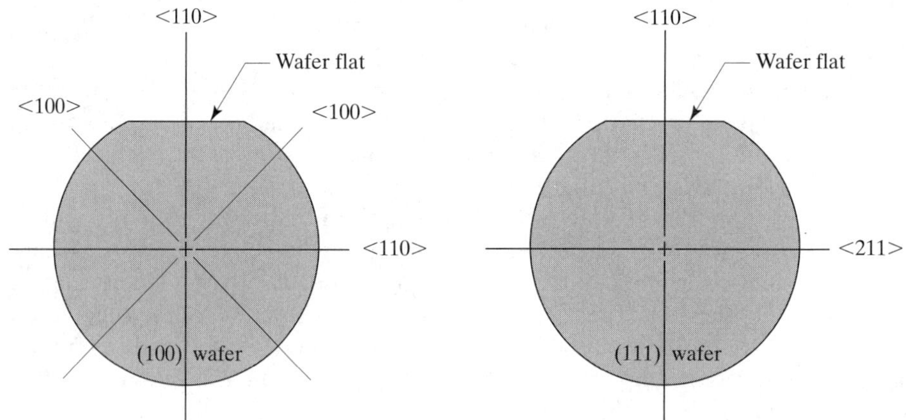

FIGURE 7.13 Identification of directions on (100) and (111) wafers.

A P-type (100) silicon wafer exhibits maximum piezoresistivity along <110> axes and minimum piezoresistivity along <100> axes. P-type diffused and implanted resistors therefore exhibit the least stress sensitivity if they lie along <100> axes. The <100> axes of a (100) wafer are rotated 45° to the wafer flat (Figure 7.13). P-type monocrystalline resistors therefore exhibit the least stress sensitivity if they are placed at 45° to the X- and Y-axes of the layout. When so oriented, the piezoresistivity

---

22 Y. Kanda, "A Graphical Representation of the Piezoresistance Coefficients in Silicon," *IEEE Trans. on Electron Devices*, Vol. ED-29, #1, 1982, pp. 64–70.

of P-type monocrystalline resistors actually falls to zero. This does not occur with N-type resistors, which still retain some degree of piezoresistivity even when oriented in the optimum directions. This difference constitutes one reason for preferring P-type monocrystalline resistors to their N-type counterparts.

The piezoresistivity of a (111) wafer does not vary with direction. Although no reason exists to prefer one orientation over another, most resistors on (111) silicon are placed either vertically or horizontally to simplify packing and interconnection.

The piezoresistivity of monocrystalline silicon exhibits little dependence on doping concentrations as long as these do not exceed about 10^{18} atoms/cm^3. Almost all matched resistors use significantly lower dopant concentrations, so low-sheet and high-sheet materials exhibit no significant differences in piezoresistivity.

Polycrystalline silicon is an isotropic material, so its piezoresistivity is the same in all directions. The magnitude of this piezoresistivity drops as the resistivity of the poly increases.[23] Lightly doped poly of the sort normally used to make resistors has a relatively small (but nonzero) stress dependency. The <100>-oriented P-type diffused resistors on (100) silicon will have lower stress sensitivities than poly resistors, but the poly resistors may have better overall matching because they do not exhibit the voltage modulation problems that plague most types of diffused and implanted resistors (Section 7.2.12).

Gradients and Centroids

Figure 7.14 graphically illustrates the stress distribution across a typical integrated circuit, ignoring localized variations caused by filler-induced stresses. The drawing at the lower left is called an *isobaric contour plot.* The curved lines (called *isobars*) indicate the stress levels at various points on the die surface. Each isobar passes through a series of points of equal stress intensity. The stress intensity rises from a broad minimum in the middle of the die to maxima at the four corners. The graph

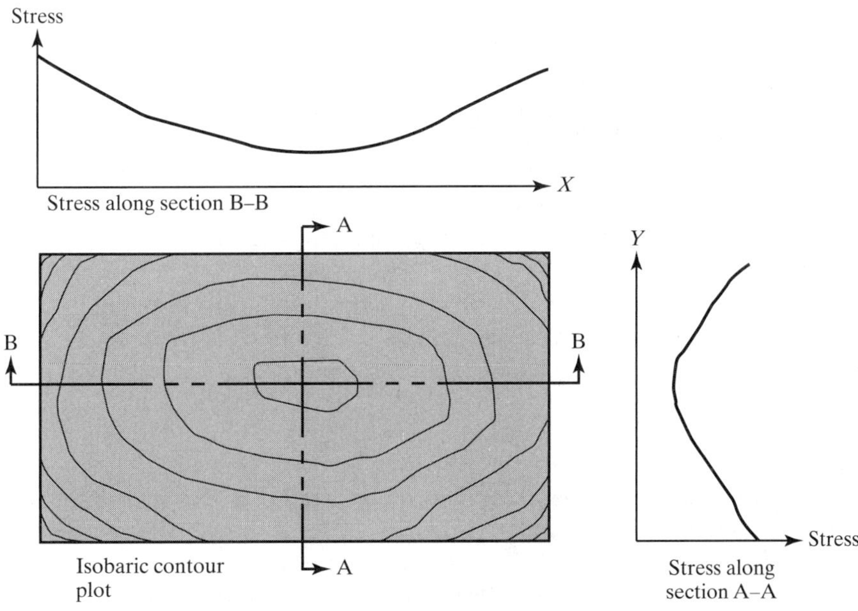

FIGURE 7.14 Isobaric contour plot of the stress distribution across the surface of a typical epoxy-mounted plastic-packaged (100) silicon die, together with two graphs showing the stress along section lines A–A and B–B.

[23] H. Mikoshiba, "Stress-Sensitive Properties of Silicon-Gate MOS Devices," *Solid-State Electronics*, Vol. 24, 1981, pp. 221–232.

above the contour plot shows the stress intensity along a line bisecting the die horizontally, while the graph at the right shows the stress intensity along a line bisecting the die vertically. By comparing the two graphs to the contour plot from which they were generated, the general nature of the latter should become apparent. Contour plots resembling this one are used in topographic mapping to display the three-dimensional shapes of hills and valleys. The distribution of stress on a die resembles a depression: It is lowest in the middle of the die and highest at the four corners.

The spacing of the isobars provides additional information about the stress distribution. The stress intensity changes rapidly where the isobars are spaced closely together, and slowly where they are spaced far apart. The rate of change of the stress intensity is called the *stress gradient.* This gradient is usually smallest in the middle of the die and slowly increases as one moves out toward the edges. The stress gradient is usually much greater at the extreme corners of the die than at any other point (Figure 7.15).

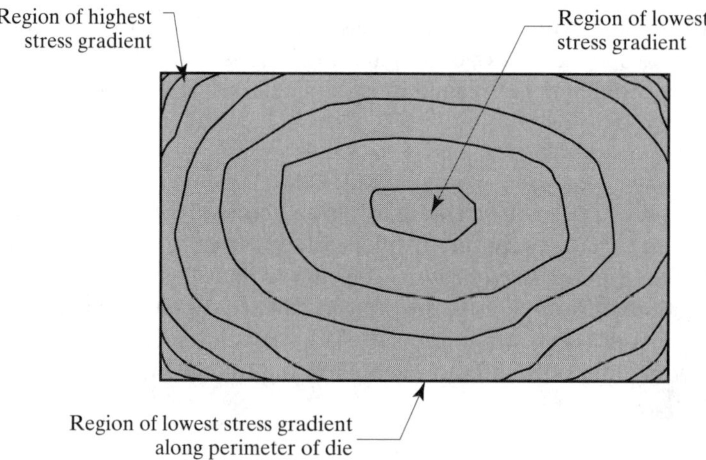

FIGURE 7.15 Isobaric contour plot showing regions of highest and lowest stress gradient.

Matched devices should reside as close to one another as possible to minimize the difference in stresses between them. Although the finite size of the devices might seem to limit how closely they can be placed, certain layout techniques can produce remarkably small effective separations. The following analysis assumes that the stress gradient is approximately constant in the region between the matched devices. This is generally a reasonable assumption, provided that the matched devices are placed to form as compact a structure as possible.

The stress difference between two matched devices is proportional to the product of the stress gradient and the separation between them. For the purposes of this calculation, the location of each device is computed by averaging the contribution of each portion of the device to the whole.[24] The resulting location is called the *centroid* of the device. The centroid of a rectangular device lies in its exact center. The centroids of other geometries can often be located by applying the *principle of centroidal symmetry,* which states that the centroid of a geometry must lie on any axis of symmetry of that geometry. Figure 7.16 shows how this principle can determine the centroids of a rectangle and a dogbone resistor. The centroid of practically any geometry used in layout can be determined in a similar manner.

---

[24] Most texts on statics include a discussion of centroids and their relationship to the well-known principle of moments, *e.g.*, R. C. Hibbeler, *Engineering Mechanics: Statics*, 4th ed. (New York: Macmillian Publishing Co., 1998), p. 435.

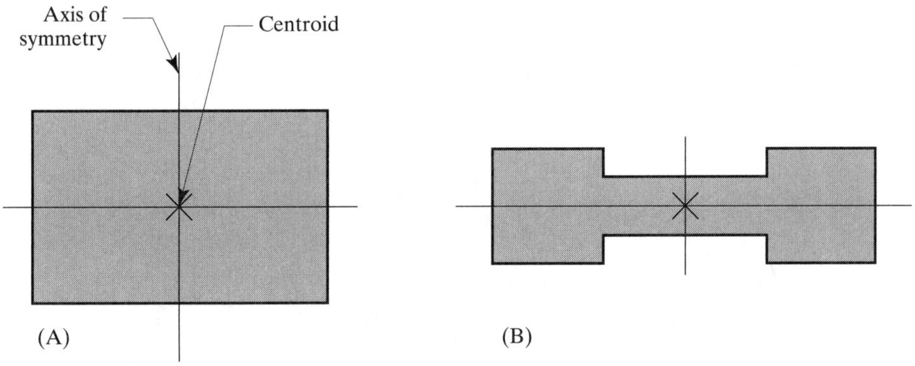

FIGURE 7.16 Locating (A) the centroids of a rectangle and (B) a dogbone resistor.

The effects of stress on resistors can be quantified in terms of piezoresistivity, centroid locations, and stress gradients. The magnitude of the stress-induced mismatch between two resistors equals

$$\delta_s = \pi_{cc} d_{cc} \nabla S_{cc} \tag{7.16}$$

where π_{cc} is the piezoresistivity along a line connecting the centroids of the two matched devices, ∇S_{cc} is the stress gradient along this same line, and d_{cc} equals the distance between the centroids. The higher order components of the stress gradient are typically much smaller than the linear component, and they are therefore ignored in the formulation of Equation 7.16. This formula reveals several ways to minimize stress sensitivity. First, the designer can reduce the piezoresistivity, π_{cc}, by choosing a suitable resistance material or by orienting the resistors in the direction of minimum piezoresistivity. Second, the designer can reduce the magnitude of the stress gradient, ∇S_{cc}, by proper location of the devices and by selecting low-stress packaging materials. Third, the designer can reduce the separation between the centroids of the device, d_{cc}. The first two options have already been discussed, so we will now focus on reducing the separation of the centroids.

Common-Centroid Layout

Suppose that a matched device is divided into sections. If these sections are all identical, and if they are arranged to form a symmetric pattern, then the centroid of the device will lie at the intersection of the axes of symmetry passing through the array. It is actually possible to arrange two arrayed devices so that they share common axes of symmetry. If this can be achieved, then the principle of centroidal symmetry ensures that the centroids of the two devices coincide. Figure 7.17A shows an example of such a *common-centroid* layout. The two devices are marked A and B, their axes of symmetry are shown as dotted lines, and their centroids are denoted by an "X" where the two axes of symmetry intersect.

Equation 7.16 predicts that the stress-induced mismatch of a common-centroid layout equals zero because the separation between the centroids equals zero. This does not actually occur, because this analysis ignores the higher-order components of the stress gradient. Despite this limitation, common-centroid layout is still the single most powerful technique available for minimizing large-scale stress-induced mismatches. Unfortunately, common-centroid layout has little effect upon filler-induced mismatches, as these are too highly localized to permit the use of simple linear approximations of the stress gradient.

Figure 7.17 shows three examples of common-centroid layouts produced by arraying segments of matched devices along one dimension. These types of layouts are usually called *interdigitated arrays* because the sections of one device interpenetrate

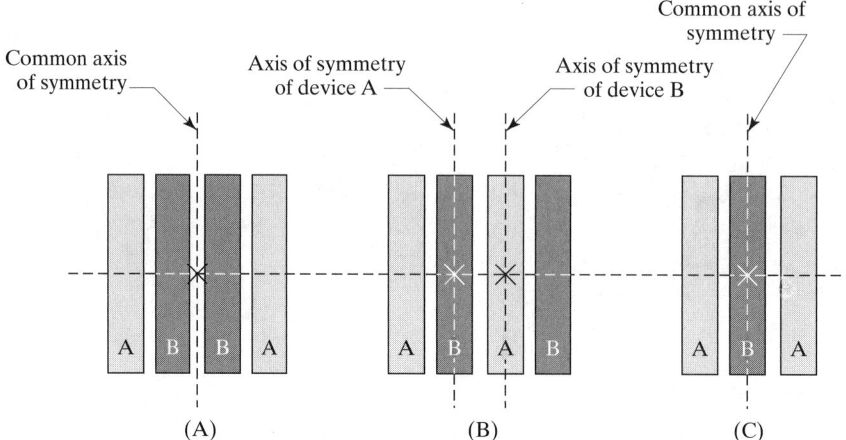

FIGURE 7.17 Examples of one-dimensional common-centroid arrays.

the sections of the other like the intermeshed fingers of two hands. Figure 7.17A shows an interdigitated array consisting of two devices, each composed of two segments. If the devices are denoted by the letters A and B, then the arrangement of the segments follows the *interdigitation pattern* (or *weave*) ABBA. This pattern has an axis of symmetry that divides it into two mirror-image halves (AB and BA). A second axis of symmetry passes horizontally through the array, but this axis results from the symmetries of the individual segments rather than the symmetry of the interdigitation pattern.

Arrays that use the interdigitation pattern ABBA require dummies because the segments of one device occupy both ends of the array. Some designers prefer to use arrays that follow the pattern ABAB (Figure 7.17B) because they mistakenly believe that this pattern eliminates the need for dummies. This belief is founded on a misunderstanding of the function of the dummy devices. Etch variations and diffusion interactions depend on the arrangement of adjacent geometries. The addition of dummy devices ensures that each segment faces an identical arrangement of geometries. If the dummies are omitted, then the segments on the ends of the array face whatever geometries happen to reside nearby. The geometries adjacent to one end of the array are likely to differ from those next to the other. If this happens, then mismatches will result if dummies are omitted regardless of whether the array follows the pattern ABBA or ABAB. The pattern ABAB should be avoided because it does not completely align the centroids of the two devices, and the resulting separation of the centroids leaves the devices vulnerable to stress-induced mismatches.

Common-centroid layouts can also consist of devices of different sizes. Figure 7.17C shows an example that implements a 2:1 ratio, using the pattern ABA. If the two devices on the end of the array are resistors, then they can connect either in series or in parallel. If they are capacitors, they can only connect in parallel because a series connection would introduce mismatches caused by the differences in parasitic capacitances between the upper and lower plates. More complicated patterns offer a larger number of possible ratios, especially if one considers that resistors can connect in series as well as in parallel. Table 7.2 lists a number of additional interdigitation patterns. The patterns marked by asterisks do not provide complete alignment of the centroids. In all such cases, one can achieve full alignment by using a larger number of segments.

The process of designing an interdigitated array begins with the identification of all of the components comprising the array. Matched devices occur in groups, and all of the devices in any one group must reside in the same array. One cannot identify

A	AA	AAA	AAAA
AB*	ABBA	ABBAAB*	ABABBABA
ABC*	ABCCBA	ABCBACBCA*	ABCABCCBACBA
ABCD*	ABCDDCBA	ABCDBCADBCDA*	ABCDDCBAABCDDCBA
ABA	ABAABA	ABAABAABA	ABAABAABAABA
ABABA	ABABAABABA	ABABAABABAABABA	ABABAABABAABABAABABA
AABA*	AABAABAA	AABAAABAAABA*	AABAABAAAABAABAA
AABAA	AABAAAABAA	AABAAAABAAAABAA	AABAAAABAAAABAAAABAA

TABLE 7.2 Sample interdigitation patterns for arrays having one axis of symmetry.

the groups of matched components in a circuit without fully understanding how the circuit operates. The circuit designer must therefore identify the groups of matched components and pass this information to the layout designer.

Once the components making up an array have been identified, they must be divided into segments. This process is not always a simple one. The designer should first check to see if all of the values have a *greatest common factor*. For example, two resistors having values of 10 kΩ and 25 kΩ have a greatest common factor of 5 kΩ. The array can then consist of a number of segments each equal to the greatest common factor. For example, an array of a 10 kΩ and a 25 kΩ resistor could be laid out using seven 5 kΩ segments.

In cases where a large common factor does not exist, try using the value of the smallest device as the value of a segment. Based on this value, determine the number of segments in the other devices. If any device requires a partial segment with a value less than about 70% of a complete segment, try dividing the smallest device value by increasingly larger integer numbers (2, 3, 4...) until a value is found that does not require a small partial segment. For example, suppose that a 39.7 kΩ and a 144.5 kΩ resistor must be arrayed. If we choose a segment resistance of 39.7 kΩ, then the 144.5 kΩ resistor would require 3.638 segments. This requires a 63.8% partial segment, so we divide the smaller resistance by two and try a segment resistance of 19.85 kΩ. The larger resistor would need 7.280 segments, which would require a 28.0% partial segment. Dividing the smaller value by three produces a segment resistance of 13.233 kΩ. The larger resistor would require 10.920 segments, so the array contains no partial segment that is less than 70% of a complete segment. This array could therefore consist of thirteen segments of 13.233 kΩ and one segment of 12.174 kΩ. In a few cases, this procedure produces a very small segment value. The designer should try larger segment values to see if one can be found that will allow all of the matched resistors to be constructed without using any partial segment that is less than about 70% of a complete segment. In certain cases, the designer may have to tolerate the presence of a small partial segment.

Once the segment value has been found, make sure that it is not so small that it precludes reasonable matching. Matched resistor segments should never contain less than five squares, and preferably they should contain at least ten. If the array seems to require short segments, consider using sections connected in parallel as well as in series. Capacitor segments (unit capacitors) should not have dimensions of much less than 100 μm^2. Capacitors are always connected in parallel because a series connection inserts parasitic capacitances that disturb matching. There are a few cases in which series-connected capacitors may form part of a set of matched capacitors, but such instances should stem only from the necessities of circuit design, not from the expediencies of layout.

Partial resistor segments are best implemented using sliding contacts to enable all of the segments to have exactly the same geometry. This precaution ensures that etch variations and diffusion interactions do not cause mismatches between the partial segment and the remainder of the array. The sliding contact can also be used to adjust the value of the partial segment should the initial setting prove incorrect. Partial unit capacitors should reside at one end of the array, where they will not unduly disturb the other unit capacitors. The partial unit capacitor should be sized using Equations 7.15A and 7.15B to ensure that its area-to-periphery ratio equals that of the other unit capacitors.

Once the array has been segmented, a suitable interdigitation pattern must be chosen. The best interdigitation patterns obey all four rules of common-centroid layout listed in Table 7.3. The *rule of coincidence* states that the centroids of the matched devices should coincide at least approximately. Patterns that do not achieve coincidence exhibit larger stress sensitivities than those that do. The *rule of symmetry* states that the array should be symmetric around both axes. A one-dimensional array should derive one of its axes of symmetry from its interdigitation pattern. For example, an array using the pattern ABBA will have an axis of symmetry dividing it into two mirror-image halves (AB and BA). A one-dimensional array must rely on the symmetries of the individual segments to produce its second axis of symmetry. In the case of resistors and capacitors, this should not present a problem, because all of the segments should have symmetric shapes.

TABLE 7.3 The four rules of common-centroid layout.

1. ***Coincidence:*** The centroids of the matched devices should coincide at least approximately. Ideally, the centroids should exactly coincide.
2. ***Symmetry:*** The array should be symmetric around both the X- and Y-axes. Ideally, this symmetry should arise from the placement of segments in the array and not from the symmetry of the individual segments.
3. ***Dispersion:*** The array should exhibit the highest possible degree of dispersion; in other words, the segments of each device should be distributed throughout the array as uniformly as possible.
4. ***Compactness:*** The array should be as compact as possible. Ideally, it should be nearly square.

The *rule of dispersion* states that the segments of each device should be distributed throughout the array as uniformly as possible. The degree of dispersion is often evident to the eye, but it can be partially quantified by counting the number of repeated segments (*runs*). For example, the pattern ABBAABBA contains three runs of two segments each, while the pattern ABABBABA contains only one run of two segments. The latter pattern is therefore more disperse than the former. Dispersion helps reduce the sensitivity of a common-centroid array to higher order gradients (nonlinearities). Dispersion is therefore especially important for arrays subject to large stress gradients, or those which spread across large distances.

The *rule of compactness* states that the array should be as compact as possible. Ideally, it should be square, but in practice many arrays can have an aspect ratio of 2:1 or even 3:1 without introducing any significant vulnerability. If the aspect ratio of the array exceeds 2:1, then consider breaking the array into a larger or smaller number of segments. If the array consists of a few long segments, try doubling the number of segments and halving the value of each. Arrays consisting of many short (or small) segments are excellent candidates for the two-dimensional arrays discussed next.

All of the common-centroid layouts discussed so far array the devices in only one dimension. Such a *one-dimensional array* derives one of its axes of symmetry from its interdigitation pattern and one of its axes of symmetry from the symmetry of its segments. The segments can also be arranged to form a *two-dimensional array* deriving both of its axes of symmetry from its interdigitation pattern. This type of arrangement generally provides better cancellation of gradients than one-dimensional arrays, primarily because of the superior compactness and dispersion possible within a two-dimensional array. Some designers believe that two-dimensional arrays are always superior to their one-dimensional counterparts, but this is not always true. If the dimensions of the arrays are small, then random mismatches will outweigh systematic mismatches caused by gradients, and little or no difference will be observed between the two styles of arrays.

Figure 7.18A shows two matched devices, each composed of two segments arranged in an array of two rows and two columns. This arrangement is often called a *cross-coupled pair.* Resistors are rarely laid out as cross-coupled pairs because the resulting arrays usually have unwieldy aspect ratios. Capacitors often produce very compact cross-coupled pairs, as do diodes and transistors. If the matched devices are large enough to segment into more than two pieces, then the cross-coupled pair can be further subdivided as shown in Figure 7.18B. This array exhibits more dispersion than a cross-coupled pair and is therefore less susceptible to higher order gradients. This two-dimensional interdigitation pattern, or *tiling,* can be indefinitely extended in both dimensions.

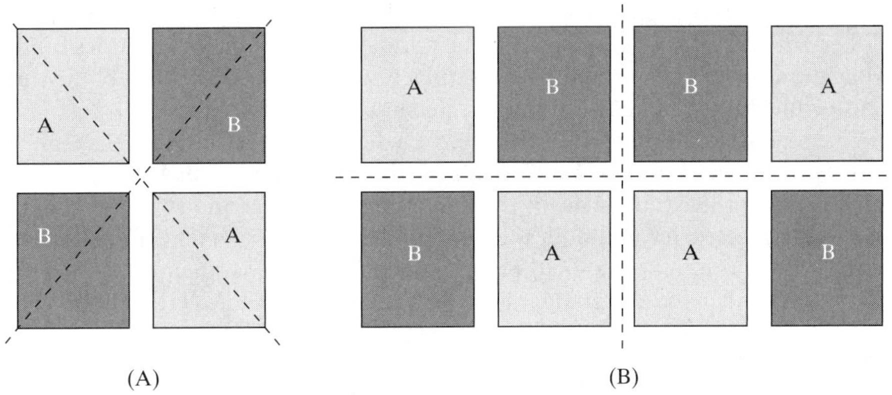

FIGURE 7.18 Examples of two-dimensional common-centroid arrays.

The rules for creating one-dimensional arrays also apply to two-dimensional arrays. The sections should be arranged so that the array has two or more axes of symmetry intersecting at the point where the centroids of the matched devices coincide. Table 7.4 lists a number of sample interdigitation patterns for two-dimensional arrays. Each row in the table consists of four examples of a given pattern. These include the simplest possible array, an extension of the array in one dimension, and extensions of the array in two dimensions. Although more complex variations are possible, most two-dimensional arrays are relatively simple because capacitors and transistors—the devices most often arrayed in two dimensions—do not lend themselves to subdivision into a large number of segments.

Location and Orientation

Any residual stress sensitivities not canceled by common-centroid layout remain proportional to the magnitude of the stress gradient. Matched devices should therefore occupy areas of the die where the stress gradients are smallest. As Figure 7.15 suggests, the stress gradient typically falls to a minimum in a broad region near the

TABLE 7.4 Sample interdigitation patterns for two-dimensional common-centroid arrays.

ABBA BAAB	ABBAABBA BAABBAAB	ABBAABBA BAABBAAB ABBAABBA	ABBAABBA BAABBAAB BAABBAAB ABBAABBA
ABA BAB	ABAABA BABBAB	ABAABA BABBAB ABAABA	ABAABAABA BABBABBAB BABBABBAB ABAABAABA
ABCCBA CBAABC	ABCCBAABC CBAABCCBA	ABCCBAABC CBAABCCBA ABCCBAABC	ABCCBAABC CBAABCCBA CBAABCCBA ABCCBAABC
AAB BAA	AABBAA BAAAAB	AABBAA BAAAAB AABBAA	AABBAA BAAAAB BAAAAB AABBAA

center of the die. The best locations for matched devices are therefore near the middle of the die. The stress gradient along the periphery of the die reaches a similar broad minimum in the middle of each side, with the lowest values usually appearing in the middle of the longer sides. If matched devices must reside along the die periphery, then they are best located in the middle of one of the longer sides of the die. Matched components should never be placed near corners because both stress intensity and stress gradients reach their maximum values here.

The stress distribution on the surface of a die exhibits symmetries that can be exploited to further improve matching. Most dice exhibit a symmetric stress distribution around at least one axis. In the case of (100) silicon, the stress distribution is usually symmetric around both the horizontal and vertical axes. Critically matched common-centroid arrays should be oriented so that one of their axes of symmetry aligns with either the horizontal or the vertical axis of the die (Figure 7.19A).

The picture is less clear in the case of (111) silicon because one axis of symmetry of the die lies along a <110> axis while the other lies along a <211> axis (Figure 7.19B). Some authors have suggested that the stress distribution is more symmetric around the <211> axis than around the <110> axis,[25] so critically

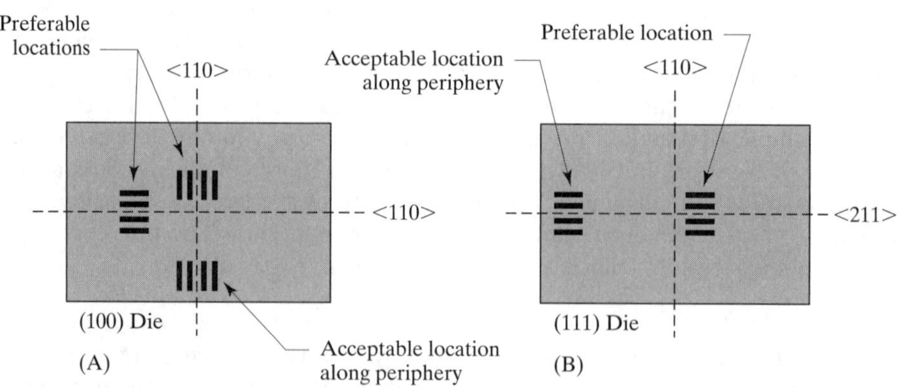

FIGURE 7.19 Locations for placing common centroid arrays on (100) and (111) dice, in the latter case assuming an axis of symmetry in the stress distribution around the <211> axis.

[25] W. F. Davis, *Layout Considerations*, unpublished manuscript, 1981, pp. 66–67.

matched common-centroid arrays on (111) silicon should be oriented so that one of the axes of symmetry of the array aligns with the <211> axis of the die (Figure 7.19B).

The placement of the common-centroid array on an axis of symmetry of the stress distribution helps reduce residual mismatches by minimizing the stress gradient. If the stress distribution is symmetric around the chosen axis of symmetry, whatever stress-induced effects do occur will have opposite polarities on either side of this axis. Providing that the matched devices are also placed symmetrically around this same axis, the effects of stress on one half of the device will cancel the effects of stress on the other half. Whenever possible, critically matched devices should be placed to take advantage of this phenomenon.

The stress distribution on a die also depends on its size and shape. Larger dice generally exhibit higher levels of stress than small ones. Stress also tends to increase with aspect ratio, so elongated dice exhibit higher stress levels than square dice having similar areas. As previously mentioned, packaging also plays a major role in determining stress levels. Epoxy mounting provides mechanical compliance, which allows stresses to dissipate. Epoxy-mounted dice in metal cans or ceramic packages exhibit relatively little stress, regardless of die size or shape. The die area and aspect ratio become more important for parts encapsulated in plastic, or mounted with solder or gold eutectic. Table 7.5 offers some general guidelines for die aspect ratios in various types of packaging.

TABLE 7.5 Suggested die aspect ratios for analog layouts.

Package Type	Die Size	Suggested Aspect Ratio	Maximum Aspect Ratio
Metal can/epoxy mount	Any	2:1 or less	Any
Plastic/epoxy mount	$<10\ mm^2$	1.5:1 or less	3:1 or less
	$>10\ mm^2$	1.5:1 or less	2:1 or less
Plastic/solder mount	$<10\ mm^2$	1.5:1 or less	2:1 or less
	$>10\ mm^2$	1.3:1 or less	1.5:1 or less

7.2.11. Temperature Gradients and Thermoelectrics

The electrical properties of many integrated components depend strongly on temperature. Most integrated resistors have temperature coefficients of 1000 ppm/°C or more (Table 5.4). Assuming a temperature coefficient of 2500 ppm/°C, a 1° temperature difference between two matched resistors produces a 0.25% mismatch. Thermal gradients of 0.1°C/µm can exist near a large power device.[26] To better understand how these thermal variations arise, we will briefly examine the concept of *thermal impedance*.

All electrical circuits dissipate some amount of power in the form of heat. This heat flows through the encapsulation out into the ambient environment. The average *junction temperature* of the die T_j equals

$$T_j = T_a + P_d\theta_{ja} \quad [7.17]$$

where T_a is the ambient temperature of the environment, P_d is the power dissipated in the package, and θ_{ja} is a constant called the *junction-to-ambient thermal impedance.* The θ_{ja} for most plastic packages exceeds 100°C/W, limiting their power

[26] See Figure 1 in R. J. Widlar and M. Yamatake, "Dynamic Safe-Area Protection for Power Transistors Employs Peak-Temperature Limiting," *IEEE J. Solid-State Circuits,* Vol. SC-22, #1, 1987, pp. 77–84.

TABLE 7.6 Typical thermal impedances for several common types of packages.[27]

Type of Package	θ_{ja} (°C/W)	θ_{jc} (°C/W)
16-pin plastic dual in-line package (DIP)	110	
16-pin plastic surface-mount package (SOIC)	131	
3-lead plastic TO-220 power package		4.2
3-lead metal TO-3 can power package		2.7

dissipation to about a watt. Specially constructed power packages exist that offer much lower thermal impedances (Table 7.6). These packages usually incorporate a metal tab or plate intended for mounting onto an external metal surface called a *heat sink.* Power packages are usually specified in terms of a *junction-to-case thermal impedance,* θ_{jc}. In this case, the average junction temperature, T_j, equals

$$T_j = T_c + P_d\theta_{jc} \qquad [7.18]$$

where T_c represents the case temperature of the package measured at a specified point on the heat tab or plate. Power packages often have remarkably low thermal impedances because of their special construction. Various manufacturers cite slightly different values due to variations in materials and manufacturing processes, but the values listed in Table 7.6 are representative of the industry.

One might expect the packages with the lowest thermal impedances to have the smallest thermal gradients, but in practice the exact opposite occurs. Power packages achieve their low thermal impedances by mounting the die on a heat sink. Heat flows vertically down to the heat sink and out of the package, rather than laterally across the die. Temperatures rise only where power is dissipated; other portions of the die remain at approximately the same temperature as the heat sink. Temperature differentials of up to 50°C can appear across the surface of a die mounted in a power package; the thermal gradients are correspondingly large.

A package lacking a heat sink presents a very different picture. Silicon is a far better conductor of heat than epoxy, so heat flows laterally across the die until the entire die rises to a high temperature. Heat then percolates out from the die through the plastic to the outside environment. The plastic package acts as a thermal insulation blanket that minimizes the magnitude of thermal gradients. Ordinary plastic packages rarely experience significant thermal differentials, except in the immediate vicinity of large heat sources.

Figure 7.20 shows an *isothermal contour plot* for a die containing one large heat source mounted in a power package. The curved lines on the surface of the die, called

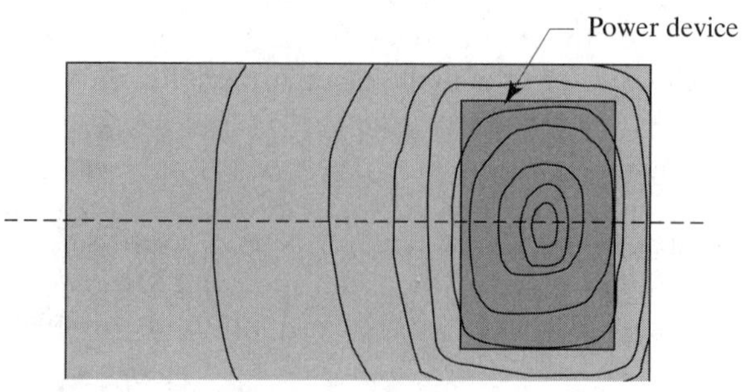

FIGURE 7.20 Isothermal contour plot of a die having only one major heat source. The axis of symmetry of the thermal distribution is marked by a dotted line.

[27] Values for 16-pin dip, 16-pin SOIC: *Power Supply Circuits Databook;* Texas Instruments #SLVD002, 1996, pp. 4–22. Values for TO-3, TO-220: *Power Products Data Book,* Texas Instruments #DB-029, 1990, pp. 4–72, 5–8.

isotherms, represent adjacent points of equal temperature. Each isotherm represents a relatively large change in temperature, perhaps 10 degrees. The same general distribution of isotherms would also occur on a die in an ordinary plastic package, but the average die temperature would be much higher and the isotherms would represent smaller changes in temperature—perhaps five degrees per isotherm.

The thermal gradients are largest within the power device and gradually decrease in magnitude as one moves away from it. Because the heat source has been placed symmetrically around the horizontal axis of the die, the heat distribution is also symmetric about this axis. The presence of this axis of symmetry can be used to improve the thermal matching of other components on the die.

Thermal Gradients

The relative spacing of the isotherms reflects the thermal gradient at each point on the die. The thermal gradient is large where the isotherms are spaced closely together, and it is small where they are spaced far apart. Thermal gradients are exactly analogous to the stress gradients discussed in the previous section. Assuming that the thermal gradient remains approximately constant in the vicinity of a pair of matched devices, then the thermally induced mismatch δ_T between the two devices equals

$$\delta_T = TC_1 d_{cc} \nabla T_{cc} \qquad [7.19]$$

where TC_1 is the linear temperature coefficient of the resistance material, d_{cc} is the distance between the centroids of the resistors, and ∇T_{cc} is the thermal gradient along a line connecting the centroids of the resistors.

Although common-centroid layouts are used to combat both stress and thermal gradients, their position and orientation differ depending on the application. The axes of symmetry of stress distributions are determined entirely by the packaging and therefore present rigid constraints on the layout. The axes of symmetry of thermal distributions are determined by the position and orientation of the power devices. The magnitude of thermally induced variations can be minimized by proper placement of the matched devices relative to the power devices.

Most dice contain only a few major heat sources, which are usually large bipolar or MOS power transistors. Whenever possible, these devices should lie upon an axis of the die in order to produce a symmetric thermal distribution. They should also lie as far as possible from critical matched devices. Consider the case of a die containing one power device (Figure 7.21A). This power device would ideally occupy one end of the die and align around an axis of symmetry of the die. This arrangement is preferable to placing the power device centrally, because it allows more separation between the power device and critical matched devices. The placement of the matched devices is a compromise between stress effects—which favor a central location—and thermal effects—which favor placement as far away from the power device as possible. The optimal arrangement places the matched devices about halfway across the portion of the die not occupied by the power transistor (Figure 7.21A). A larger separation between the power device and the matched pair can sometimes be achieved by elongating the die to an aspect ratio of 1.3 or even 1.5. Although elongation increases stress levels, the improvement in matching due to increased separation may actually outweigh the effect of increased stresses. If matched devices must reside along the periphery of the die, then they should occupy the middle of the side opposite the power device. In this case, the aspect ratio of the die should be moderated to limit the effects of stress on the less-optimal location of the matched devices. Although the effects previously discussed are difficult to quantify, the arrangements advocated here have been successfully used on many designs.

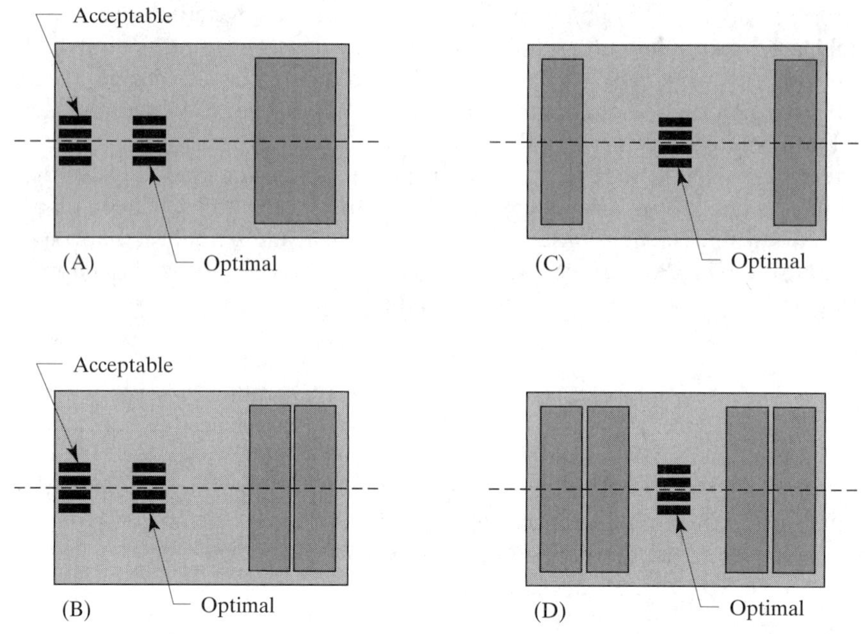

FIGURE 7.21 Various arrangements of one, two, and four power devices for optimal thermal matching. The power devices are shown as dark gray rectangles, and the axes of symmetry created by their placement are shown as dotted lines.

The layout of Figure 7.21A can easily be extended to incorporate two power transistors instead of one. Ideally, the transistors would be placed one behind another, both at one end of the die and both on the same axis of symmetry (Figure 7.21B). This yields a heat distribution resembling that of Figure 7.21A. Unfortunately, this arrangement often produces difficulties in routing the leads from the power devices to their respective pins. Many designs place the two power devices in adjacent corners of the die located on the opposite end from the matched devices. This arrangement has the advantage of separating the heat sources from the matched devices by the largest possible distance. However, the disadvantage is that the heat distribution is now asymmetric, except in the unlikely event that the devices operate at identical power levels. Another possible layout for two heat sources is shown in Figure 7.21C. This arrangement places one power device on each end of the die and the matched devices in the middle. This arrangement will probably prove satisfactory as long as a separation of at least 0.5 mm can be achieved. It has the advantage of preserving a thermal axis of symmetry regardless of the dissipation levels in the individual power devices, and it locates the matched devices in the center of the die where the stress gradients are lowest.

The layout of Figure 7.21C can be extended to include additional devices. Figure 7.21D shows an example containing four power devices. This arrangement usually suffers from a lack of separation between the power devices and the matched devices. This problem can be partially remedied by increasing the aspect ratio of the die to 1.5:1 or even 2:1. Even a large aspect ratio will not necessarily affect matching because the power devices occupy the ends of the die where the stresses are greatest. Aspect ratios larger than 1.5:1 are somewhat risky for solder or gold eutectic mounting if the die's longer dimension exceeds 3-4 mm. The stresses accumulating in the corners of such a long die may actually cause mechanical damage to the metal system or the bond wires. Epoxy die mounting provides additional mechanical compliance and therefore allows somewhat larger aspect ratios.

A variety of considerations often constrain the placement of power devices. These include the location of bondpads, the routing of power buses, and the placement of control circuitry. The compromises required to satisfy all of these constraints usually lead to a less-than-optimal layout. This does not necessarily constitute an insurmountable problem because common-centroid layout techniques can greatly reduce the impact of the remaining thermal mismatches.

Thermoelectric Effects

Resistors display two distinct types of thermal variation. One is caused by the temperature coefficient of the resistance material. Common-centroid layout techniques can ensure that the average temperatures of two resistors track one another, so even materials with large temperature coefficients will match quite precisely. The other source of thermal variation is the *Seebeck effect*, also called the *thermoelectric effect*. As discussed in Section 1.2.5, a voltage differential called the *contact potential* arises whenever two dissimilar materials come in contact with one another. The contact potentials of metal/semiconductor junctions are strong functions of temperature, so if the contacts are held at different temperatures, a net voltage difference will appear across the resistor. This *thermoelectric potential* E_T equals

$$E_T = S\Delta T_c \qquad [7.20]$$

where S is the *Seebeck coefficient* (typically about 0.4 mV/°C), and ΔT_c is the temperature difference between the two contacts of the resistor. A temperature difference of 1°C across a resistor will thus generate a voltage differential of about 0.4 mV between its contacts. This may appear inconsequentially small, but certain types of circuitry are extremely vulnerable to small voltage offsets. For example, a 0.4 mV offset in a bipolar current mirror produces a 1.5% mismatch in the currents.

Common-centroid layout cannot eliminate thermoelectrics because they arise from differences in temperature between the ends of each resistor segment. Improperly arraying the device actually compounds the problem. The individual thermoelectric potentials generated by each segment of the resistor array of Figure 7.22A add to produce an overall thermoelectric potential far larger than that of any one segment. The thermoelectric potentials of the individual segments can be canceled by reconnecting them as shown in Figure 7.22B.

In order to obtain complete cancellation, the resistor should consist of an even number of segments, half connected in one direction and half connected in the other (Figure 7.22B). If the resistor has an odd number of segments, then one segment cannot be paired. Critically matched resistors should, if possible, consist of an even number of segments, but less-sensitive resistors can tolerate the presence of an unpaired segment.

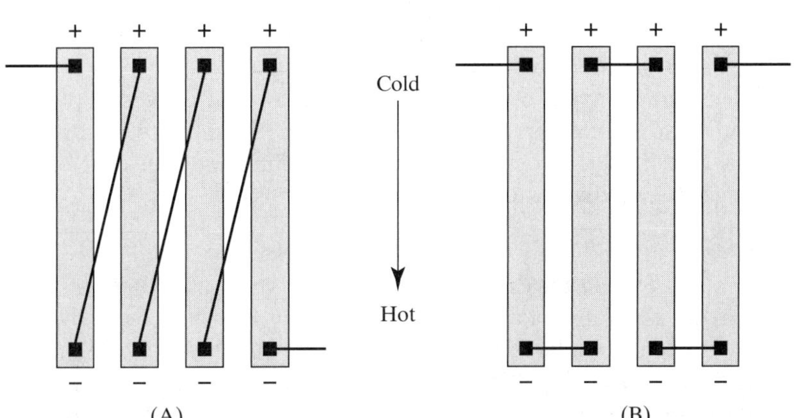

FIGURE 7.22 (A) Improper connection of resistor segments causes their thermoelectric potentials to add, while (B) proper connection of the segments cancels the thermoelectrics.

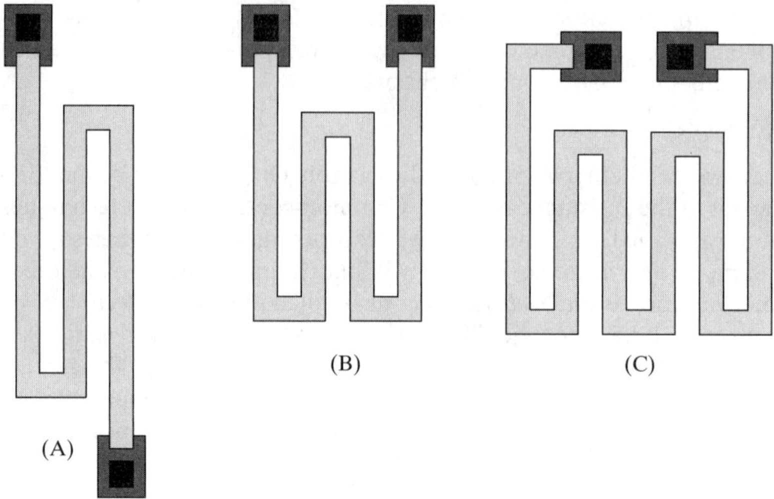

FIGURE 7.23 An HSR resistor with widely separated contacts (A) is prone to thermoelectric-induced offsets. Placing the contacts close together (B) minimizes thermoelectrics, but may increase variation due to misalignment. Placing the contacts close together and oriented in opposite directions (C) fixes both problems.

The two contacts of a serpentine resistor should reside as close to one another as possible to minimize the impact of thermoelectrics. The serpentine resistor of Figure 7.23A will have unnecessarily large thermal variations due to an excessive separation between its contacts. The layout of Figure 7.23B reduces thermal variability and improves matching by bringing the resistor heads into closer proximity. However, this layout is vulnerable to misalignment errors. If the resistor body shifts downward relative to the resistor heads, then the length of the resistor increases by twice the misalignment. This vulnerability can be eliminated by orienting the resistor heads in opposite directions, so that any shift that increases the length of the resistor protruding from one head reduces the length protruding from the other. The layout of Figure 7.23C eliminates the misalignment vulnerability, but it places the base heads adjacent to stretches of the resistor body, which can lead to diffusion interactions. It is difficult to eliminate this minor defect without introducing more serious problems in the process.

7.2.12. Electrostatic Interactions

Electrostatic fields can influence the value of both resistors and capacitors. An electrostatic field can cause depletion or accumulation of carriers within a resistive material. Most integrated resistors consist of lightly doped silicon and are therefore quite susceptible to voltage modulation. The electrostatic coupling of a capacitor to surrounding circuitry can cause unexpected variations in capacitance. Electrostatic fields can also couple noise into sensitive high-impedance nodes that often exist within arrays of matched resistors and capacitors.

The principal types of electrostatic interactions observed in resistors are voltage modulation, charge spreading, and dielectric polarization. The major electrostatic interactions in capacitors are capacitive coupling and dielectric relaxation. The following sections cover each of these mechanisms.

Voltage Modulation

The value of a resistor can be affected by the voltages present on adjacent nodes of the circuit. The value of a diffused resistor may vary with the voltage differential between the tank and the resistor body, an effect called *tank modulation* (Section 5.3.3). The mismatches caused by tank modulation in two or more diffused resistors of equal value cancel as long as the tank-to-body voltages of the resistors equal one another. Matched resistors can safely reside in a common tank as long as they are of equal values and they experience the same bias. Otherwise, each resistor should occupy its own tank. The individual tanks must connect so that each resistor sees the same tank-to-body differential voltage. This is most easily accomplished by tying the positive end of each resistor to its respective tank. If the resistors are of different values, they all must be divided into segments of equal (or approximately equal) value, and each segment must reside in its own independently biased tank. Similarly, each section of an interdigitated resistor requires its own tank (Figure 7.24).

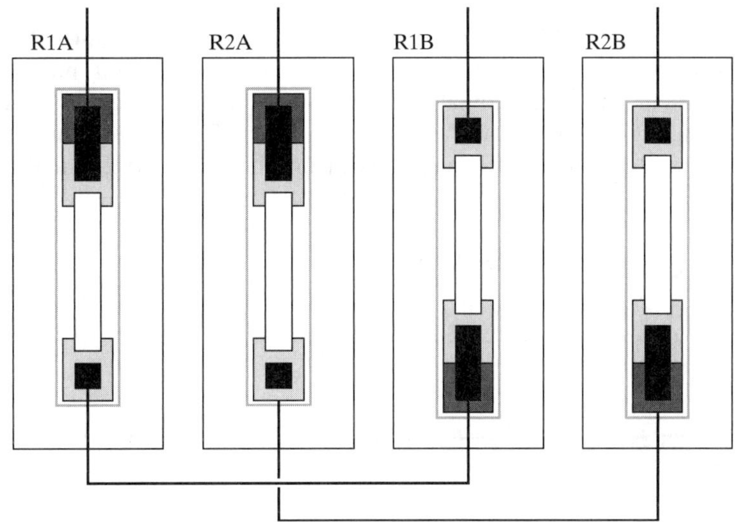

FIGURE 7.24 Two HSR resistors connected in order to cancel both thermoelectrics and tank modulation effects.

Separate tanks require enormous amounts of die area. Poly or thin-film resistors will surely provide a more compact solution. If deposited resistors are not available, consider whether the circuit can tolerate a small amount of voltage modulation. If the voltage across the matched resistors is directly (or even indirectly) trimmed, this may also at least partially compensate for voltage modulation. This is often the case for voltage references and regulators. The matched resistors should still be interdigitated to prevent thermal and stress gradients from causing package shifts and thermal drifts.

Many applications can tolerate a small amount of voltage modulation. The magnitude of the resulting systematic mismatches can be minimized by using relatively low-sheet diffusions, such as base, rather than high-sheet diffusions, such as HSR. For example, the voltage modulation of 160 Ω/□ base equals about 0.1%/V, while that of 2 kΩ/□ HSR can approach 1%/V. Matched base resistors are usually merged into a common tank to save space, while matched HSR resistors frequently require separate tanks.

The proper determination of tank biasing requires a complete understanding of circuit specifications and design, making it a job for the circuit designer rather than the layout designer. The tank connections should appear on the schematic along with the type of resistance material required, the width of the resistor, and any special matching requirements. In cases where the connections of the resistors are not obvious, or appear to be in error, the layout designer should verify them with the circuit designer.

Leads routed over resistors can also affect their operation. As a rule, leads that do not connect to matched resistors should not cross them. Not only may these leads capacitively couple noise into the resistor, but the electric field between the lead and the resistor can actually modulate the conductivity of the resistance material (a phenomenon called *conductivity modulation*). Emitter, base, and low-sheet poly (R_s < 200 Ω/□) resistors rarely experience significant conductivity modulation. High-sheet resistors are more problematic; for example, metal-1 over 2 kΩ/□ HSR can produce 0.1%/V of conductivity modulation. The impact of conductivity modulation depends on three factors: (1) the voltage difference between the lead and the underlying resistor, (2) the thickness of the intervening oxide, and (3) the area of intersection. A lead connected to an HSR resistor can safely route across the end of the resistor next to the head, while one routed entirely across the resistor array may cause problems because it has a larger area of intersection. Wires rarely need to route across resistors in a double-level metal process, but jumpers through resistor arrays are often unavoidable in single-level metal processes. For example, the interdigitated HSR resistor array shown in Figure 7.25 uses a jumper between segments R_{1A} and R_{1B} to allow a lead to exit from the left-hand terminal of R_2.

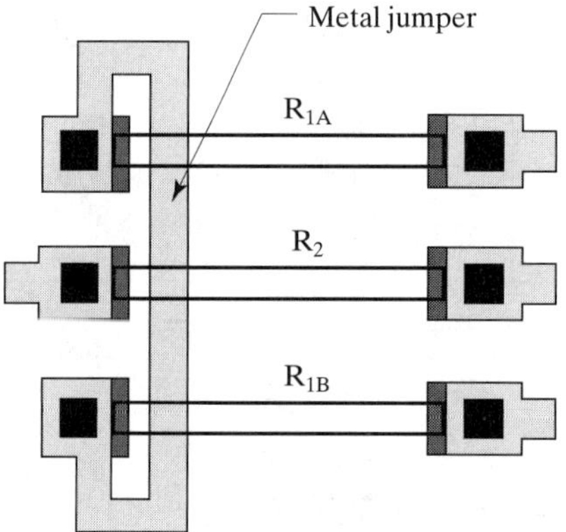

FIGURE 7.25 Portion of an interdigitated HSR array implemented in a single-level metal process showing the placement of a jumper between segments.

The jumper of Figure 7.25 is constructed so that it intersects each resistor segment in exactly the same manner. This precaution helps to minimize the mismatches produced by stress and hydrogen incorporation. Still, the mere presence of a lead running over the resistor segments will introduce some mismatch. Whenever possible, one should not route any leads across critically matched resistors.

A technique called *electrostatic shielding* (or *Faraday shielding*) can isolate a resistor from the influence of overlying leads. Electrostatic shielding not only prevents conductivity modulation but also provides considerable shielding against capacitive coupling.

Figure 7.26A illustrates the basic concept of an electrostatic shield. The shield is interposed between the two conductors—in this case the resistor and an overlying lead. The electrostatic interactions between the shield and the conductors on either side can be modeled as a pair of series-connected capacitors (Figure 7.26B). The AC voltage source, V_N, represents the time-varying voltage present on the overlying lead. Noise injected by V_N is attenuated by the RC filter formed by C_{P1} and R_1, where C_{P1} represents the capacitance between the overlying lead and the shield, and R_1 represents the resistance of the connection between the shield and AC ground. Whatever noise passes through C_{P1}–R_1 will be injected through C_{P2} onto R_2, where C_{P2} represents the capacitance between the shield and the sensitive node, and R_2 represents the resistance between the sensitive node and AC ground. The attenuation provided by the shield falls away with increasing frequency. Providing

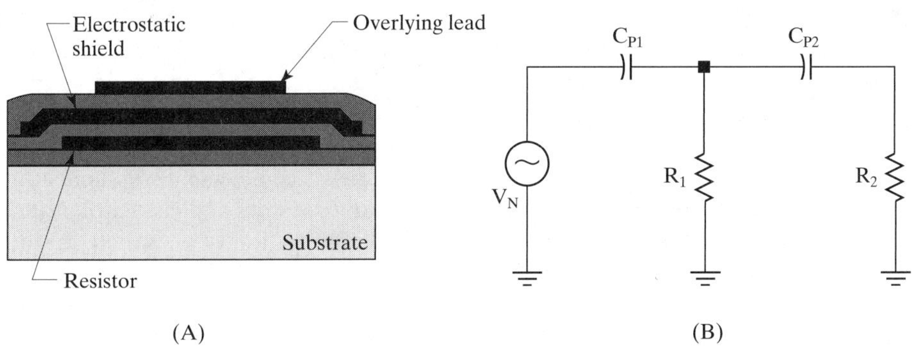

FIGURE 7.26 The concept of an electrostatic shield: (A) cross section and (B) equivalent circuit.

that the shield connects to a clean low-impedance node (such as signal ground), substantial attenuation may remain into the low RF region (1 to 10 MHz). At higher frequencies, it becomes increasingly difficult to guarantee a sufficiently low-impedance shield connection, so high-speed digital signals should not route across sensitive circuitry even if an electrostatic shield is present.

Figure 7.27 shows a practical example of an electrostatic shield. The poly resistor array includes dummies at either end, and both the shield and the dummies are connected to ground. Leads can cross the shielded resistors without modulating the conductivity of the poly resistors or injecting noise into them. Note that a fringing field exists around any conductor. In order to prevent this fringing field from coupling around the shielding, all leads should reside well away from the edges of the shield. An overhang of 3–5 μm (shield-over-conductor) will intercept the majority of the fringing fields. Since metal can elastically deform, the electrostatic shield also helps minimize stresses generated in the poly resistors by the presence of the second-metal leads running across them.

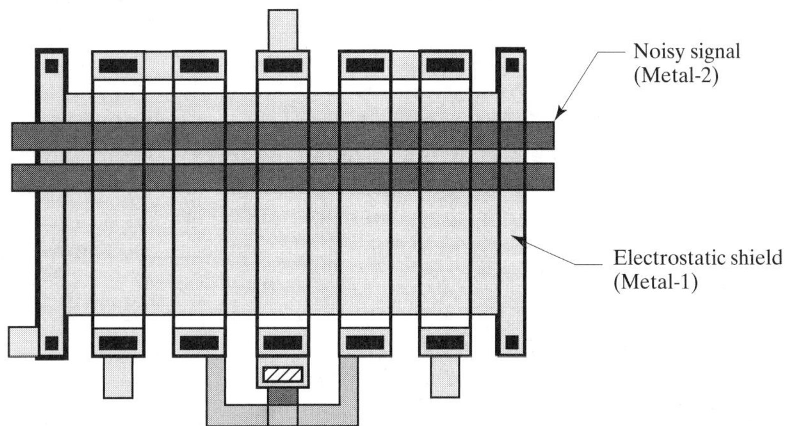

FIGURE 7.27 Example of electrostatic shielding applied to a matched poly resistor array.

Note that the electrostatic shield of Figure 7.27 covers the entire resistor array. Each resistor segment in the array sees a slightly different voltage. As long as the voltage difference across the array remains small and the sheet resistance of the resistors is less than about 500 Ω/□, a common shield will suffice. If the resistors consist of high-sheet material, or if the voltage differential across the array exceeds a few volts, then the common electrostatic shield itself will cause objectionable conductivity modulation! In such cases, the shield should be divided into individual sections placed over each resistor segment. Each shield must overlap its respective resistor by several microns (after accounting for misalignment and outdiffusion) to ensure that fringing fields do not degrade its effectiveness. Sectioned shielding requires substantially more space than common shielding.

The substrate can also inject noise into deposited resistors and capacitors. One way to minimize this source of noise coupling consists of placing a well beneath the devices and connecting this well to an AC ground. Resistors and capacitors that are especially sensitive to noise can benefit from a combination of electrostatic shields above and below them.[28] A deep-N+ sinker placed beneath deposited devices and connected to an AC ground can provide superior electrostatic shielding because of its lower series resistance. The use of deep-N+ underneath a deposited component can also minimize parasitic capacitance by thickening the field oxide through dopant-enhanced oxidation.

Charge Spreading

The mechanisms behind charge spreading were discussed at length in Section 4.3.5. Briefly, circuit operation injects electrons into the oxide overlying the die. Although most of these electrons eventually return to the silicon, a few become trapped at the interface between the interlevel oxide and the protective overcoat, or between the protective overcoat and the mold compound. These electrons constitute a mobile charge capable of varying resistor values through conductivity modulation. The electric field required to cause a fractional-percent variation of a high-sheet resistor is an order of magnitude smaller than that required to invert the surface of the silicon. Matched high-sheet resistors are therefore extremely susceptible to long-term drifts caused by charge spreading. Base resistors are much less susceptible to charge spreading due to their lower sheet resistance, and only precisely matched base resistors are significantly affected. High voltages, moisture, and mobile ion contamination amplify the effects of charge spreading. Designs that operate at voltages exceeding half of the thick-field threshold or that are fabricated on standard bipolar should be examined for potential charge spreading vulnerabilities.

Electrostatic shielding can minimize or even eliminate the effects of charge spreading on matching. The electrostatic shield also serves as a field plate, counteracting surface inversion across high-voltage tanks. The field plate must connect to a potential not greatly different from the tank bias, as described in Section 4.3.5. It is often connected to the more positive end of the resistor. Field plates may actually increase noise coupling into high-impedance resistors due to leads running above the field plate or due to fringing fields from adjacent components. Noisy signals must not be routed across field plates unless the field plates connect to low-impedance nodes that are relatively insensitive to capacitive coupling.

Figure 7.28 shows the same resistor array as Figure 7.24. Each resistor segment has been provided with an individual field plate protecting it against the effects of charge spreading. The field plates flange over the bodies of the resistors far enough to prevent channel formation. The gaps in the field plates have not been channel-stopped because the proximity of an adjacent diffusion could cause diffusion interactions. If channel stops are required, they should be carefully replicated on each section so that all of the resistors experience the same interactions. In most cases, channel stops are unnecessary as long as the field plates are properly flanged. Although diffused HSR resistors can rival the matching of deposited resistors, the area required for separate tanks and field plates makes them rather uneconomical.

---

[28] K. Yamakido, T. Suzuki, H. Shirasu, M. Tanaka, K. Yasunari, J. Sakaguchi, and S. Hagiwara, "A Single-Chip CMOS Filter/Codec," *IEEE J. Solid-State Circuits*, Vol. SC-16, #4, 1981, pp. 302–307.

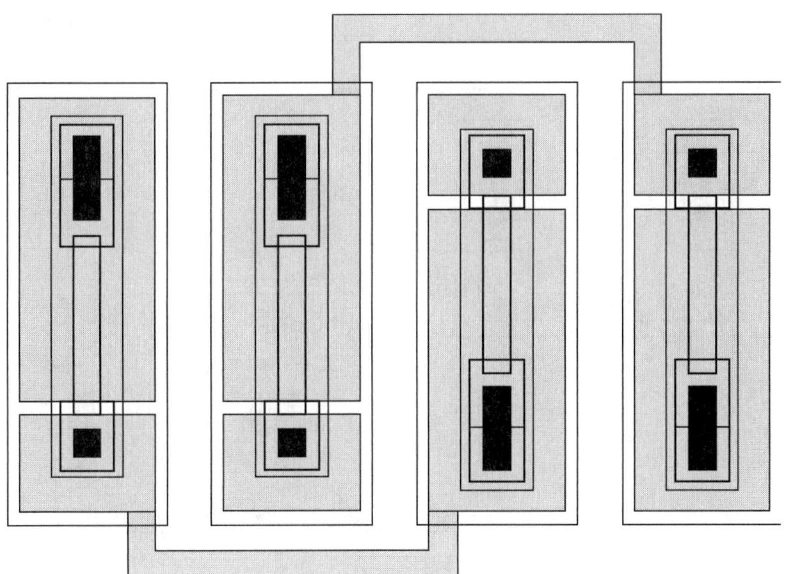

FIGURE 7.28 HSR resistor array, field-plated to minimize charge spreading. With the exception of its metallization pattern, this array matches the one in Figure 7.24.

Dielectric Polarization

Electrostatic fields can also arise due to the movement of charges within an insulator, a phenomenon called *dielectric polarization*. Oxides contaminated by alkali metal ions such as sodium or potassium exhibit a large degree of polarizability due to the mobility of these ions within the oxide. As discussed in Section 4.2.2, these *mobile ions* slowly redistribute themselves under the influence of external electrical fields. The electric field seen at the surface of the silicon shifts over time as the mobile ions gradually assume a new configuration. If the external field suddenly vanishes, the new distribution of mobile ions generates a weak residual electric field oriented in the opposite direction to the original. This residual field gradually relaxes as the mobile ions return to their original distribution. The slow variation in field intensity seen beneath the oxide can modulate the value of a high-sheet resistor. The resulting *long-term drifts* are extremely undesirable.

The addition of phosphorus to silicon dioxide effectively immobilizes alkali metal ions, presumably through a sequestration mechanism involving phosphate groups. Phosphosilicate glass (PSG) thus exhibits much less polarization when contaminated with alkali metal ions than does pure oxide. Unfortunately, the phosphate groups themselves are slightly polarizable.[29] Phosphosilicate and borophosphosilicate glasses are subject to low levels of dielectric polarization in the presence of external electrical fields, and this in turn gives rise to hysteretic voltage modulation.

Dielectric polarization normally affects only high-sheet resistors. The polarization rates are usually far too slow to affect capacitors, and the resulting fields are too weak to affect resistors with sheets of less than about 500 Ω/□. A dielectric polarization of approximately 0.1% has been observed in 2 kΩ/□ high-sheet resistors constructed on a standard bipolar process using heavily phosphorus-doped BPSG.[30]

As the foregoing example suggests, field plating is not a panacea for dielectric polarization. Field plates may actually intensify the phenomenon because they induce

---

[29] E. H. Snow and B. E. Deal, "Polarization Phenomena and Other Properties of Phosphosilicate Glass Films on Silicon," *J. Electrochem. Soc.*, Vol. 113, #3, 1966, pp. 263–269.

[30] F. W. Trafton, private communication.

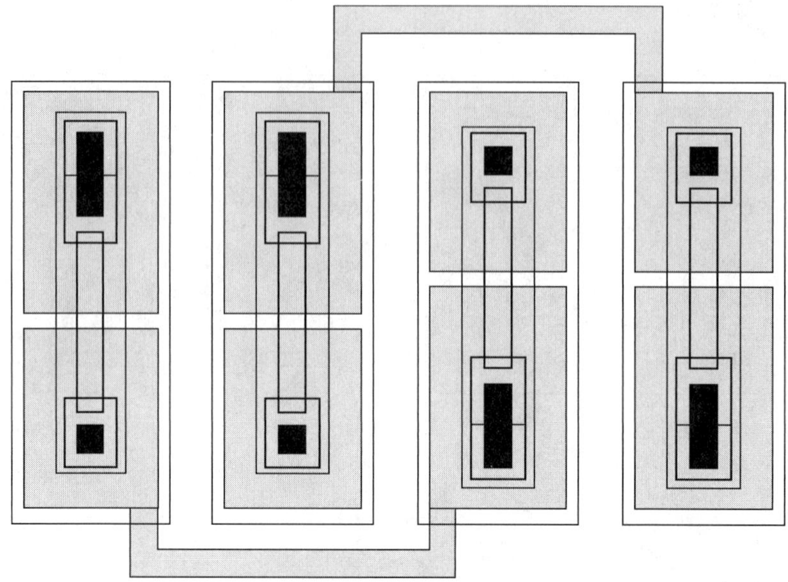

FIGURE 7.29 HSR resistor array field-plated to minimize dielectric polarization as well as to charge spreading and thermoelectrics. (Compare with Figure 7.28.)

a deliberate electric field across the interlevel oxide. On the other hand, leaving the field plates off the resistors renders them vulnerable to charge spreading. This dilemma is best solved by avoiding the use of high-sheet resistors entirely. If this is impractical, then *split field plates* should be used. Figure 7.29 shows an example of a split field plate applied to a pair of matched high-sheet resistors similar to those shown in Figure 7.28.

Split field plates differ from conventional field plates in the location of their gaps. In a conventional field plate, as much of the plate as possible connects to the positive end of the resistor to offer maximum protection against surface inversion. In a split field plate, each resistor segment requires two field plates, one connecting to either end of the resistor. The gap between the two plates falls in the exact center of the resistor. The split field plate subjects half of the resistor to an electric field equal and opposite to that seen by the other half. The dielectric polarization in each half of the resistor exactly cancels the polarization in the other. Split field plates also reduce voltage nonlinearity effects caused by the field plates, because the accumulation effects beneath one plate balance the depletion effects under the other.[31] Split field plates can be applied to resistors occupying a common tank, but each segment must have its own split field plate.

Split field plates are recommended for all applications where resistors having sheet resistances in excess of 1 kΩ/□ must match to better than ±0.5%. Split field plates are not necessary for base resistors since the lower sheet resistance of this diffusion generally renders dielectric polarization negligible. They are likewise not necessary for the majority of polysilicon resistors because these usually exhibit sheet resistances less than or equal to 1 kΩ/□.

Dielectric Relaxation

Capacitors are also susceptible to dielectric polarization in the form of a hysteretic effect called *dielectric relaxation* or *soakage*. Suppose that the capacitor is suddenly

[31] J. Victory, C. C. McAndrew, J. Hall, and M. Zunino, "A Four-Terminal Compact Model for High Voltage Diffused Resistors with Field Plates," *IEEE J. Solid-State Circuits*, Vol. 33, #9, 1998, pp. 1453–1458.

charged to produce a corresponding electric field between its electrodes. As the dielectric polarizes, the electric field intensity gradually decreases (or *relaxes*) to a lower value. The relaxation of the electric field causes a corresponding droop in capacitor voltage. After the capacitor is suddenly discharged and then disconnected, the polarization dissipates and a reverse bias gradually accumulates across the capacitor. Charge storage capacitors (such as those used in timers and sample-and-hold circuits) are especially intolerant of dielectric relaxation errors.

Charge spreading can affect capacitors in much the same manner as dielectric polarization. When the capacitor is charged, the shifting electrostatic fields cause a gradual redistribution of charges along insulating interfaces. This results in a change in the electric field intensity that is similar to what occurs in dielectric relaxation. Charge spreading can occur along the interfaces of a composite dielectric, such as the oxide-nitride-oxide sandwich used to fashion ONO capacitors. The movement of charges along the interfaces of the capacitor dielectric can occur very rapidly, so variations in capacitor value may occur at frequencies in excess of 1 MHz.

High-quality oxide dielectrics are preferred for critically matched capacitors, particularly for those that operate at high frequencies. The use of a grown oxide dielectric usually reduces dielectric relaxation to negligible levels.[32] Charge spreading and polarization phenomena arising outside the capacitor structure can be eliminated using electrostatic shielding. Figure 7.30 shows one method of shielding a poly-poly capacitor. The sensitive high-impedance node connects to the upper electrode of the capacitor. An electrostatic shield constructed from second-level metal completely covers the upper electrode. This electrostatic shield connects to the lower electrode of the capacitor to form a sandwich-capacitor structure. The sensitive node is now entirely enclosed within an electrostatic shield. Both the lower capacitor electrode and the electrostatic shield should overhang the upper capacitor electrode by 3–5 μm to suppress fringing fields. This structure is relatively insensitive to dielectric polarization and charge spreading, provided that both the capacitor oxide and the interlevel oxide above the upper electrode are not phosphorus-doped.

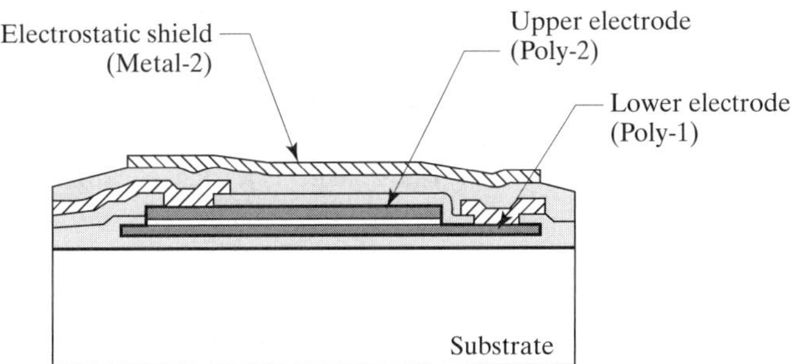

FIGURE 7.30 Cross section of a poly-poly capacitor incorporating an electrostatic shield plate. Note the overlap of the electrostatic plate over the upper electrode.

7.3 RULES FOR DEVICE MATCHING

The previous sections have discussed the various mechanisms responsible for causing mismatches. This information would ideally be used to formulate a set of quantitative rules for each process. The layout designer could then apply these rules to

[32] LPCVD oxides also exhibit little dielectric relaxation, but the same is not true of TEOS oxides: J. W. Fattaruso, M. De Wit, G. Warwar, K. S. Tan, and R. K. Hester, "The Effect of Dielectric Relaxation on Charge-Redistribution A/D Converters," *IEEE J. Solid-State Circuits*, Vol. 25, #6, 1990, pp. 1550–1561.

obtain matched devices of any desired degree of accuracy. In practice, the time and effort required to generate quantitative matching rules are usually prohibitive. This section therefore presents a set of qualitative rules applicable to many processes. In the absence of more precise matching data, these rules can be used to lay out structures with at least a modest degree of confidence.

The following rules employ the terms *minimal, moderate*, and *precise* to denote increasingly precise degrees of matching. These terms have the following meanings:

- **Minimal Matching:** Approximately ±1% mismatch, or 6 to 7 bits of resolution. Suitable for general-purpose use, such as for degenerating current mirrors in biasing circuitry.
- **Moderate Matching:** Approximately ±0.1% mismatch, or 9 to 10 bits of resolution. Suitable for ±1% bandgap references, op-amp and comparator input stages, and most other analog applications.
- **Precise Matching:** Approximately ±0.01% mismatch, or 13 to 14 bits of resolution. Suitable for precision A/D and D/A converters and for all other applications requiring extreme precision. Capacitors can obtain this level of matching more easily than resistors can.

7.3.1. Rules for Resistor Matching

Minimal matching can be obtained without much difficulty, and moderate matching can be reliably obtained using interdigitation. Precisely matched resistors are difficult to construct due to variations in contact resistance and the presence of thermal and stress gradients. The following rules summarize the most important principles of resistor design:[33]

1. Construct matched resistors from a single material.
   Resistors constructed from different materials do not even approximately match one another. Process variations will cause unpredictable shifts in the value of one resistor relative to the other, and the differing temperature coefficients of the two materials prevent them from tracking over temperature.
   Do not construct matched resistors from different materials!
2. Make matched resistors the same width.
   Uncorrected process biases will cause systematic mismatches in resistors of different widths. If for some reason one resistor must be made wider than another, consider constructing the wider resistor from a number of sections connected in parallel. Width effects are not entirely independent of temperature and stress, so even if moderately or precisely matched resistors are trimmed, they should still consist of sections of a uniform width.
3. Make the resistors sufficiently large.
   Random mismatches scale inversely with the square root of the area of the resistors. If two matched resistors are of unequal values, then the smaller resistor will contribute the majority of the mismatch. If two matched resistors are of very unequal values, consider constructing the smaller resistor from multiple segments placed in parallel.
4. Make matched resistors sufficiently wide.
   In the absence of experimental data, assume that minimal matching of resistors containing 30 or more squares requires 150% of the minimum allowed width of

---

[33] Tsividis provides a list of 10 general matching rules, entries in which correspond to numbers 1–3, 5, 6, and 13 in the above list: Tsividis, pp. 234–236.

the deposition or diffusion, moderate matching requires 200%, and precise matching requires 400%. For example, if the minimum drawn width of a poly line equals 2 μm, then minimal matching requires 3 μm, moderate matching 4 μm, and precise matching 6 μm. If the smaller of the matched resistors contains less than 20 to 30 squares, consider increasing the width of the resistors. If the smallest resistor contains more than 100 squares, consider reducing the resistor width. In no case, however, should one reduce the width of matched deposited resistors below about 1 μm because of the potential for excessive variability caused by granularity effects.

5. Where practical, use identical geometries for resistors.
The existence of corner and end effects precludes precise matching of resistors having different geometries. Resistors having the same width, but different lengths or shapes, can easily experience mismatches of ±1% or more. Matched resistors should be divided into sections as discussed in Section 7.2.10. Sectioning is not necessary for minimal (and even moderate) matching if the parameters controlled by the resistors are trimmed at wafer probe. Wafer-level trimming compensates for most (but not all) variations in resistors caused by geometric factors. Allow an extra 1 to 2% of trim range for unexpected mismatches between resistors of different geometries.

6. Orient matched resistors in the same direction.
Resistors oriented in different directions may vary by several percent. Diffused resistors show the largest orientation mismatches, but even polysilicon resistors are affected to some degree. Most resistors should be oriented vertically or horizontally. P-type diffused resistors on (100)-oriented silicon may experience less stress-induced variation if they are oriented at 45° to the X- and Y-axes.

7. Place matched resistors in close proximity.
Mismatches increase with separation. Minimally matched resistors can be spaced 50–100 μm apart, moderately matched resistors should be placed adjacent to each other, and precisely matched resistors must be interdigitated. Dice larger than about 15 mm^2, dice that dissipate more than 250 mW, dice in heat-sunk packages, designs that experience junction-to-ambient temperature rises in excess of 20°C, and dice mounted using solder or gold eutectics will experience larger mismatches. If any of these conditions apply, then minimally matched resistors should be placed adjacent to each other, and moderately or precisely matched resistors should be arrayed and interdigitated. All resistors in high-stress locations (corners or edges) or within 250 μm of a power device dissipating more than 100mW should be interdigitated.

8. Interdigitate arrayed resistors.
Arrayed resistors should be interdigitated to produce a common centroid layout. The resulting array should have an aspect ratio no greater than about 3:1, and each resistor segment should be at least five times (and preferably ten times) longer than it is wide. The interdigitation pattern should obey the rules of common-centroid layout (Table 7.3). Arrangements having an even number of segments per resistor are preferable to those having an odd number because even numbers of segments offer better rejection of thermoelectrics. Consider connecting some segments in parallel if this will reduce the total area of the array. If the array requires a large number of segments, consider arranging these into multiple banks connected to form a two-dimensional array.

9. Place dummies on either end of a resistor array.
Arrayed resistors should include dummies. Poly dummies need not have the same width as the other segments. Diffused dummies should always have the

same geometry as adjacent segments. Remember to keep the spacing between adjacent segments constant, and equal to the spacing between the dummies and their neighboring resistor segments. Whenever possible, connect diffused dummy resistors to a quiet low-impedance node.

10. Avoid short resistor segments.
    Very short resistor segments may introduce considerable variation due to contact resistance. Moderately matched resistor segments should contain no fewer than five squares, and precisely matched resistors should contain no fewer than ten. Precisely matched poly resistors should have a total length of no less than 50 μm to minimize nonlinearities caused by granularity.

11. Connect matched resistors in order to cancel thermoelectrics.
    Arrayed resistors should always be connected so that equal numbers of segments are oriented in either direction. If the array contains an odd number of segments, then one must remain unpaired. A single unpaired segment does not produce much mismatch, but arrays should preferably contain no unpaired segments. Serpentine resistors should be constructed so that their heads lie near one another, to improve thermoelectric cancellation. The heads should be oriented in opposite directions to minimize systematic errors generated by misalignment errors (Figure 7.23).

12. If possible, place matched resistors in low-stress areas.
    The stress distribution reaches a broad minimum in the middle of the die. Any location ranging from the center halfway out to the edges will lie within this broad minimum. If precisely matched resistors must reside close to an edge, then they should be placed near the middle of one side of the die, preferably a longer side. The stress distribution reaches a maximum in the die corners, so avoid placing matched devices anywhere nearby.

13. Place matched resistors well away from power devices.
    For purposes of discussion, any device dissipating more than 50 mW is a power device, and any device dissipating more than 250 mW is a major power device. Precisely matched resistors should reside on an axis of symmetry of the major power devices using one of the optimal symmetry arrangements of Section 7.2.10. Such resistors should also reside no less than 200 to 300 μm away from the closest power device. P-type diffused resistors may exhibit less package shift if placed diagonally on (100)-oriented silicon, but this arrangement precludes placement on an axis of symmetry of the power device. If large thermal gradients are expected, then these resistors should be arrayed vertically or horizontally to allow symmetrical placement because the thermal gradients are likely to produce more mismatch than stress gradients. For moderate matching, the matched devices need not lie on an axis of symmetry of the major power devices. They should, however, reside no less than 200 to 300 μm away from major power devices and at least 100 μm away from smaller power devices. Minimal matching can be achieved anywhere on the die, but if the devices must reside next to a major power device, then they must be interdigitated.

14. Place precisely matched resistors on axes of symmetry of the die.
    On (100) silicon, precisely matched resistors should be placed so that the axis of symmetry of the resistor array aligns with one of the two axes of symmetry of the die. P-type diffused resistors may benefit from a diagonal orientation that minimizes their stress sensitivity. On (111) silicon, the axis of symmetry of the array should align to the <211> axis of symmetry of the die. If large numbers of matched devices exist, reserve the optimal locations for the most critical devices.

15. Consider tank modulation effects.
    Tank modulation becomes significant for precisely matched resistors having sheet resistances of 100 Ω/□ or more, moderately matched resistors having sheet resistances of 500 Ω/□ or more, and minimally matched resistors having sheet resistances of 1 kΩ/□ or more. Substitute poly resistors for diffused resistors where possible. If diffused resistors must be employed, then consider whether use of a lower sheet material will allow merging the matched resistors into a common tank. For example, moderately matched resistors can be constructed from 160 Ω/□ base diffusions placed in the same tank. Trimmed resistors subject to known and controlled voltage biases can usually occupy a common tank regardless of their sheet resistance because trimming largely compensates for the effects of tank modulation.

16. Sectioned resistors are superior to serpentines.
    Serpentine resistors are suitable for constructing large, minimally matched resistors or for constructing minimally matched resistors that will be trimmed. All other resistors should use arrays of sections.

17. Use poly resistors in preference to diffused ones.
    Polysilicon resistors can be made much narrower than most types of diffused resistors, and, providing that they are sufficiently long, their small widths will not cause any significant increase in mismatch.

18. Place deposited resistors over field oxide.
    Deposited materials, including polysilicon, experience increased variation when crossing oxide steps. Even minimally matched deposited resistors should not cross oxide steps or other surface discontinuities.

19. Do not allow the NBL shadow to intersect matched diffused resistors.
    The NBL shadow should not intersect any precisely matched diffused resistor, or any moderately matched shallow diffused resistor (such as HSR). If the direction of NBL shift is unknown, allow adequate overlap of NBL over the resistor on all applicable sides. If the magnitude of the NBL shift is unknown, then overlap NBL over the resistor by at least 150% of the maximum epi thickness.

20. Consider field plating and electrostatic shielding.
    Field plate any matched resistor operating above 50% of the thick-field threshold. Field plate all moderately matched diffused resistors having sheet resistances of 500 Ω/□ or more. Consider split field plates for precisely matched diffused resistors having sheet resistances of 500 Ω/□ or more. Precisely matched polysilicon resistors with sheet resistances of 500 Ω/□ or more should have electrostatic shields placed over them where possible.

21. Avoid routing unconnected leads over matched resistors.
    Whenever possible, leads that do not connect to resistors should not run over them to avoid introducing stress- and hydrogenation-induced mismatches. Unconnected leads can run across minimally matched resistors having sheet resistances of less than 500 Ω/□ or moderately matched resistors having sheet resistances of less than 100 Ω/□, but the designer should carefully scrutinize all such layouts for potential noise coupling. Metal leads should not run over precisely matched resistors. Leads can run over electrostatic shields and field plates, providing that the latter connect to low-impedance nodes that can absorb the anticipated levels of noise injection. Beware of running high-speed digital signals across matched resistors regardless of whether field plates or electrostatic shielding exists.

22. Avoid excessive power dissipation in matched resistors.
Power dissipated within matched resistors can generate thermal gradients that can degrade matching. As a guideline, one should avoid dissipating more than 1–2 $\mu W/\mu m^2$ in precisely matched resistors. Moderately matched resistors can tolerate several times this level of power dissipation. Resistors that dissipate larger amounts of power should be interdigitated. High currents in narrow resistors may also induce velocity saturation nonlinearities (Section 5.3.3).

7.3.2. Rules for Capacitor Matching

Properly constructed capacitors can obtain a degree of matching unequaled by any other integrated component. Matched capacitors form the basis of most data conversion products such as analog-to-digital (A/D) and digital-to-analog (D/A) converters. Untrimmed oxide-dielectric capacitors packaged in plastic can achieve ±0.01% matching. This suffices to allow construction of 14-bit and perhaps even 15-bit converters. Beyond this, some type of wafer-level trimming is required to maintain accuracy. Matching of ±0.001% can be obtained using trimmed oxide-dielectric capacitors packaged in plastic, making possible 16- to 18-bit monolithic converters. Higher precision products usually employ hybrid assemblies rather than single-die circuits.

Precisely matched capacitors usually employ a thick oxide dielectric in conjunction with deposited electrodes. Junction capacitors have difficulty maintaining even minimal matching due to their extreme temperature dependence and the effects of outdiffusion. Composite dielectrics are inferior to pure oxide dielectrics because the multiple steps required to produce the composite increase its variability. Dielectric relaxation can also degrade matching in composite dielectrics at higher frequencies. Thick oxide dielectrics are favored over thin ones because they are more tolerant of changes in oxide thickness. A ±10 Å variation in a 100 Å oxide represents ±10% of the oxide thickness, while the same variation in a 500 Å oxide represents only ±2% of its thickness. Silicided polysilicon (*polycide*) is sometimes used for constructing the lower plates of matched capacitors because its low resistance minimizes depletion effects and because this material can withstand the high temperatures required to densify the deposited-oxide dielectric. Aluminum is the material of choice for the upper plate due to the absence of depletion effects in this material. Capacitors constructed using unsilicided poly electrodes experience significant surface depletion, even when the poly is heavily doped. These capacitors therefore exhibit temperature coefficients several times as large as metal-polycide capacitors. Despite these problems, poly-poly capacitors can still obtain adequate matching for all but the most precise applications.

The following rules summarize the most important principles of constructing matched deposited-electrode capacitors:[34]

1. Use identical geometries for matched capacitors.
Capacitors of different sizes or shapes match poorly, so matched capacitors should always use identical geometries. If the capacitors are not the same size, then each should consist of a number of segments (or *unit capacitors*), all of which are copies of the same geometry. The larger capacitor should consist of multiple segments in parallel, while the smaller capacitor should have fewer segments in parallel. Unit capacitors should not connect in series because differences between the parasitic capacitances of the upper and lower plates will produce systematic mismatches. If the required ratio does not lend itself to division into an integer number of unit capacitors, then one nonunitary capacitor

---

[34] Some of these rules follow guidelines proposed by M. J. McNutt, et al., p. 615.

should be inserted into the larger of the matched capacitors. The aspect ratio of this nonunitary capacitor should not exceed 1.5:1 (Section 7.2.2).

2. Use square geometries for precisely matched capacitors.
Peripheral variations are one of many sources of mismatch in capacitors. The smaller the periphery-to-area ratio, the higher the obtainable degree of matching. The square has the lowest periphery-to-area ratio of any rectangular geometry and therefore yields the best matching. Rectangular capacitors with moderate aspect ratios (2:1 or 3:1) can be used to construct moderately matched capacitors, but precisely matched capacitors should always be square. Oddly shaped geometries should be avoided because it is difficult to predict the magnitude of their peripheral variations.

3. Make matched capacitors as large as practical.
The random mismatch of capacitors varies inversely with the square root of their areas. However, an optimal capacitor size exists, beyond which gradient effects cause increasing variability. The optimum dimensions of square capacitors in several CMOS processes are reported to lie between 20×20 μm and 50×50 μm.[35] Capacitors larger than about 1000 μm^2 should be divided into multiple unit capacitors, as proper cross-coupling will minimize gradient effects and improve overall matching.

4. Place matched capacitors adjacent to one another.
Matched capacitors should always reside next to one another. If a large number of capacitors are involved, they should be arranged to form a rectangular array having as small an aspect ratio as possible. For example, if 32 matched capacitors are required, then consider using a 4×8 array. Alternatively, a 5×7 array could be constructed and the three unused capacitors connected as dummies. Adjacent rows of unit capacitors should have equal spacings between them, as should adjacent columns of unit capacitors. The row and column spacings need not be the same.

5. Place matched capacitors over field oxide.
Any surface discontinuities in the thick-field oxide will cause corresponding variations in the topography of the capacitor dielectric. Matched capacitors should always reside over field oxide well away from the edges of moat regions and diffusions.

6. Connect the upper electrode of a matched capacitor to the higher impedance node.
The higher impedance node of the circuit usually connects to the upper electrode, because this generally exhibits less parasitic capacitance than the lower electrode. Substrate noise also couples more strongly to the lower electrode than to the upper electrode. Some arrays may require the high-impedance node of the circuit to connect to the lower plate in order to allow an array of unit capacitors to share a common lower plate. If substrate noise coupling is a concern, consider placing a well under the entire array. This well should connect to a clean analog reference voltage, such as signal ground, so that it can serve as an electrostatic shield for the lower electrode(s) of the capacitor array.

7. Place dummy capacitors around the outer edge of the array.
Dummy capacitors will shield the matched capacitors from lateral electrostatic fields and will eliminate variations in etch rates. The dummy capacitors need not

---

[35] Shyu, Temes, and Yao, p. 1075.

have the same width as the capacitors of the array as long as an electrostatic shield covers the array. Otherwise, fringing fields can easily extend 10 to 30 μm, and arrays of identical dummy capacitors must extend at least this far to ensure precise matching. Moderate matching generally requires only a minimum-width ring of dummy capacitors, and minimal matching does not require dummy capacitors at all. Both electrodes of each dummy capacitor should be connected to prevent static charges from accumulating on the dummy electrodes. The spacing between the dummy capacitors and the adjacent unit capacitors should equal the spacing between rows of unit capacitors.

8. Electrostatically shield matched capacitors.
   Electrostatic shielding provides several benefits. First, it contains fringing fields to the capacitor array, thereby eliminating the need for wide arrays of dummy capacitors. Second, it allows leads to route over the capacitors without causing mismatch or noise injection. Third, it prevents electrostatic fields from adjacent circuitry from interfering with the matched capacitors. All precisely matched capacitors should be electrostatically shielded, and this shielding should extend over the dummies placed around the matched capacitors to seal the array against the entry of electrostatic fields. Even minimally matched capacitors can benefit from electrostatic shielding; if no dummy capacitors are used, then the shield should overlap the capacitors by at least 3 to 5 μm.

9. Cross-couple arrayed capacitors.
   Capacitor arrays lend themselves to cross-coupling because the unit capacitor forms a compact square rather than an elongated rectangle. The array typically consists of several rows and columns of capacitors. Even in the case of two matched capacitors of equal values, a very compact cross-coupled array can be constructed by dividing each capacitor into two halves. Cross-coupling minimizes the effects of oxide gradients on capacitor matching and provides protection against stress and thermal gradients. The centroids of the matched capacitors should precisely align. In practice, this is difficult to achieve with larger capacitor arrays, so the designer must often settle for less-than-optimal interdigitation patterns.

10. Consider the capacitance of leads connecting to the capacitor.
    The leads that connect a matched capacitor into the circuit will contribute some capacitance of their own. This capacitance becomes of concern when one tries to construct moderately or precisely matched arrays of capacitors. Each unit capacitor should have two minimum-width leads connecting to its top electrode, so that each capacitor will have equal overall lead capacitance. If an array contains a nonunitary capacitor, ideally its number of leads should be equal to twice the ratio of its capacitance to the unit capacitance, but this is often difficult to achieve in practice. The total lead area on each capacitor should be computed, and additional leads should be inserted until the ratio of the lead capacitance equals the ratio of the intended capacitors.

11. Do not run leads over matched capacitors unless they are electrostatically shielded.
    The capacitance between the overlying lead and the upper plate of the capacitor will induce a mismatch between the matched capacitors unless the area of the lead overlying each capacitor is identical. Even then, fringing fields and electrostatic noise coupling will degrade the performance of the matched capacitors. If leads must run over matched capacitors, an electrostatic shield should be inserted between the capacitors and the leads.

12. Use thick-oxide dielectrics in preference to thin-oxide or composite dielectrics.
Thick-oxide dielectrics exhibit less mismatch due to dimensional variations, so they are preferable to thin-oxide dielectrics. Composite dielectrics, such as oxide-nitride-oxide sandwich dielectrics, have more mismatch than homogeneous dielectrics because multiple processing steps all affect their final capacitance per unit area. While minimal and moderate matching can be obtained using thin-oxide or composite dielectrics, precise matching usually requires thick-oxide dielectrics.

13. If possible, place capacitors in areas of low stress gradients.
The stress distribution reaches a broad minimum in the middle of the die. Any location ranging from the center of the die halfway out to the edges lies within this broad minimum. Avoid placing capacitors along the edges of the die, and especially in the corners of the die, as the stress in these regions is substantially greater than elsewhere.

14. Place matched capacitors well away from power devices.
The temperature coefficient of heavily doped poly-poly capacitors is relatively low (perhaps 50 ppm/°C), and the temperature coefficient of metal-polycide or metal-metal capacitors is even lower. The direct impact of temperature on matched capacitors is much smaller than on resistors. Matched capacitors should still reside at least 200 to 300 μm away from power devices that dissipate 250 mW or more.

15. Place precisely matched capacitors on axes of symmetry of the die.
Although capacitors are far less sensitive to stress than are resistors, some degree of stress-induced mismatches may still occur. On (100) silicon, precisely matched capacitor arrays should be placed so that the axis of symmetry of the array aligns with one of the two axes of symmetry of the die. On (111) silicon, the axis of symmetry of the array should preferably lie on the <211> axis of symmetry of the die. If large numbers of matched devices are required, reserve the optimal locations for the most critically matched devices.

7.4 SUMMARY

Most integrated circuits contain numerous matched resistors and capacitors. The layout designer should determine which components must match, and with what degree of precision. Using this information, a die floor plan can be constructed showing the relative locations of power devices and matched components. The most critically matched devices should occupy locations near the middle of the die on an axis of symmetry of the power devices. Even if no power devices are present on the die, stress considerations still dictate the placement of the most sensitive matched components near the middle of the die. Preliminary die planning before layout begins will pay dividends in the form of easier circuit construction and interconnection as well as better performance.

Matched devices should never reside in the corners of the die or near a major heat source. Sometimes matched resistors must be placed along a die edge to facilitate the placement of trimpads for fuses or Zener zaps. In this case, the resistors are best placed near the middle of one of the sides of the die.

The matching of resistors and capacitors is very much a function of how well these components are laid out. Two large-value resistors haphazardly laid some distance away from one another using different sizes and shapes oriented in different directions can easily mismatch by several percent. If the same resistors are properly constructed as an array of interdigitated sections, then they will certainly match to better than ±0.1%.

The exact degree of matching achievable from any given layout is difficult to determine. The hard data required to quantitatively evaluate matching performance is rarely available in a manufacturing environment, so the layout designer must make decisions based on limited information. Although this is a difficult and sometimes frustrating process, the designer can greatly improve the performance of many circuits by following the principles discussed in this chapter.

7.5 EXERCISES

Refer to Appendix C for layout rules and process specifications.

7.1. A pair of capacitors were designed to have values of 5 pF and 2.5 pF. Measurements on 10 units provide the following pairs of values: (5.19 pF, 2.66 pF), (5.21 pF, 2.67 pF), (5.19 pF, 2.65 pF), (5.23 pF, 2.66 pF), (5.21 pF, 2.68 pF), (5.12 pF, 2.67 pF), (5.25 pF, 2.68 pF), (5.15 pF, 2.63 pF), (5.21 pF, 2.61 pF), (5.28 pF, 2.61 pF). What is the three-sigma worst-case mismatch between these two capacitors?

7.2. A 12-wafer lot of wafers numbered 1, 2, 3 ... 12 is available for use as samples in an experimental determination of mismatch. Provide detailed instructions for selecting a sample of 30 units from this lot.

7.3. A pair of 3 pF capacitors have a measured standard deviation of mismatch of 0.17%. How large must the capacitors be made to ensure that they will achieve an estimated six-sigma worst-case mismatch of ±0.5%? Assume that systematic mismatches are negligible.

7.4. A certain design contains a pair of 3 μm-wide resistors that have a measured standard deviation of mismatch of 0.32% and a measured systematic mismatch of +0.10%. Assuming that the systematic mismatch does not vary with width, how wide would the resistors have to be made to achieve an estimated three-sigma mismatch of ±0.5%?

7.5. Divide the following matched resistances into segments following the rules of Section 7.2.2, assuming a sheet resistance of 500 Ω/□. In each case, state the number of segments in each resistor, and the segment resistances.
a. 10 kΩ and 15 kΩ.
b. 7.5 kΩ and 11 kΩ.
c. 3.66 kΩ and 11.21 kΩ.
d. 75.3 kΩ and 116.7 kΩ.

7.6. Divide the following matched capacitors into unit capacitances following the rules of Section 7.2.2, assuming an areal capacitance of 1.7 fF/μm^2. State the number of unit capacitors in each device, the unit capacitor values, and their dimensions.
a. 4.0 pF and 8.0 pF.
b. 1.8 pF and 4.2 pF.
c. 3.7 pF and 5.1 pF.
d. 25 pF and 25 pF.

7.7. Choose an optimal orientation for each of the following types of resistors:
a. Standard bipolar HSR resistors.
b. Analog BiCMOS N-well resistors.
c. P-type polysilicon resistors.
d. Nichrome thin-film resistors. (Nichrome is a polycrystalline metal alloy.)

7.8. Devise one-dimensional interdigitation patterns for each of the following cases:
a. Two resistors with a ratio of 4:5.
b. Two resistors with a ratio of 2:7.
c. Three resistors with a ratio of 1:3:5.
d. Four resistors with a ratio of 1:2:4:8.

7.9. Lay out a resistor divider consisting of two 3 kΩ, 8 μm-wide base resistors placed in a common tank. Include all necessary metallization, including a separate lead for the tank contact.

7.10. Construct a resistor divider consisting of two 25 kΩ, 8 μm-wide HSR resistors.

7.11. The divider of Exercise 7.10 forms part of a die having dimensions of 2150 μm by 1760 μm. These dimensions do not include scribe streets and seals, which may be ignored for the purposes of this exercise. Draw a rectangle having these dimensions and place the divider in the best possible location for optimal matching. Assume the design contains no major heat sources.

7.12. Repeat Exercise 5.13, laying out the resistors to obtain precise matching. One of the resistors would normally contain a single segment; replace this resistor with a series-parallel network containing an even number of 1 kΩ segments.

7.13. Suppose the trim network in Exercise 7.12 forms part of a die having an active area of 5.3 mm^2, of which a power transistor consumes 3.6 mm^2. This area does not include scribe streets and seals, which need not be considered for this exercise. Choose an aspect ratio for the die and construct a rectangle having the requisite area. Place another rectangle in the layout to denote the location of the power device. Now locate the trim network in an optimal location along one side of the die. Place the trimpads as close to the edge of the die as possible. Space all metallization at least 8 μm from the edge of the die.

7.14. Construct a capacitor array using analog CMOS poly-poly capacitors. The capacitors should have the following values: 0.5 pF, 1 pF, 2 pF, 4 pF, and 8 pF. Assume that all capacitors share a common poly-1 plate, and bring a lead from each capacitor's poly-2 plate out to the edge of the array. Use copies of the unit capacitors as dummies and cover the array with a metal-2 shield. Take whatever other measures are required for precise matching.

7.15. An integrated circuit experiences a package shift in a certain parameter when packaged in plastic. The mean value of the package shift equals −1.6%, and the standard deviation of the package shift equals 0.07%. Would a polyimide overcoat significantly reduce the overall package shift? Why?

7.16. Given a rectangular area totalling 5000 μm^2, lay out the best possible resistor divider consisting of poly-2 high-sheet resistors, one having a value of 10 kΩ and the other having a value of 100 kΩ.

8 Bipolar Transistors

The *bipolar junction transistor* (BJT) is among the most versatile of all semiconductor devices. In addition to its obvious applications as a voltage or current amplifier, it can also serve as the basis of voltage and current references, oscillators, timers, pulse shapers, amplitude limiters, nonlinear signal processors, power switches, transient protectors, and many other types of circuits. There are also certain applications for which bipolar transistors are ill suited, the most important of these being low-power digital logic. Most logic is now constructed using *complementary metal-oxide-semiconductor* (CMOS) circuitry. Bipolar transistors remain important for constructing analog circuits, although many of these circuits now contain CMOS elements as well.

Much of the information required to understand the operation and construction of bipolar transistors does not appear in elementary texts. This chapter opens with a review of several of these topics, including beta rolloff, avalanche breakdown, thermal runaway, and device saturation. The remainder of the chapter covers the design of small-signal bipolar transistors. This information lays the foundation for the more specialized topics covered in Chapter 9.

8.1 TOPICS IN BIPOLAR TRANSISTOR OPERATION

Figure 8.1 shows a simple model of an NPN transistor. Diode D_1 represents the base-emitter junction of the transistor. Current-controlled current source I_1 models the minority carrier current flowing across the reverse-biased base-collector junction. The current through I_1 equals the current through D_1 multiplied by the transistor's *forward active current gain* β_F. In terms of terminal currents, this relationship becomes

$$I_c = \beta_F I_B \qquad [8.1]$$

Unlike a MOS transistor, the BJT requires a constant base current to sustain the flow of collector current. This base current represents unavoidable losses due to recombination in the neutral base and carrier injection from base to emitter. Because conduction requires constant base current, the bipolar transistor is often called a

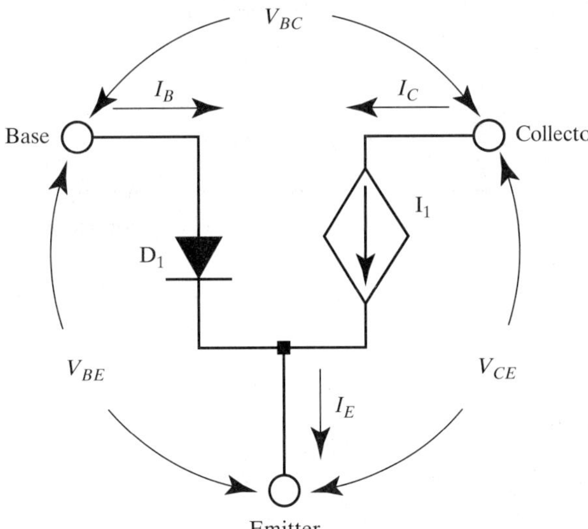

FIGURE 8.1 Simplified three-terminal model of an NPN transistor.

current-controlled device. This is somewhat misleading because the transistor can also be driven by a base-emitter voltage that forward-biases diode D_1 and provides base current to the transistor.

The base-emitter junction of a bipolar transistor is, for all intents and purposes, identical to the silicon junction diode discussed in Section 1.2.2. The base current of the transistor depends exponentially upon the base-emitter bias V_{BE}. If the forward beta remains constant and the collector-to-emitter bias V_{CE} suffices to maintain the transistor in the normal active region, then the collector current I_C also becomes an exponential function of V_{BE}. These relationships can be expressed by the following formulas:[1]

$$I_c = I_s e^{(V_{BE}/V_T)} \qquad [8.2]$$

$$V_{BE} = V_T \ln\left(\frac{I_c}{I_s}\right) \qquad [8.3]$$

The *emitter saturation current* I_s depends on several factors, including the doping profiles of the base and emitter diffusions and the effective area of the base-emitter junction. The *thermal voltage* V_T scales linearly with absolute temperature, and at 298 K (25°C) it equals 26 mV. The base-emitter voltage V_{BE} exhibits a negative temperature coefficient[2] of about −2 mV/°C. This may seem small, but since collector current depends exponentially on base-emitter voltage, an increase of only 18 mV in V_{BE} doubles the collector current. A 1°C temperature difference between two bipolar transistors will cause an 8% mismatch between their collector currents, corresponding to a temperature coefficient of 80,000 ppm/°C! This enormous temperature

---

1 These formulas are slightly simplified; the actual Ebers-Moll equations include a term accounting for reverse conduction: $I_C = I_s[\exp(V_{BE}/kT) - 1]$. The presence of the "−1" term has no significant effect upon conduction in the forward active region at any reasonable bias level and thus has been omitted. This simplified equation also neglects the ideality factor (or emission coefficient) η, which is very near unity for an NPN transistor operating at moderate currents.

2 The negative temperature coefficient of V_{BE} is mostly due to the exponential increase of I_S with temperature, which is sufficient to overwhelm the positive temperature coefficient of V_T.

coefficient has profound implications for the design of matched devices and power transistors.

8.1.1. Beta Rolloff

Elementary textbooks often assume that beta remains constant, but it actually varies greatly depending on collector current. Figure 8.2 shows typical beta curves for small signal NPN and lateral PNP transistors constructed in a standard bipolar process. The NPN beta remains relatively constant over a wide range of collector currents, which to some extent justifies the assumption of constant beta. At high current levels, typically beyond 10 μA/μm² of emitter area, the NPN beta begins to roll off. A similar but more gradual rolloff occurs at very low current levels, usually below 10 pA/μm². High-current beta rolloff is caused by high-level injection, while low-current beta rolloff results from several mechanisms, including recombination within the depletion regions and at the oxide-silicon interface, and short emitter effects (Section 8.3.1).

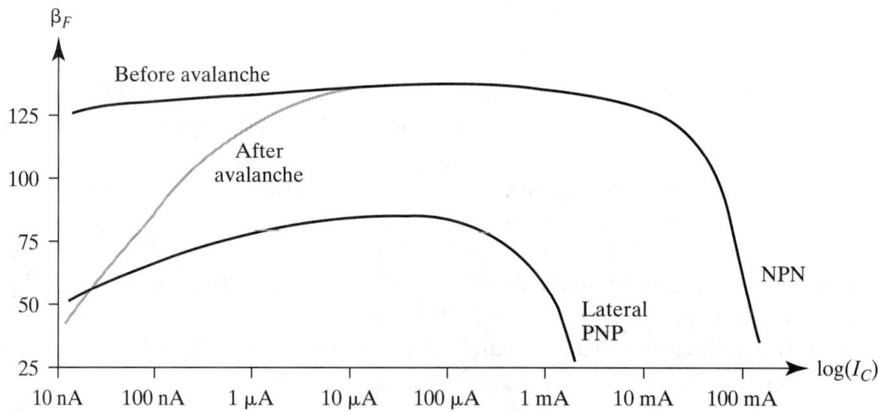

FIGURE 8.2 Beta versus collector current plots for small-signal NPN and lateral PNP transistors. The curve marked in gray shows the effect of emitter-base avalanche on NPN beta.

The beta curve of the lateral PNP differs considerably from that of the NPN. Not only does the lateral PNP have a lower peak beta, but it also exhibits more pronounced high-current and low-current beta rolloffs. Several factors account for the differences between the beta curves of the vertical NPN and the lateral PNP. The emitter of the PNP is much more lightly doped than that of the NPN, so the PNP exhibits a lower emitter injection efficiency that reduces its peak beta. The flow of carriers near the surface of the lateral transistor increases surface recombination and exacerbates low-current beta rolloff. The lightly doped base of the lateral PNP causes high-level injection to begin at relatively low current levels and thus accentuates high-current beta rolloff. The regions of high-current beta rolloff and low-current beta rolloff often overlap, causing a pronounced peaking of the beta curve.[3] Transistors that exhibit such peaking are actually operating in high-level injection at the point of peak beta, complicating the design of certain types of circuits.

8.1.2. Avalanche Breakdown

The maximum operating voltages of a bipolar transistor are determined by the breakdown voltages of the base-emitter and base-collector junctions. Depending

[3] I. Getreu, *Modeling the Bipolar Transistor* (Beaverton, Oregon: Tektronix, 1976), p. 48ff.

upon biasing conditions, several different breakdown voltages may be observed. The three most important of these are denoted V_{EBO}, V_{CBO}, and V_{CEO}. Each of these is measured between two of the transistor terminals with the third terminal left unconnected (*open*).

The breakdown voltage of the transistor base to emitter with collector held open is represented by V_{EBO}. For an NPN transistor constructed on a standard bipolar process, V_{EBO} equals approximately 7 V. This breakdown voltage remains relatively constant over process and temperature, so NPN transistors biased into emitter-base avalanche form useful Zener diodes. Base-emitter avalanche rapidly degrades the beta of NPN transistors because hot carriers generated by the avalanche process induce the formation of recombination centers along the oxide-silicon interface (Section 4.3.3). These recombination centers increase the recombination current within the depletion regions, which in turn drastically decreases the low-current beta of the device (Figure 8.2). Although the high-current beta is not affected to the same degree, one should avoid avalanching any NPN transistor not specifically intended to operate as a Zener diode.

The breakdown of the transistor collector to base with emitter left open is represented by V_{CBO}. This breakdown voltage depends on several factors relating to the base-collector junction, all of which are discussed in greater detail in Section 8.2. Both the base and the collector of most bipolar transistors are lightly doped, so V_{CBO} is usually quite large. For NPN transistors constructed on standard bipolar processes, it can range from 20 V to 120 V or more. Since collector-base breakdown largely occurs beneath the surface, it does not generate surface recombination centers and therefore does not affect beta. Both the V_{EBO} and the V_{CBO} of the lateral PNP transistor depend on the breakdown of the base-epi junction, so the beta of lateral transistors is largely unaffected by any form of avalanche.

The breakdown of the transistor collector to emitter with base held open is represented by V_{CEO}. This is substantially smaller than V_{CBO} due to an effect called *beta multiplication*. Low levels of impact ionization begin to occur at voltages well below the nominal breakdown voltage. Since the base terminal of the transistor is left unconnected, any avalanche injection into the base produces a corresponding (and larger) increase in collector current. The additional carriers transiting across the base-collector junction increase impact ionization and generate additional base drive. At a voltage equal to V_{CEO}, this positive feedback mechanism becomes self-sustaining and the transistor avalanches. The V_{CEO} of an NPN transistor usually equals about 60% of its V_{CBO} (Section 8.2.4).

Beta multiplication can be suppressed by connecting the base terminal to the emitter, holding the transistor in cutoff and preventing amplification of the collector-base leakage. The breakdown voltage collector to emitter with base shorted to emitter is denoted V_{CES}. This breakdown voltage can approach V_{CBO} as long as the base resistance of the transistor is relatively small. If for any reason current begins to pass through the transistor, the breakdown voltage will suddenly decrease to nearly V_{CEO}. This phenomenon, called *snapback,* occurs even if the extrinsic base and emitter terminals are shorted, since the necessary bias develops across the internal base resistance of the transistor. Figure 8.3 shows idealized curve tracer plots illustrating this phenomenon. As soon as the trace labeled V_{CES} exceeds 60 V, it immediately snaps back to 43 V. As the current through the transistor increases, the avalanche voltage begins increasing again. This results partially from extrinsic collector resistance and partially from high-current beta rolloff decreasing the beta multiplication effect. The V_{CEO} of many transistors also shows a small amount of snapback due to low-current beta rolloff. For example, the transistor in Figure 8.3 registers an initial V_{CEO} breakdown of 43 V and snaps back to a sustained V_{CEO} (sometimes called

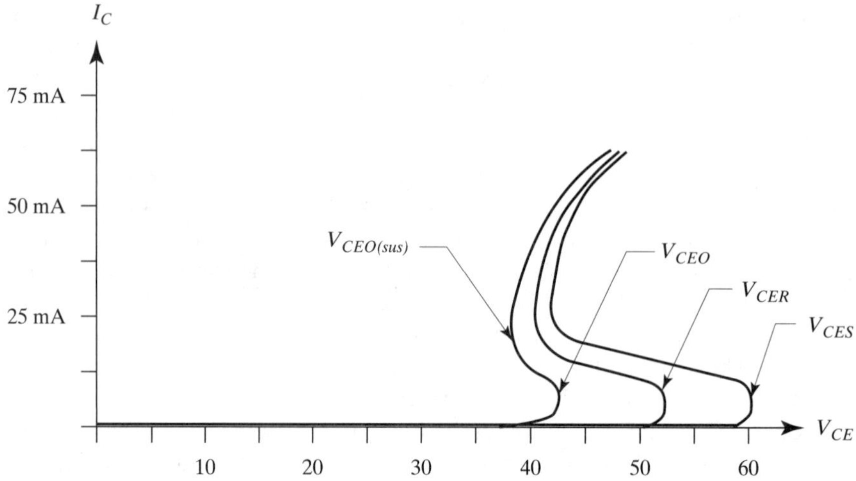

FIGURE 8.3 Idealized curve tracer plots of V_{CEO}, V_{CER}, and V_{CES} in an NPN transistor.

$V_{CEO(sus)}$) of 38 V. Allowing for a small safety margin, this device would rate a V_{CEO} of approximately 36 V.

The middle trace of Figure 8.3 represents the collector-to-emitter breakdown voltage of the transistor with a resistor connected between the base and emitter terminals (V_{CER}).[4] This condition lies between V_{CEO} and V_{CES} and shows the expected intermediate breakdown voltage along with a pronounced snapback.

8.1.3. Thermal Runaway and Secondary Breakdown

Bipolar transistors operating at relatively high power levels fall prey to a failure mechanism called *thermal runaway*. To illustrate how thermal runaway occurs, suppose that a large bipolar transistor suddenly begins to dissipate significant power. The center of the transistor quickly becomes warmer than its outer edges, causing the V_{BE} of the center to drop slightly. Due to the exponential character of the current-voltage relationship, a small change in base-emitter voltage produces a large change in collector current. Increased power dissipation occurs in the hotter portions of the transistor, leading to a further decline in V_{BE}. The region of the transistor that conducts current steadily shrinks as it grows hotter, until practically all of the current funnels through a very small, very hot area called a *hot spot*.

If the temperature in the hot spot reaches some 350 to 450°C, junction leakages become so large that the transistor essentially shorts out. Catastrophic failure occurs either when metallization is drawn through the contacts (as in a Zener zap structure) or when the silicon melts, cracks, or vaporizes. This type of self-destructive runaway does not always occur. Sometimes the increased current density causes beta to roll off far enough to stabilize the hot spot at a high, but not immediately destructive, temperature.[5] The presence of a "stable" hot spot dangerously overstresses a transistor and renders it vulnerable to electromigration, thermally accelerated corrosion, and various other long-term failure mechanisms.

A transistor that contains a stable hot spot often self-destructs during turnoff. Failure occurs due to an apparent avalanche of the collector-base junction, often at

---

[4] V_{CER} has also been used to refer to the collector-to-emitter breakdown voltage with a reverse bias applied to the base-emitter junction. This mode of biasing is sometimes used in power circuits to raise the V_{CE} breakdown of the transistor above V_{CES}.

[5] P. L. Hower, D. L. Blackburn, F. F. Oettinger, and S. Rubin, "Stable Hot Spots and Second Breakdown in Power Transistors," *National Bureau of Standards,* PB-259 746, Oct. 1976.

a voltage substantially below the rated V_{CEO} of the transistor. This unexpected reduction in avalanche voltage is called *secondary breakdown*.[6] It is a consequence of extremely high current densities within the transistor, due in this case to the presence of a stable hot spot. The velocity of carriers in the lightly doped collector increases in order to support the increasing current flow through this zone. Eventually the carrier velocity reaches its maximum limit (the *carrier saturation velocity*). Once this occurs, the electric field across the neutral collector shifts and the avalanche voltage of the transistor snaps back to a lower value, V_{CEO2}, (Section 9.1.1). If the voltage across the collector exceeds V_{CEO2}, the transistor avalanches and the resulting power dissipation quickly destroys the device.[7]

Secondary breakdown can also occur in transistors that have not experienced thermal runaway. During turnoff, the base lead withdraws charge from the neutral base. Charge removal begins in the portion of the base adjacent to the emitter periphery, and progresses inward toward the center of the transistor. As turnoff proceeds, conduction in the transistor collapses into an ever-shrinking area. This *emitter current focusing* causes the momentary appearance of extremely high current densities in small portions of the transistor. These current densities may become large enough to trigger secondary breakdown, especially if the transistor is conducting a large current at the time it is turned off.[8]

Thermal runaway and secondary breakdown can be avoided by restricting the operating conditions of the transistor. Figure 8.4 shows a graph illustrating the *forward-bias safe operating area* (FBSOA) of a typical bipolar transistor. The safe operating region is bounded by four separate curves.[9] The horizontal line represents the maximum current that the metallization and bondwires can safely carry without eventual electromigration failure. The vertical line represents the maximum voltage that can be placed across the transistor without fear of avalanche (usually assumed to equal $V_{CEO(sus)}$). A line passing diagonally across the plot represents the maximum power

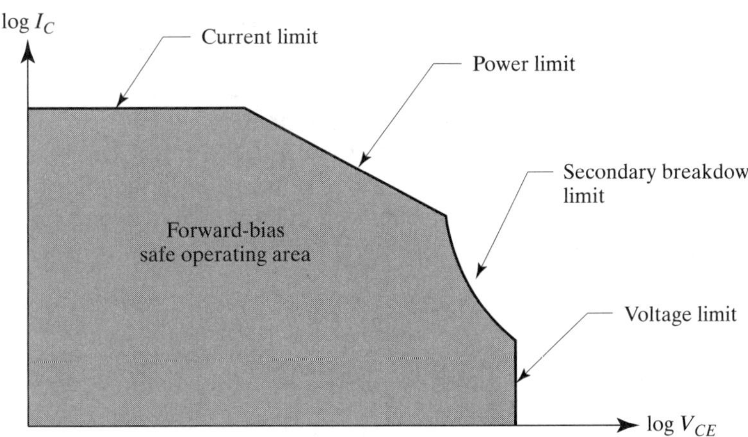

FIGURE 8.4 Forward-bias safe operating area (FBSOA) plot of a typical NPN power transistor.

[6] B. A. Beatty, S. Krishna, and M. S. Adler, "Second Breakdown in Power Transistors Due to Avalanche Injection," *Trans. on Electron Dev.,* Vol ED-23, #8, 1976, pp. 851–857.

[7] J. G. Kassakian, M. F. Schlecnt, and G. C. Verghese, *Principles of Power Electronics* (Reading, MA: Addison-Wesley, 1992), pp. 522–525.

[8] Kassakian, *et al.,* pp. 521–522. P. L. Hower and W. G. Einthoven, "Emitter Current-Crowding in High-Voltage Transistors," *IEEE Trans. on Electron Devices,* Vol. ED-25, #4, 1978, pp. 465–471.

[9] F. F. Oettinger, D. L. Blackburn, and S. Rubin, "Thermal Characterization of Power Transistors," *IEEE Trans. on Electron Devices,* Vol. ED-23, #8, 1976, pp. 831–838.

that the device can dissipate without producing excessive temperatures within the package. A fourth and final curve clips off the portion of the safe operating area where secondary breakdown may occur. A very robust transistor may not exhibit any FBSOA reduction due to secondary breakdown. A properly heat-sunk, but poorly designed power transistor may lose a substantial fraction of its potential FBSOA to secondary breakdown. Section 9.1.2 discusses ways to increase the safe operating area of bipolar transistors.

8.1.4. Saturation in NPN Transistors

An NPN transistor enters saturation when both its base-emitter and base-collector junctions simultaneously forward-bias. Power transistors are often intentionally operated in saturation to reduce the collector-to-emitter saturation voltage $V_{CE(sat)}$ and to minimize power dissipation. Unfortunately, saturation also produces a whole host of problems. The unintentional saturation of bipolar transistors has probably caused more circuit malfunctions than any other device-related design flaw.

Saturation affects discrete and integrated transistors in different ways. The saturation of a discrete transistor merely prolongs its *turnoff time* (also called its *reverse recovery time*). As soon as the base-collector junction begins to forward-bias, minority carriers flow across it in both directions. In an NPN, holes flow into the collector and electrons into the base. A substantial population of excess minority carriers soon accumulates on both sides of the collector-base junction. Now suppose that the external circuit reduces the external base-emitter bias to zero. The transistor does not turn off immediately because the minority carriers require time to recombine, and the resulting majority carrier currents must subside. Saturation increases the turnoff time by at least an order of magnitude. Fast bipolar logic usually incorporates antisaturation clamps (as in LSTTL logic) or uses circuit topologies that are immune to saturation (as in ECL and DCML logic).

Saturation has additional consequences for junction-isolated bipolar transistors. Figure 8.5 shows a cross section of a typical NPN transistor fabricated on a standard bipolar process. The arguments presented for this structure apply equally to any other junction-isolated, vertical NPN transistor, including the CDI NPN of analog BiCMOS (Figure 3.49). These arguments do not apply to fully oxide-isolated transistors, which behave as if they were discrete devices.

The presence of junction isolation introduces a fourth terminal consisting of the P− substrate and the P+ isolation. This *substrate* terminal must always be biased to a lower voltage than the collector terminal to avoid forward biasing the isolation junction. This reverse-biased junction isolates the transistor from the remainder of

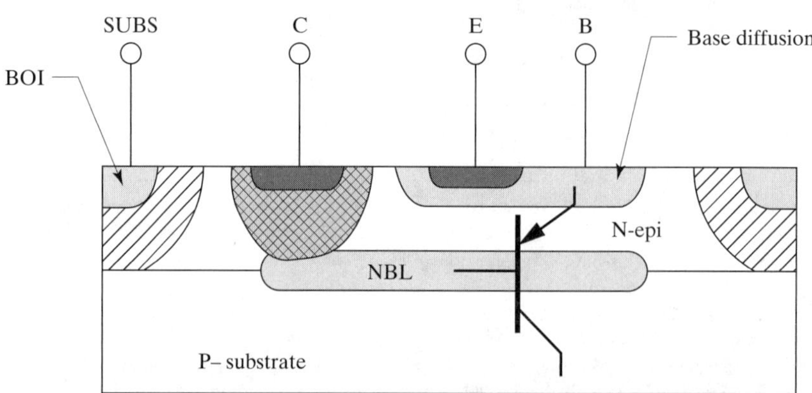

FIGURE 8.5 Cross section of an NPN transistor fabricated on a standard bipolar process, showing the parasitic PNP transistor.

the integrated circuit as long as there are few minority carriers in the neutral collector. Junction isolation fails as soon as the base-collector junction begins to forward-bias. Many of the holes injected across the base-collector junction eventually diffuse across the collector to the collector-substrate junction. The electric field across this reverse-biased junction draws these holes into the substrate, where they become majority carriers.

Saturation can also be explained by imagining a PNP transistor superimposed upon the cross section of the NPN transistor (Figure 8.5). The collector-base junction of the vertical NPN also forms the base-emitter junction of this *parasitic PNP*. When the NPN saturates and forward-biases its collector-base junction, the parasitic PNP transistor turns on and diverts excess base drive into the substrate. This unexpected diversion of base drive has several unpleasant consequences. For one, an integrated NPN transistor cannot be driven as deeply into saturation as a discrete transistor can because the parasitic PNP steals its base drive as soon as it begins to saturate. Integrated NPN transistors therefore tend to have higher $V_{CE(sat)}$ voltages than equivalently constructed discrete transistors.

A saturating NPN also represents an unexpected source of substrate current that can potentially lead to substrate debiasing (Section 4.4.1). If the base drive to the transistor exceeds a few milliamps (as is often the case for power transistors), then guard rings may be added to prevent the holes in the collector from reaching the substrate, or the base-drive circuit can be designed to reduce the base-drive once the transistor begins to saturate (Section 9.1.4).

Saturation also causes a failure mechanism called *current hogging*. When a transistor saturates, some of its base current flows through the base-collector junction rather than the base-emitter junction. This diversion of current reduces the base-emitter voltage.[10] Many circuits connect the base-emitter junctions of several transistors in parallel and expect the collector currents of these devices to track the areas of the respective base-emitter junctions. This relationship ceases to apply if one of the transistors saturates because this transistor experiences a drop in V_{BE} relative to the remaining transistors. The base current of the saturating transistor therefore increases at the expense of the other transistors. In more colloquial terms, the saturating transistor *hogs* the base drive.

Figure 8.6 shows a simple current mirror constructed from three NPN transistors. Current source I_{BIAS} feeds diode-connected transistor Q_1. All three transistors see the same base-emitter voltage and therefore draw the same collector currents, *providing that all three transistors operate in the normal active region*. Now suppose transistor Q_3 saturates. Much of its base current diverts into parasitic PNP transistor Q_P, causing the base-emitter voltage of Q_3 to decrease. Eventually, Q_3's V_{BE} drops to a point just sufficient to satisfy the *intrinsic collector current* I_3, restoring equilibrium. In this case, the intrinsic collector current equals the sum of the extrinsic collector current, I_{t3}, and the base current of Q_P. Transistor Q_2 sees the same base-emitter bias as Q_3, so its collector current, I_2, equals the intrinsic collector current, I_3. In summary, when one transistor in a mirror saturates, the extrinsic collector currents of all the other transistors decrease to equal the intrinsic collector current of the saturated transistor. This disturbs the balance of the circuit and often leads to serious malfunctions.

Circuit designers have developed several cures for current hogging, including base-side ballasting and Schottky clamps. *Base-side ballasting* requires the insertion of matched resistors into the base leads of each transistor (Figure 8.7).

---

[10] A vertical transistor is particularly susceptible to current hogging, because the built-in potential across the base-collector junction is smaller than that across the base-emitter junction, causing current to flow preferentially through the base-collector junction.

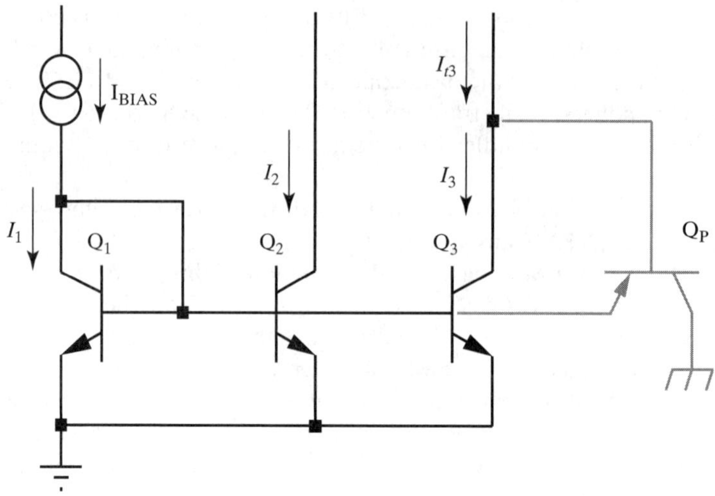

FIGURE 8.6 An example of a circuit that exhibits current hogging. Q_P represents the parasitic PNP transistor present in the structure of vertical NPN Q_3.

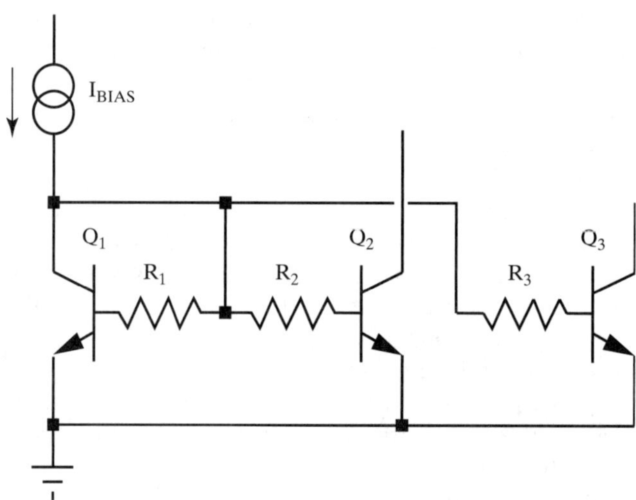

FIGURE 8.7 Base-side ballasting applied to the circuit shown in Figure 8.6.

The base ballasting resistors must ratio inversely to the emitter areas of their respective transistors. For example, if Q_2 has twice the emitter area of Q_1, then R_2 must equal half the resistance of R_1. The base ballasting resistors must match in order to maintain collector current matching. The base-side ballasting prevents current hogging by introducing localized negative feedback. If any one of the transistors begins to saturate, its base current will increase slightly. This causes a corresponding increase in the voltage drop across its base ballasting resistor, forcing a drop in the V_{BE} of this transistor. Typically, no more than 50 to 100 mV appears across the base ballasting resistor of a saturating transistor, producing a correspondingly small disturbance in the base-bias currents.

A diffused base ballasting resistor must not occupy the tank of the NPN transistor it protects because then the ballasting resistor can forward-bias into the tank (Figure 8.8). In effect, the parasitic PNP transistor simply moves from one point in the structure to another.

A clamping diode connected across the base-collector junction of a transistor will prevent it from entering saturation. In order for the clamping diode to function properly, it must have a lower forward voltage than the base-collector junction.

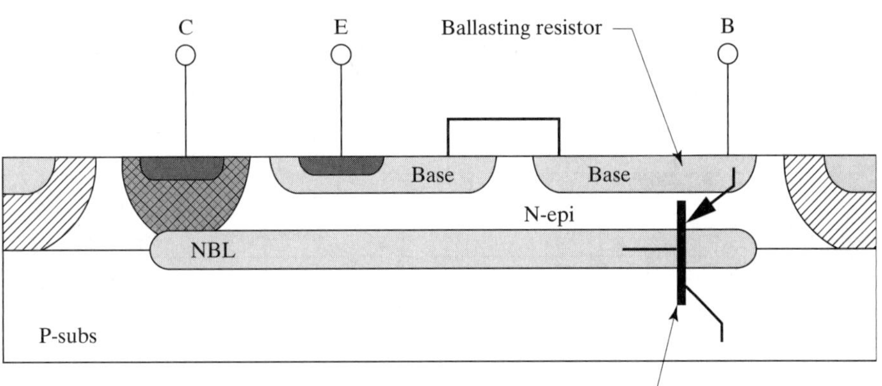

FIGURE 8.8 A base-side ballasting resistor becomes ineffective when merged into the same tank as the NPN it protects.

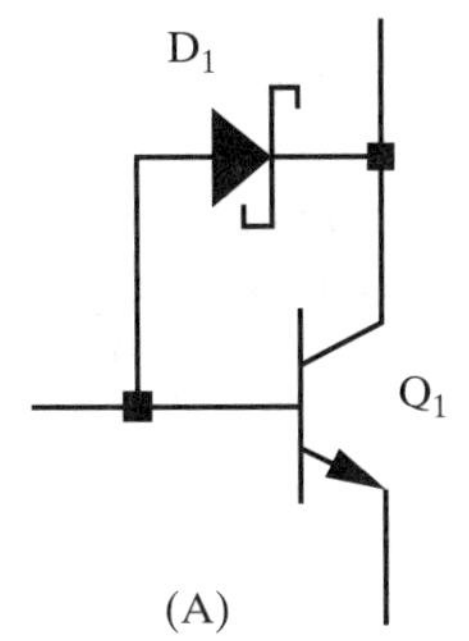

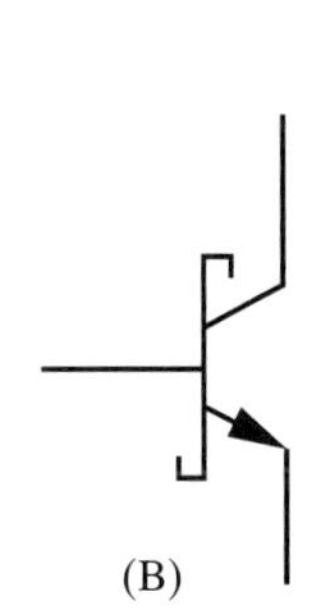

FIGURE 8.9 (A) Schottky-clamped NPN transistor and (B) its conventional schematic symbol.

Only a few types of Schottky diodes, most notably those constructed using platinum or palladium silicides, have the necessary characteristics. Figure 8.9A shows the connection of a Schottky clamp diode in parallel with the base-collector junction of an NPN transistor. The resulting *Schottky-clamped NPN* is often represented by the symbol of Figure 8.9B. The Schottky clamp works by providing an alternative path for current that would otherwise flow through the forward-biased base-collector junction. Because the diode prevents the base-collector junction from conducting, a Schottky-clamped NPN neither injects appreciable minority carriers nor experiences the prolonged turnoff times characteristic of saturation. The Schottky clamp does not prevent the base current from increasing in saturation, but it does prevent it from exceeding the collector current the transistor would normally conduct. Schottky-clamped transistors are used extensively in switching circuitry and bipolar logic to eliminate saturation-induced propagation delays. Section 10.1.3 discusses the layout of Schottky-clamped NPN transistors in further detail.

Saturating NPN transistors can also cause problems when merged with other devices. When a transistor saturates, its forward-biased base-collector junction injects minority carriers into its collector. These carriers can be collected by the reverse-biased junctions of other devices merged into the same tank. Sneak currents associated with minority conduction may cause circuit malfunctions or even catastrophic latchup. Section 13.1 discusses these difficulties in greater detail.

8.1.5. Saturation in Lateral PNP Transistors

Figure 8.10 shows a cross section of a lateral PNP transistor constructed in a standard bipolar process. An epi tank forms the base of the device, while a small plug of

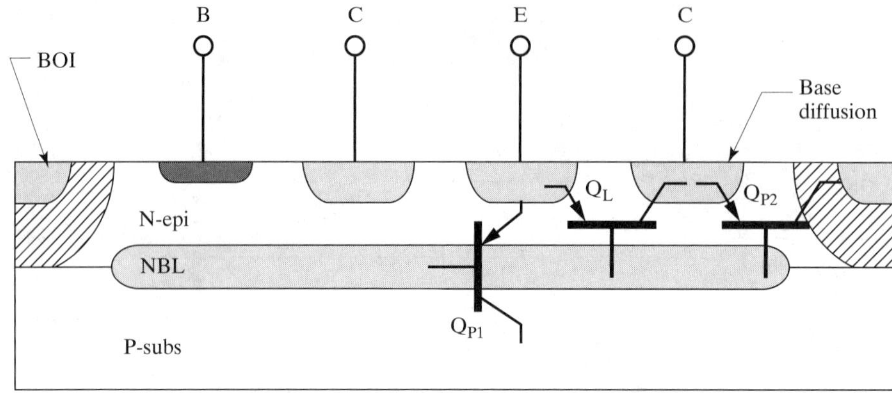

FIGURE 8.10 Cross section of a lateral PNP transistor constructed on a standard bipolar process showing parasitic substrate PNP transistors Q_{P1} and Q_{P2}.

base diffusion placed in the center of the tank acts as its emitter. The collector consists of an annular base diffusion surrounding the emitter plug. Some holes are injected across the sidewalls of the emitter and move laterally to the sidewalls of the collector. However, holes are also injected across the bottom surface of the emitter, and some of the holes injected across the emitter sidewalls diffuse downwards. These errant minority carriers can flow across the reverse-biased junctions isolating the transistor from the isolation and substrate. Parasitic PNP transistor Q_{P1} represents the undesired flow of holes down to the substrate. Parasitic PNP transistor Q_{P2} represents the lateral flow of holes to the isolation sidewalls that occurs when the transistor saturates. Figure 8.10 shows both of these parasitics superimposed upon the cross section of the transistor, along with the desired lateral PNP transistor Q_L.

A significant fraction of the total emitter current will be lost if measures are not taken to block the flow of holes to the substrate. The *collector efficiency* η_C of a lateral PNP equals the ratio of the current collected by lateral transistor Q_L to the sum of the collector currents of Q_L, Q_{P1}, and Q_{P2}. In terms of collector, emitter, and base terminal currents I_C, I_E, and I_B, the collector efficiency η_C equals

$$\eta_c = \frac{I_c}{I_E - I_B} \qquad [8.4]$$

For reasons explained next, the addition of NBL causes the collector efficiency of a forward-active lateral PNP to rise from less than 0.1 to near unity. Lateral PNP transistors constructed in CMOS processes lack NBL and thus rarely achieve high collector efficiencies, despite the use of far narrower basewidths than are possible with standard bipolar. Standard bipolar and analog BiCMOS processes both incorporate NBL and generally provide excellent lateral PNP transistors.

NBL improves the collection efficiency of a lateral PNP by repelling minority carriers moving toward it. This repulsion occurs because of the presence of an opposing electric field at the interface between the heavily doped NBL and the lightly doped N-epi. The presence of this electric field can be explained as follows: The difference in doping concentrations between N-epi and NBL causes a similar difference in majority carrier concentrations. Electrons diffuse from the NBL (where they are plentiful) to the N-epi (where they are scarce). This process produces a net negative charge on the N-epi due to an excess of electrons, and a net positive charge on the NBL due to positively charged dopant atoms. The resulting electric field generates a drift current of electrons flowing back into the NBL from the N-epi. Under equilibrium conditions, the diffusion and drift currents produced by these processes are equal and opposite. Positively charged holes entering the NBL/N-epi interface encounter the electric field present here, which forces them back up into the N-epi.[11] Most of these carriers are eventually collected by the lateral transistor, leading to a large increase in collector efficiency. A small fraction of the downward-moving holes have sufficiently large instantaneous velocities to surmount the opposing electric field and to actually enter the NBL. Most of the holes that enter the NBL recombine within it due to the large population of available majority carriers.

Most of the holes diffusing down to the NBL/N-epi interface possess energies of less than 100 meV, so a built-in potential of as little as 0.2 V represents an almost insurmountable barrier. However, if the built-in potential drops below about 0.1 V, then the interface will become somewhat permeable to holes. This rarely happens in standard bipolar processes because the N-epi is sufficiently lightly doped to provide the requisite built-in potential. However, some analog BiCMOS processes that use relatively heavily doped N-wells have exhibited high degrees of NBL permeability. This situation is extremely undesirable, as it not only degrades the performance of lateral PNP transistors, but it also prevents the construction of effective hole-blocking guard rings.

A few of the holes scattered from the NBL interface move laterally through the epi until they reach the isolation sidewalls. Although this may seem a serious problem, the dimensions of the transistor actually preclude any significant loss of current through lateral parasitic conduction. The cross section in Figure 8.10 has been exaggerated vertically to create a relatively compact drawing. The lateral distance to the sidewall is roughly an order of magnitude larger than the vertical separation between the upper edge of the NBL interface and the lower edge of the collector-base depletion region. At higher collector voltages, the depletion region extends down to the N+/N− interface, closing off the path for lateral parasitic conduction to the sidewalls. Even at low collector voltages, this path is so long and so narrow that few minority carriers can traverse its entire length without coming into contact with the collector-base depletion region.

Lateral parasitic conduction increases dramatically during saturation. The collector possesses a large sidewall area facing the isolation across a relatively narrow gap. This geometry forms an efficient PNP transistor (Q_{P2} in Figure 8.10) that activates as soon as the collector forward-biases, as happens in saturated transistors and those operated in reverse-active mode. When a lateral PNP transistor saturates, the emitter current remains unchanged. Whatever injected holes are not collected by the lateral transistor travel to the substrate instead. Therefore, lateral PNP current mirrors continue to operate properly even if one or more of their transistors in the mirror saturate. All that happens is that the unused collector current flows to the substrate. This does not present a problem as long as the collector currents do not exceed a few milliamps. If substrate injection cannot be tolerated, then a lateral PNP transistor can be fitted with a Schottky clamp. Section 9.1.4 discusses two alternative methods of limiting saturation in lateral PNP transistors.

Split-collector lateral PNP transistors are seriously affected by the saturation of any one of their several collectors. The saturating collector segment reinjects holes from all of its junction surfaces. Some of these are collected by the isolation sidewall, but significant portions of the reinjected carriers are collected by the adjacent collector segments. Consequently, the saturation of one collector in a split-collector lateral PNP causes the current drawn by the adjacent collectors to increase.

---

[11] NBL is often said to "reflect" minority carriers. This choice of wording is somewhat misleading because the carriers do not actually move in straight lines.

8.1.6. Parasitics of Bipolar Transistors

An integrated bipolar transistor behaves quite differently from an idealized textbook device due to the many parasitic elements it contains. Perhaps the most important of these is the PN junction, which isolates the transistor from the rest of the die. As discussed in the previous section, this junction forms part of a parasitic transistor. Even if this transistor does not conduct, leakages and capacitive coupling can occur across the substrate junction. The substrate connection must therefore be counted as one of the terminals of the integrated bipolar transistor. Figure 8.11B shows one conventional method of representing an NPN transistor with an explicit substrate connection. This symbol is often called a *four-terminal* NPN to distinguish it from the *three-terminal* NPN in Figure 8.11A. PNP transistors can also have both three-terminal and four-terminal symbols (Figure 8.11C and D). Although the three-terminal symbols do not show a substrate connection, one still exists as long as the process employs junction isolation. This implicit substrate connection can cause considerable trouble if the designer forgets its presence and assumes that the transistor is truly isolated from the remainder of the die.

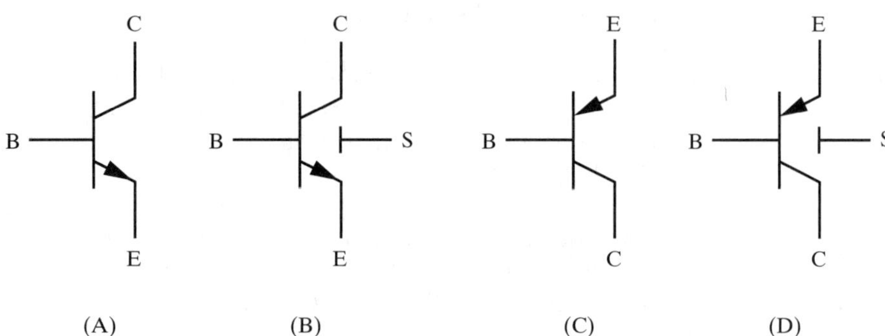

FIGURE 8.11 Symbols for three-terminal and four-terminal transistors (E: emitter, B: base, C: collector, S: substrate).

A complete model of the parasitics of a bipolar transistor includes a number of distributed effects, the discussion of which lies beyond the scope of this text. Figure 8.12 shows simplified parasitic models for a vertical NPN and a lateral PNP formed in a standard bipolar process. These same models also apply to most other junction-isolated transistors, including the vertical NPN and lateral PNP of analog BiCMOS.

The vertical NPN transistor model contains an ideal three-terminal NPN transistor Q_1 that models the intended functionality of the device. Transistor Q_2 represents the parasitic substrate PNP transistor that forward-biases when the vertical NPN transistor saturates (Section 8.1.4). Zener diodes D_{BE}, D_{BC}, and D_{CS} are not intended to model the behavior of the forward-biased junctions, as these are already subsumed into the behavior of transistors Q_1 and Q_2. The Zener diodes instead model the avalanche breakdown and capacitance associated with the respective junctions. For example, diode D_{BE} models the base-emitter capacitance C_{BE} of the transistor as well as the base-emitter breakdown voltage V_{EBO}. Diode D_{BC} models the base-collector capacitance C_{BC} as well as the base-collector breakdown voltage V_{CBO}. Diode D_{CS} models the collector-substrate capacitance C_{CS} and the collector-substrate breakdown voltage. The collector-base capacitance C_{BC} and the collector-substrate capacitance C_{CS} are of concern because they limit the operating frequency of the transistor. The transistor switches faster if these junctions are made smaller. The avalanche voltages of the diodes D_{BE}, D_{BC}, and D_{CS} also set the operating voltage of the transistor, as discussed in Section 8.1.2.

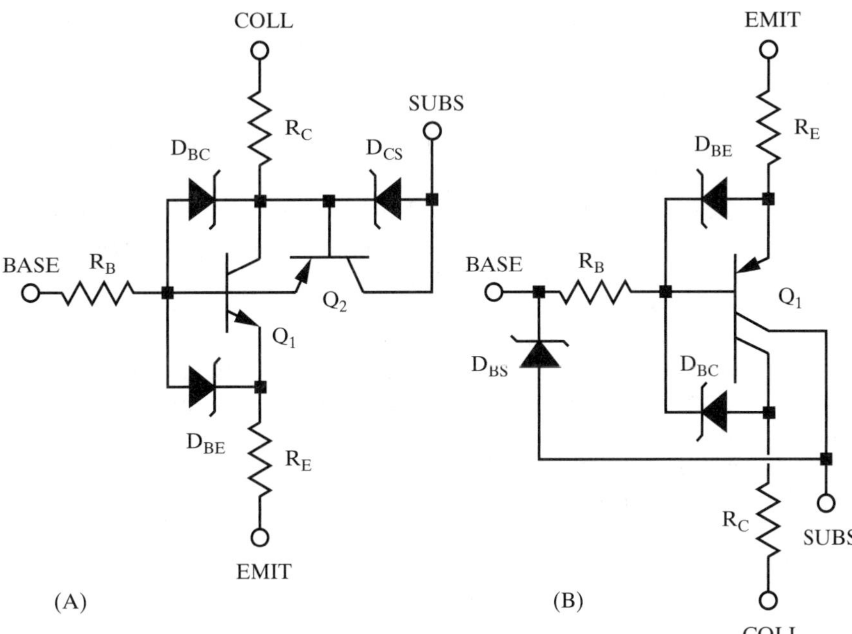

FIGURE 8.12 Subcircuit models for (A) vertical NPN transistor and (B) lateral PNP transistor.

Resistances R_E, R_B, and R_C model the Ohmic resistances of the emitter, base, and collector diffusions, respectively. Resistor R_E is usually quite small and is frequently neglected. Resistor R_B has a profound effect upon the switching speed of the transistor, as the base current must flow through a low-pass RC filter formed by R_B in series with capacitance C_{BE} of diode D_{BE} and capacitance C_{BC} of diode D_{BC}. The resulting low-frequency rolloff limits the maximum operating frequency of the transistor. Transistors with lower base resistances switch faster than those with high base resistances. The base resistance of a typical vertical NPN consists of the sum of the pinched base resistance beneath the emitter and the resistance of the extrinsic base underneath the contact. The presence of the pinched resistance causes the behavior of the base resistance to vary with bias in a very complex manner.[12] The collector resistance R_C limits the *saturation voltage* $V_{CE(sat)}$ that the transistor can achieve in saturation. Power transistors must have low collector resistances in order to minimize their saturation voltages at high currents. Also, the collector resistance of a transistor must not become too great, or the transistor may intrinsically saturate even when the terminal voltages appear to indicate operation in the normal active region.

The lateral PNP transistor model of Figure 8.12B uses an ideal dual-collector PNP Q_1 to model the desired PNP transistor and its substrate PNP parasitic (Section 8.2.3).[13] Some portion of the collector current flows through each of the two collectors. If the collector of the four-terminal transistor saturates, then most of the current instead flows to the substrate terminal (Section 8.1.5). Zener diodes D_{BE}, D_{BC}, and D_{BS} model the avalanche breakdown and capacitance effects of the three junctions of the lateral PNP transistor. Diode D_{BE} models the base-emitter

---

[12] J. R. Hauser, "The Effects of Distributed Base Potential on Emitter-Current Injection Density and Effective Base Resistance for Stripe Transistor Geometries," *IEEE. Trans. on Electron Devices*, Vol. ED-11, #5, 1964, pp. 238–242.

[13] A dual-collector transistor is electrically equivalent to a pair of ordinary single-collector transistors whose base-emitter junctions are connected in parallel. (See Figure 8.24.)

junction's capacitance C_{BE} and its breakdown voltage V_{EBO}. Diode D_{BC} models the base-collector junction's capacitance C_{BC} and its breakdown voltage V_{CBO}. The two breakdown voltages V_{EBO} and V_{CBO} are about equal because both junctions consist of the same diffusions. Capacitance C_{BC} is usually quite large due to the construction of the lateral PNP transistor, partly accounting for its poor frequency response. Diode D_{BS} models the base-substrate junction's avalanche voltage and its capacitance C_{BS}. This capacitance, which is also rather large, does not exist in a vertical transistor. The base-substrate capacitance of a lateral PNP also accounts for some of this transistor's slow frequency response.

Resistors R_E, R_B, and R_C model the Ohmic resistances of the emitter, base, and collector, respectively. Although both the emitter and the collector resistances may equal a few hundred Ohms, neither greatly affects the operation of the transistor. Resistor R_B is quite large because of the light doping of the N-epi tank. The presence of NBL does not completely counteract the effect of light doping because conduction in the lateral PNP transistor occurs near the surface, and the base current must consequently flow through virtually the entire thickness of the epi. This large base resistance forms an RC filter with capacitances C_{BC} and C_{BS}, accounting for the notoriously sluggish frequency response of lateral transistors.

8.2 STANDARD BIPOLAR SMALL-SIGNAL TRANSISTORS

Standard bipolar was originally used to construct digital logic circuits. Designers soon realized that this process could also fabricate analog integrated circuits such as voltage references, operational amplifiers, and comparators. None of these circuits operates at particularly high voltages or currents. Most of their transistors conduct, at most, a couple of milliamps. The design of these *small-signal transistors* emphasizes small size, high gain, and high speed at the expense of power-handling capability. Although designers may differ on exact definitions, most would probably agree that small-signal transistors handle currents of less than 10 mA and power levels of less than 100 mW. Transistors that exceed these limits begin to resemble power transistors more than small-signal devices (Section 9.1.2).

The standard bipolar process was optimized to fabricate NPN transistors, but its various diffusions can also create substrate and lateral PNP transistors. The design principles of small-signal transistors remain much the same regardless of the details of the process. The structure of any optimized bipolar transistor resembles that of a standard bipolar NPN. Most nonoptimized transistors resemble either the substrate or the lateral PNP. By studying the transistors available in standard bipolar, one can gain insight into how transistors are implemented in other processes.

8.2.1. The Standard Bipolar NPN Transistor

The standard bipolar NPN contains several features intended to optimize its performance. These include a heavily doped emitter; a precisely tailored base profile; a thick, lightly doped N-epi; a heavily doped NBL; and a deep-N+ sinker (Figure 8.13).

The emitter diffusion is heavily doped with phosphorus to maximize its emitter injection efficiency. The solid solubility of phosphorus in silicon allows doping levels exceeding 10^{20} atoms/cm^3 allowing the construction of a highly efficient emitter that takes full advantage of a carefully tailored base profile. Arsenic is sometimes added to the emitter diffusion to compensate for lattice strains induced by the heavy phosphorus doping. This precaution eliminates defects that might otherwise migrate into the neutral base and degrade the beta of the transistor.

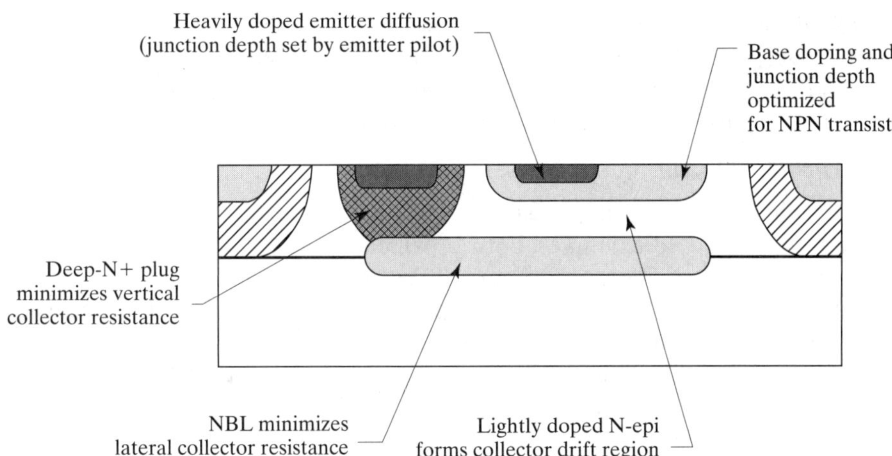

FIGURE 8.13 Key features of the standard bipolar vertical NPN transistor.

The standard bipolar base diffusion has been tailored to provide a combination of high beta, high Early voltage, and moderate V_{CEO}. Light doping aids in maintaining beta control because it allows the use of a wider, and therefore more controllable, neutral base region. A lightly doped base also increases the planar V_{CBO}, and by extension, the V_{CEO} of the vertical NPN. The standard bipolar base diffusion must still contain sufficient dopant to allow direct Ohmic contact since no shallow P+ diffusions exist in this process. Too low a surface doping concentration can also cause surface inversion and parasitic channel formation that can reduce the low-current beta of the transistor. Compromises between these conflicting requirements usually result in a base sheet resistance of 100 to 200 Ω/□.

The NPN collector consists of three separate regions: a lightly doped N-epi, an N+ buried layer, and a deep-N+ sinker. The inclusion of a lightly doped layer adjacent to the collector-base junction increases V_{CEO} and Early voltage by allowing the formation of a wide depletion region extending primarily into the collector. This lightly doped *drift region* lies sandwiched between the base and a heavily doped *extrinsic collector*. In standard bipolar, the drift region consists of the remaining lightly doped N-epi beneath the base diffusion and above the NBL, while the extrinsic collector consists of the NBL and the deep-N+ sinker. The drift region, although lightly doped, does not impede the flow of collector current as long as it remains entirely depleted. The drift region depletes through at higher currents due to velocity saturation and at higher collector-to-emitter voltages due to the extension of the base-collector depletion region. Between these two extremes lies a range of collector voltages and currents that cannot entirely deplete the drift region, causing the effective resistance of the neutral collector to increase (an effect sometimes called *quasisaturation*[14]). Quasisaturation can be minimized by keeping the drift region as thin as possible, consistent with the V_{CEO} rating of the process.

NBL creates a low-resistance path across the bottom of the transistor, but the current must still flow upward to reach the collector contact. The lightly doped epi separating the contact from the underlying NBL is highly resistive, so the inclusion of a deep-N+ sinker can reduce the total collector resistance by as much as an order of magnitude. For example, a typical minimum-size NPN transistor with NBL but

[14] G. Massobrio and P. Antognetti, *Semiconductor Device Modeling with SPICE*, 2d ed. (New York: McGraw-Hill, 1993), pp. 111–115.

without deep-N+ exhibits a collector resistance of about 1 kΩ, while the same structure with the addition of a deep-N+ sinker has a collector resistance of about 100 Ω. Power transistors always include deep-N+, but small-signal devices conducting no more than a few hundred microamps often omit it to save space.

Construction of Small-Signal NPN Transistors

Small-signal NPN transistors usually employ square or rectangular emitters. Figure 8.14 shows two examples, both of which include the features discussed in the previous section. These transistors differ only in the placement of their base and emitter contacts. The structure of Figure 8.14A places the emitter between the collector and base contacts, forming a *collector-emitter-base* (CEB) layout. The structure of Figure 8.14B places the base contact between the collector contact and the emitter, forming a *collector-base-emitter* (CBE) layout. The CEB layout places the emitter and collector contacts closer together and therefore slightly reduces the collector resistance. All other factors remaining equal, the CEB layout will outperform the CBE layout. The difference is so small, however, that many designers use the two layouts interchangeably. The substitution of one style of transistor for the other often simplifies the lead routing of single-level-metal designs.

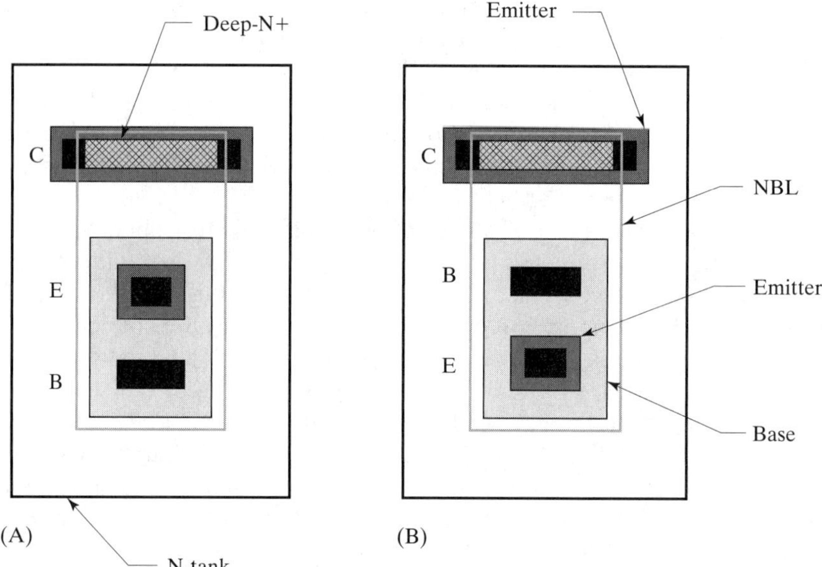

FIGURE 8.14 Two styles of NPN transistors: (A) collector-emitter-base (CEB), and (B) collector-base-emitter (CBE).

The size of an NPN transistor (meaning the magnitude of its saturation current I_S) scales approximately linearly with drawn emitter area. Other factors that exert a lesser degree of influence on scaling include lateral conduction, current crowding, and emitter push. Probably the most important of these is lateral conduction. One might simplistically try to account for lateral conduction by making the transistor size a function of both drawn area and drawn periphery, but such efforts neglect differences in doping. A more accurate, yet still imperfect, approach models the lateral and vertical conduction paths as a pair of transistors placed in parallel, one scaling with drawn area and the other with drawn periphery. Carriers injected laterally from the emitter sidewalls suffer more recombination than those injected vertically, because the laterally injected carriers must cross a wider and more heavily doped base region than the vertically injected ones. Surface states along the oxide-silicon interface also increase lateral recombination. Because of these effects, transistors with large area-to-periphery ratios usually have higher betas than those with smaller ratios. Measurements of two NPN transistors fabricated on a standard bipolar op-amp, one with a 1.0×1.0 mil emitter and the other with a 1.5×3.5 mil emitter,[15] showed that the smaller transistor exhibited a peak beta of 290 and the larger one a beta of 520. Most small NPN transistors employ either square or slightly elongated rectangular emitters to make efficient use of the available space while retaining a high area-to-periphery ratio (Section 9.2.1). Larger emitters also drive deeper into the base diffusion and thus reduce the neutral base width.

All of the other geometries that form the NPN transistor are placed relative to the emitter geometry, beginning with the emitter contact. This contact usually occupies as much of the emitter area as possible in order to minimize emitter resistance. The emitter diffusion should overlap the emitter contact equally on all sides to ensure uniform lateral current flow. The base diffusion must overlap the emitter on all sides sufficiently to prevent lateral punchthrough. To save space, the base is usually contacted along only one side of the emitter. The base contact can be elongated without enlarging the transistor, significantly reducing base resistance.

The outdiffusion of the P+ isolation determines the minimum tank overlap of base diffusion, which is among the largest spacings of the standard bipolar process. Up-down isolation can reduce this spacing by perhaps a third (Section 3.1.4). The base region occupies a tank contacted at one end by a deep-N+ sinker. Deep-N+ outdiffuses and must therefore reside some distance away from both the base and the isolation diffusions. Emitter coded over the deep-N+ increases the surface doping to ensure reliable Ohmic contact. The emitter oxide removal step also thins the thick field oxide sufficiently to allow simultaneous etching of all contacts. Both the deep-N+ sinker and the emitter diffusion used to contact it can be elongated without increasing the size of the transistor. The deep-N+ sinker is often omitted from low-current devices to save space. Regardless of whether a sinker is used, the NBL should fill as much of the tank as possible to minimize the overall collector resistance. Minimum-geometry transistors often exhibit little overlap of NBL and deep-N+. As long as the drawn geometries touch one another, the collector resistance will be reduced sufficiently to allow low-current operation without significant collector voltage drops. If the transistor must conduct more than a few milliamps, then the transistor should be enlarged to allow the NBL to completely enclose the deep-N+ sinker.

Many variations of the standard bipolar NPN layout exist, especially for single-level-metal processes. The lack of a second metal forces the designer to route leads through the transistor. Selective use of CEB and CBE layouts allows some rearrangement of the terminal ordering and often eliminates the need to jumper one or more signals. Transistors can also be stretched to allow one or more leads to route between their terminals. Stretched transistors exhibit increased base and collector resistances and capacitances that could interfere with proper circuit operation, so designers should avoid stretching devices whenever possible.

Figure 8.15A shows a typical *stretched-collector transistor*. The collector and base contacts have been moved apart to allow two leads to pass between them. This modification increases collector resistance and the collector-to-substrate capacitance. Nothing can be done to eliminate the added capacitance, but the increased resistance

[15] B. A. Wooley, S.-Y.J. Wong, and D. O. Pederson, "A Computer-Aided Evaluation of the 741 Amplifier," *IEEE J. Solid-State Circuits*, Vol. SC-6, 1971, pp. 357–365. Note that 1 mil = 25.4 μm.

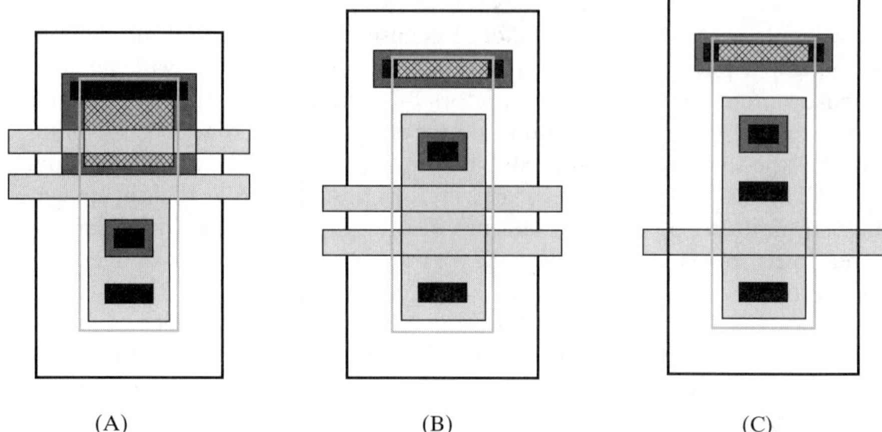

FIGURE 8.15 Three examples of stretched NPN transistors.

can be partially offset by elongating the deep-N+ and emitter diffusions surrounding the collector contact. In processes that employ a thin emitter oxide, leads routed over an extended emitter diffusion may represent an ESD vulnerability (Section 4.1.1), and only the deep-N+ (and not the enclosing emitter diffusion) should be elongated.

Figure 8.15B shows a *stretched-base transistor*. The base region of this transistor has been elongated to allow two leads to pass between the base and emitter contacts. The elongated base geometry causes an increase in base resistance and collector-base capacitance and a corresponding reduction in the frequency response. Stretched bases usually have a greater effect on circuit performance than stretched collectors, so they should be employed only when absolutely necessary, especially in high-speed signal paths. Figure 8.15C shows a different type of stretched-base transistor in which a lead tunnels through the base diffusion. This arrangement not only increases collector-base capacitance, but also inserts considerable resistance between the two base contacts. Assuming a base sheet of 160 Ω/□, the resistance between the two base contacts of the illustrated transistor equals about 200 Ω.

Double-level metal (DLM) virtually eliminates the need for stretched transistors. Not only does it eliminate stretched devices and tunnels, but it also reduces the die area because metal jumpers and vias can reside on top of other devices. DLM also allows the use of a small number of standardized layouts that can be fully characterized and modeled to enable more accurate simulation. The widespread implementation of multilevel metal systems has almost eliminated the use of stretched devices, but the technique still has merit for certain applications; for example, in photodevices where metal-2 must act as a light shield, or in power devices where the upper metal layers are devoted to power routing.

Several approaches exist for scaling up NPN transistors, all of which seek to increase emitter area without reducing device performance. This turns out to be an elusive goal that requires different strategies for different applications. The naïve approach consists of enlarging the emitter while retaining the same overall geometry (Figure 8.16A). This yields a high beta due to an increased emitter area-to-periphery ratio, but it also entails several serious disadvantages. The lightly doped pinched base region beneath the emitter has a sheet resistance of about 2 to 10 kΩ/□. This resistance introduces unwanted phase shifts and greatly slows transistor switching. More subtly, it causes nonuniform current flow at higher current levels. The portions of the emitter lying closest to the base contact experience the highest base-emitter bias and, consequently, inject more carriers than portions of the emitter far

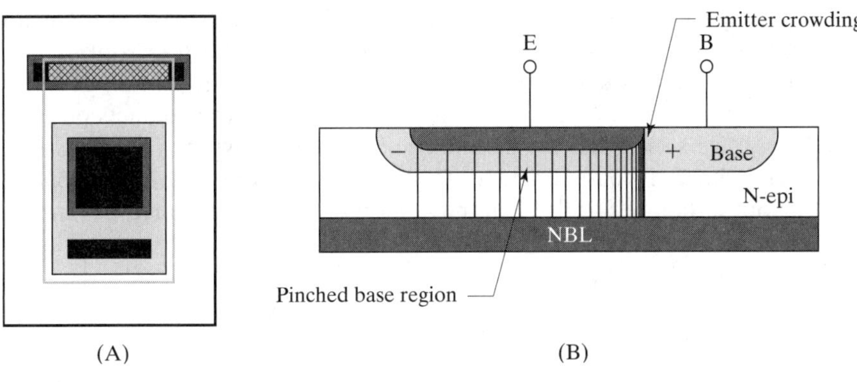

FIGURE 8.16 (A) Layout of a compact emitter NPN transistor and (B) a cross section of the active region of this device that illustrates the effects of emitter crowding.

from the base contact. Figure 8.16B illustrates this effect, called *current crowding* or *emitter crowding.* Its severity can be appreciated by remembering that a mere 18mV of debiasing doubles the emitter current.[16]

Current crowding reduces the active area of the emitter by forcing most of the conduction to occur near the base contact. This effect complicates device matching, reduces beta, and makes the transistor more susceptible to secondary breakdown. While current crowding is undesirable in small-signal transistors, it can actually enhance the robustness of properly designed power devices (Section 9.1.2).

An alternative style of layout uses long, narrow emitter stripes, or *fingers*.[17] Base contacts placed along both sides of each emitter finger help minimize base resistance, increasing switching speed and enhancing immunity to secondary breakdown (Figure 8.17A). The relatively small area-to-periphery ratio of this transistor yields a lower beta than the structure of Figure 8.16. This *narrow-emitter transistor* also becomes vulnerable to thermal runaway at emitter current densities of more than a few $\mu A/\mu m^2$ of emitter (Section 9.1.2).

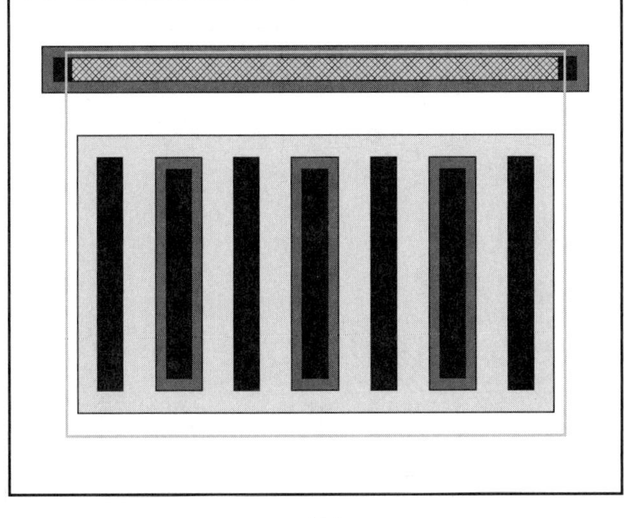

FIGURE 8.17 (A) The narrow-emitter transistor and (B) its equivalent minimum-size layout, the double-base transistor.

[16] R. J. Whittier and D. A. Tremere, "Current Gain and Cutoff Frequency Falloff at High Currents," *IEEE Trans. on Electron Devices*, Vol. ED-16, #1, 1969, pp. 39–57.

[17] A. B. Grebene, *Bipolar and MOS Analog Integrated Circuit Design* (New York: John Wiley and Sons, 1984), p. 76.

The compact emitter transistor in Figure 8.16 functions best in applications requiring high beta and moderate speeds. The narrow emitter transistor in Figure 8.17 provides superior frequency response, but inferior beta. A minimum-geometry transistor can also employ an emitter stripe paralleled by base contacts on either side, producing the somewhat misnamed *double-base transistor* (Figure 8.17B). This structure reduces the base resistance to approximately one-quarter that of the single-base layout of Figure 8.15.[18] Both the large-emitter and the narrow-emitter transistor function poorly at high emitter current densities, so neither is suitable for use as a power device (Section 9.1.1).

8.2.2. The Standard Bipolar Substrate PNP Transistor

Since NPN and PNP transistors differ only in doping polarities, one could theoretically create a PNP transistor by inverting all of the doping polarities of the standard bipolar process. The collector would then consist of a lightly doped P-type epi with the addition of a *P-buried layer* (PBL) and a deep-P+ sinker. The base would consist of a lightly doped N-type diffusion, the emitter of a heavily doped P-type one. The limitations of boron doping make it difficult to realize this structure. Buried layers should consist of slow-diffusing dopants such as arsenic or antimony. Boron diffuses relatively rapidly, so any attempt to fabricate a P-buried layer requires a radical process redesign to eliminate subsequent high-temperature drives. Standard bipolar NPN transistors also employ a heavily doped emitter to maximize emitter injection efficiency. The solid solubility of boron in silicon is only one-third that of phosphorus,[19] so the high-current performance of the PNP suffers accordingly. Even if these problems are somehow overcome, the mobility of holes in silicon is only about one-third that of electrons.

Standard bipolar cannot fabricate a fully isolated vertical PNP transistor. Although some processes do offer both vertical NPN and vertical PNP transistors, these *complementary bipolar* processes require several additional processing steps. As a compromise, standard bipolar offers a type of vertical PNP transistor called a *substrate PNP*. This transistor lacks full isolation because it employs the P-type substrate as its collector. Its base consists of N-epi, and its emitter of base diffusion (Figure 8.18). The performance of the transistor suffers because these materials are not tailored to their

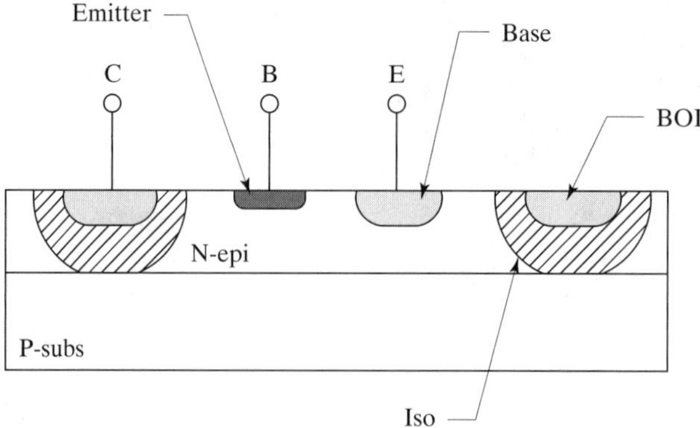

FIGURE 8.18 Cross section of a typical substrate PNP transistor fabricated in standard bipolar.

[18] A. B. Glaser and G. E. Subak-Sharpe, *Integrated Circuit Engineering* (Reading, MA: Addison-Wesley, 1977), p. 52.

[19] F.A. Trumbore, "Solid Solubilities of Impurity Elements in Si and Ge," *Bell System Technical Journal*, Vol. 39, No. 1: 1960, pp. 205–233. Datapoints are taken at 1100°C.

respective tasks. On the other hand, the substrate PNP requires no additional process steps, and it costs nothing to add one to a standard bipolar design.

The various diffusions of the standard bipolar process take their names from the functions they perform in a vertical NPN transistor. These same diffusions play very different roles in a substrate PNP transistor. The *emitter* of a substrate PNP consists of *base* diffusion, and the *base* consists of an N-tank contacted by means of *emitter* diffusion. As this example shows, one must pay close attention to the difference between the names of diffusions and the roles they play in a given structure.

The standard bipolar emitter diffusion typically has a dopant concentration in excess of 10^{20} atoms/cm^3, while the base diffusion rarely exceeds 10^{17} atoms/cm^3. The lighter doping greatly reduces the emitter injection efficiency of the substrate PNP. The still-lighter doping of the N-epi causes premature beta rolloff due to high-level injection. A substrate PNP typically retains a beta of 30 out to about 1 μA/μm^2, while a vertical NPN transistor retains a beta of 150 out to perhaps 30 μA/μm^2. The vertical resistance of the base diffusion exceeds that of the emitter diffusion by an order of magnitude, so a minimum-emitter substrate PNP typically exhibits 100 Ω of emitter resistance, compared with the 10 Ω of a comparable NPN transistor. The emitter resistance causes few problems at the low current densities typically used with substrate PNP transistors. The increased emitter resistance even provides emitter ballasting, and this combined with high-current beta rolloff ensures a high degree of immunity to thermal runaway.

Standard bipolar employs a relatively lightly doped N-epi. In the absence of NBL, the epi-substrate junction diffuses upward during the long isolation drive. The base region of the substrate PNP is therefore both thin and lightly doped—much more so than the epi thickness might suggest. Most of the carriers flow vertically from emitter to substrate, rather than laterally from emitter to isolation. Substrate PNP transistors often have peak betas of 100 or more, but high-level injection causes their betas to peak at current densities of less than 2 μA/μm^2.

The collector of the substrate PNP consists of a series combination of the P-substrate and the P+ isolation diffusion. The lightly doped substrate increases both the Early voltage and the V_{CEO} rating of the transistor, but strictly speaking, it does not act as a drift region because it is not bounded by an adjacent P+ layer. Collector voltage drops rarely have much effect upon the substrate PNP itself, but they can cause debiasing of adjacent circuitry at higher current levels. Standard bipolar designs rarely experience objectionable levels of debiasing as long as the collector current in each substrate PNP does not exceed 1 to 2 mA and the total collector current of all substrate PNP transistors does not exceed 10 mA. A standard P+ isolation diffusion supplemented by base-over-iso can cope with current levels of this magnitude as long as substrate contacts are located near the substrate transistors. Any substrate PNP transistor that injects 1 mA or more should have additional substrate contacts surrounding it (Section 4.4.1). Substrate PNP transistors become progressively more impractical as current levels increase; circuit designers may wish to consider substituting lateral PNP transistors for high-current substrate devices to minimize substrate debiasing. Alternatively, backside substrate contact can be employed to remove almost any amount of substrate current without debiasing.

Substrate PNP transistors operate primarily by vertical conduction and therefore scale with drawn emitter area. Several other factors influence the precise scaling of these devices, including outdiffusion, lateral conduction, and surface effects. As with vertical NPN transistors, precisely matched substrate PNP transistors must use identical emitter geometries.

Construction of Small-Signal Substrate PNP Transistors

The simplest style of substrate PNP consists of an N-tank containing emitter and base regions (Figure 8.19A). The tank serves as the transistor's base, and the base diffusion as its emitter. The emitter diffusion beneath the base terminal of the transistor allows Ohmic contact to the tank. The gain of this transistor depends somewhat on the area-to-periphery ratio of the emitter. This occurs in part because laterally emitted carriers must travel farther to reach the isolation than vertically emitted carriers must travel to reach the substrate. Laterally emitted carriers experience additional recombination due to defect sites along the oxide-silicon interface. Large base diffusions also drive deeper into the epi and thus narrow the base width of the substrate PNP. In summary, larger and more compact emitter structures produce higher gains.

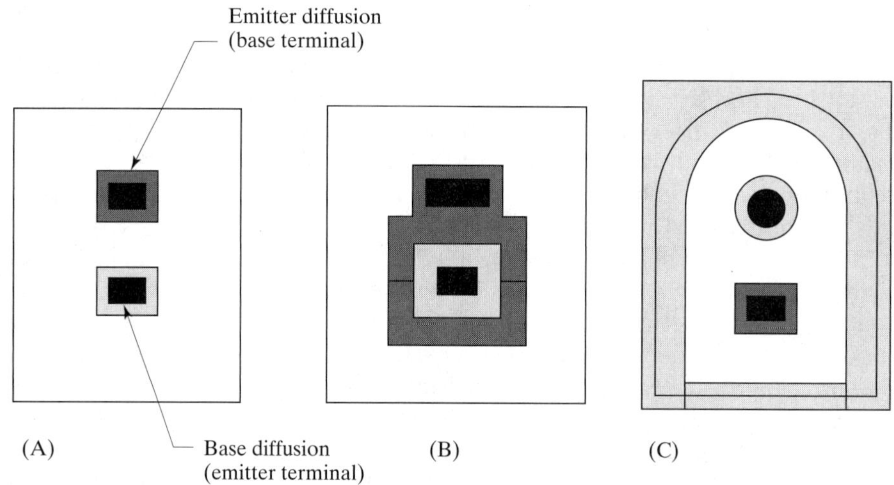

FIGURE 8.19 Examples of (A) standard, (B) emitter-ringed, and (C) verti-lat styles of substrate PNP transistor.

The transistor in Figure 8.19A is constructed by first laying out an emitter of the desired size and shape, using base diffusion. The base contact consists of a rectangle of emitter diffusion placed on one side of the emitter geometry and spaced away from it by the minimum allowed base-to-emitter spacing. The emitter diffusion need only be wide enough to contain a minimum-width contact. At least one substrate contact should reside near each substrate PNP. Ideally, this contact should abut the substrate PNP, but wiring constraints frequently force it to reside several tens of microns away. Some separation between the substrate contact and the transistor can be tolerated as long as the collector current of the substrate PNP does not exceed a milliamp or two. Higher collector currents will probably debias the substrate unless substrate contacts are placed adjacent to the transistor.

Figure 8.19B illustrates another type of substrate PNP layout, where the emitter consists of a rectangle of base diffusion surrounded on all sides by a thin ring of emitter diffusion. This emitter ring adjoins but does not overlap the drawn base diffusion. The emitter ring helps discourage lateral conduction by raising the voltage required to forward-bias the emitter sidewalls. The emitter-ringed transistor exhibits a higher beta at low current densities than the transistor in Figure 8.19A does. This advantage disappears at higher current densities because the base forward-biases into the emitter ring. Although this structure has some merit, most layouts do not employ it because it requires more room than the simple layout of Figure 8.19A, and because the latter structure has sufficient gain for most applications. If an emitter ring is used, then the spacing between the ring and the base diffusion's contact must equal the spacing between the emitter of a vertical NPN and its base contact. The other dimensions of this transistor follow much the same rules as apply to the structure of Figure 8.19A.

The device of Figure 8.19C is sometimes called a *tombstone PNP* or a *cathedral PNP* because of its distinctive tank outline. The same transistor is also called a *vertilat PNP* because it attempts to capitalize upon both vertical and lateral conduction in the same device. The emitter of this transistor consists of a circular plug of base diffusion. The tank geometry passes around the emitter in a semicircular arc digitized around the same center as the emitter. A ring of base diffusion also surrounds the tank, overlapping into it as far as layout rules allow. This base ring helps counteract the high sheet resistance of the periphery of the isolation caused by its extensive outdiffusion. Lateral conduction in the transistor occurs outward from the plug of base diffusion serving as the emitter, to the partial ring of base diffusion serving as the collector. Conduction also occurs vertically downward from the emitter plug to the underlying substrate. This style of transistor requires much the same field plating as does a lateral PNP. The field plate should entirely cover the exposed surface of the N-epi at the semicircular end of the transistor, and it should extend toward the base contact as far as the metal spacing rules allow. Failure to properly field-plate the transistor may cause unexpected leakage phenomena as well as a degradation of low-current beta (Section 8.2.3).

Theoretically, a verti-lat PNP should have a higher beta than a standard substrate PNP, because its lateral conduction pathway has been optimized by minimization of the neutral base width and by the incorporation of a base field plate. In practice, the structure of Figure 8.19B generally outperforms the verti-lat transistor because it suppresses lateral conduction and relies instead on the more efficient vertical conduction.

Figure 8.20 shows the layout of a larger substrate PNP transistor. Like the large vertical NPN transistor in Figure 8.17A, this device employs interdigitated emitter

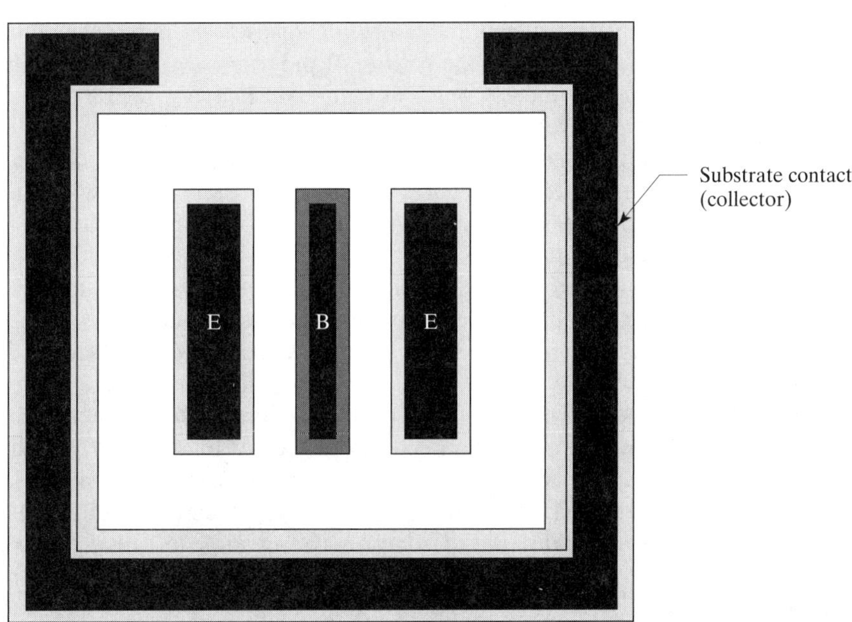

FIGURE 8.20 Higher current substrate PNP transistor employing two wide emitter stripes and a single narrow base stripe.

and base stripes. Each emitter consists of a fairly wide (~25 μm) strip of emitter diffusion. Since PNP transistors exhibit beta rolloff at relatively low current densities, hot spotting and secondary breakdown generally do not occur, and the emitter width is limited only by the resistance of the pinched region beneath the base diffusion. Because this pinched sheet may exceed 10 kΩ/□, emitter stripes more than 25 to 50 μm wide are not advisable. The base contacts consist of thin stripes of emitter diffusion placed between adjacent emitter fingers. If desired, two additional stripes of emitter diffusion can be placed on the ends of the transistor to further reduce base resistance. The large substrate contact encircling the transistor helps limit substrate debiasing. The illustrated contact cannot deal with more than a few milliamps of substrate current without saturating; a much wider strip of contact would be required to deal with the maximum collector current of which this structure is capable. The substrate contact has been interrupted along the top of the transistor to allow the emitter and base leads to emerge on first-level metal. If double-level metallization is available, then the contact should form an unbroken ring around the transistor to minimize collector resistance.

8.2.3. The Standard Bipolar Lateral PNP Transistor

Although standard bipolar cannot fabricate an isolated vertical PNP, it does offer an isolated *lateral PNP* consisting of two separate base diffusions placed in a common tank. One of these diffusions serves as the emitter and the other as the collector. When the emitter forward-biases, holes flow into the tank and travel laterally to the collector. Lateral transistors generally have slower switching speeds and lower betas than vertical devices. Although little can be done to boost switching speeds, proper design can substantially improve the beta. The layout designer can also vary the base width of a lateral PNP by moving the emitter and collector nearer together or farther apart. A narrower basewidth produces a higher beta and a lower Early voltage, while a wider basewidth produces the opposite effect. The product of beta and Early voltage remains approximately constant regardless of basewidth.

The beta of a lateral PNP depends on at least five different factors: emitter injection efficiency, base doping, base recombination rate, base width, and collector efficiency. All but the last two of these are beyond the control of the layout designer. The emitter injection efficiency of the lateral PNP suffers from the use of a relatively lightly doped base diffusion as the emitter of the transistor. The even lighter doping of the N-epi causes the early onset of high-level injection and a corresponding beta rolloff beyond a current density of about 100 μA per minimum emitter. Some processes incorporate a deep-P+ diffusion intended specifically to construct lateral PNP transistors having better high-current characteristics, but even with this modification, beta rolloff still begins at or below 500 μA per minimum emitter.

The standard bipolar lateral PNP transistor suffers from elevated recombination rates caused by the use of (111) silicon. The same surface states that increase the thick-field threshold also act as recombination centers and decrease the lifetime of minority carriers traveling near the surface. The field oxidation also spawns oxidation-induced lattice defects that migrate a short distance into the silicon. Both of these factors reduce the beta of the lateral PNP to a fraction of that of the corresponding substrate PNP. Annealing in a reducing atmosphere (such as the nitrogen-hydrogen mixture called *forming gas*) reduces the surface state concentration and increases the beta of the lateral PNP transistor at the cost of reducing the thick-field threshold. Historically, the introduction of compressive nitride protective overcoats produced an increase in lateral PNP beta, presumably because of the reducing conditions prevailing during nitride deposition. Together with smaller feature sizes, nitride protective overcoats increased the peak beta of standard bipolar lateral PNP transistors from less than 10 to more than 50.

The *drawn* basewidth of a lateral PNP equals the separation between the drawn base diffusions serving as its emitter and collector (Figure 8.21, dimension W_{B1}). The *actual* base width of the transistor is more difficult to determine because it depends on two-dimensional current flow. Near the surface, the actual base width W_{B2} is considerably less than the drawn base width W_{B1} due to outdiffusion and intrusion of the depletion regions into the neutral base. As one proceeds deeper into the silicon, the sidewalls of the emitter and collector curve away from one another. Carriers do not move in straight lines, and the greater distance traveled along alternative pathways also increases the *effective* basewidth of the transistor (W_{B3}). The effective base width of the transistor consists of a weighted average of all possible paths by which carriers might traverse the neutral base. The complexity of this problem precludes simple closed-form solution, and the solutions that exist provide little insight into transistor design.[20,21] Three general observations can still be made. First, the effective base width for purposes of computing punchthrough consists of the drawn basewidth minus twice the outdiffusion distance, or W_{B1}. Only the smallest drawn basewidths exhibit any significant decrease in operating voltage due to punchthrough. Second, the effective base width for purposes of computing beta (W_{B3}) substantially exceeds the actual base width at the surface (W_{B2}) due to the contribution of subsurface conduction. This causes beta to scale more weakly with drawn base width than one might expect. If a transistor with a drawn basewidth of 8 μm has a peak beta of 80, then one with a drawn basewidth of 16 μm will have a peak beta greater than 40. Third, the Early voltage of a lateral PNP is inversely proportional to peak beta. Suppose a transistor with a peak beta of 80 has an Early voltage of 70 V; a wider-base transistor exhibiting a beta of 60 should have an Early voltage of approximately $(80/60) \cdot 70 = 93$ V.

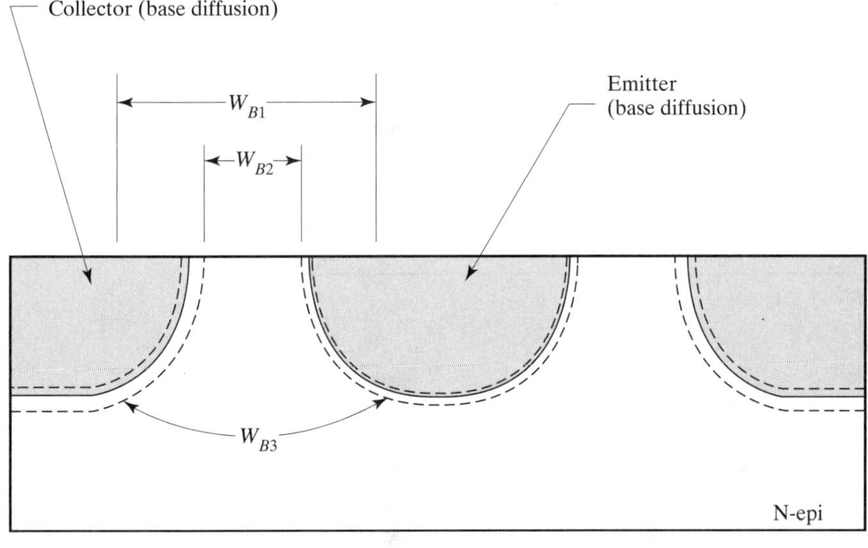

FIGURE 8.21 Cross section of a lateral PNP transistor depicting three different measures of neutral base width discussed in the text (drawn base width, W_{B1}, actual base width at the surface, W_{B2}, and effective base width beneath the surface, W_{B3}).

[20] D. E. Fulkerson, "A Two-Dimensional Model for the Calculation of Common-Emitter Current Gains of Lateral p-n-p Transistors," *Solid-State Electronics*, Vol. 11, 1968, pp. 821–826.

[21] K. N. Bhat and M. K. Achuthan, "Current-Gain Enhancement in Lateral p-n-p Transistors by an Optimized Gap in the n+ Buried Layer," *IEEE Trans. on Electron Devices*, Vol. ED-24, #3, 1977, pp. 205–214.

Designers have long known that the gain of a lateral PNP appears to increase when NBL is placed underneath the emitter. This increase occurs because the presence of the NBL improves the collector efficiency and therefore increases the collector current obtained for a given base current. This effect overwhelms any increase in base current caused by recombination within the NBL, as long as the difference in doping between the NBL and the N-epi suffices to repel minority carriers from the interface.

One might also expect some percentage of the emitted carriers to escape by traveling laterally beneath the collector diffusion. Experience shows that this does not occur to any appreciable extent while the transistor remains in the normal active region. The NBL updiffuses through the epi, establishing a retrograde doping profile that tends to drive the minority carriers upward along the built-in potential gradient. At the same time, the depletion region surrounding the collector reaches deep into the lightly doped N-epi and therefore intercepts a large fraction of minority carriers attempting to pass underneath it. Only a small percentage (usually less than 1%) of emitted carriers successfully escape the collector to reach the isolation sidewall. Transistors with shallow collector regions, such as analog BiCMOS lateral PNP transistors constructed from PMoat, may experience greater sidewall losses.

Construction of Small-Signal Lateral PNP Transistors

The lateral PNP transistor traditionally consists of a small plug of base diffusion surrounded by a larger annular base region (Figure 8.22). The central base plug serves as the emitter of the transistor, while the surrounding base ring serves as its collector. This structure ensures that almost all of the carriers injected by the emitter are intercepted by the intended collector before they can reach the isolation. The apparent reverse beta of this style of lateral PNP transistor is always far smaller than its forward beta. In the reverse mode, the outer ring of base becomes the emitter, and the majority of the injected carriers are injected toward the isolation sidewalls rather than to the small base plug in the center of the transistor. Thus, despite identical doping levels in emitter and collector, the lateral PNP transistor remains a profoundly asymmetrical device.

The circles and arcs used in constructing lateral PNP transistors are actually polygonal approximations. The number of segments in these polygons determines

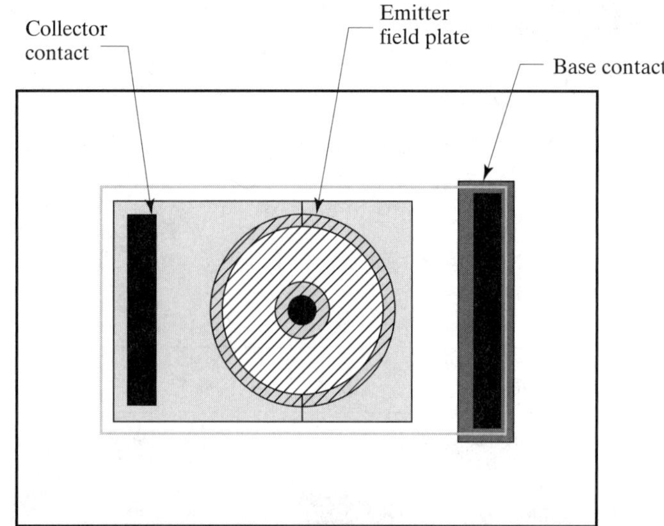

FIGURE 8.22 Layout of a lateral PNP transistor showing emitter field plate.

how closely they resemble ideal circles and arcs. The number of segments in the complete circle should be evenly divisible by four to ensure symmetry around both axes. Circular approximations with sixty-four sides are recommended for most applications. Annular geometries, such as the collector in Figure 8.22, usually consist of two matched halves to eliminate certain difficulties sometimes encountered during photomask generation.[22]

As stated previously, the peak beta of a lateral PNP depends inversely upon its effective base width. Carriers injected from the emitter sidewalls travel a shorter path than carriers injected from the bottom surface. Smaller-diameter emitters therefore have shorter effective base widths. Some authors prefer to quantify beta in terms of the emitter periphery-to-area ratio, rather than in terms of emitter diameter. Since these quantities are linearly related to one another, the two approaches are essentially equivalent.

Practical lateral PNP transistors usually employ a circular emitter geometry just large enough to contain a minimum-size contact (Figure 8.22). The emitter contact should be made circular and concentric with the emitter to ensure that all portions of the emitter periphery are equidistant from the contact edge. If they are not, some portions of the emitter periphery will inject an undue share of the carriers. This nonuniform current flow can cause subtle mismatches, especially in split-collector transistors. The collector of the lateral PNP generally consists of a square of base diffusion having a circular hole in its center. The emitter occupies the center of this hole. The drawn base width equals the difference between the radius of the emitter and the radius of the collector opening. One end of the collector extends sufficiently to allow placement of a contact inside it. Although debiasing occurs along the collector periphery, this only results in a slight increase in the effective saturation voltage of the transistor due to premature saturation of the debiased portion of the collector. If this increase in saturation voltage causes concern, then additional collector contacts can be added to reduce collector debiasing. For most applications, the additional contacts are not worth the space they consume.

The collector of the lateral PNP resides in the tank that forms its base. A strip of emitter diffusion added to one end of the tank serves as a base contact. A deep-N+ sinker is not normally required because the base current in the lateral PNP rarely exceeds a few tens of microamps. If space is at a premium, even a minimum-size base contact will suffice. On the other hand, a lateral PNP should always contain as much NBL as possible to minimize unwanted substrate injection. At an absolute minimum, the drawn NBL should entirely cover the drawn opening in the collector.

Figure 8.22 shows a metal plate covering the exposed portions of the N-tank between the emitter and the collector. This *field plate* prevents unwanted surface inversion or accumulation from occurring due to charge spreading and field interactions with the adjacent collector. Without the field plate, the beta of the lateral PNP fluctuates depending on surface potentials.[23] If a negative charge accumulates at the surface, then the minority carriers move toward the oxide interface and the beta of the transistor decreases. Similarly, a positive potential repels the carriers from the surface and increases the beta. The presence of mobile ions greatly increases the magnitude of the instabilities caused by these surface charges. The use of a field plate connected to the emitter not only stabilizes the beta of the transistor, but also helps to increase it by repelling carriers from the oxide-silicon interface.

---

[22] The boundary of a so-called *semisimple figure* becomes coincident with itself at one or more points, while the boundary of a *simple figure* does not. Some pattern generation algorithms have problems with semisimple figures, so they are often avoided.

[23] R. O. Jones, "P-N-P Transistor Stability," *Microelectronics and Reliability*, Vol. 6, 1967, pp. 277–283.

Lateral PNP transistors that lack field plates often exhibit collector-to-emitter leakage at voltages well below the thick-field threshold. These leakages result from parasitic channel formation. The geometry of the lateral PNP ensures that the parasitic channel will have a large W/L ratio, while the voltage differential present between collector and emitter attracts mobile ions and surface charges. Leakages have been observed in standard bipolar lateral PNP transistors at collector-to-emitter voltages of only 5 to 10 V on a process claiming a thick-field threshold in excess of 40 V. Properly designed field plates will completely eliminate these leakage currents. The field plate should connect to the emitter of the transistor and should entirely cover the exposed N-epi between the emitter and the collector. The field plate should overlap the periphery of the collector so that misalignment between the metal and the base diffusion cannot expose the epi surface. An overlap of 2 to 3 μm is more than enough, because the outdiffusion of the base helps shrink the size of the collector opening that the field plate must cover.

A minimum-emitter lateral PNP occupies considerably more space than a minimum-emitter NPN. A single lateral PNP can, however, be subdivided to form several smaller transistors sharing a common base and emitter. Figure 8.23A shows a simple example of such a *split-collector lateral PNP.* This transistor contains two partial collector segments, each occupying half of the emitter periphery. Since the emitter injects carriers uniformly in all directions, each collector receives half of the total injected current. This device therefore behaves as if it were actually two separate transistors, each having an effective emitter size one-half that of a normal lateral PNP. The transistor in Figure 8.23B carries the process of splitting the collector still farther. Instead of two half-sized collectors, this transistor contains four quarter-sized ones. One can even construct a split-collector transistor containing segments of different sizes. For example, the transistor in Figure 8.23C includes three 1/6 collectors and two 1/4 collectors.

The multiple collectors will match one another within about ±1% as long as they all possess identical geometries placed symmetrically.[24] The split collectors in Figures 8.23A and 8.23B meet these criteria and therefore match fairly precisely. The split

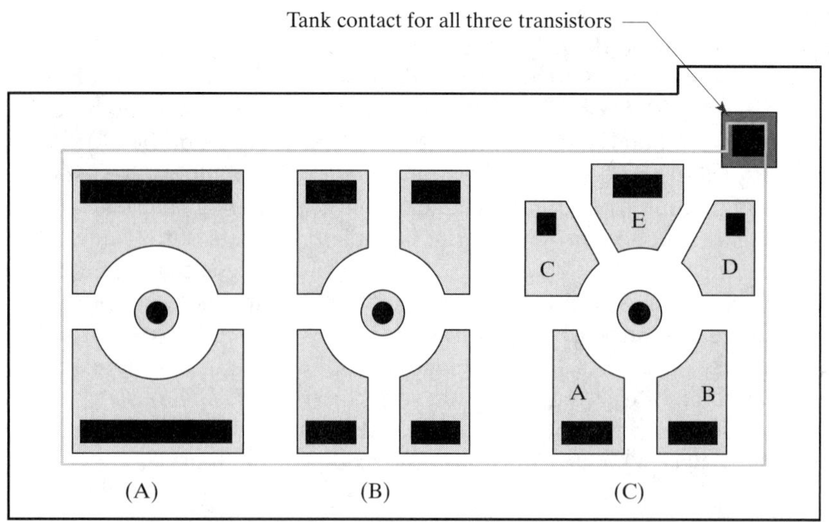

FIGURE 8.23 Examples of split collector transistors: (A) 1/2-1/2, (B) 1/4-1/4-1/4-1/4, (C) 1/6-1/6-1/6-1/4-1/4. The field plates have been omitted for clarity.

[24] B. Gilbert, "Bipolar Current Mirrors," in C. Toumazou, F. J. Lidgey, and D. G. Haigh, *Analogue IC Design: The Current-Mode Approach* (London: Peter Peregrinus, 1990), p. 250.

collectors in Figure 8.23C do not all have the same geometries and therefore do not all precisely match one another. Collectors A and B will match, as will collectors C and D. Collector E will not match collectors C and D because the former possesses a different geometry than the latter. Similarly, the ratio between collectors A and C will not exactly equal 2:3 because these segments have different geometries.

Split-collector laterals are frequently employed to construct current mirrors. A simple 1:1 current mirror can be constructed from a split collector transistor with two half-sized collectors. One collector connects back to the common base to form reference transistor Q_{1A} (Figure 8.24A). The other collector serves as the output transistor Q_{1B}. This arrangement saves considerable space, but it probably does not match as precisely as two separate lateral PNP transistors placed next to one another. Emitter degeneration cannot improve the matching of the split-collector mirror because the two transistors share a common emitter. Many schematic diagrams denote split collectors by placing multiple collector leads on a single base bar (Figure 8.24B). The ratio of the split collectors is indicated by values placed next to each collector lead. The passage of a base lead through the transistor does not indicate the presence of multiple base terminals, but rather indicates two separate connections to the same base terminal. Using this style of schematic representation, the circuit of Figure 8.24A reduces to the considerably more compact, if less familiar, schematic of Figure 8.24B.

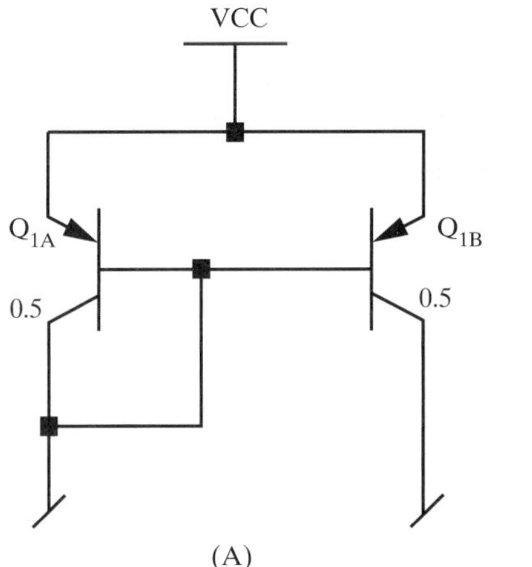

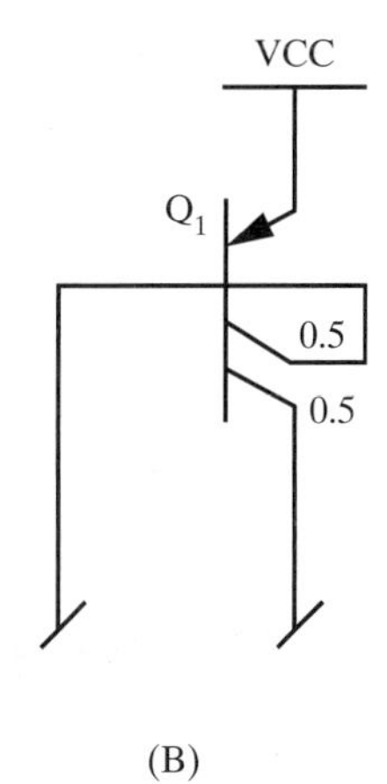

FIGURE 8.24 Schematic diagrams for 1:1 current mirrors constructed using split collector lateral PNP transistors: (A) conventional schematic diagram and (B) simplified schematic diagram.

Some designers prefer to digitize lateral PNP transistors by using a square emitter surrounded by a square collector ring. This style of lateral PNP is easier to digitize than the circular geometries previously discussed, but its base width increases slightly due to the greater length of the diagonal conduction paths and the poorer area-to-periphery ratio of the square emitter. Some designers fillet the four corners of the opening in the annular collector geometry so that the base width does not increase at the corners. Figure 8.25 shows examples of two types of square lateral PNP transistors.

The lack of radial symmetry in a square split-collector transistor prevents the use of anything except half and quarter collectors. Apart from this limitation, the same rules apply to the construction of square split-collector PNP transistors as to circular ones. In all cases, the emitter should be kept as small as possible and a field plate should entirely cover the exposed N-epi between the emitter and the collector.

Some designers claim that lateral PNP transistors scale with drawn emitter area, while others believe that they scale with drawn emitter periphery. Actually, the truth

FIGURE 8.25 Square geometry lateral PNP transistors: (A) minimum emitter and (B) 1/2-1/2 split collector. The field plates have been omitted for clarity.

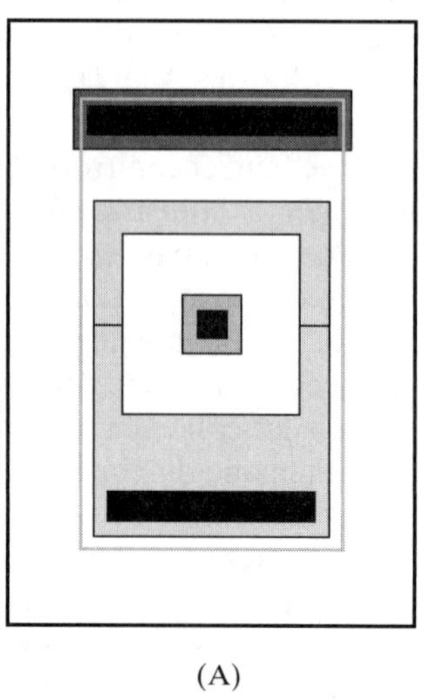
(A)

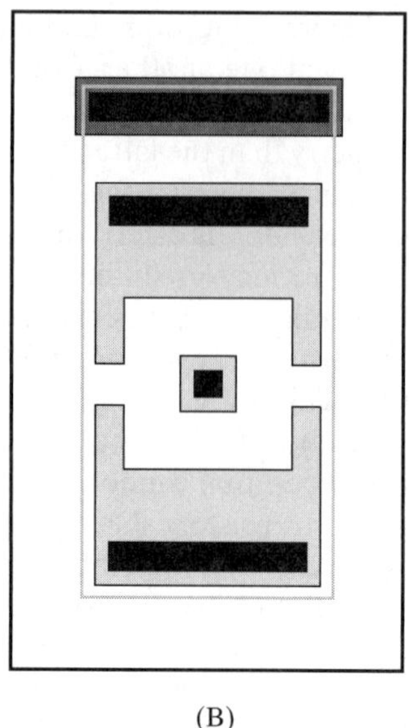
(B)

lies somewhere between these extremes, as some of the carriers are emitted from the emitter sidewalls and some are emitted from its bottom surface. The use of large square or circular emitters drastically reduces the beta of the transistor. An emitter elongated into a thin stripe can possess a very large area without much increase in effective base width (Figure 8.26A). The *elongated-emitter lateral PNP* has an emitter geometry reminiscent of the outline of a hot dog, so it is sometimes called a *hot-dog transistor.*

The area and periphery of the elongated-emitter transistor both increase at about the same rate, so arguments about whether the transistor scales with area or periphery are largely academic. An elongated-emitter transistor will not match a circular-emitter transistor because of junction sidewall effects, current crowding, high-level injection,

FIGURE 8.26 Higher current lateral PNP transistors: (A) elongated emitter or hot-dog transistor and (B) a small arrayed-emitter transistor.

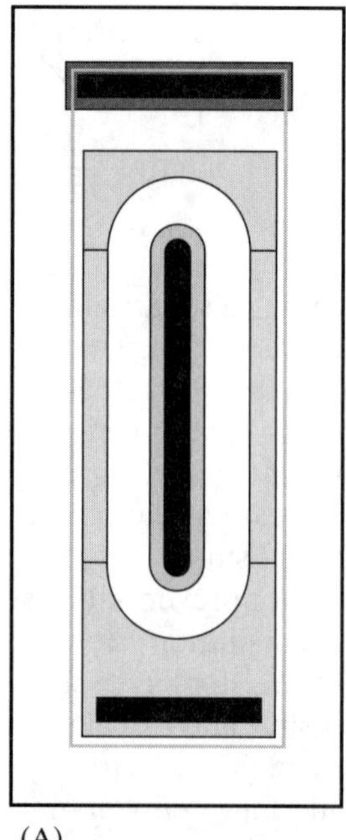
(A)

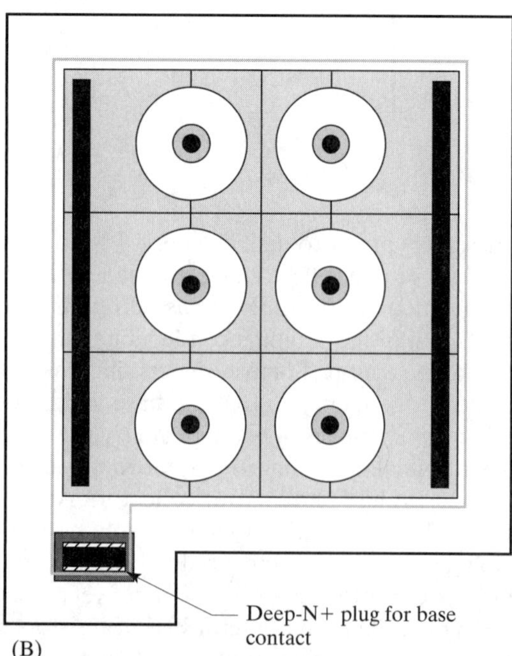

(B)

and various other sources of mismatch. The beta of the elongated-emitter transistor decreases slightly as its emitter lengthens, although this decrease is much smaller than it would be if the emitter geometry were, for example, an enlarged square. This decrease results from the contribution to the effective base width from carriers moving down the length of the emitter stripe. The transistor in Figure 8.26A has a relatively large collector resistance that may disturb transistor matching and reduce beta at low collector-to-emitter voltages. If this causes concern, then the collector contacts can be placed along the longer sides of the collector to reduce the collector resistance.

Larger lateral PNP transistors can also be produced by arraying a multitude of minimum-size emitters (Figure 8.26B). Because the geometry of each emitter cell is exactly the same as that of a minimum-emitter device, the current-handling capability of this *arrayed-emitter transistor* is directly proportional to the number of emitters it contains. Large devices of this type often stagger alternating columns of emitters to achieve a tighter hexagonal packing arrangement (Figure 9.12).

Although scaling by area and scaling by periphery both give approximately the same results when applied to the lateral PNP transistors in Figure 8.26, scaling by periphery extends more naturally to the case of split-collector laterals. Each split collector functions as a separate transistor whose size depends on the fraction of the emitter periphery it subtends. Split-collector transistors employing elongated emitters also follow the same principle. Most designers scale lateral PNP transistors by drawn-emitter periphery.

Assuming that lateral PNP transistors are scaled by periphery, they can either be assigned values in microns, or they can be assigned values that are normalized so that a minimum-sized emitter receives a size of one. Many designers prefer the latter scheme because it provides instantly recognizable values. A further complication arises when dealing with split-collector transistors: Does one assign a split collector a size based upon the fraction of the full circle it subtends, or does one assign it a size based upon the fraction of the drawn collector periphery facing the emitter that it represents? Most designers prefer the second approach because it usually produces simple integer fractions such as those given in Figures 8.23 and 8.24. This approach can be summarized by the formula

$$A_C = \frac{P_E}{P_{EU}} \cdot \frac{P_C}{\Sigma P_C} \qquad [8.5]$$

where A_C is the size assigned to a lateral PNP collector; P_{EU} is the perimeter of a minimum-sized, or unit, emitter; P_E is the perimeter of the actual emitter or emitters; ΣP_C is the sum of the perimeter of all collectors facing the emitter; and P_C is the perimeter of the collector under consideration that faces the emitter.

8.2.4. High-Voltage Bipolar Transistors

The maximum operating voltage of a process cannot exceed that of its weakest device. In standard bipolar processes, the vertical NPN transistor usually breaks down before either the lateral or the substrate PNP. The NPN V_{CEO} thus determines the maximum operating voltage. The V_{CEO} of a vertical NPN transistor is related to the avalanche voltage of the planar base-collector junction V_{CBOP} by the equation[25]

$$V_{CEO} = \frac{V_{CBOP}}{\sqrt[n]{\beta_{max}}} \qquad [8.6]$$

[25] A. Grove, *Physics and Technology of Semiconductor Devices* (New York: John Wiley and Sons, 1967), pp. 230–234.

where β_{max} represents the peak beta of the device and *avalanche multiplication factor n* usually lies in the range $3 < n < 6$. Low-beta devices have higher V_{CEO} ratings, but nominal betas below about 50 begin to restrict the usefulness of the device. Most processes therefore rely on a high base-collector planar breakdown voltage to obtain adequate V_{CEO} ratings. The width of the drift region determines V_{CBOP}, and thicker epi layers produce correspondingly higher breakdown voltages. The depth of the isolation diffusion must increase to keep pace with the epi thickness, but the other steps in the process remain the same. Manufacturers of standard bipolar processes usually offer several epi thicknesses corresponding to convenient operating voltages, such as 20 V, 40 V, and 60 V. If a process offers a choice of voltage ratings, use the lowest one possible because the higher voltages require larger isolation spacings.

Parasitic channel formation and charge spreading become increasingly serious concerns at higher voltages. Charge spreading may cause low levels of leakage at operating voltages somewhat below the thick-field threshold V_{TF}. Circuits operating at low currents or employing precision matching are especially prone to leakage problems and therefore require careful field plating and channel stopping. At voltages exceeding the thick-field threshold, metallization can directly induce parasitic channel formation, and complete field plates and channel stops become mandatory for proper circuit operation (Section 4.3.5). The exposed surface of the base region between the emitter and the collector of a lateral PNP transistor should always be field-plated regardless of operating voltages.

The breakdown voltage of any junction depends on its curvature: The sharper the curvature, the lower the breakdown voltage will be. All diffused and implanted junctions have a characteristic sidewall curvature. Deeper junctions have less sidewall curvature and therefore exhibit higher breakdown voltages. The effects of sidewall curvature can be quite dramatic. The observed breakdown voltage of a base-collector junction may equal only 60 V, even though the planar avalanche voltage exceeds 120 V. The observed breakdown voltages of shallow junctions often depend on geometry because the curvature of the corners of the diffusion exceeds the curvature of the sidewalls.[26] The observed breakdown voltages of such diffusions decrease slightly when the geometries contain acute angles. Deeper diffusions are less prone to this effect because outdiffusion rounds the corners off. Junctions with depths greater than 3 μm rarely show any significant reduction in operating voltage due to the presence of 90° vertices. Base diffusions with a junction depth of about 2 μm may experience a reduction of 5 to 10 V in breakdown voltage due to 90° vertices, and relatively shallow HSR diffusions may experience even larger reductions.

Designers must sometimes push the operating voltage of base or HSR diffusions close to their respective limits. Under such circumstances, the 90° vertices of rectangular geometries become liabilities. The designer must round off each such corner with a small arc, or *fillet*, to achieve the full operating voltage. The radius of these fillets should exceed the junction depth of the diffusion. Fillets having a radius of 5 μm serve admirably for both base and HSR diffusions. Both outside and inside corners should receive fillets. Circular geometries, such as fillets, sometimes produce false diagnostics during verification, so some designers use chamfers instead. A *chamfer* is a small diagonal facet drawn perpendicular to a line bisecting the vertex. Chamfers are not quite as effective as fillets because they still contain discernible vertices, although these are less acute than the original corner. If chamfers are used, their lengths should exceed the junction depth of the diffusion.

---

[26] C. Basavanagoud and K. N. Bhat, "Effect of Lateral Curvature on the Breakdown Voltage of Planar Diodes," *IEEE Electron Device Letters*, Vol. EDL-6, #6, 1985, pp. 276–278.

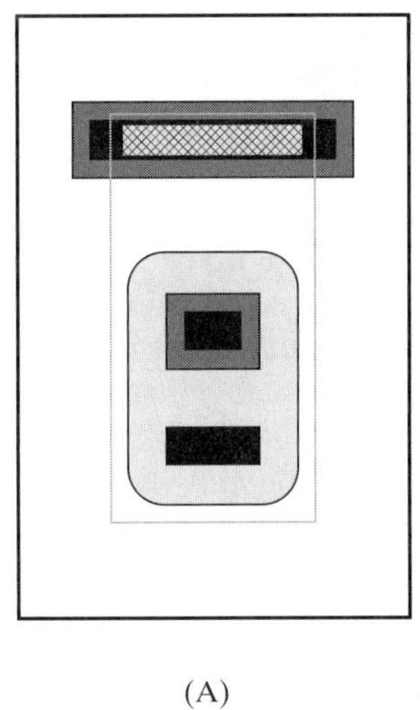

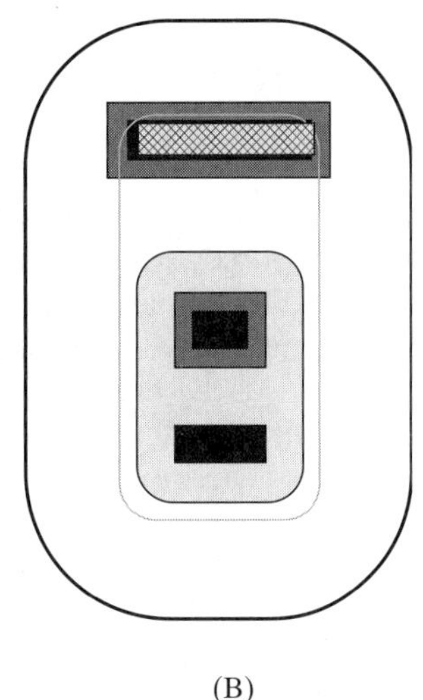

FIGURE 8.27 NPN transistors incorporating high-voltage fillets (A) on base only and (B) on base, NBL, and isolation.

Figure 8.27 shows two examples of NPN transistors that incorporate fillets. The transistor of Figure 8.27A incorporates fillets only on the corners of the base diffusion. These suffice to obtain the full V_{CBO} and are all that are necessary. Some designers also fillet the transistor tanks, although the isolation diffusion is so deep that these fillets have little or no effect. Up-down isolation may occasionally benefit from fillets because it uses shallower isolation diffusions. Figure 8.27B shows an NPN transistor incorporating fillets on base, NBL, and tank geometries. The larger fillets are customarily drawn concentric with the base fillets to maintain constant spacings.

High-voltage layouts often require increased spacings between certain diffusions to accommodate wider depletion region widths. Spacings that frequently require modification include base–base, HSR–HSR, base–HSR, base–iso, and HSR–iso. Other spacings that might require adjustment include collector–base, collector–iso, NBL–iso, deep-N+-iso, and deep-N+-base (where the collector is defined as the emitter region around the collector contact). The need for increased spacings for higher voltage diffusions causes some difficulties in verification. The simplest procedure, and one that will certainly produce a functional design, consists of applying the larger spacing rules to all diffusions. On the other hand, a considerable amount of space can be saved by applying the larger rules only to devices operating at higher voltages. This requires some means of distinguishing between high- and low-voltage geometries during design rule verification. One technique consists of drawing a geometry on a special layer around low-voltage devices to distinguish them from high-voltage ones. Some designers prefer to code a figure around high-voltage devices rather than around low-voltage ones, but this practice is not recommended. The inadvertent omission of a low-voltage marker will produce errors that become apparent during verification, while the omission of a high-voltage marker may go undetected because it does not produce any design rule violations.

Designers sometimes attempt to push the voltage ratings of devices beyond their specified limits by using special circuit topologies. The most common example of this practice consists of pushing the V_{CEO} of a vertical NPN beyond its rated maximum by ensuring the device's base terminal never sees a high-impedance state. In effect, the circuit designer relies on the V_{CER} rating of the transistor rather than the V_{CEO}. This practice can cause yield problems because V_{CER} ratings are rarely specified or controlled by the wafer fab. Transistors that are pushed beyond their rated V_{CEO} may also latch up during turnoff due to the snap-back of V_{CER} to $V_{CEO(sus)}$

during conduction. Whenever possible, one should use a higher voltage process rather than attempting to push the ratings of a lower voltage one.

8.2.5. Super-Beta NPN Transistors

The vertical NPN transistor fabricated in standard bipolar typically exhibits a peak beta of about 200. Although this represents a substantial current gain, it still proves inadequate for certain purposes. Consider the case of a low-input-current operational amplifier. If this amplifier uses an NPN input stage whose transistors operate at 50 μA each, then the input bias currents will equal 250 nA. Allowing for process and temperature variation, these bias currents could reach 500 nA. Currents of this magnitude are objectionable in the context of a low-input-current circuit. Techniques for canceling base currents exist, but they have proven less than completely successful due to mismatches and leakages.

As mentioned in Section 1.3.1, the beta of a bipolar transistor is inversely proportional to the Gummel number, which is defined as the integral of base doping along the line of carrier flow across the neutral base. In the case of an idealized vertical transistor having planar junctions and constant base doping, the Gummel number equals the product of base doping and neutral base width. Unfortunately, the Early voltage of a bipolar transistor is also proportional to the Gummel number, so any increase in beta comes only at the expense of a corresponding decrease in Early voltage. This implies that a transistor with an exceptionally high beta will have a miserably low Early voltage. Furthermore, since Early voltage merely quantifies the degree to which depletion regions intrude upon the neutral base, a transistor with a very low Early voltage will suffer punchthrough breakdown at a relatively low voltage.

Some standard bipolar processes offer an extension that produces an NPN transistor having an extremely high beta. Such devices are usually referred to as *super-beta transistors*. The process extension either employs an alternative base mask that produces a shallower base, or (more commonly) an alternative emitter mask that produces a deeper emitter. Since base doping diminishes with depth, both approaches reduce not only base width, but also base doping. Betas in excess of 5000 are possible,[27] but such devices exhibit Early voltages of a couple of volts and punchthrough voltages on the same order of magnitude. Therefore, super-beta transistors are useful only under very limited circumstances. Any circuit that uses them will also require normal NPN transistors. Super-beta transistors have now largely been supplanted by BiFET and BiCMOS alternatives, but less-extreme examples of high-gain bipolar transistors can be found in many modern processes.

Figure 8.28 shows cross sections of a standard NPN transistor and a super-beta transistor built by using a deeper emitter diffusion. As these cross sections illustrate,

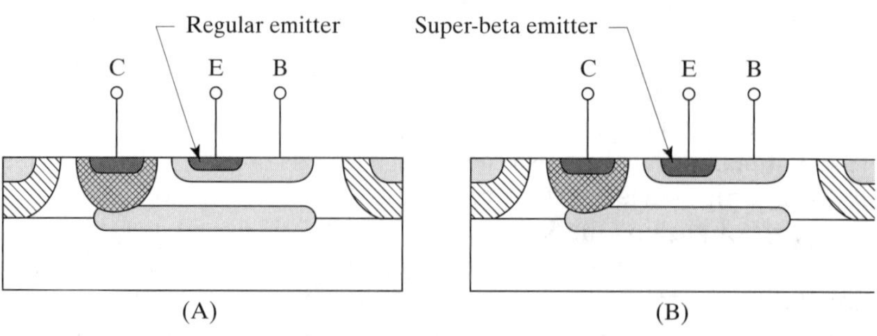

FIGURE 8.28 Comparison of representative cross sections of (A) a standard bipolar NPN and (B) a super-beta NPN.

[27] W. M. Gegg, J. L. Saltich, R. M. Roop, and W. L. George, "Ion-Implanted Super-Gain Transistors," *IEEE J. Solid-State Circuits*, Vol. SC-11, #4, 1976, pp. 485–491.

the only difference between these devices is the choice of emitter diffusion. The modified emitter results in an effective neutral base width of less than 0.1 μm. This very narrow base width is difficult to control in practice, so the beta, Early voltage, and punchthrough voltage of the transistor will all exhibit large variations. An emitter pilot will probably prove necessary to maintain adequate control of these parameters.

8.3 CMOS AND BICMOS SMALL-SIGNAL BIPOLAR TRANSISTORS

CMOS processes are optimized for constructing MOS circuits and therefore support only parasitic bipolar transistors. The performance of these devices generally leaves much to be desired. BiCMOS processes can produce transistors that rival or even surpass the performance of standard bipolar, but only at considerable cost. The advantages and disadvantages of various process options have been endlessly debated. Most current-generation analog processes fall into one of three categories: analog CMOS, power BiCMOS, and fast BiCMOS.

Analog CMOS processes are variants on digital CMOS that include minimal extensions, such as a silicide block mask, to support analog design. These processes do not include any features to support bipolar transistors, but they can always offer at least one parasitic bipolar transistor, typically a substrate PNP. Some processes also offer lateral PNP transistors and shallow-well NPN transistors. Analog CMOS processes are popular for applications that include large amounts of digital logic, and they have also been used to good effect for RF design.

Power BiCMOS processes must handle either high voltages or high currents (or both). These processes generally feature some form of DMOS transistor. They also include general-purpose CMOS transistors, but these devices are neither as small nor as fast as those fabricated by analog CMOS. Process extensions such as NBL and deep-N+ are routinely added to help control minority carrier injection. These features are also useful for constructing bipolar transistors. Most power BiCMOS processes offer reasonably adequate (if somewhat slow) collector-diffused-isolation NPN transistors. Many can also fabricate lateral PNP transistors. Some power BiCMOS processes offer other bipolar devices, such as extended-base transistors or DMOS transistors.

Fast BiCMOS processes focus upon constructing highly optimized bipolar transistors that combine extreme switching speeds with exceptional parametric performance. These processes usually support *complementary* bipolar transistors, which is to say that their NPN and PNP transistors have roughly comparable structures and performance. These transistors invariably use poly emitters, and many now employ at least partial oxide isolation. The latest generation of fast BiCMOS relies upon SiGe technology to further boost transistor speeds. These processes are used to achieve a combination of extreme speed and low signal distortion—for example, in high-speed amplifiers and certain types of line drivers.

8.3.1. CMOS PNP Transistors

Any N-well CMOS process can fabricate a substrate PNP transistor similar to that of Figure 3.31. This device uses an emitter fabricated of PMoat placed inside a base consisting of N-well contacted by an NMoat plug. The P-type substrate acts as the collector. A substrate PNP constructed on an older 10 V poly-gate CMOS process may achieve a beta in excess of 100. More advanced processes often have difficulty obtaining a beta of more than 5 or 10. Three factors account for this dismal performance: the use of shallow retrograde wells, the extreme thinness of modern source/drain implants, and the use of silicided moats.

The N-well regions of 10 V poly-gate CMOS were both deep and lightly doped. As the operating voltages of CMOS processes decreased, their wells became shallower and more heavily doped. The shallower wells were introduced to improve lateral spacings, while the heavier doping proved necessary to control channel length modulation and punchthrough. Although the shallower wells actually help minimize the vertical PNP's Gummel number, any benefit gained in this direction has been more than offset by increases in well doping. Many modern CMOS processes now employ retrograde wells that only compound this problem. These retrograde wells were introduced partly to reduce debiasing, partly to restrain downwards depletion, and partly to improve latchup immunity by killing substrate PNP beta.

Another prominent trend in CMOS development has been towards the use of thinner and thinner source-drain regions. This same trend has produced a corresponding decrease in bipolar transistor gains. This problem was first noticed in transistors that used large areas of contacts. Many designers therefore leapt to the conclusion that silicidation was the source of the beta degradation. In actuality, the problem stems from the presence of a contact (whether silicided or not) in close proximity to the emitter-base junction. Some carriers are injected from the base into the emitter. If the emitter is sufficiently thin, then these carriers will travel across it before they recombine. Any carriers that reach the contact interface will recombine almost instantaneously at this point. This drop in minority carrier concentration steepens the carrier gradient that drives diffusion and therefore increases the diffusion current of minority carriers flowing from base to emitter. Thus, the presence of a contact in close proximity to the base-emitter junction reduces the emitter injection efficiency. This *short-emitter effect* can drastically reduce the gain of a bipolar transistor.[28]

The short-emitter effect can be suppressed to a certain extent by minimizing the area of emitter contacts incorporated into the transistor. Two mechanisms are responsible for this improvement. First, silicidation actually consumes silicon as a reactant. Given the extreme thinness of modern source/drain implants, the depth of silicidation equals a large fraction of the emitter junction depth. Therefore, minimizing silicidation increases the average emitter junction depth. Second, the oxide-silicon interface has a much lower recombination rate than the silicide-silicon interface. Because of the short emitter effect, one usually obtains the highest gain from the BiCMOS transistor that incorporates the fewest possible emitter contacts. Larger transistors may require more than one contact to minimize emitter debiasing and contact electromigration issues.

Many advanced CMOS processes silicide the entire surface of the source/drain regions. These clad-moat processes typically exhibit extremely low substrate betas that are largely due to the short emitter effect. Betas of less than unity have been observed. If a clad-moat process includes a silicide block mask, then this layer can be employed to remove silicide from as much of the emitter as possible. This precaution can produce dramatic improvements in substrate PNP beta.

Processes that do not use retrograde wells may experience significant lateral base resistance, especially beneath the emitter. The low gain of these transistors greatly magnifies the impact of base resistance. Devices that exhibit near-unity gain are affected worst of all, as variations in beta produce large fluctuations in base current

[28] The author suggests this name as an analogy to the short-base diode. See R. S. Muller and T. I. Kamins, *Device Electronics for Integrated Circuits*, 2d ed. (New York: John Wiley and Sons, 1986), p. 238.

and therefore in base-emitter biasing. Drops of as little as a few hundred microvolts are of concern in matched devices. Transistors that may experience significant base debiasing should be operated with the lowest possible collector currents. Current densities of 1–10 nA/μm^2 are often used, but the designer must pay close attention to the impact of device leakage upon such small currents, especially at higher temperatures. The interdigitated layout of Figure 8.29A may also prove useful. This transistor uses long minimum-width strips of PMoat as its emitters, and interdigitates these between strips of NMoat. As few emitter contacts as possible are incorporated into the device, consistent with limiting emitter debiasing to a few hundred microvolts. A silicide block mask—if available—should be coded around each emitter to help minimize the short-emitter effect. Even with these precautions, collector current mismatches of ±3 to 5% are not uncommon in low-gain CMOS substrate transistors.

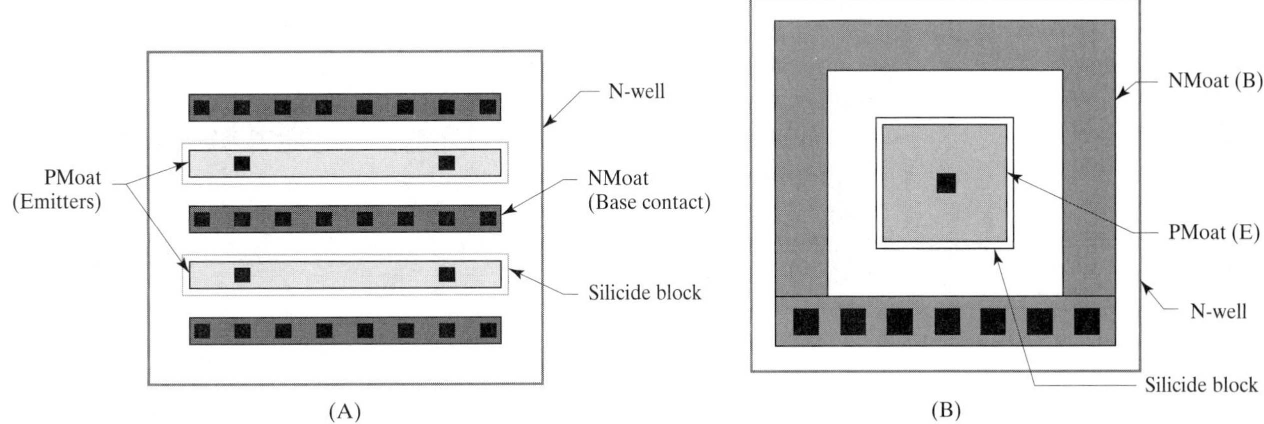

FIGURE 8.29 Layout of CMOS substrate PNP transistors using (A) interdigitated emitters and (B) a large emitter with a small contact.

If base debiasing is not a significant concern, then better performance may be achieved by using a large square emitter with a single small contact placed at its center (Figure 8.29B). The larger emitter encourages vertical injection over lateral injection and therefore increases the beta of the transistor. A silicide block geometry (if available) should be coded about the emitter. The base contact of the illustrated transistor consists of a ring of NMoat placed around the emitter. This precaution helps to minimize base resistance. If space is a concern, then either a single base contact or a pair of base contacts on opposite sides of the emitter may be employed instead.

The choice of whether to use an interdigitated emitter or a large square emitter will depend upon many factors, including the well doping profile and the impact of the short emitter effect. The only completely reliable way to access these factors consists of laying out both devices, fabricating them, and measuring their respective performances. However, as a general rule, the large emitter transistor will perform better if the process can achieve high gains, and the interdigitated emitter transistor will perform better at near-unity gains.

CMOS processes can also create a structure resembling the lateral PNP transistor of Figure 8.22. This structure will exhibit a low collector efficiency due to the lack of NBL. A typical device fabricated on a 10 V CMOS process might exhibit a collector efficiency of 0.1. This value is so low that the device cannot truly be called a lateral PNP. In fact, the gain exhibited by vertical conduction is likely to be greater than that exhibited by lateral conduction. This in turn suggests that the transistor could be improved by discarding the collector ring entirely, and by enlarging the emitter to encourage vertical carrier flow at the expense of lateral current flow. Unfortunately, the substrate PNP lacks versatility because its collector must always connect to substrate potential. Designers have therefore sought for better lateral PNP layouts.

Figure 8.30 shows an example of an improved lateral PNP transistor that uses an annular polysilicon gate as a field plate for the neutral base region. The gate is tied to the emitter so that the PMOS transistor remains off. Both the collector and the emitter of this device self-align to the field plate, producing a very short base width.[29] The emitter is contacted by a single contact at the center of a minimum-size plug of PMoat, over which a silicide block geometry has been coded. The short-emitter effect is further minimized by the fact that the lateral overlap of PSD over the contact greatly exceeds the vertical distance from the silicide to the source/drain junction.

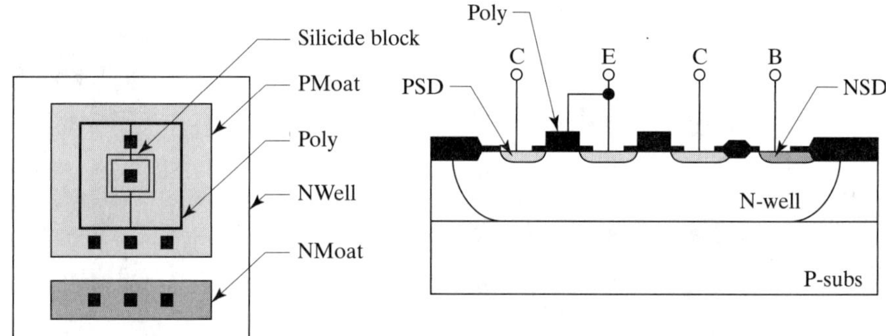

FIGURE 8.30 Layout and cross section of a lateral PNP transistor constructed in an N-well CMOS process.

Lateral PNP transistors constructed in a modern CMOS process can easily achieve gains of 50–100. These high gains are caused by a combination of high lateral beta due to the narrow base width and a dramatic improvement in collector efficiency. Much of the improvement in collector efficiency stems from the reduced dimensions of the emitter plug and the narrower spacing between emitter and collector, both of which favor lateral conduction over vertical conduction. Processes that employ heavy retrograde wells can benefit from minority carrier reflection from the retrograde interface. Collection efficiencies of greater than 0.9 are possible, providing that the retrograde well profile is suitably tailored. Processes that incorporate a retrograde shallow P-well that can be isolated from the substrate can fabricate lateral NPN transistors as well as lateral PNP transistors. Lateral NPN transistors with betas of up to 1000 and collector efficiencies of 0.99 have been reported.[30]

The narrow base width that allows the poly-plated lateral PNP to obtain its high beta also causes it to exhibit a relatively low Early voltage. Any increase in base width is inadvisable because the collector efficiency will drop off rapidly. This will in turn cause beta to drop much more quickly than Early voltage increases. Instead, the designer must rely upon circuit techniques—such as the use of cascodes—to balance the collector-to-emitter voltages seen by matched transistors and to limit the maximum collector-to-emitter voltage applied to any given transistor.

---

[29] E. A. Vittoz, "MOS Transistors Operated in the Lateral Bipolar Mode and Their Application in CMOS Technology," *IEEE J. Solid-State Circuits*, Vol. SC-18, #3, 1983, pp. 273–279.

[30] S. Verdonckt-Vandebroek, S. S. Wong, and P. K. Ko, "High Gain Lateral Bipolar Transistor," *International Electron Devices Meeting*, 1988, pp. 406–409.

8.3.2. Shallow-Well Transistors

Many processes fabricate two types of CMOS transistors: lower voltage transistors for core logic and higher voltage transistors for interface and analog circuits. A typical example of such a process might include 5 V and 15 V CMOS. Many such processes use multiple wells to accommodate the different transistors. The lower voltage transistors require shallower and more heavily doped wells than the higher voltage transistors. This usually leads to the use of three wells: a *deep N-well* for the higher voltage PMOS, a *shallow N-well* for the lower voltage PMOS, and a *shallow P-well* for the lower voltage NMOS.

Some processes offer a collector-diffused-isolation (CDI) NPN transistor that uses this shallow P-well implant as its base. The emitter of this transistor consists of a plug of NMoat contacted by a single contact at its center. This emitter resides inside a base that is contacted by a plug of PMoat. The collector consists of deep N-well and is contacted by a plug of NMoat. Figure 8.31 shows a transistor of this sort.

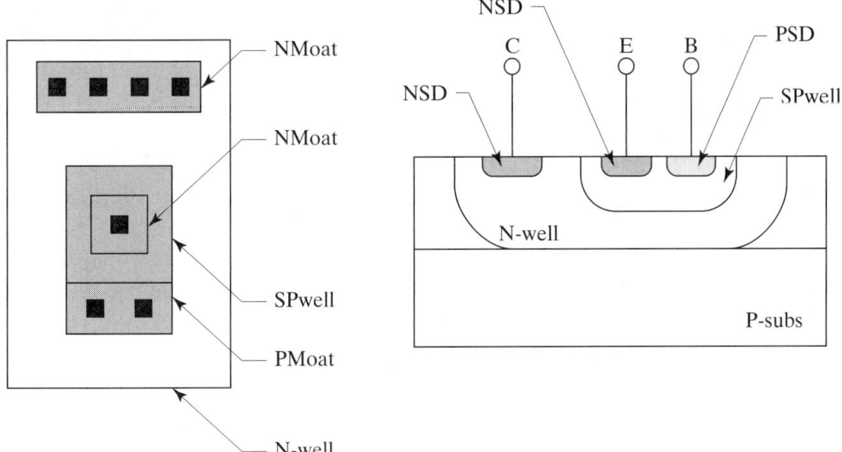

FIGURE 8.31 Cross section and layout of a shallow-well NPN transistor fabricated in a CMOS process.

The shallow-well transistor has a relatively thin and lightly doped active base region that consists of the area of deep N-well that has been counterdoped by the shallower and more heavily doped shallow P-well. This transistor therefore exhibits a relatively large beta. Practical examples of this structure on a 15 V analog BiCMOS process have achieved nominal gains well in excess of 100. The Early voltage of the shallow-well transistor is typically a little lower than that of an optimized CDI NPN that uses a separate base implant, and the base resistance is somewhat larger, but neither of these effects is sufficiently pronounced to impair the usefulness of the transistor.

Shallow-well transistors exhibit two major problems: base punchthrough and surface channel formation. The punchthrough issue arises because the lower portions of the deep N-well are usually quite lightly doped. As the bias applied across the collector-base junction increases, the depletion region rapidly depletes through the collector and reaches the underlying isolation. Punchthrough can be inhibited on BiCMOS processes by simply placing NBL beneath the transistor. A retrograde well would offer similar benefits, but most deep wells do not have retrograde profiles. Another possible means of limiting punchthrough consists of total oxide isolation; the buried oxide (BOX) will halt downwards depletion of the base-collector junction. Unfortunately, the effective base width of the transistor greatly increases when it fully depletes out. Whether or not punchthrough will cause problems on a given process depends upon the relative junction depths of the shallow P-well and the deep N-well, and their respective doping profiles. Some processes that include shallow wells cannot construct shallow well transistors.

Surface channel formation can also cause problems. The surface of the base consists of deep N-well counterdoped by shallow P-well. Depending upon the exact layout of the transistor, the base region may also include N-type channel stop. Add in the effects of boron suckup and phosphorus plow, and the surface of the base can become very lightly doped. Transistors with lightly doped base regions frequently exhibit parasitic MOS channel formation even in the absence of any obvious gate electrode. These parasitic channels cause collector-to-emitter leakage at higher collector-to-emitter biases. These parasitic channels can be suppressed by adding a channel stop, a field plate, or both. The channel stop consists of a ring of PMoat that also acts as the extrinsic base contact. The field plate consists of a plate of first level metal contacting the emitter. This metal plate should extend 1–2 μm over the PMoat ring wherever room exists. Even if a channel stop is added, the field plate must also be present in order to ensure that the area of the emitter remains constant. Without the field plate, the emitter might actually grow in size due to the inversion of the surface of the surrounding neutral base region.

Shallow-well transistors also tend to exhibit much larger collector resistances than vertical bipolar transistors constructed in standard bipolar or analog BiCMOS. The lack of both NBL and deep-N+ can easily cause the collector resistance to mount into the tens of kilohms. Such large resistances can produce a very soft transition from saturation to forward active operation (Figure 8.32). This effect not only causes high saturation voltages, but it also makes the transistor difficult to accurately model. The simple lumped collector resistance of standard SPICE models generally proves inadequate to cope with three-dimensional current flow in the collector. Collector resistance problems can be minimized by operating shallow-well transistors at low currents, typically no more than a few microamps. Shallow-well transistors constructed in analog BiCMOS should always incorporate NBL, and a deep-N+ plug should be added if the device must handle more than a few hundred microamps of current.

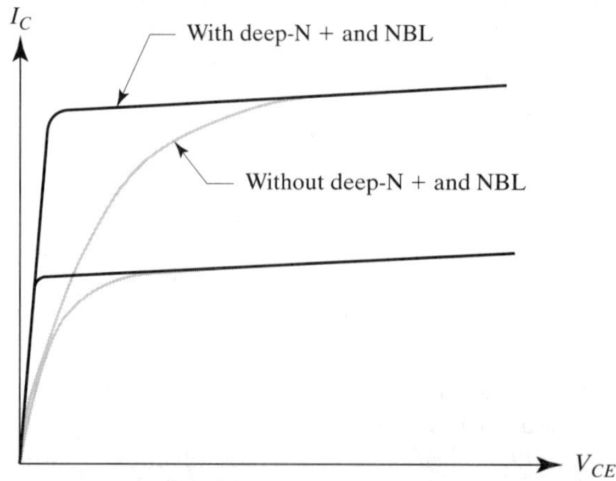

FIGURE 8.32 Comparison of saturation characteristics of a CDI NPN transistor with and without the addition of deep-N+ sinker and NBL.

8.3.3. Analog BiCMOS Bipolar Transistors

The relatively shallow junction depth of the N-well limits the operating voltage of the CDI NPN transistor to a typical value of 15 to 20 V. An alternative structure can provide higher voltages at the cost of reduced safe operating area and poorer Early voltage. Figure 8.33 illustrates the layout and cross section of this *extended-base NPN transistor.*

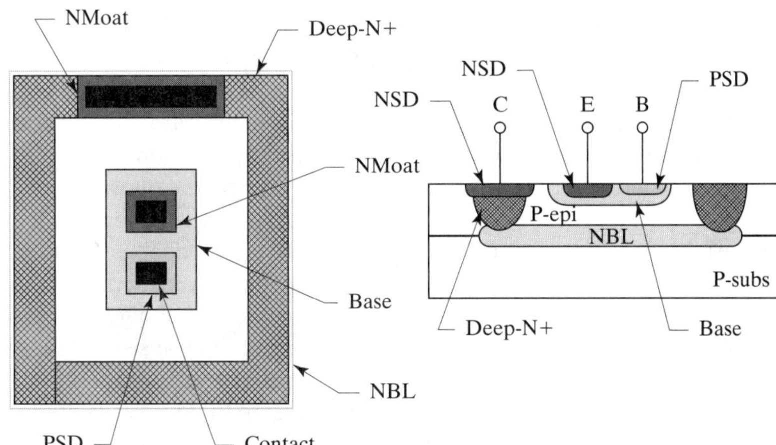

FIGURE 8.33 Layout and cross section of an extended-base NPN transistor.

Unlike the CDI NPN, the extended-base transistor does not employ N-well in its construction. The base of this transistor consists of a combination of a regular base diffusion and an isolated P-epi region, the latter lying sandwiched between the base diffusion and the underlying NBL. A ring of deep-N+ isolates the base from the surrounding P-epi and simultaneously allows contact to the NBL. The collector of this transistor consists of the NBL beneath its extended base and the deep-N+ ring surrounding it. The P-epi portion of the extended base acts as a drift region that greatly increases the operating voltage of this structure. The collector-base depletion region cannot penetrate far into the heavily doped NBL and instead moves upward through the lightly doped P-epi. Because the depletion region intrudes primarily into the neutral base rather than into the collector, the Early voltage of this structure is somewhat lower than that of the CDI NPN. Although the additional dopant in the P-epi increases the Gummel number of the extended base, the increase is smaller than one might expect because the P-epi is very lightly doped in comparison with the base diffusion. The base diffusion terminates the drift region and restricts the upward growth of the collector-base depletion region, preventing base punchthrough. Multiple-well processes often use P-well as a substitute for the base diffusion in these devices.

The extended-base structure has a higher planar base-collector breakdown voltage than the CDI NPN and therefore also exhibits a higher V_{CEO} rating, typically ranging from 40 to 60 V. The base diffusion can be eliminated entirely, resulting in an *epi-base transistor.* The elimination of the base diffusion greatly reduces the Gummel number of the device. The epi-base transistor offers a beta of several hundred at the cost of reduced operating voltage (due to punchthrough) and increased base resistance. One advantage of the epi-base transistor is that it does not require a separate base diffusion.

If the deep-N+ isolation ring is replaced by a similar N-well ring, the resulting device requires only one mask step that does not normally appear in the analog

CMOS process flow. The resulting transistors can handle only relatively small currents due to their high collector resistances, but they still offer much better low-current beta characteristics and device matching than any other bipolar transistors compatible with CMOS processing (Section 8.3.1). Unfortunately, pure CMOS processes rarely offer an NBL layer suitable for the construction of an epi-base device.

The analog BiCMOS substrate PNP (Figure 3.50) employs an emitter constructed of PSD rather than base, because the larger junction depth of the base diffusion reduces the punchthrough voltage of the transistor. The performance of the PSD substrate PNP roughly equals that of a substrate constructed in standard bipolar. Since substrate PNP transistors inject current into the substrate, the designer must take precautions to avoid substrate debiasing (Section 4.4.1).

Lateral PNP transistors constructed in analog BiCMOS exhibit surprisingly high peak betas. The relative shallowness of the base diffusion allows the emitter and collector of the transistor to be placed in close proximity to one another. Simultaneously, the graded nature of the well (aided by the presence of a phosphorus channel stop implant) helps increase the punchthrough voltage near the surface where the base is narrowest. Most of the minority-carrier injection occurs deeper in the transistor, where the built-in potential of the base-emitter junction decreases due to the graded nature of both the well and the base diffusion, so the presence of higher surface doping levels does not unduly increase the Gummel number of the transistor. The graded nature of the well—again aided by the phosphorus channel stop—generates an electric field that forces minority carriers down and away from the oxide-silicon interface. Carriers overcoming this electric field still experience relatively low levels of surface recombination due to the low surface state charge of (100) silicon. Finally, the small feature sizes possible with the shallow diffusions and superior photolithography of analog BiCMOS allow the construction of very small emitters (typically 5 μm in diameter). This increases the proportion of minority carriers injected from the emitter sidewall, while at the same time reducing the distance traveled by those carriers. All of these factors act in concert to raise the peak beta of the lateral PNP so much that it may exceed that of the CDI NPN. This high peak beta also helps extend the usable current range, allowing each minimum emitter to conduct as much as 100 μA while retaining a beta of 20. The smaller cell size of the analog BiCMOS lateral PNP enables the construction of very area-efficient power lateral PNP transistors even without the aid of a deep-P+ diffusion.

The size of the emitter has a strong effect on the beta of an analog BiCMOS lateral PNP, primarily because carriers emitted from the bottom surface of the emitter must travel farther than those emitted from the sidewalls. Additionally, the graded well doping generates a weak electric field that causes minority carriers to drift toward the NBL/N-well interface. This field prevents the interface from reflecting minority carriers toward the collector as efficiently as would otherwise occur. Worse yet, the NBL layers used in analog BiCMOS processes are often relatively lightly doped to minimize lateral autodoping, while the wells in low-voltage processes are more heavily doped to prevent punchthrough. The reduced doping difference decreases the built-in field at the NBL/N-well interface, allowing carriers to penetrate into the NBL, where they recombine or travel to the substrate. Analog BiCMOS lateral PNP transistors should always employ minimum-size emitters to ensure the highest possible gain. Larger transistors should use arrayed emitters rather than elongated ones for the same reason.

PSD implants can also form the emitter and collector of a lateral PNP transistor. The shallowness of the PSD implant diminishes the size of the transistor, but it also reduces the collector efficiency. The source/drain implants are so shallow that a substantial percentage of the minority carriers travel underneath them and escape to the sidewalls. This problem is exacerbated by the recessed thick-field oxide and N-type channel stop implants that shadow the PSD collector, and by the graded nature of the well that imposes a downward drift on minority carriers. If PSD laterals must be used, their collector efficiency can be increased by widening the collector or by ringing the transistor with deep-N+.

Lateral PNP transistors constructed in analog BiCMOS do not require base field-plating unless the operating voltage of the transistor exceeds the thick-field threshold. The presence of the phosphorus channel stop implant produces a built-in potential that repels minority carriers from the surface, and the use of (100) silicon minimizes surface recombination experienced by minority carriers that do reach the surface. Leakages and beta variations are therefore much reduced in analog BiCMOS. The elimination of base field plates helps to reduce the overall size of the transistor, making the lateral PNP a more attractive component.

8.3.4. Fast Bipolar Transistors

Many modern applications operate at gigahertz-level frequencies. Examples include RF transmitters and receivers such as those used in cellphones and satellite systems, wideband signal processing amplifiers such as those found in oscilloscopes, and digital line drivers. Bipolar transistors can theoretically switch at speeds of several tens of gigahertz. Unfortunately, none of the bipolar transistors so far discussed can achieve even a small fraction of these theoretical switching speeds. Four factors limit the performance of practical bipolar transistors: saturation, junction capacitance, base resistance, and base transit time. Each of these factors must be addressed in order to construct a truly fast bipolar transistor.

Saturation has the greatest impact upon bipolar switching speeds and therefore was historically the first factor to be addressed. Once a bipolar transistor saturates, the neutral base and collector become flooded with minority carriers. These carriers cannot be withdrawn from the transistor until they recombine or transit across the base-collector junction. The average time to recombination, known as the *minority carrier lifetime*, depends primarily upon doping. The collector drift region of a bipolar transistor is particularly wide and lightly doped. The lifetime of minority carriers in the collector typically exceeds a microsecond. This effectively limits the switching speed of a saturated bipolar transistor to a few megahertz.

Early bipolar logic used saturating bipolar transistors and therefore suffered from extremely sluggish switching speeds. Attempts were made to correct this problem by deliberately introducing recombination centers in the form of gold atoms. The additional recombination centers reduced minority carrier lifetimes and therefore improved switching speeds. Unfortunately, the reduced lifetimes also decreased beta and increased junction leakages. Furthermore, any furnaces used to process gold-doped material became contaminated with gold and were thereafter virtually useless for any other purpose. After the limitations of gold doping became painfully apparent, circuit designers began to alter the topologies of their circuits to eliminate saturating transistors. One of the earliest approaches placed Schottky diodes across the base-collector junctions (Section 8.1.4). So many other techniques have been developed that saturation no longer plays any significant role in restricting the speed of bipolar circuits.

Junction capacitance also has a very detrimental effect upon switching speed. The base-emitter, base-collector, and collector-substrate junctions of a vertical bipolar transistor all have their characteristic depletion capacitances. Each of these capacitances must be charged and discharged, at least to a certain extent, in order for the transistor to switch states. The base-collector capacitance is generally the most injurious because the charge it injects into the relatively high-resistance base circuit is magnified by the voltage swing on the collector (a principle known to circuit designers as the *Miller effect*).

Junction capacitances can be reduced by minimizing junction areas. Improved photolithography allows smaller drawn geometries, which translate directly into smaller junction capacitances and faster switching speeds. Further improvements can be achieved by redesigning the transistor to eliminate unnecessary overlaps and spacings. For example, the conventional transistor of Figure 8.34A requires the emitter diffusion to overlap the emitter contact to allow for misalignment. The *washed-emitter transistor* of Figure 8.33B eliminates this overlap, substantially reducing the size of the emitter. This in turn reduces the size of the base geometry.

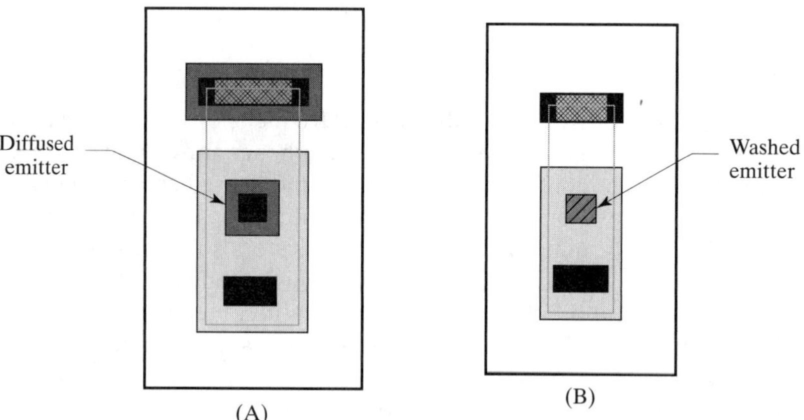

FIGURE 8.34 Comparison between (A) a conventional diffused emitter and (B) a washed emitter.

A *washed emitter* consists of an emitter diffusion self-aligned to a contact opening by means of a special etching technique. The emitter is deposited using a conventional deposition and drive during which a thin emitter oxide forms. The emitter regions remain covered by photoresist during the contact oxide removal. After all of the other contacts have been opened and the photoresist has been removed, a brief, carefully timed etch strips the thin emitter oxide and forms emitter contacts. Outdiffusion of the emitter under the original oxide opening provides just sufficient overlap of emitter over contact to prevent base-emitter shorts.[31] Although the washed emitter transistor is now primarily of historical significance, the same principles can be seen in modern polysilicon emitter transistors (Section 8.3.5). Oxide isolation (Section 8.3.6) also plays an important role in reducing junction capacitances in modern bipolar transistors.

Base resistance represents another significant obstacle to constructing high-speed bipolar transistors. The base-collector capacitance cannot be entirely eliminated, and any current that it injects into the intrinsic base must flow across the base resistance before it can be removed from the transistor by the external base drive circuit. The portion of the base resistance that lies beneath the emitter contributes the majority of the base resistance. One can therefore minimize base resistance most effectively by employing the smallest possible emitter. A washed emitter represents one method of achieving this goal. Transistors that include base contacts on both sides of the emitter, or even base contacts ringing the emitter, also

[31] Muller, *et al.*, pp. 306–307.

help to minimize the base resistance. The double-base transistor of Figure 8.17B is representative of these devices.

Once the problems posed by saturation, junction capacitance, and base resistance have been overcome, the speed of a bipolar transistor becomes a function of the time that minority carriers require to cross from the emitter to the collector. This quantity, called the *base transit time*, sets an upper limit on the speed that a bipolar transistor can hope to achieve. One can reduce base transit time by decreasing the thickness of the neutral base region. The Gummel number will remain unchanged if the base doping is increased in inverse proportion to neutral base width. The higher base doping also largely cancels the impact of neutral base width upon base pinch resistance. High-speed bipolar transistors therefore use the thinnest possible base regions, limited only by considerations of emitter injection efficiency, breakdown voltages, and dimensional control.

Once the base has been made as thin as possible, base transit times become a function of the minority carrier velocity. This velocity is in turn proportional to the mobility of the carriers in question. The mobility of electrons in silicon is almost three times the mobility of holes. Therefore, all other factors being equal, NPN transistors will be almost three times as fast as PNP transistors. Historically, this fact accounted for the emphasis placed upon NPN transistors in standard bipolar processing. Modern high-speed circuit design techniques require the use of both NPN and PNP transistors, so most fast bipolar processes are optimized to provide roughly equal performance from both types of transistors. The compromises implicit within a complementary process limit the performance of the NPN transistor, but the versatility provided by the PNP transistor more than compensates for any shortcomings of the NPN.

Once the base region has been made as thin as possible, further improvements in base transit time can only be achieved by increasing the velocity of the carriers themselves. Certain semiconductor materials have higher mobilities than silicon and can therefore achieve higher carrier velocities. The most famous of these materials is gallium arsenide, GaAs (popularly pronounced as "gas"). Gallium arsenide circuits have historically achieved several times the switching speeds of equivalent-generation silicon circuits. Advocates of GaAs design have therefore predicted that silicon would eventually be rendered obsolete as the material of choice for high-speed integrated circuit design. This never happened, because silicon processing technology advanced so rapidly and GaAs experienced so many difficulties in adapting silicon technology to its own peculiar requirements. Recently, silicon-germanium compound semiconductors have provided another way to increase carrier velocities (Section 8.3.7).

The quest for fast bipolar transistors has led to three major breakthroughs in transistor design: the polysilicon emitter, oxide isolation, and silicon-germanium processing. The next three sections will examine each of these three concepts in further detail.

8.3.5. Polysilicon-Emitter Transistors

Modern high-speed bipolar transistors require emitters with very small lateral and vertical dimensions. The lateral extent of the emitter sets the values of base pinch resistance, base-emitter capacitance, and base-collector capacitance, all of which are critical factors in determining switching speed. In order to use a thin base, one must also employ a shallow emitter to minimize variations in junction depth that would otherwise cause excessive variation in beta and breakdown voltage. The washed emitter represents an early attempt to fabricate a small shallow emitter.

Several factors conspire to limit the shallowness of a conventional emitter diffusion. First, silicidation consumes some thickness of silicon. If the emitter is made too shallow, then the silicide will punch through it. Second, the short emitter effect degrades the beta of a shallow emitter transistor unless the emitter doping is raised to compensate for the decreased emitter thickness. Solid solubility limits the maximum possible emitter doping and therefore places a lower bound upon emitter junction depth. Third, the lateral outdiffusion beneath a shallow washed emitter may become insufficient to provide the necessary process margin for the washed etch.

The problems posed by conventional emitter diffusions ultimately led to the development of the *polysilicon emitter*, or *poly emitter*. Figure 8.35 shows an example of a CDI NPN transistor fitted with a polysilicon emitter. This transistor is processed normally up through the completion of the base drive. Next, an oxide removal defines the extent of the drawn emitter. A layer of arsenic-doped polysilicon is deposited and patterned over the exposed emitter opening. A brief period of heating causes arsenic to diffuse from the polysilicon into the exposed monocrystalline silicon, producing an extremely thin and heavily doped emitter diffusion that self-aligns to the emitter oxide removal. The poly thus serves as both doping source and contact to the true emitter diffusion that lies just beneath it.

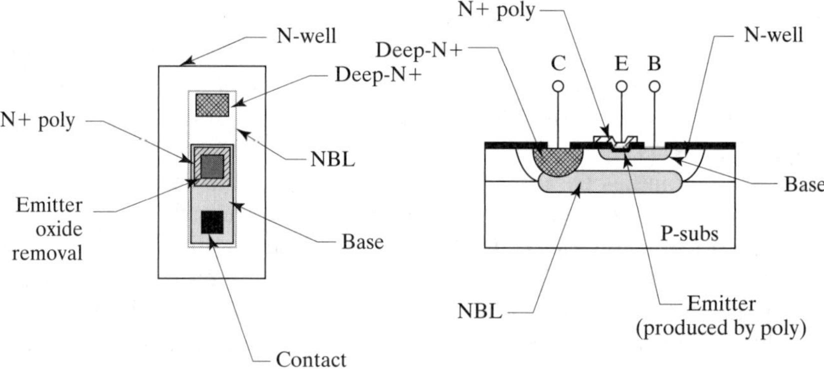

FIGURE 8.35 Layout and cross section of a CDI NPN transistor with a polysilicon emitter.

Polysilicon emitters provide all of the advantages of washed emitters with none of the corresponding disadvantages. The poly emitter self-aligns to the contact opening in the same manner as a washed emitter. However, the poly emitter does not suffer any subsequent etching; therefore, the emitter overlap of contact does not diminish. A shallow poly emitter is thus much less prone to lateral punchthrough than a washed emitter. Similarly, a poly emitter does not suffer any erosion of the silicon surface because silicide does not contact this surface. Therefore, the poly emitter also does not suffer from silicide punchthrough. Best of all, the interface between monocrystalline and polycrystalline silicon does not experience the greatly heightened recombination rates that cause the short emitter effect. In fact, the poly interface seems to actually impede the passage of minority carriers. This has the unexpected benefit of raising the emitter injection efficiency of the poly emitter above that of a corresponding deep conventional emitter. Poly emitter transistors thus exhibit betas of up to six times those of equivalent deep-emitter transistors.

Various mechanisms have been proposed to explain the impermeability of the poly interface to minority carriers. The presence of grain boundaries within the polysilicon may play a role in reducing minority carrier mobility. However, a more critical factor appears to be the nature of the interface itself. A thin layer of native oxide will form on any exposed silicon surface due to a room-temperature reaction between oxygen and silicon. Unless special precautions are taken to remove this *native oxide* before poly deposition, a layer of oxide will remain sandwiched between the poly and the monocrystalline silicon. This interfacial oxide is incredibly thin, typically consisting of no more than a couple of atomic layers. This interfacial oxide apparently reacts with the polysilicon. The exact nature of the resulting interfacial oxide appears to have a critical effect upon poly emitter injection efficiency.[32] Whatever the nature of the mechanisms at work at the poly interface, they are effective at containing both polarities of minority carriers. Therefore, both NPN and PNP transistors can benefit from the use of poly emitters.[33]

The increased emitter injection efficiency of a polysilicon emitter transistor enables the use of a more heavily doped base region to reduce base resistance. Furthermore, the depth of the emitter junction produced by the polysilicon emitter process can be controlled with great precision. These factors allow the use of a much thinner base than might otherwise be possible, and the resulting reduction in neutral base width translates into a smaller base transit time and a much faster transistor. The thinner base and emitter also allow use of a thinner epi, drastically reducing the outdiffusion of deep-N+ and N-well and therefore greatly shrinking the overall size of the transistor.

Polysilicon emitters do have certain drawbacks. Dangling bonds at the poly interface are partially passivated by hydrogen incorporation during fabrication. If the base-emitter junction of the transistor is operated at even a fraction of the avalanche voltage, then hot carriers may desorb these hydrogen atoms and regenerate the dangling bonds. These dangling bonds increase emitter recombination and decrease beta. The extremely close proximity of the poly interface to the base-emitter junction explains the heightened vulnerability of poly emitter transistors to this effect, known as avalanche-induced beta degradation (Section 4.3.3). Some poly transistors also exhibit permanent increases in beta when operated at medium-to-high forward currents, apparently because these currents mobilize hydrogen atoms to move through the silicon lattice to the emitter interface, where they passivate dangling bonds. Avalanche-induced beta degradation can be avoided by ensuring that the poly emitter transistors do not experience any significant reverse bias across the emitter-base junction. Forward-conduction beta instabilities can be minimized by increasing the doping of the polysilicon, which decreases the concentration of dangling bonds at the interface.[34] The poly interface is relatively fragile and can be readily damaged by overheating, such as may occur during electrical overstress or electrostatic discharge events. This vulnerability can be minimized by proper circuit design.

The advantages of polysilicon emitters clearly outweigh their disadvantages. Virtually all modern high-speed transistors employ some form of polysilicon emitter to provide the smallest possible emitter geometry while simultaneously enabling the use of an extremely thin and relatively heavily doped neutral base region. The oxide-isolated transistors and SiGe transistors discussed in the next two sections build upon the advantages provided by poly emitters.

---

[32] Z. Yu, B. Ricco, and R. Dutton, "A Comprehensive Analytical and Numerical Model of Polysilicon Emitter Contacts in Bipolar Transistors," *IEEE Trans. on Electron Devices*, Vol. ED-31, 1984, pp. 773–784.

[33] C. M. Maritan and N. G. Tarr, "Polysilicon Emitter p-n-p Transistors," *IEEE Trans. on Electron Devices*, Vol. ED-36, #6, 1989, pp. 1139–1143.

[34] J. Zhao, G. P. Li, K. Y. Liao, M.-R. Chin, J. Y.-C. Sun, and A. La Duca, "Resolving the Mechanisms of Current Gain Increase Under Forward Current Stress in Poly Emitter n-p-n Transistors," *IEEE Electron Device Letters*, Vol. 14, #5, 1993, pp. 252–254.

8.3.6. Oxide-Isolated Transistors

Partial or complete oxide isolation can significantly reduce junction capacitance, which in turn enables faster transistor switching. If the epi layer is sufficiently thin, then a conventional LOCOS oxide can punch entirely through it to separate adjacent tanks without the need for isolation diffusions. The elimination of isolation diffusions not only reduces collector-substrate capacitance by eliminating the isolation sidewall capacitance, but also shrinks tank dimensions. Figure 8.36A shows the layout and cross section of an NPN transistor in which the base diffusion abuts the LOCOS isolation. Figure 8.36B shows a more radical structure in which the emitter also abuts the oxide termination. Such *walled-emitter* structures generally require a more abrupt termination of the oxide region than conventional LOCOS processing can provide. Various modifications of LOCOS have been proposed that reduce the width of the bird's beak.[35] Modern processes generally replace LOCOS with *shallow trench isolation* (STI), which can produce narrow isolation regions with nearly vertical sidewalls. STI processes can produce extremely small transistor structures similar to those of Figure 8.36.

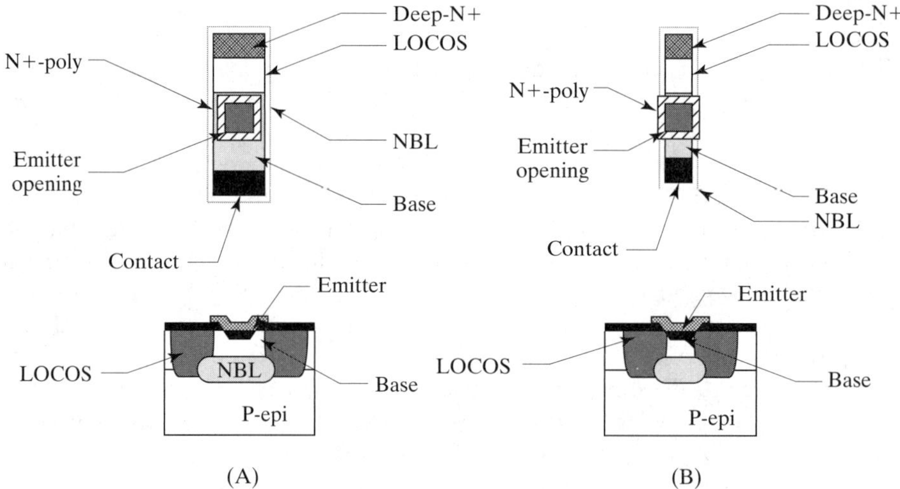

FIGURE 8.36 Partial oxide-isolated NPN transistors: (A) conventional, and (B) walled-emitter.

The walled emitter transistor of Figure 8.36B has not been widely accepted because of difficulties with the fabrication of the base sidewalls. The enhanced recombination rates caused by dangling bonds at the oxide/silicon interface have proved troublesome. Some transistors have also experienced channel formation down this sidewall region. Thus the somewhat less radical structure of Figure 8.36A has become the standard for most modern high-speed transistors.

The principal defect of the transistors of Figure 8.36 is their relatively large separation between base contact and emitter. This separation increases both the base resistance (because the base current must flow farther to reach the contact) and the base-collector capacitance (because of the increased area of the base-collector junction). These defects can be remedied by using a second layer of polysilicon to form a self-aligned base contact. The resulting structure is called a *double polysilicon self-aligned transistor*. Figure 8.37 shows a representative device of this type.

---

[35] K. Y. Chiu, J. L. Moll, and J. Manoliu, "A Bird's Beak Free Local Oxidation Technology Feasible for VLSI Circuits Fabrication," *IEEE Trans. on Electron Devices*, Vol. ED-29, #4, 1982, pp. 536–540.

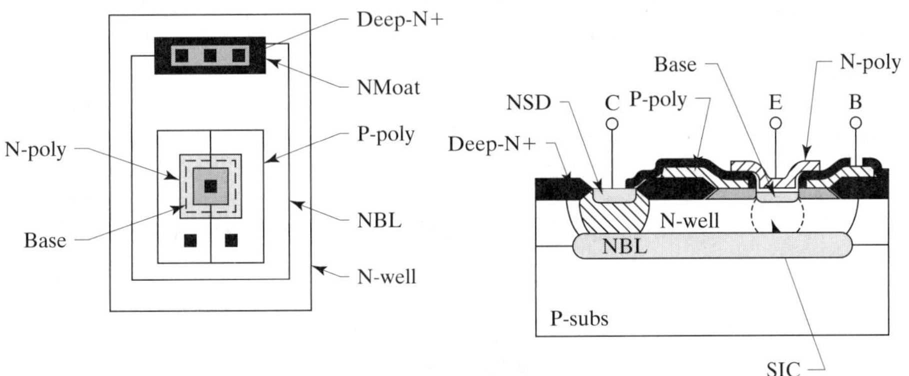

FIGURE 8.37 Layout and cross section of a double polysilicon self-aligned transistor[36].

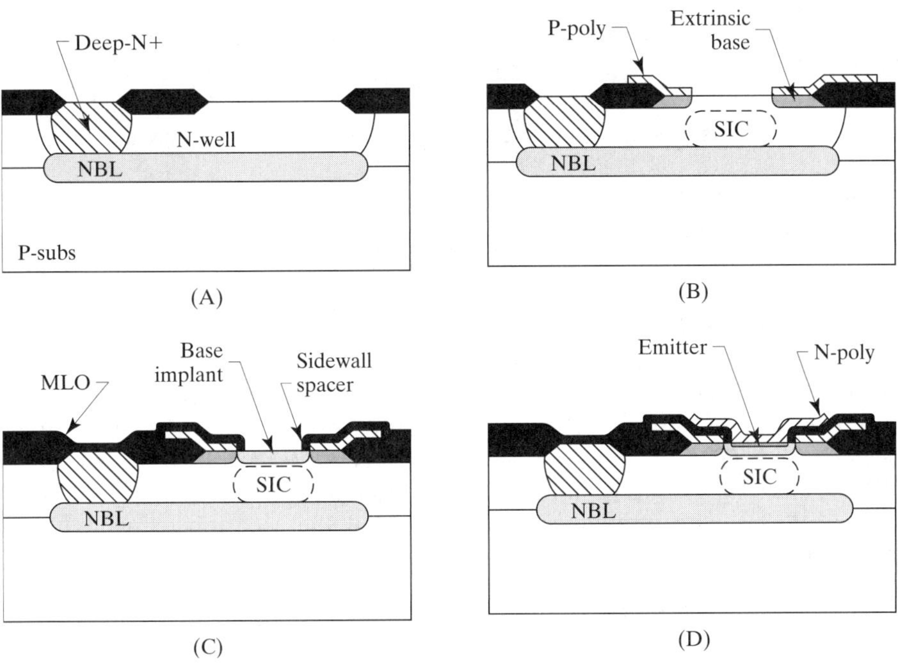

FIGURE 8.38 Key steps in fabricating a dual-poly self-aligned transistor.

This transistor begins with the formation of an N-type buried layer and the growth of an epitaxial layer. An N-well is implanted, followed by a deep-N+ plug. LOCOS oxidation then forms the field oxide. (Or alternatively, one can use trench isolation for this purpose.) Figure 8.38A shows a cross section of the resulting structure. A layer of P+ polysilicon is then deposited and etched to form the base contact. This P+ poly touches the monocrystalline silicon about the edges of what will later become the base region. A high-energy N-type implant is now conducted through the poly opening to form the *self-aligned implanted subcollector* (SIC). This implant produces an increase in collector doping below the actual base of the transistor to help minimize a phenomenon called *base push-out* (Section 9.1.1). Base push-out would otherwise widen the effective neutral base width at high currents, leading to an increase in base transit time and slower switching speeds. Next,

[36] T. H. Ning, "History and Future Perspective of the Modern Silicon Bipolar Transistor," *IEEE Trans. on Electron Devices*, Vol. 48, #11, 2001, pp. 2485–2491.

a brief high-temperature anneal causes boron to diffuse from the poly into the underlying monocrystalline silicon to form a heavily doped extrinsic base region (Figure 8.38B).

A shallow P-type implant performed through the P+ poly opening creates the intrinsic base. This implant sets the doping in the neutral base region of the transistor. The intrinsic base implant overlaps with the P+ extrinsic base that outdiffuses from beneath the surrounding P+ poly ring. Once the base implant is complete, a sidewall spacer must be fabricated along the inside edge of the hole in the P+ poly (Figure 8.38C). This sidewall spacer separates the P+ poly from the N+ poly that is deposited next. The N+ poly forms the polysilicon emitter of the transistor. A brief heat treatment causes arsenic to diffuse from the N+ poly into the underlying monocrystalline silicon. The resulting shallow N+ region acts as the true emitter of the device (Figure 8.38D). A conventional metal system is then deposited atop the completed transistor. Many variations of this process have been described in the literature.[37]

The transistor of Figure 8.37 can readily be adapted to a fully isolated process. All that must be done is to fabricate the initial buried layer within a thin silicon layer placed atop a buried oxide. This step will reduce the collector/substrate capacitance, but otherwise has no impact upon the functioning of the overall device.

8.3.7. Silicon-Germanium Transistors

The development of the high-speed silicon bipolar transistor reached its apparent conclusion in the double-poly self-aligned transistor described in the previous section. Further increases in speed would require some means of increasing the velocity of carriers crossing the neutral base. The only obvious way to achieve this goal was the use of compound semiconductors, but all known compound semiconductors were incompatible with conventional silicon processing. This difficulty was eventually resolved by the development of *ultrahigh-vacuum chemical vapor deposition* (UHVCVD) in the late 1980s.[38] UHVCVD can deposit one semiconductor material on top of another without introducing objectionable contaminants between the two. This process enabled the creation of a novel IV-IV compound semiconductor consisting of a mixture of silicon and germanium, called SiGe (pronounced "siggy").

Suppose that two layers of SiGe have the same doping but different germanium concentrations. A net contact potential will appear between these two layers due to differences in their lattice structures (or more specifically, differences in their bandgap energies). UHVCVD can create layers of SiGe that have continuously varying germanium concentrations. The germanium gradients produced in this way can be used to build electric fields into the base of a bipolar transistor. These fields can either be used to increase the velocity of minority carriers traversing the base, or to minimize the back injection of majority carriers into the emitter, and so increase emitter injection efficiency. By properly tailoring the germanium profile, one can construct a bipolar transistor that has a combination of desirable properties: faster base transit times, higher betas, and higher Early voltages. Many of the most advanced bipolar transistors therefore use deposited SiGe base layers.

---

[37] S. Konaka, Y. Yamamoto, and T. Sakai, "A 30-ps Si Bipolar IC Using Super Self-Aligned Process Technology," *IEEE. Trans. on Electron Devices,* Vol. ED-33, 1986, pp. 526–531. See also T. Y. Chiu, G. M. Chin, M. Y. Lau, R. C. Hanson, M. D. Morris, K. F. Lee, M. T. Y. Liu, A. M. Voschenkov, R. G. Swartz, V. D. Archer, S. N. Finegan, and M. D. Feuer, "The Design and Characterization of Nonoverlapping Super Self-Aligned BiCMOS Technology," *IEEE Trans. on Electron Devices,* Vol. 38, #1, 1991, pp. 141–150.

[38] D. L. Harame, and B. S. Meyerson, "The Early History of IBM's SiGe Mixed Signal Technology," *IEEE Trans. on Electron Devices*, Vol. 48, #11, 2001, pp. 2555–2567.

Figure 8.39 shows the principle steps in the fabrication of an exemplary SiGe NPN transistor.[39] This structure employs LOCOS and diffused isolation rather than the more advanced trench isolation characteristic of most advanced processes, but it is illustrative of the techniques used to fabricate SiGe. Fabrication begins with the formation of an N-buried layer and growth of epitaxial silicon. A deep-N+ sinker provides a low-resistance connection to the collector. LOCOS field oxidation defines a window where the base and emitter of the transistor will form, plus a similar window over the collector contact. Polysilicon is deposited over the entire transistor and is etched away to form a base window. A high-energy implant performed through this window forms a self-aligned implanted subcollector. The thin oxide beneath the base window is then stripped to expose bare silicon (Figure 8.39A). UHVCVD deposition forms a layer of boron-doped SiGe that is subsequently stack-etched along with the underlying polysilicon. A layer of oxide followed by a layer of nitride are deposited on top of the SiGe (Figure 8.39B). The emitter window is etched through these two layers, and arsenic-doped polysilicon is deposited to form the poly emitter. This polysilicon layer is etched away, stopping on the silicon nitride deposited underneath (Figure 8.39C). Sidewall spacers are then added and contacts are formed to the base SiGe and emitter polysilicon. Figure 8.39D shows the completed structure, minus the conventional metal system.

A SiGe transistor such as the one illustrated in Figure 8.39 is sometimes called a *heterojunction bipolar transistor* (HBT). This term is something of a misnomer when applied to conventional SiGe transistors because it implies that their most important feature is a step change in composition at the emitter-base junction. Such a step change would be primarily useful in improving the emitter injection efficiency.

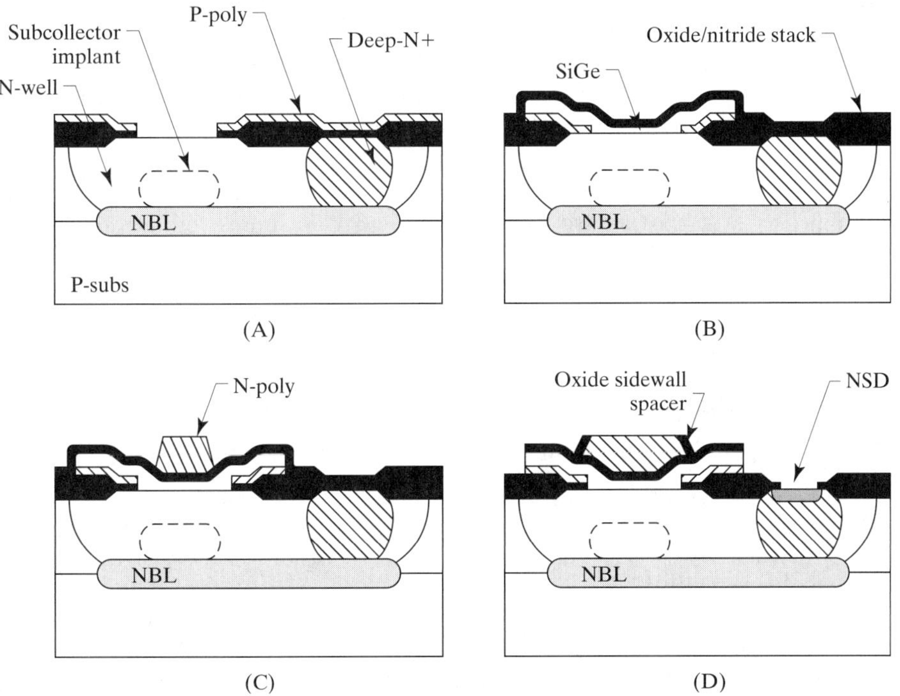

FIGURE 8.39 Steps followed in fabricating a SiGe bipolar transistor.

[39] A. Chantre, M. Marty, J. L. Regolini, M. Mouis, J. de Pontcharra, D. Dutartre, C. Morin, D. Gloria, S. Jouan, R. Pantel, M. Laurens, and A. Monroy, "A high performance low complexity SiGe HBT for BiCMOS integration," *Proc. IEEE Bipolar/ BiCMOS Circuits and Technologies Meeting*, 1998, pp. 93–96.

Instead, conventional SiGe transistors normally concentrate upon decreasing the base transit time by imposing a drift force upon minority carriers within the base. This choice implies that SiGe transistors would be better termed graded-base transistors than heterojunction transistors.

Current-generation SiGe transistors have exhibited peak switching speeds of over 50 GHz. These speeds are still somewhat short of what other compound materials such as GaAs can provide, but SiGe provides two crucial advantages over all other compound materials. First, SiGe is fully compatible with conventional silicon processing. The transistor of Figure 8.39 illustrates this compatibility, as it very closely resembles the double-polysilicon self-aligned transistor of Figure 8.37. Second, SiGe adds relatively little additional cost to the existing advanced bipolar transistor flow. These two advantages make SiGe bipolar a strong contender for high-speed large-scale integrated circuits.

8.4 SUMMARY

Bipolar transistors are extremely versatile devices, but they have their limitations. Improperly designed bipolar transistors may suddenly fail under heavy loads due to thermal runaway or secondary breakdown. Saturating bipolar transistors can also inject current into the substrate, debiasing adjacent circuitry and causing catastrophic latchup failures. These problems have caused circuit designers much anguish, but they can almost always be overcome by proper circuit design and device layout.

Bipolar transistors fall into one of two general categories: small-signal transistors and power transistors. Small-signal transistors are optimized for dense packing rather than for power handling capability. These devices are primarily used in analog signal processing and control circuitry, where they are particularly prized for their high transconductance and superior device matching. Standard bipolar processes offer relatively high-performance vertical NPN transistors plus several varieties of somewhat less useful PNP transistors. BiCMOS processes generally incorporate a similar selection of bipolar devices. The increasing popularity of analog BiCMOS ensures that bipolar transistors will remain an important part of analog circuit design for the foreseeable future.

The layout of bipolar transistors can be tailored to suit specific applications. Bipolar power transistors are frequently designed for improved immunity to thermal runaway and secondary breakdown. Small-signal transistors are frequently laid out to minimize device mismatches. The following chapter examines the techniques used to create these optimized bipolar transistor layouts.

8.5 EXERCISES

Refer to Appendix C for layout rules and process specifications.

8.1. What is the magnitude of the thermal voltage V_T at −55°C? At 125°C?

8.2. A lateral PNP transistor exhibits an emitter current of 110 μA, a collector current of 98 μA, and a base current of 7 μA. What is the transistor's current gain and its collection efficiency?

8.3. Select base-ballasting resistors for the circuit in Figure 8.7. Assume $I_{BIAS} = 100$ μA, and design the base-ballasting resistors so that a transistor in deep saturation will not consume more than 10% of I_{BIAS}. Assume that the V_{BE} of a NPN drops by a maximum of 100 mV when it enters deep saturation.

8.4. Lay out the circuit in Figure 8.7 using the resistor values computed in Exercise 8.3 and the standard bipolar layout rules in Appendix C. Assume that all three transistors have minimum emitter areas. Use 6 μm HSR resistors for R_1, R_2, and R_3. Place Q_1, Q_2, and

Q_3 alongside one another and interdigitate the base ballasting resistors for proper matching. Justify your choice of tank biasing for R_1, R_2, and R_3.

8.5. Lay out a minimum-size standard-bipolar NPN transistor, including a deep-N+ sinker. Construct a similar device, but omit the sinker. Allow space for all necessary metallization. Assuming that the areas of the two devices equal the areas of their respective tanks, what percentage area reduction does the omission of deep-N+ produce?

8.6. Lay out a standard-bipolar, stretched-collector CEB transistor that allows one minimum-width lead to pass between its collector and emitter. The transistor should have a minimum-size emitter. Minimize the collector resistance, assuming the process uses a thick emitter oxide.

8.7. Lay out a standard-bipolar, narrow-emitter transistor having four minimum-width emitter fingers, each 100 μm long. Place a deep-N+ sinker along one side of the transistor, and make sure that NBL fully encloses the sinker to minimize collector resistance. Include all necessary metallization to allow connection of the transistor into a circuit.

8.8. Construct minimum-size, standard-bipolar substrate PNP transistors using the standard, emitter-ringed, and verti-lat layout styles. Allow room for all necessary metallization. For the verti-lat transistor only, provide an emitter field plate overlapping the collector by 2 μm.

8.9. Construct a split-collector lateral PNP transistor containing four quarter-sized collectors arranged around a minimum circular emitter. Allow room for all necessary metallization, including an emitter field plate overlapping the collectors by 2 μm.

8.10. Construct a set of merged PNP transistors occupying the same tank. One transistor should have an elongated emitter 25 μm long, while the second transistor should consist of two half-size collectors arranged around a minimum-size circular emitter. Allow room for all necessary metallization, including emitter field plates overlapping the collectors by 2 μm.

8.11. Modify the layout in Exercise 8.10 to field plate the collector of the elongated transistor and one of the two collectors of the split-collector transistor. Overlap the field plate over the collectors by at least 6 μm. Use flanges to elongate all channels as far as possible without enlarging the tank.

8.12. Construct a high-voltage NPN transistor, using the standard bipolar layout rules. Assume the transistor uses a deep-N+ sinker and that it has a minimum-size emitter. Use 4 μm fillets on the base diffusion and include all necessary field plates and channel stops.

8.13. Lay out an extended-base NPN transistor, using the analog BiCMOS layout rules. Overlap the NBL 5 μm into the deep-N+ isolation ring to ensure an adequate seal between the two. Construct a 64 μm^2 emitter and allow for all necessary metallization.

8.14. Construct a CMOS substrate PNP transistor containing two emitter fingers that are each 30 μm long. Assume that the process only silicides contacts and therefore no silicide block mask is required. Use two emitter contacts per finger.

8.15. Construct a CMOS lateral PNP transistor, using a minimum-width poly field plate. Connect this field plate to the emitter.

8.16. Construct a minimum-size poly-emitter NPN transistor, using the analog BiCMOS process. Draw the emitter contact on a layer named ECONT and use the following rules:

1. ECONT width 2.0 μm
2. POLYI overlap ECONT 1.0 μm

9 Applications of Bipolar Transistors

The bipolar transistor possesses two key advantages over its MOS counterpart: higher transconductance and superior device matching. These advantages translate into faster circuits that consume less power and offer higher precision. Many high-performance operational amplifiers and comparators use bipolar circuitry to minimize input offset and maximize output drive. Some families of high-speed logic also employ bipolar transistors as output drivers. Voltage regulators and references almost always use bipolar circuitry to obtain precise temperature-invariant voltages. The majority of the highest-speed and highest-accuracy integrated circuits rely on bipolar circuitry in one form or another.

The *transconductance* of a bipolar transistor equals the ratio of the change in collector current to the change in base-emitter voltage. A high transconductance produces a large change in collector current for a small change in base-emitter voltage. The transconductance of a bipolar transistor is directly proportional to emitter current and is independent of emitter area, so even a small bipolar transistor can provide a large transconductance if it receives enough current. MOS circuitry dominates low-power design because MOS transistors retain moderate transconductances at very low currents. As the current levels increase, bipolar transistors become increasingly attractive. A micropower amplifier probably uses all-CMOS circuitry to conserve power, but high-drive amplifiers frequently incorporate bipolar output stages to reduce output impedance and minimize standby currents. The bipolar transistors in these output stages must handle high currents while dissipating large amounts of power. Small-signal transistors, even ones with enlarged emitters, perform poorly in power applications, so a variety of specialized layouts have been developed for this purpose.

The high transconductance of bipolar transistors also improves their base-emitter voltage matching. An untrimmed differential input stage constructed using bipolar transistors can routinely achieve three-sigma input offset voltages of less than ±1 mV over temperature. Only relatively large and well constructed MOS input stages can hope to rival this performance.[1] Ratioed bipolar transistors can

[1] H. C. Lin, "Comparison of Input Offset Voltage of Differential Amplifiers Using Bipolar Transistors and Field-Effect Transistors," *IEEE J. of Solid-State Circuits*, Vol. SC-5, #3, 1970, pp. 126–129.

also generate very accurate voltage differentials, which form the basis of most voltage and current references. MOS references, even carefully constructed ones, rarely perform as well as their bipolar counterparts.

Although bipolar transistors offer distinct advantages over MOS, many designers are reluctant to use them. Bipolar transistors can fall prey to a number of failure mechanisms that rarely affect MOS designs. The problem of saturation in bipolar transistors has no direct equivalent in MOS design. Improperly constructed bipolar transistors frequently self-destruct under heavy loads, while MOS transistors rarely do so. Carelessly matched bipolar transistors are far more vulnerable to thermal gradients than similar MOS devices. This chapter explains how to retain the unique advantages of bipolar transistors while avoiding their many pitfalls.

9.1 POWER BIPOLAR TRANSISTORS

The previous chapter discussed the layout of small-signal transistors. These devices usually employ minimum-area emitters to conserve space. These small emitters are acceptable because small-signal transistors rarely conduct more than a fraction of a milliamp. Transistors conducting larger currents experience beta rolloff unless their emitter areas increase in proportion to their emitter currents in order to maintain a constant emitter current density. The beta of a typical vertical NPN transistor begins to roll off at current densities of about 1 $\mu A/\mu m^2$ of emitter. To conserve space, power transistors generally operate at betas lower than their small-signal counterparts. A beta of 10 is often chosen as a minimum acceptable limit for high-current operation. Power NPN transistors can usually handle 10–20 $\mu A/\mu m^2$ before their beta drops below 10. Few PNP transistors can handle more than a small fraction of this current density. Although substrate PNPs may retain a beta of 10 up to current densities of 1 $\mu A/\mu m^2$, substrate injection usually limits them to maximum currents of a few milliamps. Lateral PNP transistors rarely achieve more than 250 μA/minimum emitter. Most high-current circuits avoid the use of PNP transistors entirely, even if this eliminates otherwise-attractive circuit topologies.

Small-signal transistors can handle up to about 10 mA and 100 mW without any precautions. Beyond this point they become increasingly vulnerable to failure mechanisms caused by high currents and high power dissipation. These problems become especially acute in transistors handling currents in excess of 100 mA or dissipating power in excess of 500 mW. Such transistors require specialized layouts to protect them from thermal runaway and secondary breakdown. With careful layout, one can successfully integrate transistors capable of conducting 10 A and dissipating 100 W. Power transistors of this magnitude require so much die area that they completely dominate the layout of the integrated circuit. The cost of constructing a large integrated power device greatly exceeds the cost of purchasing an equivalent discrete device. Very high-power or high-current devices also require special, and often costly, packaging. Most integrated power transistors conduct less than 2 A and dissipate less than 10 W. Power lateral PNP transistors require enormous amounts of die area, and few designs incorporate PNP transistors conducting more than 500 mA. The vast majority of power bipolar transistors are therefore power NPN devices.

Many different power NPN layouts have been proposed. Each offers its own unique combination of advantages and disadvantages. No single structure outperforms all the others in all applications. In order to make an intelligent choice, the designer must understand the mechanisms that cause power transistors to fail.

9.1.1. Failure Mechanisms of NPN Power Transistors

The three most common problems encountered in the design of power bipolar transistors are emitter debiasing, thermal runaway, and secondary breakdown. These three problems all result from the large currents and the high power dissipations typical of power devices. None of these mechanisms cause much trouble in small signal transistors, but all impose significant constraints on power transistor design.

Emitter Debiasing

The term *emitter debiasing* refers to a nonuniform current distribution that may develop in a power bipolar transistor due to voltage drops in the extrinsic base and emitter, and in their respective leads. The high transconductance of bipolar transistors makes these devices very susceptible to changes in base-emitter bias. Small voltage drops down the base or emitter leads can radically redistribute current flow through the transistor. Some portions of the transistor may conduct little or no current, while others conduct far more current than they were designed to handle. The overloaded portions of the transistor become vulnerable to thermal runaway and secondary breakdown.

Figure 9.1 shows an example of emitter debiasing occurring between the separate emitter fingers of a power transistor. In the accompanying schematic, transistors Q_1 to Q_4 represent the four emitter fingers and resistors R_1 to R_3 represent the resistance of the metal leads connecting the fingers together. Assume that each emitter finger conducts 50 mA, and that each resistor consists of one square of 20 kÅ aluminum with a sheet resistance of 12 mΩ/□. The total drop across the three resistors

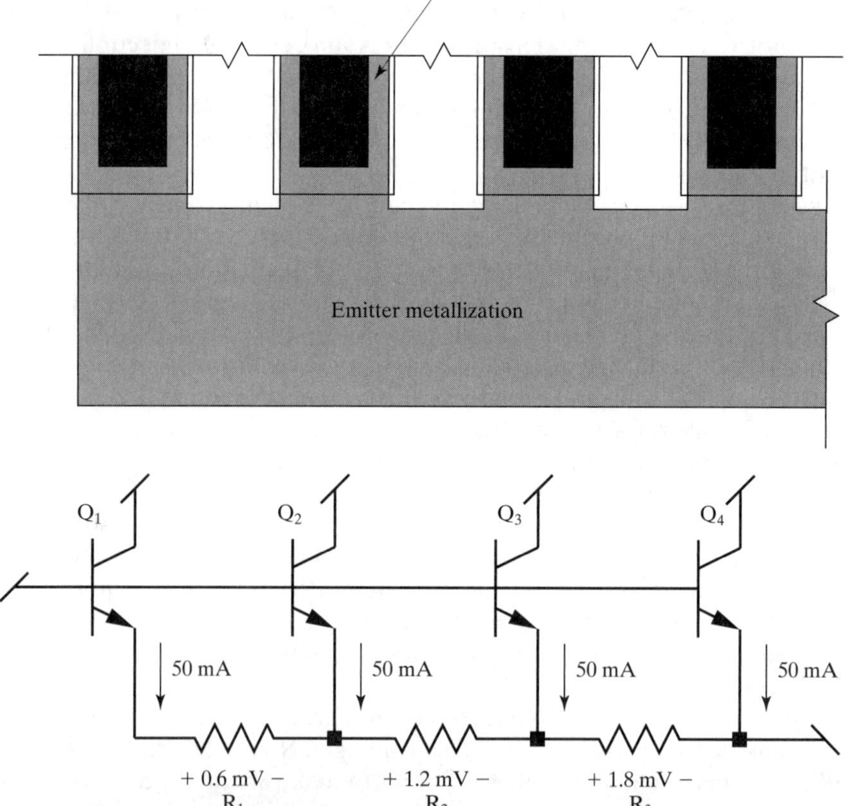

FIGURE 9.1 The layout and equivalent schematic of a power transistor having four emitter fingers. The values listed on the schematic follow the computations described in the text.

equals 3.6 mV. The ratio of emitter currents η between two transistors whose base-emitter voltages differ by a voltage ΔV_{BE} equals

$$\eta = e^{\Delta V_{BE}/V_T} \qquad [9.1]$$

where V_T represents the thermal voltage of silicon (approximately 26 mV at room temperature). In this example, the ratio of currents equals 1.15, so the rightmost finger Q_4 would conduct about 15% more current than the leftmost finger Q_1. Analog BiCMOS processes encounter even more severe debiasing problems because they use thinner metallization (typically 10 kÅ).

The previous example illustrates the severity of emitter debiasing—relatively small currents flowing through short, wide leads still cause 3.6 mV of debiasing. A technique called *emitter ballasting* can greatly reduce the impact of debiasing. Emitter ballasting requires the insertion of resistors into each emitter lead (Figure 9.2). These resistors are typically sized to provide a voltage drop of 50 to 75 mV at full rated current. For example, emitter fingers conducting 50 mA each might employ 1 Ω ballasting resistors. The addition of these ballasting resistors forces the emitter current to redistribute about equally between the emitter fingers. If any emitter finger attempts to draw more than its fair share of current, then the voltage drop across its ballasting resistor increases. This limits the amount of current that can flow through this emitter finger. Voltage drops between ballasted emitters appear primarily across the ballasting resistors rather than across the base-emitter junctions of the transistors. Thus, 3.6 mV of debiasing between two emitters ballasted with 1 Ω resistors would result in a 1.8 mA current increase in one emitter and a 1.8 mA current decrease in the other. These numbers are only approximations, but they serve to show how much benefit emitter ballasting provides. A ballasted transistor can generally tolerate debiasing equal to 25% of the voltage drop across the ballasting resistors. A transistor that drops 50 mV across each ballasting resistor can easily tolerate 10 mV of debiasing in its emitter leads. If the layout would produce more debiasing, the size of the ballasting resistors can be increased to compensate. Remember, however, that the voltage drop across the ballasting resistors adds to the saturation voltage of the transistor, lowers its effective transconductance, and increases its power dissipation. If the design requires more than about 100 mV of ballasting, consider altering the metallization pattern or the aspect ratio of the transistor.

Emitter debiasing can also develop within a single emitter finger (*intrafinger debiasing*). Voltage drops accumulate as the current flows along the finger. One end of the emitter finger sees a larger base-emitter voltage and therefore conducts more current than the other. Debiasing along a long emitter finger can actually become a

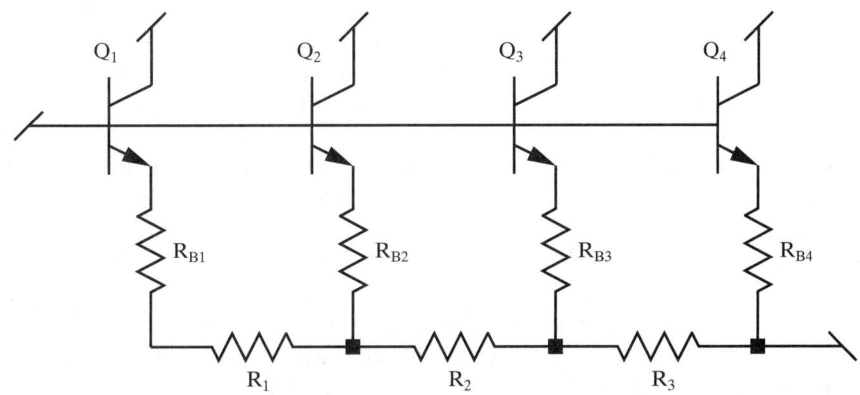

FIGURE 9.2 Connection of ballasting resistors to the segmented power transistor in Figure 9.1; R_{B1} to R_{B4} are the ballasting resistors for the emitter fingers Q_1 to Q_4.

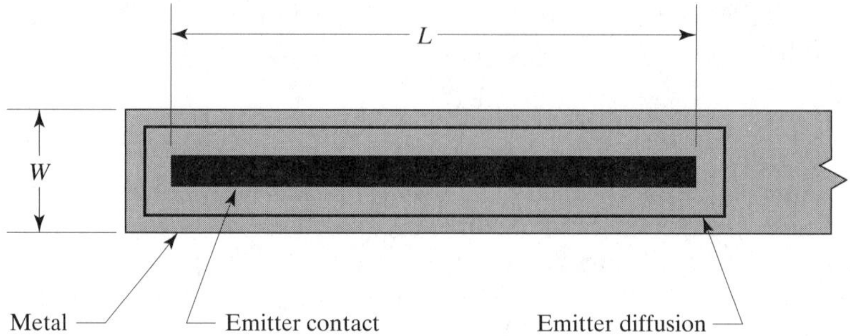

FIGURE 9.3 A sample emitter finger layout showing measurements L and W used in Equation 9.2.

more serious problem than debiasing between separate fingers. For a narrow emitter finger like that in Figure 9.3, the voltage drop from one end to the other should not exceed about 5 mV. Assuming that an emitter lead of constant width runs down the emitter finger, and assuming that equal currents flow into the emitter lead along each increment of its length, then the total voltage drop from one end of the emitter contact to the other end equals

$$\Delta V_{BE} = \frac{LR_sI_E}{2W} \qquad [9.2]$$

where R_s represents the sheet resistance of the metallization, W is the width of the emitter lead, L is the length of the emitter contact, and I_E equals the total current flowing out of the entire emitter finger (Figure 9.3). For example, suppose an emitter finger conducts 50 mA along a lead 300 μm long by 30 μm wide constructed from 12 mΩ/□ aluminum. Equation 9.2 indicates that the debiasing along this emitter finger equals 9 mV, which exceeds the maximum suggested debiasing of 5 mV. Although this computation does not consider the redistribution of emitter current in response to debiasing, it still demonstrates the severity of the debiasing problem.

Several options exist for reducing intrafinger debiasing. The emitter fingers may be shortened and widened. This not only minimizes the finger length but also allows the use of wider metal leads. Alternatively, the transistor may employ a larger number of shorter emitter fingers of the same width as the originals. A ballasting technique also exists that applies to individual fingers (Section 9.1.2), but it can only provide a limited amount of ballasting, which may not suffice to compensate for poor emitter finger design.

Thermal Runaway and Secondary Breakdown

Both thermal runaway and secondary breakdown result from an intensification of current flow through portions of the power transistor. In the case of thermal runaway, the current flow localizes in response to increasing temperature. Suppose that one portion of the power transistor becomes slightly warmer than the rest. The V_{BE} required to maintain constant collector current drops by 2 mV/°C, so a temperature rise of only a few degrees results in significant emitter debiasing. Almost all of the current flows through the hottest portion of the transistor, raising its temperature still further. In a matter of milliseconds, the region of conduction collapses to a tiny *hot spot* comprising only a few percent of the transistor's area. Perhaps beta rolloff can limit the collapse of the hot spot sufficiently to prevent catastrophic device failure, or perhaps not. Even if the hot spot stabilizes, the transistor will be so severely overstressed that it will become vulnerable to other failure mechanisms such as secondary breakdown, electromigration, and thermally accelerated corrosion.

Since thermal runaway involves emitter debiasing, ballasting resistors can provide some measure of protection against it. If each finger of a multi-emitter transistor has its own ballasting resistor, then a hot spot that develops in one finger cannot steal current from the other fingers. Even in a worst-case scenario in which hot spots develop in all of the fingers, each hot spot absorbs only a fraction of the total current. Usually 50 mV of ballasting suffices to control thermal runaway, but more ballasting is sometimes necessary to offset voltage drops in the emitter metallization system.

Hot spots can still develop in individual emitter fingers even if all of the fingers have ballasting resistors. If each finger has its own ballasting resistor, then the current drawn by any one hot spot will decrease as the number of emitter fingers increases. Distributed emitter ballasting (Section 9.1.2) also provides some measure of protection against hot spot formation. Extremely demanding applications may require a combination of distributed emitter ballasting and individual ballasting resistors for each emitter finger.

Secondary breakdown occurs when the emitter current density in a transistor exceeds a *critical current density* J_{crit}. Beyond this point, the sustained collector-to-emitter breakdown voltage $V_{CEO(sus)}$ snaps back to a new, lower value called the *secondary breakdown voltage* V_{CEO2}. A transistor is most vulnerable to secondary breakdown when it is in the process of turning off. The collector-to-emitter voltage across the transistor rises as the emitter current through the transistor decreases. Secondary breakdown occurs if the collector-to-emitter voltage exceeds V_{CEO2} while the emitter current density exceeds J_{crit}. Once avalanche begins, the base drive circuit can no longer turn the transistor off, and the transistor is soon destroyed by overheating or metallization failure.

Transistors driving inductive loads are extremely vulnerable to secondary breakdown. Consider a power transistor Q_1 driving a high-side inductive load L_1 (Figure 9.4A). As soon as Q_1 begins to turn off, the inductive kick-back of L_1 drives the collector voltage V_{CE} upward until recirculation diode D_1 begins to conduct (Figure 9.4B). The collector voltage reaches its maximum value almost immediately, long before the emitter current drops to zero. Secondary breakdown will occur if the collector voltage exceeds V_{CEO2} while the emitter current density exceeds J_{crit}.

Conservative design rules dictate that power transistors operate at an emitter current density of no more than 10–15 μA/μm². These current densities lie well below those required to trigger secondary breakdown in order to provide a safety

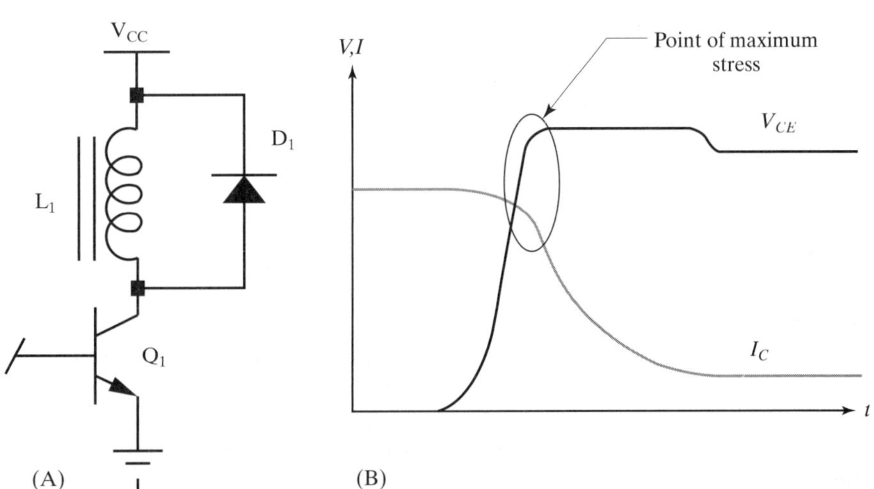

FIGURE 9.4 An example of (A) a bipolar transistor driving an inductive load and (B) the waveforms associated with the turnoff interval of the transistor.

margin for emitter debiasing, hot spot formation, emitter current focusing, and thermal gradient formation.

Kirk Effect

As previously discussed, secondary breakdown occurs when the current density through a transistor exceeds a critical current density, J_{crit}. When this occurs, the transistor's avalanche voltage rating suddenly and dramatically decreases due to a mechanism called the *Kirk effect.* Although the Kirk effect is generally considered an advanced topic in device physics, it can still be explained in terms of the simplified device physics introduced in Chapter 1.

Consider the collector-base depletion region of a NPN transistor. The strong electric field that exists across this depletion region quickly sweeps carriers from the base to the collector. This junction normally operates under *low-level injection*, which means that the concentration of carriers within the depletion region is far smaller than the concentration of dopant atoms (Figure 9.5A). Because there are so few carriers in the depletion region, we can ignore them when computing the space charge. The depletion region extends a distance x_{mb} from the metallurgical junction into the base, uncovering negatively charged acceptors. Similarly, the depletion region extends a distance x_{mc} from the metallurgical junction into the collector, uncovering positively charged donors. The total donor charge must equal the total acceptor charge, so the depletion region will extend farther into a lightly doped region than into a heavily doped one. The collector of an optimized bipolar transistor is usually more lightly doped than its base, so $x_{mc} > x_{mb}$.

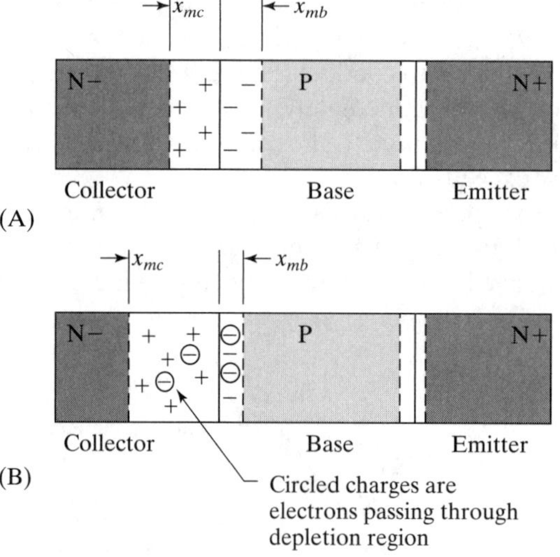

FIGURE 9.5 Cross-section of an idealized NPN transistor operating (A) under low-level injection, and (B) under high-level injection, showing the movement of the depletion region caused by the Kirk effect.

As the current density across the collector-base junction increases, so does the number of carriers in the depletion region. Eventually, the number of carriers becomes so great that their contribution to the total charge within the depletion region can no longer be ignored. The junction is now said to be operating in *high-level injection* (Figure 9.5B). The flow of electrons across the collector-base junction causes a negative charge to appear on both sides of the junction. This negative charge adds to the negative charge to the exposed acceptors and thus decreases x_{mb}. This negative charge cancels a portion of the position charge of the exposed donors and thus increases x_{mc}.

Because x_{mb} decreases, the width of the neutral base of the transistor increases. This effect, called *base push out*, causes an increase in base transit time and thus a reduction in operating frequency. Kirk's original paper[2] explains the mechanism behind base push out, and consequently this mechanism has been dubbed the *Kirk effect*.

Returning to an analysis of Figure 9.5B, if the current density across the collector-base junction rises further, then x_{mc} will increase and x_{mb} will decrease. This process continues until the concentration of electrons equals the concentration of donors on the collector side of the metallurgical junction. When this happens, the negative charge of the electrons exactly counterbalances the positive charge of the donors, and this portion of the depletion region has an effective charge density of zero. This implies that x_{mc} increases to infinity. The current density at which this occurs is called the *critical current density*, J_{crit}.

The transistor of Figure 9.5 was highly idealized. An actual NPN transistor will have a collector consisting of two distinct regions: a lightly doped drift region next to the collector-base junction, and a heavily doped extrinsic collector that serves as a low-resistance path to the collector contact. The NBL of a vertical NPN transistor serves as its extrinsic collector. Figure 9.6A shows what happens when this structure operates at the critical current density. The entire drift region depletes out, and the collector-base region now touches the extrinsic collector. Figure 9.6B shows what happens when the current density increases yet further. The drift region now becomes negatively charged, and the depletion region must extend into the extrinsic collector to counterbalance this negative charge. The peak electric field intensity $E_{\max}$ occurs at the point where the depletion region charge switches from positive to negative. At or below the critical current density, the peak field intensity occurs at the metallurgical collector-base junction. Above the critical current density, the peak field intensity abruptly shifts to the boundary between the extrinsic collector and the drift region. Further increase in the current causes more negative charge to accumulate within the drift region, and consequently more donors must be uncovered within the extrinsic collector. This implies that the peak field intensity increases with current density. The increase in peak field intensity in turn implies an increase in impact ionization and a reduction in avalanche voltage. Thus,

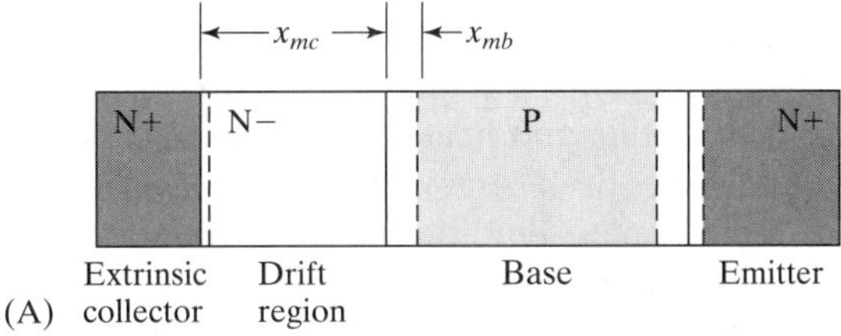

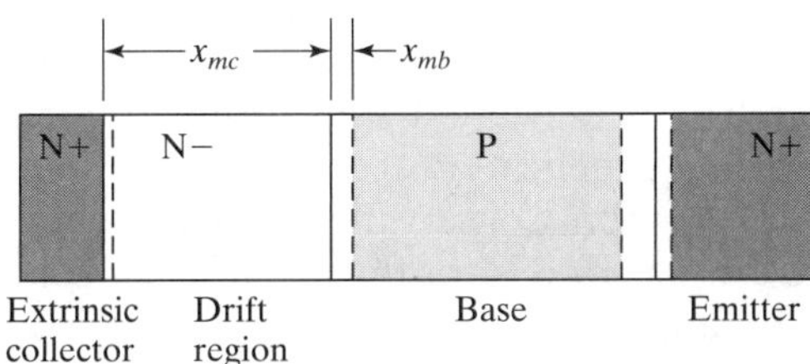

FIGURE 9.6 Cross section through an idealized NPN transistor incorporating an extrinsic collector region, showing the device operating (A) at the critical current density J_{crit} and (B) above this current density.

2 C. T. Kirk, Jr., "A Theory of Transistor Cut off Frequency (f_T) Falloff at High Current Densities," *IRE Trans. on Electron Devices*, Vol. ED-9, 1962, pp. 164–174.

the avalanche voltage of the transistor decreases with increasing current density. This effect becomes apparent only at current densities approaching J_{crit}, and it is the driving force behind secondary breakdown.

The critical current density J_{crit} can easily be computed. If we assume that the peak electric field E_{max} has increased to the point where the carrier drift velocity has saturated (as is usually the case when a transistor approaches avalanche breakdown), then the carrier concentration n_c equals

$$n_c = \frac{J}{qv_{lim}} \qquad [9.3]$$

where J is the current density, q is the charge on the electron, and v_{lim} is the carrier saturation velocity, which for electrons equals roughly $1 \cdot 10^7$ cm/s. The critical current density J_{crit} occurs when the carrier concentration n_c equals the drift region doping N_C. Therefore,

$$J_{crit} = qN_C v_{lim} \cong 1.6 \cdot 10^{-12} N_C \qquad [9.4]$$

where N_C has units of cm^{-3}. For a typical drift region doping $N_C = 1 \cdot 10^{15}\ cm^{-3}$, $J_{crit} = 80\ \mu A/\mu m^2$. This calculation is only an approximation because actual junctions are not abrupt, but the analysis does lend credence to the rule that emitter current densities in vertical NPN power transistors should not exceed 10–20 $\mu A/\mu m^2$.

9.1.2. Layout of Power NPN Transistors[3]

Over the years, a number of alternative layouts have been proposed for NPN power transistors. Each layout has certain strengths and weaknesses, so knowledge of several different types of layouts will aid the designer in choosing the best style for any given application. Any layout can be scaled by adding or removing emitter sections or by connecting several power devices in parallel.

A transistor used in a *linear-mode* application remains in the forward active region for long periods of time. Linear transistors must withstand large collector-to-emitter differentials while simultaneously conducting large collector currents. Such a transistor must cover enough area to allow for heat dissipation. As a general rule, linear-mode transistors should not dissipate more than 150 $\mu W/\mu m^2$ of emitter nor conduct more than 10 $\mu A/\mu m^2$. These are conservative guidelines, and with sufficient ballasting and heatsinking, it is possible to successfully operate transistors at several times these stress levels. Still, one should generally follow these conservative guidelines unless empirical measurements show that a transistor can safely operate at higher stress levels.

Transistors used in *switched-mode* applications operate either in cutoff where no current flows, or in saturation where collector-to-emitter differentials remain small. Switching transistors dissipate power only during brief switching intervals. The average power dissipation of switching transistors remains relatively small, so they rarely experience hot spot formation or thermal runaway. On the other hand, switching applications generate many opportunities for emitter current focusing during turnoff. Conservative designs should not exceed an emitter current density of 20 $\mu A/\mu m^2$ to ensure that emitter current focusing does not trigger secondary breakdown.

Transistors that drive purely capacitive loads such as MOS gates conduct current only as infrequent short-duration pulses. Such *pulsed-mode* transistors typically

---

[3] F. W. Trafton, "High Current Transistor Layout," unpublished manuscript, 1988.

conduct large currents for a few hundred nanoseconds, then rest for a few microseconds before conducting again. Pulsed-mode transistors are unharmed by emitter-current focusing because the external capacitive load will quench conduction regardless of whether secondary breakdown occurs. Pulsed-mode transistors are also immune to thermal runaway because hot spots cannot form and localize in less than a few microseconds.[4] Most pulsed-mode applications rely on high-current beta rolloff and collector resistance to limit conduction. This practice is acceptable as long as the pulse duration does not exceed 1 μS, the intervals between pulses are no shorter than 250 nS, and the *average* emitter current density does not exceed 20 $\mu A/\mu m^2$. The metallization for pulsed-power transistors should be designed following the electromigration rules for intermittent currents described in Section 14.3.3.

The Interdigitated-Emitter Transistor

The oldest style of power transistor, the *interdigitated-emitter transistor*, remains in use because it can operate at higher speeds than any other style of bipolar power transistor. Figure 9.7 shows an interdigitated-emitter transistor constructed using a single-level-metal standard-bipolar process.

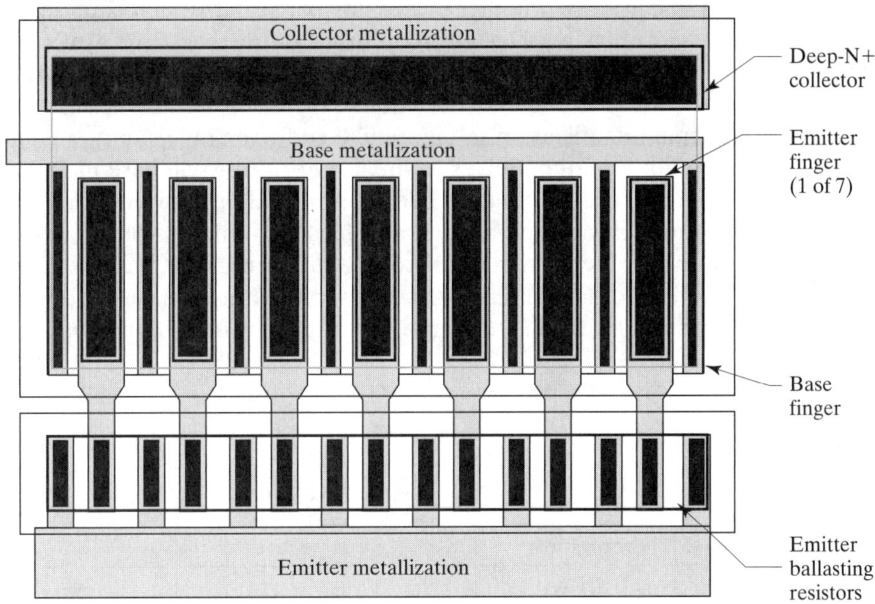

FIGURE 9.7 An example of an interdigitated-emitter power transistor. Each emitter finger has a separate ballasting resistor. Metallization is shown in gray for emphasis.

This transistor consists of a number of emitter fingers, each having its own dedicated emitter ballasting resistor. The ballasting resistors are all formed from a single strip of emitter diffusion placed in a separate tank. The emitter diffusion is not isolated from the tank because small current leakages from one finger to another cause no harm. Each emitter finger connects to two ballasting resistors placed in parallel, each consisting of about one square of emitter. Assuming a minimum emitter sheet resistance of 5 Ω/□, this provides a ballasting resistance of 2.5 Ω per finger. This resistance will provide 50 mV of ballasting at 20 mA of emitter current.

The interdigitated-emitter transistor is extremely vulnerable to intrafinger debiasing. The voltage drop down each emitter finger computed using Equation 9.2

[4] H. Melchior and M. J. O. Strutt, "Secondary Breakdown in Transistors," *Proc. IEEE*, Vol. 52, 1964, p. 439.

should not exceed 5 mV. A large number of short emitter fingers are preferable to a small number of long fingers. The width of the emitter fingers also affects performance. Widening the emitter fingers widens the pinched base regions underneath them. The resulting increase in base resistance causes the transistor to exhibit slower switching and increased emitter focusing. The fastest and most robust designs incorporate minimum-width emitter fingers, but it is difficult to place enough metal on narrow emitter fingers to prevent them from debiasing. Double-level metal helps, but narrow fingers still do not make efficient use of available area. Most designers compromise on an emitter width of 8 to 25 μm. In this style of transistor, the emitter contacts are always made as large as possible to reduce emitter resistance.

Base contacts along either side of each emitter finger reduce the base resistance and enable faster switching. Base contacts placed on either end of the emitter array ensure that the end fingers turn off as quickly as the others. If these end contacts were omitted, the end fingers would turn off more slowly than the others. This could lead to emitter current focusing and secondary breakdown during turnoff. Minimum-width base contacts help conserve space, and relatively few designs require more base metallization. Power transistors operating at high current densities may, however, experience enough beta rolloff to necessitate wider base metallization. Computation can show whether or not any particular design experiences significant base-lead debiasing. Designs exhibiting more than 2 to 4 mV of base debiasing should be redesigned to reduce base metallization resistance. The comb-style base metallization in Figure 9.7 exhibits much less metallization resistance than the serpentine metallization in Figure 9.8. Unfortunately, many single-level-metal designs do not lend themselves to the use of comb-style base metallization.

The transistor in Figure 9.7 contains deep-N+ only along one side. This may suffice for a linear-mode device operating at a collector-to-emitter voltage differential of a half-volt or more. A switching transistor is a different matter, as its efficiency is determined by its collector-to-emitter voltage drop in saturation (its *saturation voltage*). If the saturation voltage is too large, then the transistor dissipates too much

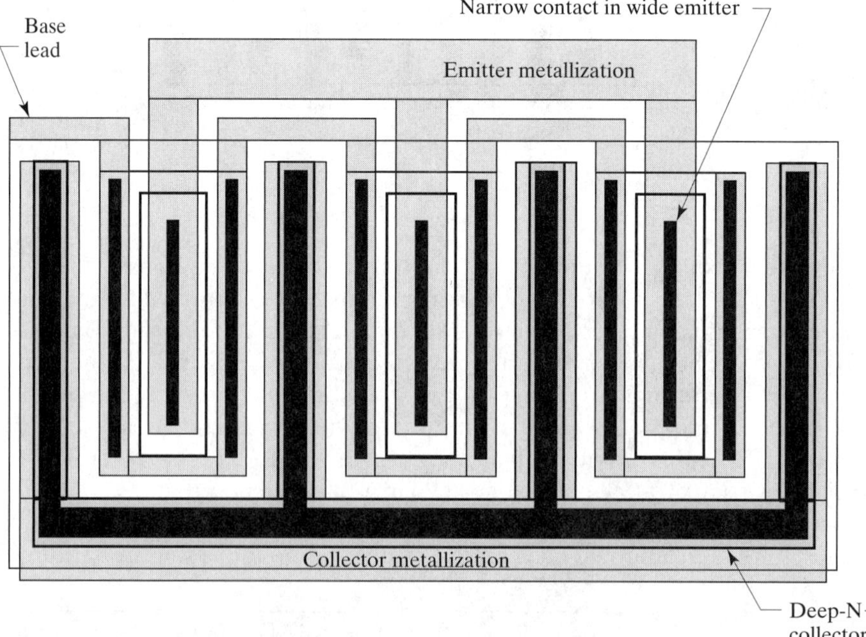

FIGURE 9.8 An example of wide-emitter narrow-contact power transistor. Emitter ballasting resistors, although not shown, can easily be added. Metallization is shown in gray for emphasis.

power. At high currents, the collector resistance of a switching transistor equals the sum of its vertical deep-N+ resistance and its lateral NBL resistance.[5] The vertical resistance of the deep-N+ sinker can be reduced by increasing its area. The sinker should not be less than 10 μm wide to ensure that outdiffusion does not dilute its doping and increase its vertical resistance. The lateral NBL resistance can be reduced by contacting the NBL along a longer periphery or by decreasing the distance between the active portions of the device and the sinkers. Placing sinkers along both sides of the transistor reduces the NBL resistance by a factor of four, and an unbroken ring of deep-N+ around the transistor reduces it still further. The NBL should extend to the outer edge of the deep-N+ sinker to ensure a low-resistance connection between the two.

An unbroken ring of deep-N+ around a power transistor also forms a hole-blocking guard ring that helps control substrate injection during saturation. When an NPN transistor saturates, all of its unused base drive flows to the substrate. The guard ring does not reduce the amount of base drive consumed by the transistor, but it does prevent the majority of it from flowing to the substrate. Section 9.1.4 discusses several techniques for limiting the base current consumed during saturation.

The Wide-Emitter Narrow-Contact Transistor

The interdigitated-emitter transistor uses relatively narrow emitter fingers to reduce base resistance and to control emitter crowding. This structure's low base resistance allows it to operate at higher frequencies than any other. Unfortunately, the narrow emitters are quite prone to emitter crowding. Emitter debiasing causes conduction to concentrate at the exit end of each finger, while thermal gradients focus conduction into the middle of the transistor. In either case, current tends to localize at one point in each emitter finger. Ballasting resistors can help ensure that the fingers conduct equal currents, but they cannot prevent intrafinger debiasing. Even well-ballasted interdigitated-emitter transistors tend to develop hot spots at higher current densities.

If each emitter finger is divided into a large number of individually ballasted sections, then no one portion of the emitter finger can contact more current than any other. Although it is generally not feasible to segment an emitter finger in this manner, placing a narrow emitter contact in a wide emitter finger provides similar benefits.[6] Figure 9.8 shows the resulting *wide-emitter narrow-contact transistor.*

The use of a wide emitter finger and a narrow contact produces the equivalent of a distributed network of ballasting resistors. This network consists partly of emitter resistance and partly of pinched base resistance. The emitter resistance is largest at the periphery of the emitter and smallest in the center directly beneath the narrow contact. Conversely, the base resistance is smallest at the periphery and is largest in the center directly under the emitter contact. These two forms of ballasting complement one another. At low currents, the base resistance is relatively insignificant, and current distributes uniformly across the width of the emitter finger. As the current increases, debiasing in the pinched base region causes conduction to move out toward the periphery of the emitter finger. The current must now flow through a larger emitter resistance. The resulting emitter voltage drops counteract the movement of current toward the emitter periphery. Together, the base-side and emitter-side distributed ballasting ensure that conduction occurs relatively uniformly across the

[5] The drift region usually contributes little or no resistance since it depletes through under the influence of reverse bias or velocity saturation.

[6] A. B. Grebene, *Bipolar and MOS Analog Integrated Circuit Design* (New York: John Wiley and Sons, 1984), p. 510.

entire width of the emitter finger. This type of emitter ballasting is distributed along the length of the emitter finger, so it protects all portions of the device against emitter debiasing and the formation of hot spots.

The emitter must overlap the contact by a distance sufficient to provide adequate ballasting. Typical wide-emitter narrow-contact structures employ emitter overlaps of 12 to 25 μm. Larger overlaps unnecessarily slow the frequency response of the transistor, while smaller overlaps may not provide enough distributed ballasting to fully protect against thermal runaway and secondary breakdown. Transistors operating under extreme conditions often benefit from additional ballasting resistors inserted into the leads of each emitter finger, as illustrated in the interdigitated-emitter transistor in Figure 9.7.

Some designers employ a trapezoidal emitter contact tapering from a wide low-current end to a narrow high-current end. This design provides additional ballasting at the high-current end of the emitter finger to offset the effect of voltage drops in the emitter metallization. The lack of matching between metal and emitter sheet resistances makes this type of design problematic, and most wide-emitter narrow-contact transistors employ minimum-width contacts instead. One must not stretch the contact to the ends of the emitter finger, because this eliminates the ballasting at these points and renders them vulnerable to hot spot formation.

The transistor in Figure 9.8 employs multiple base regions with fingers of deep-N+ interdigitated between them. This structure minimizes collector resistance at the cost of increasing area and complicating lead routing. In single-level-metal layouts, the base lead must serpentine through the transistor between the collector and emitter metallization. The added length of the serpentined base lead can cause significant debiasing in the base metallization. Even with the distributed ballasting, the transistor should not contain metallization drops of more than a few millivolts. Base debiasing can be reduced by a factor of roughly four by connecting both ends of the serpentined emitter lead. Layouts employing double-level metal frequently use comb or grid arrangements to combat base metallization debiasing.

The wide-emitter narrow-contact structure is remarkably robust. The distributed emitter ballasting helps prevent thermal runaway and secondary breakdown within individual fingers, allowing the device to operate at higher current densities than comparable interdigitated structures do. Wide-emitter narrow-contact transistors that must operate under especially harsh conditions may benefit from the insertion of an additional 50 to 75 mV of emitter ballasting in the leads of the individual emitter fingers. This structure does not switch as quickly as the interdigitated-emitter transistor, but the degradation is not as large as one might expect since considerable high-current conduction occurs along the emitter periphery.

The Christmas-Tree Device

Another type of power transistor layout is nicknamed the *christmas-tree device* because of the peculiar shape of its emitter geometry (shown in dark gray in Figure 9.9). Historically, this structure was widely used in linear applications because of its exceptional resistance to thermal runaway. It is rarely used for switching applications, because the same features that improve its immunity to thermal runaway degrade its ability to withstand emitter current focusing during turnoff.

The emitter of this transistor consists of a central spine surrounded by a complex branching structure of triangular prongs that give the transistor its picturesque name. Most of the conduction occurs in the triangular prongs along the emitter periphery. These connect to the central spine of the emitter through narrow emitter strips that act as ballasting resistors. At low currents, all portions of the emitter conduct. As the current increases, emitter crowding forces conduction

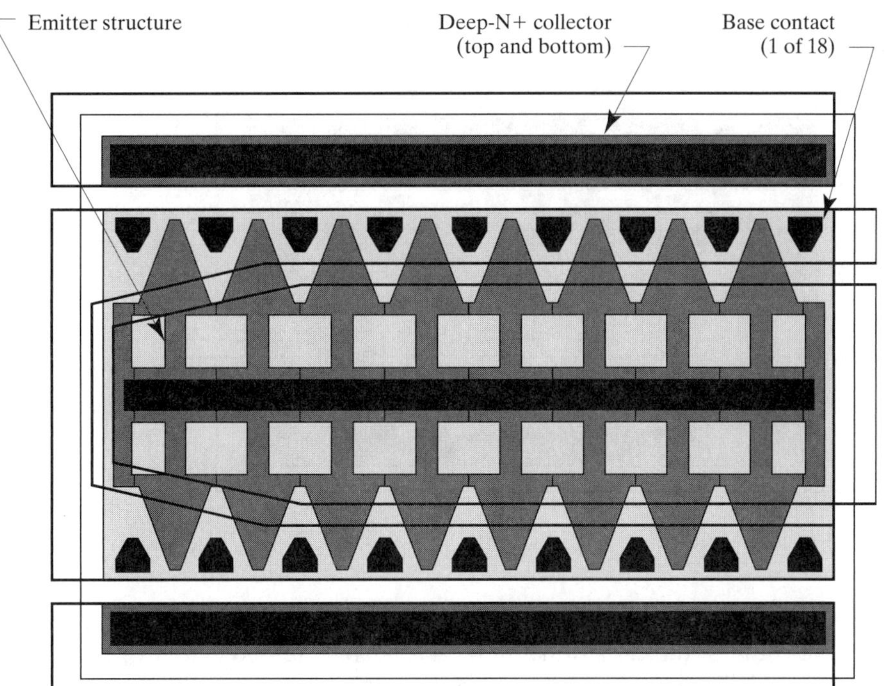

FIGURE 9.9 An example of a christmas-tree power transistor. The base is shown in light gray and the emitter in dark gray in order to highlight the peculiar structure of the emitter.

out toward the periphery, causing current to flow through the ballasting resistors incorporated into the emitter structure. This device gains its resistance to thermal runaway from a large amount of distributed ballasting. Unfortunately, the great width of the emitter structure renders it vulnerable to emitter current focusing. As the transistor begins to turn off, the area of conduction retreats from the periphery toward the central spine. Because the spine represents only a small portion of the total emitter area, the emitter current density increases dramatically during the final stages of turnoff. This concentration of current flow can (and often does) trigger secondary breakdown. The wide-emitter narrow-contact structure exhibits superior immunity to secondary breakdown because the distance from the periphery to the center of the emitter is not as great and the effects of emitter focusing are not as dramatic.

The christmas-tree device serves best in applications that dissipate large amounts of power, but where abrupt turnoff transitions never occur. Historically, this style of device was frequently chosen for the series-pass devices of linear voltage regulators and the output stages of audio power amplifiers. A number of variations on the christmas-tree device have been developed in an attempt to minimize its vulnerability to emitter focusing while retaining its immunity to hot spot formation. None of these variants are as robust as the wide-emitter narrow-contact transistor, or its descendent, the cruciform-emitter transistor.

The Cruciform-Emitter Transistor

The *cruciform-emitter transistor* represents an evolutionary development of the wide-emitter narrow-contact structure that seeks to incorporate additional emitter ballasting without rendering the device vulnerable to secondary breakdown. The emitter of this device consists of a series of cross-shaped (*cruciform*) sections stacked end-to-end to form a continuous emitter finger (Figure 9.10). The base contacts occupy the small notches between the arms of the crosses.

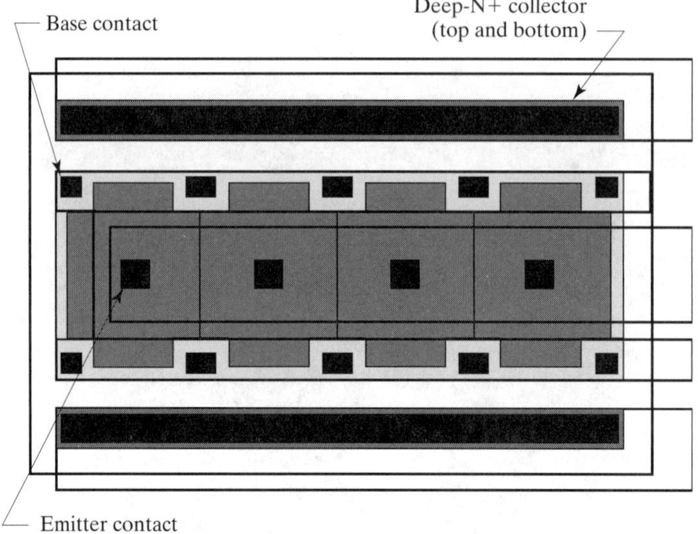

FIGURE 9.10 An example of a cruciform-emitter transistor. The base is shown in light gray and the emitter in dark gray for emphasis.

The width of the cruciform emitter has been increased to 75 to 125 μm to obtain additional ballasting. The narrow emitter contact has also been replaced by a series of small, square or circular contacts occupying the center of each cross. All of the emitter current must flow through these contacts, producing a distributed three-dimensional ballasting effect considerably more efficient than the two-dimensional ballasting generated by the wide-emitter narrow-contact structure. Consequently, the cruciform emitter combines the best features of the wide-emitter narrow-contact transistor and the christmas-tree device. The cruciform transistor does not have quite the immunity to secondary breakdown that the wide-emitter narrow-contact transistor does, but it vastly outperforms the christmas-tree device in this respect. The cruciform emitter transistor also makes extremely efficient use of space.

The cruciform structure suffers from two drawbacks. First, the small size of the emitter contacts renders them vulnerable to electromigration. All of the emitter current must cross the sidewalls of these contacts, and this produces very high localized current densities in the metallization. Even refractory barrier metal has its limits, which this transistor may exceed. Some designers replace the single contact in the center of each cruciform with an array of minimum contacts to increase the sidewall perimeter. Second, the compact design of the cruciform emitter can cause extreme localized heating at high power levels. Less area-efficient transistors are actually preferable to more compact ones from the standpoint of heat dissipation. If the heat produced by the transistor spreads over a wider area, then the thermal impedance between the transistor and the package decreases and the transistor can handle more power before it overheats. The cruciform structure is best suited for switching applications, as these are more strongly constrained by current-handling capability than by power dissipation.[7]

Power Transistor Layout in Analog BiCMOS

Any of the power transistors discussed up to now can also be implemented in analog BiCMOS. Figure 9.11 shows a BiCMOS version of a wide-emitter narrow-contact

[7] A related structure called the *H-emitter transistor* is described in F. F. Villa, "Improved Second Breakdown of Integrated Bipolar Power Transistors," *IEEE. Trans. on Electron Devices*, Vol. ED-33, #12, 1986.

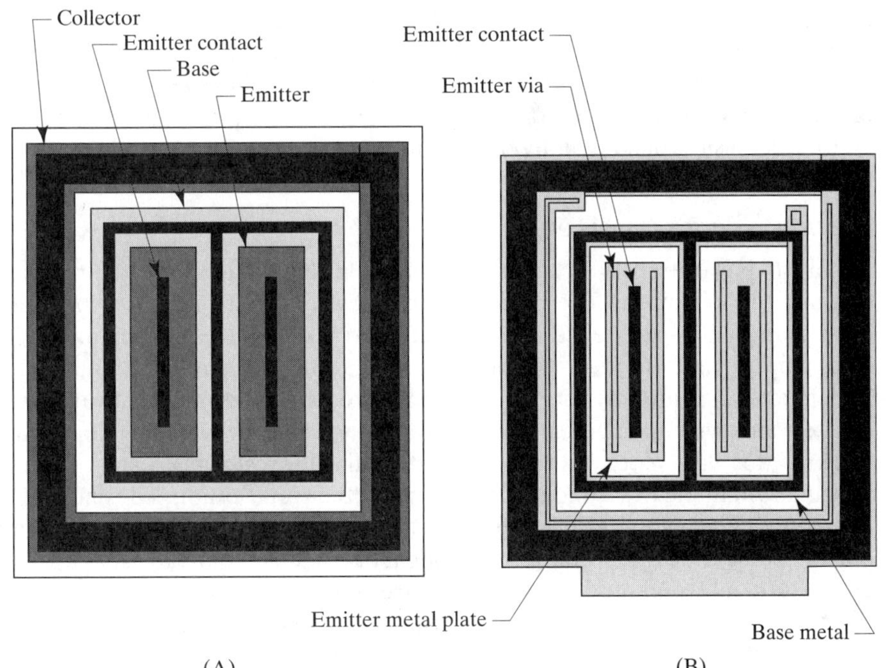

FIGURE 9.11 A wide-emitter narrow-contact transistor constructed in analog BiCMOS using double-level metal: (A) diffusions and (B) metal-1 pattern. The metal-2 pattern is not shown.

transistor. Double-level metallization allows the base contacts to completely encircle each emitter finger, whereas in a single-level-metal design they can only reach two or three sides of each finger. The complete ring of base contact helps ensure that all portions of the emitter periphery are equally active. An unbroken ring of deep-N+ sinker minimizes collector resistance and blocks substrate injection during saturation.

Figure 9.11B shows further details of the metallization system. Emitter current flows from the narrow emitter contacts to vias placed parallel to them. Passing up through these vias, the current reaches a metal-2 plate covering the top of the transistor. This plate minimizes emitter debiasing by reducing the metal-2 resistance to an absolute minimum. The resistance in the metal-1 plates actually serves as emitter ballasting and therefore is unobjectionable. The base metallization consists of a grid of first-level metal covering the base contacts. The base current exits the transistor through a metal-2 jumper placed between the emitter metal-2 plate and the encircling collector metal-2. If necessary, a second base lead can exit on the other side of the emitter plate. The collector metallization consists of a complete ring of metal-1 covering the collector contact and a U-shaped metal-2 plate covering the collector on three sides of the transistor. Vias along the inner edges of the collector contact allow current to flow through both levels of metallization. The collector lead can exit any side of the transistor except the side where the emitter lead exits. The best arrangement places the collector lead diametrically opposite of the emitter lead. This minimizes the resistance of the collector metallization by ensuring that half of the current flows through the metal on either side of the transistor.

The structure in Figure 9.11 has been used to fabricate pulse-power transistors capable of operating at emitter current densities of more than 150 $\mu A/\mu m^2$. This structure uses an overlap of emitter over contact of 8 to 12 μm and a continuous ring of deep-N+ sinker at least 8 μm wide. The NBL should completely overlap the deep-N+ sinker to minimize resistance and to ensure that no minority carriers can escape through a lightly doped portion of the extrinsic collector. This structure can continuously conduct in excess of 15 μA/μm^2 and can operate as either a linear dissipative device or as a switching element. The distributed emitter ballasting inherent in the wide emitter fingers prevents hot spots from forming even at very high power levels. The use of a solid metal-2 plate to terminate the emitters helps minimize emitter debiasing, making individual emitter ballasting resistors unnecessary for all but the most demanding applications.

Selecting a Power Transistor Layout

All of the power transistor layouts presented in this section have their advantages and their disadvantages. The Christmas-tree device is best suited for linear applications that do not experience rapid switching transients. The interdigitated emitter transistor provides the best switching speeds and frequency response, but it requires individual ballasting resistors on each finger to avoid thermal runaway. The wide-emitter narrow-contact and cruciform transistors excel in switching applications. A wide-emitter narrow-contact transistor with ballasting resistors in each emitter finger is virtually immune to secondary breakdown at the voltages normally encountered in integrated circuit applications (10 to 40 V). Some applications require that a small portion of the transistor's emitter be brought out independently to act as a sensing element. The interdigitated emitter structure offers the easiest insertion of a sense emitter and the best matching of the sense emitter to the remainder of the transistor. Table 9.1 summarizes these advantages and disadvantages.

	Interdigitated Emitter	Wide-emitter Narrow-contact	Christmas-tree Device	Cruciform Transistor
Thermal runaway	Good*	Good	Excellent	Excellent
Secondary breakdown	Fair	Excellent	Poor	Good
Frequency response	Excellent	Good	Fair	Fair
Compactness of layout	Poor	Good	Good	Excellent
Ease of emitter sensing	Excellent	Fair	Poor	Poor

*Assumes individually ballasted emitter fingers; otherwise *poor*.

TABLE 9.1 Comparison of four types of power NPN layouts.

9.1.3. Power PNP Transistors

Most processes do not fabricate isolated vertical PNP transistors, and even those that do seldom create a device capable of handling significant power. The two candidates for a power PNP are therefore the substrate PNP and the lateral PNP. Both of these devices have significant drawbacks that prevent their general adoption as power devices.

Substrate PNP transistors are generally unable to handle more than a few tens of milliamps without debiasing the substrate contacts. This disadvantage can be circumvented by contacting the backside of the die, using a conductive die attach. Conductive epoxies are generally considered unsatisfactory for this purpose, so either a gold eutectic bond or a solder mount must be performed in order to obtain a backside contact. The mount pad must then be connected to a pin, either by means of a down bond, or through use of a *fused leadframe* in which one of the lead fingers is connected to the mount pad. Even if a low-resistance backside contact can be established, substrate PNP transistors have limited functionality because their collector connects to ground. This configuration rules out many circuit techniques that could otherwise take advantage of the existence of a power PNP transistor.

Lateral PNP transistors make relatively poor power devices because they cannot handle high current densities. The lateral PNP emitter cannot be increased in size without severely degrading the beta of the device, so a power lateral PNP typically employs numerous minimum-geometry emitters arranged in either a square grid (Figure 9.12A) or a hexagonal grid (Figure 9.12B). The hexagonal-array layout exhibits slightly denser packing than the square-array layout. Most high-power lateral PNP transistors use a continuous ring of deep-N+ around the edge of the transistor to contact the base. This deep-N+ ring serves not only as a base contact, but also as a hole-blocking guard ring that minimizes substrate injection in the event that the transistor saturates.

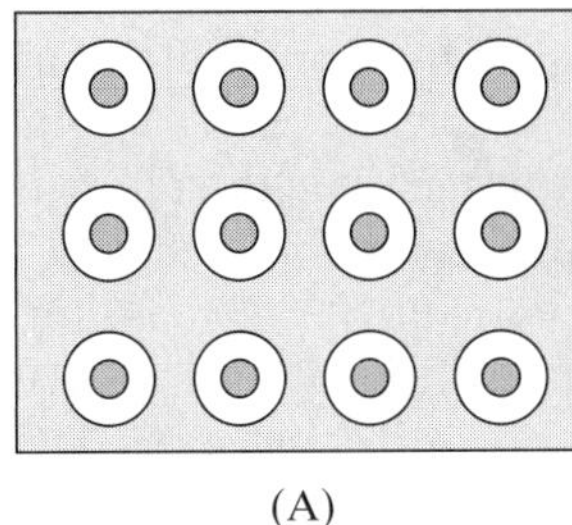
(A)

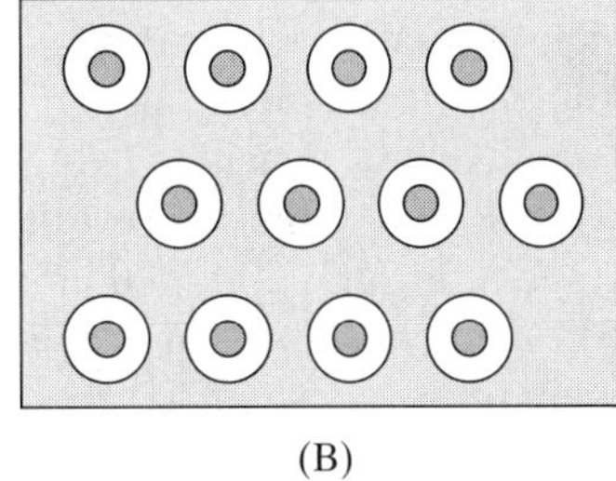
(B)

FIGURE 9.12 Base patterns for power lateral PNP transistors laid out using (A) a square grid layout, and (B) a hexagonal grid layout.

A typical minimum-emitter lateral PNP can handle only a few hundred microamps before high-level injection into the lightly doped base begins to degrade its beta. A single minimum emitter cannot handle more than 0.25–1 mA before the beta degrades to unusable levels. Lateral PNP power transistors therefore consist of many hundreds of individual emitters. Even the largest such devices are seldom rated for more than an amp of collector current.

Ironically, the limitations of the power lateral PNP transistor are responsible for one of its greatest strengths: its remarkable ruggedness. The lateral's pronounced high-current beta rolloff acts as a form of unintentional ballasting. If any portion of the transistor begins to conduct too much current, then its beta will begin to rapidly decrease. The transistor simply does not have the high-current beta required to generate dangerous current densities through hot spotting or current focusing. Similarly, the low current density of the lateral transistor makes it very difficult to destroy through overheating. Add to these advantages the lack of a fragile base-emitter junction, and the resulting device proves nearly indestructible.

Power lateral PNP transistors were widely used in the 1970s and 1980s to construct *low-dropout* (LDO) voltage regulators. A PNP LDO exhibits a very objectionable increase in ground return current at low input-to-output differential voltages. This effect is due to saturation within the PNP and the resulting increase in base current. A new generation of LDO regulators based on PMOS transistors were developed in the 1990s. These regulators exhibit low ground currents at all differential voltages. However, the lateral PNP LDO regulators still retain one major advantage over their PMOS counterparts: The lateral PNP maintains a high output impedance down to very low collector-to-emitter voltages, while a PMOS exhibits severe degradation of output impedance under these conditions due to the movement of its operating point from saturation to the linear region. Because of this limitation, power lateral PNP transistors still find application in LDO regulators even today.

Many of the older-generation PNP LDO regulators use a variation of the standard bipolar process that incorporates a special *deep-P+* diffusion that is both

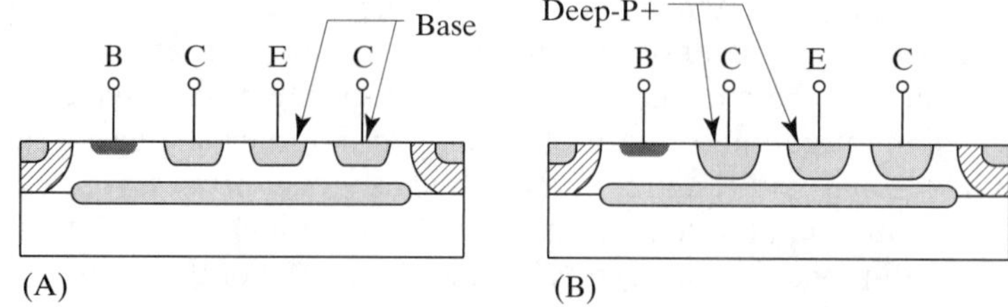

FIGURE 9.13 Comparison of representative cross sections of (A) a standard bipolar lateral PNP and (B) a deep-P+ lateral PNP.

deeper and more heavily doped than the regular base diffusion (Figure 9.13). The increase in dopant concentration improves the emitter injection efficiency of the lateral PNP, while the deeper junction ensures that a larger percentage of emitter injection occurs from the sidewalls.[8] The high-current beta of a deep-P+ lateral does not roll off as quickly as that of a base lateral. Deep-P+ laterals can therefore operate at current densities two or three times greater than base laterals. The beta for a typical 10 μm-diameter emitter constructed using deep-P+ falls to half its peak value at around 200–500 μA, compared with 100–200 μA for an equivalent device constructed solely from base diffusion. Although this increase in performance may seem relatively small, it translates directly into device area. The deep-P+ process extension requires only a single mask and thus is relatively cheap.

9.1.4. Saturation Detection and Limiting

Both lateral PNP and vertical NPN transistors inject current into the substrate when they saturate. Substrate injection wastes supply current and may cause substrate debiasing and device latchup. Several techniques have been developed to suppress substrate injection, either by intercepting minority carriers before they reach the substrate or by preventing the transistor from saturating in the first place. Most of these techniques require specialized layouts.

The emitter of a lateral PNP continuously injects minority carriers into its tank. When the transistor saturates, most of this current flows to the substrate. Small-signal transistors rarely inject enough current to warrant concern, but a few designs incorporate large lateral PNP transistors that conduct tens or even hundreds of milliamps. Currents of this magnitude can easily produce enough debiasing to trigger latchup.

Figure 9.14A shows one way to prevent minority carriers from reaching the substrate. This transistor incorporates a continuous, unbroken ring of deep-N+ around the outside edge of its tank. This ring merges with the underlying NBL and completely encloses the base region of the lateral PNP within a hole-blocking guard ring (Section 4.4.2).

Figure 9.14B shows another method for preventing substrate injection. The illustrated device incorporates a ring of base diffusion completely encircling its primary collector. This ring acts as a *secondary collector.* As long as the primary collector does not saturate, few carriers can reach the secondary collector and it conducts little current. When the primary collector saturates, the carriers begin to flow to the secondary collector. So long as the secondary collector does not simultaneously saturate, it collects most of the carriers and prevents them from reaching the isolation sidewalls. The secondary collector is sometimes called a *ring collector* because it often takes the form of an unbroken ring enclosing the primary collector.

---

[8] B. Murari, "Power Integrated Circuits: Problems, Tradeoffs, and Solutions," *IEEE J. Solid-State Circuits*, Vol. SC-13, #3, 1978, pp. 307–319.

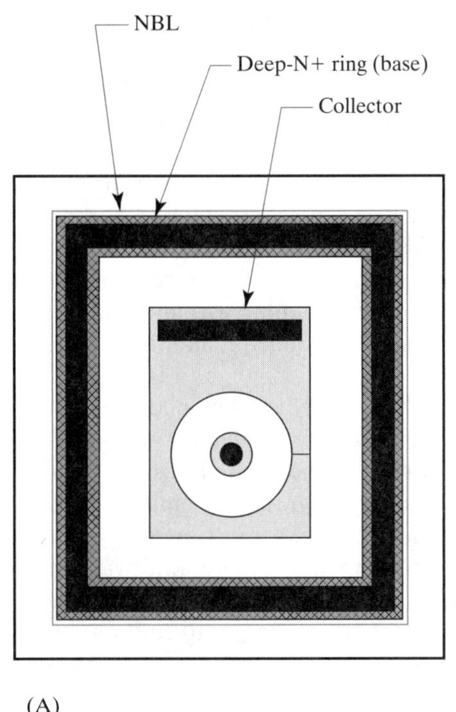

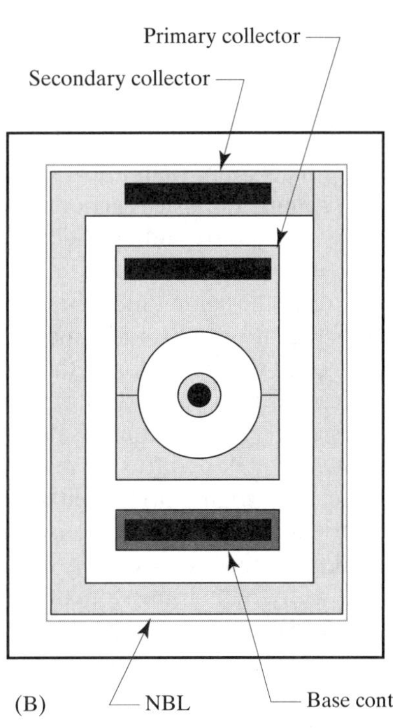

FIGURE 9.14 Two examples of lateral PNP transistors modified to minimize saturation: (A) transistor ringed with deep-N+ and (B) transistor with secondary collector.

The secondary collector can perform one of several functions depending on how it is connected. If it connects to ground, then it returns any carriers it collects to the ground return line. The secondary collector then behaves as a hole-collecting guard ring. Alternatively, the secondary collector can connect to the base lead. When the transistor saturates, the carriers collected by the secondary collector add to the base current and cause the apparent beta to rapidly decline. This connection provides the same functionality as a deep-N+ ring, while consuming considerably less space. If the designer wishes to increase the efficiency of the ringed collector still further, then the base ring can be supplemented by a deep-N+ ring placed outside it.

Secondary collectors can also be constructed in analog BiCMOS processes. These should consist of base diffusion (or shallow P-well) rather than PSD, because the source/drain implant is frequently too shallow to act as an efficient collector. Analog BiCMOS secondary collectors are generally less effective than their standard bipolar counterparts due to the downward drift of minority carriers produced by the well doping gradient.

A secondary collector can also function as a saturation detector. Current begins to flow through the secondary collector as soon as the primary collector saturates, and the current stops as soon as the primary collector ceases to saturate. The secondary collector can be used to dynamically control the base drive to prevent the primary collector from saturating. Instead of dumping the unwanted current to ground, a *dynamic antisaturation circuit* throttles the base drive back to reduce the emitter current. The negative feedback loop required to control the base drive may become unstable unless properly compensated, and the phase shift across the secondary collector is difficult to model. Secondary collectors used for saturation detection do not have to encircle the entire transistor because they need only intercept a small fraction of the minority carriers to generate the necessary control signal. Since the dynamic antisaturation circuit must incorporate a relatively large signal delay to ensure stability, a deep-N+ ring may be required to suppress transient substrate injection.

Saturating NPN transistors also inject current into the substrate. Small-signal transistors rarely inject enough substrate current to necessitate antisaturation rings, but power transistors are another matter entirely. Any saturating NPN transistor that conducts more than a few milliamps of base drive requires some form of protection against minority carrier injection. The first (and best) solution consists of a deep-N+ ring surrounding the periphery of the collector. This ring not only functions as a hole-blocking guard ring, but also reduces the collector resistance of the transistor at the same time. Minority carriers that recombine within the tank or the deep-N+ guard ring become majority carriers in the collector, and from there they pass through the transistor to the emitter.

A base diffusion placed within the collector of a NPN transistor collects minority carriers and can also act as a saturation detector. Power switching transistors often include a small base diffusion in the collector tank connected to a dynamic antisaturation circuit (Figure 9.15). Antisaturation circuits for grounded-emitter transistors are difficult to construct because the secondary collector must operate at or near ground potential. A knowledgeable circuit designer can usually find ways around this difficulty. NPN transistors fitted with antisaturation circuitry should still employ a complete deep-N+ ring around the periphery of the collector to contain minority carriers injected during transients, as well as to reduce the collector resistance of the power transistor.

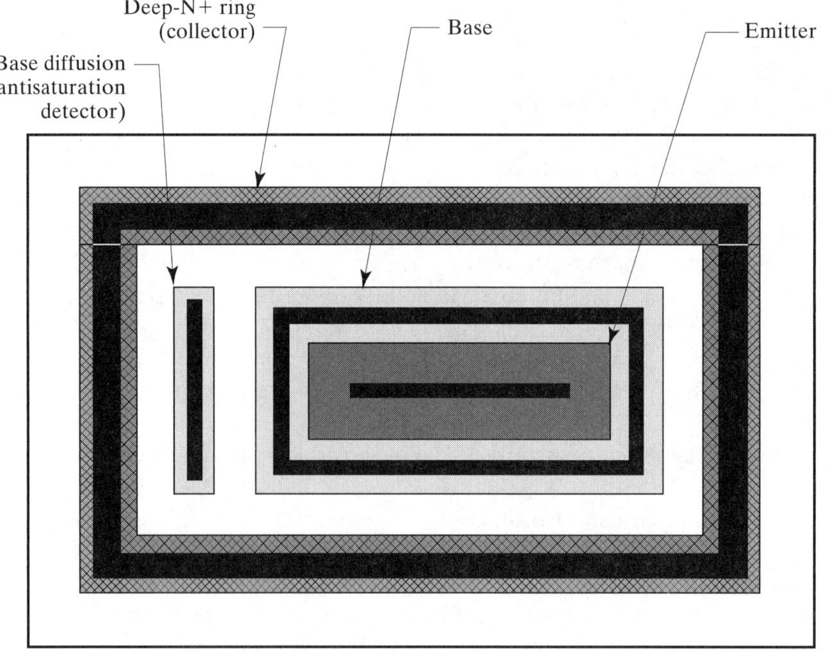

FIGURE 9.15 A NPN switching transistor incorporating both a complete deep-N+ ring and a secondary collector that functions as a saturation detector.

No generally accepted symbols exist for transistors with secondary collectors or deep-N+ guard rings. Figure 9.16 shows a set of symbols that have achieved some degree of industry recognition. A thick base bar denotes a power transistor, or more generally, any transistor requiring special layout (A). A diagonal slash across the collector lead of a NPN indicates the presence of a deep-N+ sinker (B), as does the

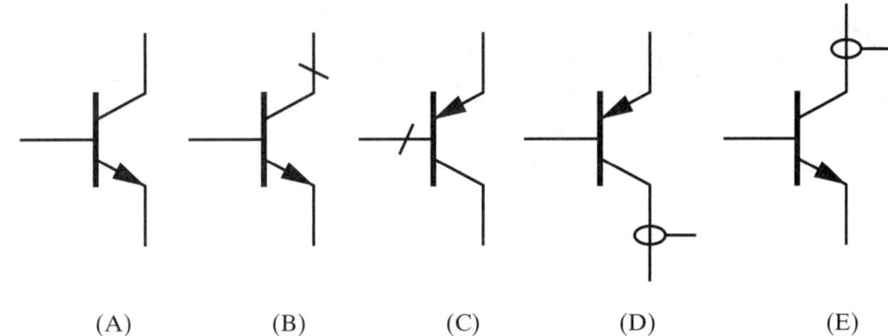

FIGURE 9.16 Proposed symbols for (A) a power NPN, (B) a power NPN with a deep-N+ collector, (C) a lateral PNP with deep-N+ in base, (D) a lateral PNP with a secondary collector, and (E) a NPN transistor with saturation detection.

presence of a similar slash across the base lead of a lateral PNP (C). The addition of a small ring encircling the collector lead of a lateral PNP denotes the addition of a secondary collector[9] (D), while the addition of a similar ring around the collector lead of an NPN transistor denotes a base diffusion placed in the tank as a saturation detector (E). The majority of these circuits use NPN transistors because of their superior device characteristics.

9.2 MATCHING BIPOLAR TRANSISTORS

Many analog circuits require matched bipolar transistors. Current mirrors and current conveyors use them to replicate currents; amplifiers and comparators use them to construct differential input stages; references use them to produce known voltages and currents; and Gilbert translinear circuits use them to perform analog computations. All of these applications depend on precise matching of collector currents and base-emitter voltages, sometimes from transistors of the same size, and sometimes from ones of different sizes.

NPN collector currents scale approximately with drawn emitter area, but no model can precisely predict the influence of emitter geometry on matching. It is therefore very difficult to match transistors with different sizes and shapes of emitters. Most bipolar circuits employ simple integer ratios such as 1:1, 2:1, 4:1, or 8:1. These ratios are easily obtained by assembling multiple copies of an identical unit device. The same techniques can obtain virtually any ratio of small integer numbers, such as 2:3, 3:4, and 2:5. Ratios requiring more than eight or ten unit devices become increasingly impractical due to area requirements and the sensitivity of large devices to temperature gradients.

Two transistors having identical dimensions and operating at equal collector currents should theoretically develop exactly the same base-emitter voltage. In practice, small differences in the emitter saturation currents cause the two base-emitter voltages to vary slightly. The difference between the base-emitter voltages of two transistors operating at equal currents is called the *offset voltage* ΔV_{BE} and can be computed from the equation

$$\Delta V_{BE} = V_T \ln\left(\frac{I_{s1}}{I_{s2}}\right) \qquad [9.5]$$

---

[9] A different symbol for the ringed-collector PNP, as well as a unique application for this device, are found in H. Lehning, "Current Hogging Logic (CHL)—A New Bipolar Logic for LSI," *IEEE J. Solid-State Circuits,* Vol. SC-9, #5, 1974, pp. 228–233.

where I_{s1} and I_{s2} are the emitter saturation currents of the two transistors. Given that the thermal voltage V_T equals 26 mV at room temperature, a 1% mismatch in emitter saturation currents produces an offset voltage of 0.25 mV. Saturation currents scale approximately with drawn emitter area, so a 1% variation in emitter area also produces a 0.25 mV offset.

9.2.1. Random Variations

Random fluctuations in base doping and emitter junction area set the ultimate limits of vertical bipolar transistor matching. Other significant sources of random variation include recombination in the emitter-base depletion region and lateral injection across the base diffusion, both of which scale inversely with the emitter area-to-periphery ratio. Matched bipolar transistors therefore employ relatively compact emitter geometries. The three geometries favored for constructing matched vertical NPN transistors are squares, octagons, and circles (Figure 9.17). Each style of emitter has its proponents. The circle has the largest area-to-periphery ratio and therefore theoretically provides the best possible matching. However, circles are approximated as many-sided polygons during pattern generation. Squares require no such approximations, so many designers believe that they are rendered more precisely on the photomask. Octagons also require no approximations, and they possess slightly larger area-to-periphery ratios than do squares.

FIGURE 9.17 Examples of NPN transistors designed with square, octagonal, and circular emitters.

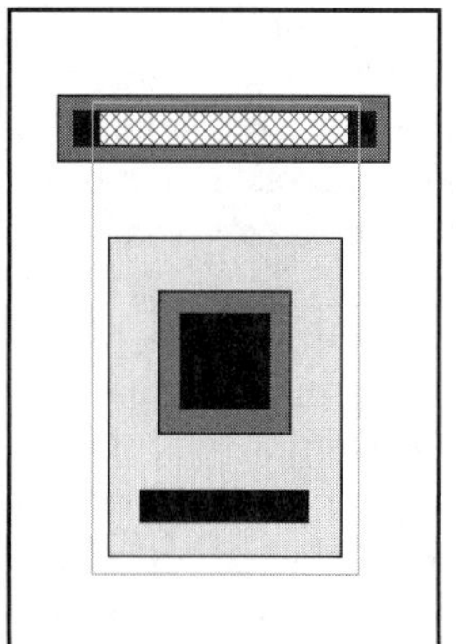
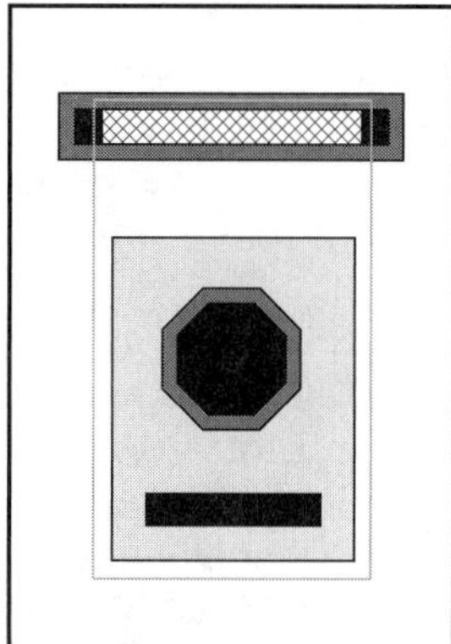
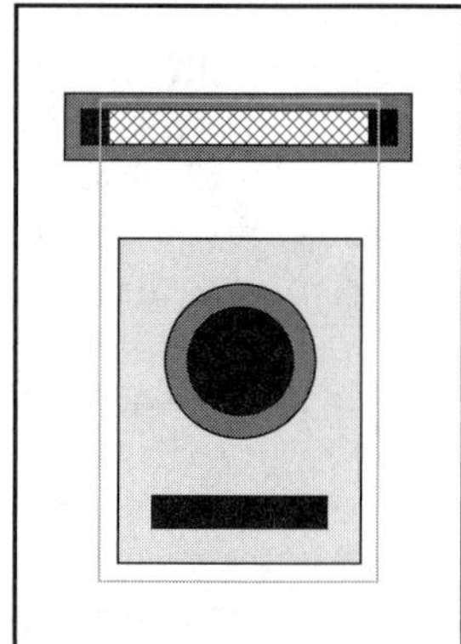

In practice, all three styles of emitters provide excellent matching. Although circles are approximated as polygons, identical circles produce identical polygons. The approximations involved in generating circular emitters therefore have little impact on their matching. Furthermore, the differences in area-to-periphery ratios among squares, octagons, and circles are relatively insignificant. The area-to-periphery ratio R_{AP} of any geometry can be determined using the equation

$$R_{AP} = k_r\sqrt{A_e} \qquad [9.6]$$

where A_e represents the emitter area and k_r is a dimensionless constant equal to 0.250 for squares, 0.274 for octagons, and 0.282 for circles. Note that the area-to-periphery ratio is not itself a dimensionless quantity. For example, if the emitter area is measured in square microns, then R_{AP} will have dimensions of microns. Equation 9.6 shows that the reduction in peripheral effects gained by using a circular emitter can be equaled by simply increasing the area of a square emitter by 25%.

As with most components, areal fluctuations are responsible for the majority of bipolar mismatch.[10] The standard deviation of the mismatch in emitter areas, s equals

$$s = \sqrt{\frac{k_B}{A_e}} \qquad [9.7]$$

where k_B is a constant of proportionality that quantifies the matching achievable with a specific type of transistor in a specific process.

Although large emitters exhibit less random mismatch than small ones, there are other factors to consider. Any increase in emitter size increases the spacing between the devices and therefore renders them more vulnerable to thermal and stress gradients. Large emitters also exhibit increased base pinch resistance. Because of these problems, one must avoid making matched emitters either too large or too small. As a general rule, the diameter of the emitter of a matched NPN transistor should not be less than twice nor more than 10 times the minimum possible diameter. For example, a minimum contact width of 2 μm and a minimum overlap of emitter over contact of 1 μm produce a minimum emitter diameter of 4 μm. Matched emitters in this process should have diameters of no less than 8 μm and no greater than 40 μm. More accurate guidelines require actual data that rarely exist for a production process.

The choices between circular, square, and octagonal emitters are usually of little consequence, but there are exceptional cases where one type of emitter may confer a specific advantage. In the case of lateral PNP transistors, the emitter area must remain small to conserve beta. Circular emitters increase the area-to-periphery ratio of the emitter without increasing its diameter, and thus help not only to improve matching but also to raise beta. Therefore, matched lateral PNP transistors often employ minimum-diameter circular emitters such as those in Figures 8.22 and 8.26B. The emitter should also overlap its contact equally on all sides to ensure an even distribution of emitter current. Consequently, circular emitters should contain circular contacts, and octagonal emitters should contain octagonal contacts. These arrangements are not possible in processes that allow only minimum-dimension square contacts. In such cases, square emitters should be used rather than circular or octagonal ones.

Many circuits require matched transistors having unequal device areas. Although it is possible to connect identical unit transistors in parallel, the large amounts of die area required by collector isolation actually degrade matching by exacerbating the effect of thermal and stress gradients. Matched NPN transistors can occupy a common tank because the geometry of the collector has almost no effect on their matching (Figure 9.18A). The geometry of the base-collector junction also has relatively little impact on matching since most of the conduction occurs either directly underneath the emitter or immediately adjacent to it. Several emitters can therefore occupy the same base region (Figure 9.18B). The emitters must be placed far enough apart to prevent minority carriers that are injected by one from being collected by another. Similarly, the emitters should reside far enough inside the base diffusion to minimize lateral conduction that would otherwise produce mismatches between transistors having different numbers of emitters. These requirements can be met by increasing both the emitter-to-emitter spacing and the base overlap of emitter by 1 to 2 μm. These increased spacings ensure that the individual emitters will not interact with one another or with the collector-base junction.

---

[10] One study has shown that the area parameter alone provides a good fit to as small as 16 μm^2: H.-Y. To and M. Ismail, "Mismatch Modeling and Characterization of Bipolar Transistors for Statistical CAD," *IEEE Trans. on Circuits and Systems—I: Fundamental Theory and Applications*, Vol. 43, #7, 1996, pp. 608–610.

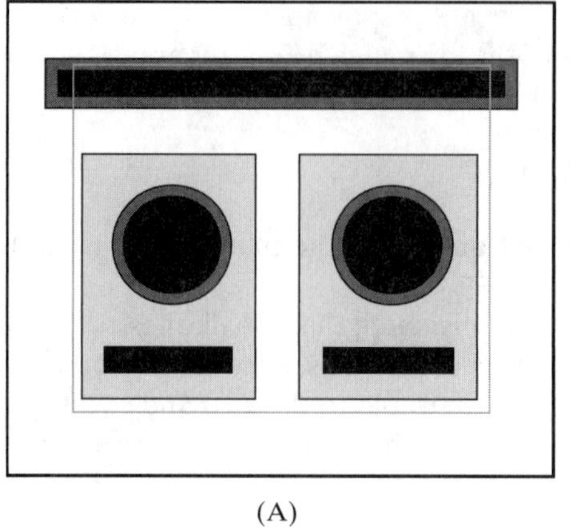

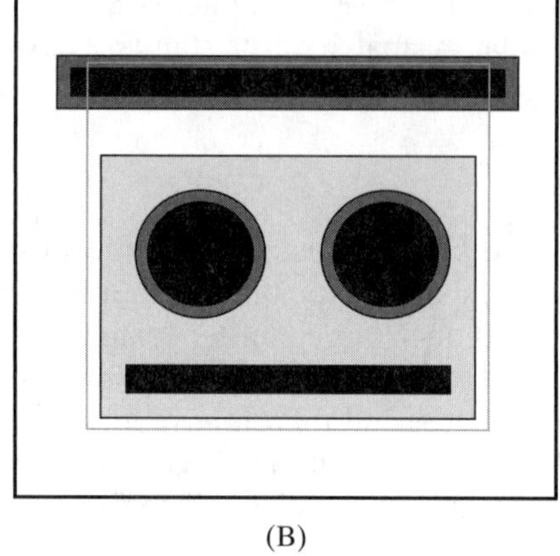

(A) (B)

FIGURE 9.18 Two styles of multiple-emitter NPN transistors: (A) separate base regions in a common tank and (B) separate emitters in a common base region.

Matched lateral PNP transistors can also occupy a common tank to save area. Multiple emitters cannot occupy a single opening in a collector geometry because each emitter would interfere with the flow of minority carriers from the others. Instead, each emitter must occupy its own collector opening, and all of these openings must have identical dimensions. The outside dimensions of the collector geometry and the size and shape of the tank have little impact on matching. Therefore, matched lateral transistors usually consist of rectangular arrays of minimum emitters placed in a common collector region (Figure 8.26B).

9.2.2. Emitter Degeneration

Regardless of the care taken in their construction, some types of bipolar transistors simply do not match very well. A technique called *emitter degeneration* can transfer the burden of matching from a set of bipolar transistors to a set of associated resistors. This technique will improve the overall matching of the circuit as long as the resistors match more precisely than the bipolar transistors. Emitter degeneration also increases the output resistance of bipolar transistors and therefore reduces the systematic errors due to finite Early voltages. The systematic mismatch in collector currents between two matched bipolar transistors operating at different base-collector voltages equals

$$\frac{I_{C1}}{I_{C2}} \cong 1 + \left(\frac{\Delta V_{BC}}{V_A}\right)\left(\frac{V_T}{V_T + V_d}\right) \qquad [9.8]$$

where $\Delta V_{BC} = V_{BC1} - V_{BC2}$, V_A is the Early voltage of the transistors, V_T is the thermal voltage (26 mV at 25°C), and V_d is the voltage developed across the degeneration resistors. Equation 9.8 is only valid as long as V_{BC1} and V_{BC2} are both much smaller than V_A. This equation indicates that 50 mV of degeneration reduces the Early error by approximately a factor of three.

Figure 9.19 shows a lateral PNP current mirror consisting of three lateral PNP transistors Q_1 to Q_3. Each of these transistors has an associated emitter degeneration resistor R_1 to R_3. The mirror also uses a *beta helper* transistor Q_4 to minimize the effect of low betas on matching. Transistor Q_4 does not need to match any of the other transistors, and it does not normally require emitter degeneration.[11]

[11] Sometimes a small resistor is added in series with the emitter of the beta helper as part of a frequency compensation network; this resistor does not take part in the matching of the mirror, and its value is noncritical.

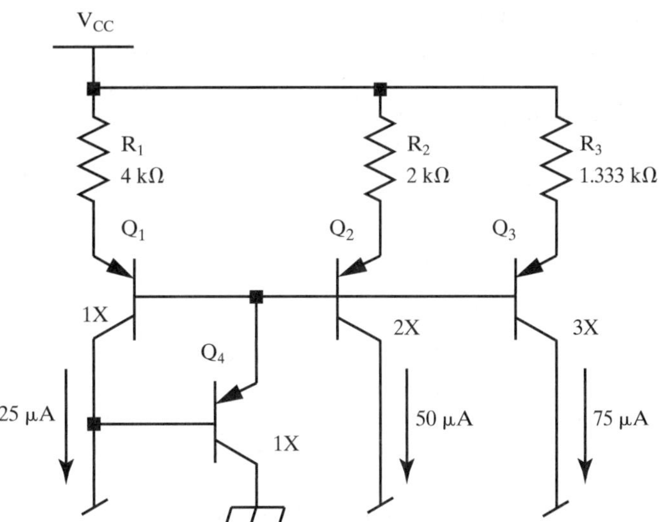

FIGURE 9.19 Lateral PNP current mirror that incorporates emitter degeneration resistors.

In this example, transistors Q_1 to Q_3 have device sizes of one, two, and three, respectively. These sizes represent the number of unit emitters in each transistor. This dimensionless notation avoids any possible confusion between emitter periphery and emitter area, and simultaneously frees the circuit designer from worrying about exact layout dimensions. The values of resistors R_1 to R_3 are inversely proportional to the sizes of transistors Q_1 to Q_3. Each resistor therefore generates the same voltage differential, which in this case equals 100 mV. This voltage represents the amount of degeneration applied to the transistors. Approximately 50 to 75 mV of degeneration suffices to ensure that the resistors determine the matching of the mirror rather than the transistors. Few circuits require more than 100 mV of degeneration as long as the ratio of the emitter areas of the transistors lies within ±10% of the desired value.

The improvement in matching obtained through emitter degeneration depends on how well the resistors match and on the nature of the mismatches between the bipolar transistors. Well-matched resistors vary no more than ±0.1%, while the currents of well-matched minimum-area emitters typically vary by ±1% or more. The matching of the degenerated transistors will approximately equal the matching of the emitter degeneration resistors. On the other hand, the area consumed by the resistors could also be used to increase the emitter areas. Increasing the emitter areas of matched NPN transistors usually proves more area-effective than adding emitter degeneration resistors, since well-matched resistors are not small. Emitter degeneration may sometimes provide better matching in the presence of large thermal gradients because resistors are less susceptible to these gradients than bipolar transistors.

Lateral PNP transistors might seem to benefit less from emitter degeneration than vertical NPN transistors do, since much of the mismatch between lateral transistors stems from beta variations that remain unaffected by degeneration. Despite this, lateral PNP transistors are frequently degenerated because these devices use minimum emitters to maintain acceptable betas. Lateral PNP transistors also benefit from increased output resistance caused by emitter degeneration because these devices often have rather low Early voltages.[12] Emitter degeneration cannot improve the

[12] Gray, *et al.* conclude that NPN transistors derive more benefit from emitter degeneration than PNP transistors, but their arguments are flawed because they ignore the other factors discussed in the text; P. R. Gray and R. G. Meyer, *Analysis and Design of Analog Integrated Circuits,* 3d ed. (John Wiley and Sons, New York: 1993), pp. 317–320.

matching of split collector transistors because it is not possible to provide a separate emitter degeneration resistor for each split collector.

Large amounts of emitter degeneration are sometimes used to obtain noninteger ratios between transistors. The size of the transistors becomes relatively unimportant in the presence of 250 to 500 mV of degeneration. For example, a 3.4:1 ratio can be obtained by ratioing a 3X transistor and a 10 kΩ resistor with a 1X transistor and a 34 kΩ resistor. This technique works equally well with both NPN and PNP transistors.

9.2.3. NBL Shadow

Surface discontinuities caused by oxidation during the NBL anneal propagate upward during epitaxial deposition to produce a surface discontinuity called the *NBL shadow.* A mechanism called *pattern shift* can displace the NBL shadow laterally by a distance of up to twice the epi thickness (Section 7.2.4). Mismatches can occur if the NBL shadow intersects the emitter of a vertical NPN transistor. Arrays of multiple-emitter NPN transistors are particularly vulnerable to pattern shifts perpendicular to their axis of symmetry. Consider the two transistors in Figure 9.20A. Pattern shift has displaced the NBL shadow toward the right, causing it to intersect the leftmost emitter of each device. Suppose that the affected emitters experience a 1% reduction in emitter area. Only one of Q_A's two emitters is affected, so its emitter area becomes 1.99. The sole emitter of Q_B is also affected, so its emitter area becomes 0.99. The new ratio between the two devices is 1.99:0.99, or about 2.01:1. This represents a mismatch of 0.5%, or an offset voltage of about 0.13 mV.

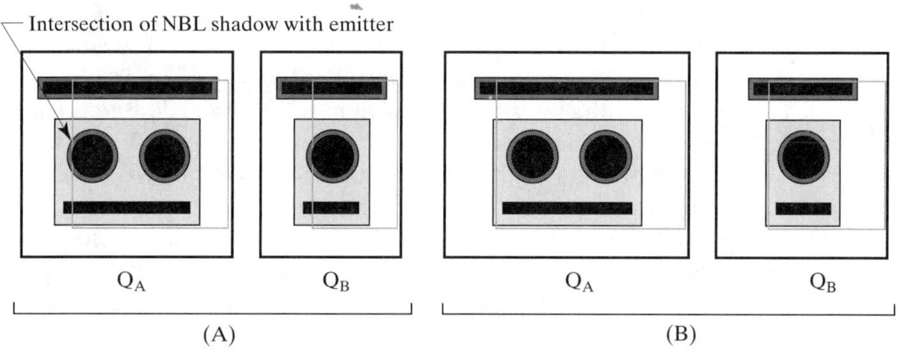

FIGURE 9.20 (A) NBL shadow causes mismatch between two transistors. (B) This mismatch is eliminated by oversizing NBL to prevent intersection of NBL and emitter.

Several ways exist to prevent pattern shift from causing mismatches. One approach consists of replacing the multiple-emitter transistor Q_A with two single-emitter transistors that are identical to Q_B. The NBL shadow will now intersect all three emitters in exactly the same manner, and the resulting systematic variations should cancel one another. Unfortunately, pattern distortion can also produce random variations in the NBL shadow that do not cancel one another. These can only be avoided by ensuring that the NBL shadow does not intersect the active area of the transistor, which, in the case of NPN transistors, is defined by the emitter diffusion.

If the direction of pattern shift is known, then the transistors can be laid out in a CEB array in which the main axis of symmetry lies parallel to the direction of pattern shift. In this way, the shift will displace the NBL shadow into either the collector contact or the base contact. The space required by these contacts usually suffices to prevent the NBL shadow from reaching the emitter. Pattern shift in (111) silicon usually occurs along the <211> axis. The direction of pattern shift in tilted (100) silicon depends on the direction of the tilt, which may vary from one manufacturer to the next. In order to properly orient the transistor array, the designer must determine the relationship between the X–Y coordinates of the layout and the wafer orientation. This information can usually be obtained by microscopic examination of a die. Planarization will obscure the NBL shadow, but a wafer can be removed prior to planarization for examination. Remember that the reticle array may differ from one device to another depending on the choices made during pattern generation, and that rotated or reflected reticle arrays will alter the apparent direction of pattern shift.

Sometimes the direction of pattern shift is not known, or layout considerations preclude the use of a specific orientation. In such cases, the overlap of NBL over emitter can be increased to prevent the NBL shadow from intersecting the emitter (Figure 9.20B). If no data exists on the magnitude of the pattern shift, then the designer should overlap the NBL over the emitter by 150% of the epi thickness.

Some designers eliminate the NBL shadow by omitting NBL from the matched transistors. Without NBL, the collector resistance of an NPN transistor can reach several kilohms. The V_{CEO} of the transistor may also diminish due to punchthrough of the lightly doped collector. CDI NPN transistors are particularly vulnerable to punchthrough due to the extremely light doping of the lowest portions of the N-well. One should not remove NBL from transistors unless characterization data indicates that the resulting device will function properly without it.

Lateral PNP transistors generally do not suffer from mismatches caused by the NBL shadow. The width of the collector usually suffices to prevent the shadow from intruding into the base region. If the NBL shadow does fall across the base region, it can cause mismatches by disturbing the flow of carriers from emitter to collector. These mismatches are easily eliminated by reorienting the transistors or by enlarging the NBL region, as has been discussed. One must never eliminate NBL from a lateral PNP, because this would greatly reduce its collection efficiency.

9.2.4. Thermal Gradients

Bipolar transistors are extremely sensitive to thermal gradients. The base-emitter voltage V_{BE} exhibits a temperature coefficient of about −2 mV/°C, corresponding to a collector current temperature coefficient of about 80,000 ppm/°C. Matched bipolar transistors are routinely expected to achieve offset voltages of less than ±1 mV, corresponding to a temperature difference of only ±0.5°C. Temperature variations of this magnitude can occur in almost any integrated circuit.

Matched bipolar transistors are often used to construct differential pairs, ratioed pairs, and ratioed quads. A *differential pair* (also called a *diff* pair, an *emitter-coupled pair,* or a *long-tailed pair*) consists of two matched bipolar transistors whose emitters are connected as in Figure 9.21A. The input stages of amplifiers and comparators often consist of differential pairs whose collectors terminate into matched resistors or current mirrors. The input offset voltage of a bipolar amplifier or comparator depends largely upon the matching of the input differential pair. Various trimming schemes can reduce the random component of the offset voltage to a fraction of a millivolt. Trimming also minimizes the temperature coefficient of the offset voltage as well as its absolute value,[13] so it is often used to minimize the vulnerability of high-gain amplifiers to a phenomenon called *thermal feedback.*

---

[13] Mismatch in a differential pair has the same impact as deliberate ratioing of emitter areas; The ΔV_{BE} voltage so developed has a large positive temperature coefficient. Therefore, minimizing the offset at one temperature also tends to minimize temperature variability.

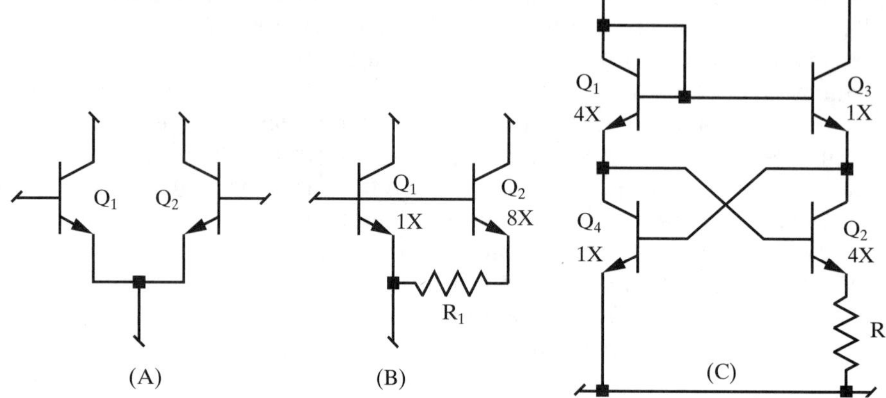

FIGURE 9.21 Three circuits containing matched NPN transistors: (A) differential pair, (B) ratioed pair, and (C) ratioed quad.

Thermal feedback occurs when one portion of a circuit influences another through thermal interactions rather than electrical ones. Changes in voltage or current within relatively high-power circuits (such as the output stages of an amplifier) produce localized temperature fluctuations that, in turn, generate small offsets between devices in the input stages of the circuit. The circuit then amplifies these offsets as if they were an electrical signal. The amplified offsets can produce further temperature variations, possibly even leading to oscillations. Since many amplifiers have voltage gains in excess of 10,000, even a very weak thermal interaction can cause significant thermal feedback. The frequency response of many commercial operational amplifiers contains low-frequency poles and zeros caused by this mechanism.[14] Thermal feedback can be minimized by increasing the separation of the input and output stages and by reducing the thermal sensitivity of the input stage. Many operational amplifiers place the input circuit on one side of the die and the output circuit on the other. Even so, thermal coupling remains a serious problem. The input differential pair of a high-gain amplifier should always be located and constructed to achieve the highest possible degree of matching in order to minimize its sensitivity to thermal variations.

A *ratioed pair* consists of two bipolar transistors whose emitter areas are in integer ratio. Assuming that the two transistors conduct equal currents, then their base-emitter voltages will differ by an amount ΔV_{BE} equal to

$$\Delta V_{BE} = V_T \ln\left(\frac{A_1}{A_2}\right) \qquad [9.9]$$

where V_T equals the thermal voltage (26 mV at 25°C), and A_1 and A_2 represent the emitter areas of transistors Q_1 and Q_2.[15] The thermal voltage scales linearly with absolute temperature;[16] therefore, ΔV_{BE} is a *voltage proportional to absolute temperature* (VPTAT). The VPTAT produced by a ratioed pair of NPN transistors remains linear with temperature and independent of current over a remarkably wide range

[14] J. E. Solomon, "The Monolithic Op Amp: A Tutorial Study," *IEEE J. Solid-State Circuits,* Vol. SC-9, #6, 1974, pp. 314–332.

[15] This derivation ignores the ideality factor (or emission coefficient) η, which is usually very near unity for NPN transistors operating at moderate current levels.

[16] An *absolute temperature* is one measured with respect to absolute zero. The SI unit of absolute temperature is the Kelvin degree (K), which has the same magnitude as the Celsius degree (°C); 0°C ≅ 273 K and 25°C ≅ 298 K. The thermal voltage V_T equals kT/q, where k is Boltzmann's constant ($1.38 \cdot 10^{23}$ J/K), T is the absolute temperature (in K), and q is the charge on the electron ($1.60 \cdot 10^{-19}$ C).

of operating conditions. A resistor connected between the emitters of a ratioed pair (as in Figure 9.21B) can transform this VPTAT into a *current proportional to absolute temperature,* or IPTAT.[17] IPTAT circuits form the basis of many precision voltage and current references.

In order for VPTAT and IPTAT circuits to operate properly, the ratioed pair must match very precisely. The most common ratio used in VPTAT and IPTAT circuits is probably 8:1, which produces a ΔV_{BE} of 54 mV. A 1 mV mismatch in such a circuit would produce approximately a 2% error in the voltage or current. A typical trimmed voltage reference must produce a voltage that varies no more than ±1% over all possible operating conditions. Poor layout often leads to excessive output voltage variation with input voltage (poor *line regulation*) or to excessive output voltage variation with output current (poor *load regulation*). Both of these problems often stem, at least in part, from thermal feedback.

A *ratioed quad* is essentially a variation on the ratioed pair. The four transistors of the quad produce a VPTAT voltage that is usually imposed across a resistor to produce an IPTAT (Figure 9.21C). The VPTAT voltage ΔV_{BE} equals

$$\Delta V_{BE} = V_T \ln\left(\frac{A_1 A_2}{A_3 A_4}\right) \qquad [9.10]$$

where A_1 to A_4 are the emitter areas of transistors Q_1 to Q_4, respectively. The ratioed quad can provide a much larger ΔV_{BE} than a simple ratioed pair because the sizes of Q_1 and Q_2 multiply together. Two 4X transistors can produce a ΔV_{BE} of 72 mV, while two 8X transistors can generate 108 mV.

The extreme thermal sensitivity of bipolar transistors requires that matched devices be laid out to cancel thermal gradients. Critical matched devices almost always employ common-centroid layout techniques similar to those discussed in Section 7.2.10. Differential pairs usually employ the two-dimensional common-centroid layout shown in Figure 9.22. This configuration is popularly called a *cross-coupled quad.*[18] Common-centroid layouts help minimize the impact of thermal variations, but they cannot completely cancel nonlinear thermal variations. The exponential nature of the I_C-vs.-V_{BE} relationship sharply limits the cancellation—an extremely important criterion for common-centroidal layouts. Compactness is thus a property of matched bipolar arrays. Most of the more complex common-centroidal arrangements lack compactness and are therefore inferior to the simple cross-coupled quad.

The VPTAT voltage developed by a ratioed pair increases as the logarithm of the area ratio, while offsets produced by thermal and stress gradients increase roughly as the square root of the ratio. As the area ratio increases, a point is eventually reached beyond which mismatch increases more rapidly than VPTAT. For any given design, there exists a ratio that will provide optimal matching. This optimal ratio depends on many factors, but in most cases it probably lies somewhere between 8:1 and 16:1. Even-number ratios greatly simplify the task of constructing common-centroid layouts, and smaller ratios consume less space, so 8:1 is probably the most popular choice of ratio. Similar arguments lead to ratioed quads of 4:1:1:4.

The layouts of ratioed pairs and ratioed quads should provide a very compact and symmetric layout. In a ratioed pair, one device usually possesses a single emitter, the other multiple emitters. One simple layout for a ratioed pair places the single-emitter

---

[17] In actuality, the temperature coefficient of the resistor will distort the linearity of the IPTAT. Most circuits use the IPTAT current to regenerate a VPTAT across another resistor. If the two resistors have the same temperature coefficient, then this can be (and is) neglected.

[18] Grebene, pp. 348–349, 365.

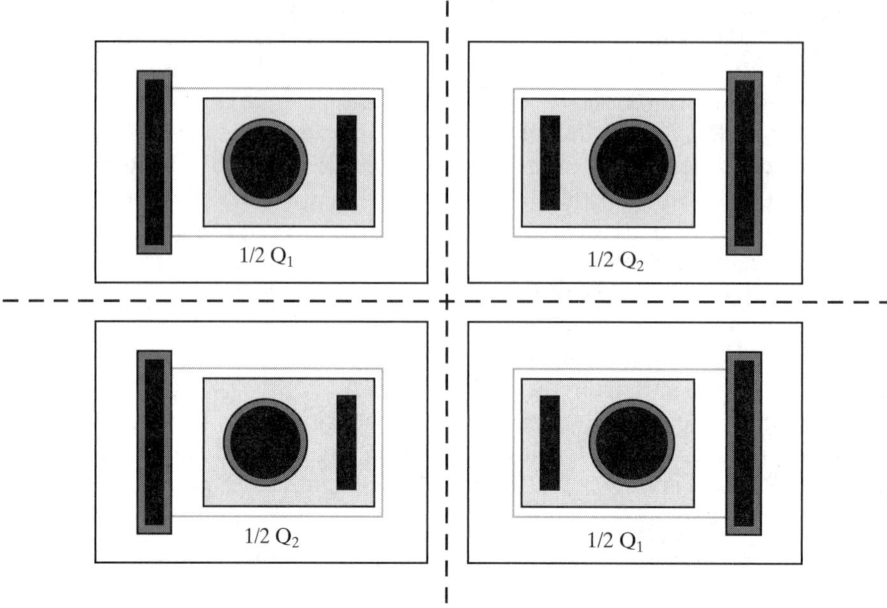

FIGURE 9.22 Example of cross-coupled bipolar transistors.

device in the middle of two halves of the multiple-emitter device. The simplest structure embodying this arrangement is a one-dimensional common-centroid layout following the pattern ABA (Figure 9.23A). The placement of all of the emitters in a line generates a secondary axis of symmetry S_2 that enables the structure to reject thermal gradients perpendicular to the row of transistors, while the primary axis of symmetry S_1 rejects thermal gradients parallel to the line. The elongated shape of the array makes it more difficult to cancel gradients around S_1. Therefore, arrays of this sort should be oriented so that the secondary axis S_2 lies parallel to the expected isotherms.

A more compact arrangement can be achieved for ratios that are multiples of 4:1. The emitters of the larger transistor Q_2 can then be arranged in two rows around an axis of symmetry S_2 (Figure 9.23B). Axis S_2 should also pass through the center of the single-emitter transistor Q_1. This arrangement is particularly beneficial for large ratios, such as 16:1, that would otherwise produce very elongated layouts. As before, the array should be oriented so that the secondary axis of symmetry S_2 lies parallel to the expected isotherms.

Ratioed quads are laid out as if they consisted of a pair of ratioed mirrors. Both the upper pair (Q_1 and Q_3 in Figure 9.21C) and the lower pair (Q_2 and Q_4) can employ layouts similar to those in Figure 9.23. Ideally, the two pairs should lie one above the other so that their primary axes of symmetry (S_1) coincide. This converts the entire arrangement into a two-dimensional common-centroid array. If for some reason this arrangement is not feasible, then each of the two pairs can be treated as an independent ratioed pair. A reasonable degree of matching will be achieved even if the two ratioed pairs reside some distance apart.

Some designers advocate laying out ratioed pairs by using a circular array in which the smaller device occupies the center of a ring-shaped array of emitters forming the larger device. If the larger device contains a multiple of four emitters, then this arrangement will possess both horizontal and vertical axes of symmetry as well as a number of subsidiary axes dependent on the total number of emitters. Although this arrangement has a high degree of symmetry, symmetry is less important then compactness.

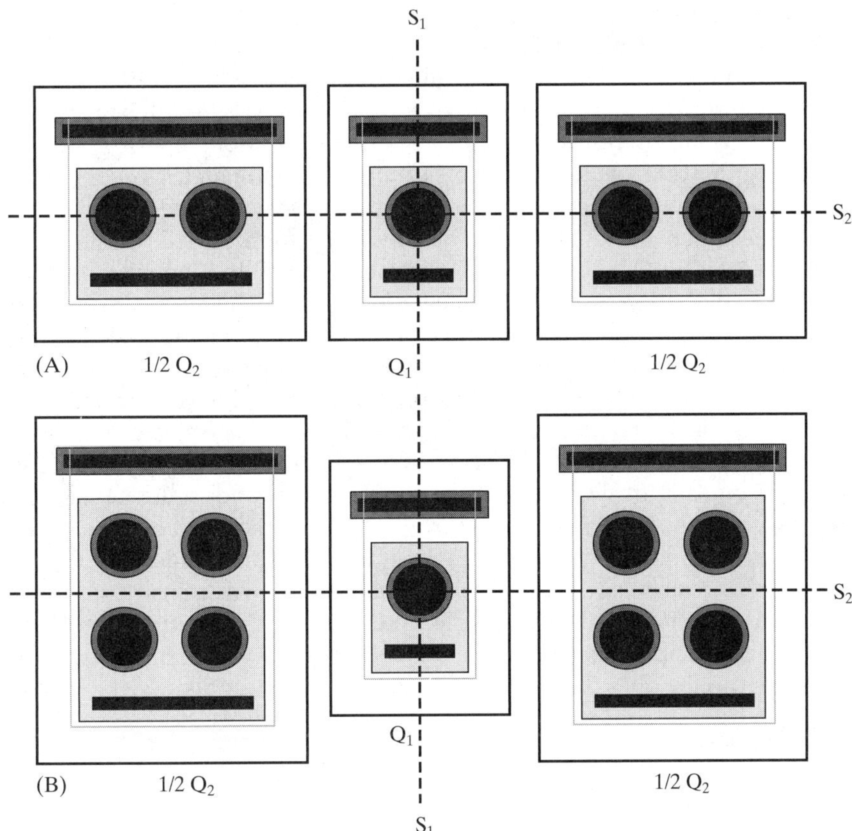

FIGURE 9.23 Two common-centroid layouts frequently used for constructing ratioed pairs.

Therefore, the layouts of Figure 9.23 are generally recommended over circular arrangements.

9.2.5. Stress Gradients

Mechanical stress can induce mismatches between bipolar transistors by altering their base-emitter voltages or by reducing their betas. The base-emitter voltage of a transistor depends on the bandgap voltage of silicon, which varies slightly under stress.[19] The degradation in beta under mechanical stress is principally due to mobility variations induced by piezoresistivity.[20] Together these effects can easily produce several millivolts of offset.

Assembled integrated circuits almost always experience a certain amount of stress. Plastic-molded devices are cured at elevated temperatures, and the mold compound elastically deforms as the assembled units cool. The resulting package stresses cause the base-emitter voltages of bipolar transistors to shift and can produce offsets between matched pairs of devices. These *package shifts* cannot be entirely trimmed out at wafer probe because they only appear after packaging. Packaging stresses vary with temperature, so even postpackage trimming cannot entirely counteract them, stress-relieving overcoats may help minimize the impact

---

[19] J. J. Wortman, J. R. Hauser, and R. M. Burger, "Effects of Mechanical Stress on p-n Junction Device Characteristics," *J. Applied Physics*, Vol. 35, #7, 1964, pp. 2122–2131.

[20] H. Mikoshiba and Y. Tomita, "Piezoresistance as the Source of Stress-induced Changes of Current Gain in Bipolar Transistors," *Solid State Electronics*, Vol. 25, #3, 1982, pp. 197–199.

of package shifts, provided that proper common-centroid design techniques have been employed (Section 9.2.6).

Common-centroid layout techniques can greatly reduce the impact of stress on matched transistors. These techniques effectively move the matched transistors closer together and therefore equalize the stresses on them. Unfortunately, even the best common-centroid arrangements cannot cancel the higher order components of the stress gradient. Matched transistors should therefore reside in low-stress regions of the die. Figure 9.24A shows the best locations for matched transistor arrays on a die fabricated in (100) silicon. These layouts assume that no significant sources of heat exist in the vicinity of the matched transistors. The best locations lie near the center of the die where the magnitude of the stresses reaches a broad minimum. The major axis of symmetry of the array S_1 should lie along one of the axes of symmetry of the die. This helps ensure that the isobars lie parallel to the secondary axis of the array S_2. If the matched transistors must reside along the side of the die, then they should be placed in the center of one side so that the primary axis of the array S_1 aligns to one axis of symmetry of the die. If the die is not square, then a location in the middle of the longer side is preferable to one in the middle of the shorter side. If possible, the transistors should reside at least 250 μm inside the edge of the die. Under no circumstances should critical matched bipolar transistors be placed near the corners of a die, as these experience excessively large stress gradients. In summary, the principles that guide the placement of matched bipolar transistors are analogous to the ones that apply to matched resistors, as discussed in Section 7.2.10.

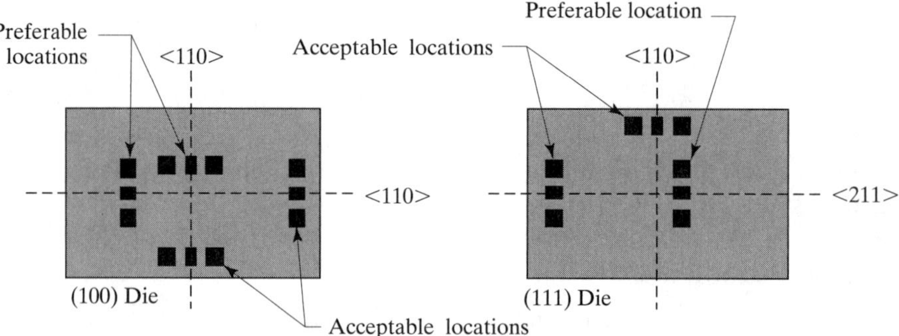

FIGURE 9.24 Locations for placing common centroid bipolar transistor arrays on (100) and (111) dice, in the latter case assuming an axis of symmetry exists in the stress distribution around the <211> axis. (Compare with Figure 7.19.)

Figure 9.24B shows the best locations for critical matched transistors on a die fabricated in (111) silicon. Since isobars tend to lie symmetrically around the <211> axis, matched transistors ideally should reside around this axis instead of around the <110> axis. Locations near the center of the die will again give the lowest overall stress, although locations near either end of the die will provide acceptable matching. Less-critical matched transistors can also be placed on the <110> axis of symmetry. Again, matched transistors should reside at least 250 μm inside the die edge to avoid the increased stress gradients along the edges of the die, and matched transistors should never reside near the corners of the die because of the high stresses there.

Figure 9.25 shows the compromises required when the die contains both a significant heat source and critical matched transistors. Bipolar transistors are more susceptible to thermal gradients than to stress gradients, so the larger the distance between the heat source and the matched transistors, the better the matching. The matched transistors should reside at least 125 to 250 μm away from the edge of the

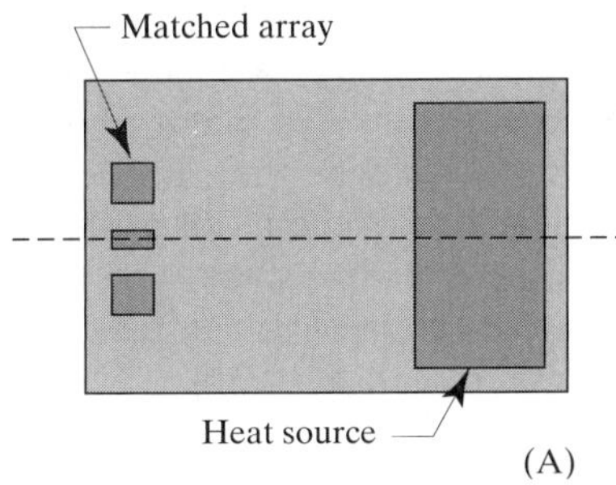

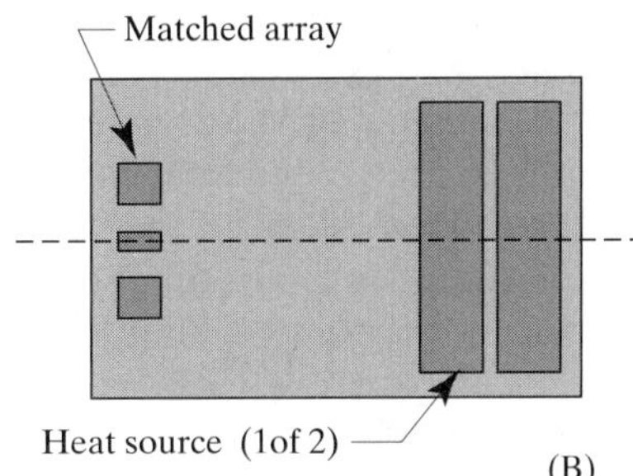

FIGURE 9.25 Layouts for matched transistor arrays on dice that contain one or more power devices.

die opposite the heat source. This location should provide better overall matching than a location near the center of the die, even though the stresses are greater near the edges than at the center.

High-power integrated circuits may benefit from judicious application of emitter degeneration to transistors that would normally have sufficient area to match quite precisely on their own. High levels of power dissipation produce correspondingly large thermal gradients, especially in dice mounted on heat-sunk leadframes. These gradients have a much stronger effect on bipolar transistors than on passive components, so the matching of degenerated transistors usually proves superior to the matching of undegenerated ones. Emitter degeneration is particularly beneficial for matched transistors residing next to a large heat source; without degeneration, such transistors often exhibit mismatches of many millivolts. Whenever emitter degeneration resistors are added to critically matched bipolar transistors, the resistors must be carefully laid out to ensure that they actually improve matching rather than degrade it. Poorly constructed degeneration resistors can easily ruin the matching of an otherwise well-laid-out circuit.

9.2.6. Filler-Induced Stress

The mechanical stresses that arise during packaging stem from two separate mechanisms, both of which are driven by unequal coefficients of thermal expansion between the die and the packaging materials. Stress gradients arise from forces applied to the die as a whole. The resulting pattern of forces varies gradually and predictably across the die. Filler-induced stresses are caused by irregular particles of silica added to the plastic encapsulant. The contraction of the plastic forces the particles against the surface of the die, creating highly localized stresses on underlying components (Section 7.2.9). Filler-induced stresses are essentially random, while stress gradients are largely systematic in character.

The impact of stresses upon the die can be quantified by measuring the package shifts that they generate. The package shift Δx in a parameter x for a single unit equals the difference between the value measured after assembly x_a and the value measured before assembly at wafer probe x_w:

$$\Delta x = x_a - x_w \qquad [9.11]$$

In practice, one normally measures a large number of units at wafer probe and after assembly. The mean and standard deviation of the package shift are then computed using the equations

$$m_{\Delta x} = m_{xa} - m_{xw} \qquad [9.12A]$$

$$s_{\Delta x} = \sqrt{s_{xa}^2 - s_{xw}^2} \qquad [9.12B]$$

where m_{xw} and s_{xw} are the mean and standard deviation of the wafer probe data, and m_{xa} and s_{xa} are the mean and standard deviation of the postassembly data.

One can quickly determine if filler particles generate significant mismatch by comparing the mean and standard deviation of the package shift. If filler-induced stresses are not significant, then the standard deviation of the package shift will be far smaller than the mean. If the standard deviation is comparable to the mean, or if it exceeds the mean, then filler-induced stresses play a major role in generating the package shift.

Some data have been reported for package shifts in bandgap voltages. A *bandgap reference* uses bipolar transistors to generate a voltage that remains approximately constant over temperature. Figure 9.26 shows the essential components of a Brokaw bandgap circuit.[21] Transistors Q_1 and Q_2 form a ratioed pair. Amplifier X_1 adjusts voltage V_{bg} until the collector currents I_{C1} and I_{C2} equal one another. Under these conditions, the *bandgap voltage* V_{bg} equals

$$V_{bg} = V_{Be1} + \frac{2R_1}{R_2}\ln(N) \tag{9.13}$$

When the resistors are adjusted to minimize the temperature coefficient of V_{bg}, this voltage equals about 1.25 V. In one family of bandgap references, assembled in SOT-23 packages, the 25°C package shift exhibited a mean of −1.1 mV and a standard deviation of 2.3 mV.[22] The majority of this package shift stems from the base emitter voltage V_{be1}. Compressive stress upon a base-emitter junction causes its forward-bias voltage to drop.[23] As filler-induced stresses are largely compressive, one would expect that they would produce a downward shift in bandgap voltage. This observation accounts for most of the −1.1 mV mean package shift observed in the preceding example. The package shift in these SOT-23 references was probably primarily (if not solely) due to filler-induced stress.

Layout has little effect upon mismatches generated by filler-induced stresses. Filler particles are typically no more than a few hundred microns in diameter. The area of contact between a filler particle and the underlying die is even smaller. The

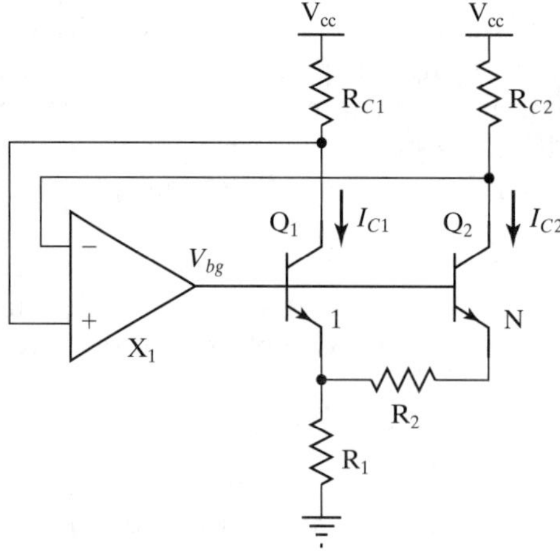

FIGURE 9.26 Simplified schematic of the critical matched components of a Brokaw bandgap reference.

---

[21] A. P. Brokaw, "A Simple Three-Terminal IC Bandgap Reference," *IEEE J. Solid-state Circuits*, Vol. SC-9, #6, 1974, pp. 388–393.

[22] Author, unpublished data.

[23] J. J. Wortman, et. al.

stresses generated by fillers are thus so localized that common-centroid layout techniques are of little or no value. Instead, one must resort to the use of alternative packaging materials or mechanically compliant overcoats, as discussed in Section 7.2.9. Thick copper metallization has proven to be a valuable resource in this regard. The package shift in one particular bandgap was altered from $m_{\Delta Vbg} = -5.06$ mV, $s_{\Delta Vbg} = 2.64$ mV to $m_{\Delta Vbg} = -2.26$ mV, $s_{\Delta Vbg} = 1.38$ mV by the addition of a 15 μm-thick layer of copper on top of the nitride protective overcoat.[24] The thick copper effectively halved the package shift in this design. Again, the relatively large value of the standard deviation suggests that the majority of the package shift was due to filler-induced stress.

9.2.7. Other Causes of Systematic Mismatch

Many other mechanisms can potentially cause mismatches between bipolar transistors. For example, imbalanced collector-emitter voltages can generate systematic mismatches due to Early voltage or collector efficiency limitations of the devices. Cross injection in merged lateral PNP transistors and depletion region intrusion into the base of a multiple-emitter NPN transistor can also cause serious systematic mismatches. This section will briefly discuss each of these mechanisms.

A difference in the collector-emitter voltages of two bipolar transistors can generate a systematic mismatch between them. These mismatches can arise from either of two mechanisms, one caused by Early voltage and the other by collector efficiency. The *Early voltage* V_A of a bipolar transistor quantifies the influence of collector-to-emitter voltage V_{CE} upon collector current I_C:

$$I_C = I_S e^{V_{BE}/V_T}\left(1 + \frac{V_{CE}}{V_A}\right) \qquad [9.14]$$

The Early effect occurs because the majority of the collector-to-emitter voltage appears across the reverse-biased collector-base junction. As the voltage across this junction increases, the depletion region intrudes farther into the neutral base of the transistor. The narrowing of the base increases the beta of the transistor, and this in turn increases the collector current.[25] A vertical NPN transistor typically has an Early voltage of 100 V. A 1 V change in V_{CE} would thus produce a 1% change in collector current. If two matched bipolar transistors operate at different collector-to-emitter voltages, then the Early effect will produce a systematic error between their collector currents. This error can be eliminated simply by operating the two transistors at the same collector-to-emitter voltages. A variety of circuit techniques have been developed that allow circuit designers to achieve this goal.

Collector efficiency in lateral PNP transistors also varies with collector-to-emitter voltage. A saturating lateral PNP will obviously have a dramatically reduced collector efficiency, so one cannot expect a saturating lateral transistor to match one operating in the forward active region. More subtly, collector efficiency also varies in the active region of operation due to variations in the depth of the base-collector depletion region. Higher collector-to-emitter voltages drive the collector-base depletion region further into the collector and thus improve the collector efficiency. Variations in collector efficiency translate into variations in collector current. This effect can cause systematic mismatches in collector currents unless one operates matched devices at equal collector-to-emitter voltages.

---

[24] B. Abesingha, G. A. Rincón-Mora, and D. Briggs, "Voltage Shift in Plastic-Packaged Bandgap References," *IEEE Trans. on Circuits and Systems—II: Analog and Digital Signal Processing*, Vol. 49, #10, 2002, pp. 681–685.

[25] J. M. Early, "Effects of Space-Charge Layer Widening in Junction Transistors," *Proc. IRE*, Vol. 40, 1952, pp. 1401–1406.

Cross injection between merged lateral PNP transistors can create differences in collector efficiency between devices that result in collector current mismatches. The most extreme example of cross injection involves multiple lateral PNP transistors merged into a common tank, one of which saturates. The saturating transistor injects carriers into the tank that are collected by the other transistors. This type of cross injection can cause extremely large mismatches. Subtler mismatches can occur due to cross injection between devices operating in the forward active region. Some carriers will escape from beneath the collector of each device, and these carriers can potentially flow to other devices merged in the same tank. The simplest and surest way of eliminating cross injection is to place each lateral PNP in its own tank. Some designers have used N-bars (Section 4.4.2) to achieve the same goal with slightly less area.

Depletion regions are responsible for another systematic mismatch mechanism in vertical bipolar transistors that incorporate multiple emitters in a common base region (such as is shown in Figure 9.18B). The base-emitter depletion region extends primarily into the lightly doped base. If two emitters are placed at the minimum allowed distance from one another, then the depletion regions around these two emitters approach one another quite closely. The neutral base region between the two emitters becomes very thin. Voltage drops in this base region cause the portions of the emitters that face one another to become debiased and to thus conduct less current. A similar effect can occur if the base overlap of the emitter is insufficiently large. These mismatches can be avoided by increasing the emitter-to-emitter spacing and the base overlap of the emitter by several microns beyond the minimums specified by the layout rules.

9.3 RULES FOR BIPOLAR TRANSISTOR MATCHING

The previous section explained the mechanisms that cause mismatch in bipolar transistors. This section attempts to condense this information into a set of qualitative rules that will allow a designer to construct matched bipolar transistors with some degree of confidence, even if (as is usually the case) quantitative matching data are not available.

The following rules use the terms *minimal, moderate*, and *precise* to denote increasingly precise degrees of matching. These terms should be interpreted as follows:

- **Minimal matching:** Offset voltages of ±1 mV or collector current mismatches of ±4%. This is suitable for constructing input stages of op-amps and comparators that must achieve offsets of ±3 to 5 mV without trim. It is also suitable for use in current mirrors for biasing noncritical circuitry.
- **Moderate matching:** Offset voltages of ±0.25 mV or collector current mismatches of ±1%. This level is suitable for use in ±1% bandgap references and in op-amps and comparators that must achieve ±1 to 2 mV without trim. Since lateral transistors have difficulty maintaining this degree of matching, most untrimmed, moderately matched circuits use vertical NPN transistors instead.
- **Precise matching:** Offset voltages of ±0.1 mV or collector current mismatches of ±0.5%. This level of matching usually requires trimming or the addition of precisely matched degeneration resistors. Proper layout is still important because degeneration and trimming cannot entirely eliminate the effects of thermal gradients or package shifts. Lateral transistors cannot obtain this degree of matching unless they are heavily degenerated and the circuitry incorporates

some means of base current cancellation. Circuits requiring precise matching usually employ heavily degenerated vertical NPN transistors.

9.3.1. Rules for Matching Vertical Transistors

Vertical transistors inherently match better than lateral transistors because they are not subject to the vagaries of surface conduction. Most processes optimize the performance of their vertical NPN transistors at the expense of their lateral PNP transistors, which only strengthens the case for using NPN transistors. The following rules summarize the principles of designing matched vertical transistors:

1. Use identical emitter geometries.
   Transistors with different sizes or shapes of emitters match *very* poorly. Even minimal matching requires the use of identical emitter geometries. Matched transistors are therefore restricted to ratios of small integer numbers. The geometry of the base and collector regions matters much less than the geometry of the emitter region. Multiple emitters can thus reside in a common base region.

2. The emitter diameter should equal 2 to 10 times the minimum allowed diameter.
   The minimum diameter of the emitter equals the minimum contact width plus twice the minimum emitter overlap of contact. For example, a process having a minimum contact width of 2 μm and a minimum overlap of 1 μm has a minimum emitter diameter of 4 μm. Matched emitters in this process should have diameters of 8 to 40 μm. Emitter areas at the lower end of this range suffice for minimal matching. Moderate and precise matching generally require the use of larger emitters, but the presence of power devices may justify the use of smaller emitters to produce a more compact structure that is less susceptible to thermal gradients.

3. Maximize the emitter area-to-periphery ratio.
   For a given emitter area, the transistor with the largest area-to-periphery ratio produces the best possible matching. Circular geometries provide the highest area-to-periphery ratios, but octagonal and square emitters are almost as good.

4. Place matched transistors in close proximity.
   Bipolar transistors are very sensitive to thermal gradients. Even minimally matched transistors should reside within a few hundred microns of one another. Moderately or precisely matched transistors should use common-centroid layout techniques to minimize the separation between the transistors.

5. Keep the layout of matched transistors as compact as possible.
   The use of common base and collector regions may cause slight mismatches, but the increase in compactness usually more than compensates. Layouts that arrange the emitters in tight clusters generally provide better matching than layouts that place them in a line. A pair of matched transistors of equal sizes should employ a cross-coupled layout.

6. Construct ratioed pairs and quads using even integer ratios between 4:1 and 16:1.
   Ratios that are too small or too large will match less well than those that lie within a certain range, typically between 4:1 and 16:1 for ratioed pairs and between 4:1:1:4 and 8:1:1:8 for ratioed quads. The ratios used for quads tend to be smaller than those used for pairs, because quads develop larger VPTAT voltages for a given number of unit emitters.

7. Place matched transistors far away from power devices.
   Power devices represent a significant threat to bipolar transistor matching. Minimally matched transistors should lie at least 250 μm away from major

power devices (those dissipating 250 mW or more) and should not reside adjacent to any power device dissipating more than 50 mW. Moderately matched devices should lie at least 100 to 250 μm from any device dissipating more than 50 mW and should be placed at the opposite end of the die from major power devices. Precisely matched devices should be separated as far as possible from any power device. Consider elongating the die to a 1.5:1 or even a 2:1 aspect ratio to increase the separation between precisely matched transistors and major power devices. Power dissipations of a watt or more generally preclude precise matching unless the transistors are heavily degenerated.

8. Place matched transistors in low-stress areas.
   The presence of any significant heat source on the die precludes placing the matched transistors in the center, because they would lie too close to the heat source. In this case, moderately matched transistors should occupy the middle of the opposite end of the die from the heat source. Moderately matched transistors should not reside within about 250 μm of an edge of the die because stress levels increase near edges. Similarly, moderately matched transistors should be kept well away from the corners of the die where the stresses are greatest. Precise matching is very difficult to maintain in the presence of large thermal gradients.

9. Place moderately or precisely matched transistors on axes of symmetry.
   Moderately or precisely matched transistor arrays should be oriented so that their major axis of symmetry, S_1, coincides with one of the axes of symmetry of the die. If possible, matched arrays should reside around the <211> axis of a (111) die rather than the <110> axis.

10. Do not allow the NBL shadow to intersect matched emitters.
    The NBL region of a moderately or precisely matched transistor should overlap its emitters by a distance sufficient to ensure that it does not intersect them. If the direction of NBL shift is unknown, allow adequate overlap of NBL on all sides of the emitter. If the magnitude of the shift is unknown, then overlap NBL over the emitter by at least 150% of the maximum epi thickness. Minimally matched transistors can forgo this precaution because the impact of the NBL shadow is relatively small.

11. Place emitters far enough apart to avoid interactions.
    If multiple emitters must occupy a common base region, then space them far enough apart to prevent their depletion regions from intersecting. If the layout rules specify the spacing between unconnected emitters, use this rule for matched emitters regardless of how they are connected. If no such rule exists, then the spacing between the matched emitters should exceed the minimum spacing by 2 to 3 μm.

12. Increase the base overlap of moderately or precisely matched emitters.
    If the base barely overlaps the emitter, misalignment can cause the lateral beta of a portion of the emitter periphery to increase enough to produce minor mismatches. The base region of a moderately or precisely matched transistor should overlap its emitter by 1 to 2 μm more than minimum.

13. Operate matched transistors on the flat portion of the beta curve.
    The β-*vs.*-I_C plot of an NPN transistor usually exhibits a plateau across a relatively broad range of currents. Whenever possible, matched transistors should operate on this plateau. Transistors operating at higher currents will suffer from high-level injection that may induce mismatches. In ratioed pairs and quads,

high-level injection can cause deviations from the theoretical VPTAT voltages. Most NPN transistors do not enter high-level injection except at relatively high currents that can also cause undesirable self-heating. Transistors operated at a current level that is too low are prone to variations in beta due to surface effects. The low-current beta rolloff of most vertical NPN transistors occurs only at extremely low current densities, so this effect rarely interferes with device matching.

14. The contact geometry should match the emitter geometry.
A circular emitter should contain a concentric circular contact. Similarly, an octagonal emitter should contain an octagonal contact and a square emitter should contain a square contact. If the process allows only minimum-size square contacts, then use square emitters and square arrays of minimum contacts. The emitter contacts should fill as much of the emitter area as possible, except in cases where silicidation must be minimized to prevent beta degradation. These precautions help prevent interactions between the contact and the edge of the emitter from distorting the flow of emitter current.

15. Consider using emitter degeneration.
Minimally matched transistors will not normally benefit from emitter degeneration. Moderately matched transistors may benefit from degeneration in the presence of large thermal gradients. Precisely matched transistors are often degenerated, if for no other reason than to allow adjustment of their offset voltage by trimming the degeneration resistors. The degenerating resistors should develop at least 50 mV for moderate matching and 100 mV for precise matching. Emitter degeneration can also be used to match transistors with different emitter sizes or geometries. In this case, at least 200 mV of degeneration should be employed for minimal matching and 500 mV for moderate matching. This technique can achieve noninteger ratios between matched transistors. For example, a 1.64:1 ratio can be constructed from two transistors having equal emitter areas and a pair of emitter degeneration resistors with a ratio of 1:1.64.

16. Operate moderately or precisely matched transistors at equal collector-to-emitter voltages.
The Early effect can induce systematic collector current mismatches between devices operating at different collector-to-emitter voltages. The magnitude of this mismatch depends upon the Early voltage of the devices. Vertical transistors typically have Early voltages of 100–300 V, which correspond to mismatches of 0.3–1%/V. Various circuit design techniques—such as the insertion of cascode devices—can ensure that matched transistors operate at equal voltages.

17. Do not allow the base-emitter junctions of matched devices to avalanche.
Avalanche breakdown of the base-emitter junction reduces the beta of a vertical transistor (Section 4.3.3). The impact ionization responsible for this degradation mechanism actually begins at voltages somewhat below the apparent breakdown voltage. The resulting mismatches in beta may indirectly produce differences in collector currents through interaction with resistances in the base circuit of the devices. The input differential pairs of bipolar amplifiers and comparators are particularly susceptible to this problem because they are directly exposed to pins, and they can therefore suffer degradation from electrical overstress and ESD events. Matched transistors should therefore not operate with reverse biases across their base-emitter junctions that exceed 50% of the emitter-base breakdown voltage V_{EBO}. Devices connected to external pins should include clamps or other protective structures that limit the reverse bias placed across their base-emitter junctions.

9.3.2. Rules for Matching Lateral Transistors

Lateral transistors generally do not match as well as vertical transistors. Their poorer matching is due partly to surface effects and partly to an inability to use large emitters. Emitter degeneration is frequently used to improve the matching of lateral PNP current mirrors and whatever other circuits can tolerate its presence. The following rules summarize the principles of designing matched lateral transistors:

1. Use identical emitter and collector geometries.
   Both the emitter and the collector geometries affect conduction in lateral transistors. Transistors with different emitter or collector geometries match very poorly. For minimal matching, only the size and shape of the inner periphery of the collector facing the emitter matters. For higher degrees of precision, the entire collector geometry should be duplicated. The shape and size of the base region are unimportant as long as none of the transistors saturate. If a transistor can saturate, it is safest to place it in its own tank. P-bar or N-bar isolation schemes (Section 4.4.2) should not be counted on to ensure complete isolation between matched devices. Shallow-collector transistors (such as analog BiCMOS devices constructed from PSD implants) should be placed in separate tanks or wells to minimize cross injection caused by carriers passing underneath the shallow collectors.

2. Use minimum-size emitters for matched transistors.
   Larger emitters will degrade the beta of the transistor, and this effect usually hurts matching more than the increased area helps. Ratioed transistors should employ multiple copies of a minimum-emitter cell (Figure 8.26B).

3. Field-plate the base region of matched lateral PNP transistors.
   Field-plating ensures that electrostatic charges do not interfere with the flow of current across the neutral base. Improperly field-plated transistors are susceptible to long-term drifts that can play havoc with matching. Lateral PNP transistors constructed in analog BiCMOS processes that incorporate a channel stop implant across the neutral base generally do not require field-plating, because the channel stop performs this function. Still, the addition of field plates never hurts.

4. Split-collector lateral PNP transistors can achieve moderate matching.
   Moderate matching can be achieved only as long as all of the split collectors are identical copies of one another, and none of the collectors saturates. The presence of gaps between the collectors makes it impossible to accurately predict the division of current between split collectors of different sizes. The saturation of any split collector destroys the matching between the remaining split collectors. Split-collector laterals can be used to form very compact cross-coupled transistors that exhibit surprisingly precise matching.[26]

5. Place matched transistors in close proximity.
   Even minimally matched lateral PNP transistors should reside near one another to minimize the impact of thermal gradients. Moderately or precisely matched transistors may benefit from placement in a common base tank. If this is done, make sure that none of the transistors in the tank can saturate.

6. If possible, avoid constructing VPTAT circuits from ratioed lateral PNP transistors.
   An ideality factor ignored in the derivation of Equations 9.9 and 9.10 becomes significant in high-level injection, where lateral PNP transistors usually operate.

---

[26] Gilbert reports ±0.1% typical matching from cross-coupled split-collector lateral PNP transistors; this is presumably a one-sigma value. See B. Gilbert, "Bipolar Current Mirrors," in C. Toumazou, F. J. Lidgey, and D. G. Haigh, *Analogue IC Design: The Current-Mode Approach* (London: Peter Perigrinus, 1990), pp. 249–250.

The VPTAT voltages developed by ratioed mirrors and quads often exhibit significant deviations from the values predicted by the equations due to the contribution of the ideality factor.

7. Place matched transistors far away from power devices.
Minimally matched transistors should reside at least 250 μm away from major power devices and should not be placed adjacent to any device dissipating more than 50 mW. Moderately matched devices should reside at least 100 to 250 μm away from any device dissipating more than 50 mW, and they should be placed at the opposite end of the die from major power devices. Precisely matched devices should be separated as far as possible from any power device. Devices that dissipate a watt or more generally preclude precise matching unless the matched transistors are heavily degenerated. Consider elongating the die to a 1.5:1 or even a 2:1 aspect ratio to increase the separation between precisely matched transistors and major power devices.

8. Place matched transistors in low-stress areas.
Precisely matched transistors should occupy the center of the die, but the presence of any significant heat source generally precludes placing the matched transistors in the center of the die. Moderately matched transistors should instead occupy the middle of the end of the die opposite the heat source. They should not reside within about 250 μm of an edge of the die, and they should be kept well away from the corners of the die.

9. Place moderately or precisely matched transistors on axes of symmetry of the die.
Moderately or precisely matched transistor arrays should be oriented so that their major axis of symmetry, S_1, coincides with one of the axes of symmetry of the die. If possible, matched arrays should be placed on the <211> axis of a (111)-oriented die.

10. Do not allow the NBL shadow to intersect the base region of a lateral PNP.
The presence of the surface discontinuity that causes the NBL shadow distorts the flow of current across the neutral base of the transistor. If the direction of NBL shift is unknown, allow adequate overlap of NBL on all sides of the base region. If the magnitude of the shift is unknown, then overlap NBL over the base region by at least 150% of the maximum epi thickness. The NBL shadow will have little or no effect on matching if it merely intersects the collector of the transistor.

11. Operate matched lateral PNP transistors near peak beta.
The β *vs.* I_C of a lateral PNP transistor usually exhibits a pronounced peak. Matched transistors should operate at or slightly below this peak in order to minimize base current errors. Operating the transistor at either lower or higher current densities causes the beta to roll off and increases base current errors. Also, the nonidealities mentioned in Rule 6 become increasingly important away from the point of maximum beta.

12. The contact geometry should match the emitter geometry.
A circular emitter should contain a concentric circular contact. Similarly, an octagonal emitter should contain an octagonal contact and a square emitter should contain a square contact. These precautions help prevent interactions between the contact and the edge of the emitter from distorting the flow of emitter current.

13. Consider using emitter degeneration.
Lateral PNP transistors usually benefit more from emitter degeneration than do vertical NPN transistors because of their lower Early voltages and

the inadvisability of increasing their emitter areas. The degenerating resistors should develop at least 50 mV for moderate matching and 100 mV for precise matching. Emitter degeneration can also be used to match transistors with different emitter sizes or geometries. In this case, 200 mV of degeneration should be employed for minimal matching and 500 mV for moderate matching. This technique can also achieve noninteger ratios between matched transistors. Split collectors cannot be degenerated relative to one another because they share a common emitter.

14. Operate moderately or precisely matched transistors at equal collector-to-emitter voltages.
The Early effect can induce systematic collector current mismatches between devices operating at different collector-to-emitter voltages. The magnitude of this mismatch depends upon the Early voltage of the devices. Lateral transistors typically have Early voltages of 50-200 V, which correspond to mismatches of 0.5-2%/V. In addition to the Early effect, variations in collector efficiency with collector-to-emitter bias can also induce mismatches between transistors. The magnitude of these mismatches increases as the collector-to-emitter voltages drop, and the effect becomes extremely severe when one or both devices begin to saturate. Various circuit design techniques—such as the insertion of cascode devices—can ensure that matched transistors operate at equal collector-to-emitter voltages.

9.4 SUMMARY

Bipolar power transistors are substantially more difficult to design than MOS power transistors. The negative temperature coefficient of V_{BE} makes bipolar transistors susceptible to thermal runaway, and current focusing during turnoff can destroy an otherwise robust transistor through secondary breakdown. Proper design can minimize these vulnerabilities. Bipolar transistors offer several unique advantages over MOS transistors: Their high transconductance does not depend on large device areas or small channel lengths, and they also exhibit superior transient power handling capability due to the larger volume of silicon available to dissipate heat. Bipolar transistors perform exceptionally well as MOS gate drivers and as ESD protection devices (Section 13.5). Large lateral PNP transistors are incredibly robust. The large die area required to construct such a device spreads the heat dissipation over a corresponding volume of silicon, and the pronounced beta rolloff of the lateral PNP makes it almost impossible to destroy the transistor by excessive current conduction.

Bipolar transistors also exhibit better voltage matching characteristics than MOS transistors. The high transconductance of the bipolar transistor allows a single stage to generate higher gains and thus minimizes the number of matching transistors required. The emitter area of a vertical NPN transistor can be increased without impairing transconductance, while increasing the channel length of a MOS transistor rapidly decreases its transconductance and increases the required area. Matched MOS transistors almost always consume more area than matched bipolar transistors of similar precision. Properly ratioed bipolar transistors develop extremely accurate VPTAT voltages that form the basis of many voltage and current references. MOS transistors only develop VPTAT voltages when operated in subthreshold, a mode incompatible with high-temperature operation.

Although MOS transistors have supplanted their bipolar counterparts in many applications, bipolar transistors still have their advantages. The future of analog design appears to lie with analog BiCMOS processes that merge high-density CMOS with high-performance bipolar. These processes can provide high-density submicron CMOS logic in combination with precision analog functions currently achievable only through the use of bipolar transistors.

9.5 EXERCISES

Refer to Appendix C for layout rules and process specifications. For all power transistors, assume a minimum beta at full rated current of 10. Do not exceed an emitter current density of 8 μA/μm^2 for linear-mode devices and 15 μA/μm^2 for switched-mode devices.

9.1. What is the maximum current that can flow through a 100 μm-long emitter finger with a metallization width of 12 μm? Assume that the metallization consists of 10 kÅ of aluminum/copper/silicon alloy and that emitter debiasing must not exceed 5 mV.

9.2. Construct an interdigitated-emitter power transistor, using standard bipolar layout rules. The transistor is intended as a 500 mA series-pass transistor for a linear regulator. Construct the transistor around a central spine of deep-N+ 20 μm wide, and place banks of emitter fingers on both sides of this spine. Make the emitter fingers 20 μm wide. Do not allow intrafinger debiasing to exceed 5 mV, or base debiasing to exceed 3 mV. Use emitter ballasting resistors that develop 50 mV at full rated current. Include all necessary metallization.

9.3. Construct a wide-emitter narrow-contact power transistor for a lamp driver, using standard bipolar layout rules. This switching transistor must handle 150 mA of collector current and should contain as much deep-N+ as possible. Make all deep-N+ sinkers 16 μm wide and all emitter fingers 24 μm wide. Assume only one end of the base serpentine can be connected. Do not allow the sum of base metallization debiasing and emitter metallization debiasing to exceed 10 mV.

9.4. Construct a cruciform-emitter power transistor for a relay driver, using standard bipolar layout rules. This switching transistor must handle 700 mA of collector current and should be ringed with a deep-N+ sinker 20 μm wide. Maximize connection to the collector. Make the cruciform emitter sections 75 μm wide and contact them with 10 μm-diameter circular emitter contacts. Assume only one end of the base lead can be connected. Do not allow the sum of base metallization debiasing and emitter metallization debiasing to exceed 10 mV.

9.5. Construct a wide-emitter narrow-contact transistor, using analog BiCMOS layout rules. This gate driver transistor must conduct 500 mA pulses. Assume that the transistor operates at a peak emitter current density of 100 μA/μm^2. Use an emitter overlap of emitter contact of 8 μm and completely ring the transistor with a deep-N+ sinker no less than 10 μm wide. Maximize metallization of both emitter and collector. Since the layout rules do not allow strip contacts, use rows of minimum-width contacts instead. Make the narrow contact for the emitter out of two rows of minimum contacts. Include all necessary metallization.

9.6. Construct the relay driver transistor from Exercise 9.4, using analog BiCMOS layout rules. Maximize metallization to both emitter and collector. Devise a suitable replacement for the circular emitter contacts.

9.7. Construct the circuit shown in Figure 9.19, using standard bipolar layout rules. Transistors Q_1, Q_2, and Q_3 are minimum-area lateral PNP transistors having one, two, and three emitters, respectively, and transistor Q_4 is a minimum-area substrate PNP. Resistors R_1, R_2, and R_3 consist of 6 μm-wide base resistors placed in a tank connected to V_{CC}.

9.8. Construct the circuit shown in Figure 9.21C, using analog BiCMOS layout rules. Transistors Q_1, Q_2, Q_3, and Q_4 should use square emitters having a width of 10 μm Include as many emitter contacts as possible. Compute the value of resistor R_1 necessary to produce a current of 10 μA, and lay this resistor out using PSD-doped poly-2 6 μm wide. Take all necessary precautions to obtain optimal matching.

9.9. Lay out the Brokaw bandgap cell shown in Figure 9.27A using standard bipolar layout rules. Transistors Q_1 and Q_2 should employ circular emitters with diameters of 10 μm. Resistors R_1 and R_2 should be constructed as an interdigitated array of base resistors in a common tank. This tank should connect to the base of transistors Q_1 and Q_2. Include all necessary interconnection and label all devices.

9.10. Lay out the simple operational amplifier shown in Figure 9.27B, using analog BiCMOS layout rules. Transistors Q_1 to Q_5 should employ 5×5 μm square emitters. Transistors Q_6, Q_7, and Q_8 should use 8×8 μm square emitters. Cross-couple Q_4 and Q_5. Include all necessary interconnection and label all devices. The values on this schematic represent the areas, in μm^2, of the respective emitters.

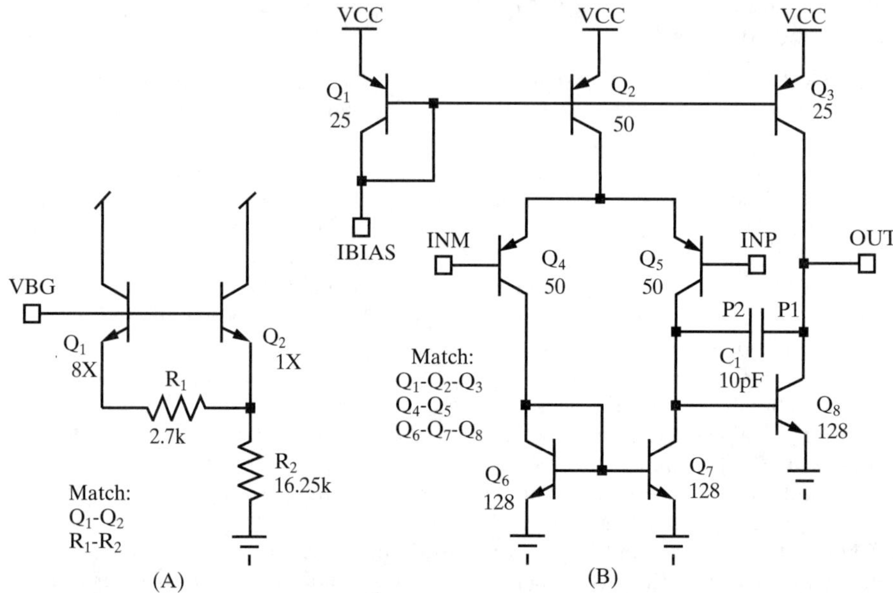

FIGURE 9.27 (A) Brokaw bandgap cell and (B) Simple operational amplifier for Exercises 9.9 and 9.10.

9.11. Lay out the Gilbert multiplier core shown in Figure 9.28, using analog BiCMOS layout rules. Transistors Q_1 to Q_4, Q_6, Q_7, Q_9, and Q_{10} use 8×8 μm emitters. Transistors Q_5, Q_8, and Q_{11} use 6×6 μm emitters. Lay out all transistors for optimal matching. Transistors sharing a common collector connection can occupy the same tank. Include

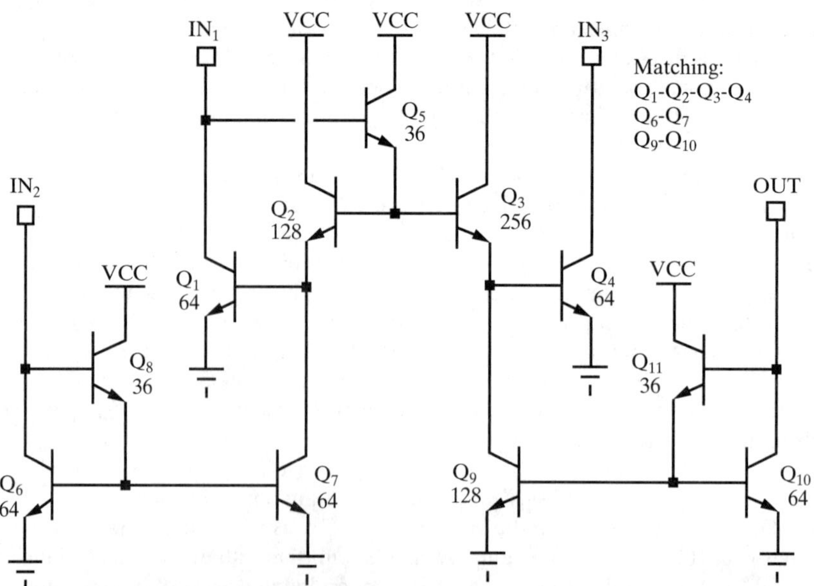

FIGURE 9.28 Gilbert multiplier core for Exercise 9.11.

all necessary interconnection and label all devices. The values on this schematic represent the areas, in μm^2, of the respective emitters.

9.12. Suppose the Gilbert multiplier core in Exercise 9.11 forms part of a die with an area of 7.6 mm^2 (not including scribe streets and seals), which also includes a power NPN transistor with an area of 4.3 mm^2. Select an aspect ratio for the die, and place rectangles representing the outline of the die and power transistor. Place the multiplier core at an optimal location for best matching.

9.13. A pair of matched NPN transistors of equal emitter areas have Early voltages of 175 V. Suppose the two transistors operate at collector-to-emitter voltages of 1 V and 2 V, respectively. The collector currents through the two transistors are equal. What systematic mismatch, in mV, exists between the transistors' base-to-emitter voltages?

9.14. A lot of 1000 bandgap references are measured at wafer probe; their bandgap voltages exhibit a mean of 1.233 V and a standard deviation of 1.8 mV. After packaging, the same devices exhibit a mean of 1.231 V and a standard deviation of 3.1 mV. What are the mean and standard deviation of the package shift?

10 *Diodes*

The device now called a *diode* was invented in the late nineteenth century, but it first saw widespread use in the galena crystal detector of 1907. This device was actually a Schottky diode formed between a metallic cat's whisker and semiconducting lead sulfide (galena). The copper oxide rectifiers and selenium stacks of the vacuum-tube era were also primitive Schottkies. Modern semiconductor diodes emerged from a different line of development that began with germanium point-contact diodes developed for military and computer applications. These were replaced in the mid-1960s by silicon PN-junction diodes similar to those in use today.

Diodes have found a number of applications in modern integrated circuits. Schottky diodes are often used as antisaturation clamps for the collector-base junctions of NPN transistors. PN junction diodes form part of current mirrors and biasing networks. Junction diodes operated in reverse breakdown can also serve as voltage references and clamping devices. This chapter examines these and other applications of integrated diodes.

10.1 DIODES IN STANDARD BIPOLAR

The standard bipolar process can construct a wide variety of diodes. Of these, the most popular are the diode-connected transistor, the base-emitter Zener, and the Schottky diode. The first two are both variations of the bipolar NPN transistor, while the Schottky diode relies on the formation of a rectifying contact to lightly doped silicon. Not all versions of standard bipolar offer Schottky diodes, because they require the formation of platinum or palladium silicides and the addition of a special masking step to allow contact through the thick-field oxide. This section also discusses several additional types of Zener diodes sometimes available in standard bipolar.

10.1.1. Diode-Connected Transistors

An NPN transistor consists of two back-to-back PN junctions, either of which could theoretically serve as a PN junction diode. In practice, the parasitic transistors associated with these junctions generally render them unsuitable for use as diodes. The collector-base diode loses most of its current to substrate due to parasitic PNP action. NBL does not stop this parasitic conduction, because the carriers can still flow to the isolation sidewalls. Ringing the tank with deep-N+ contains the minority carriers and minimizes the current loss, but only at the price of greatly enlarging the structure. The base-emitter diode loses the vast majority of its current to the enclosing tank due to parasitic NPN action. Most of the carriers injected by the emitter travel across the base and into the tank. The accumulation of electrons in the collector causes the base-collector junction to forward-bias, turning on the parasitic PNP and diverting current to the substrate.

A more useful type of diode is created by tying the collector and the base of an NPN transistor together (Figure 10.1). The resulting device is often called a *diode-connected transistor.* Most of the current flows through a diode-connected transistor from collector to emitter by means of transistor action. Only a small current flows through the base terminal, so the base resistance has little effect on the device's forward voltage. The forward voltage also remains independent of collector resistance as long as the transistor does not fully saturate. A typical diode-connected transistor can tolerate about 400 mV of debiasing at 25°C, or about 200 mV at 150°C. If the collector debiasing exceeds these limits, then the diode begins losing current to the substrate due to parasitic PNP action. Diode-connected transistors usually incorporate NBL to minimize collector series resistance. Diodes that must conduct more than a few hundred microamps should also contain a deep-N+ sinker. Diodes conducting 10 mA or more should be laid out as power devices and should incorporate deep-N+ rings to minimize substrate injection during transients.

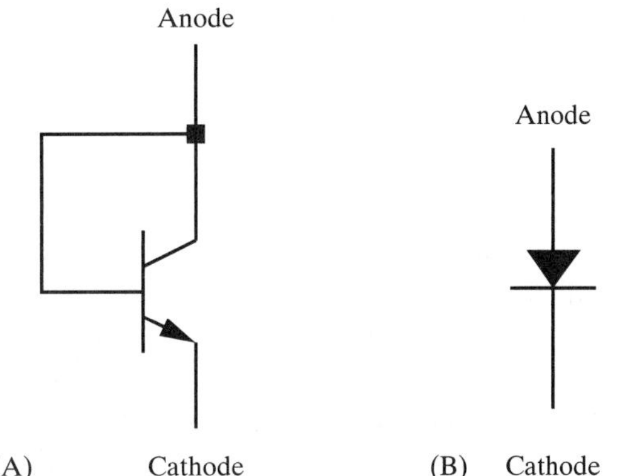

FIGURE 10.1 (A) Schematic and (B) symbol for a diode-connected transistor.

The diode-connected transistor will not suffer any loss of current to the substrate as long as collector debiasing does not exceed the limits given above. It also has much less series resistance than either the base-emitter diode or the base-collector diode. A minimum-size diode-connected transistor typically exhibits no more than 10 to 20 Ω of series resistance. Its only serious drawback lies in its relatively low reverse breakdown voltage, which is limited by the V_{EBO} of the NPN transistor to 6 to 8 V.

Diode-connected transistors usually employ the CBE configuration rather than the CEB configuration (Figure 8.14). Although the CBE configuration has slightly more collector resistance, it allows first-level metal to connect the collector and base contacts. Many processes allow merged collector-base contacts, which save additional space (Figure 10.2). In this structure, the emitter diffusion surrounding the collector

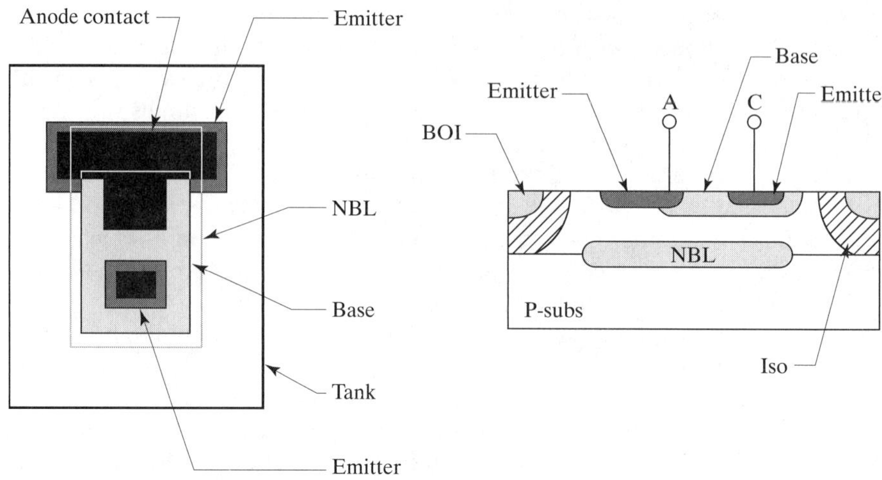

FIGURE 10.2 Layout and cross section of a standard bipolar diode-connected transistor.

contact overlaps the base diffusion so that a single contact can touch both.[1] This contact must extend far enough into the base diffusion to account for misalignment and outdiffusion while still allowing sufficient base contact area for conduction. The contact must also extend into the collector far enough to counter misalignment while allowing adequate collector contact. Even with these overlaps, the merged structure is still considerably smaller than a traditional NPN layout.

The diode-connected transistor can serve as a convenient voltage reference. A base-emitter junction exhibits a forward voltage of about 0.65 V at a current density of 1 μA/μm^2 and a temperature of 25°C. A typical layout has an emitter junction area of 100 μm^2 and requires a current of about 100 μA to develop a forward voltage of 0.65 V. The forward voltage of a diode is relatively insensitive to small fluctuations in current. Even if the current through the diode were to double, the forward voltage would increase by only 18 mV. The forward voltage also exhibits a temperature coefficient of about −2 mV/°C. A stack of several diodes connected in series can develop larger voltages, but temperature and current variability increase proportionally.

A substrate PNP transistor can also serve as a diode, but the collector current of this device flows directly into the substrate. Currents that are much in excess of 1 mA may debias the substrate enough to saturate the transistor. Diode-connected substrate transistors are sometimes used in CMOS processes that cannot fabricate other bipolar components (Section 10.2.1). They rarely see much use in processes that can fabricate isolated bipolar transistors.

Lateral PNP transistors make relatively poor diodes. They require large tanks that not only consume die area, but also contribute unwanted parasitic capacitance. Some portion of the collector current always flows to the substrate regardless of how thoroughly the device has been guard ringed. The cathode current of a diode-connected lateral PNP is always less than its anode current. This loss prevents the use of lateral PNP transistors in applications where current matching is critical. Still, diode-connected lateral PNP transistors are occasionally inserted into circuits to balance other PNP base-emitter voltage drops. An NPN transistor would not serve as well because its base-emitter voltage does not exactly match that of a PNP. The base and collector of the lateral PNP form the cathode of a diode-connected PNP,

---

1 "Diodes," *Semiconductor Reliability News*, Vol. III, #7, 1991, p. 9.

while the emitter acts as its anode. The layout sometimes uses a merged cathode contact that is analogous to the merged anode contact of the diode-connected NPN transistor shown in Figure 10.2.

10.1.2. Zener Diodes

A reverse-biased diode conducts very little current until the voltage across it exceeds a certain value. Beyond this point, the current through the diode increases exponentially until it eventually approaches an asymptote defined by the series resistance of the diode (Figure 10.3). The breakdown curve usually shows a fairly definite inflection point or *knee* corresponding to the *breakdown voltage* of the diode. The magnitude of the breakdown voltage depends on the width and curvature of the depletion region of the diode (Section 1.2.4). Carriers can move across a very thin depletion region by Fowler-Nordheim tunneling. Diodes with breakdown voltages of less than 6 V conduct primarily by tunneling and are called *Zener diodes* after the individual who first predicted their behavior.[2] Diodes with breakdown voltages in excess of 6 V are properly called *avalanche diodes* because they conduct primarily by avalanche multiplication instead of by tunneling. Some authors use the term *breakdown diode* to refer to both Zener diodes and avalanche diodes, but the engineering community has not widely adopted this term. Instead, designers use the term *Zener diode* to describe all junction diodes operated in reverse breakdown regardless of conduction mechanism.

Fowler-Nordheim tunneling increases at higher temperatures because the thermally agitated carriers are more energetic and thus do not have to tunnel as far through the depletion region to gain the energy needed to free them from the lattice. A lower reverse bias therefore suffices to maintain conduction at a given current. The breakdown voltage of diodes that conduct by tunneling exhibits a negative temperature coefficient that increases in magnitude as the breakdown voltage diminishes. Avalanche conduction decreases at higher temperatures because the

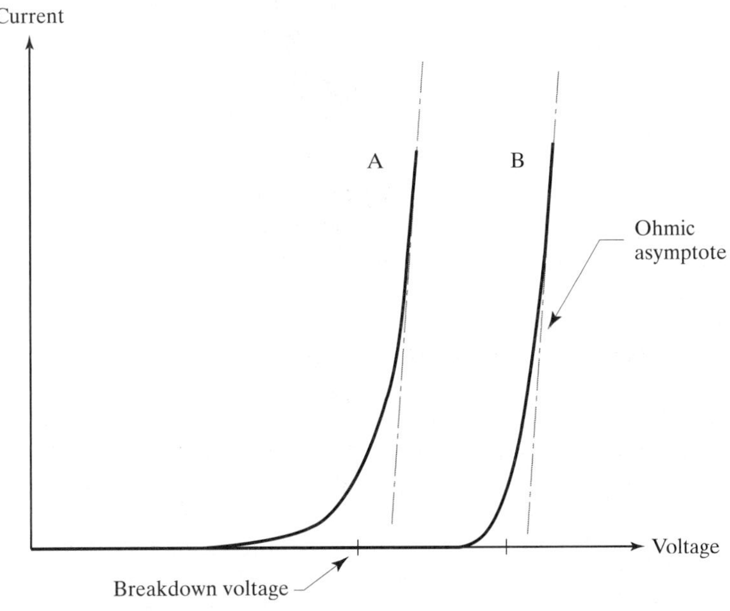

FIGURE 10.3 A comparison of the reverse breakdown characteristics of (A) a Zener diode and (B) an avalanche diode.

[2] C. Zener, *Proc. Roy. Soc.* A145, London: 1934, p. 523.

thermal vibration of the lattice enhances scattering and therefore limits the mobility of the hot carriers. A higher reverse bias is therefore necessary to maintain conduction at a given current. The breakdown voltage of avalanche diodes exhibits a positive temperature coefficient that increases in magnitude with increasing breakdown voltage. These two competing mechanisms cause breakdown voltages of less than about 5.6 V to have negative temperature coefficients and those of more than about 5.6 V to have positive temperature coefficients. The familiar emitter-base Zener has a breakdown voltage of 6 to 8 V with a positive temperature coefficient of 2 to 4 mV/°C. A 40 V base-collector Zener exhibits a much larger temperature coefficient, perhaps 35 to 40 mV/°C.

Zener diodes with breakdown voltages of 5 to 6 V have very small temperature coefficients. These diodes are sometimes used to construct temperature-independent voltage references, but they have several drawbacks that severely restrict their usefulness. Zener walkout (Section 4.3.2) causes the reference voltage to drift over time unless the device is specifically constructed to ensure subsurface breakdown. Zener references also require a supply voltage of at least 6 V. Because of these disadvantages, Zener references have largely been replaced with lower voltage alternatives such as bandgap references. Zener references are less affected by package stresses than bandgap references, so the former are still preferred for applications requiring extreme precision.

Zener diodes with breakdown voltages below 5 V exhibit a rather gradual onset of reverse conduction (Figure 10.3A). This *soft breakdown* becomes more pronounced as the breakdown voltage decreases. Designers often blame soft breakdown on leakage, but it is actually an unavoidable characteristic of Zener tunneling. In any event, Zener diodes with breakdown voltages of much less than 5 V have little practical application because they do not exhibit well-defined reverse characteristics.

Surface Zener Diodes

The emitter-base junction of an NPN transistor forms a convenient Zener diode. Its breakdown voltage, V_{EBO}, depends on base doping and emitter junction depth. Most standard bipolar processes provide an emitter-base breakdown of about 6.8 V. Advanced bipolar and BiCMOS processes often use lightly doped bases that exhibit emitter-base breakdowns of as much as 10 V. Breakdown proceeds primarily by avalanche rather than by tunneling, so the temperature coefficient of the breakdown voltage is positive. A typical 6.8 V emitter-base Zener exhibits a temperature coefficient of +3 to 4 mV/°C.

Emitter-base Zeners have historically exhibited large amounts of process variation and long-term drift. Older bipolar processes used an emitter pilot step to control NPN beta, allowing the use of poorly controlled dopant sources such as boron nitride disks. The resulting variation in base doping and the compensatory changes in emitter junction depth caused V_{EBO} to vary by as much as ±1 V. Ion implantation has dramatically improved base doping control, and most modern processes guarantee no more than ±0.25 V initial variation. Process improvements have also reduced long-term drift due to Zener walkout, which now rarely exceeds 0.1 V.

Emitter-base Zeners use essentially the same layout as NPN transistors (Figure 10.4). The emitter acts as the *cathode* of the Zener and the base acts as the *anode.* The tank serves only to isolate the Zener from the surrounding isolation. It should connect either to the cathode of the Zener or to some equal or higher voltage. The tank must never connect to the anode of the Zener lest the transistor become biased into the reverse-active region. If this occurs, then V_{ECO} breakdown will produce a snapback phenomenon similar to that exhibited by V_{CEO} breakdown (Section 8.1.2). Some designers leave the tank unconnected, but this practice is not

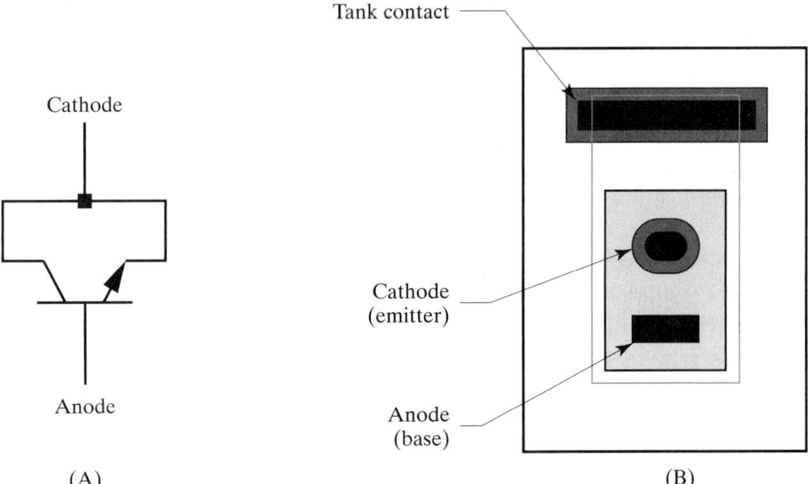

FIGURE 10.4 (A) Schematic and (B) layout of a typical base-emitter Zener diode.

recommended because it can amplify leakage currents. The floating tank acts as the base of a parasitic substrate PNP. The leakage current across the tank-isolation junction exceeds that across the smaller base-tank junction, and the difference between these currents forms the base drive of the substrate PNP. This mechanism can cause substantial loss of anode current to the substrate at elevated temperatures. Connecting the tank to the cathode prevents the parasitic PNP from amplifying leakage currents in this manner. The tank contact does not require a deep-N+ sinker because it only conducts leakage currents. NBL has little effect on a Zener and can therefore be omitted if the designer so chooses.

Emitter-base Zeners usually have circular or oval emitters like the one shown in Figure 10.4. These round geometries are intended to prevent electric field intensification at the corners of the emitter. Not all processes exhibit this phenomenon, but if it occurs it increases the variability of the breakdown voltage. One can determine whether a given process exhibits this effect by examining an avalanching rectangular emitter under a microscope in a darkened room. The avalanching junction will emit a faint white light.[3] If this light appears brightest at the corners, then electric field intensification is occurring at these points, and the device will benefit from the use of round emitters. Many designers routinely use round emitters because, even if they do not provide any benefit, they do no harm.

The base doping profile causes the base-emitter depletion region to narrow near the surface. The intensification of the electric field across the narrowed depletion region ensures that avalanche breakdown occurs at this point. Hot carriers produced by the intense electric field sometimes penetrate into the overlying oxide, where a small fraction becomes trapped. The gradual accumulation of this trapped oxide charge causes *Zener walkout* (Section 4.3.2). Some designers have attempted to suppress surface breakdown by flanging metal over the base-emitter junction to form a field plate. The thickness of the emitter oxide and the relatively low voltages placed across it prevent this field plate from having much effect. The avalanche process continues to occur near the surface and the field plate cannot prevent Zener walkout.

Emitter-base Zeners are relatively fragile devices because they dissipate energy in the small volume of the emitter-base depletion region. The heat generated by this

---

[3] "Junction Breakdown Characteristics," *Semiconductor Reliability News,* Vol. II, #1, January 1990, p. 8.

process can damage the junction or the adjacent contacts. Extreme overloads usually induce metal migration that causes permanent short-circuit failures. This mechanism has been employed as a replacement for fuses (Section 5.6.2). Base-emitter Zeners used as voltage references or clamp devices should not conduct more than about 10 μA per micron of emitter periphery. For example, a 5 × 5 μm emitter can safely conduct some 200 μA. Zeners can tolerate much higher currents for brief periods of time, but long-term operation at elevated currents can cause shifts in breakdown voltage due to junction damage. If an application requires more current than a small Zener diode can safely handle, consider using a power transistor to amplify the current conducted by the Zener, since this circuit takes much less area than a large Zener (Figure 13.22).

Some circuits connect the anode of the base-emitter Zener to the substrate. Since the base diffusion of the Zener operates at the same potential as the surrounding isolation, these two diffusions can overlap one another. This practice saves considerable area because it eliminates the large spacing required to isolate the base diffusion from the isolation (Figure 10.5). A tank is placed beneath the emitter to prevent the isolation diffusion from reducing the Zener voltage. Although this tank remains unconnected, the diffusions around it are all biased to the same potential. The parasitic PNP transistor inherent in this structure cannot amplify leakage currents and is therefore harmless.

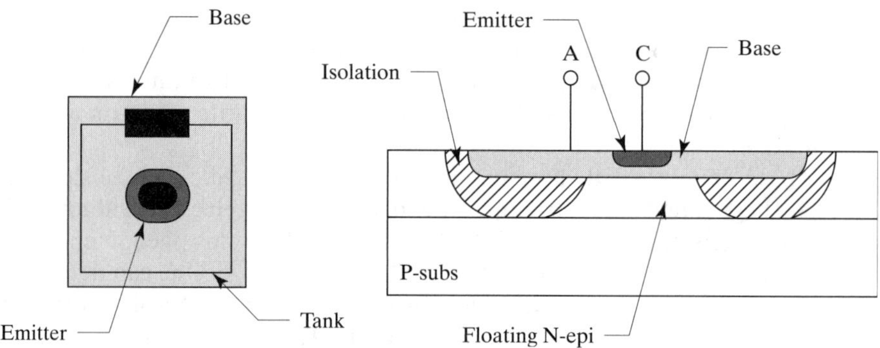

FIGURE 10.5 Layout and cross section of a nonisolated base-emitter Zener diode.

The nonisolated base-emitter Zener is vulnerable to both substrate debiasing and noise coupling. Nonisolated Zeners should never reside near structures that inject substrate currents of more than a few hundred microamps. Conservative designers often avoid using nonisolated Zeners, preferring to accept the extra die area of a conventional base-emitter Zener rather than risk unexpected debiasing or noise coupling.

Buried Zeners

If the avalanche region lies several microns beneath the oxide, hot carriers will scatter off the lattice and lose energy before they can reach the oxide interface. Zeners that avalanche beneath the surface are called *subsurface Zeners,* or more colloquially, *buried Zeners.* The breakdown voltages of buried Zeners remain constant throughout their operational lifetime, making these devices ideal for use as precision voltage references. This section presents several common varieties of buried Zeners that are compatible with standard bipolar processing.

A buried Zener can be constructed from the emitter and isolation diffusions of certain standard bipolar processes. This diode consists of a plug of P+ isolation covered

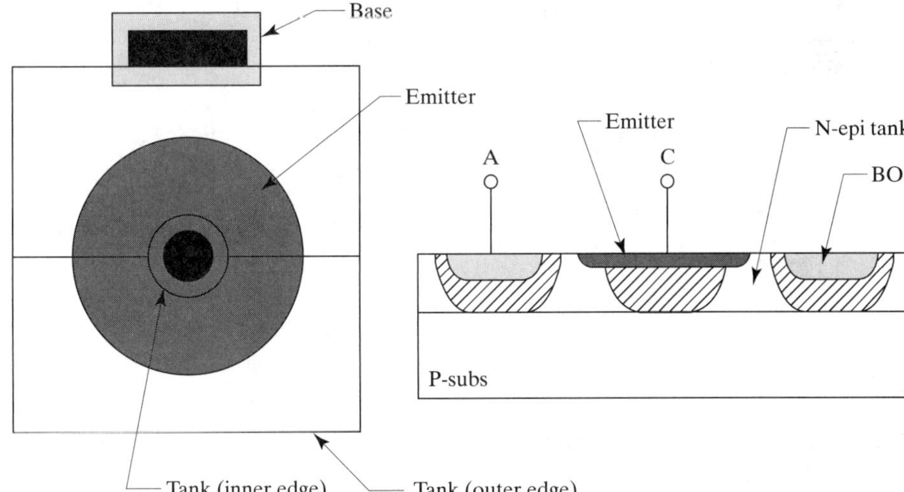

FIGURE 10.6 Layout and cross section of an emitter-in-isolation Zener diode.

by an emitter diffusion that overlaps into a surrounding tank (Figure 10.6).[4] The emitter counterdopes the surface of the isolation to form the cathode of the Zener. The anode consists of the portion of the isolation plug beneath the emitter. The N-epi surrounding the isolation plug prevents the emitter sidewall radius from further reducing the already-low breakdown voltage of the structure. The *emitter-in-iso* buried Zener usually exhibits a breakdown voltage of 5 to 6 V and a temperature coefficient of about +1 mV/°C.

Some processes use base over isolation to prevent channel formation. The presence of base in the isolation plug further reduces the breakdown voltage of the emitter-in-iso Zener. Although this may help minimize the residual temperature coefficient of the Zener, it also produces a softer breakdown characteristic. Breakdown voltages of less than 5 V become rather indefinite and are therefore unsuited for use as voltage references. If the addition of base to the emitter-in-iso Zener causes its breakdown voltage to drop below 5 V, then the designer should omit it from the area around the emitter plug.

The emitter-in-iso Zener shares most of the same drawbacks as the nonisolated base-emitter Zener. Its anode is electrically common to the substrate. This severely limits its range of applications and raises the possibility of substrate debiasing and noise coupling. Emitter-in-iso Zeners do not experience any appreciable long-term drift, but their initial voltage varies considerably because of the heavy deposition and long drive time required to fabricate the P+ isolation. A typical emitter-in-iso Zener exhibits a breakdown voltage of 5.4 ± 0.4 V.

Several alternative styles of buried Zeners eliminate the drawbacks of the emitter-in-iso device at the cost of introducing additional processing steps. Figure 10.7 shows a buried Zener that requires the use of a deep-P+ diffusion inserted after isolation and before base.[5] The doping concentration in this deep-P+ diffusion significantly exceeds that of the base diffusion but falls short of that of the emitter. The active region of the Zener consists of a plug of deep-P+ diffusion covered by emitter. The emitter diffusion is in turn enclosed by a base diffusion. The anode of the diode

[4] A. B. Grebene, *Bipolar and MOS Analog Integrated Circuit Design* (New York: John Wiley and Sons, 1984), pp. 133–134.

[5] Grebene, p. 134.

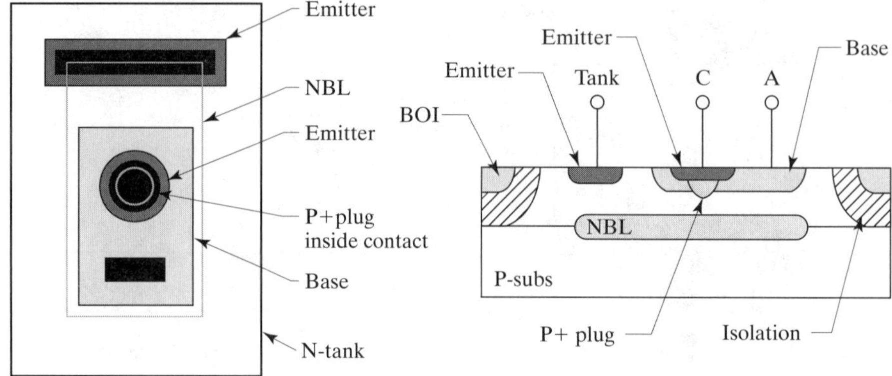

FIGURE 10.7 Buried Zener using a special deep-P+ diffusion in combination with emitter diffusion.

consists of the deep-P+ plug and is contacted by means of the surrounding base diffusion. The cathode of the diode consists of the emitter diffusion. The tank of this Zener is usually connected to the cathode to prevent beta multiplication of leakage currents. The structure in Figure 10.7 contains NBL, but it actually plays no role in the operation of the device. It can be omitted as long as the deep-P+ is sufficiently shallow to prevent punchthrough breakdown to the substrate.

The breakdown voltage of this structure can be tailored to suit a specific application by adjusting the profile of the deep-P+ diffusion. A lightly doped diffusion will have a higher breakdown voltage, while a heavily doped diffusion will have a lower one. The breakdown voltage cannot exceed that of the base-emitter junction, nor should it drop so low as to produce a soft breakdown characteristic. In practice, the breakdown voltage ranges from 5 to 6.5 V. A typical device has a breakdown voltage of 6.3 ± 0.2 V with a temperature coefficient of about +2 mV/°C.

Another style of buried Zener substitutes a high-energy implant for the emitter diffusion of the standard base-emitter structure (Figure 10.8).[6] At high implant energies, the peak of the dopant distribution actually lies beneath the surface of the silicon. The high-energy implant therefore creates a shallow N-buried layer. The intersection of this layer with a plug of base diffusion forms a buried Zener diode. The breakdown voltage is set by adjusting the implant energy and dose used to fabricate the implanted NBL.

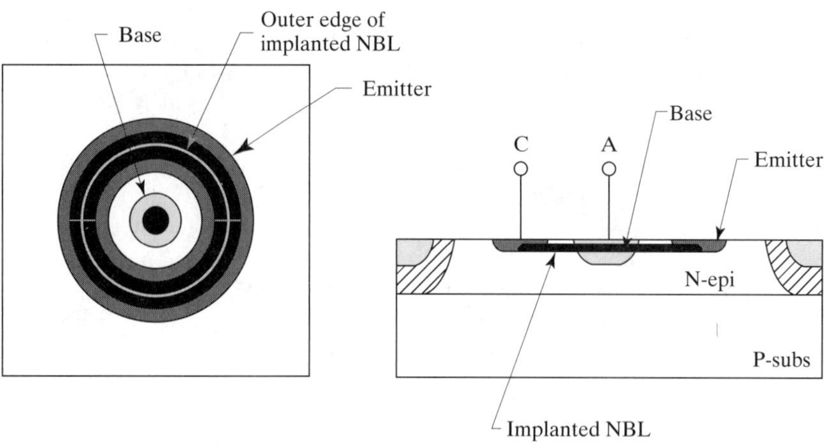

FIGURE 10.8 Buried Zener using an implanted NBL in combination with base diffusion.

[6] Grebene, p. 134.

All three of the buried Zeners just discussed provide breakdown voltages of 5 to 7V with relatively low temperature coefficients. Certain applications, such as ESD protection, require higher voltage Zeners. These are usually constructed from stacks of series-connected base-emitter Zeners, possibly supplemented by diode-connected transistors. In addition to helping adjust the voltage of the stack, the negative temperature coefficient of the diode-connected transistors can offset part or all of the temperature coefficient of the Zeners. A few specialized high-voltage Zener structures take advantage of the large breakdown voltages between deep, lightly doped diffusions. Of these, the isolation-NBL Zener deserves special mention. This Zener consists of a plug of isolation diffused into a plate of NBL. The NBL plate is enclosed in a tank contacted by means of an emitter diffusion (Figure 10.9). The voltage of the isolation-NBL Zener can vary by a few volts because it depends on a number of factors, including NBL and isolation doping, isolation drive time, and epi thickness. The isolation-NBL junction lies far beneath the surface and is therefore immune to Zener walkout. The relatively large area of the depletion regions and their depth beneath the surface also allow this structure to safely dissipate several times the power density of a base-emitter Zener.

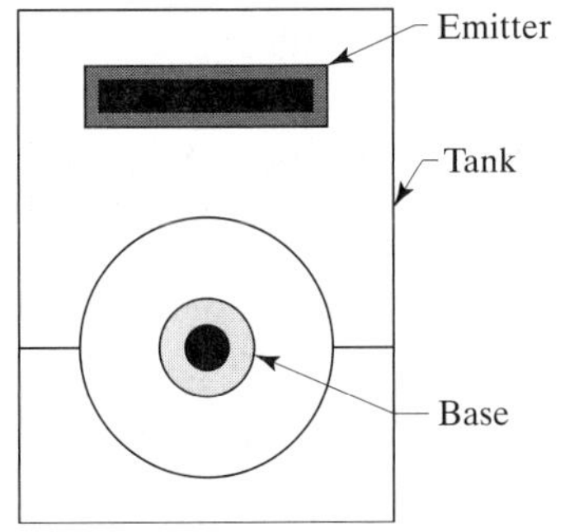

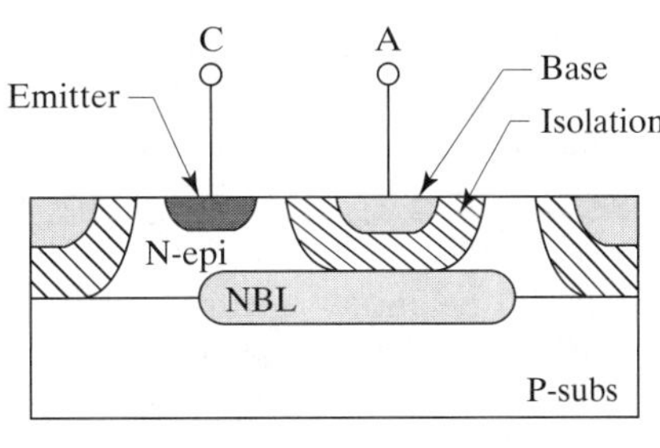

FIGURE 10.9 Layout and cross section of an isolation-NBL Zener.

10.1.3. Schottky Diodes

Schottky diodes depend on the formation of a rectifying Schottky barrier between a conductor and a semiconductor. Both P-type and N-type silicon form rectifying Schottky barriers with numerous metals and metal silicides. The forward voltages of the resulting diodes depend on the composition of the conductor and the polarity of the silicon (Table 10.1). Schottky diodes with forward voltages of less than

Material	N-Type Silicon	P-Type Silicon
Aluminum	0.54 V	0.40 V
Gold	0.62 V	0.16 V
Molybdenum	0.50 V	0.24 V
Palladium silicide (Pd_2Si)	0.57 V	
Platinum silicide (PtSi)	0.66 V	
Titanium silicide ($TiSi_2$)	0.42 V	

TABLE 10.1 Typical forward voltages of Schottky diodes constructed from selected metals and silicides[7] (25°C, 1 μA/μm^2).

[7] These values are derived from the barrier potential ϕ_B, assuming current density of 1 μA/μm^2 and a Richardson constant of 120 A/cm^2/K^2. Under these circumstances, the forward voltage is 180 mV less than ϕ_B. Barrier potentials from S. M. Sze, *Physics of Semiconductor Devices,* 2d ed. (New York: John Wiley and Sons, 1981), pp. 290–291.

0.5 V usually exhibit large junction leakages, especially at higher temperatures. This limitation generally prevents the use of P-type silicon in Schottky diodes.

The concentration of acceptors or donors in the silicon determines whether a given Schottky barrier exhibits rectifying or Ohmic behavior. Dopant concentrations exceeding 10^{17} atoms/cm^3 reduce the width of the depletion region to the point at which carriers can successfully surmount it by tunneling. The resulting tunneling current increases almost exponentially with doping.[8] When the tunneling current exceeds the forward conduction current, the Schottky barrier begins to resemble a resistor. Practical Schottky diodes normally employ surface doping concentrations of no more than 10^{16} atoms/cm^3 to minimize tunneling effects.

Many applications require Schottky diodes with forward voltages significantly lower than those of PN junction diodes. Aluminum would appear to be an ideal material for constructing such diodes. Its forward voltage lies comfortably below that of a PN junction diode, yet not so low as to cause excessive junction leakage. Unfortunately, sintering can drastically alter the properties of an aluminum-silicon Schottky. The aluminum dissolves a small amount of silicon during sintering. As the wafer cools, some of the dissolved silicon redeposits at the metal-semiconductor interface as an aluminum-doped P-type semiconductor. This deposit constricts the Schottky contact area and, in extreme cases, may entirely cover the contact opening. This mechanism can cause the forward voltage of a Schottky to increase by several hundred millivolts.[9]

Few other metals exhibit the properties required to construct practical Schottky diodes. Most are difficult to sinter and exhibit excessive forward voltage variation. Modern integrated Schottky diodes employ the silicides of certain noble metals, most notably platinum and palladium. These *noble silicides* provide extremely stable and repeatable forward voltages lying in the desired range of 0.5 to 0.7 V. The inability of noble silicides to withstand the temperatures required for source/drain annealing limits their application in CMOS processes. The forward voltages of the *refractory silicides* (such as titanium silicide) are rarely sufficient to prevent leakage, so processes that use these silicides rarely offer Schottky diodes.

Most processes indiscriminately silicide all contact openings. Those lying over heavily doped silicon become Ohmic contacts, while those residing over lightly doped N-type silicon become Schottky diodes. These processes generally prohibit the opening of contacts over lightly doped P-type silicon, because the resulting Schottky barrier has a contact resistance that is too high to function as an Ohmic contact and a forward voltage that is too low to function as a Schottky diode.

Figure 10.10 shows the layout and cross section of a Schottky diode constructed in a standard bipolar process. The metal system consists of a sandwich of platinum silicide, refractory barrier metal, and copper-doped aluminum. The Schottky barrier forms between the platinum silicide and the lightly doped N-type epi. In order to reach the epi, the Schottky contact must penetrate the thick-field oxide. If this contact opening were etched simultaneously with the base and emitter contacts, the latter would suffer severe overetching before the former cleared. Most standard bipolar processes include an additional etching step to thin the oxide over the Schottky contact openings prior to the regular contact oxide removal. This step uses the so-called *Schottky contact* mask. The process extension required to form Schottky contacts therefore consists of a single masking step and a single oxide removal.

---

[8] Actually, theory predicts that the tunneling current varies exponentially with respect to the square root of the dopant concentration and linearly with respect to the applied voltage; See W. R. Runyan and K. E. Bean, *Semiconductor Integrated Circuit Processing Technology* (Reading, MA: Addison-Wesley, 1994), p. 524.

[9] M. Mori, "Resistance Increase in Small-Area Si-Doped Al-n-Si Contacts," *IEEE Trans. on Electron Devices*, Vol. ED-30, #2, 1983, pp. 81–86.

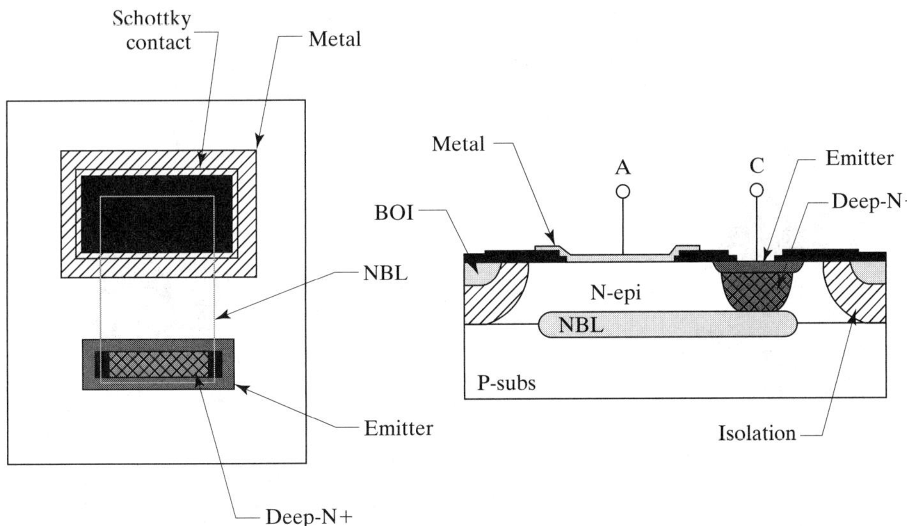

FIGURE 10.10 Layout and cross section of a field-plated Schottky diode.

The cathode of the Schottky includes NBL and deep-N+ to minimize its series resistance. A small plug of deep-N+ suffices to extract the cathode current as long as this does not exceed a milliamp or two. High-current Schottky diodes usually employ rings of deep-N+ around the periphery of their tanks to further reduce the series resistance. Low-current Schottkies may omit the deep-N+ plug entirely. Large diodes can occupy empty spaces between other components, because the shape of the Schottky contact has little effect on its performance.

The planar breakdown voltage of a Schottky diode usually exceeds the V_{CEO} rating of the corresponding NPN transistor by at least a factor of two. Practical Schottky diodes obtain only a fraction of this theoretical breakdown voltage because the sharp edges of the Schottky contact greatly intensify the electric field. An unprotected Schottky usually begins to avalanche at a reverse bias of only a few volts. Several techniques have been developed that diminish the electric field intensity at the edges of the Schottky contact. The simplest of these consists of flanging the metallization over the Schottky contact to form a field plate. The full reverse-bias voltage appears between the field plate and the underlying silicon. The resulting vertical electrical field repels electrons from the surface of the silicon and makes it appear to be more lightly doped. This delays the onset of avalanche and increases the reverse-bias voltage rating by a few volts. The field plate need only overlap the contact by 3 to 5 μm. This relatively simple and compact arrangement suffices for many low-voltage applications.

Figure 10.11 shows an alternative style of Schottky diode that can withstand much higher reverse voltages. This structure encloses the edge of the Schottky contact within a thin strip of base diffusion called a *field relief guard ring.*[10] The presence of the guard ring completely eliminates the lateral electric field at the edge of the Schottky contact and raises the breakdown voltage of the structure to equal the V_{CBO} of the base diffusion. These field-relief guard rings are completely unrelated to the minority carrier guard rings of Section 4.4.2.

The inclusion of a field-relief guard ring inevitably enlarges the Schottky diode. The spacing between the tank and the Schottky contact must increase to avoid

---

[10] M. P. Lepselter and S. M. Sze, "Silicon Schottky Barrier Diode with Near-Ideal I-V Characteristics," *Bell Sys. Tech. J.,* Vol. 47, #2 1968, pp. 195–208.

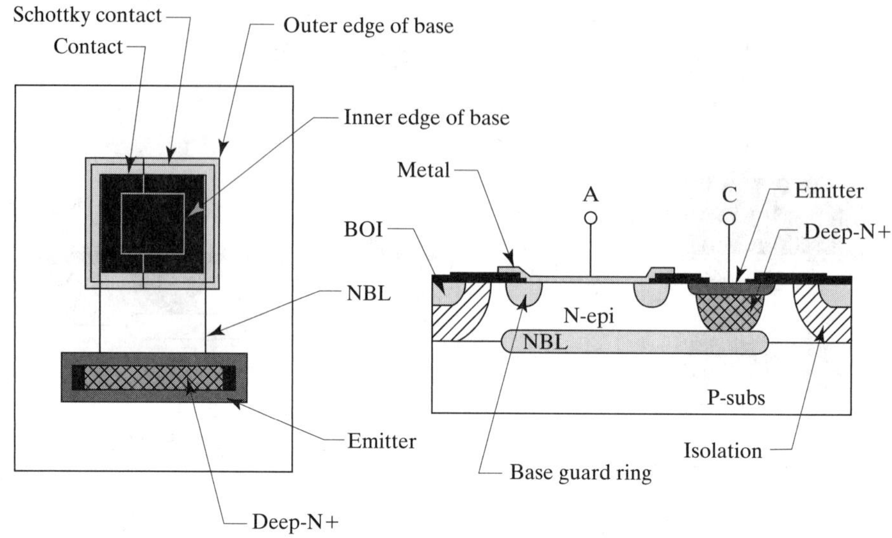

FIGURE 10.11 Layout and cross section of a base guard-ring Schottky diode.

punchthrough between the guard ring and the isolation. Outdiffusion also constricts the contact opening by several microns on all sides. These considerations affect the area of small Schottky diodes much more severely than they affect the area of large ones. Many designers routinely add guard rings to large Schottkies because they render the breakdown characteristics of the diode much more predictable and repeatable. On the other hand, these same designers frequently use field plates on small Schottkies to conserve area.

Schottky diodes are often used as antisaturation clamps for NPN transistors (Section 8.1.4). An antisaturation diode can occupy the same tank as the transistor it protects. A field-plated diode is easily created by extending the base contact out into the surrounding tank (Figure 10.12A). A guard-ringed diode requires the addition of a narrow strip of base around the periphery of the Schottky contact opening (Figure 10.12B). Minimum-size devices usually forgo the addition of the guard ring

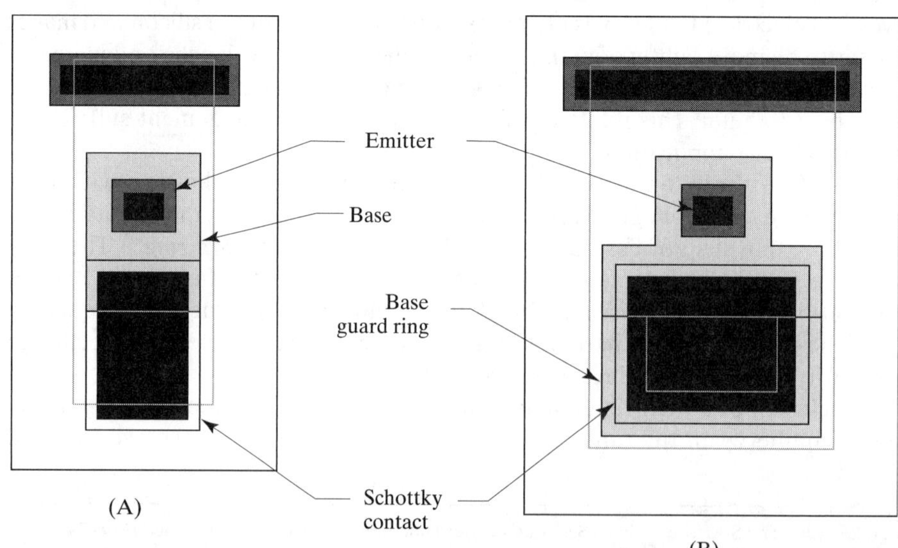

FIGURE 10.12 Schottky-clamped NPN transistors using (A) field plates and (B) base guard rings.

to conserve area. Devices that must tolerate higher collector-base voltages may require guard rings, as may those that cannot tolerate any base-collector leakage.

In order for a Schottky diode to serve as an antisaturation clamp, it must have a lower forward voltage than the base-collector junction of the protected transistor. The size of the required contact opening depends on three factors: the materials used to fabricate the Schottky, the maximum operating temperature, and the current it must conduct. Palladium silicide Schottky diodes make excellent antisaturation clamps because they have relatively small forward voltages. Platinum silicide Schottky diodes have larger forward voltages and therefore require careful attention to contact areas. Devices that must operate at high temperatures require larger contact areas because the Schottky forward voltage has a smaller temperature coefficient than the base-collector forward voltage. This phenomenon prevents platinum silicide Schottky antisaturation clamps from operating at temperatures much higher than 150°C. The same effect can cause guard-ringed Schottky diodes to begin to inject minority carriers to the substrate at temperatures in excess of 130 to 140°C. Junction leakage imposes similar temperature limitations on palladium silicide Schottky diodes.

The area of the Schottky contact required to protect a transistor from saturation must be determined by empirical measurement. In order to perform this measurement, current is forced through the base-emitter junction of a transistor whose collector remains unconnected. The base drive is increased until substrate injection exceeds a predetermined percentage of the base drive—perhaps 5%. This measurement should be performed at the maximum operating junction temperature. The amount of base drive that the Schottky clamp can successfully absorb scales with the ratio of Schottky area divided by base-collector junction area. Based on a single experiment, one can determine approximately how much Schottky contact area is required to protect any given transistor. Outdiffusion may also affect the Schottky contact area of a guard-ringed device. Assuming the Schottky contact is rectangular, then the effective Schottky area, A_S, can be computed from the drawn dimensions of the base opening X_D and Y_D, using the equation

$$A_s = (X_D - 2\delta)(Y_D - 2\delta) \qquad [10.1]$$

where δ is a correction factor that accounts for base outdiffusion, photolithographic size adjusts, and so forth. Since forward current scales linearly with Schottky area, the correction factor can be determined by measuring the currents required to obtain a specified forward voltage for two Schottky diodes of different dimensions.

Schottky diodes also see occasional use as power devices. High current densities tend to activate parasitic mechanisms that would otherwise remain dormant. Figure 10.13 shows the parasitics associated with a typical Schottky diode. D_1 represents the Schottky diode itself, R_1 models the resistance of the NBL and

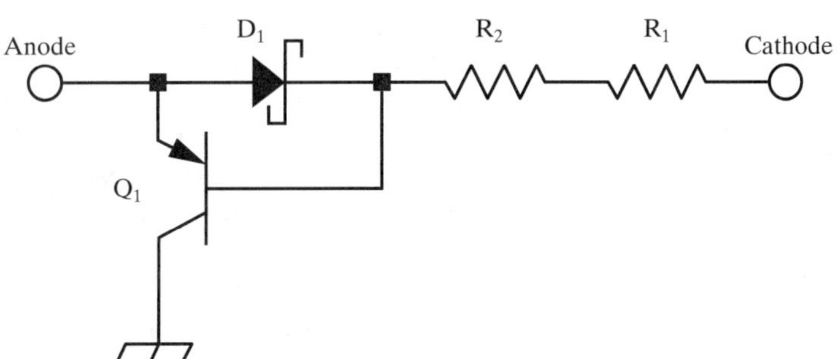

FIGURE 10.13 A simplified parasitic model of a Schottky diode.

deep-N+, and R_2 models the resistance of the N-epi beneath the Schottky contact. The epi contributes more resistance to a Schottky diode than to an NPN transistor because current in the Schottky must cross the entire thickness of the epi. This resistance is not necessarily undesirable because it acts as distributed ballasting and prevents thermal runaway under all but the most extreme conditions. Power Schottky diodes can therefore consist of a single contact opening of any desired shape.

Transistor Q_P in Figure 10.13 models minority carrier injection within the Schottky diode. Although Schottky diodes are majority carrier devices, they do experience some minority carrier injection at high current. Schottky field-relief guard rings will inject minority carriers into the cathode tank when the voltage across the diode exceeds the forward voltage of the junction between the guard ring and the N-epi. Field-plated devices lack this parasitic PN junction, but the Schottky barrier itself emits small numbers of minority carriers at high current densities. Unless these minority carriers are blocked, they will flow across the N-epi into the substrate. The presence of NBL in the Schottky helps minimize substrate injection, but power Schottky diodes should also contain a continuous ring of deep-N+ around their perimeter. This ring serves as both a cathode contact and as a minority carrier guard ring.

10.1.4. Power Diodes

Although most junction diodes are made from the base-emitter junctions of NPN transistors, one can also fabricate useful diodes from collector-base junctions. The collector-base junction of an NPN transistor has a much higher reverse breakdown voltage than the base-emitter junction. Depending upon the process, collector-base breakdown voltages range from about 20 V to 120 V or more. Collector-base diodes are therefore often found in high-voltage circuits. The collector-base junction is also more rugged than the base-emitter junction because it is deeper and more lightly doped. The heat generated in a collector-base junction spreads through a larger volume of silicon and is situated further away from thermally fragile contacts. Collector-base diodes sometimes form part of ESD structures. They also find use as inductive "catch" diodes for relay and solenoid drivers.

The simplest conceivable structure for a standard bipolar collector-base diode consists of a base region placed inside an N-tank. This structure does not function as intended because holes injected by the base into the collector travel down into the substrate. Electrically, the resulting device operates as a substrate PNP transistor. In order to force the device to function as a diode, one must block the flow of minority carriers to the substrate. Vertical carrier flow can be blocked by placing NBL beneath the diode. Lateral current flow can be collected using a ring of base diffusion tied back to the cathode. This structure is nothing more than a diode-connected lateral PNP transistor. Although it functions admirably at low currents, it begins to inject substrate current as soon as it saturates. Saturation typically occurs at relatively modest current levels because of the substantial resistance of the base diffusion, combined with the small reverse bias across the base-collector junction of the lateral. Diode-connected lateral PNP transistors are thus unsuited for use as power devices.

An alternative type of collector-base diode consists of a base diffusion placed inside an N-tank that is floored with NBL and ringed with deep-N+. This structure is essentially a collector-base junction surrounded by a hole-blocking guard ring (HBGR). Holes flood into the N-tank when the collector-base junction forward biases. The hole-blocking guard ring keeps most of these carriers confined within the tank until they recombine. The device thus functions as a diode rather than as a transistor. It is capable of handling high reverse voltages, high forward currents, and significant power dissipation. Many designers therefore call this structure a *power diode*.

Power diodes have two prominent problems, both of which are caused by the flood of minority carriers injected into the collector. First, the large minority carrier population represents a significant stored charge that must be removed in order to turn the diode off. Power diodes switch relatively slowly and are unsuited to high-speed switching applications. Second, the effectiveness of the HBGR diminishes at higher current levels.[11] This effect is probably due to high-level injection eroding the built-in potential at the N-epi/NBL interface. As large numbers of holes are injected into the collector, the population of electrons in the N-epi must increase to maintain charge neutrality. This in turn reduces the built-in potential and therefore allows more holes to surmount the N-epi/NBL interface.

The power diode can be improved by the addition of a hole-collecting guard ring (HCGR) inside the deep-N+ ring. The HCGR will collect some of the holes injected into the N-tank, reducing stored charge and substrate injection. The HCGR typically consists of a ring of base diffusion that acts as the collector of a lateral PNP. Since the HCGR connects to the cathode, it can be butted against the deep-N+ to save space. A single merged contact can run over both base and deep-N+ (Figure 10.14). This structure, or some modification of it, forms the classic power diode used by many relay and solenoid drivers.

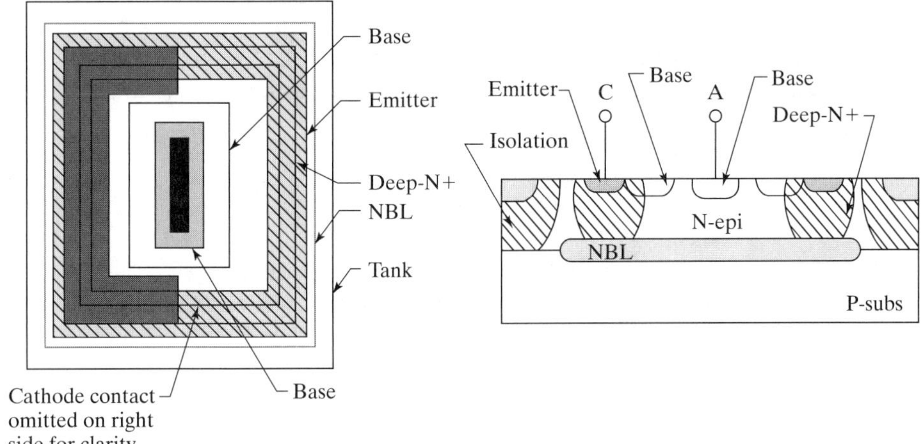

FIGURE 10.14 Layout and cross section of a power diode incorporating merged hole-collecting and hole-blocking guard rings.

Power diodes exhibit several advantages over Schottky diodes. First, the layers required to construct the power diode are almost always available in a power process. By contrast, noble silicides are by no means universally available. Standard bipolar Schottky diodes also require an additional masking step that adds processing cost. Second, the power diode is more rugged than a Schottky diode, as its junction lies far below the surface of the silicon, whereas the Schottky contact lies immediately adjacent to the thermally sensitive metallization. A power diode can therefore dissipate far more power than a Schottky diode before it fails. This observation is particularly true if the diode is forced into avalanche breakdown, as Schottky diodes are typically very fragile under these conditions, whereas power diodes are remarkably rugged.

Power diodes also have their share of disadvantages. They typically exhibit higher forward voltage drops at low currents than Schottky diodes. This disadvantage

[11] B. Murari, "Power Integrated Circuits: Problems, Tradeoffs, and Solutions," *IEEE J. Solid-state Circuits*, Vol. SC-13, #3, 1978, pp. 307–319.

fades at higher currents, where the series resistances of both power diodes and Schottkies dominate their respective forward drops. Power diodes are also very large devices because they must incorporate deep-N+ rings. This disadvantage is also less serious than it might at first appear, as high-current Schottky diodes must also incorporate full deep-N+ rings to prevent holes injected by their guard rings from reaching the substrate. Finally, and most tellingly, power diodes are relatively slow devices. Applications that require rapid reverse recovery typically use large-area Schottky diodes rather than power diodes.

10.2 DIODES IN CMOS AND BiCMOS PROCESSES

Standard bipolar integrated circuits can employ either the collector-base or the base-emitter junction of an NPN transistor as junction diodes. Some versions of standard bipolar also offer a process extension that fabricates Schottky diodes. Diodes constructed in analog BiCMOS closely resemble those fabricated in standard bipolar in both appearance and performance. These devices therefore require no further discussion.

CMOS processes present quite a different picture. Unlike their bipolar counterparts, CMOS transistors do not operate their PN junctions under forward bias. These processes therefore make no provision to contain the flow of minority carriers or to optimize the behavior of forward-biased junctions. Diodes fabricated in CMOS processes usually exhibit much larger parasitics than those constructed in either standard bipolar or analog BiCMOS. The following section examines the various types of diodes that can be fabricated in a CMOS process, along with their capabilities and limitations.

10.2.1. CMOS Junction Diodes

Although one could theoretically construct PN junction diodes from any of the three junctions available in a CMOS process, only one of these junctions can be biased into conduction under normal operating conditions. In an N-well CMOS process, the anodes of the NSD/P-epi and the N-well/P-epi diodes both connect to the substrate. These diodes can only be forward-biased by pulling their cathodes below the substrate. Not only do these diodes require a special negative power supply, but they also pose a latchup hazard because they inject minority carriers into the substrate. The PSD/N-well junction does not exhibit these problems, but it contains a parasitic PNP that diverts a significant fraction of the diode current to the substrate. Without the addition of a buried layer, the resulting device functions as a substrate PNP transistor (Section 8.3.1).

Some modern CMOS processes use a high-energy implant to generate a retrograde well. The heavily doped lower portion of the well functions much like a buried layer. If the doping profile generates a sufficiently large built-in potential, then it will retard the flow of minority carriers to the substrate. Even if this is not the case, the high doping level within the lower portion of the well enhances recombination and therefore lowers the substrate PNP beta. If the substrate beta drops below one, then the resulting device is better described as a junction diode than as a substrate PNP transistor.

An N-well CMOS process can construct Zener diodes by using either the PSD/N-well junction or the NSD/P-epi junction. Both of these Zener diodes exhibit breakdown voltages in excess of the operating voltage of the CMOS transistors. This limitation severely restricts the circuit applications for either type of diode. However, both diodes are useful ESD protection structures.

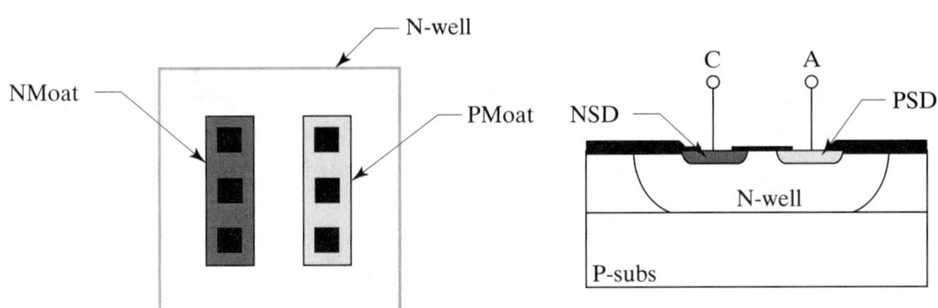

FIGURE 10.15 Layout and cross section of a PSD/N-well Zener diode.

Figure 10.15 shows a typical layout for a PSD/N-well Zener diode. Avalanche breakdown occurs across the depletion region surrounding the PSD implant. The high sheet resistance of the well tends to focus conduction at the edges of the PSD implant facing the NSD contact. The effective area of conduction can be increased by using long narrow strips of PSD interdigitated with strips of NSD. Even with these improvements, the current-handling capability of the PSD/N-well Zener is substantially inferior to that of a base-emitter Zener because of the relatively shallow nature of the source-drain implant.

Processes that silicide the source/drain regions have especially fragile junctions because the silicide eliminates ballasting that would otherwise exist between the contact and the edges of the avalanching diffusion. If the process includes a silicide block mask, then a silicide block geometry should be coded around the avalanching diffusion. For the transistor of Figure 10.15, the silicide block geometry should surround the PMoat region. Even in the absence of silicide, the extremely small overlap of PSD over contact may not provide enough ballasting to survive an ESD strike. This problem can be mitigated by increasing the PSD overlap of contact to 1–2 μm.

An inventive designer can find other options for creating PN junction diodes in a CMOS process. A P-type channel stop region placed inside an N-well creates a lightly doped PN junction. Similarly, an N-type channel stop region placed in the P-epi also forms a junction diode. Many CMOS processes use a combination of deep, lightly doped wells for higher voltage CMOS and shallow, heavily doped wells for lower voltage CMOS. A shallow well placed inside a deep well of the opposite polarity forms a junction diode. For example, on a P-epi process, a shallow P-well (SPWell) placed inside a deep N-well (DNWell) will form a SPWell/DNWell diode. Neither this diode nor the channel stop diodes have much practical application because they offer no compelling advantages over the other devices already discussed.

10.2.2. CMOS and BiCMOS Schottky Diodes

Any process that employs a noble silicide can theoretically fabricate functional Schottky diodes. Standard bipolar processes require an additional masking step to thin the field oxide for Schottky contacts. CMOS and BiCMOS processes do not require this additional processing step because they employ a selective field oxidation.

Unfortunately, few CMOS processes employ platinum or palladium silicides because neither of these materials can withstand the temperatures required to anneal the source/drain implants. Titanium silicide has found widespread application because of its ability to reduce the thin layer of native oxide that appears on the surface of silicon after even momentary exposure to atmospheric oxygen.[12] The low contact potential of titanium silicide renders it completely unsuited to the fabrication of Schottky diodes. More recently, nickel and cobalt silicides have begun to replace titanium silicide because they offer lower sheet resistances at very narrow linewidths. These materials have larger contact potentials that might allow some limited application for Schottky diodes. Neither material is likely to prove suitable for applications that must operate at junction temperatures much beyond 100°C. Therefore, the majority of practical Schottky diodes will continue to employ noble silicides.

Figure 10.16 shows a typical Schottky diode formed in a CMOS process between a noble silicide and the N-well. The moat geometry coded across the contact opening ensures that the contact resides over thin oxide. A ring of PSD diffusion placed around the edge of the contact opening serves as a field-relief guard ring. The reverse breakdown voltage of the Schottky is limited by the avalanche voltage of the PSD/N-well junction rather than by the planar breakdown voltage of the Schottky. A simple field plate can replace the PSD guard ring, but since field-plated Schottky diodes exhibit larger leakages than guard-ringed Schottkies, most designers opt to use guard rings whenever possible.

Schottky diodes fabricated in CMOS processes have relatively high series resistances due to the absence of NBL and deep-N+. This resistance can be minimized by elongating the Schottky contact and by surrounding it on all sides with cathode contacts. Larger Schottky diodes may employ interdigitated anode and cathode contacts. These measures cannot reduce the series resistance as effectively as NBL and deep-N+ so the resulting Schottky diodes are useful only for relatively low-current applications.

Advanced CMOS processes often require the use of arrays of minimum-sized contacts instead of a single large contact. Arrays of small contacts do not lend themselves to the construction of Schottky diodes because they cannot be guard-ringed. Unless some structure analogous to a clad moat can be fabricated over the lightly doped well, these processes cannot support Schottky diodes. A large silicied area, if permitted, would take the place of the single large contact shown in the Schottky diode of Figure 10.16.

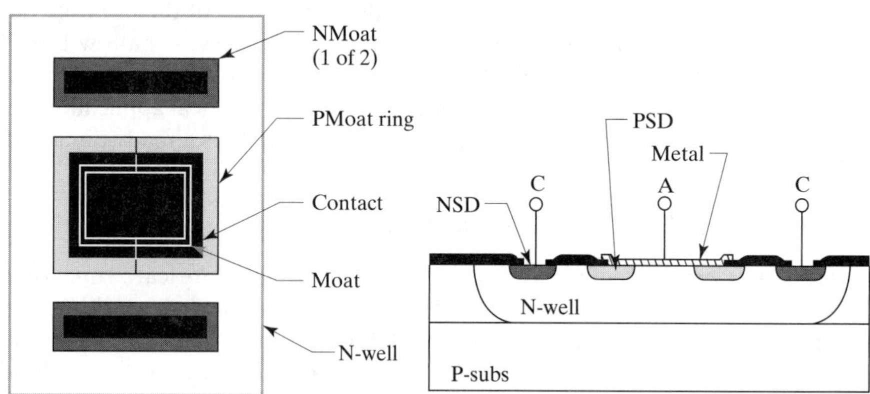

FIGURE 10.16 Layout and cross section of a PSD guard-ring Schottky diode in a CMOS process.

[12] D. Lévy, P. Delpech, M. Paoli, C. Masurel, M. Vernet, N. Brun, J.-P. Jeanne, J.-P. Gonchond, M. Ada-Hanifi, M. Haond, T.T. D'ouville, and H. Mingam, "Optimization of a Self-Aligned Titanium Silicide Process for Submicron Technology," *IEEE Trans. Semiconductor Manufacturing*, Vol. 3, #4, 1990, pp. 168–175.

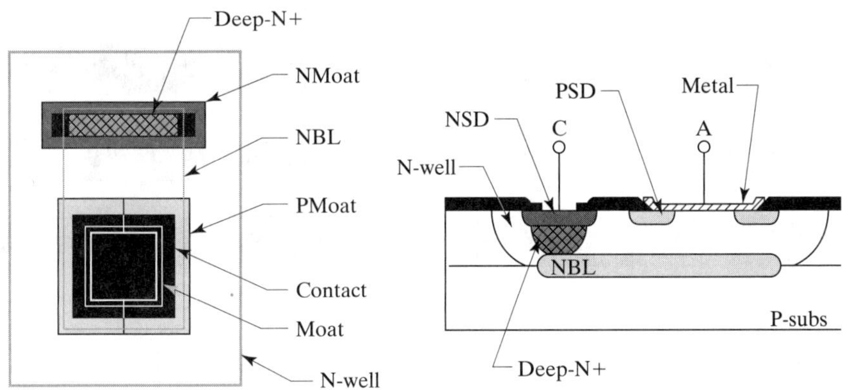

FIGURE 10.17 Layout and cross section of a PSD guard-ring Schottky diode in a BiCMOS process.

Analog BiCMOS processes that employ platinum or palladium silicides can also fabricate Schottky diodes. The structure of these diodes resembles that of their standard bipolar counterparts. A moat geometry coded around the anode contact takes the place of the Schottky contact geometry (Figure 10.17). This substitution allows Schottky diodes to be constructed without requiring any additional masking steps. Many BiCMOS designers use these devices as replacements for diode-connected transistors.

10.3 MATCHING DIODES

The devices discussed in this chapter fall into three broad categories: PN junction diodes, Zener diodes, and Schottky diodes. Different types of diodes do not match one another because they obey different principles of operation. Diodes of the same type can match, but only if they are properly constructed. This section briefly discusses the advantages and disadvantages of each type of matched diode and presents some guidelines to aid the designer in constructing matched diodes.

10.3.1. Matching PN Junction Diodes

Most junction diodes used in bipolar and BiCMOS processes are actually diode-connected transistors. As such, their layouts closely resemble those of conventional bipolar transistors. The only unique feature found in diode-connected transistors is the merged collector-base contact (Figure 10.2). The techniques used to match diode-connected transistors are otherwise identical to those used for other bipolar transistors (Sections 9.2 and 9.3).

Pure CMOS processes do not support the construction of diode-connected transistors, but they can still produce PN junction diodes. An N-well CMOS process can construct a PSD/N-well diode, and a P-well process can construct an NSD/P-well diode. Both of these devices contain parasitic vertical bipolar transistors that divert a significant fraction of their forward current to the substrate. The beta of these bipolar transistors can range from less than 0.1 to more than 10. A typical device has a beta of 2. The resulting base current generates a voltage drop across the well resistance that adds to the forward voltage of the diode. Any variation in beta produces a corresponding variation in the voltage drop across the well resistance. The matching of junction diodes therefore depends on the matching of parasitic bipolar betas and well resistances as well as forward voltages.

The layout of a CMOS PN junction diode is identical to that of a CMOS substrate transistor (Section 8.3.1). This device may be viewed either as a PN junction diode containing a parasitic bipolar transistor, or as a bipolar transistor having so low a gain that it resembles a diode. Either way, one must maximize beta and minimize well

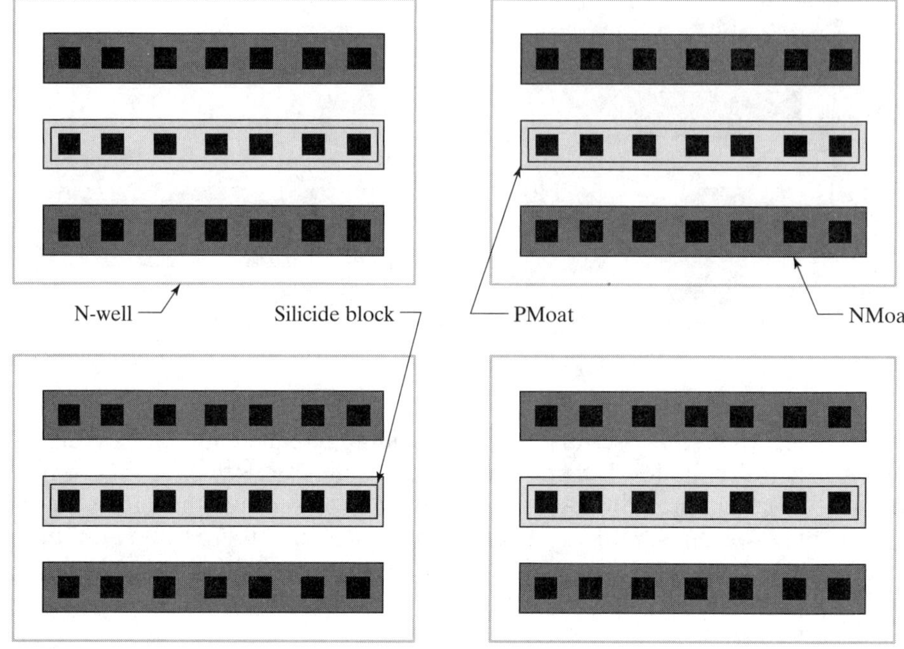

FIGURE 10.18 Layout of a pair of matched PSD/N-well diodes.

resistance to obtain any semblance of matching. These factors have such a strong impact that a device with a small junction area-to-periphery ratio will actually exhibit better matching than one that has a large area-to-periphery ratio. The best layout consists of an array of minimum-width junctions interdigitated with well contacts. Matched devices should be interdigitated or cross coupled to produce a common-centroid array (Section 7.2.10). Two diodes should not be interdigitated in the same well because current flow through the well debiases the diodes to different degrees, depending upon their position and relation to one another. Instead, each diode should be divided into several identical sections, each placed in its own well, and these wells interdigitated or cross coupled. Figure 10.18 shows a pair of matched PSD/N-well diodes.

Many CMOS processes use silicided moats (*clad moats*), but the emitter of a bipolar transistor should not be silicided because this reduces its beta. The matching of PN junction diodes actually improves when they incorporate high-gain parasitic bipolar transistors, because the high gain reduces the currents that must flow through the well resistance. If a silicide block mask is available, then the designer should code this layer around the fingers of the diodes that form its PN junctions (Figure 10.18).

The matching of PN junction diodes also improves at low current densities, because the voltage drop across the well resistance decreases. The current density can be decreased either by increasing the area of the devices or by operating them at lower currents. CMOS diodes typically operate at current densities of 5–50 nA/μm^2.

Regardless of the care taken in their construction, PSD/N-well and NSD/P-well diodes will usually exhibit residual mismatches of several millivolts because of the large emitter periphery-to-area ratios and the influence of well resistance and beta variation.

10.3.2. Matching Zener Diodes

Zener diodes are difficult to match because their breakdown voltages depend so strongly on electric field intensity. Any curvature in the junction geometry intensifies the electric field and produces a localized reduction in breakdown voltage. The portion of the junction that has the largest curvature conducts the majority of the current and therefore sets the breakdown voltage of the device. Localized breakdown degrades matching because it reduces the effective area of the junction and magnifies the effects of outdiffusion. Matched Zeners should employ circular junction geometries to avoid introducing unnecessary corner curvature. Unfortunately, even circular junctions seldom break down uniformly. Linewidth variations produce minute irregularities that have larger curvatures than the remainder of the junction sidewall. The resulting variations in conduction are often visible under a microscope in a darkened room. The faint glow of the avalanching junction usually appears at only a few spots around the perimeter of the junction. Almost all of the current flows through these spots, which represent only a small fraction of the junction periphery. Devices that have identical layouts often show radically different patterns of conduction. These variations highlight the essentially random distribution of defects and linewidth variations.

Large devices generally exhibit a more uniform distribution of defects. Higher current densities also minimize variability by increasing the voltage drop across the lightly doped diffusion adjacent to the junction. The resulting voltage drop provides ballasting and helps distribute conduction across a larger area, but it also represents another source of variability.

Matched Zeners usually employ large circular geometries to minimize random variations and to eliminate edge effects. Both the anode and the cathode contacts should be circularly symmetric, or nearly so. If necessary, the amount of resistive ballasting in the device can be increased by spacing the contacts farther away from the junction. Figure 10.19 shows a cross-coupled pair of matched base-emitter Zeners constructed following these guidelines. The shape of the anode contacts resembles a four-leaf clover, or *quatrefoil,* to allow leads to connect the cathodes to one another without stacking vias over contacts. The four individual Zeners occupy a common tank to minimize the separation between them. The tank isolates the Zeners from the substrate, but it does not include NBL and deep-N+ because it does not have to conduct any significant current.

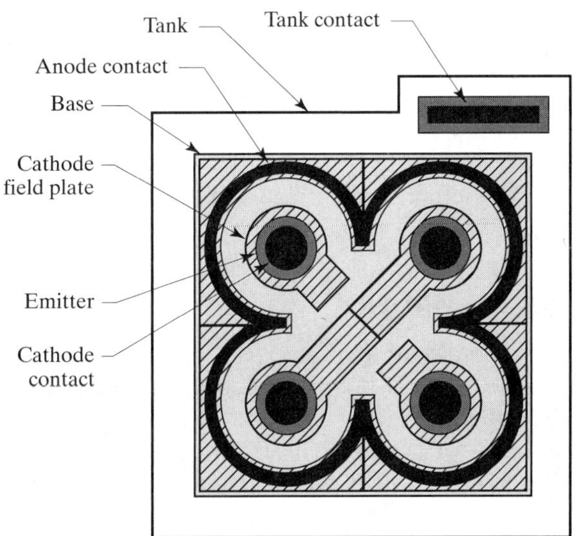

FIGURE 10.19 Quatrefoil layout of cross-coupled base-emitter Zeners. This layout assumes the use of double-level metal.

Even the elaborate quatrefoil layout of Figure 10.19 may not provide precise matching between surface Zeners, because these devices are susceptible to Zener walkout. The magnitude of this walkout depends on the total charge that has passed through the device, the magnitude and direction of the electric fields through the overlying oxide, and the concentration of mobile ions in the oxide. The layout in Figure 10.19 flanges the emitter metallization over the junction. This flange functions as a field plate that ensures that each Zener sees exactly the same voltage across its junction. The overlap of the field plate over the base-emitter junction should account for misalignment, outdiffusion, and fringing fields. An overlap of 5 to 8 μm usually suffices. The field plate cannot stop Zener walkout, but it may help minimize its variability. Buried Zeners provide much better matching because they do not exhibit Zener walkout.

10.3.3. Matching Schottky Diodes

Like Zeners, Schottky diodes are inherently difficult to match. The characteristics of the Schottky barrier depend on several factors, including metal composition, silicon doping, edge effects, annealing conditions, and the presence or absence of surface contaminants. Most of these factors are difficult to control, so Schottky diodes usually exhibit larger mismatches than junction diodes.

Matched Schottky diodes should always incorporate diffused guard rings rather than field plates because field-plated structures often exhibit leakages that could interfere with low-current matching. The contact opening should have a large area-to-periphery ratio to minimize mismatches due to linewidth variations. If possible, the diodes should incorporate NBL and deep-N+ to minimize the portion of the cathode resistance not directly associated with the Schottky contact. If NBL and deep-N+ are used, then several matched Schottky diodes can reside in the same tank or well. If deep-N+ and NBL are not used, then each Schottky should occupy its own well or tank, all of which should have exactly the same dimensions to ensure that the cathode resistances of the diodes match. Ratioed Schottky diodes should always consist of arrays of identical unit contacts, in much the same way (and for much the same reasons) as ratioed NPN transistors use arrays of identical unit emitters. Because Schottky diodes are very sensitive to thermal gradients, they should always use interdigitated or cross-coupled layouts similar to those employed for bipolar transistors.

Schottky diodes will not reliably match PN junction diodes or bipolar transistors. The voltage differential between a diode forward voltage and a Schottky forward voltage may vary by a few percent because of variations in surface conditions, anneal times, and other factors. Even the voltage differential between two Schottky diodes operated at different current densities may depend on processing conditions due to the nonideality factor in the diode equation, which typically plays a larger role in Schottky diodes than it does in junction diodes and bipolar transistors.

10.4 SUMMARY

Every semiconductor process offers one or more types of diodes. Standard bipolar and analog BiCMOS processes can fabricate excellent diode-connected NPN transistors that exhibit mismatches of only a few millivolts and can handle large amounts of current. These transistors can also function as emitter-base Zener diodes. Pure CMOS processes offer fewer options, but they may still be able to construct Schottky diodes if their metallization system includes a noble silicide.

10.5 EXERCISES

Refer to Appendix C for layout rules and process specifications.

10.1. Lay out a standard-bipolar diode-connected transistor, using a minimum-size emitter. Compare the area of this device with the area of a minimum-size NPN transistor that does not contain deep-N+. Additional rules for constructing the collector-base contact of the diode are as follows:

1. CONT extends into EMIT 4 μm
2. CONT overhang EMIT 6 μm
3. BASE overhang EMIT 2 μm

10.2. Lay out a minimum-size emitter-in-isolation Zener, using standard bipolar rules. The emitter should overlap the isolation plug by 18 μm.

10.3. What is the approximate temperature coefficient of a series combination of a 6.8 V emitter-base Zener and two diode-connected NPN transistors?

10.4. Construct a standard bipolar field-plated Schottky diode with an area of 200 μm^2. Overlap the field plate over the contact by 4 μm. Include a deep-N+ sinker along one end of the device. Include all necessary metallization.

10.5. Construct a standard bipolar base guard-ringed Schottky diode having an effective area of 200 μm^2. Assume that the correction factor δ equals 2.0 μm. The layout rules for the base guard ring are as follows:

1. BASE overhang CONT 4 μm
2. BASE extends into CONT 4 μm

10.6. Construct a standard bipolar power Schottky diode with an effective area of 15000 μm^2. Include a base guard ring constructed according to the rules given in Exercise 10.5. Ring the cathode tank with deep-N+ and provide as much metallization as possible, leaving space for a 12 μm-wide cathode lead. The anode metallization should be at least 10 μm wide at all points and should exit the device through a 12 μm-wide anode lead opposite the cathode lead.

10.7. Why does the analog BiCMOS process of Appendix C not support Schottky diodes? What modification would allow their construction?

10.8. Lay out a pair of CMOS PSD/N-well diodes for optimal matching. The diodes should have drawn areas of 60 and 120 μm^2, respectively. Draw silicide block (SBLOCK) around the anodes so that SBLOCK overlaps PMOAT by 1 μm.

10.9. Lay out a quatrefoil Zener structure consisting of four individual diodes sharing a common anode, using standard bipolar rules. Use a diameter of 12 μm for each of the four emitters. Include all necessary metallization, showing how the cathode lead exits from the device and how the tank contact is connected.

11 *Field-Effect Transistors*

The *field-effect transistor*, or FET, has a long and complicated history. Its initial invention preceded that of the bipolar transistor by some 17 years, but all of the early attempts to manufacture field-effect transistors failed because of processing problems.[1] Many of these problems were associated with the growth of thin, high-quality dielectric films. By the time these problems were finally overcome, Bardeen and Brattain had already developed the bipolar transistor.

Since the growth of thin dielectric films proved to be so difficult, the first practical field-effect transistors used reverse-biased junctions in place of dielectrics. The resulting devices were called *junction field-effect transistors* (JFETs). Although JFETs were relatively cumbersome devices, they offered much lower input currents than bipolar transistors could achieve. Certain types of operational amplifiers were designed with JFET input stages to reduce their input currents. These devices have become quite successful and are still being produced today.

Thin insulating films suitable for gate dielectrics were finally produced in 1960.[2] This achievement made possible the manufacture of the *metal-oxide-semiconductor field-effect transistor* (MOSFET), often simply called the *MOS transistor*. The early MOS devices still had their share of problems. Their threshold voltages were notoriously unstable, and their thin gate oxides were exceedingly vulnerable to electrostatic discharge (ESD). Once these problems were overcome, MOS transistors began to seriously challenge established bipolar technologies. MOS integrated circuits proved especially useful for low-power digital devices such as digital watches and pocket calculators.

The earliest MOS processes offered only PMOS transistors. These were soon superseded by processes that could produce both enhancement and depletion NMOS transistors. Demands for lower current consumption and greater design flexibility

---

1 D. Kahng, "A Historical Perspective on the Development of MOS Transistors and Related Devices," *IEEE Trans. on Electron Devices*, Vol. ED-23, #7, 1976, pp. 655–657.

2 D. Kahng and M. M. Atalla, "Silicon-silicon dioxide field-induced devices," *Solid-State Device Research Conference*, Pittsburgh, 1960.

led to the introduction of processes that could simultaneously fabricate both NMOS and PMOS transistors. Although originally intended for digital applications, these *complementary metal-oxide-semiconductor* (CMOS) processes could also fabricate a variety of analog integrated circuits. These soon began to replace bipolar integrated circuits in selected applications, but CMOS transistors were not able to duplicate all of the capabilities of bipolar. Many newer processes now merge both bipolar and CMOS transistors onto a common substrate.

Another interesting class of field-effect transistors consists of devices whose gates are completely surrounded by an insulating dielectric. The behavior of these *floating-gate* transistors depends upon the quantity of charge stored upon their floating gates. Various techniques have been developed to either inject charge onto floating gates or to remove it. The resulting devices behave as nonvolatile memory elements. They have many important applications in both analog and digital integrated circuits.

This chapter describes the operation and construction of MOS transistors, particularly those employing self-aligned polysilicon gate technology. This chapter also covers the operation and construction of JFETs and floating-gate transistors. Chapter 12 covers high-voltage and power MOS devices, as well as the matching of MOS transistors.

11.1 TOPICS IN MOS TRANSISTOR OPERATION

This section reviews the basic operating principles of MOS transistors, and then focuses upon several topics of interest to layout designers. These include the effects of process and layout on device parameters, the behavior of MOS transistors operating in breakdown, and causes of leakage in MOS transistors.

11.1.1. Modeling the MOS Transistor

Figure 11.1 shows a simplified three-terminal circuit model of an NMOS transistor. No DC current flows through the gate terminal because of the insulating layer between it and the rest of the transistor. Capacitors C_{GS} and C_{GD} represent the gate-to-source and gate-to-drain capacitances produced by this *gate dielectric*. The slashes through these capacitors signify that their values depend on biasing. Voltage-controlled current

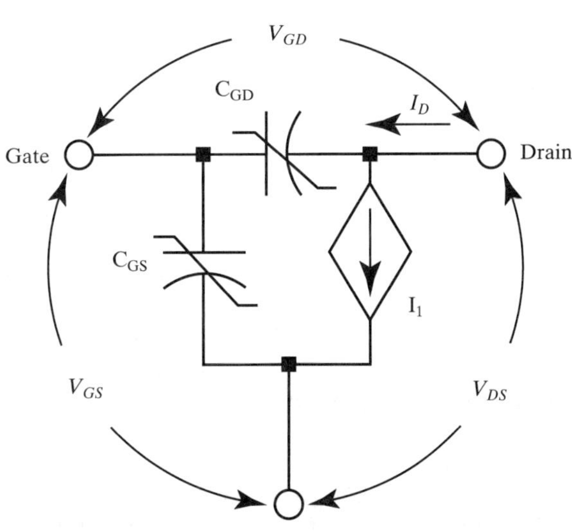

FIGURE 11.1 Simplified three-terminal model of a NMOS transistor.

source I_1 models the current flowing from drain to source through the channel beneath the gate.

The magnitude of the drain current I_D depends on the gate-to-source voltage V_{GS} and the gate-to-drain voltage V_{DS}. If the gate-to-source voltage is less than the *threshold voltage* V_t, then no channel forms and the transistor remains in *cutoff* and conducts little or no current. A channel begins to form as soon as the gate-to-source voltage V_{GS} exceeds the threshold voltage V_t. The difference between these two quantities is sometimes called the *effective gate voltage* V_{gst}:

$$V_{gst} = V_{GS} - V_t \qquad [11.1]$$

A larger V_{gst} generates a stronger channel that can conduct more current. The drain current also depends on the drain-to-source voltage V_{DS}. If the drain-to-source voltage is less than the effective gate voltage, then the drain current varies approximately linearly with drain-to-source voltage and the transistor is said to operate in the *linear region* (also called the *triode region*). If the drain-to-source voltage exceeds the effective gate voltage, then the drain current becomes essentially independent of drain-to-source voltage and the transistor is said to operate in the *saturation region*. The relationship between I_D, V_{GS}, and V_{DS} can be described by a pair of equations—one for the linear region and the other for saturation:

$$If\ 0 \leq V_{DS} < V_{gst},\ I_D = k\left(V_{gst} - \frac{V_{DS}}{2}\right)V_{DS} \qquad [11.2A]$$

$$If\ V_{DS} \geq V_{gst},\ I_D = \frac{k}{2}V_{gst}^2 \qquad [11.2B]$$

These are the *Shichman-Hodges equations* for the NMOS transistor.[3] The parameter k is called the *device transconductance*, which is something of a misnomer because it has units of A/V^2 rather than A/V. The equations for the PMOS transistor differ in only one respect: Equation 11.2A applies when $0 \geq V_{DS} > V_{gst}$, and Equation 11.2B applies when $V_{DS} \leq V_{gst}$. For an enhancement-mode PMOS, V_{DS}, V_{GS}, V_t, k, and I_D must all be negative quantities for the equations to yield the proper results.

The Shichman-Hodges equations for the NMOS do not cover the case where $V_{DS} < 0$. Strictly speaking, this condition cannot occur because the source and drain of a MOS transistor are determined by electrical biasing rather than by arbitrary terminal markings. If one attempts to make the drain-to-source voltage less than zero, then the source and drain simply swap roles. The drain terminal now plays the role of the source, and *vice versa*. If one insists on retaining terminal names that no longer correspond to the roles that the terminals actually play, then the terminal conditions must be transformed into actual biasing conditions before applying the Shichman-Hodges equations. (See Exercise 11.2.)

Device Transconductance

The *device transconductance*, k, determines the amount of drain current that flows through a MOS transistor in response to a given V_{gst}. The device transconductance thus specifies the size of a MOS transistor in much the same way that the emitter saturation current I_S quantifies the size of a bipolar transistor. The device transconductance has units of A/V^2 or (more commonly) μA/V^2. It is related to the layout

---

[3] The equations given here ignore channel length modulation and the body effect. For further details see H. Shichman and D. A. Hodges, "Modeling and Simulation of Insulated-Gate Field-Effect Transistor Switching Circuits," *IEEE J. Solid-State Circuits*, SC-3, 3, 1968, 285-289.

dimensions of the transistor by the equation

$$k = k'\left(\frac{W}{L}\right) \tag{11.3}$$

where W and L represent the width and length of the MOS channel and k' is a constant called the *process transconductance*, which equals

$$k' = \frac{\mu\varepsilon_o\varepsilon_r}{t_{ox}} \tag{11.4}$$

where the quantity μ represents the *effective mobility* of the carriers (electrons in an NMOS, holes in a PMOS). Surface scattering reduces the mobility of carriers confined within a MOS channel, so the effective mobilities appearing in Equation 11.4 are considerably smaller than the bulk mobilities discussed in Section 1.1.1. The effective mobility of electrons and holes in silicon are about 675 $cm^2/V \cdot s$ and 240 $cm^2/V \cdot s$, respectively. The constant ε_o denotes a universal physical constant called the *permittivity of free space*, which equals $8.85 \cdot 10^{-12}$ F/m. The constant ε_r represents the relative permittivity of the gate dielectric, which for pure silicon dioxide equals about 3.9. The actual permittivity of oxide may vary slightly from this theoretical value. (See Table 6.1.) The quantity t_{ox} represents the thickness of the gate dielectric. Substituting these values into the previous equation yields the following simplified formulas for the process transconductance of an NMOS transistor k'_n and of a PMOS transistor k'_p:

$$k'_n \cong \frac{23000}{t_{ox}} \mu A/V^2 \tag{11.5A}$$

$$k'_p \cong \frac{8200}{t_{ox}} \mu A/V^2 \tag{11.5B}$$

Here, t_{ox} is measured in Angstroms (Å). MOS processes use the thinnest possible gate oxides to produce the largest possible device transconductances. The dielectric strength of gate oxide equals about 10^7 V/cm, or about 0.1 V/Å. In practice, the gate dielectric is restricted to substantially lower field intensities to prevent a delayed breakdown mechanism called *time-dependent dielectric breakdown* (TDDB). Transistors with gate oxides thicker than 500 Å are generally restricted to field intensities of no more than $3 \cdot 10^6$ V/cm (30 mV/Å). Thinner gate oxides can withstand somewhat higher electric field intensities,[4] so transistors with a gate oxide only 100 to 200 Å thick can safely operate at fields in excess of $5 \cdot 10^6$ V/cm (50 mV/Å). Assuming a conservative limit of 30 mV/Å, the maximum possible process transconductance for an operating voltage V_{op} equals

$$k'_n \cong \frac{690}{V_{op}} \mu A/V^2 \tag{11.6A}$$

$$k'_p \cong \frac{240}{V_{op}} \mu A/V^2 \tag{11.6B}$$

These formulas indicate that a 5 V CMOS process can achieve an NMOS transconductance of about 140 $\mu A/V^2$ and a PMOS transconductance of about 50 $\mu A/V^2$. In

[4] C. M. Osburn and D. W. Ormond, "Dielectric Breakdown in Silicon Dioxide Films on Silicon, II. Influence of Processing and Materials," *J. Electrochem. Soc.*, Vol. 119, #5, 1972, pp. 597–603.

practice, short-channel transistors often experience additional transconductance reductions caused by velocity saturation and other high-field effects. In order to construct a PMOS transistor with the same transconductance as a given NMOS, the W/L ratio of the PMOS must equal almost three times the W/L ratio of the NMOS. Assuming equal lengths, the PMOS transistor requires nearly three times the area of the NMOS. This disparity is most noticeable in power transistors, but even minimum-size logic gates often increase the size of PMOS transistors to compensate for their low process transconductances.

The device transconductance of MOS transistors decreases with temperature. This variation is primarily due to the temperature coefficient of the carrier mobility. As the temperature rises, lattice vibrations become more energetic and cause increased carrier scattering. Consequently, carrier mobilities are approximately proportional to the inverse square of absolute temperature.[5] The device transconductance at 150°C equals about half of its value at 25°C. The drain current for a given gate-to-source voltage scales somewhat similarly. Since an increase in temperature causes a decrease in drain current, many designers believe that MOS transistors are immune to thermal runaway, and that they do not require ballasting. Neither of these assumptions is necessarily true. Every MOS transistor contains a bipolar parasitic transistor that can conduct current under certain conditions, potentially leading to thermal runaway (Section 11.1.2). Although ballasting is generally ineffectual in MOS power transistors, it finds widespread application in constructing ESD devices and transient suppressors (Section 13.5.2).

The device transconductance given in Equation 11.4 corresponds to that found in most engineering texts,[6] as well as that used in the level-1 MOS model of the simulation program SPICE. Since SPICE was written at the University of California at Berkeley, this definition of device transconductance is often called the *Berkeley* k. A few authors use an alternative definition equal to one-half of the Berkeley k and adjust the Shichman-Hodges equations accordingly.

Threshold Voltage

The *threshold voltage* V_t equals the gate-to-source voltage required to just establish a channel beneath the gate dielectric when the backgate is connected to the source. An *enhancement-mode* MOS transistor requires the application of a nonzero gate-to-source voltage in order to form a channel. The channel of an enhancement-mode NMOS consists of electrons attracted to the surface of the P-type backgate by the positively charged gate electrode (Figure 11.2A). The threshold voltage of the enhancement NMOS is therefore positive. The channel of an enhancement PMOS consists of holes attracted to the surface of an N-type backgate by the negatively charged gate electrode (Figure 11.2B). The threshold voltage of an enhancement PMOS is therefore negative. Another type of MOS transistor exhibits a channel even at a gate-to-source voltage of zero. These *depletion-mode* transistors are normally conducting and require the application of an external gate-to-source voltage in order to turn them off (Figures 11.2C, D). The threshold voltage of a depletion NMOS is negative and the threshold of a depletion PMOS is positive.

MOS transistors are often described as electrically actuated switches. An enhancement MOS resembles a normally open switch because it is normally off and it requires the application of an external gate bias to turn it on. A depletion MOS resembles a normally closed switch because it is on by default and it requires the

---

5 Electron mobility varies as $T^{-2.42}$ and hole mobility varies as $T^{-2.20}$; see S.M. Sze, *Physics of Semiconductor Devices*, 2d ed. (New York: John Wiley and Sons, 1981), p. 29.

6 For instance, see R. S. Muller and T. I. Kamins, *Device Electronics for Integrated Circuits*, 2d ed. (New York: John Wiley & Sons, 1986), p. 430.

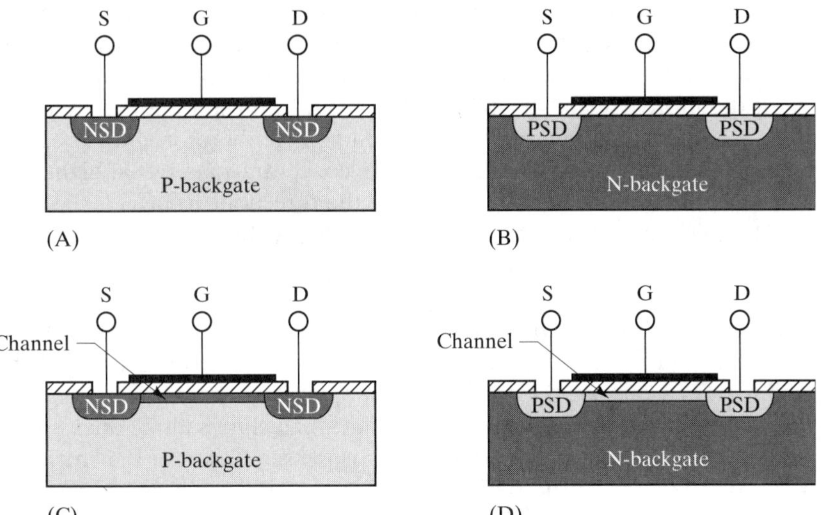

FIGURE 11.2 The four types of MOS transistors: (A) enhancement NMOS, (B) enhancement PMOS, (C) depletion NMOS, and (D) depletion PMOS.

application of an external gate bias to turn it off. Most processes are optimized to fabricate enhancement-mode transistors because these devices are much easier to use than their depletion-mode counterparts. Some processes also offer depletion-mode devices as process extensions.

The threshold voltage of a MOS transistor depends on several factors, including the gate electrode material, the doping of the backgate, the thickness of the gate oxide, the surface state charge density, and the oxide charge density (both fixed and mobile). Each of these factors will be considered in turn.

If the materials used to form the gate and the backgate are not identical, then a nonzero contact potential develops between them. This contact potential represents a net voltage difference between the two materials even if they are separated by an insulating layer. Any change in the contact potential produces a corresponding change in the threshold voltage. Modern MOS transistors almost always use heavily doped polysilicon gate electrodes. There are two choices of gate material: N+ poly or P+ poly. The substitution of a P+ gate for an N+ gate causes the contact potential to increase by about 1.2 V, regardless of backgate doping (Table 11.1). Similarly, the substitution of an N+ gate for a P+ gate causes the contact potential to decrease by about 1.2 V. These variations in contact potential will produce equal variations in threshold voltage.

A few examples may help to explain the effects of swapping gate materials. Suppose that a certain NMOS transistor develops a threshold voltage of 0.7 V, using an N+ gate. If this same transistor were to use a P+ gate, it would have a threshold of about 1.9 V. Suppose that the corresponding enhancement PMOS develops a threshold voltage of −0.7 V, using a P+ gate. If this transistor were to use an N+ gate, it would become a depletion device with a threshold of about +0.5 V.

Backgate Material	N+ Poly Gate	P+ Poly Gate
N-type, $N_D = 10^{14}\ \text{cm}^{-3}$	−0.36 V	0.82 V
N-type, $N_D = 10^{16}\ \text{cm}^{-3}$	−0.24 V	0.94 V
N-type, $N_D = 10^{18}\ \text{cm}^{-3}$	−0.12 V	1.06 V
P-type, $N_A = 10^{14}\ \text{cm}^{-3}$	−0.82 V	0.36 V
P-type, $N_A = 10^{16}\ \text{cm}^{-3}$	−0.94 V	0.24 V
P-type, $N_A = 10^{18}\ \text{cm}^{-3}$	−1.06 V	0.12 V

TABLE 11.1 Computed poly gate-to-backgate contact potentials ($10^{20}\ \text{cm}^{-3}$ poly).

The backgate doping concentration also has a strong impact on the threshold voltage. In order to form a channel, enough carriers must be attracted to the surface to invert the silicon. A heavily doped backgate is difficult to invert, and the gate must therefore exert a stronger electric field to muster enough carriers to create a channel. The magnitude of the threshold voltage therefore increases with backgate doping. The effect is small at low doping concentrations, but at higher doping levels it actually dominates the expression for the threshold voltage.

The thickness of the gate oxide can also play an important role in determining the threshold voltage of a transistor. A given gate-to-source voltage produces a weaker electric field across a thick gate oxide than across a thin one. Transistors with thick gate oxides are more difficult to invert than ones with thin gate oxides, so increasing the thickness of the gate oxide increases the magnitude of the threshold voltage. For example, the oxide in the field regions is made as thick as possible to raise the thick-field threshold. Unfortunately, thickness alone does not guarantee an adequate thick-field threshold for most processes. Consider the entries in Table 11.2 for a 10 kÅ oxide. If the background doping equals 10^{15} cm^{-3}, then the thick-field threshold only equals about 4 V. This is obviously inadequate. Most processes use channel stop implants to raise the doping in the field regions.[8] If the channel stop implant raises the doping under a 10 kÅ field oxide to 10^{17} cm^{-3}, then the thick-field threshold will approach 50 volts. Since both the NMOS and the PMOS backgate doping levels are relatively low, most processes must employ a combination of boron and phosphorus channel stops to ensure that both thick-field thresholds lie well above the nominal operating voltages.

TABLE 11.2 Typical NMOS threshold voltages as a function of backgate doping and gate oxide thickness (N-type poly gate).[7]

Backgate Doping	100 Å	250 Å	10k Å
10^{14} cm^{-3}	−0.23 V	−0.21 V	0.89 V
10^{15} cm^{-3}	−0.08 V	−0.02 V	3.91 V
10^{16} cm^{-3}	0.14 V	0.35 V	13.9 V
10^{17} cm^{-3}	0.60 V	1.31 V	47.9 V
10^{18} cm^{-3}	1.86 V	4.28 V	162 V
10^{20} cm^{-3}	7.94 V	19.1 V	747 V

The threshold voltage is also affected by the presence of residual charges within the gate oxide and along the oxide-silicon interface. These residual charges can be divided into three types: fixed oxide charge, mobile oxide charge, and surface state charge. The *fixed oxide charge* Q_f consists of defect sites scattered randomly throughout the oxide film. Gate oxides grown in dry oxygen at relatively low temperatures have very small fixed oxide charges. The fixed oxide charge can drastically increase if holes are injected into the oxide, as happens during oxide breakdown, hot carrier injection, and exposure to ionizing radiation.

The *mobile oxide charge* Q_m consists of positively charged mobile ions such as sodium and potassium.[9] The threshold voltage shift that they produce also depends on their location within the oxide film, and this in turn depends on gate biasing. Consider the case of an NMOS transistor whose gate oxide is contaminated by

---

7 The values in this table assume $Q_{ss} = 0$ and $\Phi_{MS} = -0.7$ V.

8 J. D. Sansbury, "MOS Field Threshold Increase by Phosphorus-Implanted Field," *IEEE Trans. on Electron Devices*, Vol. ED-20, #5, 1973, pp. 473–476.

9 Sodium is the primary mobile ion encountered in silicon processing; lesser contributors include potassium and hydrogen ions. See B. E. Deal, "The Current Understanding of Charges in the Thermally Oxidized Silicon Structure," *J. Electrochem. Soc.*, Vol. 121, # 6, 1974, pp. 198C–205C.

sodium. When a positive gate-to-source voltage is applied, the mobile ions move away from the positively charged gate electrode and move closer to the negatively charged backgate. As the mobile ions shift closer to the channel region, they exert an increasing effect on it. Thus the movement of the mobile ions causes the NMOS threshold to decrease. Any shift in the threshold voltage of MOS transistors can cause offsets. In order to obtain accurate matching, mobile ions must either be excluded from the process, or they must somehow be immobilized (Section 4.2.2). Modern processing techniques and purer chemicals have reduced the magnitude of the mobile oxide charge to negligible proportions.

The *surface state charge* Q_{ss} is concentrated in a thin layer near the oxide/silicon interface. It is generally positive, but its magnitude depends on silicon crystal orientation and annealing conditions. The precise mechanisms responsible for generating the surface state charge are not fully understood, but they are believed to involve mismatches between the molecular structure of the silicon lattice and that of the oxide macromolecule. A large component of the surface state charge stems from unfilled valence shells, also called *dangling bonds*, that appear along the oxide interface.[10] The number of dangling bonds at the interface depends upon the crystal orientation of the silicon surface. Oxides grown on (100) silicon have less than half the surface state charge density of oxides grown on (111) silicon.[11] All modern MOS processes employ (100) silicon to minimize the impact of surface state charge on the threshold voltages of MOS transistors. Standard bipolar uses (111) silicon to deliberately raise the NMOS thick-field threshold. The surface state charge can also be reduced by annealing the wafer in a reducing atmosphere such as hydrogen or forming gas (a mixture of nitrogen and hydrogen). During the anneal, hydrogen atoms tie off dangling bonds at the surface and thus eliminate the surface state charge these dangling bonds would otherwise create. Although it is not possible to entirely eliminate the surface state charge, the use of (100) silicon in combination with proper annealing can minimize threshold voltage variation.

Each of the mechanisms just discussed contributes a small amount of variability to the threshold voltage. With care, manufacturing variations in threshold voltages can be held to within approximately ±0.1 V. Threshold voltages also vary with temperature. The magnitude of this variation depends on the backgate doping and oxide thickness, but it typically ranges from −2 mV/°C to −4 mV/°C.[12,13] A temperature variation of −2 mV/°C over a temperature range of −55 to 125°C produces a threshold variation of about ±0.2 V. Combining this with the process variation yields a total variation of about ±0.3 V. A transistor with a nominal threshold voltage of 0.7 V might actually have a V_t as low as 0.4 V or as high as 1.0 V. Although it might seem that the minimum threshold could safely be decreased to 0.3 V or less, this is actually not the case. MOS transistors continue to conduct small amounts of current when the gate-to-source voltage is less than the threshold voltage due to a mechanism called *subthreshold conduction* (Section 11.1.2). The magnitude of the subthreshold current decreases exponentially, and the gate-to-source voltage must drop at least 0.3 V below the threshold in order to reduce the drain current to negligible levels. The magnitude of the nominal threshold voltage must therefore equal

---

[10] This discussion is somewhat oversimplified, as several types of charges reside along the interface, including a component of the fixed oxide charge and a variety of interface traps (the so-called *fast surface states* and *slow surface states*). The component due to fixed oxide charge is always positive, while the interface traps may contribute either positive or negative charges. (See Muller and Kamins, p. 152ff.)

[11] Deal's data shows Q_{SS} values on (100) silicon equal to 20 to 30% of those on (111) silicon: Deal, *ibid.*

[12] R. Wang, J. Dunkley, T. A. DeMassa, and L. F. Jelsma, "Threshold Voltage Variations with Temperature in MOS Transistors," *IEEE Trans. on Electron Devices*, Vol. ED-18, #6, 1971, pp. 386–388.

[13] F. M. Klaasen and W. Hes, "On the Temperature Coefficient of the MOSFET Threshold Voltage," *Solid-state Electronics*, Vol. 29, #8, 1986, pp. 787–789.

at least 0.6 V. Transistors with smaller threshold voltages are useful in certain applications, but they cannot be used as switching devices. Transistors used in very low current applications may require nominal threshold voltages of 0.8 V or more to prevent objectionable subthreshold conduction. Alternatively, special circuit design techniques may be used to apply a reverse gate-to-source voltage to ensure proper cutoff.

11.1.2. Parasitics of MOS Transistors

A real-world MOS transistor contains a number of parasitic elements that affect its operation. Perhaps the most important of these are the junctions that isolate the source and drain regions from the backgate. These junctions remain reverse-biased during normal operation, but either or both of them may begin to conduct under certain circumstances. The forward-biased junctions will inject minority carriers into the backgate, at best causing unexpected leakage and at worst triggering latchup.

A complete model of all of the parasitics contained in an MOS transistor would include a number of distributed effects, the discussion of which lies beyond the scope of this text. Figure 11.3 shows a simplified parasitic model of an NMOS transistor constructed in an N-well CMOS process. The circuit contains a three-terminal NMOS transistor, M_1, that models the intended functionality of the device. Capacitors C_{GD}, C_{GS}, and C_{GB} represent the gate-to-drain, gate-to-source, and gate-to-backgate capacitances, respectively. The gate-to-backgate capacitance, C_{GB}, models the capacitance across the thin gate oxide separating the gate electrode from the backgate diffusion. This capacitance decreases as the transistor approaches inversion, and drops to zero when a channel forms. When the transistor is in cutoff, the gate-to-drain and gate-to-source capacitances, C_{GD} and C_{GS}, consist primarily of the overlap capacitances between the gate electrode and the respective diffusions. These have been greatly reduced by the introduction of self-aligned polysilicon gates, and they now consist mainly of fringing capacitances. The gate-to-source and gate-to-drain capacitances abruptly increase when a channel forms, because of the sudden addition of the capacitance between the gate electrode and the channel.[14]

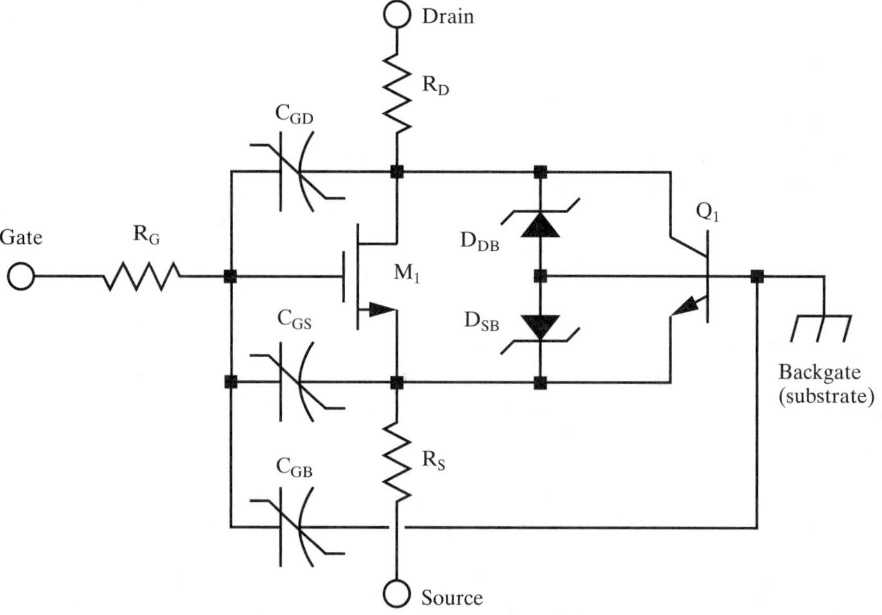

FIGURE 11.3 Simplified parasitic model of an NMOS transistor constructed in N-well CMOS.

[14] J. E. Meyer, "MOS Models and Circuit Simulation", *RCA Rev.*, Vol. 32, 1971, pp. 42–63.

The channel behaves as a resistor when the device operates in the linear region. The gate capacitance is distributed uniformly along this resistance, but it can be modelled as a π-network consisting of two equal capacitances (C_{GS} and C_{GD}). When the device saturates, the channel pinches off and the capacitance between the gate and the channel appears entirely on the source side of the device. C_{GS} thus greatly exceeds C_{GD} in saturation. The slashes drawn through C_{GD}, C_{GS}, and C_{GB} denote the voltage dependence of these devices.

Resistors R_G, R_S, and R_D represent the resistance of the gate, source, and drain terminals, respectively. The gate resistance actually forms a distributed network with the three gate capacitances, but this simplified model depicts it as a lumped quantity. The gate resistance has no effect on the DC performance of the transistor, but it does slow the switching speed because it limits the current that is available to charge and discharge capacitances C_{GS}, C_{GD}, and C_{GB}. The gate poly is often silicided to minimize R_G. The drain and source resistances R_D and R_S consist of the Ohmic resistances between the contact and the edge of the source/drain diffusions abutting the channel. These resistances can be minimized by siliciding the surfaces of the source/drain diffusions. A silicided poly gate electrode is sometimes called a *clad gate*, and silicided source/drain regions are also called *clad moats*. Many processes use clad gates, but generally only submicron processes have sufficient transconductance to merit clad moats.

Diodes D_{DB} and D_{SB} represent the drain-backgate and source-backgate junctions, respectively. The diodes model the junction capacitance added to the drain and source terminals by these junctions, and they also model their avalanche breakdown characteristics. Lateral NPN transistor Q_1 represents one possible path for minority carriers to travel from drain to source (or *vice versa*). Since the NMOS transistor resides in the epi, the minority carriers can also flow to adjacent NMOS source/drain regions, or to adjacent wells. The flow of carriers to adjacent wells can lead to CMOS latchup, as discussed shortly.

Figure 11.4 shows a simplified parasitic model for a PMOS transistor constructed in an N-well process. Capacitors C_{GS}, C_{GD}, and C_{GB} play the same roles in this device as their counterparts do in an NMOS. Similarly, resistors R_G, R_S, and R_D represent the same terminal resistances in the PMOS as they do in the NMOS. Diodes D_{SB} and D_{DB} represent the junctions between the source/drain regions and

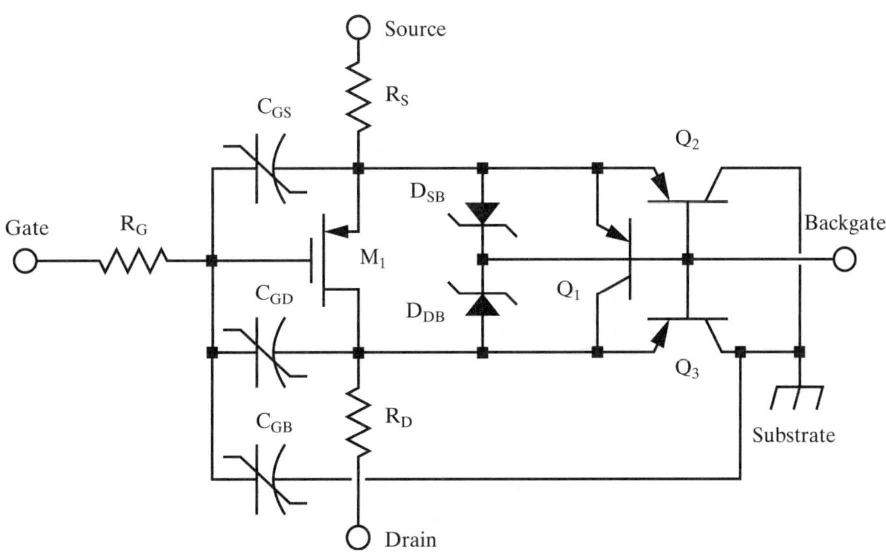

FIGURE 11.4 Simplified parasitic model of a PMOS transistor constructed in N-well CMOS.

the backgate, in this case consisting of an N-well diffusion contacted through a fourth terminal. If either of these junctions forward-biases into the well, then the minority carriers will travel either to the other source/drain diffusion or to the substrate. Lateral PNP Q_1 represents the minority carrier conduction path from drain to source (or *vice versa*). PNP transistors Q_2 and Q_3 represent the minority carrier paths from the source/drain diffusions to the substrate.

Breakdown Mechanisms

Several different mechanisms limit the operating voltage of MOS transistors. One of these corresponds to the V_{CER} breakdown mechanism seen in bipolar transistors (Section 8.1.2). For purposes of discussion, assume that the gate and backgate electrodes both connect to the source. As the drain-to-source voltage rises, it eventually reaches a point where the drain-backgate junction begins to break down. Avalanche multiplication injects large numbers of majority carriers into the lightly doped backgate, causing it to debias. The source-backgate junction begins to inject minority carriers into the backgate as soon as it forward-biases. Most of these minority carriers flow across to the drain, where they stimulate further avalanche multiplication. This beta multiplication process causes the breakdown voltage of the MOS transistor to snap back from the initial *trigger voltage* to a lower *sustain voltage* (Figure 11.5). Short-channel transistors have somewhat lower breakdown voltages because the narrower basewidth of their parasitic lateral bipolar transistor raises its gain and enhances the beta multiplication process.

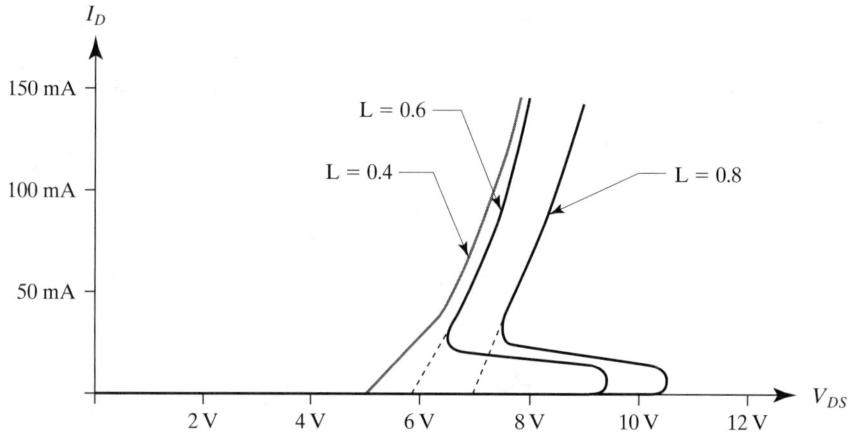

FIGURE 11.5 Breakdown characteristics of short-channel NMOS transistors. The solid curves show the breakdown characteristics in cutoff, and the dotted lines show the change in characteristics when in conduction.

Once a MOS transistor avalanches, most of its drain current flows through the parasitic bipolar transistor rather than through the MOS channel.[15] The device therefore becomes vulnerable to hot spot formation, current focusing, and secondary breakdown. Section 12.2.1 discusses the impact of these mechanisms upon MOS power transistors and their safe operating areas.

Short-channel transistors sometimes experience another form of breakdown called *punchthrough*. The source/drain regions are heavily doped, so the depletion regions surrounding them extend primarily into the lightly doped backgate. The drain depletion region widens as the drain-to-source voltage increases. If the drain/backgate junction does not avalanche first, the drain depletion region will

[15] F.-C. Hsu, P.-K. Ko, S. Tam, C. Hu, and R. S. Muller, "An Analytical Breakdown Model for Short-Channel MOSFETs," *IEEE Trans. on Electron Devices*, ED-29, #11, 1982, pp. 1735–1740.

eventually extend entirely across the channel to touch the source depletion region. Punchthrough breakdown does not exhibit snapback because it does not activate the parasitic lateral bipolar transistor. The curves of Figure 11.5 highlight the differences between punchthrough (in the 0.4 μm device) and avalanche breakdown (in the 0.6 and 0.8 μm devices). The snapback characteristic of avalanche breakdown appears only if the transistor is in cutoff. Even low levels of subthreshold conduction produce enough beta multiplication to obscure the snapback characteristic (as shown by the dotted lines in Figure 11.5). Low-voltage isolated MOS transistors may also exhibit punchthrough breakdown between the drain and the substrate. In N-well CMOS processes, only the PMOS transistor normally experiences this form of punchthrough. This problem can be solved by using a deeper well diffusion, but this will result in larger spacings between components. Some low-voltage processes use high-energy implants called *punchthrough stops* to increase the doping at the bottom of the well to prevent vertical punchthrough. Alternatively, the well can be formed using a high-energy implant to produce a peak doping concentration deep beneath the surface of the silicon. Such a *retrograde well* has characteristics similar to those of a regular well augmented by a punchthrough stop.[16]

The thin gate oxide is vulnerable to dielectric breakdown. Fowler-Nordheim tunneling injects electrons into the oxide when the electric field across it exceeds a critical field intensity of about 11 MV/cm. The hot electrons spawn holes through collisions with the oxide macromolecule. The movement of the holes back through the oxide generates additional traps that enhance the rate of tunneling (Section 4.1.3). Even the most uniform oxides vary slightly in thickness from one location to another. The thinnest spots experience the largest electric fields and consequently the largest rates of hot electron injection. The appearance of additional traps further weakens these spots. Eventually, the weakest spot will catastrophically fail and the gate will short-circuit to the underlying silicon. The speed at which dielectric breakdown occurs depends upon the degree of electrical overstress. A voltage that generates an electric field greatly exceeding the critical field intensity causes almost instantaneous failure. Voltages near the critical field cause delayed failures, a process called *time-dependent dielectric breakdown* (TDDB). The amount of charge required to induce catastrophic breakdown of a given area of oxide remains relatively independent of the magnitude of the field or the time to breakdown. The charge-to-breakdown, or Q_{BD}, thus serves as an empirical measure of the quality of a gate dielectric.[17]

MOS transistors are usually rated to operate at only a small fraction of the critical field intensity in order to ensure adequate safety margin against TDDB. Manufacturers typically quote a maximum electric field intensity E_{max} of 3.5–4 MV/cm (0.035–0.04 V/Å) for gate oxides 300–500 Å thick, and a maximum of 4–4.5 MV/cm (0.04–0.045 V/Å) for thinner gate oxides. The maximum gate-to-source voltage rating, $V_{GS(max)}$, can be computed from the maximum electric field intensity E_{max} and the oxide thickness t_{ox} using the equation

$$V_{GS(\max)} = \frac{E_{\max}}{t_{ox}} \qquad [11.7]$$

Thus, a 200 Å dielectric will provide a maximum gate-to-source voltage of at least 8 V. The maximum gate-to-drain voltage $V_{GD(max)}$ of a self-aligned poly-gate transistor

---

16 R. D. Rung, C. J. Dell'oca, and L. G. Walker, "A Retrograde p-well for Higher Density CMOS," *IEEE Trans. on Electron Devices*, Vol. ED-28, #10, 1981, pp. 1115–1119.

17 G. A. Swartz, "Gate Oxide Integrity of NMOS Transistor Arrays," *IEEE Trans. on Electron Devices*, Vol. ED-33, #11, 1986, pp. 1826–1829.

equals the maximum gate-to-source voltage. Higher gate-to-drain voltages can be obtained by altering the geometry of the drain (Section 12.1).

MOS transistors are also subject to a fourth breakdown mechanism. The electric field across the pinched-off portion of an MOS channel can become very intense. The carriers flowing across the pinched-off region accelerate to very high velocities and become so-called *hot carriers*. Some of these carriers collide with the lattice and recoil out of the channel. Most of them travel into the backgate and eventually contribute to the backgate current. Some of the hot carriers also travel into the gate oxide. The reducing anneal used to minimize surface state charge ties off dangling bonds along the oxide interface with hydrogen atoms. Hot carriers can break the relatively weak silicon-hydrogen bonds, regenerating the dangling bonds and the surface state charge they represent. As the transistor continues to operate, the accumulating charge causes the threshold voltage to gradually shift. This effect can easily disrupt matching between MOS transistors operating under different biases. If the threshold voltage shifts too far, the transistor may not even be able to switch on and off. Short-channel transistors are especially susceptible to hot-carrier generation because the high backgate doping levels required to prevent punchthrough shorten the pinched-off portion of the channel. The resulting electric fields produce hot carriers at lower voltages than would occur in a transistor with a lightly doped backgate. Transistors acting as switches are less susceptible to hot carrier generation because they normally operate either in the linear region or in cutoff. MOS transistors used in analog circuitry are more vulnerable, as these devices often operate continuously in the saturation region.

Any given MOS transistor will not necessarily experience all of these breakdown mechanisms. Long-channel transistors are usually limited by avalanche breakdown and dielectric breakdown. Short-channel transistors are usually limited by punchthrough and dielectric breakdown. Transistors operating for long periods of time under high drain-to-source voltages may also experience hot carrier-induced threshold voltage shifts.

CMOS Latchup

When a source/drain diffusion forward-biases into the backgate, it injects minority carriers that can flow to the reverse-biased junctions of adjacent devices. The exchange of minority carriers between adjacent NMOS and PMOS transistors can trigger *CMOS latchup* (Section 4.4.2). Minority-carrier guard rings can prevent latchup, but they are not necessarily easy to construct in CMOS processes.

The absence of NBL and deep-N+ makes it impossible to construct effective blocking guard rings in a pure CMOS process. One can still construct hole- and electron-collecting guard rings by using PMoat and NMoat, although the collection efficiency of these shallow diffusions usually leaves much to be desired. Any PMOS transistor that can inject minority carriers into its well should be surrounded by a hole-collecting guard ring constructed of PMoat (Figure 11.6A). This guard ring should connect to substrate potential to reverse-bias the PMoat/N-well junction as strongly as possible. The guard ring collects a percentage of the minority carriers injected laterally by the enclosed PMOS transistor. Although this reduces the lateral flow of carriers toward adjacent devices, it does not stop holes from traveling downward to the substrate. CMOS processes usually employ a P+ substrate to minimize debiasing.

Any NMOS transistor that can forward-bias into the substrate should be surrounded by an electron-collecting guard ring. These guard rings are usually constructed from NMoat rather than from N-well (Figure 11.6B). The deeper well diffusion intercepts a larger fraction of the carriers, but its large vertical resistance makes it prone to debiasing unless it connects to a relatively high-voltage supply.

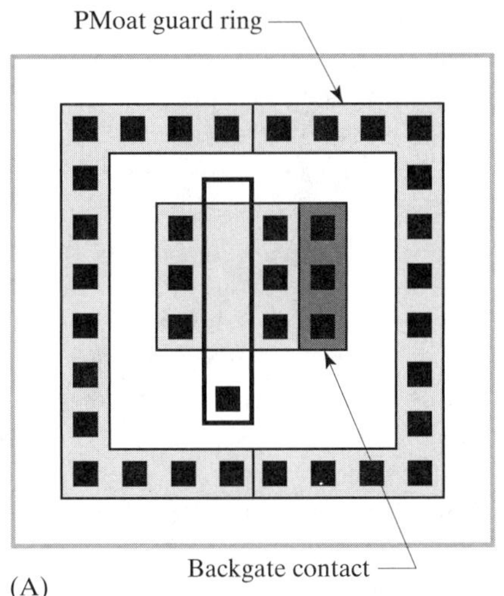

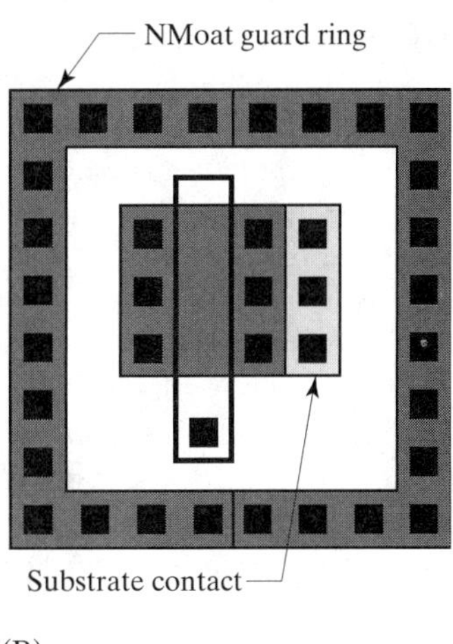

FIGURE 11.6 Examples of guard rings for protecting N-well CMOS transistors: (A) PMOS with a hole-collecting guard ring and (B) NMOS with an electron-collecting guard ring.

NMoat has a lower collection efficiency, but it is virtually immune to debiasing. NMoat guard rings are sometimes connected to a positive power supply to help drive the depletion region deeper into the substrate to enhance collection efficiency. In low-voltage processes, NMoat guard rings should be connected to substrate potential to minimize hot-carrier generation in their depletion regions (Section 13.2.3).

Guard rings, by themselves, cannot provide total latchup immunity. The flow of even a few minority carriers around the guard rings will trigger parasitic bipolar conduction if not for the presence of backgate contacts. Backgate contacts remove the collected carriers and prevent them from biasing the parasitic lateral bipolar transistors into conduction. Section 11.2.7 discusses the design of backgate contacts in greater detail.

Leakage Mechanisms

Strictly speaking, the term *leakage current* refers specifically to currents flowing across reverse-biased junctions. However, most designers use the term to describe any small current that appears where it is not wanted. This generalization covers a host of mechanisms that affect MOS transistors, including the eponymous junction leakage, subthreshold conduction, minority carrier injection, hot carrier injection, stress-induced leakage current, and gate-induced drain leakage. This section briefly describes each of these mechanisms and discusses the impact that each has upon MOS transistor design and operation.

Junction leakage typically becomes significant only when transistors operate in cutoff. Consider the case of the sample-and-hold circuit of Figure 11.7. NMOS transistor M_1 and PMOS transistor M_2 together form a switch called a *transmission gate*.

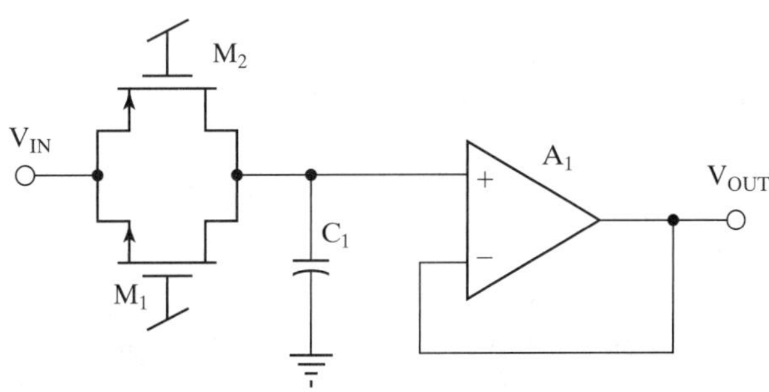

FIGURE 11.7 Schematic diagram of a simple analog sample-and-hold circuit.

When these transistors are enabled, capacitor C_1 quickly charges to voltage V_{IN}. This phase of operation is called *sampling*. When M_1 and M_2 are disabled, capacitor C_1 holds its charge, and CMOS operational amplifier A_1 replicates the voltage on the capacitor at the output of the circuit. This phase of operation is called *holding*. Any leakage on or off capacitor C_1 will cause the voltage to drift during the holding phase of the sample-and-hold operation.

The magnitude of junction leakage in a MOS transistor operating in cutoff depends upon two factors: the junction area of the source/drain diffusion in question, and the junction temperature. The area of the source/drain diffusion plays the most obvious role. A smaller diffusion will leak less current. The area of the source and drain regions scales linearly with the width of the transistor, so one should use the smallest widths possible for low-leakage switches. A further reduction in source/drain area can sometimes be achieved by placing two transistors back-to-back so that they share a common source or drain connection. One should always avoid connecting the well of a PMOS transistor to a charge storage node because the large well/substrate junction has an equally large leakage current.

The magnitude of junction leakage currents increases exponentially with temperature. The junction leakage current doubles approximately every 8°C. Whenever possible, one should place low-leakage circuitry well away from power transistors. Unfortunately, many integrated circuits must operate at elevated ambient temperatures. No method exists for shielding the internal circuitry of an integrated circuit from the ambient temperature of its environment.

Circuit designers sometimes attempt to further reduce junction leakages by balancing the leakage contributions of PMOS and NMOS transistors. In the circuit of Figure 11.7, PMOS transistor M_2 injects a leakage current onto capacitor C_1, while NMOS transistor M_1 draws a leakage current off of C_1. If these two currents were equal, then the capacitor would experience no net gain or loss of charge. Unfortunately, the cancellation of NMOS and PMOS junction leakages depends upon maintaining a precise doping ratio between the N-well and P-epi. This balance proves impossible to maintain in practice, so one cannot rely upon NMOS and PMOS leakages to compensate one another.

Subthreshold conduction can also cause small leakage currents to flow through transistors that are supposedly operating in cutoff. Contrary to the simple Shichman-Hodges equations, the drain current I_D does not drop to zero when the gate-to-source voltage V_{GS} equals the threshold voltage V_t. Instead, it transitions from a linear function of V_{GS} to an exponential function of V_{GS}. Figure 11.8 shows a plot of drain current versus gate-to-source voltage for a typical NMOS transistor. The drain current is plotted on a logarithmic scale. The slope of the drain current in the subthreshold region typically ranges from 80 to 100 mV per decade of current.

The subthreshold current typically drops to insignificant levels about 0.3 V below the threshold voltage. Subthreshold leakage can be eliminated by ensuring that the transistors are biased at least 300 mV into cutoff. This can be achieved by either using devices with sufficiently large threshold voltages, or providing an external reverse bias from source to gate. Consider the case of the sample-and-hold circuit of Figure 11.7. Assuming that the gate of M_1 operates at 0 V in cutoff, then this transistor experiences a reverse gate-to-source bias equal to the voltage stored on the capacitor.

Minority carrier injection can sometimes cause unexpected currents to flow across reverse-biased junctions. If the magnitude of these currents is small, then they will appear much like leakages caused by other mechanisms. NMOS transistors residing in the P-epi are particularly vulnerable to minority carrier injection because so many other devices also reside in the P-epi. Guard rings can reduce the magnitude

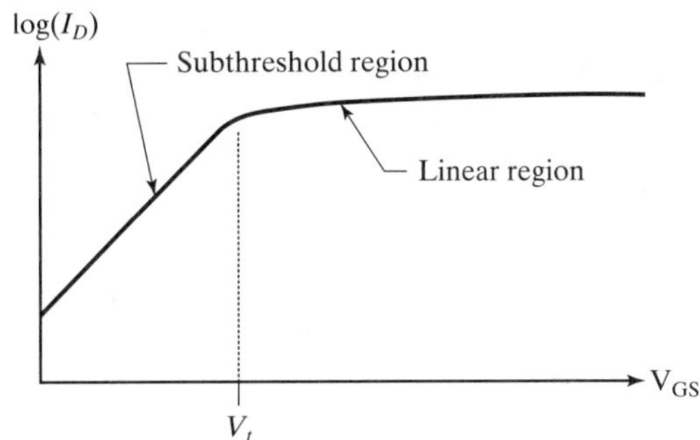

FIGURE 11.8 Plot of NMOS drain current I_D versus gate-to-source voltage V_{GS}, showing the linear and subthreshold regions of operation.

of minority carrier injection, but they cannot eliminate it entirely. They are thus of only limited utility in guarding low-current circuitry. Sometimes circuits can be rearranged to eliminate transistors that share backgates with other devices. In the circuit of Figure 11.7, NMOS transistor M_1 resides in the epi, while PMOS transistor M_2 occupies its own well. Transistor M_1 is thus susceptible to minority carrier injection, while M_2 is not. Transistor M_1 can be eliminated if the circuit can tolerate a restricted range of input voltages. Some BiCMOS processes offer the possibility of isolating NMOS transistors, using NBL (Section 11.2.2). Dielectrically isolated processes provide another sure solution for minority carrier injection in the form of oxide isolation rings placed around the critical components. Consider using these techniques if they are available, as they greatly increase the robustness of low-current circuits without greatly increasing their area.

Hot carrier injection causes unexpected currents to flow through both the backgates and the gates of saturated MOS transistors operating at large drain-to-source voltages. Backgate currents rarely cause much trouble in low-current circuitry because designers take care not to connect backgates to critical nodes. The gate currents are usually so small that they can safely be ignored. Floating-gate devices (Section 11.3) are exceptions to this generalization, as their gates are completely insulated from the rest of the circuit. If hot carrier injection causes troublesome leakages, these can be minimized—or even entirely eliminated—by reducing the drain-to-source voltage across the offending transistors.

Stress-induced leakage current (SILC) can also cause gate leakage. The injection of hot carriers into the gate oxide not only causes gate leakage, but also generates traps within the oxide. If these traps become sufficiently numerous, then trap-assisted tunneling can occur. This form of tunneling causes a leakage current to flow across the trap-filled oxide even at very low voltages (Section 4.1.3). Ordinary MOS transistors should not experience appreciable stress-induced leakage currents within their normal operating lifetime unless the circuit is improperly designed, or if it is subjected to electrical overstress. On the other hand, floating-gate circuits rely upon either hot-carrier injection or Fowler-Nordheim tunneling to transfer charge to and from the floating gates. Both of these mechanisms produce oxide traps. Floating-gate devices are therefore vulnerable to SILC (Section 11.3).

Gate-induced drain leakage (GIDL) represents yet another potential leakage mechanism in MOS transistors. Dangling bonds along the oxide-silicon interface can act as traps for Fowler-Nordheim tunneling. The vertical component of the electric field generated by the gate electrode adds to the horizontal component generated by the drain-to-source voltage differential. If the total electric field intensity exceeds the value required for Fowler-Nordheim tunneling (which is itself dependent upon trap density), then a leakage current begins to flow across the drain-to-backgate depletion region.[18] This current increases exponentially with the total electric field, which means that it responds exponentially to increases in either drain-to-gate voltage or drain-to-source voltage. As with all forms of tunneling, GIDL current also increases exponentially with temperature.

In practice, GIDL usually occurs when a transistor operates in cutoff under high drain-to-source voltages at elevated temperatures. Turning the transistor on reduces the drain-to-gate voltage differential and therefore reduces the magnitude of GIDL. Similarly, reducing the temperature lessens the rate of tunneling and therefore minimizes GIDL. NMOS transistors are less susceptible to GIDL than PMOS transistors because hot-carrier injection concerns limit the electric field intensity allowed across NMOS transistors.

GIDL also depends upon the density of traps along the oxide interface, especially in the vicinity of the drain. Any mechanism that generates additional trap states can therefore enhance the rate of GIDL. In particular, hot-carrier injection and charge injection due to the antenna effect have been shown to increase GIDL currents.[19] Incomplete hydrogen annealing will also increase GIDL.

Gate-induced drain leakage can be eliminated by reducing the electric field stresses upon the affected transistors. This goal can be achieved by reducing the operating voltage, by increasing the channel length to reduce the horizontal component of the electric field, or by using a thicker gate oxide to reduce the vertical component of the electric field. The use of a field relief structure (Section 12.1) may also reduce the vertical component of the field and thus minimize GIDL.

11.2 CONSTRUCTING CMOS TRANSISTORS

Most modern CMOS and BiCMOS processes are designed to produce self-aligned poly-gate transistors. Figure 11.9 shows a layout and cross section of a simple self-aligned poly-gate NMOS. The backgate of this transistor consists of a P− epitaxial layer grown on a P+ substrate. The areas between adjacent transistors are called field regions. LOCOS oxidation covers these with a thick-field oxide that helps suppress parasitic channel formation. The nitride oxidation mask prevents thick oxide from growing in the moat regions where transistors will eventually reside. After the removal of the nitride, the moat regions are re-oxidized to form the thin *gate oxide* of the MOS transistors. Doped polysilicon is then deposited on top of the *gate oxide* to form the gate electrodes of the MOS transistors. After the poly has been patterned,

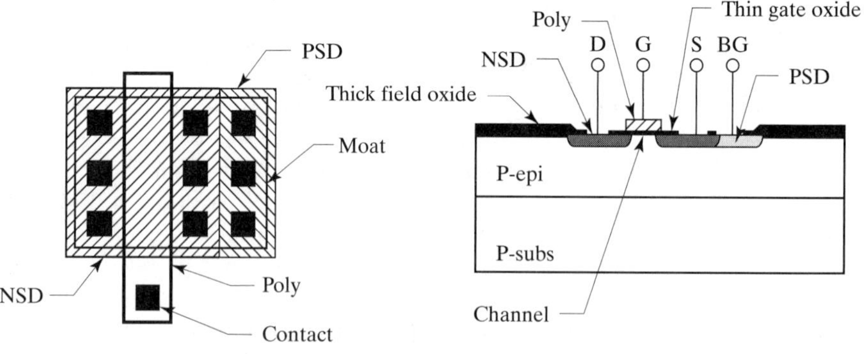

FIGURE 11.9 Layout and cross section of a simple self-aligned poly-gate NMOS transistor.

[18] V. Nathan and N. C. Das, "Gate-Induced Drain Leakage Current in MOS Devices," *IEEE Trans. on Electron Devices*, Vol. 40, #10, 1993, pp. 1888–1890.

[19] S. Ma, Y. Zhang, M. F. Li, W. Li, J. Xie, G. T. T. Sheng, A. C. Yen, and J. L. F. Wang, "Gate-Induced Drain Leakage Current Enhanced by Plasma Charging Damage," *IEEE Trans. on Electron Devices*, Vol. 48, #5, 2001, pp. 1006–1008.

a low-energy arsenic implant produces the source and drain regions of the transistors. This N-source/drain (NSD) implant does not possess enough energy to penetrate the polysilicon or the thick-field oxide. The NSD only penetrates the thin gate oxide regions bounded by the poly gate and the thick-field oxide. The resulting source/drain regions are said to self-align to the poly gate and the thick-field oxide. Next, a *P-source/drain* (PSD) implant is performed. This implant gains its name from the role it plays in the construction of PMOS transistors. The NMOS transistor uses PSD to contact the lightly doped P-epi backgate. A brief anneal activates the source/drain implants and completes the formation of the transistors.

Early MOS processes used aluminum to form the gate electrodes. Aluminum-gate processes are inferior to polysilicon-gate processes in several respects. Aluminum cannot withstand the temperatures required to anneal the source/drain implants, so it must be deposited after implantation. This precludes the self-alignment of the source/drain diffusions, so these implants must overlap the gate by an amount sufficient to account for misalignment. These overlaps greatly increase the gate-to-source capacitance, C_{GS}, and the gate-to-drain capacitance, C_{GD}, which in turn greatly reduces the switching speed of the transistor. The overlap capacitances of a poly-gate transistor are much smaller because the source and drain regions self-align to the gate.

Some attempts have been made to construct gate electrodes from refractory metals such as tungsten. These materials facilitate self-alignment because they can withstand the temperatures required to anneal the source/drain implants. Despite this advantage, refractory-metal gates have not enjoyed widespread success because they still exhibit threshold voltage variations caused by mobile ions and by variations in contact potential. Poly gates are preferred because they provide much more stable and reproducible threshold voltages (Section 4.2.2).

11.2.1. Coding the MOS Transistor

A simple N-well CMOS process requires a total of seven masks: N-well, moat, poly, NSD, PSD, contact, metal, and protective overcoat. The layout database contains the geometric information required to construct each of these seven masks. In the simplest database, the geometries for each mask are drawn upon a different layer. Figure 11.10A shows the layout of an NMOS transistor following this approach.

Figure 11.10B shows another way to draw the same transistor. Two new layers called *NMoat* and *PMoat* are used to generate the NSD, PSD, and moat masks. A geometry placed on the NMoat layer produces corresponding geometries on both the moat and the NSD masks. The NSD geometry is automatically oversized to account for misalignment.[20] A geometry drawn on the PMoat layer creates corresponding geometries on the moat and PSD masks, and the PSD geometries are again automatically oversized.

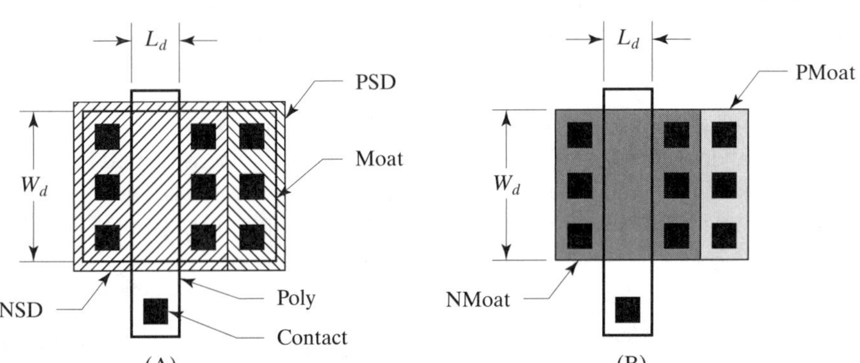

FIGURE 11.10 A NMOS transistor can be coded (A) using NSD, PSD, and moat mask layers or (B) using NMoat and PMoat coding layers.

The NSD, PSD, and moat layers used to construct the layout in Figure 11.10A are called *mask layers* because the information they contain is transferred directly to the corresponding masks without any intermediate processing. NMoat and PMoat are called *drawing layers* or *coding layers* because they are used only during the drawing (or *coding*) of the layout. The data on the coding layers must pass through a series of geometric transformations to generate the actual mask data.

Coding layers simplify the layout in a number of ways. Data entry takes less time because the layout contains fewer geometries, displays and plots become less cluttered, verification programs run more quickly, and databases consume less space. Each of these advantages may seem relatively insignificant, but together they make a strong argument for the use of coding layers.

The process of transforming coding data into mask data sometimes produces unexpected results on complicated geometries. After the NSD and PSD geometries are oversized, they must be trimmed so that the two implants do not overlap along abutting edges (Figure 11.10A). The trimming algorithm is relatively simple and straightforward as long as the NMoat and PMoat geometries do not contain bends or notches. The algorithm becomes much more complex if it must handle these special cases. The more complex the algorithm becomes, the more likely it is to produce unexpected results under circumstances that are not anticipated by its designer. The only way to entirely eliminate problems of this sort is to avoid the use of coding layers.

The choice between mask layers and coding layers is by no means an easy one. Many designers appreciate the simplifications introduced by coding layers, but they do not always understand the accompanying problems. The people responsible for writing verification and pattern generation programs must ultimately decide whether to use coding layers or to avoid them. This text uses NMoat and PMoat coding layers because they significantly simplify the illustrations.

Width and Length

The length of the transistor equals the distance between the source diffusion and the drain diffusion. The *drawn length* L_d of a self-aligned transistor equals the distance across the poly gate from source to drain as measured in the layout database. The *effective length* L_{eff} of the transistor may be slightly larger or smaller than the drawn length due to overetching, underetching, straggle, outdiffusion, and other factors.[21] These corrections remain relatively constant regardless of the dimensions of the gate, so L_{eff} can be approximated by

$$L_{\text{eff}} \cong L_d + \delta L \qquad [11.8]$$

where δL is constant for any given process. The value of δL is usually much less than 1 μm, so it primarily affects short-channel devices. Submicron transistors, in particular, exhibit substantial differences between drawn and effective channel lengths. In such cases, the channel length, L, used in the Shichman-Hodges equations 11.2A and 11.2B must equal the effective channel length L_{eff} and not the drawn channel length L_d.

The poly gate must overhang both ends of the source/drain region to prevent the source and drain from shorting together. The width of a self-aligned MOS transistor is therefore set by the moat mask rather than the poly mask. The *drawn width* W_d equals the width of the moat geometry in the layout database or, equivalently, the

---

[20] Although the moat geometries could be undersized instead, this would affect the drawn width of the devices and is therefore a less intuitive option.

[21] G. Massobrio and P. Antognetti, *Semiconductor Device Modeling with SPICE*, 2d ed. (New York: McGraw-Hill, 1993), pp. 279–283.

width of the NMoat or PMoat geometry (Figure 11.10). The *effective width* W_{eff} varies slightly due to straggle, outdiffusion, the presence of the bird's beak, and other factors. These corrections remain relatively constant regardless of the moat dimensions, so the effective width W_{eff} can be approximated by

$$W_{eff} \cong W_d + \delta W \qquad [11.9]$$

where δW is a constant for any given process. The value of δW is also usually less than 1 μm.

11.2.2. N-Well and P-Well Processes

The NMOS transistors in Figure 11.10 are fabricated in a P− epitaxial layer deposited on a P+ substrate. The heavily doped substrate improves latchup immunity, but it introduces an additional process step. Providing that other measures have been taken to prevent latchup, the epitaxial layer can be eliminated and the transistors built directly into the substrate. Many early processes used this approach to minimize fabrication costs, and some processes still do so today:[22] Most analog CMOS processes use an epitaxial layer because the epi doping can be controlled very accurately, and therefore the threshold voltages of transistors constructed in the epi vary less than the thresholds of those constructed in the substrate.

A P-epi allows the construction of NMOS transistors, and an N-epi allows the construction of PMOS transistors, but neither allows the construction of both simultaneously. In order to build complementary transistors, another diffusion must be added to counterdope the backgate region of one transistor or the other. If a P-epi is used, then a deep, lightly doped N-type diffusion must be added for PMOS transistors (Figure 11.11A). If an N-epi is used, then a deep, lightly doped P-type diffusion must be added for NMOS transistors (Figure 11.11B). These deep diffusions are commonly called wells. An N-type well is called an *N-well* and a P-type well is called a *P-well*. Many processes use either an N-well or a P-well, but not both. In these *single-well* processes, one type of transistor or the other resides in the epi. In an N-well process, the NMOS occupies the epi and the PMOS the N-well. In a P-well process, the PMOS occupies the epi and the NMOS the P-well. Some processes include both an N-well and a P-well (Figure 11.11C). In such a *twin-well process*, the NMOS is formed in the P-well and the PMOS is formed in the N-well.

Single-well processes are simpler and cheaper than twin-well processes, but submicron processes often require two wells. As the channel length of the transistor decreases, the backgate doping must increase to prevent punchthrough breakdown. The counterdoping mechanism that creates the well becomes difficult to control on heavily doped substrates. Heavy counterdoping also causes a slight reduction in carrier mobility and a more significant reduction in well-substrate breakdown voltages. Additionally, the introduction of a well offers an opportunity to tailor the dopant distribution by altering the implant energy. The use of very high energy implants can produce a so-called *retrograde well* whose lower portions are more heavily doped than the upper portions. The resulting structure resembles a buried layer. A retrograde well profile not only reduces well resistance and enhances latchup immunity, but also allows the use of shallower wells without the risk of vertical punchthrough.

---

[22] A layer of P− silicon can be created on a P+ substrate by annealing the wafer in hydrogen at high temperature. Hydrogen combines with boron at the surface and outgasses as diborane: M. Aminzadeh, K. V. Ravi, G. Sery, S. Hu, K. Wu, and C. Peng, "Pseudo Epi, Material Cost Reduction," *Int. Semiconductor Manufacturing Symp.* 2001, pp. 16–170.

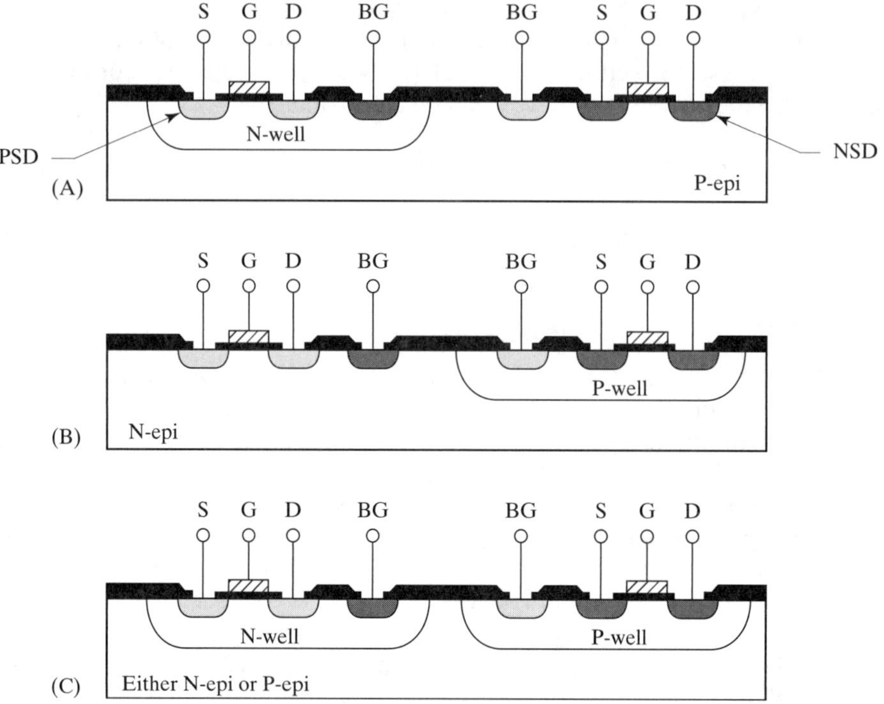

FIGURE 11.11 Three types of CMOS processes: (A) N-well, (B) P-well, and (C) twin-well.

These considerations force most submicron processes to use twin wells driven into a lightly doped epi.

The choice of epi also has several consequences. In a single-well process, transistors formed in the epi share a common backgate connection, while transistors formed in wells can be isolated from one another. Although the separate wells consume additional die area, isolation offers an extra degree of design flexibility. An N-well process produces isolated PMOS transistors, while a P-well process produces isolated PMOS transistors. A similar consideration affects the choice of the epi type for a twin-well process. If a P-epi is chosen, then this epitaxial layer shorts all of the P-wells on the die together, and all of the NMOS transistors share a common backgate connection. Similarly, if an N-epi is chosen, the epi shorts the N-wells together and all of the PMOS transistors share a common backgate connection.

N-well processes are favored over P-well processes for several reasons. Most schematics reference their power supplies to a common ground potential. If all of the power supplies deliver positive voltages with respect to ground, as is often the case, then ground becomes the most negative node in the circuit. The substrate of an N-well process can connect to this common ground, but the substrate of a P-well process must connect to the highest-voltage power supply. In multiple-supply systems, it is difficult to ensure that one power supply will always generate a higher voltage than the others, especially during start-up and shut-down. P-well processes are thus poorly suited to multiple-supply applications. One could theoretically reference multiple negative voltages to a positive ground, but this is rarely done in practice.

The mobility of carriers in the counterdoped well will be slightly less than the mobility of carriers in the epi. Since electrons are more mobile than holes, the NMOS transistor has a higher transconductance than the PMOS transistor. Many circuit designers prefer to degrade the performance of the already-inferior PMOS rather than reduce the superior transconductance of the NMOS. This consideration also favors the use of an N-well process.

BiCMOS processes generally employ a P-epi on a P-substrate because this combination simplifies the isolation of the bipolar transistors. The NPN transistor uses the lightly doped N-well as a collector region and the P-epi as isolation, a practice called *collector-diffused isolation* (CDI). Most analog BiCMOS processes are either N-well processes or twin-well processes built on a P-type epi.

N-well BiCMOS processes can construct isolated NMOS transistors as well as isolated PMOS transistors. The isolated NMOS uses a combination of NBL and deep-N+ (or N-well) to isolate the section of P-epi forming the backgate of the transistor (Figure 11.12). The NBL severs the isolated P-epi tank from the P-substrate beneath, and the deep-N+ (or N-well) ring isolates it from adjacent P-epi regions.[23] In order to ensure complete isolation, the ring must contain no gaps and the NBL must overlap it sufficiently to allow for misalignment. If deep-N+ is available, then it is frequently used instead of N-well because it produces a lower resistance connection to the NBL. The N-well/epi junction usually has a higher breakdown voltage than the deep-N+/epi junction, so devices that must operate at a high voltage relative to the substrate will normally use N-well isolation rings, either alone or surrounding a deep-N+ isolation ring.

The isolation ring must connect to a voltage equal to or greater than that applied to the isolated P-epi tank. The source/drain regions easily punch through the lightly doped tank, so most isolated NMOS transistors cannot withstand the application of more than a few volts drain to isolation or source to isolation. These operating voltages increase slightly if the isolation ring connects to a potential midway between that of the backgate and that of the source/drain diffusions. This configuration allows

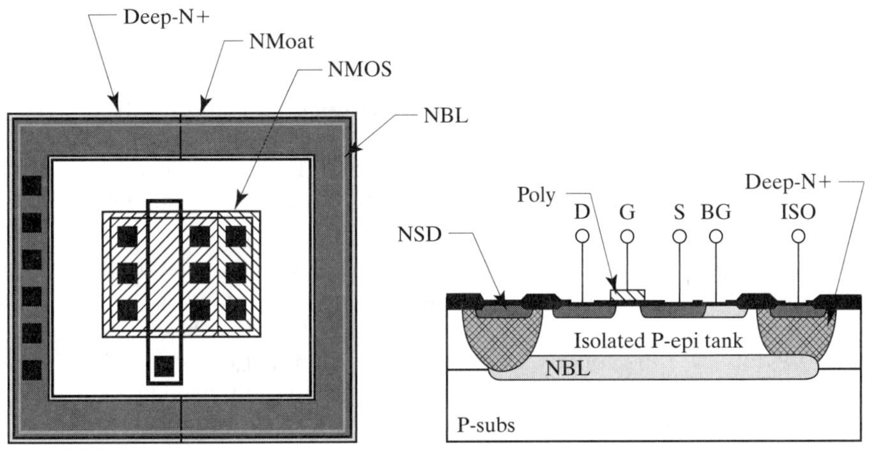

FIGURE 11.12 Layout and cross section of an isolated NMOS in an N-well analog BiCMOS process.

a portion of the depletion region surrounding the isolation/NBL region to intrude into the lightly doped outer fringes of the NBL. Because the NBL dopant diffuses very slowly, the degree of field relief offered is slight and the improvement in operating voltage amounts to only a few volts.

The backgate region of the isolated NMOS consists of a thin, lightly doped P-epi layer. This layer has much more lateral resistance than the P-epi/substrate sandwich

---

[23] E. Bayer, W. Bucksch, K. Scoones, K. Wagensohner, J. Erdeljac, and L. Hutter, "A 1.0 μm Linear BiCMOS Technology with Power DMOS Capability," *Proc. Bipolar/BiCMOS Circuits and Technology Meeting*, 1995, pp. 137–141.

constituting the backgate of the nonisolated NMOS. Rapidly slewing signals can momentarily forward-bias the source/drain regions of an isolated NMOS into its backgate. Most of the injected minority carriers flow to the isolation ring, but a few travel from source to drain (or *vice versa*). If the transistor operates at a relatively large drain-to-source potential, then minority carrier injection can trigger V_{CER} breakdown and snapback. Adequate backgate contact minimizes the magnitude of the snapback and the likelihood of triggering it (Section 11.2.7).

11.2.3. Channel Stop Implants

Self-aligned poly-gate MOS transistors form wherever poly intersects PMoat or NMoat geometries. Under certain circumstances, MOS transistors can also form underneath the thick-field oxide. These unwanted *parasitic transistors* interfere with the operation of the integrated circuit unless they are somehow suppressed.

The threshold voltages of the parasitic transistors can be raised by implanting the field regions with a suitable dopant before growing the thick-field oxide. A doping concentration of 10^{17} atoms/cm^3 beneath a 10 kÅ field oxide will produce a thick-field threshold of nearly 50 V (Table 11.2). This thick-field threshold will provide a comfortable safety margin for a 30 V process. Implants used to raise the doping of field regions are called *channel-stop implants*.

The thick-field thresholds can also be raised by deliberately introducing surface state charges. Standard bipolar processes produce large surface state charges by using (111)-oriented silicon in combination with a final oxidizing anneal. The resulting positive charge raises the magnitude of the PMOS thick-field threshold and lowers the magnitude of the NMOS thick-field threshold. Standard bipolar uses a heavily doped P+ isolation system to suppress NMOS parasitic channel formation, and relies on the surface state charge to elevate the PMOS thick-field threshold. Thick-field thresholds of 40 V can routinely be achieved by this means.

CMOS processes cannot tolerate the introduction of excess surface state charge because its effects are not limited to field regions. The magnitude of the surface state charge varies with processing conditions, and this, in turn, causes threshold voltage fluctuations. CMOS processes use (100)-oriented silicon and conduct a hydrogen anneal to minimize the residual surface state charge. This anneal usually occurs in conjunction with the deposition of the protective overcoat. Many designers do not appreciate the importance of proper annealing. If a wafer is removed from processing before nitride deposition, then an anneal must be conducted in order to stabilize the threshold voltages and to sinter the contacts. Because this anneal does not necessarily duplicate the conditions of nitride deposition, the threshold voltages of no-nitride wafers do not always correspond to those of the finished product.

Most CMOS processes use two complementary channel-stop implants to suppress both NMOS and PMOS parasitic channels. All P-type field regions receive the P-type channel-stop implant to increase the magnitude of the PMOS thick-field threshold. Similarly, all N-type field regions receive the N-type channel stop implant to increase the magnitude of the NMOS thick-field threshold. Several methods have been devised to ensure the proper alignment of these channel-stop implants. The most common techniques involve either a blanket boron channel-stop implant and a patterned phosphorus channel-stop implant, or *vice versa*. Figure 3.23 shows the steps required to produce a blanket boron channel-stop implant and a patterned phosphorus channel-stop implant in an N-well CMOS process. Figure 11.13 shows the results.

The channel-stop implants diffuse downwards during the long, high-temperature field oxidation. Lateral outdiffusion causes the two implants to intersect along the

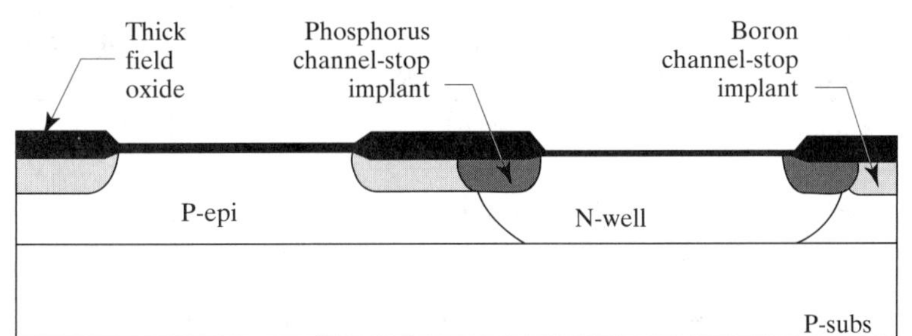

FIGURE 11.13 N-well CMOS wafer with boron and phosphorus channel-stop implants.

edges of the N-well. This zone of intersection limits the breakdown voltage of the N-well/P-epi junction. Fortunately, the channel-stop implants are sufficiently deep and lightly doped that the breakdown voltage lies well above normal operating voltages, and a 15 V CMOS process can generally obtain an N-well/P-epi breakdown voltage in excess of 30 V.

The patterned channel-stop implant requires a masking step. The locations receiving it depend on the type of process selected and which of the two implants is patterned. A patterned phosphorus channel stop in N-well CMOS goes into all N-well regions not inside moat. Although it is possible to draw the geometries, the channel-stop mask can be generated from the data present on the N-well, PMoat, and NMoat coding layers.

Submicron processes can sometimes dispense with one or both channel-stop implants. As the channel length shrinks, the backgate doping concentration rises and the operating voltage drops. The wells of a submicron process may therefore contain enough dopant to raise the thick-field threshold above the relatively low operating voltage of the process.

The channel-stop implants are designed to raise the NMOS and PMOS thick-field thresholds above the maximum operating voltage of the process. Many layout designers assume that this provides unconditional protection against parasitic channel formation, but in practice it does not. The maximum operating voltage of a process is usually set by either the gate oxide rupture voltage or by the breakdown voltage of the source/drain implants into their respective backgates. Certain components can operate at much higher voltages; for example, a poly resistor is limited only by the breakdown of the thick-field oxide, which can easily reach several hundred volts. Similarly, the well-epi junctions can usually withstand several times the operating voltage of the process. Section 4.3.5 discusses the techniques used to suppress parasitic channel formation in circuits operating at or above the thick-field threshold.

11.2.4. Threshold Adjust Implants

Ideally, the threshold voltages of enhancement transistors should lie between 0.6 and 0.8 V. The *native*, or *natural*, thresholds are determined by the doping of the gate and backgate and by the thickness of the gate oxide. Most processes dope the gate poly with phosphorus, reducing the magnitude of the NMOS threshold and increasing that of the PMOS. The natural NMOS threshold usually lies well below 0.6 V and the magnitude of the natural PMOS threshold well above 0.8 V. Over extremes of process and temperature, the NMOS goes into depletion, and the magnitude of the PMOS threshold exceeds 1.5 V (Table 11.3). These thresholds are completely unacceptable for most applications.

The threshold voltage of an MOS transistor can be altered by implanting its channel region. A P-type implant produces a positive threshold shift and an N-type implant a negative one. NMOS and PMOS transistors using phosphorus-doped gate

TABLE 11.3 Worst-case natural and adjusted threshold voltages for a typical 10 V N-well CMOS process.[24]

Worst-Case Corner	Natural NMOS	Adjusted NMOS	Natural PMOS	Adjusted PMOS
Minimum	−0.10 V	0.50 V	−1.75 V	−1.15 V
Nominal	0.20 V	0.80 V	−1.40 V	−0.80 V
Maximum	0.55 V	1.15 V	−1.10 V	−0.50 V

poly both require a positive threshold shift. Providing that the initial backgate dopant concentrations have been properly chosen, a single boron implant can adjust the thresholds of both types of transistors. This boron implant is called the *threshold adjust implant*, or simply the *threshold adjust*. Transistors receiving this implant are called *adjusted* transistors, while those not receiving it are called *native*, or *natural*, transistors. The threshold adjust implant does not necessarily require a photomask. If the implant is performed across the entire wafer immediately after stripping the LOCOS nitride, then it appears in every moat region. This blanket implant simultaneously adjusts the threshold voltage of every MOS transistor to the targeted value. This practice precludes the fabrication of natural devices.

Circuit designers can often improve the performance of their circuits if they have access to both natural and adjusted transistors. Many processes therefore offer natural transistors as a process option. This option requires a single mask, properly called the *threshold adjust implant mask*, but more often referred to as a *natural* V_t *mask*. The associated coding layer has been given many names; in this text it is called *NatVT*.[25] This layer must be coded around the gate region of each natural transistor (Figure 11.14). The NatVT figure should slightly overlap the channel region to allow for misalignment and lateral outdiffusion. If a design does not use any natural transistors, then the NatVT mask can usually be omitted. Some processes may use the NatVT mask to fabricate certain other devices, such as Schottky diodes.

Although many processes have successfully used a single boron threshold adjust implant, submicron processes often require a different strategy. The boron implant

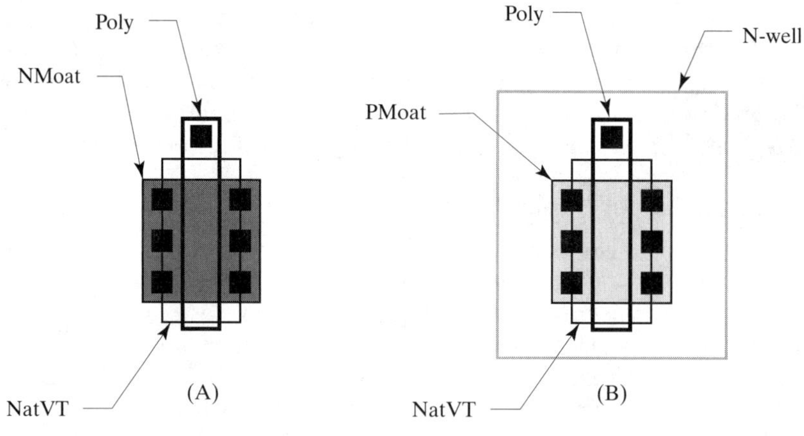

FIGURE 11.14 Layout of natural transistors using NatVT: (A) natural NMOS and (B) natural PMOS.

[24] These values assume the listed natural V_t targets, ±0.15 V threshold control, and a −2 mV/°C temperature coefficient.

[25] The natural V_t mask is also called *NVT*, but this name is also used for the N-type threshold adjust mask used in a dual-doped poly process.

reduces the magnitude of the PMOS threshold, but it also has the undesired effect of producing a *buried channel*. In order to obtain a large threshold shift, so much boron must be implanted that it actually inverts a thin layer of the backgate. The inversion region appears beneath the surface because this is where the peak doping concentration from the implant occurs. The buried channel lies so close to the surface that the electric field produced by the contact potential of the gate electrode inverts it, and it does not interfere with the normal operation of the transistor. This situation changes in a submicron transistor because the backgate doping increases as the channel length decreases. The increased doping partially shields the buried channel from the influence of the gate electrode. The gate can no longer fully invert the channel, so the buried channel begins to conduct current. Submicron buried-channel PMOS transistors are therefore somewhat leaky.

The buried channel can be eliminated by using a phosphorus channel-stop implant for the PMOS transistors. Since phosphorus induces a negative threshold shift, the PMOS transistor must begin with a relatively low threshold voltage, which can be achieved by using a boron-doped gate poly (Figure 11.15).

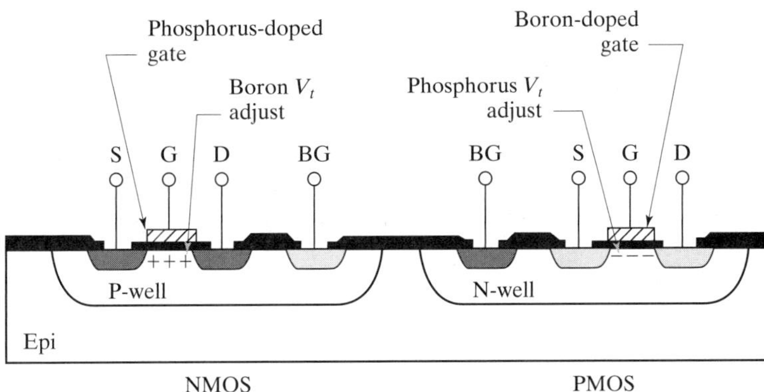

FIGURE 11.15 Cross section of dual-doped poly CMOS transistors.

The processing required to produce *dual-doped poly* CMOS transistors is as follows: After the LOCOS nitride has been stripped, the wafer is patterned using the mask for the *P-type* V_t *adjust* (PVT). This low-energy boron implant adjusts the NMOS threshold. Next, the wafer is patterned using the mask for the *N-type* V_t *adjust* (NVT). This low-energy phosphorus implant adjusts the PMOS threshold. After gate oxidation, the gate poly is deposited in a near-intrinsic state. Before etching, it is doped with a patterned boron implant using the *P-type gate poly* (PPoly) mask, followed by a patterned phosphorus implant using the *N-type gate poly* (NPoly) mask.

The most elaborate version of this process requires four new masks (PVT, NVT, PPoly, and NPoly). As long as no natural transistors are required, the P-well mask can be reused for the PVT step, and the N-well mask can be reused for the NVT step. If natural transistors are required, then the well masks cannot be reused in this manner and separate PVT and NVT masks become necessary. A well mask can also be used to define either the PPoly or the NPoly, but not both. Suppose that the N-well mask is used to pattern the NPoly. Both the P-well and the epi regions must receive the PPoly implant, and a special mask must be produced for this purpose. This process therefore requires at least one new mask, and possibly as many as four.

The number of masking steps can be reduced at the cost of compromising performance. For example, the gate poly can be doped with a blanket boron implant and a patterned phosphorus implant. The resulting poly is not quite as heavily doped as that formed by using two separate masking steps. Since most advanced CMOS processes silicide the poly, this increase in sheet resistance is less objectionable than it might seem. The process can also use a blanket V_t implant followed by a patterned V_t implant of the opposite polarity in order to save one masking step. The transistor that receives both implants has somewhat more threshold voltage variation than it would have had using one implant. If both of these modifications are adopted, then only two masking steps are required instead of four. Even so, the additional steps add cost and complexity, so they are only used for submicron processes that would otherwise exhibit unacceptable leakage due to buried channel formation. In general, operating voltages of 5 V or more require well dopings compatible with single-doped poly, while lower operating voltages require dual-doped gate poly.

Selective gate doping produces unwanted poly diodes unless the gate poly is silicided. PN junctions appear wherever the PPoly and the NPoly abut one another. They can be shorted by metal jumpers or, more conveniently, by silicidation. As long as the designer takes care not to block the silicide from locations where PPoly and NPoly abut one another, silicidation automatically shorts all of the poly diodes. Silicidation greatly increases the rate at which dopants diffuse through the poly, so the intersections between PPoly and NPoly must be spaced well away from the gate regions of adjacent MOS transistors.[26] Although the PPoly–NPoly junction does exhibit rectification, poly diodes are not recommended as circuit components because the presence of grain boundaries within the depletion regions causes substantial leakage.

Almost all CMOS processes adjust the NMOS and PMOS threshold voltages. Most analog processes offer natural NMOS and PMOS transistors either as part of the baseline process or as process extensions. A few processes offer additional threshold voltage options, such as depletion-mode transistors or low-V_t PMOS transistors. Each such option requires its own threshold adjust implant, formed through an additional masking step. Transistors using these special implants are coded much like natural transistors, except that NatVT is replaced by the layer that codes for the special implant.

11.2.5. Scaling the Transistor

Integrated circuits have become vastly more complex over the past 30 years. The first digital integrated circuits contained 10 or 20 transistors; their modern equivalents contain hundreds of millions. This remarkable increase in complexity has largely been made possible by corresponding reductions in the size of individual transistors. From 1973 to 2000, minimum channel lengths went from 8 μm to about 0.2 μm.[27] These reductions in size have also improved the performance of the transistors. A set of guidelines called *scaling laws* has been developed that dictates how the various dimensions of an MOS transistor should be reduced to obtain the best performance.

Scaling laws fall into two general categories, both of which presume that width and length are multiplied by a *scaling factor S. Constant-voltage scaling* holds the operating voltage of the transistor constant while scaling its dimensions. As the transistor shrinks further and further, it becomes increasingly difficult to avoid hot-carrier generation and punchthrough breakdown. *Constant-field scaling* avoids these problems by reducing the supply voltage to keep the electric fields in the transistor constant regardless of scale. Most modern processes use some variant of constant-field

---

[26] Y. P. Tsividis, *Operation and Modeling of the MOS Transistor* (New York: McGraw-Hill, 1988), p. 439.

[27] The 8 μm figure is from D. A. Pucknell and K. Eshraghian, *Basic VLSI Design*, 3d ed. (Sydney: Prentice-Hall Australia, 1994), p. 7. Both figures are approximations of industry practice; much smaller dimensions are possible in a research environment.

scaling. Table 11.3 shows simplified rules for both constant-voltage and constant-field scaling laws.[28]

As an example, consider a process producing 5 V transistors with minimum dimensions of 1 μm long by 2.5 μm wide, using a 250 Å gate oxide and a backgate doping concentration of 10^{16} cm^{-3}. Suppose the channel length of this process is reduced to 0.8 μm by means of constant-field scaling. The scaling factor S equals 0.8 μm/1.0 μm, or 80%. According to Table 11.4, the scaled transistor should have a minimum width of 2.0 μm, a 200 Å gate oxide, and a backgate doping of $1.25 \cdot 10^{16}$ cm^{-3}. Since processes are generally identified by gate length, the original (100%) process would be considered a 1 μm process, while the 80% shrink would be an 0.8 μm process.

TABLE 11.4 Constant-voltage and constant-field scaling laws.

Quantity	Constant-Voltage	Constant-Field
Supply voltage	1	S
Minimum channel width	S	S
Minimum channel length	S	S
Gate oxide thickness	1	S
Backgate doping	$1/S^2$	$1/S$
Gate delay	S^2	S
Power-delay product	S^2	S^3

Shrinking a transistor actually improves its performance. The smaller dimensions reduce parasitic capacitances and increase switching speeds. The *gate delay* of a CMOS process equals the time required for a digital signal to propagate through a representative CMOS gate. As the transistors scale down, the gate delay decreases and the circuit can handle faster switching speeds. Early microprocessors operated at clock speeds of 1 to 10 MHz; their modern equivalents operate at well over 1 GHz.

Not only does a smaller transistor switch faster, but it requires less power to do so. CMOS logic gates require pulses of power to charge and discharge gate capacitances each time they switch. The faster the gate switches, the more transitions occur per second, and the larger the current consumption becomes. The supply current required by a gate can be reduced at the expense of increasing its gate delay. The product of gate delay and power consumption remains approximately constant for any given process. This *power-delay product* decreases as the size of the transistor shrinks. For example, 80% constant-field scaling reduces the power-delay product to about half of its initial value. As this example suggests, even relatively minor decreases in size significantly reduce power consumption. This is fortuitous, since otherwise a microprocessor running at more than a few hundred megahertz would literally melt from its own waste heat.

Scaling laws are frequently applied to existing digital layouts to convert them for use with newer processes. Rather than laboriously recoding the layout, the designer simply runs a program that scales all of the data by a specified amount. This type of scaling is called an *optical shrink* because it produces the same results as photoreducing the existing mask set. Optical shrinks are denoted by the percentage scaling factor used to transform the data from their original, or *drawn*, dimensions to their final, or *shrunk*, dimensions. A 100% shrink indicates that the final dimensions equal the drawn dimensions, while an 80% shrink indicates that they equal 4/5 of the drawn dimensions. Figure 11.16A shows a 1 μm transistor drawn at 100%.

[28] These laws have been adapted from Pucknell *et al.*, p. 129.

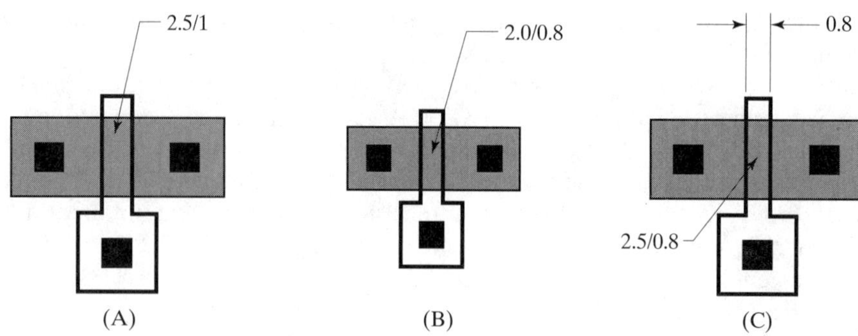

FIGURE 11.16 Examples of scaled MOS transistors: (A) drawn at 100%, (B) optically shrunk to 80%, (C) selectively shrunk to 80% of drawn gate length. The wells have been omitted for clarity.

Figure 11.16B shows the same transistor optically shrunk to 80%. The optical shrink scales both the width and the length of the device in a manner consistent with either constant-voltage or constant-field scaling. The process engineers will adjust gate oxide thickness, backgate doping, and other parameters according to the desired type of scaling.

An optical shrink affects all dimensions equally, but some are more difficult to scale than others. Multilayer metal systems have proved especially difficult to scale to submicron dimensions. Although fine-line metal systems certainly exist, they are very expensive. Many processes selectively scale channel length while retaining the previous dimensions for all other layout rules. Figure 11.16C shows a 1 μm transistor whose gate has been shrunk to 0.8 μm. The selective gate shrink requires a more complicated set of geometric transformations than a simple optical shrink, but it is still far simpler and quicker than a full relayout. The benefits of a selective gate shrink are somewhat less than those of a full optical shrink (Table 11.5), but they are still sufficient to justify selective gate shrinks for many processes.

TABLE 11.5 Scaling laws for selective gate shrinks.

Quantity	Constant-Voltage	Constant-Field
Supply voltage	1	S
Minimum channel width	1	1
Minimum channel length	S	S
Gate oxide thickness	1	S
Backgate doping	$1/S^2$	$1/S$
Gate delay	S	1
Power-delay product	S	S^2

The scaling laws were originally developed for digital processes. CMOS logic circuits respond quite predictably to scaling, but the same is not true of analog or mixed-signal circuits. No set of predetermined scaling laws can comprehend the full complexity of analog circuit design. Indiscriminate scaling usually causes analog circuits to fail parametric specifications, and in some cases it may cause outright malfunctions. For example, an 80% optical shrink reduces all capacitors to 64% of their former values. Since analog designs rely on capacitors to stabilize feedback loops, a reduction in capacitance can actually destabilize the circuit. Constant-field scaling only makes matters worse because it simultaneously increases transconductance and reduces capacitance. Selective gate shrinks do not change capacitance values, but they are still risky because they can introduce unforeseen parametric changes in short-channel transistors that frequently prove more significant for analog circuits than for digital ones. To summarize, analog and mixed-signal circuits should not be scaled without re-evaluating the performance of the resulting circuit to ensure that it still meets functional and parametric specifications.

11.2.6. Variant Structures

The simplest type of self-aligned poly-gate transistor consists of a rectangle of NMoat or PMoat bisected by a strip of poly. This type of structure serves admirably for width-to-length ratios of less than 10. Transistors with larger *W/L* ratios become increasingly unwieldy unless they are divided into multiple identical sections connected in parallel. Figure 11.17A shows the layout of a three-section transistor. The paralleled fingers not only produce a more convenient aspect ratio, but also save area because adjacent sections share source and drain fingers. The merger of adjacent source/drain fingers can also reduce parasitic junction capacitance by up to 50%.

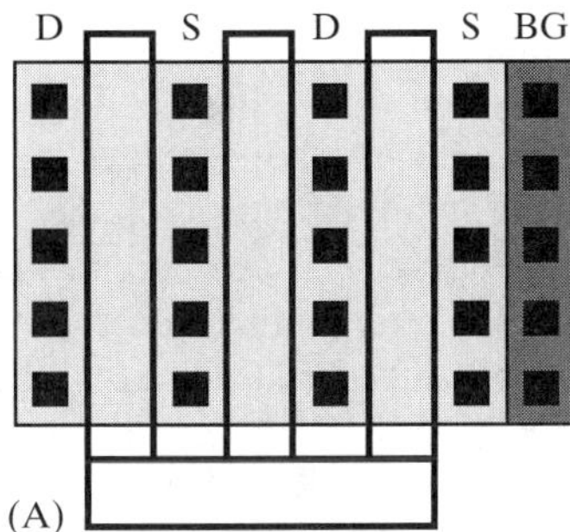

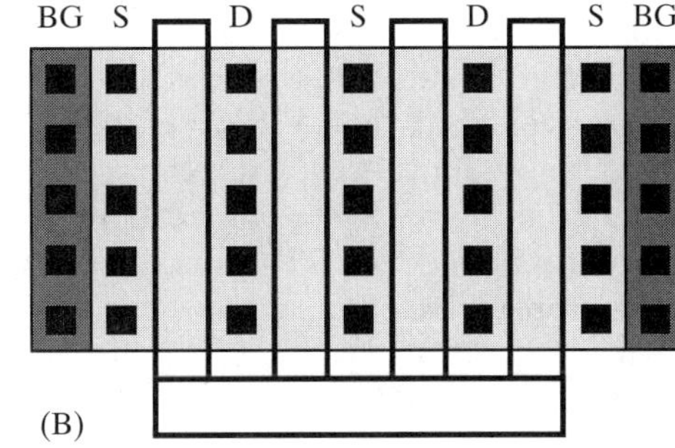

FIGURE 11.17 Sectioned transistors of (A) three and (B) four sections. The fingers are marked S (source), D (drain), and BG (backgate). The wells have been omitted for clarity.

The division of a transistor into sections can affect its matching, so circuit designers often specify the number of sections for critical transistors. The most common notation for a sectioned transistor is *N(W/L)*, which denotes *N* sections, each with a drawn width of *W* and a drawn length of *L*. Transistors specified in this manner should be laid out exactly as requested. If the transistor is specified as having dimensions *W/L*, this usually indicates that it can have any number of segments desired. If this transistor must match one having dimensions *N(W/L)*, then the former device should be laid out as a single section. Sometimes a circuit designer unwittingly specifies a large W/L ratio that results in a poor layout. One can usually multiply the number of segments in a transistor by an integer number, and divide the width of each segment by this same number, so long as all transistors that must match are partitioned in like manner.

Transistors with even numbers of sections always contain odd numbers of source/drain fingers (Figure 11.17B). Such transistors are usually constructed with source fingers at either end. Not only does this allow the use of abutting backgate contacts on either or both ends, but it also reduces the number of drain fingers by one. This arrangement minimizes parasitic drain junction capacitance at the expense of source capacitance. Drain capacitance usually has more effect upon circuit performance than source capacitance, so a reduction in drain capacitance at the expense of source capacitance usually improves circuit performance.

Transistors sharing common source or drain connections are frequently merged to save space or to minimize parasitic junction capacitance. The merger is a relatively simple matter so long as both transistors contain sections of the same width. Differing widths require the use of a notched moat (Figure 11.18). The layout rules usually prohibit the placement of polysilicon immediately adjacent to a moat edge, due to the large oxide step present at this location. The spacing between poly and

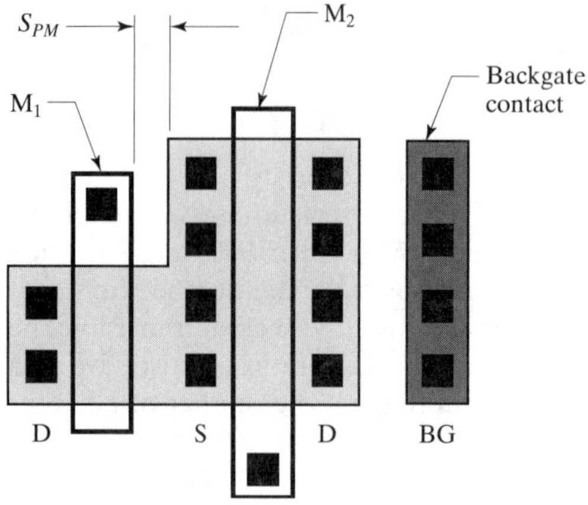

FIGURE 11.18 Merged transistors M_1 and M_2 share a common source (well omitted for clarity).

moat S_{PM} forces a slight increase in the area of the shared source/drain finger, but one shared finger still consumes less area than two separate fingers.

Transistors M_1 and M_2 in Figure 11.18 share a common source region, so the drain fingers must occupy the ends of the array. This arrangement precludes the use of an abutting backgate contact, so the backgate contact is placed some distance away from the devices. The spacing between the backgate contact and the merged transistors may seem to eliminate any area benefit produced by the merger, but this backgate contact can also serve several other devices. Transistors sharing a common drain have source fingers on either end of the array and can therefore use abutting backgate contacts.

CMOS layout makes extensive use of merged devices to save space and to minimize capacitance. Figure 11.19 shows a simple layout of a two-input NAND gate that illustrates many of the techniques in common use. PMOS transistors M_1 and M_2 occupy a common well placed at the top of the layout. These transistors share a common drain region that not only reduces the width of the cell but also minimizes the drain capacitance on the output node Z. The two PMOS transistors also share a single backgate contact at the right end of the well. NMOS transistors M_3 and M_4

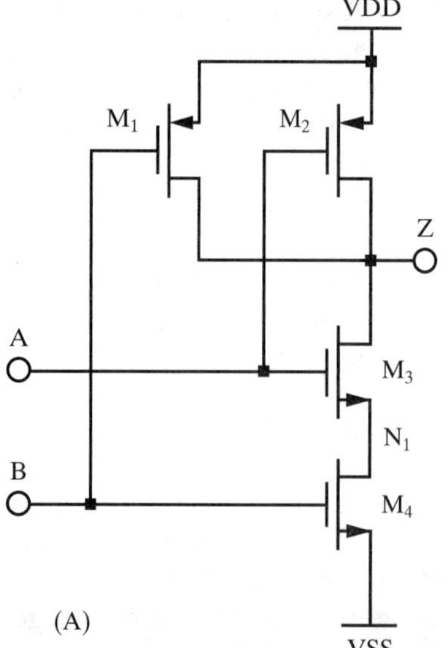

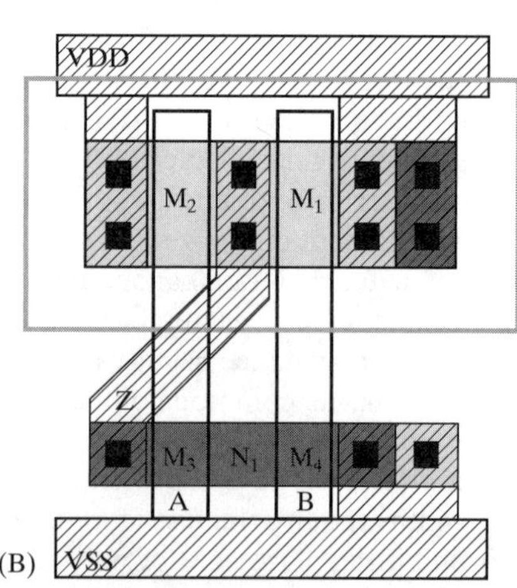

FIGURE 11.19 (A) Schematic and (B) layout of a two-input NAND gate.

reside next to one another near the bottom of the layout. These transistors have been placed in series—the drain of M_3 simultaneously acts as the source of M_4. No contacts are necessary, because the current simply flows from one channel to the next. One strip of poly forms the gates of transistors M_2 and M_3, and a second strip of poly forms the gates of M_1 and M_4. The spacing between M_3 and M_4 is slightly larger than minimum. If desired, this spacing can be minimized by angling the gate leads toward one another.

The layout in Figure 11.19 follows the general guidelines of digital standard cell design. The power and ground rails run across the top and bottom of the cell, respectively. The width and spacing of these leads should be the same for all logic cells so they can stack end-to-end. The PMOS transistors occupy a common well spanning the top of the cell. When multiple logic cells stack end-to-end, their wells overlap to form a single contiguous region running the entire length of the assembly. This arrangement avoids the well-to-well spacings that would otherwise appear between adjacent cells. The NMOS transistors reside near the bottom of the cell, either in the epi or in another common well. Each cell contains at least one substrate and one backgate contact. Larger cells should contain additional substrate and backgate contacts wherever possible. The input and output connections exit from either the top or the bottom of the cell, whichever is more convenient for a given layout. Digital standard cells frequently contain special elements called *ports* and *prels* (port relationships) required by autorouting software. Conventional autorouters cannot handle analog layout because of the severe routing constraints imposed by matching, channel formation, and noise coupling. Some advanced routers now incorporate sufficient flexibility to render them attractive tools for large-scale mixed-signal design. The operation of these autorouters is beyond the scope of this text. However, the analog designer may still wish to employ concepts such as standard cell heights and consistent power and ground rail placement to allow analog cells to stack together. The height of analog cells is usually much greater than that of digital cells to accommodate their larger components and greater interconnection complexity. Additional rails are sometimes necessary to accommodate separate analog and digital supplies or to distribute several different supplies throughout a multisupply system.

Serpentine Transistors

Some designs require transistors with very long channels. The most convenient layout for such devices consists of a strip of NMoat or PMoat placed underneath a plate of polysilicon. A very compact layout results if one folds the moat into a serpentine pattern (Figure 11.20). The total channel length is computed by a procedure analogous to that used for serpentine resistors. Each 90° bend in the channel adds one-half of the transistor's width to its total length. The channel length of the transistor in Figure 11.20 therefore equals $2L_X + L_Y + W$. Serpentine transistors will not

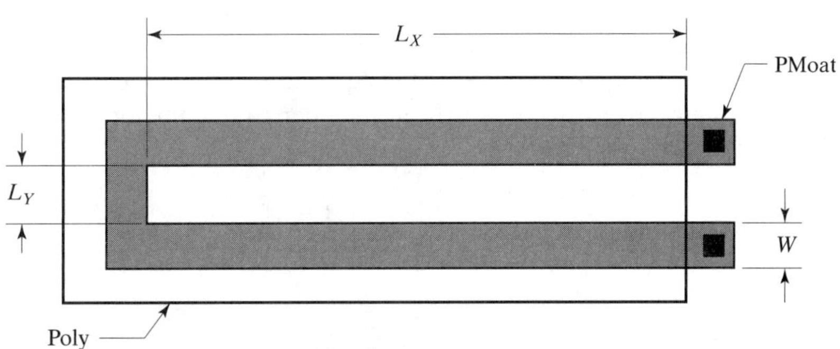

FIGURE 11.20 Serpentine PMOS transistor (well omitted for clarity).

match precisely unless they have identical geometries, but most designs do not require especially precise matching from long-channel devices.

Annular Transistors

The drain capacitance of a MOS transistor limits its switching speed and frequency response. Many circuits, analog as well as digital, can benefit from reduced drain capacitances. A smaller transistor has less capacitance, but it also provides less transconductance. These factors offset one another, so the smaller transistor is generally no faster than its larger counterparts. In order to actually increase switching speed, one must reduce the ratio of drain capacitance to transistor width C_D/W. Interdigitation reduces the C_D/W ratio by half because it surrounds each drain with two gates. This same principle can be carried still further by surrounding the drain on all four sides by an annular gate (Figure 11.21). An annular transistor will provide the smallest possible C_D/W ratio, but the decreased drain capacitance comes at the expense of increased source capacitance. The increased source capacitance is not necessarily injurious because the source often connects to a low-impedance node such as a power supply rail.

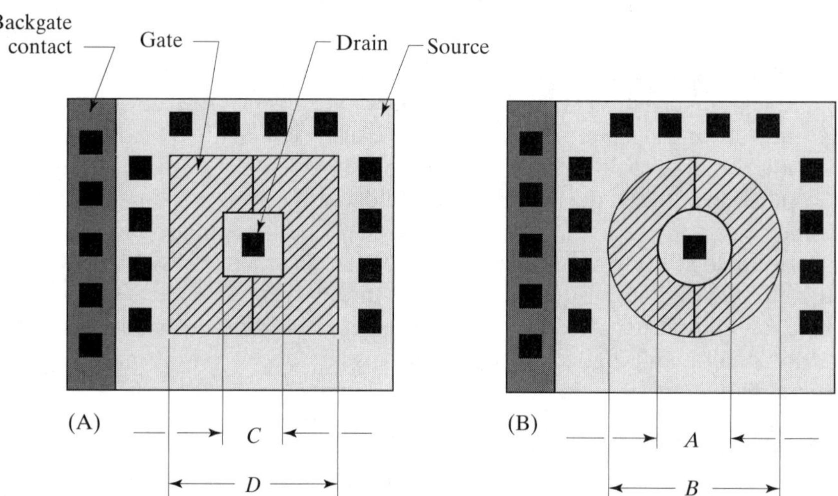

FIGURE 11.21 Annular MOS structures: (A) square and (B) circular.

Two basic types of annular transistors exist: those that use a square gate geometry (Figure 11.21A) and those that use a circular gate (Figure 11.21B). The circular gate theoretically provides the highest C_D/W ratio because it minimizes the area-to-periphery ratio of the drain. The current flow through a circular gate is quite symmetric, and the width of the transistor is easily computed. The current flow through square gates is less uniform, and the effective width of the transistor is less easily computed. Square gates also have sharp corners that can induce premature avalanche breakdown due to electric field intensification.

The following equations (derived in Appendix D) give the width and length of a circular annular transistor in terms of the inner gate diameter A and the outer gate diameter B:

$$W = \frac{\pi(B - A)}{\ln(B/A)} \quad [11.10A]$$

$$L = \frac{B - A}{2} \quad [11.10B]$$

Some designers approximate the width of a circular annular transistor as the perimeter of a circle drawn halfway between the source and drain, giving $W \cong \frac{1}{2}\pi(A + B)$. This approximation slightly overestimates the true width of the transistor. The errors caused by the approximation have little impact because precision circuits always rely on matching between identical devices rather than the properties of any one device.

The width and length of a square annular transistor are given by the following approximations, which do not correct for corner effects:

$$W \cong 2(C + D) \tag{11.11A}$$

$$L \cong \frac{D - C}{2} \tag{11.11B}$$

Annular transistors are often elongated to produce gate geometries similar to those in Figure 11.22. The C_D/W ratio of the elongated annular transistor is not much smaller than that of a conventional interdigitated transistor, so elongated structures are not recommended for minimizing drain capacitance. They are still sometimes used to produce an enclosed channel (Section 12.1.2). The W and L of an elongated circular annular transistor (Figure 11.22A) are approximately

$$W = \pi\frac{(B - A)}{\ln(B/A)} + 2U \tag{11.12A}$$

$$L = \frac{B - A}{2} \tag{11.12B}$$

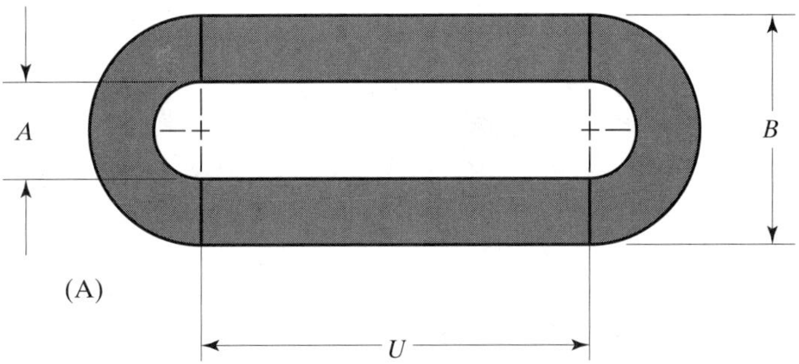

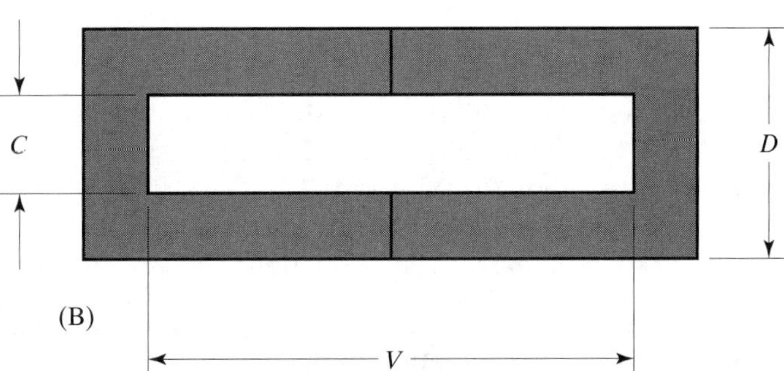

FIGURE 11.22 Gate geometries for elongated annular transistors: (A) circular and (B) square.

Similarly, the W and L of an elongated square annular transistor (Figure 11.22B) are approximately

$$W \cong 2V + C + D \qquad [11.13A]$$

$$L \cong \frac{D - C}{2} \qquad [11.13B]$$

11.2.7. Backgate Contacts

All MOS transistors require electrical connection to their backgates even though no current normally flows through these connections. MOS transistors that lack backgate contacts or exhibit excessive backgate resistance are particularly prone to latchup. Every PMOS transistor contains a parasitic lateral PNP, and every NMOS contains a parasitic lateral NPN. Together these form a parasitic SCR (Figure 4.22). The backgate contacts short the base-emitter junctions of these parasitic transistors, and the associated backgate resistances become base turnoff resistors (R_1 and R_2 in Figure 4.22). The SCR will remain off as long as the voltage across both of these resistors remains less than the forward voltage of the respective base-emitter junctions. The voltage necessary to trigger the SCR into conduction equals about 0.65 to 0.7 V at 25°C, but this falls to 0.4 to 0.45 V at 150°C due to the temperature coefficient of V_{BE}. Not only does the trigger voltage drop at high temperatures, but the betas of the parasitic transistors increase significantly. Thus, CMOS latchup is most likely to occur at high temperature.

Most CMOS products must pass a standardized test that measures their latchup susceptibility. Positive and negative test current pulses are applied to each pin, while power is applied to the part. Depending on specifications, the magnitude of these test pulses may range from as little as ±100 mA to as much as ±250 mA. The supply current is measured both before and after the application of each test pulse. If these two currents are not approximately the same, then the part fails the test.[29]

This latchup test can be modeled mathematically. Suppose that a test current I_T flows through the source/drain junction of an MOS transistor M_1. In order to prevent latchup from occurring between M_1 and a complementary MOS transistor M_2, at least one of the following inequalities must be true:

$$\beta_{12}\beta_{21}(1 - \eta_{c12})(1 - \eta_{c21}) < 1 \qquad [11.14A]$$

$$I_T R_{B2}(1 - \eta_{c12})\left(\frac{\beta_{12}}{\beta_{12} + 1}\right) < V_{\text{trig}} \qquad [11.14B]$$

β_{12} represents the beta of the parasitic bipolar formed by minority carriers flowing from the source/drain region of M_1 to the backgate of M_2 in the absence of guard rings. β_{21} represents the beta of the parasitic bipolar formed by minority carriers flowing from the source/drain region of M_2 to the backgate of M_1, again in the absence of guard rings. η_{c12} represents the fraction of minority carriers flowing from M_1 to M_2 intercepted by guard rings. Similarly, η_{c21} represents the fraction of minority carriers flowing from M_2 to M_1 intercepted by guard rings. I_T equals the test current, R_{B2} equals the backgate resistance of M_2, and V_{trig} equals the trigger voltage of the SCR (about 0.4 V at 150°C).

These equations provide some insight into the roles of guard rings and backgate contacts in suppressing latchup. Equation 11.14A represents the condition required

---

[29] For details on latchup testing, see JEDEC publication JESD-78, *IC Latchup Test*, 1997.

to avoid sustained feedback. Minimizing parasitic betas β_{12} and β_{21} and adding guard rings to improve collector efficiencies η_{c12} and η_{c21} can help prevent sustained conduction. Any device meeting this criterion is invulnerable to CMOS latchup regardless of the magnitude of the test currents applied. Unfortunately, few CMOS processes can satisfy Equation 11.14A because their transistors lie too close together and their guard rings are too inefficient. CMOS devices can still achieve conditional latchup immunity by satisfying the conditions of Equation 11.14B. The four terms in this inequality represent the magnitude of the test current and the backgate resistance, the effectiveness of the guard rings, and the magnitude of the parasitic beta, respectively. The contributions of guard rings and backgate contacts multiply one another, producing a synergistic relationship between the two. Even if neither guard rings nor backgate contacts alone can stop latchup, a combination of the two frequently can. Guard rings require so much room that they can be placed around only a few devices—usually those that may potentially inject minority carriers into the die. Backgate contacts require much less area, so each transistor can have its own backgate contact or can at least share a backgate contact with another transistor.

The backgate of an NMOS must connect to a voltage less than or equal to its source, and the backgate of a PMOS must connect to a voltage greater than or equal to its source. In many applications, the backgate can be connected to the source. However, some transistors operate under conditions in which it is difficult or impossible to distinguish source from drain. These applications must connect the backgate to a voltage that is different from the source to increase the threshold voltage, using the body effect. Some high-speed circuits also avoid connecting the source and backgate in order to minimize the capacitance appearing at the source node. All of these circuits require an independent backgate contact like that in Figure 11.23B. Transistors whose source and backgate operate at the same potential can use an abutting backgate contact (Figure 11.23A). This contact saves considerable area by eliminating the spacing between source and backgate diffusions. The intersection of two heavily doped diffusions produces a leaky and unreliable junction, but this defect can be tolerated as long as the two diffusions are connected together by metal or silicide.

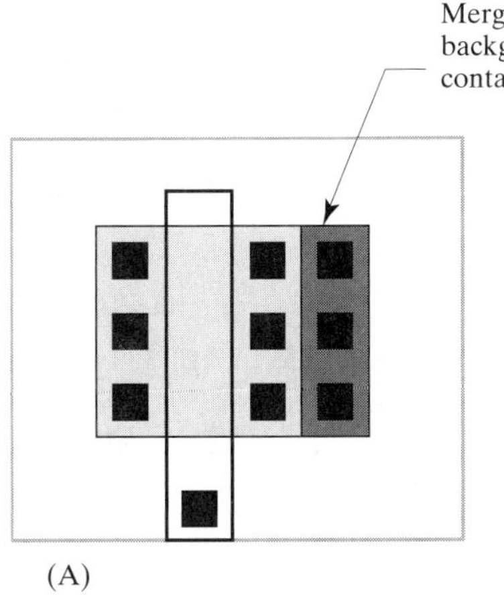

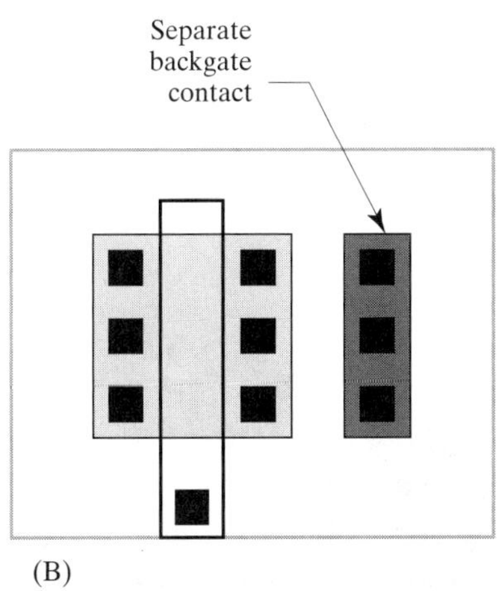

FIGURE 11.23 Examples of (A) abutting and (B) separate backgate contacts.

Relatively small transistors such as those in Figure 11.23 require only one backgate contact. The distance across the transistor grows larger as additional segments are added, and at some point this distance becomes so great that it produces an unacceptably large backgate resistance. A second backgate contact on the opposite side of the transistor reduces the distance to the backgate contacts at the cost of slightly increasing the device area (Figure 11.17B). The point at which a second backgate contact becomes necessary varies depending on the sheet resistance of the backgate. An NMOS transistor constructed in the P-epi above a P+ substrate would have a lower backgate resistance than a PMOS constructed in a shallow, lightly doped N-well. Adding a buried layer to a well likewise decreases the backgate resistance of transistors occupying it. Factors of this sort make it difficult to provide quantitative rules for backgate contact spacing. Some processes specify a maximum distance between any portion of a transistor and the nearest backgate contact. The maximum allowed distance becomes shorter as the backgate resistance increases. Typical spacings range from 25 μm to 250 μm. Transistors subject to large transients should use a smaller distance to provide additional latchup immunity. These include ones whose source/drain regions connect to pins, and those residing next to transistors whose source/drain regions connect to pins.

Large transistors with many fingers may require substrate contacts embedded within the body of the transistor itself. This is usually achieved by placing strips of backgate contact through the transistor at regular intervals (Figure 11.24A). Although these *interdigitated backgate contacts* reduce the distance to the nearest substrate contact, they also substantially increase the size of the transistor. Some processes allow another type of backgate contact, consisting of small plugs of backgate diffusion placed in holes within the source fingers of the transistor (Figure 11.24B). These *distributed backgate contacts* slightly increase the source resistance of the overall transistor, but they greatly reduce the area required by the backgate contacts. A substantial area savings can be obtained even if the transistor must be enlarged to compensate for increased source resistance. Distributed backgate contacts can be placed on every source finger, as shown, or they may be placed on only a few source fingers distributed at regular intervals across the transistor. A larger number of distributed backgate contacts further reduces backgate resistance, but not all applications necessarily require the same degree of latchup protection. The plugs of source/drain diffusion used for distributed backgate contacts must be sufficiently large to ensure contact to the backgate even after lateral outdiffusion. Careful adherence to the applicable layout rules is required to ensure proper performance.

Analog BiCMOS processes often contain additional diffusions that can help reduce the latchup susceptibility of MOS transistors. For example, many analog BiCMOS processes fabricate NPN transistors in the same N-well as PMOS transistors. The NBL that is used to reduce the collector resistance of the CDI NPN can

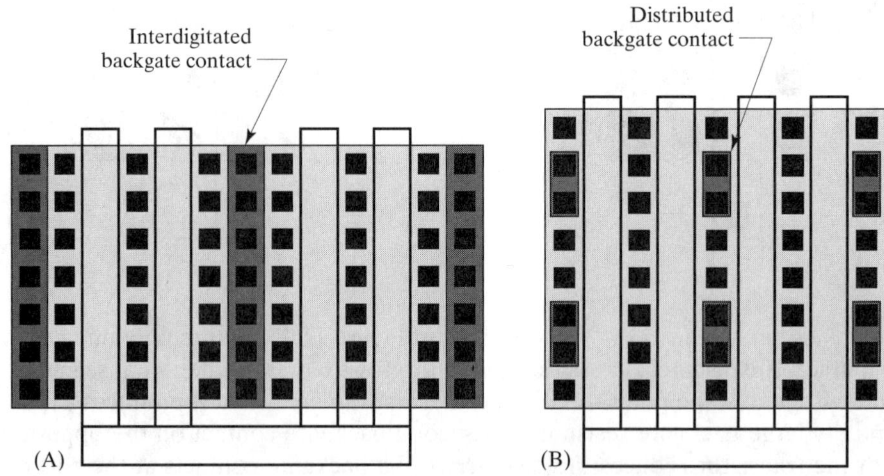

FIGURE 11.24 Additional styles of backgate contacts: (A) interdigitated backgate contact and (B) distributed backgate contact.

also reduce the backgate resistance of the PMOS. If NBL is available, it should be placed in all MOS transistors that use the same well as the CDI NPN. Very efficient hole-blocking guard rings can be produced in many processes that offer deep-N+ and NBL. Some transistors may actually operate under conditions in which their source/drain regions regularly forward-bias into the backgate. Deep-N+ guard rings can help ensure that substrate injection from these PMOS transistors does not disrupt the operation of the rest of the circuit.

NMOS transistors are somewhat more difficult to protect from latchup than PMOS transistors. An isolated NMOS structure (Section 11.2.2) offers total immunity to CMOS latchup, but at the price of greatly increased backgate resistance. Although these transistors do not require a low backgate resistance in order to suppress CMOS latchup, parasitic lateral NPN action remains a concern. If the transistors must operate at relatively high voltages, they may become susceptible to snapback breakdown due to parasitic NPN action. Even if they operate well below the $V_{CEO(sus)}$ of the NPN, minority carrier injection still causes sluggish switching due to charge storage. A system of distributed backgate contacts can minimize these problems. An NMOS transistor fabricated above a P+ substrate and surrounded by a deep-N+ electron-collecting guard ring will also prove quite resistant to latchup. This style of transistor is suited to applications where the transistor must withstand severe transients, but where its source/drain regions do not routinely forward-bias into the backgate. Transistors of the latter sort are best constructed as isolated NMOS transistors.

11.3 FLOATING-GATE TRANSISTORS

Virtually all digital circuits contain elements that remember digital information from one cycle of operation to the next. Some of these elements are nothing more than individual flip-flops embedded within digital logic. Others consist of arrays of many identical circuits arranged in densely packed structures called *memories*. The size of a memory depends upon the number of *bits* that it can store. Each bit represents a single binary value—either a zero or a one. Digital memories usually organize bits in groups of eight called *bytes*. Modern digital memories are so large that they are usually specified in *kilobytes* (KB), *megabytes* (MB), or even *gigabytes* (GB). A kilobyte equals 1024 (not 1000) bytes. Similarly, a megabyte equals 1024 kilobytes, and a gigabyte equals 1024 megabytes. Thus, a 1MB memory chip contains 1,048,576 bytes, or 8,388,608 bits of memory.

Memories are generally classified as either volatile or nonvolatile. The information stored in a *volatile memory* vanishes as soon as power is removed. By contrast, information stored in a *nonvolatile memory* (NVM) will remain intact for an indefinite period of time after loss of power. Volatile memories are typically found only in digital applications. Nonvolatile memories, on the other hand, have proved so useful that even analog designs now routinely incorporate them.

The simplest form of nonvolatile memory is called *read-only memory* (ROM). A ROM leaves the factory preprogrammed to a specific pattern of ones and zeros. The creation of a new ROM requires the development of a mask set and the fabrication of a new batch of integrated circuits. This process costs tens of thousands of dollars and takes months of time. Any error in the contents of the ROM requires another entire design cycle.

The cost and time required to introduce new versions of ROM led to the development of *programmable read-only memory* (PROM). A typical PROM circuit uses nichrome fuses as memory elements. Each fuse represents one bit. Initially, all of the fuses in the PROM are intact. A network of transistors within the PROM allows external circuitry to access and blow any given fuse. The development of a new PROM requires only a blank PROM chip, a programming machine, and a few seconds of time. Any error in the contents of a PROM can be corrected by programming a replacement.

While PROM memories enjoyed a certain degree of success, they also exhibited serious flaws. The large currents required to program the fuses made the memory cells so bulky that the average PROM could store only a few hundred bytes of memory. Reliability problems also plagued the early nichrome fuses. Researchers sought more compact and more reliable PROM memory elements. As early as 1967, researchers at Bell Labs suggested that MOS transistors incorporating floating gates could potentially serve as nonvolatile memory elements.[30] A *floating-gate transistor* has a gate electrode completely surrounded by insulating oxide. One can program such a transistor by injecting charge onto its gate by hot-carrier injection, causing the transistor to switch from a nonconducting state to a conducting state. The charge on a floating-gate transistor can be erased by exposing the device to ultraviolet light (UV).

An *erasable programmable read-only memory* (EPROM) consists of an array of floating-gate transistors in a UV-transparent package. An EPROM can be programmed by using a machine similar to those formerly employed to program PROMs. EPROMs can be erased by placing them under a high-intensity UV lamp. Each EPROM can survive several hundred programmings and erasures. EPROMs placed in ordinary plastic packages can only be programmed once, and so become PROMs.

Floating-gate transistors can also be programmed by Fowler-Nordheim tunneling. This process can both inject charge onto a floating gate and remove it. The resulting devices are called *electrically erasable programmable read-only memories* (EEPROM). An EEPROM does not need to be exposed to UV light to be erased. Not only does this eliminate expensive UV-transparent packaging, but it also allows the device to be erased without removing it from the application. Dense EEPROM arrays called *flash memories* have become very popular as replacement for bulky, fragile magnetic media such as tapes and floppy disks.

Analog integrated circuits have long used fuses and Zener zaps for trimming (Section 5.6.2). These devices are really just another form of programmable read-only memory. The analog versions of fuses and Zener zaps require programming currents of tens, if not hundreds, of milliamps. These large currents generally dictate the addition of a dedicated probe pad for each bit of memory. These pads not only severely limit the number of bits of available memory, but they also render post-package trimming impossible. Attempts to circumvent these limitations by using transistors to switch the programming currents have met with equivocal success because of the size of the required transistors. EPROM and EEPROM offer the ability to integrate far more trim memory, to program this memory after packaging, and to reprogram the memory in the event that the initial trimming proves less than optimal. These advantages make EPROM or EEPROM a highly desirable addition to many analog CMOS or BiCMOS products.

The next section discusses the theory of operation of floating-gate devices, including the means by which they are programmed and erased, and the mechanisms which limit their lifetimes. Section 11.3.2 covers the construction of simple EEPROM memory cells compatible with conventional CMOS processing. This text does not cover the construction of dense EPROM and EEPROM arrays intended specifically for digital applications.

---

[30] D. Khang and S. M. Sze, "A Floating Gate and Its Application to Memory Devices," *Bell Systems Tech. J.*, Vol. 46, 1967, p. 1283.

11.3.1. Principles of Floating-Gate Transistor Operation

A floating gate consists of poly completely surrounded by oxide. Carriers require an energy of about 3.2 eV to surmount the oxide-silicon interface. Very few carriers have such large energies, so a charge placed upon a floating gate typically requires many years to leak away. The programming and erasure of floating-gate devices requires the generation of carriers with energies in excess of 3.2 eV. Four common processes can generate the necessary energies: heating, ionizing radiation, hot-carrier injection, and Fowler-Nordheim tunneling. Each of these processes deserves further discussion.

Heating a wafer to a temperature of 400–500°C will produce a few high-energy carriers capable of surmounting the oxide-silicon interface. The relatively small charge on a floating gate will gradually leak away at these temperatures. Heating provides a simple means of removing any charge that may have accumulated on floating gates during manufacture. The final anneal often performs this function. If the final anneal proves inadequate, then a post-anneal bake can serve the same purpose. Finished devices generally cannot be discharged by baking because the required temperatures cause plastics to decompose and accelerate the formation of intermetallic compounds between gold bondwires and aluminum metallization.

Ionizing radiation can also generate high-energy carriers. A mercury vapor lamp generates short-wave ultraviolet (UV) light whose photons have an energy of about 4.9 eV. Such a lamp will erase a floating-gate device in a matter of minutes. Unfortunately, UV radiation generates photocurrents that interfere with proper device operation. Therefore, UV exposure can only serve as a means of erasing unpowered devices. Furthermore, the packaging must include a UV-transparent window constructed of fused silica, and the die must use a UV-transparent protective overcoat composed of silicon dioxide or silicon oxynitride. These considerations limit the use of UV erasure to devices specifically designed for this purpose, such as EPROM memories. Few analog circuits use UV erasure.

Intense electric fields can also generate hot carriers with sufficient energy to surmount the oxide-silicon interface. The earliest EPROMs were programmed by hot-carrier injection from an avalanching junction.[31] The floating-gate device used in these EPROMs was called a *floating-gate avalanche-injection metal-oxide-semiconductor* (FAMOS) transistor.[32] Figure 11.25 shows a cross section of this device.

The FAMOS transistor resembles a normal PMOS transistor whose gate has been left unconnected. It can be erased by heating it or by exposing it to UV

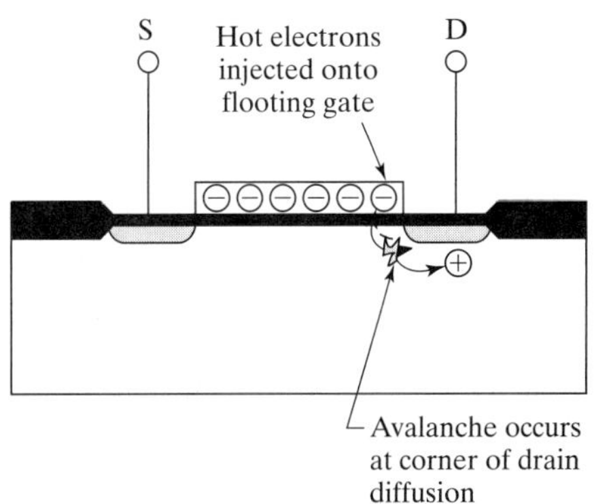

FIGURE 11.25 Cross section of a FAMOS transistor, showing programming by means of electrons injected from the avalanching drain/backgate junction.

[31] D. Frohman-Bentchkowsky, "A Fully-Decoded 2048-Bit Electrically-Programmable MOS ROM," *ISSCC Digest of Tech. Papers*, 1971, pp. 80–81.

[32] FAMOS is a trademark of Intel.

radiation. Either process causes any charge present upon the gate to leak away. The erased device exhibits a gate-to-source voltage of 0 V, and it therefore operates in cutoff. The FAMOS transistor is programmed by avalanching the drain–backgate junction. The doping gradient and sidewall curvature ensure that avalanche breakdown occurs primarily at the surface. A small percentage of the hot electrons produced by the avalanching junction are injected into the gate oxide. Some of these electrons pass through the oxide and accumulate on the floating gate. The floating gate gradually accumulates a negative charge. The resulting negative gate-to-source voltage induces the formation of a channel beneath the floating gate. The FAMOS transistor thus acts as a normally open switch that closes when programmed.

Intense electric fields can also appear at the drain end of a saturated MOS transistor. These fields can generate hot carriers capable of surmounting the oxide-silicon interface. NMOS transistors are especially prone to hot carrier injection because of the relatively high mobility of electrons. Figure 11.26A shows a cross section of a double-poly EPROM transistor programmed by hot-carrier injection.

This transistor has two gate electrodes: a lower floating gate, and an upper *control gate*. Figure 11.26B shows the equivalent electrical circuit for this device. NMOS

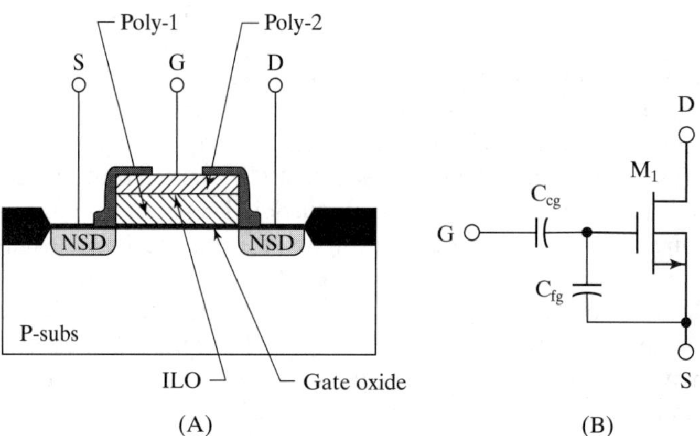

FIGURE 11.26 (A) Cross section of a double-poly EPROM transistor,[33] and (B) the equivalent circuit of the double-poly transistor.

transistor M_1 represents the device formed by the floating gate, the source, and the drain. The floating gate capacitively couples to the backgate through the gate oxide, forming capacitor C_{fg}. The floating gate also capacitively couples to the control gate through the interlevel oxide, forming capacitor C_{cg}. The voltage on the floating gate relative to the backgate, V_{fg}, equals

$$V_{fg} = \frac{C_{cg}}{C_{cg} + C_{fg}} V_{cg} + \frac{Q_{fg}}{C_{cg} + C_{cg}} \qquad [11.15]$$

where Q_{FG} equals the charge on the floating gate. This charge equals zero if the transistor has been erased by exposure to heat or UV radiation. Under these conditions, a channel will form beneath the floating gate when

$$V_{t(cg)} = \frac{C_{cg} + C_{fg}}{C_{cg}} V_{t(fg)} \qquad [11.16]$$

[33] R. Bucksch, *MSEE Thesis*, Friedrich-Alexander University, Erlangen-Nürnberg, Germany, 1997 p. 5.

where $V_{t(fg)}$ equals the voltage that must be applied to the floating gate to invert the underlying silicon. The double-poly transistor therefore initially acts as an NMOS transistor with a somewhat larger-than-expected threshold voltage.

The double-poly transistor can be programmed by operating it in saturation under a high drain-to-source differential voltage. Under these conditions, hot electrons generated in the pinched-off region travel across the gate dielectric to the floating gate. The resulting negative charge Q_{fg} effectively increases the threshold voltage of the double-poly transistor to

$$V_{t(\text{cg})} = \frac{C_{\text{cg}} + C_{\text{fg}}}{C_{\text{cg}}} V_{t(\text{fg})} + \frac{Q_{\text{fg}}}{C_{\text{cg}}} \qquad [11.17]$$

The programming process makes the double-poly transistor more difficult to turn on. The double-poly transistor thus acts as a normally closed switch that opens when programmed.

Fowler-Nordheim tunneling can either inject electrons onto a floating gate or remove them from it. Figure 11.27 shows a *floating gate tunneling oxide* (FOTOX) transistor that can be programmed and erased by means of Fowler-Nordheim tunneling. This device resembles the double-poly transistor of Figure 11.26, except that a portion of the oxide beneath the floating gate has been made particularly thin. This *tunneling oxide* resides over an extension of the transistor's drain. Holding the drain at ground while applying a high voltage to the control gate will program the transistor by placing a large positive voltage between the floating gate and the drain. Electrons tunnel from the drain across the thin tunnel oxide to the floating gate. The resulting negative charge increases the effective threshold voltage of the device in accordance with Equation 11.17. The FOTOX transistor can be erased by holding the control gate at ground while placing a high voltage on the drain. Electrons now tunnel from the floating gate to the drain, removing the negative charge upon the floating gate and diminishing the effective threshold voltage of the device.

The FOTOX transistor, or some variation of it, forms the basis for all modern EEPROM memories. However, the basic FOTOX device shown in Figure 11.27 requires several special processing steps to fabricate the extended drain, the thin tunneling oxide, and the control gate. Few analog processes incorporate these extensions. Section 11.3.2 describes a single-poly EEPROM cell compatible with

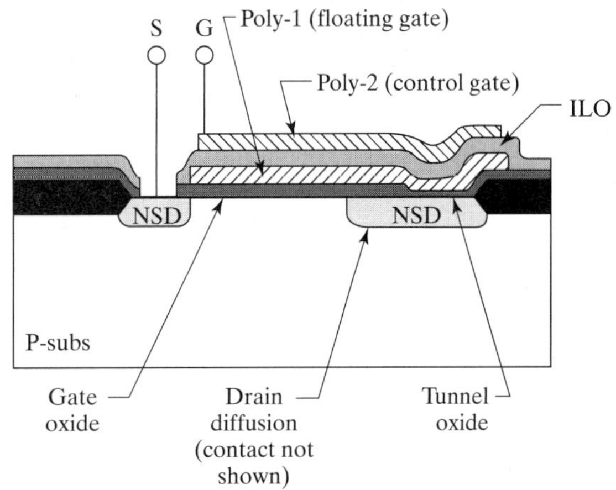

FIGURE 11.27 Cross section of a FOTOX transistor.

[34] Ibid, p. 18.

conventional CMOS processing. Many analog processes use some variation on the single-poly EEPROM transistor.

All floating-gate devices degrade after repeated programming and erasure. The passage of high-energy carriers through the oxide generates trap sites that eventually enable current to leak off of the floating gate (Section 4.1.3). The stress-induced leakage current (SILC) responsible for the failure of a floating-gate device will occur only after a certain amount of charge has passed through the oxide. This charge corresponds to a certain number of program-erase cycles. Modern double-poly EEPROM transistors can survive several hundred thousand program-erase cycles.

Floating-gate devices are also vulnerable to the accumulation of charges within the oxide. During programming and erasure, some of the electrons that pass through the oxide surrounding the floating gate can become trapped. These trapped electrons represent a gradually increasing negative charge that cannot be removed by any means short of a high-temperature bake. Properly fabricated dry oxides experience relatively low levels of electron trapping. Mobile ionic contaminants can also interfere with the proper operation of floating-gate devices by causing gradual shifts in the apparent threshold voltages of the devices. Proper manufacturing precautions and the use of impermeable nitride or oxynitride overcoats can largely eliminate problems caused by mobile ionic contaminants.

11.3.2. Single-Poly EEPROM Memory

Many analog applications require the integration of a limited amount of nonvolatile memory onto a standard CMOS or BiCMOS process. A special version of EEPROM has been developed for this purpose. This type of EEPROM uses only a single layer of polysilicon, and it requires neither a tunnel oxide nor a drain extension. Single-poly EEPROM requires much more area than conventional double-poly EEPROM, limiting its application to memories that contain at most a few hundred bits. This amount of memory more than suffices for most analog applications.

Figure 11.28A shows the layout of the critical components required to construct one bit (or *cell*) of a single-poly EEPROM memory. This device incorporates a tunnel capacitor C_T, a control capacitor C_C, and a sense transistor M_S. All three devices share a common floating gate. The tunnel capacitor and the control capacitor are both laid out as PMOS transistors, each contained within its own N-well. The tunnel capacitor is made as small as possible by minimizing the poly overlap of its PMoat. The control capacitor, on the other hand, is deliberately enlarged. The sense transistor resembles a normal NMOS transistor, except that its gate connects only to the control and tunnel capacitors. Figure 11.28B shows the equivalent schematic diagram of this circuit. The control capacitor C_C connects to the control electrode C,

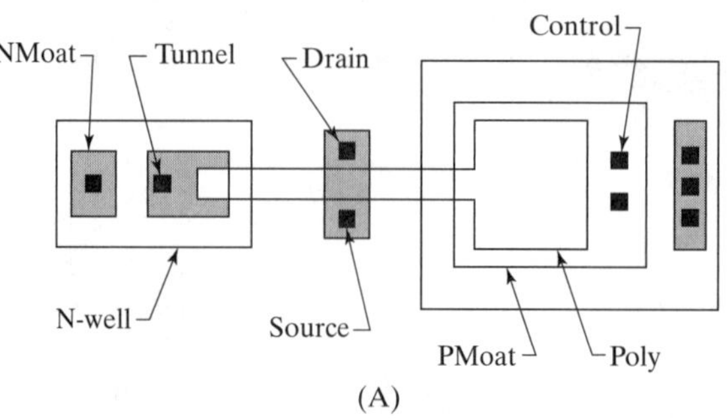

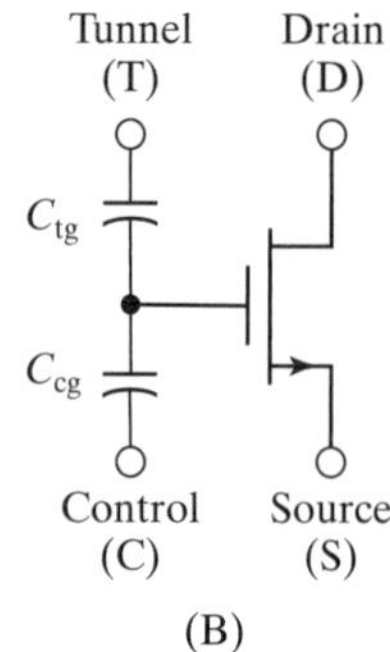

FIGURE 11.28 (A) Layout and (B) equivalent schematic of a typical single-poly EEPROM cell.[35]

[35] Ibid, p. 27.

and the tunneling capacitor C_T connects to the tunneling electrode T. The ratio of C_C to C_T typically equals at least 20.

The single-poly EEPROM cell can be programmed by placing a high voltage on the tunneling electrode while grounding the control electrode (Figure 11.29A). The source and drain of the sense transistor can be either grounded or left floating. Because the control capacitance greatly exceeds the tunneling capacitance, most of the voltage differential between the tunneling electrode and the control electrode appears across the dielectric of the tunneling capacitance. Electrons tunnel from the floating gate to the tunneling electrode. Programming is complete as soon as this tunneling current subsides.

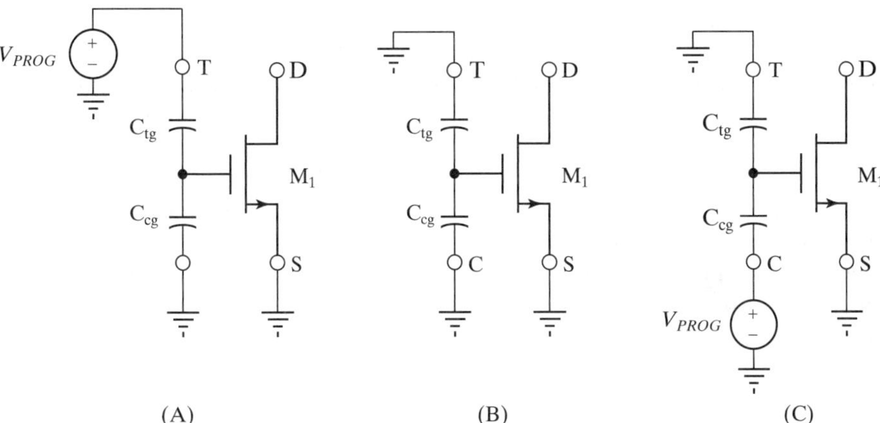

FIGURE 11.29 Connections required to (A) program, (B) read, and (C) erase the single-poly EEPROM cell.

The single-poly EEPROM cell can be read by grounding both the tunneling electrode and the control electrode. The charge injected across the tunnel oxide during programming leaves a positive charge on the floating gate, and this charge induces a channel within the sense transistor (Figure 11.29B). The sense transistor therefore conducts an electrical current. The single-poly EEPROM cell thus behaves as a normally open switch that closes during programming.

The single-poly EEPROM cell can be erased by placing a high voltage on the control electrode while grounding the tunneling electrode (Figure 11.29C). The source and drain of the sense transistor can again be either grounded or left floating. The voltage differential across the tunneling capacitor's dielectric causes electrons to tunnel to the floating gate. This process removes the positive charge that programmed the transistor and replaces it with a negative charge. When the cell is again read, this negative charge suppresses the formation of a channel in the sense transistor. The erased single-poly EEPROM cell thus again behaves as a normally open switch.

Many variations on the single-poly EEPROM cell have been developed. Some eliminate the tunnel capacitor and instead use the sense transistor as the tunneling element.[36] This variation on the structure saves space within the cell, but complicates the switching network required to program, erase, and read the device.

All versions of EEPROM require extensive testing in order to assure their reliability over the desired number of program/erase cycles. Many analog applications only program the EEPROM once, as part of a trimming operation conducted during manufacture. Such applications require a very low number of program–erase cycles. Other applications allow the end user to reprogram the EEPROM. These applications may require as many as several hundred thousand program–erase cycles. The

[36] E. Carman, P. Parris, H. Chaffai, F. Cotdeloup, S. Debortoli, E. Hemon, J. Lin-Kwang, O. Perat, and T. Sicard, "Single Poly EEPROM for Smart Power IC's," *ISPSD*, 2000, pp. 177–179.

endurance of any given EEPROM cell depends upon many factors. Higher operating temperatures degrade the endurance of the device by enhancing oxide leakage. Poorly controlled programming voltages can result in either increased charge injection, which leads to premature oxide wearout, or insufficient charge stored on the floating gate, which magnifies the impact of oxide leakage. Layout also plays a critical role in determining the endurance of the cell. EEPROM cells designed for a low number of program–erase cycles can use extremely large ratios of control capacitance to tunnel capacitance to maximize the charge stored on the floating gate. The larger charge requires more time to leak away, thereby enhancing the endurance of the cell. On the other hand, cells designed for high numbers of program–erase cycles will favor lower ratios of control capacitance to tunnel capacitance to minimize the amount of charge injected across the tunnel oxide during each program–erase cycle. This precaution allows the device to undergo more program–erase cycles before the charge injected across the oxide reaches the value needed to generate a stress-induced leakage current.

The process engineers will generally develop the layout for the critical components of the EEPROM cell for a specific process. They will then subject this layout to extensive reliability testing to ensure that the cell will remain programmed for the required lifetime (often specified as 10 years) after having undergone the maximum specified number of program–erase cycles. In order to assure that the EEPROM cell functions as specified, one must duplicate the original layout exactly—even if this means accepting less-than-optimal sizes, spacings, or aspect ratios.

11.4 THE JFET TRANSISTOR

Junction field-effect transistors (JFETs) were used throughout the 1970s and the early 1980s as substitutes for the less-reliable MOS devices of that era. JFETs were often used in the input stages of operational amplifiers to obtain input leakage currents several orders of magnitude smaller than those generated by the best bipolar circuits.[37] JFETs were also used as analog switches and as current sources.

Standard bipolar easily accommodated the steps required to construct simple JFET structures. The resulting *BiFET* processes merged bipolar and JFET transistors in much the same way that modern BiCMOS processes merge bipolar and CMOS. These BiFET processes were primarily used to construct low-input-current and low-noise operational amplifiers. The older BiFET processes have become largely obsolete because modern BiCMOS processes generally offer better performance. (Although low-noise BiMOS amplifiers can still outperform their BiCMOS counterparts.)

JFET transistors remain of interest because they can be constructed on many existing processes without requiring any additional masking steps, and the resulting devices can replace high-value resistors in startup circuits. The following sections provide a brief overview of the operation and construction of JFETs, with an emphasis on structures compatible with standard bipolar and analog BiCMOS processes.

11.4.1. Modeling the JFET

Although the I-V characteristics of the JFET broadly resemble those of the depletion-mode MOS transistor, the underlying physics of the two devices are quite different. Most textbooks derive the JFET equations from fundamental principles, but so

[37] The advantages of JFET input stages vanish at higher temperatures because the input current of a JFET increases exponentially with temperature, while the input current of a base-current-compensated bipolar circuit increases somewhat more slowly.

many assumptions are made along the way that the results have little practical value. This section discusses only those aspects of the theoretical model required to understand the sizing of JFET transistors and leaves the remaining details to other texts.[38]

The *pinchoff voltage* V_P of a JFET equals the minimum drain-to-source voltage V_{DS} required to pinch off the drain end of the channel when the gate-to-source voltage V_{GS} equals zero. In theory, the pinchoff voltage of an ideal JFET equals

$$V_P \cong 1.9 \cdot 10^{-16} N_C t^2 \qquad [11.18]$$

where N_C equals the doping concentration of the channel in atoms/cm^3 and t equals the channel thickness in microns.[39] In practice, the channel doping usually varies with depth, and the pinchoff voltage must be determined empirically. This is done by examining the I-V characteristics of the JFET for $V_{GS} = 0$. The drain current I_D remains approximately constant at high drain-to-source voltages. As V_{DS} decreases, a point is eventually reached at which the drain current begins to diminish (Figure 11.30B). The pinchoff voltage equals the drain-to-source voltage at this inflection point.

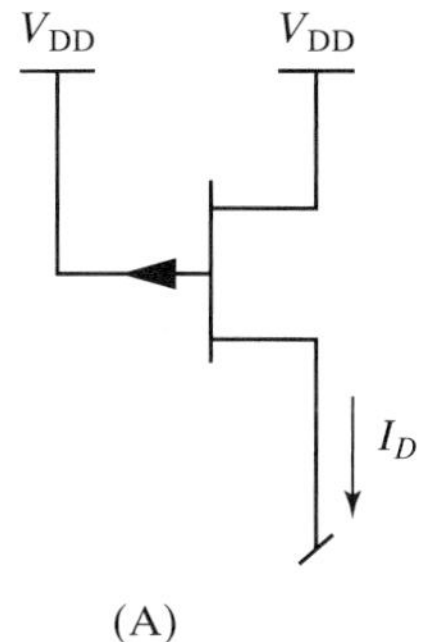

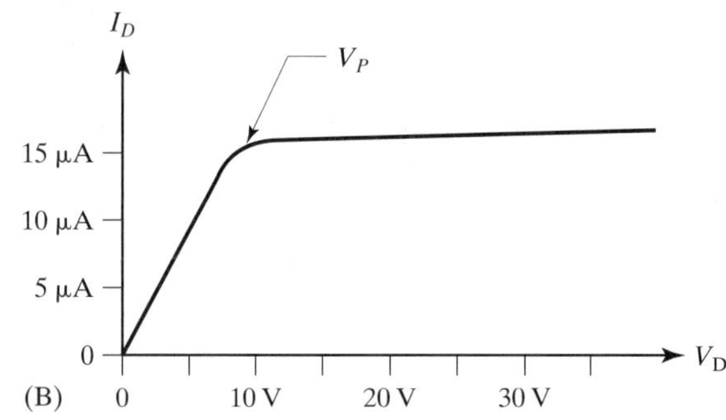

FIGURE 11.30 (A) A schematic of a P-JFET connected as a current source and (B) an I-V plot of the same device showing $V_P = 8$ V and $I_{DSS} = 16\ \mu$A.

The *saturation current* I_{DSS} of a JFET equals the drain current at $V_{GS} = 0$ and $V_{DS} = V_P$. If one assumes a uniformly doped channel with resistivity ρ, width W, length L, and thickness t, then the saturation current equals

$$I_{DSS} = \frac{V_P t}{3\rho}\left(\frac{W}{L}\right) \qquad [11.19]$$

This equation can also be used for nonuniformly doped channels, providing that one empirically determines the effective channel resistivity ρ by measuring devices of different widths and lengths and fitting these measurements to the equation. Several factors complicate the extraction of I_{DSS}. The width and length used in the equation do not exactly correspond to the drawn dimensions of the device, any more than the effective width and length of MOS transistors exactly correspond to their drawn dimensions (Section 11.2.1). Correction factors δW and δL relate the effective width W_{eff} and effective length L_{eff} to the drawn width W_{d}

---

[38] R. S. Muller and T. I. Kamins, *Device Electronics for Integrated Circuits,* 2d ed. (New York: John Wiley & Sons, 1986), p. 202ff.

[39] The full equation is $V_P = qN_C t^2/2\varepsilon$, where q is the charge on the electron and ε is the permittivity of silicon.

and drawn length L_d:

$$W_{\text{eff}} = W_d + \delta W \quad [11.20]$$

$$L_{\text{eff}} = L_d + \delta L \quad [11.21]$$

For devices with channel widths of less than 10 μm, the value of I_{DSS} can be accurately determined only by measuring a device having the desired channel width. Devices that channel lengths of less than 10 μm are better avoided because a variety of short-channel effects complicate the task of sizing the transistors.

11.4.2. JFET Layout

Practical JFET devices can be created using existing layers of a standard bipolar or an analog BiCMOS process. Figure 11.31 shows one type of N-channel JFET compatible with standard bipolar processing. This device is sometimes called an *epi-FET* because its channel consists of a portion of the N-type epitaxial layer. The epi-FET is also called an *epi pinch resistor,* particularly when it operates in its linear region (Section 5.5.5). The thickness of the channel has been greatly reduced by placing a base diffusion over the epi. The updiffusion of the underlying substrate causes substantial grading of the backgate-body junction and renders the constant-doping approximations that underlie Equations 11.18 and 11.19 of questionable validity. These devices are usually sized by interpolating between the I_{DSS} currents measured on an array of test devices.

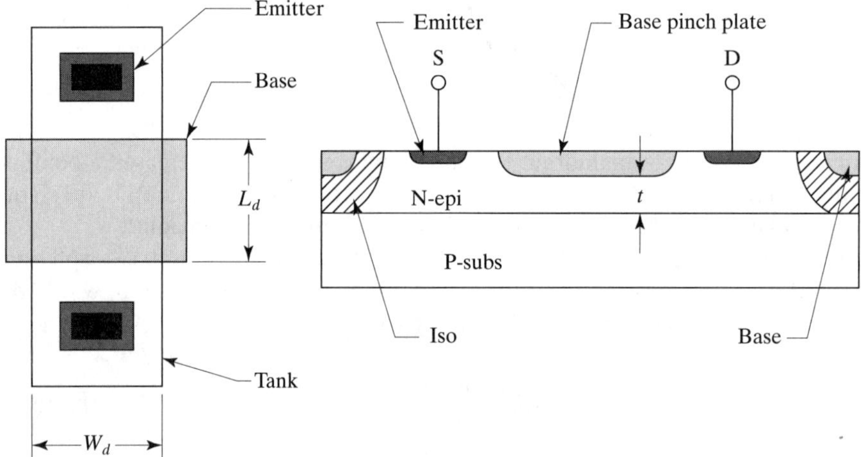

FIGURE 11.31 Layout and cross section of a N-channel JFET constructed in standard bipolar. The gate connects to the substrate and is accessed through an adjacent substrate contact.[40]

The tank geometry determines the drawn width W_d, while the base pinch plate determines the drawn length L_d. The effective width of an epi-FET is substantially smaller than its drawn width because of isolation outdiffusion. The relationship between effective width and drawn width becomes nonlinear for small widths because of diffusion interactions between the opposing sidewalls of the channel. The width correction factor δW may therefore vary with width, especially for small widths. The length correction factor δL also varies with length because of the presence of the extrinsic source/drain regions on either end of the channel, but δL has little effect upon devices having channel lengths of at least 50 μm.

---

[40] A similar device is discussed in D. J. Hamilton and W. G. Howard, *Basic Integrated Circuit Engineering* (New York: McGraw-Hill, 1975), p. 170.

The base pinch plate extends into the surrounding isolation and shorts the gate of the epi-FET to substrate. Most epi-FETs are used as startup devices in which the grounded-gate configuration is quite acceptable. If the drain-to-source voltage across the epi-FET is large enough, then its drain will draw a current equal to the saturation current I_{DSS}. In practice, most epi-FETs have such large pinchoff voltages that they do not fully saturate under normal operating conditions, and they therefore resemble pinch resistors. The main advantages of the epi-FET include high breakdown voltage and low transconductance, which together allow it to replace a much larger pinch resistor. The operating voltage of an epi-FET is limited only by the breakdown of the epi-base junction. JFETs are largely immune to hot-carrier-induced threshold shifts because they do not contain a gate dielectric. They are also immune to the parasitic channel formation and conductivity modulation because the base pinch plate serves as a field plate covering the active region of the device.

Epi-FETs are designed for compactness rather than for precision. These transistors normally use the minimum channel width, even though wider devices exhibit less variability. The channel is frequently serpentined to fit into unused areas in the layout. Contacts are usually placed over the base pinch plate and connected to substrate potential. Although not strictly necessary, these contacts help minimize variations in epi-FET current caused by substrate debiasing. Any contact to the base pinch plate also serves as a substrate contact in its own right and helps extract stray substrate currents flowing near the epi-FET. Some designers use rounded bends in serpentine epi-FETs, believing that these increase the breakdown voltage by preventing electric field intensification. Although this practice causes no harm, it provides little or no benefit because the exposed edge of the base pinch plate usually breaks down before the isolation sidewalls.

Analog BiCMOS processes can construct a *N-well JFET* analogous to the epi-FET in Figure 11.31 by substituting N-well for the tank and NMoat for emitter (Figure 11.32). The resulting device usually has a lower pinchoff voltage than its epi-FET counterpart due to the graded nature of the well. The pinchoff voltage can be reduced still further by growing field oxide over the base pinch plate, as the resulting oxidation-enhanced diffusion drives the base deeper into the N-well.

N-well JFETs and epi-FETs vary in several ways. One critical difference concerns the overlap of the base pinch plate over the channel. In the epi-FET, the isolation diffuses inward and the base pinch plate need overlap the tank only slightly, if at all. The base pinch plate of the N-well JFET must overlap the well by a much greater distance because the N-well diffuses outward rather than inward. The high

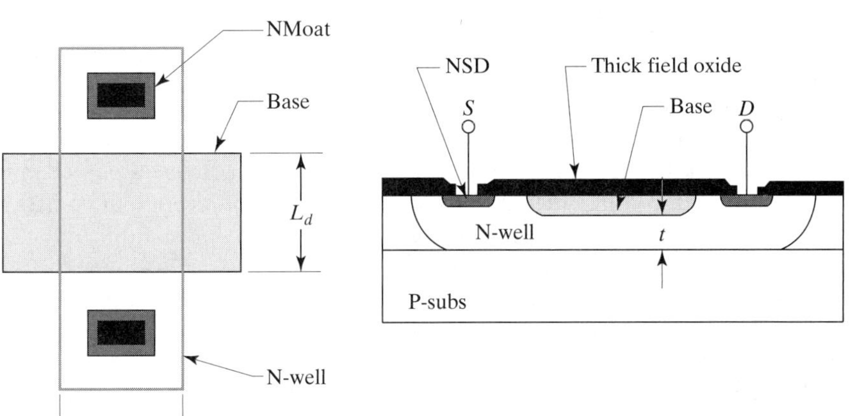

FIGURE 11.32 Layout and cross section of a N-well JFET constructed in an analog BiCMOS process. (The contacts to the base plate have been omitted for clarity.)

resistance of the P-epi also makes it desirable to add contacts directly to the base pinch plate rather than to rely on the presence of substrate contacts elsewhere on the die. These contacts should not reside over the channel of the N-well JFET because the moat region required for the contact alters the thickness of the channel. Instead, the contacts should be located next to the device and should connect to it by a strip of base or PSD diffusion.

A N-well JFET with a minimum-width channel has a much lower pinchoff voltage than one with a wider channel. If the channel is covered by thick-field oxide, the narrowest devices may even pinch off entirely and become unusable. These effects occur because the dopant in a narrow N-well diffuses laterally as well as vertically, leaving a lower overall doping concentration within the channel. The dopant in a wider well diffuses laterally near the edges, but the center of the well still retains a high dopant concentration. The wider device thus has a higher pinchoff voltage than the narrow one. If necessary, the pinchoff voltage of a N-well JFET can be increased by placing a moat region above the base or by substituting a PSD implant for the base implant. The PSD implant usually gives so high a pinchoff voltage that the device cannot saturate under normal operating conditions. Thus, it behaves as a nonlinear pinch resistor rather than as a true FET (Section 5.5.9).

P-channel JFETs can be constructed in both standard bipolar and analog BiCMOS processes, but those constructed from existing diffusions leave much to be desired. The standard bipolar device has the same structure as a base pinch resistor (Figure 3.15). The analog BiCMOS device has a similar structure, consisting of base pinched by NSD rather than by emitter. The operating voltages of these devices are limited by the avalanche of the base-emitter and base-NSD junctions, respectively. The pinchoff voltages of both devices greatly exceed their respective breakdown voltages, so neither device ever saturates. Both of these devices are really nothing more than nonlinear pinch resistors (Section 5.5.3). A special N-type implant must be added to the process to construct a true P-JFET capable of operating in saturation. This implant must have a slightly shallower junction depth than the base diffusion, and a doping concentration just sufficient to invert the base diffusion. A shallower diffusion yields too large a pinchoff voltage, and a more heavily doped one produces too low a breakdown voltage. No suitable diffusion exists in either standard bipolar or analog BiCMOS, although one can be added as a process extension. Previous-generation BiCMOS processes were generally derived from standard bipolar by the addition of just such an extension. The P-channel transistors constructed in this way are called *double-diffused JFETs* because their gates are produced by the diffusion of the N-implant into the base. The layout and cross section of the double-diffused P-JFET are essentially the same as that of the base pinch resistor in Figure 3.15, with the substitution of the new N-implant for the emitter. New processes rarely support the P-JFET extension because CMOS transistors have largely supplanted JFETs.

All of the layouts previously discussed short the gate to the backgate, which in the N-channel device consists of the substrate. In order to use the N-channel JFET in any application other than as a grounded current source, one must first separate the gate and the backgate electrodes by using an annular structure similar to that in Figure 11.33. The gate of the annular N-JFET consists of a ring-shaped P-type diffusion placed inside an N-epi tank. Tank contacts placed inside and outside this ring serve as the drain and source, respectively. This arrangement minimizes the drain capacitance at the cost of increased source capacitance.

The schematic symbol used for the annular N-JFET is exactly the same as that used for the conventional N-JFET (Figure 1.29A). The two can be differentiated by examining the connection of the gate electrode. The conventional layout in Figure 11.32

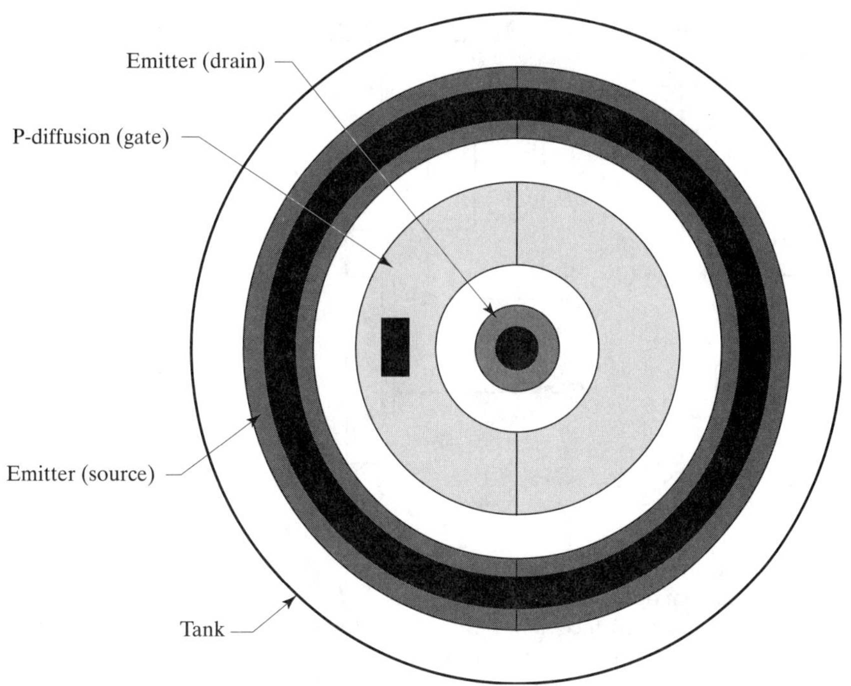

FIGURE 11.33 Annular circularly symmetric N-channel JFET.

should be used if the gate connects to substrate potential. If the gate connects to any other potential, the transistor must use the annular layout shown in Figure 11.33. The substrate forms the backgate of the annular device. The width and length of annular JFET devices are computed using the rules presented for annular MOS transistors (Section 11.2.6).

11.5 SUMMARY

This chapter has covered the construction of conventional small-signal poly-gate CMOS transistors, floating-gate devices, and JFETs. The next chapter covers a variety of more specialized types of transistors, including extended-voltage transistors, power transistors, and DMOS transistors. These transistors can fill a very wide range of applications, including many that are traditionally filled by bipolar transistors.

11.6 EXERCISES

Refer to Appendix C for layout rules and process specifications.

11.1. Suppose an enhancement NMOS has a threshold voltage of 0.7 V and a transconductance of 220 μA/V^2. Determine the region of operation and compute the drain current for each of the following biasing conditions:
 a. $V_{GS} = 1.2$ V, $V_{DS} = 2.3$ V.
 b. $V_{GS} = 1.2$ V, $V_{DS} = 0.2$ V.
 c. $V_{GS} = -1.0$ V, $V_{DS} = 4.4$ V.

11.2. Suppose the enhancement NMOS in Exercise 11.1 is subjected to the following terminal voltages: $V_{GS} = 1.2$ V, $V_{DS} = -2.3$ V. Recognizing that the source and drain have swapped roles, determine the true electrical biasing conditions, the mode of operation, and the drain terminal current.

11.3. What is the process transconductance of an NMOS transistor having a composite gate dielectric consisting of 150 Å of nitride ($\varepsilon_r = 6.8$) sandwiched between two layers of oxide, each 50 Å thick ($\varepsilon_r = 3.9$)? *Hint:* See Section 6.1.

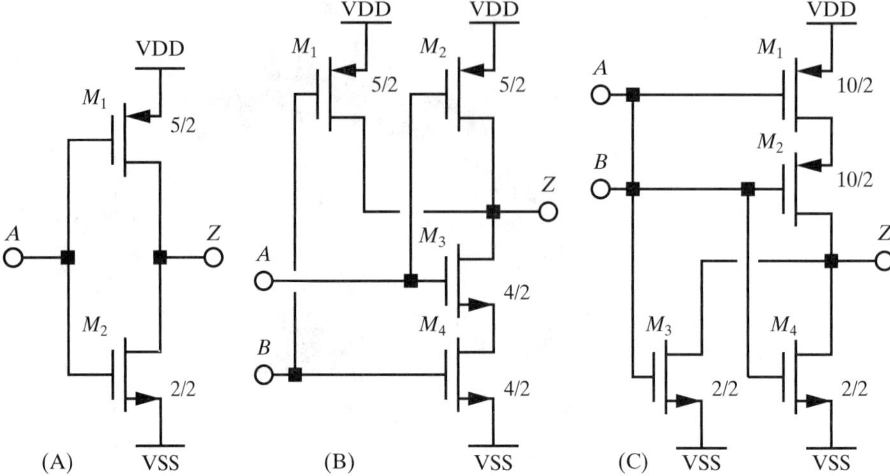

FIGURE 11.34 Three standard-cell logic gates: (A) Inverter, (B) NAND, and (C) NOR.

11.4. Estimate the process transconductances of NMOS and PMOS transistors having a maximum operating voltage of 15 V.

11.5. Suppose an enhancement PMOS transistor with an N+ poly gate electrode has a nominal threshold voltage of −0.95 V. What would the nominal threshold voltage become if the transistor used a P+ poly gate?

11.6. Would an enhancement PMOS transistor with a nominal threshold voltage of −0.4 V serve as a useful device for constructing digital logic gates? Explain.

11.7. Lay out the inverter shown in Figure 11.34A, using the poly-gate CMOS rules listed in Appendix C. Place a hole-collecting guard ring around the PMOS transistor and an electron-collecting guard ring around the NMOS. Connect the guard rings to provide the best possible protection.

11.8. Construct an isolated NMOS transistor, using the analog BiCMOS rules in Appendix C. The transistor should have a W/L ratio of 10/5. Include a single abutting substrate contact.

11.9. A 3 μm CMOS process can withstand a maximum operating voltage of 12 V and offers a gate delay of 2.3 nS. Suppose constant-field scaling is applied to produce a 2 μm process. Predict the maximum operating voltage and gate delay of the scaled process.

11.10. Lay out the following transistors, using poly-gate CMOS rules. Include abutting backgate contacts, gate interconnections, and well geometries as necessary.

a. NMOS, 3(5/15)
b. NMOS, 12(20/5)
c. PMOS, 7(10/25)
d. PMOS, 4(10/3)

11.11. Lay out a natural NMOS transistor with dimensions of 3(10/3) and a natural PMOS transistor with dimensions of 25/25. Include abutting backgate contacts, gate interconnections, and well geometries as necessary. The layout rules for the NATVT layer are as follows:

1. NATVT width	4 μm
2. NATVT overlap of GATE	2 μm
3. NATVT spacing to TOX	4 μm
4. NATVT spacing to POLY	4 μm

Note: GATE is defined as the intersection of POLY with either NMOAT or PMOAT.

11.12. Construct standard-cell layouts of each of the three logic gates in Figure 11.34. The NAND gate should resemble the layout in Figure 11.19B. The VDD and VSS leads should be 4 μm wide, and each cell should have at least one substrate contact and one well contact. Design the cells so they can be stacked together by abutting their

VDD and VSS leads. No violations of layout rules should occur regardless of the order in which the cells are stacked.

11.13. Using the poly-gate CMOS rules of Appendix C, construct a serpentine PMOS with a nominal device transconductance of 0.01 μA/V^2. Fold the gate as many times as necessary to produce an approximately square layout.

11.14. Lay out the following annular transistors, using poly-gate CMOS rules. Include abutting backgate contacts, gate interconnections, and well geometries as necessary.

a. Circular NMOS, 31.4/4

b. Square PMOS, 48/4

c. Elongated circular PMOS, 51.4/4

11.15. Construct a 5000/2 PMOS transistor, using poly-gate CMOS rules. Divide the transistor into as many sections as required to produce an approximately square layout. Include enough interdigitated backgate contacts to ensure that no part of the transistor is more than 50 μm from the nearest backgate contact.

11.16. The single-poly EEPROM of Figure 11.28 can also be erased by hot carriers injected from the channel of the sense transistor. Describe the terminal biasing required to perform this operation.

11.17. Construct a single-poly EEPROM layout similar to that of Figure 11.28, using the poly-gate CMOS rules of Appendix C. Make the tunneling capacitor and sense transistor as small as possible. Lay out the control capacitor so that its gate capacitance is 20 times greater than the gate capacitance of the tunneling capacitor. Ignore fringing capacitance.

11.18. Layout a standard bipolar epi-FET having drawn dimensions of 30/8. Assume that the base pinch plate must extend at least 2 μm into the isolation.

11.19. Construct a minimum-size circularly symmetric epi-FET. Include all necessary metallization. What are the drawn width and length of this device?

12 Applications of MOS Transistors

Some designs require MOS transistors to operate at high voltages without experiencing breakdown or parametric shifts. Although breakdown voltages can be increased by using lower dopant concentrations and thicker gate oxides, hot-carrier generation remains a concern. A variety of specialized transistor structures has been developed to minimize hot-carrier generation at high voltages. This chapter examines several structures that are compatible with ordinary CMOS processing, as well as several that require process extensions.

MOS power transistors are particularly useful for high-current, low-resistance switching. Conventional CMOS transistors can conduct large currents, but have operating voltage limitations. Many processes offer *double-diffused MOS,* or DMOS, transistors as a process extension. These devices allow the construction of compact, high-voltage, high-current transistors. With the development of lateral DMOS, and more recently, with advances in RESURF technology, DMOS transistors have become the power devices of choice for 15 V and beyond.

Finally, this chapter also discusses the matching of MOS transistors. MOS transistors used in analog circuits frequently require a high degree of matching of both transconductances and threshold voltages. The techniques used to match MOS transistors differ quite markedly from those used for bipolar transistors.

12.1 EXTENDED-VOLTAGE TRANSISTORS

Early MOS processes produced relatively long-channel transistors with lightly doped backgates. Avalanche breakdown of the source/drain regions limited these transistors to operating voltages of 10 to 15 V. Processing improvements have enabled channel lengths to decrease from about 8 μm to less than 0.3 μm. If backgate doping and operating voltages remained constant, the pinched-off portion of the channel would represent an ever-increasing percentage of the channel length. The resulting transistors would exhibit increased channel length modulation and premature punchthrough breakdown.

Channel length modulation can be minimized and punchthrough averted either by reducing the operating voltage or by increasing the backgate doping. Since the operating voltages of analog circuits are not easily reduced below 2–3 V, most processes have opted to increase backgate doping. This minimizes the width of the pinched-off region at the cost of intensifying the lateral electric field across it. Intense electric fields generate hot carriers that in turn produce undesirable long-term parametric drifts (Section 4.3.1). Holes are more difficult to accelerate than electrons, so PMOS transistors are less prone to hot-carrier generation than are NMOS transistors.

The *operating voltage* of a MOS transistor equals the maximum drain-to-source voltage allowed during saturation, while the *blocking voltage* equals the maximum drain-to-source voltage allowed during cutoff. Since the pinched-off region disappears in cutoff, hot-carrier generation ceases and the blocking voltage is limited only by drain-backgate avalanche and gate oxide rupture. Hot-carrier generation may restrict the operating voltage to a lower value than the blocking voltage. Higher voltages do not necessarily cause instant failure, but they produce gradual shifts in both threshold voltage and device transconductance.[1] Analog circuits are particularly sensitive to hot carrier degradation because they contain matched devices that continuously operate in saturation. MOS transistors in digital circuits enter saturation only during brief switching transients. Thus, not only are digital circuits more robust due to an absence of matched devices, but they also experience lower rates of deterioration, especially at lower clock speeds.

Hot carriers become an increasingly serious problem as channel lengths diminish. Specialized transistor structures offer extended operating voltage ranges at the cost of additional process complexity and larger spacings. Virtually all submicron CMOS processes use some form of these *extended-voltage transistors.*

12.1.1. LDD and DDD Transistors

All extended-voltage transistors incorporate some form of specialized drain structure that absorbs a portion of the electric field that would otherwise appear across their channel. Figure 12.1A shows a plot of the lateral electric field intensity across the drain end of a saturated MOS transistor. This example assumes constant backgate and drain doping concentrations and abrupt junctions. The field intensity rises linearly across the pinched-off region and reaches a maximum at the drain metallurgical junction. The field then drops linearly across the depletion region inside the drain. The widths of the pinched-off region x_p and the drain depletion region x_d are proportional to the inverse square root of the doping of the respective regions. The voltages sustained across the pinched-off region V_p and across the drain depletion region V_d equal the areas of the respective triangles (Figure 12.1A).

The total drain-to-source voltage V_{DS} equals the sum of the areas of both triangles:

$$V_{\mathrm{DS}} = \frac{E_{\mathrm{max}}}{2}(x_p + x_d) \qquad [12.1]$$

The maximum electric field intensity E_{max} and the width of the pinched-off region x_p are limited by hot-carrier generation and channel length modulation, respectively. No such hard-and-fast limit exists on the width of the drain depletion region x_d. The only way to increase the operating voltage V_{DS} is to increase the width of the depletion region x_d. This in turn requires a more lightly doped drain.

---

1 C. Duvvury and S. Aur, "Hot-Carrier Degradation Effects in CMOS Technologies," *TI Technical Journal,* Vol. 8, #1, 1991, pp. 56–66.

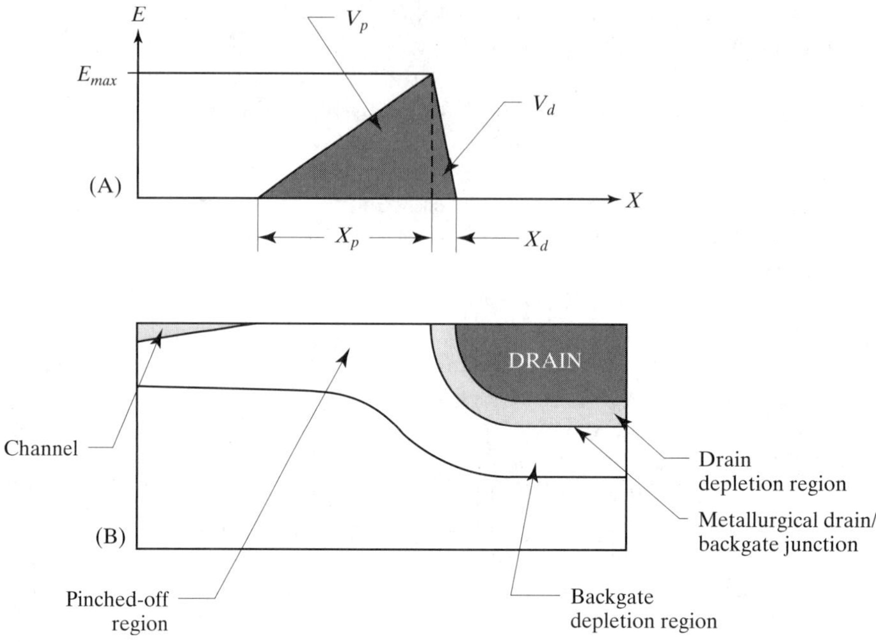

FIGURE 12.1 Diagram showing the lateral electric field across the drain end of a saturated MOS transistor.

In order to provide significant benefit, the width of the drain depletion region x_d must equal a significant fraction of the width of the pinched-off region x_p. This dictates that the drain doping not greatly exceed the backgate doping. Unfortunately, a lightly doped drain is also a highly resistive drain. Most modern processes minimize drain resistance by using a composite structure consisting of a lightly doped fringe surrounding a heavily doped core. The fringe depletes at relatively low voltages, forming a *drift region* bounded by the metallurgical junction on one side and by the heavily doped core, or *extrinsic drain,* on the other. The width of the drift region sets the width of the drain depletion region x_d. The drift region should be made just wide enough to support the desired operating voltage, and no wider. Any additional width would increase drain resistance without providing any corresponding benefit. The optimal width of the drift region usually represents no more than a small fraction of the channel length.

Several device structures have been developed that control the drift region width through various forms of self-alignment. The drift region must self-align to both the extrinsic drain and to the poly gate in order to minimize overlap capacitance. Figure 12.2 shows two structures that fulfill these requirements. Both employ a special feature called an *oxide sidewall spacer* formed by isotropically depositing and anisotropically

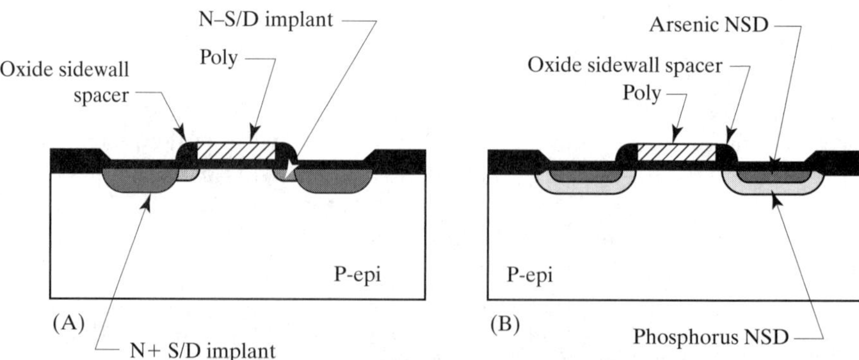

FIGURE 12.2 Extended-voltage transistors: (A) lightly doped drain and (B) double-diffused drain.

etching an oxide layer (Section 3.2.4). By its very nature, the oxide sidewall spacer self-aligns to the poly gate. The spacer is approximately as wide as the poly is thick. A limited range of spacer widths can be fabricated by adjusting poly thickness and etch conditions.

A *lightly doped drain,* or LDD, uses oxide sidewall spacers to define the width of the drift region (Figure 12.2A). The LDD structure requires two separate source/drain implants: one performed before spacer formation and one afterward. The first implant forms a lightly doped drift region aligned to the polysilicon gate, while the second forms a heavily doped extrinsic drain aligned to the oxide sidewall spacer.[2] This technique requires no additional masking steps, but it does require an oxide deposition, an oxide removal, and a second drain implant. The performance advantages gained from higher operating voltages offset the cost of the additional processing.

The process that forms the LDD structure does not discriminate between source and drain. The source resistance can be reduced if the LDD structure appears only on the drain end of the device. The resulting transistor is said to be *asymmetric* because its source and drain terminals are not interchangeable. Asymmetric LDD transistors require additional masking and processing steps to remove the sidewall spacer from the source end of the transistor. The benefits produced by this additional processing rarely justify the additional cost.

The *double-diffused drain,* or DDD, uses two implants driven through the same oxide opening to form a composite drain structure (Figure 12.2B).[3] These two implants require dopants of widely differing diffusivities, most commonly arsenic and phosphorus. A brief drive causes the phosphorus to diffuse outside the boundaries of the arsenic implant to form a lightly doped drift region. An oxide sidewall spacer minimizes overlap capacitances by preventing the drift region from diffusing underneath the poly gate.

The double-diffused drain is not readily applicable to PMOS transistors because of the absence of a slow-diffusing acceptor for silicon. NMOS transistors sometimes use the DDD structure in preference to the LDD structure because it can fabricate very narrow drift regions with great precision. The DDD structure also increases the avalanche voltage of the source/drain implants by grading their junctions. It is difficult to construct wide DDD drift regions because the drive required to force the phosphorus under a wide sidewall spacer also interferes with threshold voltage control. The LDD structure is therefore favored for wider drift regions and the DDD structure for narrower ones. The cost and complexity of the two techniques are comparable, as both require oxide sidewall spacers and both employ two drain implants.

The electric field intensity required to generate hot holes is two or three times larger than that required to generate hot electrons, so many applications that require LDD or DDD NMOS transistors can still use ordinary PMOS transistors. These PMOS devices require only a single source/drain implant, so they are called *single-diffused drain* (SDD) transistors. If the process includes an LDD or DDD NMOS, then the SDD PMOS also receives oxide sidewall spacers. The spacers can actually transform a buried-channel PMOS transistor into an LDD device. The contact potential of the doped poly gate inverts the portions of the buried channel underneath it. The portions of the buried channel protruding under the oxide sidewall spacers do not invert because they do not experience the full electric field generated by the gate

---

2 S. Ogura, P. J. Tsang, W. W. Walker, D. L. Critchlow, and J. F. Shepard, "Design and Characteristics of the Lightly Doped Drain-Source (LDD) Insulated Gate Field-Effect Transistor," *IEEE Trans. on Electron Devices,* Vol. ED-27, #8, 1980, pp. 1359–1367.

3 E. Takeda, H. Kume, T. Toyabe, and S. Asai, "Submicrometer MOSFET Structure for Minimizing Hot-carrier Generation," *IEEE Trans. on Electron Devices,* Vol. ED-29, #4, 1982, pp. 611–618.

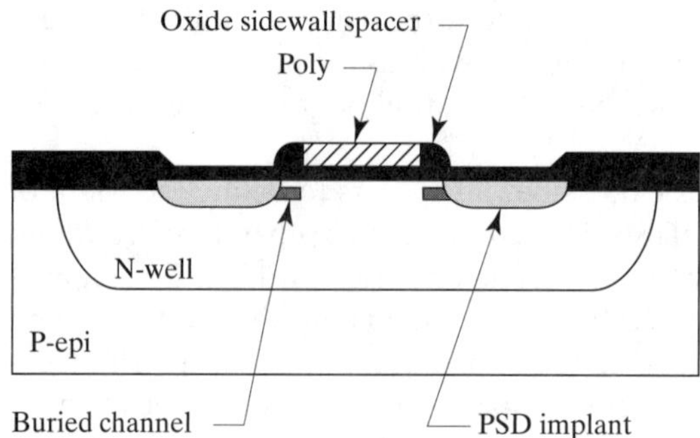

FIGURE 12.3 PMOS buried-channesl lightly doped drain (BCLDD).

electrode. The stubs of the buried channel therefore form two lightly doped P-type regions that act as lightly doped drains (Figure 12.3). This type of structure is sometimes called a *buried-channel lightly doped drain* (BCLDD).[4]

Modern submicron CMOS processes use extensions of lightly doped drain technology to help minimize the need for increased backgate doping. Instead of doping the entire backgate equally, these processes inject localized areas of heavier doping around the lightly doped drain (Figure 12.4). These regions are called *pocket implants* or *punchthrough stoppers*. Several methods exist for fabricating them. One technique relies upon an implant in which the ion beam strikes the surface of the wafer diagonally. As the wafer rotates, the diagonal beam of ions penetrates laterally beyond the edges of the lightly doped drain.[5] Pocket implants help minimize backgate modulation of the threshold voltage, which would otherwise become a serious problem in submicron devices.

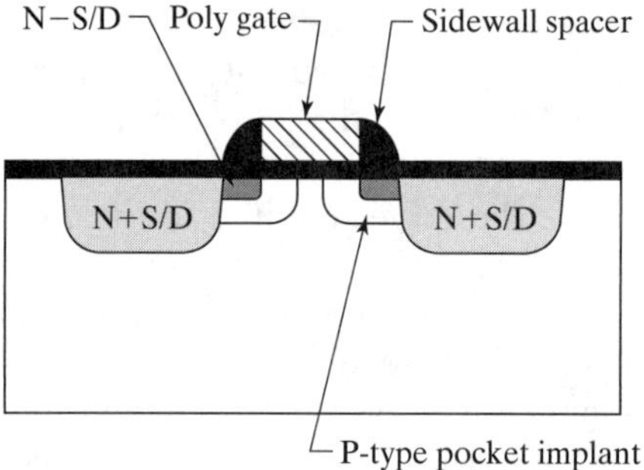

FIGURE 12.4 CMOS transistor with pocket implants formed by high-angle implantation.

12.1.2. Extended-Drain Transistors

Higher-voltage LDD and DDD transistors require thicker oxide sidewall spacers. The difficulties inherent in constructing wide spacers limit these structures to voltages of about 15 to 20 V. Much higher voltages can be achieved using non-self-aligned composite drains that do not employ sidewall spacers to delimit the drift regions. The simplest structure of this sort is called an *extended drain.* It consists of a shallow, heavily doped diffusion entirely contained within a deeper, more lightly

[4] R. H. Eklund, R. A. Haken, R. H. Havemann, and L. N. Hutter, "BiCMOS Process Technology," in *BiCMOS Technology and Applications,* 2d ed., A. R. Alvarez, ed. (Boston: Kluwer Academic Publishers, 1993), pp. 93–95.

[5] T. Hori, "A 0.1-µm CMOS Technology with Tilt-Implanted Punchthrough Stopper (TIPS), *International Electron Devices Meeting*, 1994, pp. 75–78.

doped one. The inner diffusion forms the extrinsic drain, while the fringes of the outer diffusion act as a drift region. For example, an NMOS extended drain might consist of NSD within N-well; the NSD implant then acts as the extrinsic drain, and the N-well as the drift region.

One can often create extended-drain transistors from existing diffusions. The resulting devices are usually larger than purpose-built devices such as lateral DMOS transistors (Section 12.2.3). On the other hand, the extended-drain devices do not require additional processing steps or masks. If an integrated circuit only requires a few small high-voltage transistors, then the most economical solution probably consists of extended-drain transistors constructed from existing masks. On the other hand, large, low-resistance devices are better constructed using purpose-built devices that can achieve lower specific on-resistances and overlap capacitances.

Extended-Drain NMOS Transistors

Figure 12.5A shows the cross section of a typical extended-drain NMOS transistor constructed in an N-well CMOS process. The extended drain consists of an NSD plug contained inside a larger N-well geometry. The N-well diffuses outward to produce a very lightly doped drain capable of withstanding high voltages. These voltages will rupture the thin gate oxide unless the drain incorporates a special *field-relief structure*. The structure on the illustrated transistor consists of a section of thick-field oxide placed just inside the metallurgical drain-backgate junction. As the depletion region intrudes farther into the drain, it passes underneath the gradually increasing thicknesses of oxide that comprise the bird's beak. The highest drain-to-source voltages appear across the thick-field oxide over the drift region. The field-relief structure has no effect on the transconductance or the threshold voltage of the transistor because these depend solely on the channel, all of which remains underneath the thin gate oxide. This type of field relief relies upon the gradual taper of the bird's beak. The more abrupt transition of shallow trench isolation causes electric field intensification that renders it unsuitable for use as a field relief. An alternative style of field relief leaves a gap between the gate and the drain. Some transistors combine both a drain gap and a field oxide structure. The dimensions of all field relief structures are critical to their proper operation, so one must pay close attention to the applicable layout rules and closely follow reference designs.

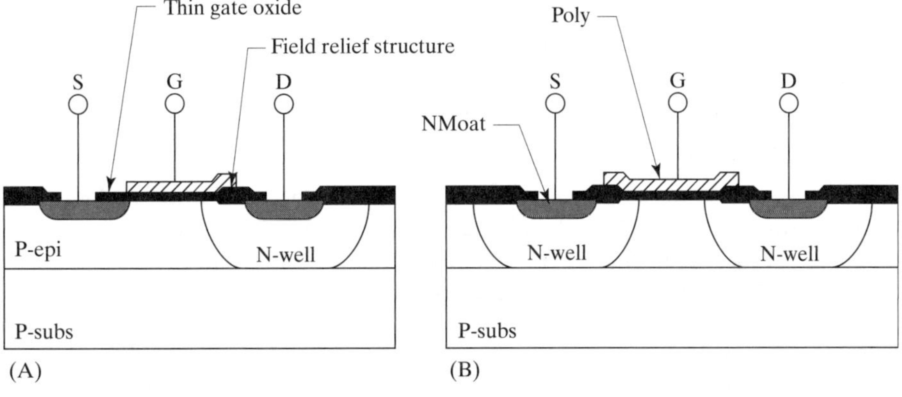

FIGURE 12.5 Cross sections of (A) asymmetric and (B) symmetric extended-drain NMOS transistors.

Figure 12.5A illustrates an *asymmetric extended-drain NMOS* transistor. Only one end of this transistor receives an extended-drain structure. This produces a relatively compact layout, but one that is not suitable for applications where either end of the transistor may see high voltages. The *symmetric extended-drain NMOS* in Figure 12.5B equips both ends of the transistor with extended drains. The symmetric transistor can withstand large drain-to-source voltages regardless of which end of the transistor acts as its drain. It cannot simultaneously withstand large voltage differentials between the gate and both source/drain regions, because one of these must serve as its source. Transistors that must withstand large gate-to-source voltages require thicker gate oxides (Section 12.1.3).

Figure 12.6 shows the layout of asymmetric and symmetric extended-drain NMOS transistors. In both cases, the drawn gate length, L_d, equals the distance across the moat beneath the gate. The minimum drawn gate lengths of these transistors are relatively large (typically 4 to 6 μm), but the effective gate lengths are much smaller because of outdiffusion of the wells. Asymmetric transistors with multiple gate fingers lend themselves to a compact layout in which source and drain fingers alternate; this allows efficient use of the N-well strips, which form the extended drains.

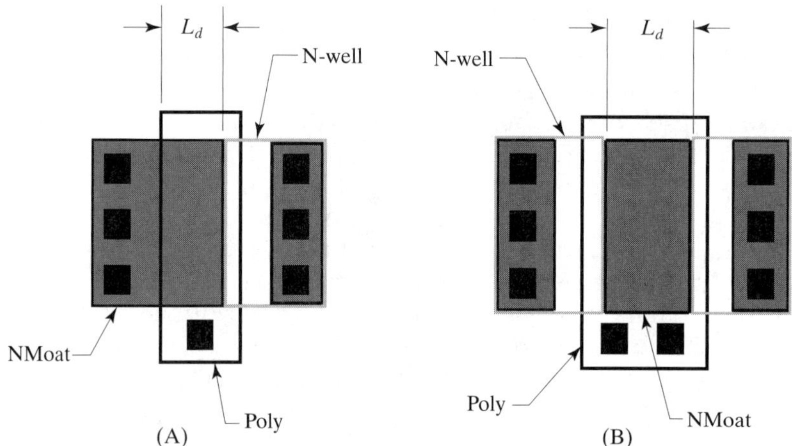

FIGURE 12.6 Layouts of (A) asymmetric and (B) symmetric extended-drain NMOS transistors.

The extremely light doping of the N-well suppresses hot electron generation, so the operating voltage ratings of properly designed extended-drain NMOS transistors are limited only by the avalanche voltage of the well-substrate junction and by the effectiveness of the field-relief structures. It is not particularly difficult to design extended-drain transistors that can withstand operating voltages two or three times larger than those of regular NMOS and PMOS transistors. Such voltages generally exceed the thick-field threshold of the process, requiring the addition of field plates or channel stops (Section 4.3.5). High-voltage extended-drain devices often suffer from electrical SOA limitations that limit allowed biasing conditions (Section 12.2.1).

Extended-Drain PMOS Transistors

Figure 12.7A shows the cross section of an *asymmetric extended-drain PMOS* transistor constructed in an N-well CMOS process. The drift region of the extended drain consists of a P-type channel-stop implant. The CMOS process flow presented in Section 3.2 uses a patterned phosphorus channel-stop implant to counterdope a blanket boron channel-stop implant. If the channel-stop mask is modified to block the phosphorus implant from the vicinity of the extended drain, then this region receives only the boron implant. The boron outdiffuses during the field oxidation to form a deep, lightly doped P-type diffusion suitable for use as a drift region. Figure 12.7B shows a *symmetric extended-drain PMOS* that uses the channel-stop implant to form drift regions for both source/drain terminations.

Extended-drain PMOS transistors use a special coding layer called *Chstop* to block the patterned phosphorus channel-stop implant. The geometries on the Chstop

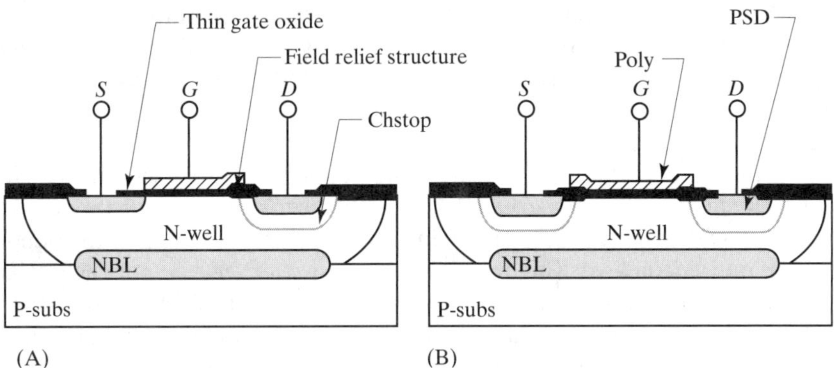

FIGURE 12.7 Cross sections of (A) asymmetric and (B) symmetric extended-drain PMOS transistors.

layer are added to the channel-stop mask during mask generation. Different processes may employ other coding techniques, but the principles remain broadly the same.

Figure 12.8 shows the layout of asymmetric and symmetric extended-drain PMOS transistors. In both cases, the drawn gate length equals the distance across the moat region beneath the gate. The transistors must contain NBL to stop vertical depletion from punching through the lightly doped bottom of the N-well and shorting the drain to the substrate.

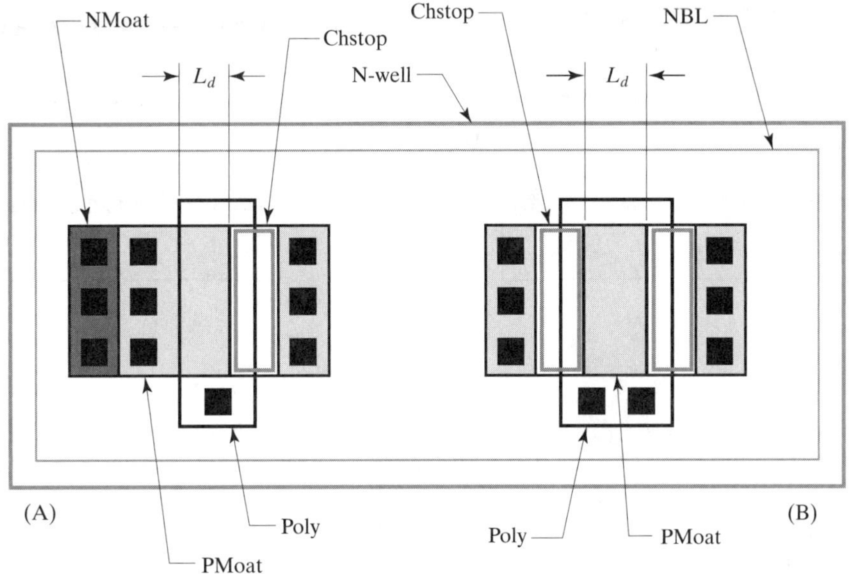

FIGURE 12.8 Layouts of (A) asymmetric and (B) symmetric extended-drain PMOS transistors.

12.1.3. Multiple Gate Oxides

The previous two sections have described transistors that can operate at large drain-to-source voltages. These transistors can also operate at large drain-to-gate voltages as long as they incorporate suitable field-relief structures. If they must operate simultaneously at large gate-to-source voltages, then no remedy exists but to increase the thickness of the gate oxide. Any increase in gate oxide thickness also reduces device transconductance. Circuit designers take an understandably dim view of increasing the gate oxide thickness of all transistors merely to accommodate increased voltages on a few. This conflict can be resolved by producing two separate thicknesses of gate oxide. The thinner gate oxide provides high transconductance for low-voltage applications, while the thicker gate oxide can withstand higher voltages. The circuit designer chooses which gate oxide each transistor receives on the basis of expected operating conditions.

Multiple gate oxides are fabricated using either staged oxidation or etch-and-regrowth techniques. *Staged oxidation* requires a separate polysilicon deposition for each gate oxide. The thinnest gate oxide is grown first, followed by the deposition of the first polysilicon layer (Figure 12.9A). Once patterned, the poly acts as an oxidation mask for a continuation of the gate oxidation (Figure 12.9B). After the gate oxidation is complete, the second poly layer is deposited and patterned (Figure 12.9C). Any transistor with a poly-1 gate receives the thin gate oxide, and any transistor with a poly-2 gate receives the thick gate oxide.

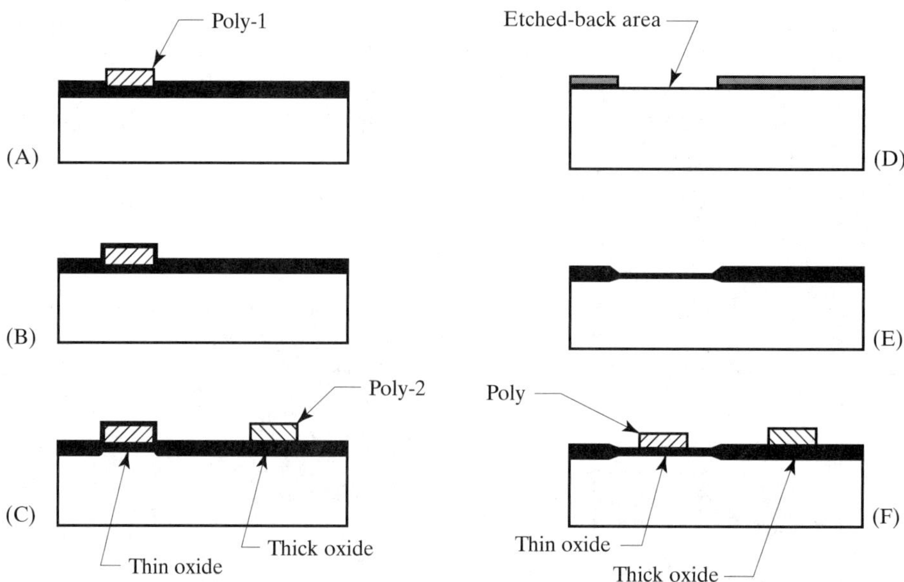

FIGURE 12.9 Process steps for growing multiple thicknesses of oxide using staged oxidation (A-B-C) and etch-and-regrowth (D-E-F) techniques.

Processes with only one layer of poly can use the *etch-and-regrowth* technique instead of staged oxidation. The etch-and-regrowth process requires an additional masking step instead of an additional poly deposition. The extra mask patterns a layer of photoresist spun on top of a thin-gate oxide. The exposed oxide regions are etched away (Figure 12.9D), after which the gate oxidation is resumed. A thin-gate oxide forms over the areas that were etched back, while a thicker layer of oxide forms over the areas where the initial oxide was left undisturbed (Figure 12.9E). A single layer of polysilicon can now form the gates of both thin-oxide and thick-oxide transistors (Figure 12.9F).

Some processes use staged oxidation, while others use etch-and-regrowth. Insofar as the layout designer is concerned, the main difference between the two lies in the number of polysilicon layers required. Figure 12.10 shows a comparison of layouts required by staged oxidation and etch-and-regrowth. Both processes use a *Moat-2* geometry coded around the gate region, but this geometry performs different functions for each process. In the staged oxidation process, Moat-2 defines the region receiving the threshold adjusts for the thick-oxide devices. In the etch-and-regrowth

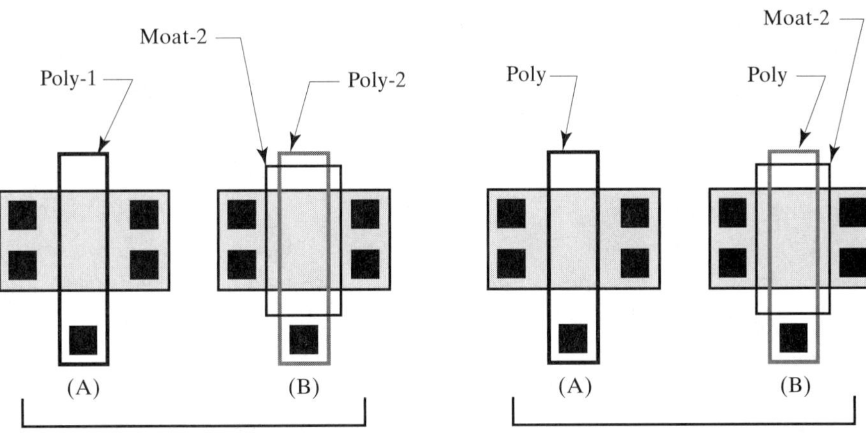

FIGURE 12.10 Comparison of layouts of (A) thin-oxide and (B) thick-oxide transistors constructed for a staged-oxidation process and an etch-and-regrowth process.

process, the Moat-2 geometry defines both the regions protected from etchback and the regions receiving the thick-oxide threshold adjusts.

12.2 POWER MOS TRANSISTORS

MOS transistors can switch or regulate large amounts of power. Devices specifically designed for such applications are called *power transistors* to distinguish them from low-power, or *small-signal*, devices. MOS power transistors have several advantages over their bipolar counterparts, including a lack of saturation delays, much simpler drive requirements, and lower forward voltages (particularly at low currents). Together these advantages make MOS transistors extremely useful power devices.

Bipolar transistors are poorly suited to high-speed switching applications because they saturate when their collector-base junctions forward-bias. Saturation greatly increases the amount of minority carrier charge stored in both the neutral base and the neutral collector. A transistor cannot turn off until these stored charges recombine or diffuse across a junction. A typical power bipolar transistor therefore exhibits a saturation delay of about a microsecond. This delay effectively places an upper limit on switching speeds of about 500 kHz. Clamping the transistor to prevent saturation enables faster operation, but only at the cost of greatly increased forward voltage drops. MOS transistors, on the other hand, are majority carrier devices. They do not exhibit any saturation delay. A power MOS transistor can therefore switch at speeds well in excess of 1 MHz.

Power MOS transistors also have an advantage over bipolar transistors in terms of the drive circuitry they require. The gate of a MOS transistor presents a purely capacitive load to its drive circuit. Although large currents flow during switching transitions, these currents quickly subside. The average current through the gate drive circuit of a typical one-amp power MOS transistor equals only a few milliamps. Bipolar transistors require constant base current. Since these transistors often have high-current betas as low as 10, a one-amp transistor may require as much as 100 mA of base drive. The simplest circuits provide full base drive to the transistor regardless of the collector current. Such circuits are obviously very inefficient at light loads. More complicated base drive circuits adjust the base drive depending upon the collector current, but even the best of these circuits cannot match the efficiency of MOS gate drive circuits.

MOS transistors can also conduct large currents at very low drain-to-source voltages. The behavior of a MOS transistor under these conditions can be derived from the Shichman-Hodges equation for the linear region. This equation is first rearranged in the form

$$I_D = k(V_{GS} - V_t)V_{DS} + \frac{kV_{DS}^2}{2} \qquad [12.2]$$

The quadratic term becomes negligible at low drain-to-source voltages ($V_{DS} \ll V_{GS} - V_t$). The preceding equation then simplifies to

$$I_D \cong k(V_{GS} - V_t)V_{DS} \qquad [12.3]$$

This equation reveals a linear relationship between the drain-to-source voltage V_{DS} and the drain current I_D. The transistor therefore behaves as if it were a resistor whose value $R_{DS(on)}$ equals

$$R_{DS(on)} \cong \frac{1}{k(V_{GS} - V_t)} \qquad [12.4]$$

The on-resistance $R_{DS(on)}$ varies inversely with device transconductance and inversely with effective gate voltage V_{gst}. In theory, the on-resistance can be reduced to arbitrarily small values by increasing the *W/L* ratio. In practice, considerations such as die size and cost, metallization resistance, and bondwire resistance place practical limitations upon $R_{DS(on)}$ Discrete MOS transistors with on-resistances of only a few milliohms are readily available, but most integrated power devices must fit in smaller areas and thus have resistances that lie between 25 mΩ and 1 Ω.

In practice, nonidealities prevent computation of $R_{DS(on)}$ by simple formulas like Equation 12.4. Not only do metallization and bondwire resistances play a large role in determining the final on-resistance, but the transconductance of power transistors varies with their effective gate voltage due to carrier velocity saturation. These and other considerations make the *a-priori* determination of on-resistance nearly impossible. Instead, designers typically construct sample devices and construct a graph of on-resistance versus device area (Section 12.2.2).

12.2.1. MOS Safe Operating Area

Although MOS transistors do have many advantages over their bipolar counterparts, they also share some of the same limitations, particularly in regards to safe operating area (SOA). The nature and severity of these limitations have become apparent only in recent years. In particular, high-density high-voltage MOS transistors often exhibit severe SOA limitations. This section discusses the nature of these limitations, their ultimate causes, and the means by which they may be minimized.

An ideal power MOS transistor would exhibit a safe operating area plot resembling that of Figure 12.11A. The transistor's breakdown voltage (whether due to avalanche or punchthrough) determines its maximum drain-to-source voltage, $V_{DS(max)}$. Electromigration limitations determine the transistor's maximum drain current, $I_{D(max)}$. The maximum operating temperature of the die, combined with the nature of the heat sink, determines the maximum steady-state power dissipation, $P_{D(max)}$. This device can operate at higher power levels for brief periods of time (typically less than 10 mS), as the die can absorb a certain amount of energy before it reaches its maximum operating temperature. Many, if not most, discrete power

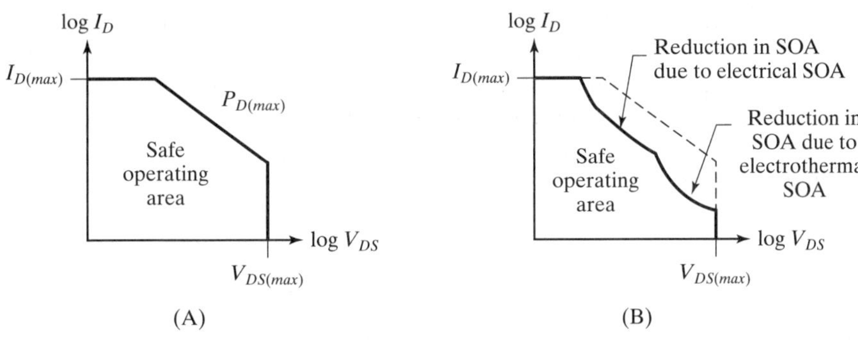

FIGURE 12.11 Comparison of safe operating area (SOA) plots of (A) an ideal power MOS transistor, and (B) a typical integrated power MOS transistor, showing the reduction in safe operating area that often occurs in integrated devices.

MOS transistors have SOA plots that resemble Figure 12.11A. These plots show no evidence of the secondary breakdown phenomenon that so often restricts the capabilities of bipolar power transistors (Section 9.1.1).

Figure 12.11B shows an SOA plot of a typical integrated power MOS transistor. The dotted outline represents the SOA boundary that would apply if the device had the ideal characteristics described in the previous paragraph. The solid outline shows the reduced SOA region that actually applies to this device. Two separate reductions are visible, each of which stems from a different mechanism. Researchers have termed these two portions of the SOA boundary the *electrical SOA* and the *electrothermal SOA*.[6] Not all MOS transistors exhibit both reductions—some show only an electrical SOA, and some show only an electrothermal SOA.

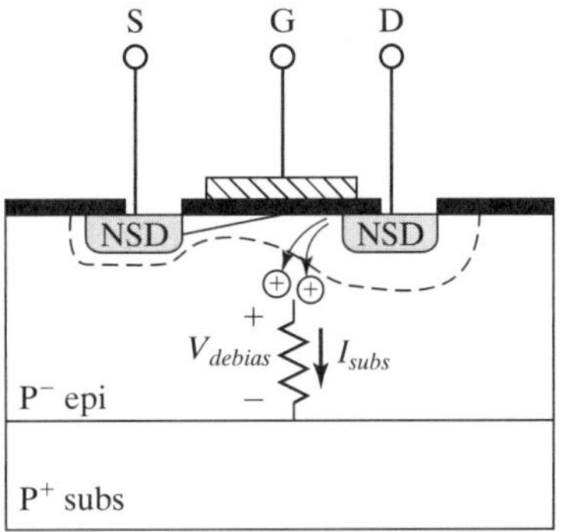

FIGURE 12.12 Cross section of a power MOS transistor showing impact ionization and backgate debiasing.

Electrical SOA

The electrical SOA of a power transistor is ultimately due to impact ionization. At high drain-to-source voltages, hot carriers injected across the pinched-off portion of the channel slam into silicon atoms and generate electron–hole pairs. Many of the resulting minority carriers flow across the drain-backgate junction. The resulting current must flow through the backgate until it reaches a contact. Power transistors typically have relatively lightly doped backgates and large spacings. Consequently, backgate debiasing can become quite significant at higher drain currents. If the backgate debiasing exceeds the forward voltage of the source/backgate junction, then this junction forward biases and begins to inject minority carriers into the backgate (Figure 12.12). These minority carriers transit across the backgate to the drain. This conduction path effectively forms a parasitic bipolar transistor in parallel with the power MOS transistor. The collector current of this parasitic bipolar transistor also

---

[6] P. L. Hower, "Safe Operating Area—a New Frontier in Ldmos Design," *Proc. 14th Int. Symposium on Power Semiconductor Devices and ICs*, 2002, pp. 1–8.

experiences impact ionization as it flows across the drain/backgate depletion region. This impact ionization in turn generates further backgate debiasing. This positive feedback mechanism causes the current through the transistor to increase dramatically. The gate voltage cannot limit the current through the parasitic bipolar transistor. The device will eventually self-destruct through overheating if the external circuit does not limit the current flow. This snapback mechanism resembles that which occurs in a bipolar transistor in V_{CER} breakdown (Section 8.1.2).

The electrical SOA limitations of a MOS transistor are often quantified in terms of its drain-to-source breakdown voltage rating. The drain-to-source breakdown voltage with gate shorted to source, BV_{DSS}, experiences no reduction in magnitude due to impact ionization because no drain current flows. (Shorting the gate to the source places the device in cutoff.) The drain-to-source breakdown voltage will diminish as soon as impact ionization reaches a level where the parasitic bipolar begins to conduct. This reduced breakdown voltage is denoted by the term BV_{DII}, where "II" stands for impact ionization. BV_{DII} diminishes as drain current increases because higher currents generate additional impact ionization.

Device designers often characterize the electrical SOA of a MOS transistor by plotting its drain-to-source breakdown voltage BV_{DS} against its gate-to-source voltage V_{GS}. Higher gate-to-source voltages correspond to higher drain currents, so BV_{DS} will diminish with increasing V_{GS} if the device has an electrical SOA limitation. Figure 12.13A shows the BV_{DS}-versus-V_{GS} plot of a transistor with no electrical SOA limitation. Such a transistor is said to possess a *square SOA* characteristic. Figure 12.13B shows a BV_{DS}-versus-V_{GS} plot of a device that has a severe electrical SOA limitation. Many designers customarily derate such devices to their lowest BV_{DII} rating, but with care one can actually operate these devices at higher breakdown

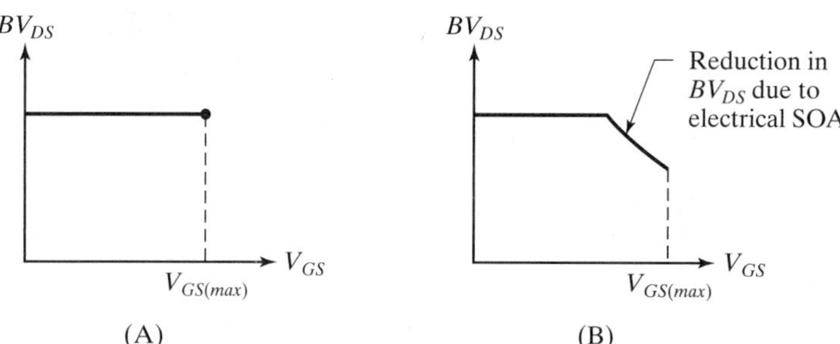

FIGURE 12.13 Plots of drain-to-source breakdown voltage BV_{DS} versus gate-to-source voltage V_{GS} for (A) a transistor with square SOA characteristics, and (B) a transistor with an electrical SOA limitation.

voltages. This feat requires the designer to pay careful attention to the exact waveforms of voltage and current that the device experiences in actual operation.

If one injects a constant current into the drain of a MOS transistor, then the drain-to-source voltage will rise until it reaches the breakdown voltage (either BV_{DSS} or BV_{DII}). The drain-to-source voltage will then snap back to a somewhat lower value called the *sustaining voltage*, $BV_{\text{DS(sus)}}$. The large current that flows through the drain/backgate depletion region generates intense localized heating. Unless some external agency interrupts the drain current, the temperature of this part of the transistor will soon reach destructive levels. The maximum amount of energy that an avalanching MOS transistor can safely absorb during a transient overload condition is called its *avalanche energy rating*.[7] The larger this rating, the more robust the transistor.

Transistors with electrical SOA limitations often have surprisingly low avalanche energy ratings. In many cases, these ratings are so small that the energy stored in the

---

[7] R. R. Stottenburg, "Boundary of Power-MOSFET, Unclamped Inductive-Switching (UIS), Avalanche-Current Capability," *Applied Power Electronics Conf. and Expo*, 1989, pp. 359–364.

drain/backgate capacitance suffices to destroy the transistor. Such a device will instantly self-destruct if its drain-to-source voltage ever exceeds its breakdown voltage (either BV_{DSS} or BV_{DII}). Researchers have attributed this type of failure to a mechanism called *filamentation*.[8] Impact ionization generates carriers that flood the drain/backgate depletion region. Larger currents generate more carriers, which can in turn carry more current. This positive feedback cycle gives the device the negative resistance characteristic responsible for snapback. This same phenomenon causes current flow to localize in a slender thread or *filament* through the high-field region. Current crowding at the entry and exit points of the filament becomes worse as the filament's diameter shrinks. Eventually the positive resistance due to current crowding equals the negative resistance of the filament itself, and the dimensions of the filament stabilize. Filamentation occurs within a couple of nanoseconds. If the dimensions of the filament are sufficiently small, temperatures within it can quickly reach destructive levels. Only a minute amount of energy is required to melt the filament, because of its tiny dimensions.

High-voltage MOS transistors typically use a composite drain consisting of a heavily doped extrinsic drain diffusion enclosed in a much wider lightly doped drain region that serves the same function as the drift region in a high-voltage bipolar transistor. The peak electric field normally occurs at the metallurgical junction between the lightly doped drain and the backgate, and sufficient ballasting exists to prevent destructive localization of the current flow. At high current levels, the Kirk effect (Section 9.1.1) causes the depletion region to collapse towards the drain contact. The electric field now peaks at the interface between the lightly doped drain and the extrinsic drain. Impact ionization causes filamentation within this region. The low resistance of the extrinsic drain provides insufficient ballasting to prevent the collapse of the filament to destructively small dimensions.

Proper layout can improve the electrical SOA rating of a MOS transistor by reducing the backgate resistance. The inclusion of interdigitated backgate contacts can provide some benefit at the cost of significantly increased device area (Section 11.2.7). A more compact, and hence more attractive, solution consists of inserting integrated backgate contacts into the source fingers of the device (Figure 11.24B). Both the size and spacing of the integrated backgate contacts strongly influence their effectiveness. Most high-voltage power devices use elongated annular gate geometries such as the one illustrated in Figure 11.22A. The ends of such a transistor usually avalanche before the rest of the device due to electric field intensification. These transistors often benefit from the inclusion of an integrated backgate contact placed at either end of each source finger so as to minimize conduction along the curved portion of the gate.

Process modifications can also improve the electrical SOA rating of a transistor. Any measure that increases backgate doping will minimize backgate debiasing and therefore increase the BV_{DII} rating of the transistor. Examples of such modifications include the use of deeper or more heavily doped backgates, the use of buried layers, and the adoption of retrograde well profiles. Many high-voltage processes include some or all of these features to ensure that the high-voltage MOS transistors obtain square (or nearly square) SOA characteristics.

Any modification that alleviates the Kirk effect will tend to suppress destructive filamentation.[9] One could simply increase the doping of the lightly doped drain. Unfortunately, this solution reduces the BV_{DSS} rating of the device because the

---

8 P. Hower, S. Pendharkar, R. Steinhoff, J. Brodsky, J. Devore, and W. Grose, "Using Two-Dimensional Structures to Model Filamentation in Semiconductor Devices," *Proc. 13th Int. Symposium on Power Semiconductor Devices and ICs*, 2001, pp. 385–388.

9 P. L. Hower, J. Lin, and S. Merchant, "Snapback and Safe Operating Area of Ldmos Transistors," *Int. Electron Device Meeting*, 1999, pp. 193–196.

drain/backgate depletion region can no longer spread as far into the lightly doped drain. A better solution widens the lightly doped drain to encourage current to spread out vertically. This solution requires that the lightly doped drain extend to a considerable depth. Transistors with gaps between the gate and the drain contact, such as are often created by field relief structures, serve this purpose. However, any increase in the width of the lightly doped drain also increases the on-resistance of the structure. A third approach involves the placement of an additional diffusion within the lightly doped drain to delay the progress of the Kirk effect towards the extrinsic drain without unduly constraining the growth of the depletion region into the lightly doped drain. This approach should minimize the increase in on-resistance while simultaneously maintaining a high BV_{DSS}. The adaptive RESURF device of Section 12.2.3 uses this technique.

Electrothermal SOA

The parasitic bipolar transistor inherent in the MOS structure suffers the same vulnerabilities as any other bipolar transistor. In particular, it can experience thermal runaway (Section 9.1.1). The resulting current localization can cause the destruction of an avalanching MOS transistor after a time delay of as much as a millisecond. This mechanism has been termed *electrothermal SOA*.

An electrothermal SOA failure begins in exactly the same manner as an electrical SOA failure (Figure 12.14). The drain-to-source voltage across the MOS transistor rises so high that impact ionization commences within the drain/backgate depletion region. Carriers injected from this depletion region into the backgate cause debiasing and subsequent triggering of the parasitic bipolar transistor. Minority carriers injected across the source/backgate junction flow through the backgate to the drain. These carriers add to the drain current flowing through the MOS transistor, causing loss of gate control and avalanche snapback. Assuming that filamentation does not occur, the transistor continues to operate in this mode of operation for some hundreds of microseconds. The drain/backgate depletion region rapidly heats due to the flow of electrical current across the large voltage drop present here. Heat spreads from the drain/backgate junction to the

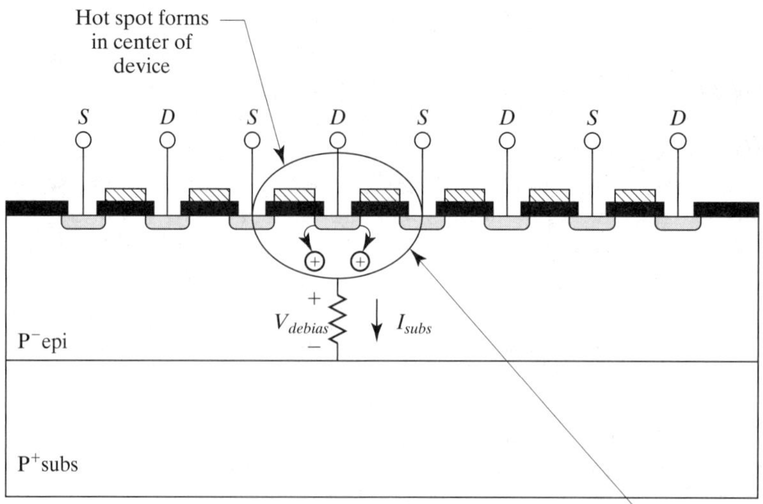

FIGURE 12.14 Cross section of a MOS transistor experiencing electrothermal SOA, illustrating the principal features of this failure mechanism.

source/backgate junction, which also serves as the base/emitter junction of the bipolar transistor. Inevitably, this heating does not occur with perfect uniformity. Whatever portion of the source/backgate junction is hotter than the rest will conduct proportionately more current. This process causes progressive localization of current flow into a hot spot. Eventually the temperature within the drain/backgate junction adjacent to the hot spot becomes so great that the silicon melts and the device self-destructs.[10]

Electrothermal SOA differs from electrical SOA in the magnitude of the time required for failure to occur. Failures due to electrical SOA typically occur in a matter of nanoseconds due to filamentation. By contrast, failures due to electrothermal SOA occur only after a delay on the order of a few hundred microseconds. A transistor can exhibit both modes of failure along different portions of its SOA boundary, as illustrated in Figure 12.11B.

Rapid Transient Overload

MOS transistors sometimes experience current focusing problems caused by debiasing in long, narrow polysilicon gate fingers. The resistance of these fingers can become quite large, especially if the poly is not silicided. This resistance forms a distributed RC network with the gate capacitance. When the gate voltage slews rapidly, the ends of the transistor nearest the gate connection turn on and off before the rest of the device. This progressive turn-on characteristic sometimes leads to localized overheating and device failure.

Figure 12.15 shows a simplified model of a single gate finger of an MOS power device. M_3 represents the portion of the gate finger nearest the gate connection, while M_2 and M_3 represent more distant portions. Resistors R_1 and R_2 model the resistance of the polysilicon gate. If the gate drive voltage V_G rises rapidly, then transistor M_1 turns on before transistors M_2 and M_3 due to the RC time delays caused by gate resistance and capacitance. The portion of the finger nearest its termination (M_1) be-

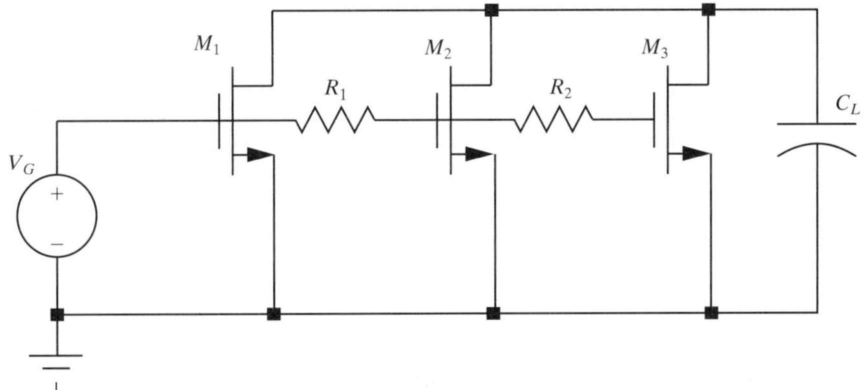

FIGURE 12.15 Model of a long MOS gate finger driving a capacitive load.

gins to discharge load capacitance C_L before the other portions of the transistor can take up their share of the load. If the voltage across C_L is large, then the current density through M_1 may become large enough to damage the device. This type of failure occurs only if the rise time of the gate drive voltage V_G is less than the gate time

[10] P. Hower, C.-Y. Tsai, S. Merchant, T. Efland, S. Pendharkar, R. Steinhoff, and J. Brodsky, "Avalanche-induced thermal instability in Ldmos transistors," *Proc. 13th Int. Symposium on Power Semiconductor Devices and ICs*, 2001, pp. 153–156.

delay, which is typically about a few nanoseconds. Rapid transient overloads of this sort often occur in ESD protection circuits and MOS gate drivers.

12.2.2. Conventional MOS Power Transistors

MOS transistors generally do not benefit from ballasting. The same finger layouts used to construct small-signal transistors therefore serve equally well for power applications.

MOS power transistors are usually specified in terms of their on-resistance $R_{DS(on)}$ measured at a specified gate voltage V_{GS} and junction temperature. The $R_{DS(on)}$ of a power MOS transistor typically increases by 50% when the junction temperature rises from 25°C to 125°C, and it typically varies about ±30% over process. The metallization resistance becomes significant for on-resistances of less than an Ohm, and the equation for $R_{DS(on)}$ then becomes

$$R_{DS(on)} \cong \frac{1}{k(V_{GS} - V_1)} + R_M \qquad [12.5]$$

where R_M is the sum of the resistance of the source and drain metallization. This metallization resistance is difficult to compute because it depends on transistor geometry. Many designers avoid the need for determining R_M by relying on measured $R_{DS(on)}$ data. This method requires that one measure the $R_{DS(on)}$ of a sample device whose layout resembles that of the proposed power device. The measured $R_{DS(on)}$ is then used to compute a figure of merit called the *specific on-resistance* R_{SP},

$$R_{SP} = A_d R_{DS(on)} \qquad [12.6]$$

where A_d represents the drawn area of the sample layout. The specific on-resistance is usually given in units of $\Omega \cdot mm^2$. Smaller values of R_{SP} indicate increasingly area-efficient layouts.

Once the specific on-resistance has been determined, one can use Equation 12.6 to compute the area required to obtain any desired on-resistance. The biggest problem with this technique lies in obtaining an accurate estimate of the specific on-resistance. This is much more difficult than it might seem. The measured value of R_{SP} should not include the resistances of bondwires and leadframe because these do not scale with device area. This is best done by providing Kelvin connections (Section 14.3.2) to the sample device. Alternatively, one can measure the resistance of the leads and bondwires of a dummy unit that contains no die. $R_{DS(on)}$ computations based on R_{SP} do not include the bondwire and leadframe resistance, so these must be added to obtain the total $R_{DS(on)}$.

The specific on-resistance also varies with device area and aspect ratio. Any one value of R_{SP} only applies to a limited range of device sizes and aspect ratios. In practice, one cannot rely on an empirical value of R_{SP} to accurately scale the $R_{DS(on)}$ of a device by more than a factor of two or three. If R_{SP} values are available for a range of device sizes, then one can interpolate between these measurements to find the R_{SP} value for a device of intermediate size. A similar process can be used to account for variations in aspect ratio.

Alternatively, one can attempt to compute the metallization resistance on the basis of an analysis of the geometry of the transistor. Hand calculations yield only approximate results because of the large number of assumptions and simplifications required to render the problem tractable. A computerized finite-element analysis provides more accurate results because it takes into account a larger number of geometric factors. In either case, the computations require a detailed knowledge of the metallization pattern. The following sections describe two of the most popular metallization patterns for MOS power transistors. Many other patterns have been proposed for specific applications, but most of these are not sufficiently general to merit further discussion.

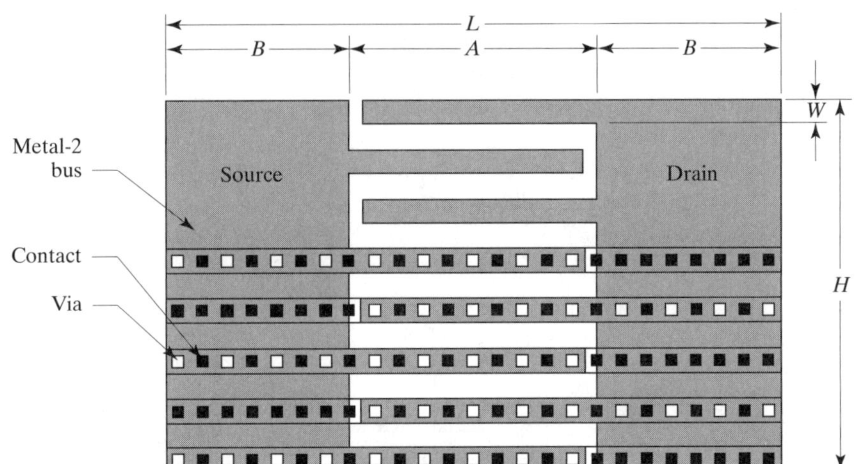

FIGURE 12.16 Metallization pattern for a rectangular MOS power device.

The Rectangular Device

Figure 12.16 shows a diagram of a simple double-level-metal pattern that produces a compact rectangular device. The top portion of the diagram shows only the interdigitated metal-2 patterns for the source and the drain fingers. The lower portion of the diagram shows these metal-2 patterns superimposed over the metal-1 fingers. Each of these fingers consists of a narrow strip of metal-1 spanning the entire width of the transistor and containing one row of contacts and vias. The metal-2 buses running up the left and right sides of the transistor collect the currents from all of the fingers and feed the source and drain terminations, which may lie at either the top or the bottom of the transistor.

The metallization pattern of Figure 12.16 provides the maximum possible amount of metallization for both the source and the drain. To understand how this has been achieved, consider a single finger in isolation from the rest of the device—for example, the bottom finger. Current flows along this finger from right to left, finally exiting into the metal-2 bus connecting to the source. No vias exist on the rightmost third of the finger. The current flowing through this portion of the finger must pass entirely through metal-1. Each contact feeds a small amount of current into the metal, so the magnitude of the current increases as it flows leftward. Metal debiasing becomes an increasingly serious concern as the magnitude of the current increases. Once the current reaches the middle third of the finger, a portion of it flows upward through vias to reach a strip of metal-2. The current now flows through a sandwich of metal-1 and metal-2. The magnitude of the current continues to increase as it flows leftward. Once the current reaches the leftmost third of the finger, it flows up into the metal-2 bus and out to the termination of the transistor.

The magnitude of the voltage drop along a source finger, V_{SM}, increases from right to left, while the magnitude of the voltage drop along a drain finger, V_{DM}, increases from left to right (Figure 12.17). The sum of these voltage drops, $V_{DM} + V_{SM}$, varies less than either of the terms that compose it. This not only ensures approximately equal conduction through all parts of the transistor, but also reduces the overall $R_{DS(on)}$. This principle can be applied to the metallization patterns for almost any type of power device, but it is particularly applicable to MOS power transistors where metallization resistance plays such an important role.[11]

[11] Krieger analyzes the distribution of currents for both parallel and antiparallel current flow in G. Krieger, "Nonuniform ESD Current Distribution Due to Improper Metal Routing," *EOS/ESD Symposium Proc.*, EOS-13, 1991, pp. 104–108.

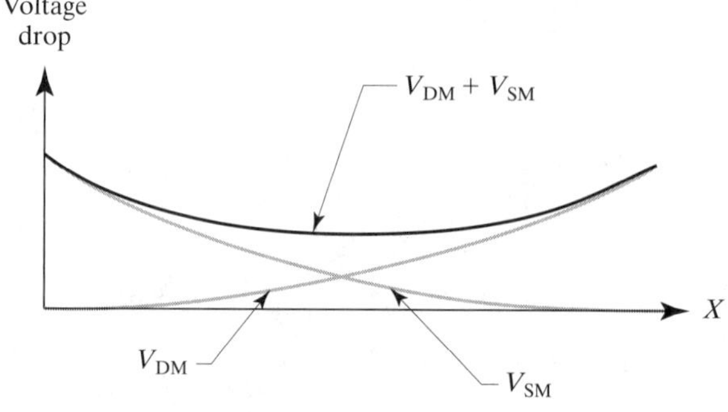

FIGURE 12.17 Graph of voltage drops across a lateral section of the power transistor of Figure 12.15.

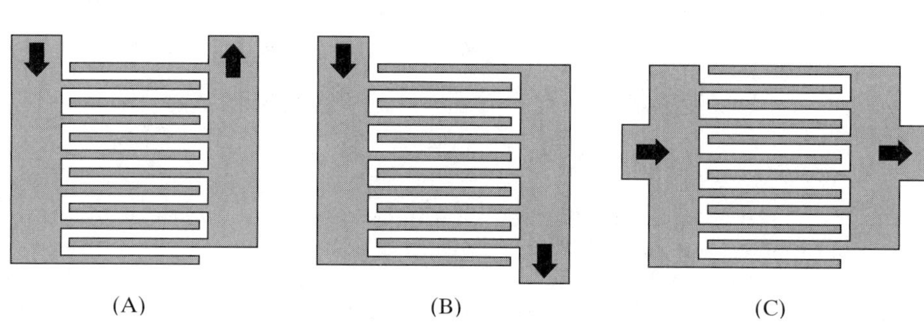

FIGURE 12.18 Three different metallization patterns for rectangular power transistors. The arrows indicate the flow of current to the termination points.

The metal-2 buses along the left and right sides of the transistor collect the currents flowing from the individual source and drain fingers. These currents flow vertically to the terminations of the device. A variety of different termination arrangements exist, some of which perform better than others. Figure 12.18A shows a common arrangement in which both terminations lie on the same end of the transistor. The paired terminations allow connections to adjacent bondpads, but they produce excessive voltage drops and an uneven distribution of current in the device. Figure 12.18B shows a better arrangement where the source and drain terminations lie at opposite ends of the transistor. This arrangement delivers a more even distribution of current and exhibits a lower total resistance than the arrangement in Figure 12.18A.

Figure 12.18C shows another method sometimes used to minimize metallization resistance. The termination points lie midway along either bus, so the current does not have to flow through the full length of the buses. This arrangement does minimize resistance, but it also produces an uneven current distribution. For most purposes, the layout in Figure 12.18B is superior to those in both Figure 12.18B and 12.18C.

The Diagonal Device

The rectangular layout in Figure 12.18 has one glaring flaw. The width of the metal-2 buses remains constant, yet the current flowing through them varies. Tapered buses can minimize debiasing and can provide a more uniform distribution of current among the fingers of the transistor. Figure 12.19 shows a layout employing tapered buses. The fingers of the transistor are arranged in a diagonal pattern that naturally produces trapezoidal metal-2 buses on either side of the device. The source and drain terminations must lie on opposite ends of the transistor. This device is more difficult to construct than the rectangular layout shown in Figure 12.16, and computer simulations are required to determine its optimum dimensions. The triangular areas beneath the metal-2 buses must be filled with circuitry in order to obtain the full packing density promised by this layout. Many designers prefer to use rectangular layouts, such as the one in Figure 12.18B, which are easier to construct and to optimize.

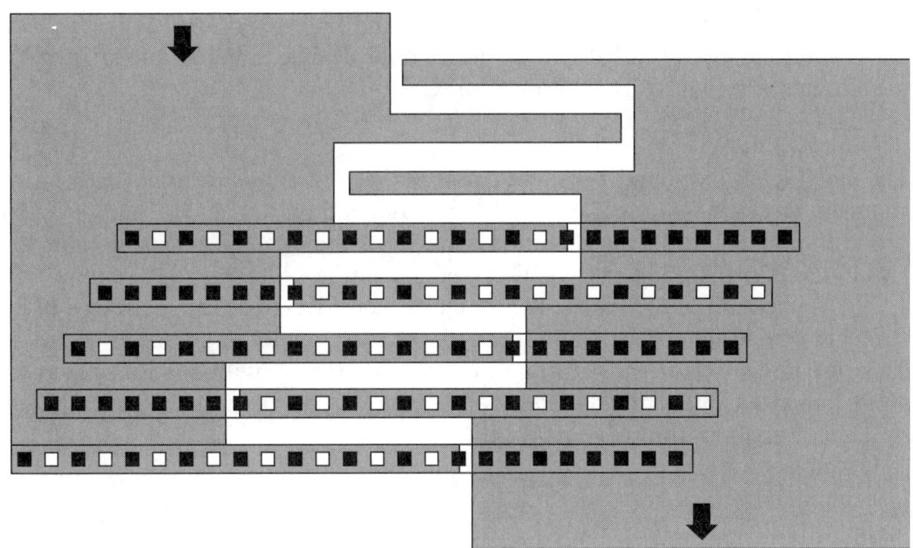

FIGURE 12.19 Metallization pattern for a diagonal MOS power device.

Computation of R_M

Accurate calculations of the metallization resistance become very complex and generally require computer modeling. The metallization pattern in Figure 12.18B represents an exception, as it is possible to estimate its metallization resistance using the formula

$$R_M = \frac{B^2 R_{S1}}{2W\, N_D L} + \frac{A R_{S12}}{2W\, N_D} + \frac{H R_{S2}}{2B} \qquad [12.7]$$

where N_D equals the number of drain fingers (or half the total number of source/drain fingers), R_{S1} equals the sheet resistance of metal-1, R_{S2} equals the sheet resistance of metal-2, and R_{S12} equals the sheet resistance of a parallel combination of metal-1 and metal-2. Figure 12.16 shows the relationship between dimensions *A, B, W, L,* and *H.* This derivation assumes that each source/drain finger conducts an equal amount of current and that the current flowing through a finger increases linearly along its length. The formula also neglects the variation in voltage across the width of the metal-2 buses.

The preceding equation can be analyzed (Appendix D) to determine the optimum width, *B,* of the metal-2 buses, which equals

$$B = \left(\frac{R_{S12}}{R_{S1}}\right)L \qquad [12.8]$$

Assuming that both metal-1 and metal-2 have the same composition, Equation 12.8 becomes

$$B = \left(\frac{t_1}{t_1 + t_2}\right)L \qquad [12.9]$$

where t_1 and t_2 are the thicknesses of metal-1 and metal-2, respectively. This equation provides a means of sizing the metal-2 buses. If the thickness of metal-1 equals or exceeds the thickness of metal-2, then the buses should each extend across half of the transistor. In this case, the interdigitated region described by dimension *A* vanishes entirely. If the thickness of metal-2 exceeds the thickness of metal-1, then the buses should not entirely cover the transistor, and an interdigitated region should exist between them. Although Equations 12.8 and 12.9 were derived specifically for the structure in Figure 12.18B, they concern only a single finger and so apply also to the structures in Figures 12.18A and 12.18C.

Other Considerations

The metal leads and bondwires that connect a power transistor to its load can substantially increase $R_{DS(on)}$. These resistances depend on the general size and shape of the transistor and its placement relative to its bondpads. The calculations form part of the floorplanning process discussed in Section 14.2.

The connection of the gate lead also merits consideration. The resistance of long stretches of polysilicon substantially slows the switching of large power transistors. This resistance can be minimized by connecting the individual gate fingers with metal jumpers. Connecting both ends of the gate further reduces the gate resistance by a factor of about four (Figure 12.20).

Other factors worth considering when laying out power transistors include the placement of backgate contacts and guard rings. Power transistors may use either interdigitated or distributed backgate contacts, depending on which technique provides the smallest overall area while maintaining acceptable SOA characteristics (Section 11.2.7). Devices requiring independent connection to the backgate must use interdigitated backgate contacts spaced away from the neighboring source fingers. PMOS transistors constructed in analog BiCMOS processes may use NBL to obtain a solid backgate connection without the need for interdigitated or distributed backgate contacts. Processes that employ retrograde wells may obtain similar benefits from the heavily doped lower regions of these wells.

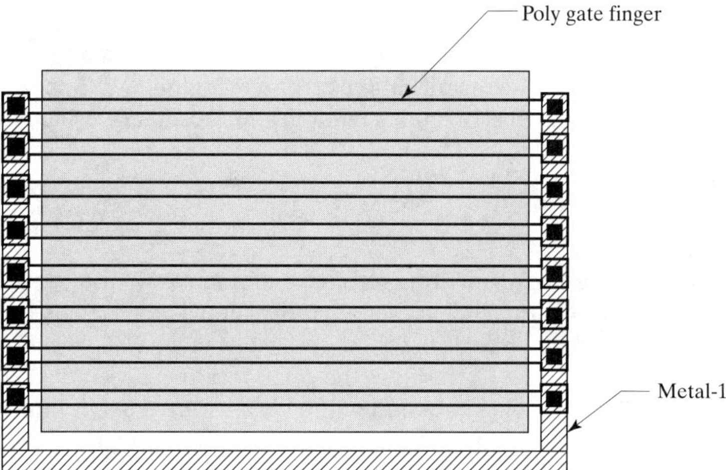

FIGURE 12.20 Gate metallization connecting both ends of the gate fingers reduces gate resistance.

Some applications momentarily forward-bias the backgate diode of a power MOS transistor. If this occurs, then the transistor not only requires an extensive network of backgate contacts to provide recombination current, but also requires an efficient system of minority carrier guard rings to prevent substrate injection and debiasing.

Transistors constructed in the epi pose a particular problem because no way exists to block the flow of minority carriers to the substrate. This situation occurs when NMOS transistors are constructed in N-well CMOS processes. In BiCMOS processes, one can sometimes isolate these NMOS transistors by using a combination of deep-N+ sinker and NBL (Section 11.2.2). An isolated NMOS cannot inject electrons into the substrate because the isolation structure acts as an electron-collecting guard ring. If an isolated structure is not feasible, then the designer must fall back on guard rings placed around the periphery of the transistor. These guard rings are often quite effective when combined with a heavily doped substrate. If the process uses a lightly doped substrate, then the guard rings should be made as wide as possible to try to prevent minority carriers from passing underneath them. Substrate contacts should be placed on the far side of the guard ring from the point of injection. Majority carrier current flowing underneath the guard ring through the lightly doped substrate generates an electric field that opposes the flow of minority carriers underneath the guard ring. One can sometimes arrange the wells of adjacent power transistors so they act as guard rings.

Some circuits use a small transistor to sense the current passing through a much larger one. Ideally, the sense transistor should consist of a number of segments scattered throughout the transistor, so the average of these segments represents the average operating conditions of the power transistor. In practice, it often proves difficult to embed sense transistors within the interior of the power transistor. Instead, these devices usually lie at the ends of gate fingers along one or two sides of the device. If only one sense transistor segment can be used, then this should occupy the center of one side of the device. Two sense segments should occupy the center of opposite sides of the device. Four sense segments should occupy pairs of sites located symmetrically around an axis passing through the centroid of the power transistor. If the sense segments can lie within the confines of the power transistor, they should occupy locations approximately half way from the center of the transistor to its periphery, and should be located in symmetric locations, as discussed previously.

Figure 12.21 shows a typical example of a single embedded sense transistor located on the end of a gate finger. In practice, the power transistor would have a much larger number of fingers, and the one chosen for the sense device would lie as near to an axis of symmetry of the device as possible. The sense device shares common gate, source, and backgate connections, but it has an independent drain connection.

When designing sense transistors, keep in mind that the smaller of two matched devices generates most of the mismatch. The short channel lengths characteristic of most power devices make it difficult to obtain sufficient gate area to minimize mismatch while keeping current through the sense resistor to desirably low levels.

Nonconventional Structures

The conventional self-aligned poly-gate transistor consists of a series of interdigitated source and drain fingers. Although this arrangement possesses the virtue of simplicity, it does not produce the densest possible layout. Other designs can achieve

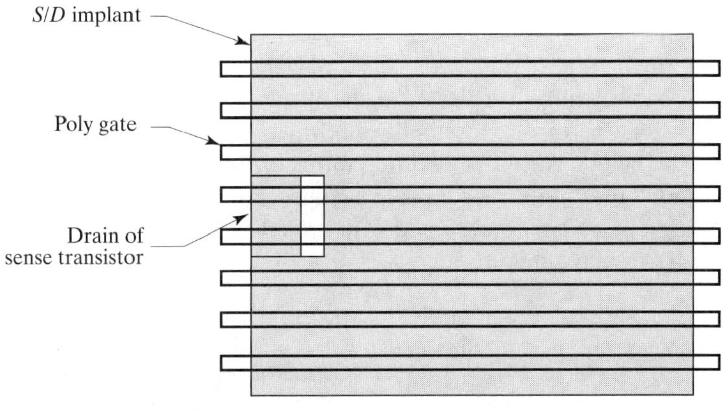

FIGURE 12.21 Construction of an embedded sense transistor (contacts and metallization not shown).

lower specific on-resistances by tightly packing arrays of cleverly shaped source and drain elements. The *waffle transistor* in Figure 12.22A exemplifies this concept. It uses a mesh of horizontal and vertical poly strips to divide the source/drain implant into an array of squares. Each square contains a single contact. By alternately connecting these contacts to the source and drain metallization, one can arrange four drains around each source and four sources around each drain. The drain and source metallization consists of a series of diagonal strips of metal-1, which are usually combined with an interdigitated metal-2 pattern similar to those used for conventional transistors.

An analysis of the W/L ratios achieved for a given device area will show that the waffle transistor provides an increase in packing density equal to

$$\frac{(W/L)_w}{(W/L)_c} \cong \frac{2S_d}{L_d + S_d} \tag{12.10}$$

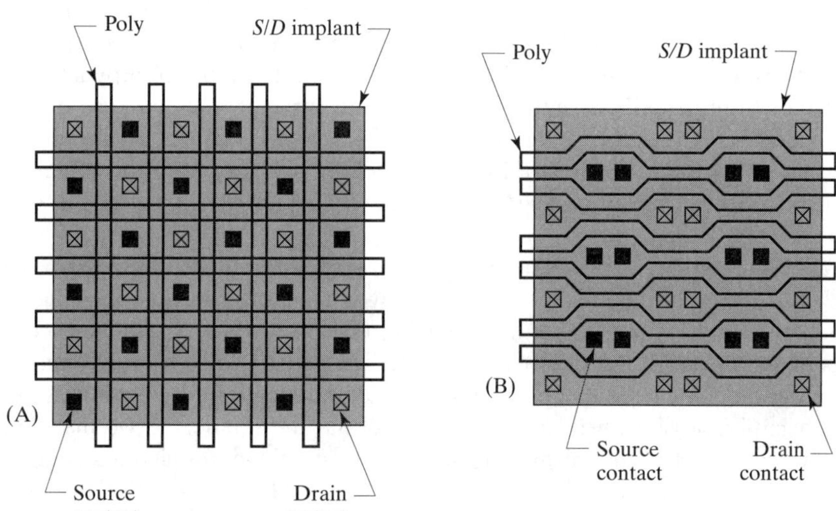

FIGURE 12.22 Nonconventional MOS transistor layouts: (A) waffle and (B) bent-gate. The drain contacts have been drawn differently than the source contacts to aid in their identification.

where $(W/L)_w$ of the waffle transistor and $(W/L)_c$ of the conventional interdigitated transistor are measured from two devices consuming equal die areas. The waffle transistor offers better packing than the conventional interdigitated transistor as long as the spacing between the gates S_d exceeds the gate length L_d. Almost all power transistors meet this requirement. Suppose the layout rules specify a minimum drawn gate length of 2 μm, a minimum contact width of 1 μm, and a minimum spacing poly-to-contact of 1.5 μm. Using these rules, Equation 12.10 indicates that the waffle transistor provides approximately 33% more transconductance than the interdigitated-finger transistor. A more precise estimate of the benefits of the waffle transistor would account for the differences between drawn gate length and effective gate length and would include corner conduction terms for the waffle transistor, neither of which are included in Equation 12.10. A hexagonal arrangement of alternating source and drain plugs within a poly grid provides an even denser layout, but one not easily constructed using most layout editors.[12]

[12] A. Van den Bosch, M. S. J. Steyaert, and W. Sansen, "A High-Density, Matched Hexagonal Structure in Standard CMOS Technology for High-Speed Applications, *IEEE Trans. Semiconductor Manufacturing*, Vol. 13, #2, 2000, pp. 167–172.

The waffle transistor has three crucial defects. First, the foregoing analysis does not consider the effects of metallization resistance. The metallization invariably contributes a significant portion of the $R_{DS(on)}$ of the transistor, and in thin CMOS metal systems it often becomes the dominant factor. If one assumes that the metallization contributes about half the total $R_{DS(on)}$, then the improvement gained by using the waffle layout drops by half, or from 33% to 16% for this example. The situation is actually even worse, because the waffle layout is difficult to properly metallize. The metal-1 fingers must repeatedly cross the gate poly, introducing significant step-induced metal thinning. Second, the waffle transistor contains a large number of bends in its channels. These bends produce sharp corners in the source/drain regions that avalanche at lower voltages than the remainder of the transistor. Localized avalanche limits the amount of energy the waffle transistor can dissipate. This limitation becomes apparent in ESD testing, where the performance of waffle transistor may fall short of that of the conventional layout. Fillets or chamfers applied to the corners of the source/drain squares will largely eliminate this problem.[13] Third, the waffle transistor makes no provision for backgate contacts. Unless the transistor is used in combination with a heavily doped substrate or a buried layer to provide backgate contact, it is quite susceptible to backgate debiasing and latchup. No simple way exists to add interdigitated or distributed backgate contacts to a waffle layout.

Figure 12.22B shows a *bent-gate transistor* that avoids most of the difficulties of the waffle transistor, while offering some unique advantages. The bent gates increase the gate width while simultaneously allowing the gate strips to pack more closely together. This layout readily accommodates distributed backgate contacts without sacrificing undue die area. It also avoids the use of 90° bends in favor of gentler 135° bends, which are less prone to localized avalanche. The diagonal arrangement of the source and drain contacts also provides additional source/drain ballasting that can improve robustness under extreme conditions, such as those encountered during ESD testing. These benefits, combined with the ease of inserting extensive networks of distributed backgate contacts, make this device attractive for applications that routinely experience transient overloads.

12.2.3. DMOS Transistors

High-voltage transistors require short, heavily doped backgates and wide, lightly doped drift regions. These are more conveniently produced by diffusing the backgate into the drift region than *vice versa*. The *double-diffused MOS*, or DMOS, uses this approach to produce short-channel, high-voltage transistors optimized for use as power devices.

Like the DDD transistor, the DMOS relies on the self-alignment of two diffusions driven through a common oxide opening. An N-channel DMOS is fabricated by diffusing boron and arsenic into lightly doped N-type silicon (Figure 12.23). Boron outdiffuses more rapidly than arsenic, producing a moderately doped P-type region enclosing a shallower and more-heavily doped N-type region. The heavily doped arsenic core forms the source of the DMOS transistor, while the surrounding, moderately doped boron diffusion forms the backgate. The channel length of the transistor equals the difference between the surface outdiffusion distances of the

---

[13] L. Baker, R. Currence, S. Law, M. Le, C. Lee, S. T. Lin, and M. Teene, "A 'Waffle' Layout Technique Strengthens the ESD Hardness of the NMOS Output Transistor," *EOS/ESD Symposium Proc.*, EOS-11, 1989, pp. 175–181.

FIGURE 12.23 Layout of (A) DMOS mask geometry and (B) the resulting pattern of diffusions.

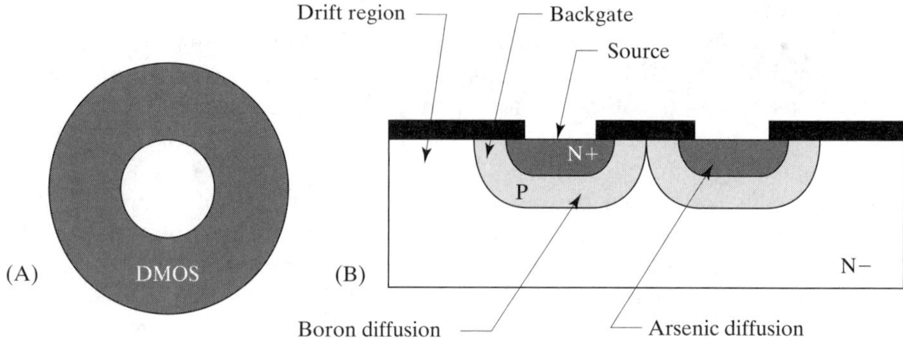

boron and arsenic implants, which depends solely on doping concentrations and drive times.

The lightly doped N-type region surrounding the backgate serves as the drift region of the DMOS transistor. This drift region must be contacted by means of a heavily doped N-type region. Discrete DMOS transistors often occupy a lightly doped N-type epi deposited on a heavily doped N-type substrate. The drain current flows down through the epi to the substrate and exits through the backside of the die. The epi thickness determines the width of the drift region and hence the maximum operating voltage of the transistor. In order to integrate this *vertical DMOS*, or VDMOS, the drain region must be isolated from the substrate. This can be achieved by placing the transistor in an N-well furnished with NBL and deep-N+. This resolves the isolation issue, but the transistor still exhibits excessive drain resistance at low forward voltages because of incomplete depletion of the drift region. Drain resistance represents a major challenge for constructing low-voltage DMOS transistors. The epi thickness cannot be reduced too far or the tail of the NBL will intersect the DMOS diffusions, so another approach must be tried.

The Lateral DMOS Transistor

Most integrated DMOS transistors use a shallow, heavily doped N-type diffusion placed next to the DMOS backgate to extract the drain current. This type of device is called a *lateral* DMOS, or LDMOS.[14] The separation of the backgate and drain contact diffusions determines the width of the lateral drift region. This drift region is designed to fully deplete through at a relatively low voltage. This type of transistor does not require NBL or deep-N+, although these are often added to minimize substrate injection in the event that the backgate forward-biases into the drain.

The DMOS backgate contact represents something of a problem. Not only is the backgate relatively lightly doped, but it is also extremely narrow. A heavily doped P-type diffusion can contact the backgate, but only if it also contacts the N+ source. The resulting P+/N+ junction may leak so badly that the backgate cannot be isolated from the source. Most DMOS transistors use an annular geometry containing a central P+ plug that serves as a backgate contact. The P+ plug shorts to the source of the transistor through a single contact opening that covers both (Figure 12.24A).

---

[14] J. D. Plummer and J. D. Meindl, "A Monolithic 200-V CMOS Analog Switch," *IEEE J. Solid-State Circuits,* Vol. SC-11, #6, 1976, pp. 809–817.

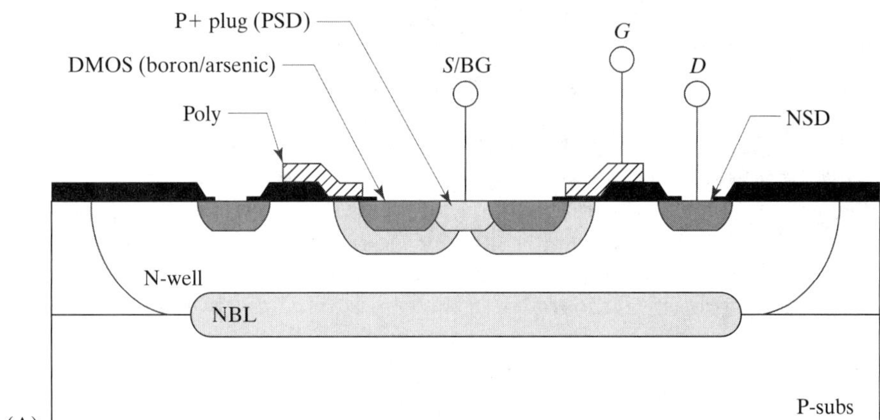

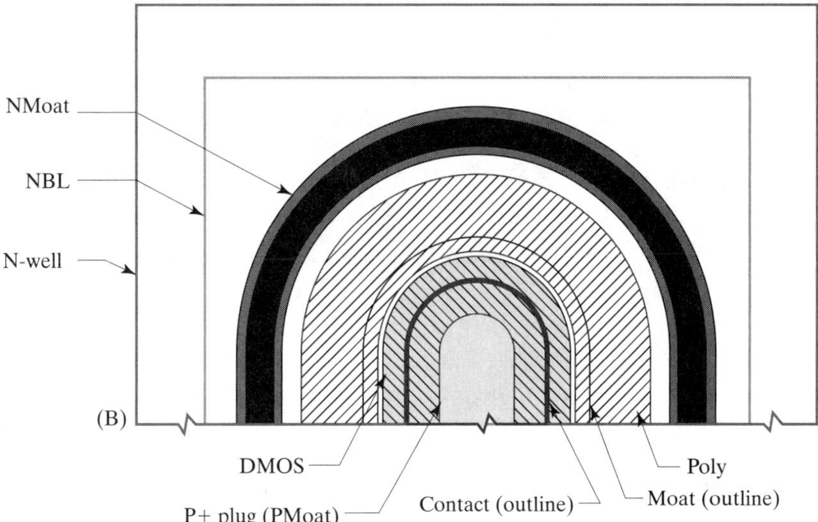

FIGURE 12.24 Layout and cross section of a lateral DMOS transistor constructed in analog BiCMOS.

DMOS transistors are considered asymmetric devices because their backgate usually connects to their source and because the diffused backgate is more lightly doped near the drain than near the source.

Figure 12.24 shows a simple LDMOS transistor constructed in an N-well analog BiCMOS process. An annular DMOS implant consisting of arsenic and boron defines the source and backgate regions of the device, and an enclosing N-well serves as its drift region. The P+ plug contacting the backgate consists of a PSD implant whose outer edge coincides with the inner edge of the DMOS implant. The extrinsic drain consists of a ring of NSD implanted around the transistor. The spacing between the NMoat and the DMOS geometries determines the width of the drift region. The poly gate overlaps the thick-field oxide over the drift region to form a field-relief structure, allowing the transistor to withstand large drain-to-gate voltage differentials without requiring a thick gate oxide. Many elaborations upon this structure exist.

The smallest DMOS transistor uses a circular ring of DMOS implant, like that shown in Figure 12.23. Larger widths can be obtained by connecting many small annular devices in parallel, but such arrays of annular transistors pack rather loosely. The transistor in Figure 12.24 uses an elongated annular geometry to improve packing density. A large power device consists of an array of annular DMOS transistors interdigitated with drain contacts. This type of structure resembles a conventional interdigitated transistor and can use any of the metallization patterns discussed in Section 12.2.2.

NBL often forms part of a DMOS transistor, but its presence does not always prove beneficial. Consider the case of a DMOS transistor without NBL. The drain-backgate and drain-substrate depletion regions both widen as the drain voltage increases. If the well is sufficiently shallow, then it will punch through from backgate to substrate long before the well-backgate junction avalanches. This punchthrough only causes problems if the source of the transistor connects to a voltage other than substrate potential. NBL can prevent punchthrough by constraining the depletion region, but in so doing it intensifies the electric field and reduces the well-backgate avalanche voltage. The DMOS transistor therefore has two separate drain-to-source voltage ratings. A device whose source connects to substrate potential can omit NBL and can obtain a higher operating voltage. Devices whose sources connect to voltages other than substrate potential require NBL and are therefore constrained to lower operating voltages. These two styles of devices are often called *low-side drive* (LSD) and *high-side drive* (HSD) layouts. The LSD layout omits NBL and therefore requires the source to connect to substrate potential. This in turn requires the transistor to reside between the load and the ground return, or on the "low side" of the load. The HSD layout includes NBL and allows source operation above substrate potential. An HSD device can therefore reside between the load and the positive rail, or on the "high side" of the load.

RESURF Transistors

High-voltage transistors require lightly doped drains to provide adequate breakdown voltages. These lightly doped drains deplete both laterally and vertically. Conventional designs use relatively deep wells or epitaxial layers to keep the drain/substrate depletion region away from the active device. Recent research has shown not only that such deep drains are unnecessary, but that the use of shallower drains actually reduces the lateral surface field and therefore allows the device to operate at higher voltages. Devices that embody this principle are called *reduced surface field* (RESURF) transistors.

Figure 12.25A shows a cross section of a portion of a RESURF device. The rectangular N–region represents a lightly doped N-type epi layer deposited upon a lightly doped P-type substrate. A moderately doped P-type isolation diffusion driven through the N-epi isolates one epi tank from another. A high voltage applied to the tank causes a depletion region to form between the tank and the surrounding P-type regions. The epi/substrate depletion region extends vertically into the epi a distance X_V. The epi/isolation depletion region would normally extend laterally into the epi a distance X_{lat} in order to uncover sufficient ionized donors to balance the ionized accepter charge within the isolation region. However, the shaded rectangle has already been depleted by the epi/substrate depletion region. The epi/isolation depletion region must therefore extend farther into the N-epi in order to uncover the requisite number of ionized donors. Figure 12.25B shows the actual lateral encroachment into the N-epi once the interaction between the vertical and lateral depletion regions has been taken into account. The electric field across a depletion region varies inversely with its width, so the increase in the lateral depletion region width implies a corresponding decrease in electric field intensity.[15]

[15] M. Imam, M. Quddus, J. Adams, and Z. Hossain, "Efficacy of Charge Sharing in Reshaping the Surface Electric Field in High-Voltage Lateral RESURF Devices," *IEEE Trans. on Electron Devices*, Vol. 51, #1, 2004, pp. 141–148.

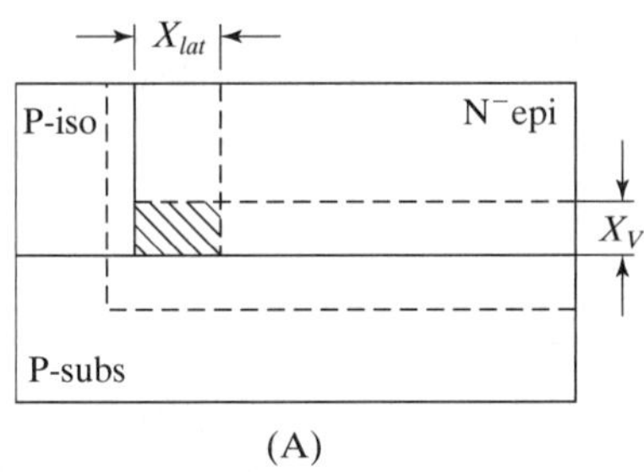

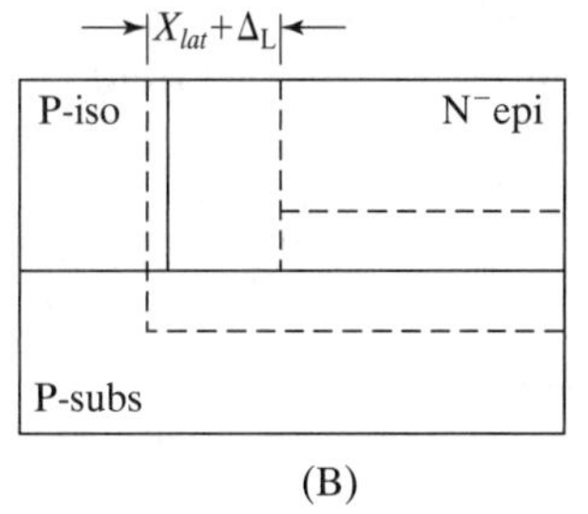

FIGURE 12.25 Cross section of an N-type tank showing vertical and lateral components of the depletion region surrounding the tank, (A) without consideration of lateral-vertical depletion region interaction, and (B) including lateral-vertical interaction.

RESURF devices make use of the lateral-vertical depletion region interaction to increase the breakdown voltage of the device. The RESURF effect requires that the vertical breakdown exceed the lateral breakdown, as happens in almost all planar devices. Consider the structure of Figure 12.25B. The P-type isolation diffusion must contain sufficient dopant to reach the epi/substrate interface. This implies that the upper portions of the isolation diffusion are relatively heavily doped. The depletion region therefore narrows as it approaches the surface, and this narrowing reduces the lateral breakdown voltage. A similar situation exists when an N-well diffuses into a P-epi region. Not only does the upper portion of the well contain more dopant than the lower portions, but the channel stop implants also increase the surface doping. The RESURF effect can therefore increase the lateral breakdown voltage of an epi tank or a well. In a properly constructed device, RESURF ensures that vertical breakdown occurs before lateral breakdown. However, this ideal will not occur if the vertical dimension is either too large or too small. A large vertical dimension reduces the amount by which the lateral depletion region expands. If the vertical dimension is too small, then the vertical depletion region reaches the surface. When this happens, the vertical electric field begins to intensify and the vertical breakdown voltage drops.

Figure 12.26 shows a cross section of a RESURF DMOS transistor. The drain consists of an N-well driven into a lightly doped P-type epi layer. The DMOS transistor resides within this well. When the drain operates at the maximum rated voltage, the N-well region beneath the P-type body depletes entirely through. This does not cause any difficulty so long as the transistor is employed as a low-side drive device, or in other words, so long as the source (and therefore also the backgate) are held at substrate potential. The lateral intrusion of the depletion region around the

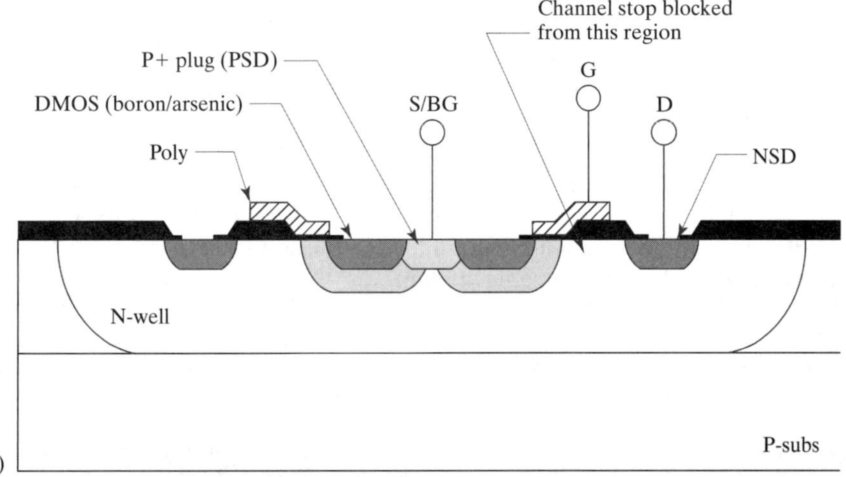

FIGURE 12.26 Cross section of a low-side-drive RESURF DMOS transistor.

well/epi junction increases considerably because of the RESURF effect. This reduces the electric field at the drain sidewall, and increases the breakdown of the transistor. A similar effect also occurs at the drain/backgate junction. Although most RESURF transistors are LSD devices, properly constructed HSD devices can also benefit from the RESURF effect.[16]

Early high-voltage RESURF devices suffered from poor safe operating area characteristics due to electrical SOA limitations. These devices were extremely fragile, and consequently designers were reluctant to adopt them. The electrical SOA limitations of these devices stemmed from the Kirk effect, which caused the point of maximum electric field intensity to move from the vicinity of the drain/backgate junction to the interface between the lightly doped drain and the extrinsic drain (Section 12.2.1). The addition of a moderately doped diffusion beneath the field relief structure halts this effect. The N-channel stop implant that normally appears beneath the field oxide is sufficiently deep and heavily doped to perform this function. Since most high-voltage transistors include a field oxide region as part of their field relief structure, this N-channel stop implant forms a natural part of the transistor. This technique has been termed *adaptive RESURF*.[17]

Lateral DMOS transistor design has matured considerably over the last several years. The best devices available today offer an extraordinary combination of high breakdown voltage, square or near-square SOA characteristics, and low specific on-resistance. Consequently, LDMOS transistors are now considered the pre eminent high-voltage power devices for integrated circuit applications.

The DMOS NPN

The DMOS structure of Figure 12.23 contains a parasitic NPN transistor. The source of the DMOS acts as the emitter of this transistor, the backgate as its base, and the drain as its collector. This parasitic NPN has a heavily doped emitter that enhances its emitter injection efficiency; a thin, moderately doped base that reduces its Gummel number; and a wide, lightly doped collector that minimizes the Early effect. The performance of the DMOS NPN can approach that of a conventional CDI NPN, making it a useful alternative to the latter device.

Figure 12.27 shows a layout and cross section of a typical DMOS NPN. This structure uses a circular DMOS implant to form the base and emitter of the transistor. The drawn emitter area equals the drawn area of the DMOS implant. The emitter is contacted by a central plug of NSD, and the base is contacted by a ring of PSD surrounding the DMOS implant. The boron DMOS implant must overlap the PSD implant sufficiently to allow for misalignment. This usually requires that the two implants practically abut one another. The extrinsic collector consists of NBL and deep-N+, just as in a conventional NPN transistor.

A P+/N+ junction appears in the conventional DMOS structure between the P+ backgate contact and the N+ DMOS implant. The potential for leakage across this junction is of no concern in a DMOS transistor because the source always shorts to the backgate. The same is not true in the DMOS NPN because these diffusions form its base and emitter. Leakage can be avoided by reducing the dosage of the arsenic DMOS implant. NSD must now be added to the source regions to allow Ohmic contact to the lightly doped arsenic implant.

---

[16] V. Khemka, V. Parthasarathy, R. Zhu, and A. Bose, "Correlation Between Static and Dynamic SOA (Energy Capability) of RESURF LDMOS Devices in Smart Power Technologies," *Proc. 14th Int. Symp. on Power Semiconductor Devices and ICs,* 2002, pp. 125–128.

[17] P. Hower, J. Lin, S. Merchant, and S. Paiva, "Using 'Adaptive RESURF' to Improve the SOA of Ldmos Transistors," *Proc. 12th Int. Symp. on Power Semiconductor Devices and ICs,* 2000, pp. 345–348.

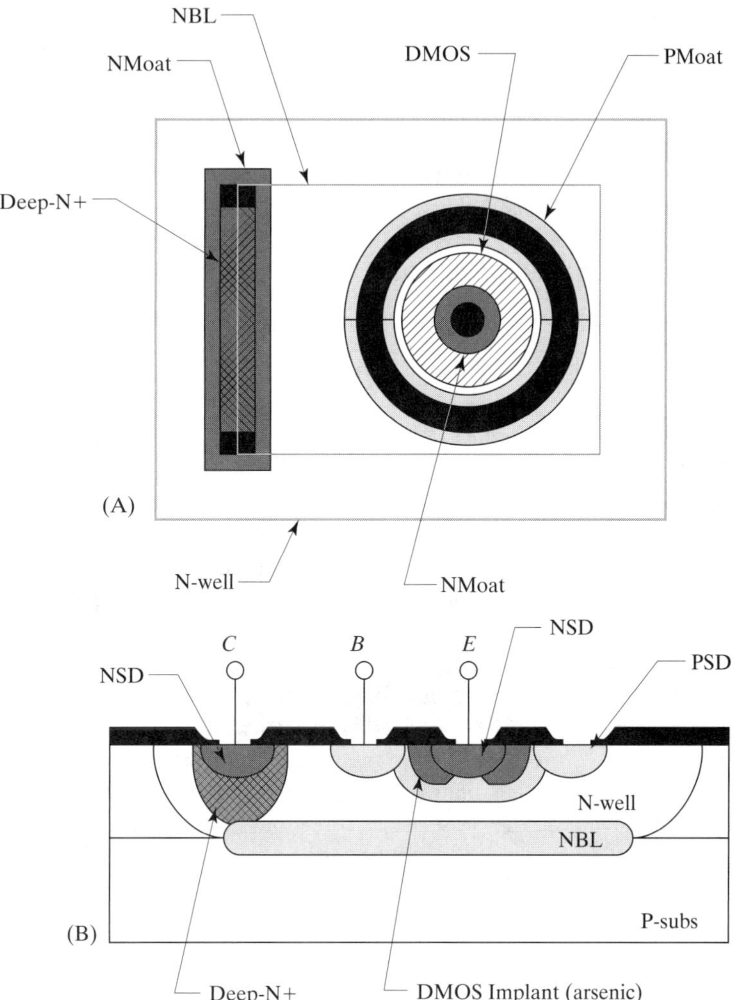

FIGURE 12.27 Layout and cross section of a DMOS NPN (omitting poly field plate).

The structure in Figure 12.27 omits the moat geometry normally covering the DMOS implant, and allows thick-field oxide to grow over it. Dopant segregation and oxidation-enhanced diffusion drive the arsenic emitter deeper into the boron base, reducing the base width of the transistor. Conducting a field oxidation over the DMOS implant thus increases the beta of the DMOS NPN.

All DMOS NPN transistors contain a parasitic DMOS transistor connected between collector and emitter. The structure in Figure 12.27 does not show the poly gate electrode required to suppress this parasitic device. The poly electrode, or *field plate,* must cover the exposed boron DMOS implant with sufficient overlap to allow for misalignment. This field plate is usually connected to the emitter, since this connection shorts the gate and source of the parasitic DMOS.

12.3 MOS TRANSISTOR MATCHING

A wide variety of analog circuits use matched MOS transistors. Some circuits, such as differential pairs, rely on matching of gate-to-source voltages, while others, such as current mirrors, rely on matching of drain currents. The biasing conditions required to optimize voltage matching differ from those required to optimize current matching. One can optimize MOS transistors either for voltage matching or for current matching, but not simultaneously for both.

The relationship between biasing and voltage matching is easily derived from the Shichman-Hodges equations (Section 11.1.1). Suppose two matched MOS transistors operate at the same drain current I_D. If the transistors were ideal devices, then they would develop exactly the same gate-to-source voltage V_{GS}. In practice, mismatches cause the gate-to-source voltages of the two transistors to differ by an amount $\Delta V_{GS} = V_{GS1} - V_{GS2}$. Assuming that the transistors operate in saturation, as is usually the case, then the offset voltage ΔV_{GS} equals

$$\Delta V_{GS} \cong \Delta V_t - V_{gst1}\left(\frac{\Delta k}{2k_2}\right) \qquad [12.14]$$

where ΔV_t equals the difference between the threshold voltages of the two transistors, Δk equals the difference between their device transconductances, V_{gst1} equals the effective gate voltage of the first transistor, and k_2 equals the device transconductance of the second (Appendix D). The offset voltage ΔV_{GS} depends on device dimensions due to the presence of the device transconductance k_2 and effective gate voltage V_{gst1} in the second term. The offset voltage also depends on biasing conditions because of presence of effective gate voltage in the equation. These dependencies are unique to MOS transistors and are not shared by bipolar transistors (Section 9.2).

The MOS designer can minimize the offset voltage ΔV_{GS} by reducing the effective gate voltage V_{gst} of the matched transistors. MOS circuits that depend on voltage matching therefore benefit from the use of large W/L ratios and low operating currents. The improvements obtainable in this manner are limited by the onset of subthreshold conduction and by the presence of threshold mismatches. As a practical matter, reducing V_{gst} below about 0.1 V produces little improvement in voltage matching.

MOS circuits relying on current matching behave quite differently. The mismatch between two drain currents, I_{D1} and I_{D2}, can be specified in terms of a ratio I_{D2}/I_{D1} equal to

$$\frac{I_{D2}}{I_{D1}} \cong \frac{k_2}{k_1}\left(1 + \frac{2\Delta V_t}{V_{gst1}}\right) \qquad [12.15]$$

The mismatch in drain currents actually increases at low effective gate voltages due to a larger contribution from the threshold mismatch ΔV_t (Appendix D). MOS circuits relying on current matching should operate at reasonably large effective gate voltages to avoid exacerbating threshold voltage variations. The optimal value of V_{gst} depends on many factors and is difficult to quantify. As a practical matter, one should endeavor to maintain a nominal V_{gst} of at least 0.3 V (and preferably 0.5 V) in MOS transistors generating matched currents. Larger effective gate voltages may provide some additional benefit, but most applications cannot spare the headroom to support a higher V_{gst}.

In summary, MOS circuits that generate matched voltages should operate at low effective gate voltages, while MOS circuits that generate matched currents should operate at high effective gate voltages. For most purposes, a nominal V_{gst} of 0.1 V or less will suffice for voltage matching, and a nominal V_{gst} of 0.3 V or more will suffice for current matching. Assuming that the circuit designer adjusts the biasing of the transistors to these values, the matching now depends almost entirely on the care taken in transistor layout. The next three sections discuss layout considerations that affect MOS matching.

12.3.1. Geometric Effects

The size, shape, and orientation of MOS transistors all affect their matching. Large transistors match more precisely than small ones because increased gate area helps minimize the impact of localized fluctuations. Long-channel transistors match more precisely than short-channel ones because longer channels reduce channel-length modulation. Transistors oriented in the same direction match better than those oriented in different directions because of the anisotropic nature of monocrystalline silicon. This section discusses the impact of these and other geometric factors on MOS transistor matching.

Gate Area

MOS mismatches have been experimentally measured for a number of processes. These measurements reveal that the magnitude of the threshold voltage mismatch varies inversely with the square root of the active gate area. This relationship can be expressed in terms of the effective channel dimensions W_{eff} and L_{eff} as

$$s_{Vt} = \frac{C_{Vt}}{\sqrt{W_{\text{eff}} L_{\text{eff}}}} \qquad [12.16]$$

where s_{Vt} is the standard deviation of the threshold voltage mismatch and C_{Vt} is a constant.[18] The value of C_{Vt} is empirically determined by measuring the random mismatch between pairs of transistors of different sizes. The results apply only to transistors closely resembling the test devices used to derive C_{Vt}. The relationships between drawn dimensions and effective dimensions are not always known, and sometimes the drawn dimensions W_d and L_d must be substituted for the effective dimensions W_{eff} and L_{eff}. This substitution will have little effect on the accuracy of the predictions as long as both dimensions of the transistor are several times greater than minimum.[19]

Strictly speaking, Equation 12.16 applies only to MOS transistors that have been carefully laid out to ensure optimal matching. Poorly matched transistors often exhibit gross defects that do not scale as predicted. Once these gross defects have been eliminated, the residual threshold mismatches usually follow Equation 12.16 quite precisely. Theoretical studies suggest that residual threshold mismatches stem mostly from statistical fluctuations in the distribution of backgate dopants.[20] Statistical fluctuations in the distribution of fixed oxide charge may also play a minor role.

Random short-range variations also appear to determine the residual transconductance mismatches observed between well-matched devices. If the transconductance mismatch is described as a normalized ratio s_k/k, then it varies with the effective dimensions, W_{eff} and L_{eff}, as

$$\frac{s_k}{k} = \frac{C_k}{\sqrt{W_{\text{eff}} L_{\text{eff}}}} \qquad [12.17]$$

---

[18] K. R. Lakshmikumar, R. A. Hadaway, and M. A. Copeland, "Characterization and Modeling of Mismatch in MOS Transistors for Precision Analog Design," *IEEE J. Solid-State Circuits*, SC-21, #6, 1986, pp. 1057–1066.

[19] Substituting drawn for effective dimensions will have very grave effects if either the width or the length of the matched devices is small; see S. J. Lovett, M. Welten, A. Mathewson, and B. Mason, "Optimizing MOS Transistor Mismatch," *IEEE J. Solid-State Circuits*, Vol. 33, #1, 1998, pp. 147–150.

[20] M. J. M. Pelgrom, A. C. J. Duinmaijer, and A. P. G. Welbers, "Matching Properties of MOS Transistors," *IEEE J. Solid-State Circuits*, Vol. SC-24, #5, 1989, pp. 1433–1439. Also see Lakshmikumar, *et al.,* p. 1059.

where C_k is a constant. Possible causes for short-range variations in transconductance include linewidth variation, gate oxide roughness, and statistical variations in mobility. The relative importance of these causes is not known, although several authors have suggested that mobility variations predominate.

Transconductance mismatches can also arise from peripheral variations. These so-called *edge effects* rarely play a significant role in devices with dimensions of 2 µm or more, but they can increase the mismatches between short-channel or narrow-channel transistors. One should generally avoid using matched transistors with minimum dimensions significantly less than 2 µm, but if such devices must be used, then their transconductance mismatch can be accurately computed with the equation

$$\frac{S_k}{k} = \sqrt{\frac{C_k^2}{W_{\text{eff}}L_{\text{eff}}} + \frac{C_{kp1}^2}{W_{\text{eff}}^2 L_{\text{eff}}} + \frac{C_{kp2}^2}{W_{\text{eff}} L_{\text{eff}}^2}} \qquad [12.18]$$

where C_k quantifies the areal-based mismatches, and C_{kp1} and C_{kp2} quantify the peripheral-based mismatches. For devices with relatively large dimensions, Equation 12.18 reduces to the same form as Equation 12.17, and the constant C_k that appears in these two equations is thus the same quantity. A study has shown that edge effects may become significant for devices with a channel length of about one micron.[21]

Gate Oxide Thickness

Transistors with thinner gate oxides generally exhibit better matching than those with thicker gate oxides. In the case of transconductance matching, the critical factor is actually backgate doping, which process designers usually increase for low-voltage thin-oxide devices in order to minimize channel length modulation and retard device punchthrough. A higher backgate doping concentration reduces variations caused by the random scattering of dopant atoms throughout the backgate. However, circuit designers do not always wish to use thin oxide transistors for current matching applications, as their higher transconductances make it more difficult to obtain adequate effective gate voltages without resorting to either excessively narrow or excessively long devices.

Thinner gate oxides benefit threshold voltage matching in several ways. Researchers have shown that the constant C_{Vt} in Equation 12.16 depends upon both oxide thickness t_{ox} and backgate doping N_b:

$$C_{Vt} = at_{\text{ox}}\sqrt{N_b} \qquad [12.19]$$

Here, a is a constant of proportionality. Although the mismatch actually increases with higher backgate doping, the effect of oxide thickness dominates the equation, and V_t mismatch consequently improves as process dimensions shrink.[22] Thinner oxides also provide higher transconductances that decrease effective threshold voltages. This indirectly improves voltage matching between MOS transistors, as shown by Equation 12.14.

Circuit designers generally favor thin-oxide devices for matching. However, thick-oxide transistors tend to have higher operating voltage ratings than their thin-oxide counterparts. Although cascodes can enable the use of thin-oxide devices at

---

[21] J. Bastos, M. Steyaert, R. Roovers, P. Kinget, W. Sansen, B. Graindourze, A. Pergoot, and E. Janssens, "Mismatch characterization of small size MOS transistors," *Proc. IEEE Conf. on Microelectronic Test Structures*, Vol. 8, 1995, pp. 271–276.

[22] M. J. M. Pelgrom, H. P. Tuinhout, and M. Vertregt, "Transistor matching in analog CMOS applications," *Int. Electron Devices Meeting Technical Digest*, 1998, pp. 915–918.

higher voltages, many circuit designers favor the simplicity of designs that employ thick-oxide devices. Thick-oxide transistors are also favored for analog circuitry in advanced submicron processes where the thin-oxide transistors suffer from severe channel length modulation and extremely limited operating voltages.

Channel Length Modulation

Channel length modulation can cause severe mismatches between short-channel transistors operating at different drain-to-source voltages. The systematic mismatch between the transistors is proportional to the difference between their drain-to-source voltages, and inversely proportional to their channel length. Drawn lengths of 10 to 20 μm are generally adequate for noncritical applications such as current distribution networks. Greater precision can be obtained by operating the matched transistors at equal drain-to-source voltages, for example, through the addition of cascodes. MOS designers rarely use source degeneration to combat channel length modulation because the low transconductance of MOS transistors makes it difficult to obtain adequate degeneration without using extremely large resistors.

Orientation

The transconductances of MOS transistors depend on carrier mobilities, and these in turn exhibit orientation-dependent stress sensitivities. MOS transistors oriented along different crystal axes will therefore exhibit different transconductances under stress. Since all packaged devices experience some stress, these mismatches can be avoided only by orienting matched transistors in the same direction. The devices in Figure 12.28A, which are oriented along the same crystal axis, match better than the devices in Figures 12.28B and 12.28C, which are not. Stress-induced mobility variations can induce current matching errors of several percent between rotated devices.[23] The use of tilted wafers may induce current matching errors of as much as 5%.[24]

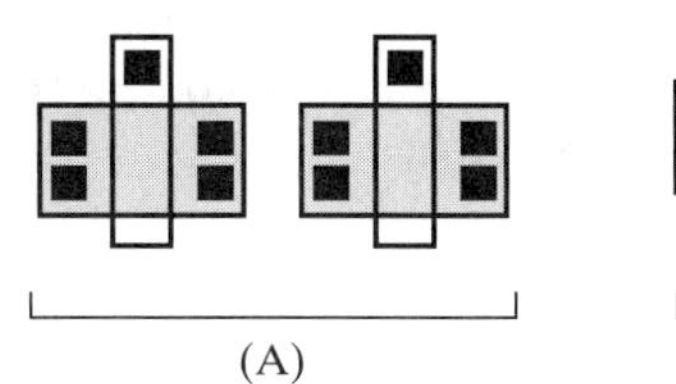

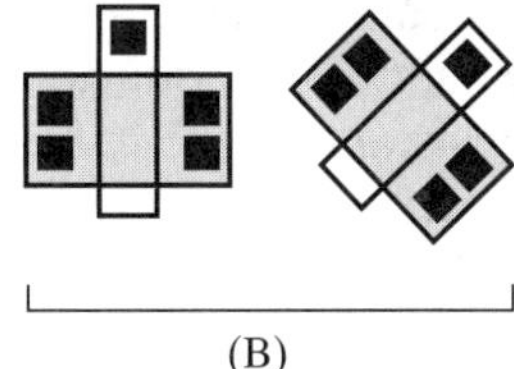

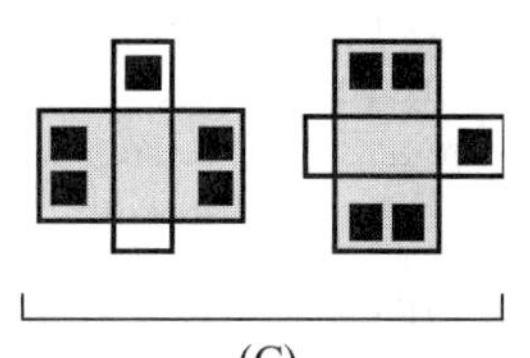

FIGURE 12.28 (A) Devices oriented in the same direction match more precisely than (B,C) those oriented in different directions.

Editing can easily introduce orientation errors if the design has not been properly partitioned. Consider a circuit that contains two matched transistors: M_1, located in cell X_1; and M_2, located in cell X_2. During top-level layout, the designer decides to rotate cell X_1 by 90°. Although this operation seems innocuous, it actually introduces a 90° difference between the orientations of M_1 and M_2. Errors of this sort can be prevented by grouping matched devices together in the same cells. This can sometimes make the schematic more difficult to comprehend, but it greatly reduces the risk of inadvertently introducing matching errors during editing.

MOS transistors that do not self-align must follow very strict orientation rules. Consider the asymmetric extended-drain NMOS transistors, M_1 and M_2, in Figure 12.29A. Each of these transistors is a mirror image of the other. The channel lengths of M_1 and M_2 are both defined by the overhang of their poly gates beyond

---

23 Pelgrom, *et al.*, p. 1436.

24 J. E. Chung, J. Chen, P.-K. Ko, C. Hu, and M. Levi, "The Effects of Low-Angle Off-Axis Substrate Orientation on MOSFET Performance and Reliability," *IEEE Trans. on Electron Devices*, Vol. 38, #3, 1991, pp. 627–633.

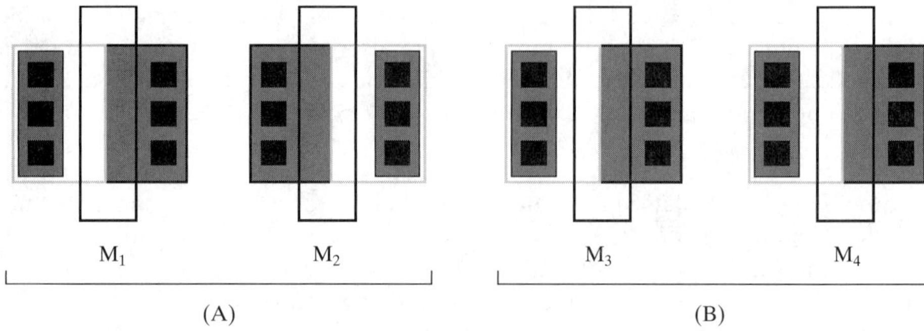

FIGURE 12.29 Extended-drain transistors that are (A) mirror images of each other experience mismatches that do not affect (B) superimposable transistors.

their respective N-well regions. Suppose that photolithographic misalignment causes the poly gates to shift to the right. This misalignment increases the channel length of M_1 and decreases the channel length of M_2. These mismatches are easily eliminated by ensuring that the matched devices are superimposable, as are M_3 and M_4 in Figure 12.29B. Even fully self-aligned transistors may experience slight orientation-dependent mismatches due to diagonal shifting of the source/drain implants (Section 12.3.5).

12.3.2. Diffusion and Etch Effects

The previous section examined sources of mismatch that depended solely upon geometry. Certain other types of mismatch are caused by the presence or absence of other structures near the matched transistors. For example, the presence of poly regions near the gate electrodes can cause slight variations in polysilicon etch rates. These variations produce mismatches in the effective widths and lengths of the matched transistors. Similarly, the placement of other diffusions near the channel may influence the backgate dopant concentration and may therefore cause variations in both threshold voltage and transconductance. The presence of contacts over the active gate region of a transistor can also induce mismatches, as can penetration of dopants through grain boundaries in the gate polysilicon.

Polysilicon Etch Rate Variations

Polysilicon does not always etch uniformly. Large poly openings clear more quickly than small ones because etchant ions have freer access to the sides and bottom of the large opening. The edges of the large opening therefore exhibit some degree of overetching by the time the smaller openings clear. This effect can cause variations in the gate lengths of poly-gate MOS transistors. Consider the layout in Figure 12.30A. The gate of transistor M_2 faces adjacent gates on both sides, but the gates of transistors M_1 and M_3 face an adjacent gate on only one side. The outside edges of the gates of M_1 and M_3 experience more erosion than the corresponding edges of the gate of M_2, so the gate lengths of M_1 and M_3 are slightly shorter than the gate length of M_2.

The etch rate variations experienced by MOS transistors are usually smaller than those experienced by poly resistors (Section 7.2.5), because poly gates do not lie as close together as poly resistor segments do. Many MOS transistors also use relatively long channel lengths. Even so, transistors that must achieve moderate or precise current matching should use dummy gates to ensure uniform etching. Failure to do so may produce current mismatches of 1% or more. Figure 12.30B shows an example of an array of MOS transistors incorporating dummies. Most designers make the dummy gates the same width as the active ones, but this precaution is not strictly necessary because the width of the poly strips is far less significant than their spacing.

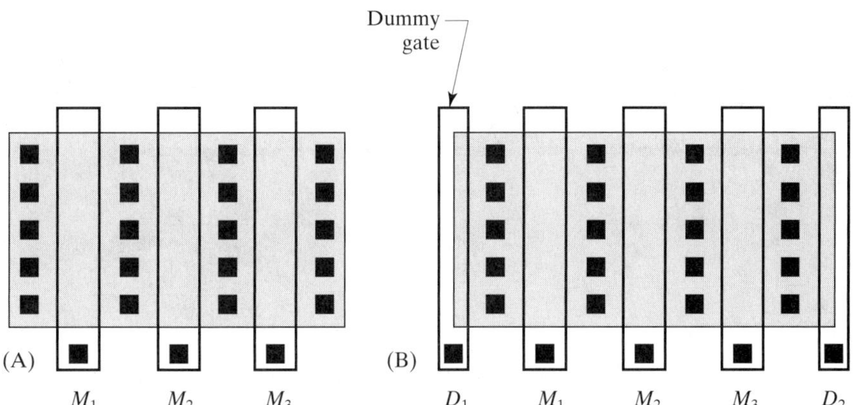

FIGURE 12.30 MOS transistor arrays (A) without dummy gates and (B) with dummy gates.

Dummies D_1 and D_2 are therefore made as narrow as possible while still allowing space for a contact. The spacing between the dummies and the actual gates must exactly equal the spacing between the actual gates themselves.

Since the dummies are not actual transistors, they do not require the presence of source/drain regions along their outside edges. The source/drain implant can therefore terminate on top of the dummies, as it does in Figure 12.30B. This should not introduce significant mismatches as long as the moat geometry extends beyond the inner edge of the dummy gate electrodes by a few microns to ensure that the edge of the dummy rests on thin gate oxide. Most sets of design rules include one that addresses this situation.

The dummy gate electrodes are usually electrically connected to either the source or the backgate of the transistor. Although this precaution is not strictly necessary, it helps ensure that the electrical characteristics of the transistors are not affected by the formation of spurious channels beneath the dummies. Some designers connect the dummies to the adjacent gate electrodes, but this practice is not recommended because it increases terminal capacitances and leakage currents.

Many designers interconnect multiple gate electrodes with a strip of polysilicon to produce a comblike gate structure. While this is undeniably convenient, it may introduce etch rate variations due to the presence of an adjacent polysilicon geometry. For the best possible matching, one should use simple rectangular strips of polysilicon connected by metal. If matched gates must be connected in poly, then increase the spacing between the connecting poly and the moat region by 1–2 μm beyond the minimum allowed by the design rules.

Diffusion Penetration of Polysilicon

Most processes deposit gate polysilicon in an intrinsic or near-intrinsic state. Dopants added by ion implantation are then redistributed throughout the poly during a subsequent annealing step. Unfortunately, dopants do not diffuse uniformly through the heterogeneous structure of polysilicon. Most dopants diffuse rapidly along grain boundaries and more slowly through the interior of individual grains. The dopant thus reaches the gate oxide first at points where grain boundaries intersect the plane of the oxide/poly interface (Figure 12.31A). If the anneal terminates at this stage in the diffusion, then the transistor will exhibit excessive threshold voltage variations due to partial depletion of incompletely doped poly grains. Continuing the anneal redistributes the dopants sufficiently to prevent depletion-induced mismatches (Figure 12.31B). However, if the anneal continues for too long

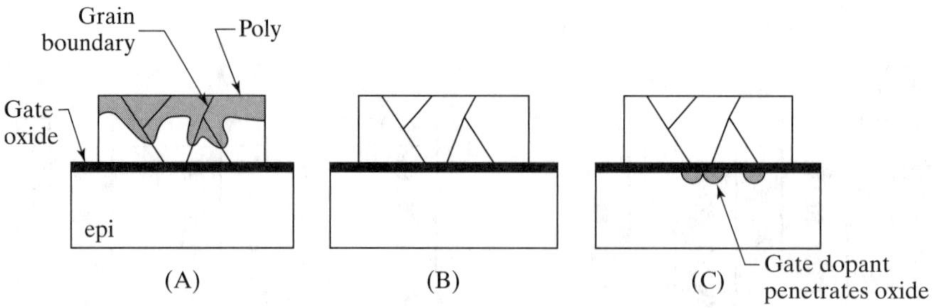

FIGURE 12.31 Cross section of polysilicon gate at three stages during diffusion of gate dopant: (A) before dopant completely redistributes, (B) after complete redistribution, and (C) after over-annealing causes dopant penetration of the gate oxide.

a time, then dopants may diffuse entirely through the thin oxide and into the backgate. Dopant penetration of the gate oxide occurs first near poly grain boundaries because the dopant contacted the gate oxide first at these points (Figure 12.31C). This mechanism can again introduce device mismatches.[25]

Contacts Over Active Gate

For reasons not well understood, the placement of contacts over the active gate regions of MOS transistors sometimes induces significant threshold voltage mismatches. One possible explanation for this effect is the presence of metal above the active gate. (See Section 12.3.3.) Another potential mechanism for contact-induced mismatches involves the localized silicidation of contacts. In processes where the gate poly is sufficiently thin, some silicide may actually penetrate entirely through the gate poly. The presence of silicide at the oxide interface drastically alters the work function of the gate electrode in the vicinity of the contact and can cause gross threshold voltage mismatches. Changes in grain size, dopant distributions, and stress patterns may also play a role in generating contact-induced mismatches. Figure 12.30 illustrates the proper placement of gate contacts in extensions of the poly gate electrodes. This precaution ensures that the contacts reside over thick-field oxide, where they cannot significantly alter transistor properties.

Annular transistors such as those in Figure 12.22 present a special problem because they require contacts to be placed over active gate regions. Matched annular transistors should be used only if absolutely necessary. If they are used, then they should incorporate identical arrangements of minimal numbers of small gate contacts. In annular extended-drain transistors, the gate contacts should reside over the field-relief regions so they rest on field oxide rather than on gate oxide. This precaution effectively locates the contacts outside the active gate region. In cases in which the field relief region is not wide enough to accommodate the contacts, they should still be located as far inside it as possible to take advantage of the zone of intermediate oxide thickness just inside the edges of the moat region (the bird's beak).

Diffusions Near the Channel

Deep diffusions can affect the matching of nearby MOS transistors. The tails of these diffusions extend a considerable distance beyond their junctions, and the excess dopants they introduce can shift the threshold voltages and alter the transconductances of nearby transistors. The deep-N+ sinker of the analog BiCMOS process

---

[25] H. P. Tuinhout, A. H. Montree, J. Schmitz, and P. A. Stolk, "Effects of Gate Depletion and Boron Penetration on Matching of Deep Submicron CMOS Transistors," *Int. Electron Devices Meeting Tech. Digest*, 1997, pp. 631–634.

represents one example of a deep diffusion. All sinkers and similar diffusions should be spaced away from matched channels by at least twice their junction depth.

Wells also qualify as deep diffusions. N-well geometries should not be placed near matched NMOS transistors, to prevent the tail of the N-well dopant distribution from intersecting the channels of the matched transistors. PMOS transistors should be placed far inside the edges of their enclosing N-well regions to prevent outdiffusion from causing variations in backgate doping. In all cases, a spacing from the active gate regions equal to or greater than twice the junction depth of the deep diffusion should limit interactions to negligible levels (Figure 12.32).

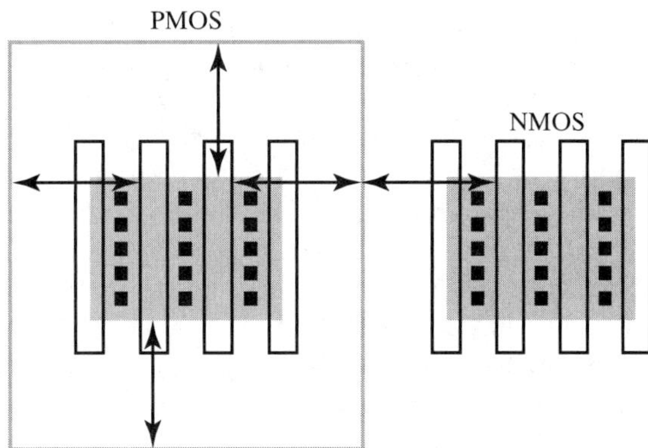

FIGURE 12.32 Spacings between drawn well boundaries and active gate regions.

Recent research has revealed another mechanism that can cause mismatches if the drawn edges of wells approach within a couple of microns of active gate regions. The high-energy ion implants used to form retrograde wells scatter within the photoresist. Some ions are deflected a few microns laterally and dope adjacent uncovered silicon regions more heavily than would otherwise be the case.[26]

Because MOS transistors are surface devices, they are vulnerable to the surface discontinuities produced by the NBL shadow. The channels of matched MOS transistors should be placed far enough away from NBL boundaries to allow for both misalignment and pattern shift. If the pattern shift has not been characterized, assume that it can displace the NBL shadow by up to 150% of the epi thickness. Thus the spacing from the active gate regions to the edge of the nearest NBL region should equal at least 150% of the epi thickness. Although this substantially increases the overlap of NBL over the matched transistor, much of this space is already required to satisfy the increased well spacings discussed previously.

PMOS versus NMOS Transistors

NMOS transistors often match more precisely than PMOS transistors. This phenomenon has been observed on a number of different processes, including both P-well and N-well variants. Several authors have reported that PMOS transistors exhibit 30 to 50% more transconductance mismatch than comparable NMOS transistors.[27] Some studies have also found increased threshold mismatches in PMOS transistors, although these do not appear to be as significant as the differences in transconductance matching.

---

[26] T. B. Hook, J. Brown, P. Cottrell, E. Adler, D. Hoyniak, J. Johnson, and R. Mann, "Lateral Ion Implant Straggle and Mask Proximity Effect," *IEEE Trans. Electron Devices*, Vol. 50, #9, 2003, pp. 1946–1951.

[27] Lakshmikumar, *et al.*, pp. 1060, 1062; Pelgrom, *et al.*, p. 1437.

The mechanisms responsible for the differences between PMOS and NMOS transistors are not well understood. Possible culprits include increased backgate doping variability, the presence of buried channels, and orientation-dependent stress effects. Several authors have suggested that the increased variability stems (at least in part) from differences in the threshold adjust implants, but this seems an unlikely explanation since so many different processes behave similarly.

12.3.3. Hydrogenation

Process designers have long used a reducing-atmosphere anneal to help stabilize the threshold voltages of MOS transistors. Researchers have shown that hydrogen seeps into the interlevel oxide during the course of the anneal. Some of the hydrogen eventually finds its way to the oxide-silicon interface, where it reacts with dangling bonds. This reaction neutralizes the positive fixed-oxide charge that the dangling bonds would otherwise introduce. The elimination of this charge helps minimize threshold voltage variations. More subtly, the random distribution of dangling bonds causes corresponding random fluctuations in threshold voltages. Hydrogen annealing therefore improves threshold voltage matching.

Differences in metallization patterns over matched MOS transistors can introduce large mismatches between devices that are otherwise identical. This effect was initially ascribed to stress variations induced by the presence (or absence) of metal above a structure. Such stress differences undoubtedly exist. However, they have little effect upon threshold voltages,[28] and only moderate effects upon transconductances. Incomplete hydrogenation appears to be the actual culprit behind metallization-induced mismatches.[29]

Several factors contribute to the prevalence of incomplete hydrogenation in modern CMOS integrated circuits. First, the presence of a nitride protective overcoat prevents hydrogen from entering the wafer from the annealing atmosphere. The small amounts of hydrogen that remain in the oxide from previous processing may not suffice to neutralize all dangling bonds. Second, metal geometries impede the redistribution of hydrogen to areas beneath them. This effect becomes particularly noticeable in metal systems that incorporate titanium silicide, or titanium-tungsten as titanium exhibits a considerable affinity for hydrogen. Much of the hydrogen left beneath metal geometries becomes adsorbed within the silicide before it can reach the dangling bonds at the oxide-silicon interface. Even if excess hydrogen exists elsewhere on the die, it must diffuse laterally beneath metal geometries. This process requires a prolonged anneal. Many modern processes do not incorporate sufficient annealing to neutralize all of the oxide charge beneath metal geometries.

Systematic drain current mismatches of up to 20% may occur between MOS transistors covered with metal and those without.[30] These mismatches persist regardless of the metal layers involved. Some designers have attempted to eliminate metallization-induced mismatches by entirely covering matched transistors with metal-2. The diffusion of hydrogen beneath the edges of the metal generates threshold voltage gradients that can produce significant mismatches between covered devices. Even if this were not the case, the presence of dangling bonds would still increase random threshold variations in the covered transistors. Therefore, no metallization should reside above the active gate regions of critical matched transistors. Less critical devices

---

[28] A. T. Bradley, R. C. Jaeger, J. C. Suhling, and K. J. O'Connor, "Piezoresistive Characteristics of Short-Channel MOSFETs on (100) Silicon," *IEEE Trans. on Electron Devices*, Vol. 48, #9, 2001, pp. 2009–2015.

[29] H. Tuinhout, M. Pelgrom, R. P. de Vries, and M. Vertregt, "Effects of Metal Coverage on MOSFET Matching," *Int. Electron Devices Meeting Tech. Digest*, 1996, pp. 735–738.

[30] Ibid.

may reside entirely under metal, or may have metal running across them, so long as each transistor sees exactly the same metal pattern. In no event should two matched transistors have different metal patterns placed above them. Similar metallization-induced mismatches have been observed in poly resistors (Section 7.2.8).

The presence of a metal geometry in close proximity to the active gate region of a MOS transistor can also induce mismatches. This effect occurs because the metal geometry (or the silicide beneath it) leaches hydrogen from surrounding areas. Only a little hydrogen remains in the oxide layers beneath the protective overcoat, and no additional hydrogen can penetrate through the nitride to replace that adsorbed by the metal. Consequently, regions that lie within several microns of a metal geometry may suffer from incomplete annealing of dangling bonds. Critical matched MOS transistors should therefore have identical metallization patterns surrounding them, at least out to a distance of several microns.

Fill Metal and MOS Matching

Modern metallization systems employ chemical-mechanical polishing (CMP) to obtain the high degree of planarity required for fine-line photolithography. CMP processing often requires the insertion of additional metal geometries to maintain a roughly constant metal pattern density (Section 2.6.4). The additional metal geometries are called *fill metal*. They often take the form of a pattern of squares or rectangles called *dummy tiles* automatically inserted during pattern generation. This process can result in the placement of metal geometries over matched MOS transistors.

Most processes that employ automatic fill metal generation allow the coding of a pseudolayer to exclude fill metal from selected regions. Alternatively, the layout designer can edit the database produced by the fill metal generator to remove tiles that lie above matched devices. In either case, the designer must heed the rules limiting the distance between fill metal areas in order to maintain adequate planarization.

Mismatches of up to 1% in drain current have been reported between transistors surrounded by different fill metal patterns, even though no fill metal actually resided over the transistors themselves.[31] These mismatches can probably be ascribed to hydrogen adsorption beneath the dummy tiles leaching hydrogen from adjacent exposed oxide regions. One should therefore surround matched transistors with custom-crafted dummy metal patterns to ensure that each transistor sees the same pattern of surrounding metal.

12.3.4. Thermal and Stress Effects

Another important category of mismatches stems from long-range variations called *gradients*. The magnitude of gradient-induced mismatches depends on the separation between the effective centers, or *centroids,* of the matched devices. Providing that the devices are placed relatively close to one another, the variation ΔP in parameter P between two matched devices equals the product of the distance d between the centroids and the gradient ∇P along a line connecting the two centroids:

$$\Delta P \cong d\nabla P \qquad [12.20]$$

The impact of the gradient on matching depends on both the magnitude of the gradient and the distance between the centroids of the matched devices. Gradients that affect MOS matching include those of oxide thickness, stress, and temperature.

---

[31] H. P. Tuinhout and M. Vertregt, "Characterization of Systematic MOSFET Current Factor Mismatch Caused by Metal CMP Dummy Structures," *IEEE Trans. Semiconductor Manufacturing*, Vol. 14, #4, 2001, pp. 302–310.

Oxide Thickness Gradients

The thickness of a grown oxide film depends on the temperature and composition of the oxidizing atmosphere used to create it. Although modern oxidation furnaces are very precisely controlled, slight variations of temperature and atmospheric composition still occur within the furnace tube. Thick oxide layers often exhibit a pattern of concentric rainbow-colored rings that betray the presence of a radial oxide thickness gradient. Gate oxides are too thin to exhibit interference colors, but they also tend to exhibit radial oxide thickness gradients. Devices placed close to one another have very similar oxide thicknesses, while devices placed farther apart exhibit greater differences in oxide thickness. These differences directly affect threshold voltage matching.

Stress Gradients

Stress affects the device transconductance of MOS transistors by causing variations in carrier mobilities. As discussed in Section 7.2.10, the effects of stress on mobility depend on orientation. In bulk <100> silicon, holes experience maximum stress dependence along the <110> axis and minimum stress dependence along the <100> axis. Similarly, electrons in bulk silicon experience maximum stress dependence along the <100> axis and minimum stress dependence along the <110> axis. Dice are oriented to the major wafer flat, which lies perpendicular to a <110> axis. Therefore, electrons experience minimum stress-induced bulk mobility variation in directions aligned with the X- and Y-axes of a (100)-oriented die, while holes experience minimum stress-induced bulk mobility variations in directions oriented 45° to these axes.

The stress dependence of bulk mobilities drops to nearly zero along the preferred orientations, but unfortunately the same is not true of the *effective mobilities* of carriers confined to a channel. The stress dependence of the effective mobilities does decrease along the directions predicted by theory, but these minima are nowhere near as pronounced as in the case of bulk mobilities. A diagonal placement of a PMOS transistor may reduce the stress dependence of its device transconductance by only 50%, rather than by the 90% or more one would expect based on bulk mobility data.[32] The randomizing effects of carrier collisions with the oxide/silicon interface probably account for the reduced orientation dependence of effective mobilities, but not all researchers agree on the details of this mechanism. Given these uncertainties, there seems little reason to diagonally orient PMOS transistors. One should instead rely on proper design of common-centroid layouts to minimize stress sensitivity.

Stress has relatively little effect on voltage matching because the threshold voltages of MOS transistors are largely independent of stress. What small stress dependencies do exist are probably caused by stress-induced changes in the bandgap voltage of silicon. The threshold voltage generally does not exhibit more than a few millivolts of stress-induced variation, which can be reduced still further by using common-centroid layout techniques.

Thermal Gradients

The voltage matching of MOS transistors depends primarily on the matching of threshold voltages. Threshold voltages decrease with temperature at roughly the same rate as base-emitter voltages of bipolar transistors—about −2 mV/°C. Most of the temperature coefficient stems from variations in the work functions of the gate

---

[32] H. Mikoshiba, "Stress-sensitive Properties of Silicon-gate MOS Devices," *Solid-State Elect.*, Vol. 24, #3, 1881, pp. 221–232.

and backgate materials with temperature, and it is therefore virtually independent of drain current.[33] Voltage-matched MOS transistors therefore exhibit about the same sensitivity to thermal gradients as bipolar transistors.

MOS and bipolar input differential pairs respond quite differently to offset trimming. In bipolar circuits, trimming the offset voltage to zero also trims its temperature dependence to zero. This happens because the equation for the offset voltage between two matched bipolar transistors contains only one significant source of temperature variation: the thermal voltage V_T. The temperature dependence therefore scales directly with ΔV_{BE}, and when ΔV_{BE} has been trimmed to zero, it vanishes. The input offset voltages of MOS transistors are trimmed by adjusting drain current densities. This operation attempts to cancel the mismatch in threshold voltages by introducing a compensating offset in device transconductance. Different mechanisms cause the temperature coefficient of threshold voltage and the temperature coefficient of transconductance, so the two are not equal, and the trimming operation cannot reduce the temperature coefficient to zero. Thus, while trimmed bipolar input differential pairs retain their very low offset voltages over temperature, trimmed MOS input differential pairs do not. Trimmed bipolar amplifiers and comparators therefore provide much better performance over temperature than do their MOS counterparts.

The current matching of MOS transistors depends primarily on the matching of device transconductances. These transconductances are directly proportional to effective carrier mobilities, which exhibit rather large temperature coefficients. At temperatures near 25°C, MOS device transconductances typically exhibit temperature coefficients of about –7000 ppm/°C. Temperature variations in threshold voltage have little effect on current matching as long as the transistors operate at a relatively large effective gate voltage V_{gst}. The low transconductance of MOS transistors makes them much less sensitive to thermal gradients than bipolar transistors, but it also makes it difficult to improve matching by source degeneration. Instead of relying on degeneration resistors, one should use common-centroid layout techniques.

12.3.5. Common-Centroid Layout of MOS Transistors

Gradient-induced mismatches can be minimized by reducing the distance between the centroids of the matched devices. Some types of layout can actually reduce the distance between the centroids to zero. These *common-centroid* layouts can entirely cancel the effects of long-range variations as long as these are linear functions of distance. Even if the variations contain a nonlinear component, they still remain approximately linear over short distances. The more compact the common-centroid layout can be made, the less susceptible it becomes to nonlinear gradients. The best layouts for MOS transistors combine exact alignment of the centroids with compactness.

The active gate region of an MOS transistor usually takes the form of a long, narrow rectangle. As in the case of resistors, MOS transistors are usually divided into segments, or *fingers,* to allow the construction of a compact array. The simplest types of arrays involve the placement of multiple device fingers in parallel. If these fingers are properly interdigitated, then the centroids of the matched devices will align at a point midway along the axis of symmetry bisecting the array. Figure 12.33 shows an example of a pair of matched MOS transistors laid out as an interdigitated array.

This layout uses the interdigitation pattern ABBA to ensure exact alignment of the centroids (Section 7.2.10). If source and drain fingers are denoted by subscripts,

---

[33] F. M. Klaassen and W. Hes, "On the Temperature Coefficient of the MOSFET Threshold Voltage," *Solid-State Elect.,* Vol. 29, #8, 1986, pp.787–789.

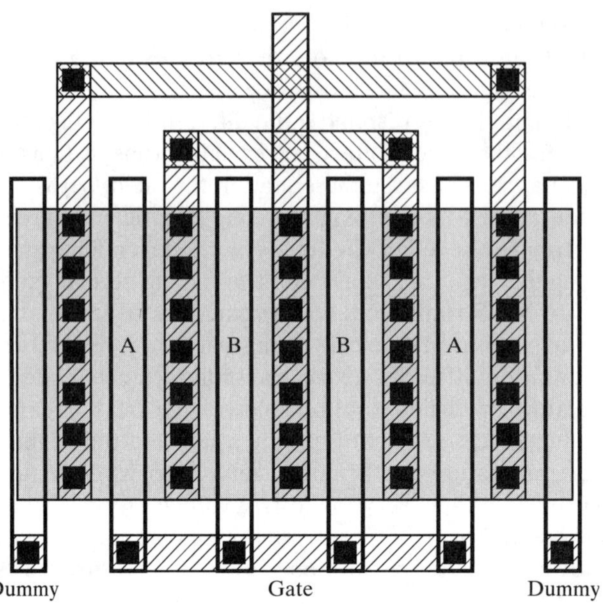

FIGURE 12.33 Interdigitated MOS transistors.

then the pattern becomes ${}_{D}A_{S}B_{D}B_{S}A_{D}$. Notice that the A-segment on the right has its drain on the right, while the A-segment on the left has its drain on the left. Similarly, the B-segment on the right has its source on the right, while the B-segment on the left has its source on the left. Each transistor thus contains one segment oriented in either direction. The reason for this precaution is rather subtle. Suppose one transistor consists entirely of segments with drains on the left, while a second transistor consists entirely of segments with drains on the right. If left-oriented and right-oriented segments differ in any way, then the two transistors will not match. If both transistors consist entirely of segments oriented in the same direction, then the effect of orientation on each transistor will be the same (Section 12.3.1). If each transistor consists of an equal number of left-oriented and right-oriented segments, then the effects of orientation will cancel and the transistors will again match.

More generally, if we define the *chirality* of a transistor as the fraction of right-oriented segments it contains minus the fraction of left-oriented segments it contains, then transistors having equal chirality will not experience orientation-dependent mismatches.[34] For example, a transistor having three right-oriented segments and one left-oriented segment has a chirality of $3/4 - 1/4 = 1/2$. Similarly, a transistor having nine right-oriented and three left-oriented segments has a chirality of $9/12 - 3/12 = 1/2$. Since these transistors have equal chirality, they do not exhibit any orientation-dependent mismatch. Most designers prefer to use transistors having chiralities of zero; in other words, transistors that consist of equal numbers of left- and right-oriented segments.

Orientation-dependent mismatches can develop in MOS transistors due to diagonal shifts in the source/drain implants. Such diagonal shifts occur when ion implantation is performed at an angle to prevent channeling.[35] Such *tilted implants* cause the source/drain regions on the left side of the gates to differ from the source/drain

---

[34] The term *chirality* refers to the asymmetry, or handedness, of an object. The term is most commonly encountered in stereochemistry.

[35] J. F. Gibbons, "Ion Implantation in Semiconductors—Part I: Range Distribution Theory and Experiment," *Proc. IEEE,* Vol. 56, #3, 1968, pp. 296–319.

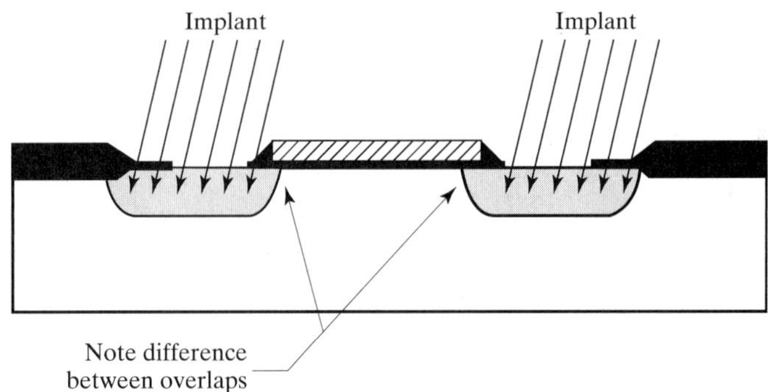

FIGURE 12.34 Diagonal shift in the source/drain regions of an implanted transistor due to the use of a tilted implant. The angle of implantation has been exaggerated for clarity.

regions on the right side (Figure 12.34). If the matched devices are arranged in a pattern such as $_DA_SB_D$, then the drain of the left-hand device differs from the drain of the right-hand device. Similarly, the source of the left-hand device differs from the source of the right-hand device. Tilted implants have little effect upon the matching of transistors operated in the linear region, but saturated devices sometimes experience small transconductance differences. These mismatches become worse as the voltage drop across the device approaches maximum because tilted implants have an especially strong impact on hot-carrier generation.[36] These orientation dependencies cancel as long as the matched transistors have equal chirality.

Some newer ion implantation systems support on-axis implantation to help minimize the problems caused by tilted implants. The addition of an oxide layer helps minimize channeling, but some still occurs. BiCMOS processes sometimes use *tilted wafers* that have been cut off-axis to minimize pattern distortion. The tilted silicon lattice channels a portion of the ion beam, and this diagonal channeling may cause slight device asymmetries despite the use of on-axis implants.

The interdigitation patterns for common-centroid MOS transistor arrays are often difficult to construct, as it is not easy to satisfy all of the rules of common-centroid layout. MOS transistors must obey not only all four rules given in Section 7.2.10, but also a fifth rule—that of *orientation*. This additional rule ensures that tilted implants (and other device asymmetries) do not affect matching. The full set of rules for MOS devices are as follows:

1. **Coincidence:** The centroids of the matched devices should at least approximately coincide. Ideally, the centroids should exactly coincide.
2. **Symmetry:** The array should be symmetric around both the X- and Y-axes. Ideally, this symmetry should arise from the placement of segments in the array and not from the symmetry of the individual segments themselves.
3. **Dispersion:** The array should exhibit the highest possible degree of dispersion; in other words, the segments of each device should be distributed throughout the array as uniformly as possible.
4. **Compactness:** The array should be as compact as possible. Ideally, it should be nearly square.
5. **Orientation:** Each matched device should consist of an equal number of segments oriented in either direction; more generally, the matched devices should possess equal chirality.

[36] F. K. Baker and J. R. Pfiester, "The Influence of Tilted Source-Drain Implants on High-Field Effects in Submicrometer MOSFETs," *IEEE Trans. on Electron Devices*, Vol. 35, #12, 1988, pp. 2119–2124.

TABLE 12.1 Sample interdigitation patterns for MOS transistor arrays.

1. $({}_S A_D A)({}_S B_D B_S B_D B)({}_S A_D A)_S$
2. $({}_D A_S B_D - {}_D B_S A_D) - ({}_D A_S B_D - {}_D B_S A_D)$
3. $({}_D A_S B_D B_S A)_D$
4. $({}_S A_D A_S B_D B)_S(B_D B_S A_D A_S)$
5. $({}_S A_D A_S B_D B_S A_D A)_S$
6. $({}_S A_D A_S B_D - {}_S A_D A_S - {}_D B_S A_D A)_S$
7. $({}_S A_D A_S B_D B_S C_D C)_S(C_D C_S B_D B_S A_D A_S)$

Table 12.1 shows a few of the simpler interdigitation patterns used for MOS transistors. Source and drain fingers are denoted by subscripts, and sequences of segments that may be repeated are enclosed in parentheses: $({}_S A_D A)$. When a pattern includes more than one repeated sequence, each portion of the sequence in parentheses must be replicated the same number of times. Certain patterns contain locations where the source/drain fingers cannot merge with one another; these are denoted by dashes. All of the entries in this table obey the rules of coincidence, symmetry, and orientation, but many of them are not as disperse nor as compact as possible. For example, consider patterns 1 to 4, all of which provide a 1:1 ratio between two matched devices. Pattern 1 lacks dispersion because it contains long runs of segments belonging to the same device. Pattern 2 contains gaps that make it less compact than the others. Patterns 3 and 4 both exhibit considerable dispersion because the segments appear in pairs throughout most parts of the array. However, the middle of pattern 4 contains a run of four segments belonging to the same device. The middle of pattern 3 contains a run of only two segments, so it provides better dispersion than pattern 4. In summary, pattern 3 should exhibit more precise matching than patterns 1, 2, and 4. The device of Figure 12.33 uses pattern 3.

Two-dimensional common-centroid arrays generally provide better matching than one-dimensional interdigitation patterns. This effect has been attributed to tighter packing of segments within the two-dimensional array. Common centroid layout cancels the linear component of a gradient-induced mismatch, but it does not cancel the higher-order components. The larger the spacing between segments, the larger the impact of these residual mismatches. Finite-element analysis suggests that a two-dimensional array AB/BA of square elements should have about 60% of the residual mismatch of a linear array ABBA of the same elements.[37] Since MOS transistors often have approximately square geometries, the use of a two-dimensional array of the form AB/BA, often called a *cross-coupled pair*, can frequently improve matching.

Cross-coupled pairs only improve matching if a significant gradient exists across the devices, such as a temperature gradient produced by a nearby power device, or a stress gradient produced by packaging. Test devices often do not reveal any difference between cross-coupled pairs and interdigitated pairs because these devices do not include heat sources. Furthermore, they are often packaged in ceramic or metal can packages rather than in the higher stress plastic packaging typical of production devices. Practical devices usually experience sufficient gradient-induced offsets to make consideration of two-dimensional arrays worthwhile for precisely matched devices.

---

[37] M. F. Lan, A. Tammineedi, and R. Geiger, "A New Current Mirror Layout Technique for Improved Matching Characteristics," *Proc. Midwestern Semiconductor Circuits and Systems Conf.*, 1999, pp. 1126–1129.

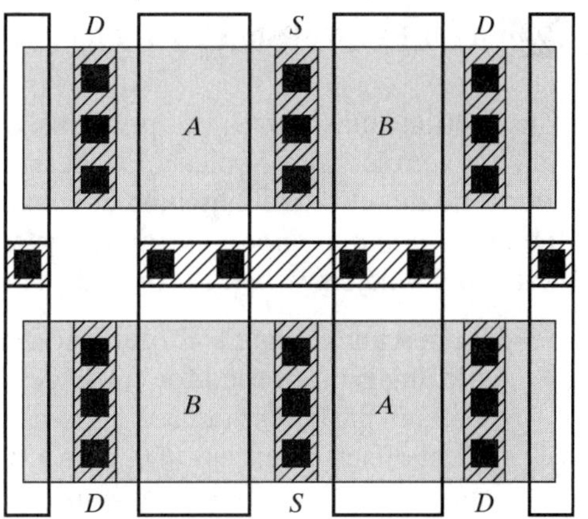

FIGURE 12.35 Cross-coupled MOS transistors.

Figure 12.35 shows the simplest possible cross-coupled pair. This layout follows the interdigitation pattern $_{D}A_{S}B_{D}/_{D}B_{S}A_{D}$, where the slash (/) separates the segments that occupy the upper two quadrants from those that occupy the lower two.[38] Not only does this produce a very compact layout, but it also satisfies the rule of orientation, because the two segments belonging to each matched device are oriented in opposite directions. This layout is especially suited for pairs of relatively small MOS transistors.

Larger cross-coupled pairs are more difficult to construct. Most designers simply divide each transistor into two equal halves and place these halves in diametrically opposite corners of the array. A layout of this sort can be represented by the pattern XY/YX, where *X* and *Y* are subarrays composed entirely of segments of transistors A and B, respectively. A typical implementation of such an array is $(_{S}A_{D}A)_{S}(B_{D}B_{S})/(_{S}B_{D}B)_{S}(A_{D}A_{S})$. While this pattern satisfies most of the rules of interdigitation, it does not provide optimal dispersion. As the array grows larger, its lack of dispersion renders it increasingly susceptible to mismatches caused by nonlinear components of the variation. A much better pattern for large cross-coupled pairs is $(_{D}A_{S}B_{D}B_{S}A)_{D}/_{D}(B_{S}A_{D}A_{S}B_{D})$. If the array becomes very large, then additional dispersion can be introduced by elaborating this array in the vertical dimension, as shown in the following examples:

$_{D}A_{S}B_{D}B_{S}A_{D}$	$_{D}A_{S}B_{D}B_{S}A_{D}$	$_{D}A_{S}B_{D}B_{S}A_{D}A_{S}B_{D}B_{S}A_{D}$
$_{D}B_{S}A_{D}A_{S}B_{D}$	$_{D}B_{S}A_{D}A_{S}B_{D}$	$_{D}B_{S}A_{D}A_{S}B_{D}B_{S}A_{D}A_{S}B_{D}$
	$_{D}A_{S}B_{D}B_{S}A_{D}$	$_{D}A_{S}B_{D}B_{S}A_{D}A_{S}B_{D}B_{S}A_{D}$
	$_{D}B_{S}A_{D}A_{S}B_{D}$	$_{D}B_{S}A_{D}A_{S}B_{D}B_{S}A_{D}A_{S}B_{D}$
		$_{D}A_{S}B_{D}B_{S}A_{D}A_{S}B_{D}B_{S}A_{D}$
		$_{D}B_{S}A_{D}A_{S}B_{D}B_{S}A_{D}A_{S}B_{D}$

The main drawback to the more elaborate patterns of this type lies in the difficulty of connecting the various segments together to form the full device. This becomes particularly difficult in cases where the gates of the two matched devices do not connect together. The simpler—and hence easier to connect—patterns generally serve unless the cross-coupled transistors are relatively large.

[38] Tsividis discusses a $_{D}A_{S}B_{D}/_{D}B_{S}A_{D}$ array, but without the dummies: Y. Tsividis, *Mixed Analog-Digital VLSI Devices and Technology* (New York: McGraw-Hill, 1997), p. 233.

12.4 RULES FOR MOS TRANSISTOR MATCHING

This section summarizes the previously given information in the form of a set of qualitative rules. These rules allow designers to construct matched MOS transistors, even if no quantitative matching data exist for the process in question. The rules use the terms *minimal, moderate,* and *precise* to denote increasingly precise degrees of matching, which may be interpreted as follows:

- **Minimal matching:** Drain current mismatches of several percent. Minimal matching is often used for constructing bias current networks that do not require any particular degree of precision. This level of matching corresponds to typical offsets in excess of ± 10 mV and is therefore generally considered inadequate for voltage matching applications.
- **Moderate matching:** Typical Offset voltages of ± 5 mV or drain current mismatches of less than $\pm 1\%$. Useful for constructing input stages of noncritical op-amps and comparators, where untrimmed offsets of ± 10 mV can be maintained.
- **Precise matching:** Typical Offset voltages of less than ± 1 mV or drain current mismatches of less than $\pm 0.1\%$. This level of matching usually involves trimming, and the resulting circuit will probably meet specification within only a limited range of temperatures due to the presence of uncompensated temperature variations.

The following rules summarize the most important principles of MOS transistor matching:

1. Use identical finger geometries.
   Transistors of different widths and lengths match very poorly. Even minimally matched devices must have identical channel lengths. Most matched transistors require relatively large widths and are usually divided into sections, or *fingers*. Each of these fingers should have the same width and length as all others. Do not attempt to match transistors of different widths and lengths, because the width and length correction factors, δW and δL, vary substantially from lot to lot.

2. Use large active areas.
   The active area of an MOS transistor equals the product of its effective channel width and length. Assuming that all other matching considerations have been addressed, the residual offset due to random fluctuations scales inversely with the square root of device area. Moderate matching usually requires active areas of several hundred square microns, while precise matching requires thousands of square microns.

3. For voltage matching, keep V_{gst} small.
   The offset voltage of a pair of matched MOS transistors contains a term dependent on device transconductance. This term scales with V_{gst}, so smaller values of V_{gst} provide better voltage matching. Reducing the V_{gst} below 0.1 V provides little additional benefit because threshold voltage variations begin to dominate the offset equation. Most designers decrease V_{gst} by using larger W/L ratios because these simultaneously increase the active area of the transistors.

4. For current matching, keep V_{gst} large.
   The current mismatch equation contains a term dependent upon threshold voltage. This term scales inversely with V_{gst}, so large values of V_{gst} minimize its impact upon current matching. Circuits relying upon current matching should

maintain a nominal V_{gst} of at least 0.3 V. Moderately matched transistors should maintain a nominal V_{gst} of at least 0.5 V whenever headroom allows. Precisely matched transistors should use the largest value of V_{gst} permitted by the configuration of the circuit, but in any event should equal at least 0.5 V.

5. Use thin-oxide devices in preference to thick-oxide devices.
Some processes offer multiple thicknesses of gate oxides. The transistors with thinner gate oxides generally exhibit better matching characteristics than thick-oxide devices. Whenever circuit configurations allow, consider using thin-oxide devices in preference to thick-oxide devices. The higher transconductance of thin-oxide devices can also improve voltage matching by reducing V_{gst}.

6. Orient transistors in the same direction.
Transistors that do not lie parallel to one another become vulnerable to stress- and tilt-induced mobility variations that can cause several percent variation in their transconductance. This effect is so severe that even minimally matched transistors should lie parallel to one another. Matched transistors, especially those that are not fully self-aligned, should have equal chirality. This condition can be met by ensuring that each transistor contains an equal number of segments oriented in each direction.

7. Place transistors in close proximity.
MOS transistors are vulnerable to gradients in temperature, stress, and oxide thickness. Even minimally matched devices should reside as close as possible to one another. Moderately or precisely matched transistors should be kept next to one another to facilitate common-centroid layout.

8. Keep the layout of the matched transistors as compact as possible.
Wide MOS transistors naturally lend themselves to long, spindly layouts that are extremely vulnerable to gradients. Each device should be divided into a number of segments to provide as compact an array as possible. Matched devices should all consist of segments having the same width and length.

9. Where practical, use common-centroid layouts.
Moderately and precisely matched MOS transistors require some form of common-centroid layout. This can often be achieved by dividing each transistor into an even number of fingers and by then arranging these fingers in an interdigitated array. Pairs of matched transistors should be laid out as cross-coupled pairs to take advantage of the superior symmetry of this arrangement.

10. Avoid using extremely short or narrow transistors.
Dimensions much less than 1 μm may cause transistors to experience increased random mismatches due to peripheral effects. Unless data exist to show the relative significance of peripheral effects, avoid submicron dimensions in precisely matched transistors.

11. Place dummy segments on the ends of arrayed transistors.
Arrayed transistors should include dummy gates at either end. These dummies need not have the same length as the actual gates, but the spacing between the dummy gates and the actual gates must equal the spacing between actual gates. The moat diffusion should extend at least several microns into the dummies to prevent the edge of the dummies from resting on the bird's beak. The dummy gates should be connected, preferably to a potential that prevents channel formation beneath them. This is most easily achieved by connecting the dummies to the backgate potential.

12. Place transistors in areas of low stress gradients.
The stress gradients reach a broad minimum in the center of the die. Any location ranging from the center of the die out halfway to the edges will fall within this broad minimum. Whenever possible, precisely matched transistors should reside within this area. Moderately and precisely matched transistors should reside at least 250 μm away from any side of the die. The stress distribution reaches a maximum in the die corners, so avoid placing any matched transistors near corners. PMOS transistors may experience slightly less stress dependence when oriented along [100] directions. This effect is not sufficiently pronounced to justify placing minimally or moderately matched transistors diagonally, but precisely matched transistors might benefit from this unconventional orientation. NMOS transistors should always be oriented horizontally and vertically.

13. Place transistors well away from power devices.
For purposes of discussion, any device dissipating more than 50mW should be considered a power device, and any device dissipating more than 250mW should be considered a major power device. Precisely matched transistors should reside on an axis of symmetry of the major power devices using an optimal symmetry arrangement. (See Section 7.2.10.) Moderately and precisely matched transistors should reside no less than 300–500 μm away from the closest power device. Minimally matched devices may be placed next to power devices, but only if they use some form of common-centroid layout.

14. Do not place contacts on top of active gate area.
Whenever possible, extend the gate poly beyond the moat and place the gate contacts over thick-field oxide. When this is not possible, minimize the number and size of the gate contacts and place them in the same location on each transistor. Consider placing the gate contacts of high-voltage annular transistors over the field-relief structure because this is not part of the active gate.

15. Do not route metal across the active gate region.
Whenever possible, avoid routing metal across the active gate region of moderately or precisely matched MOS transistors. Leads may route across minimally matched MOS transistors, but additional dummy leads should be added so that every section of the array of matched devices is crossed at the same location along its channel by an identical length of lead.

16. Keep all junctions of deep diffusions far away from active gate area.
The minimum spacing between a drawn well boundary and a precisely matched MOS transistor should equal at least twice the well junction depth. Moderately and minimally matched transistors need only obey the applicable layout rules. Similar considerations apply to deep-N+ sinkers and other deep diffusions.

17. Place precisely matched transistors on axes of symmetry of the die.
Arrays of precisely matched transistors should be placed so that the axis of symmetry of the array aligns with one of the two axes of symmetry of the die. If the design contains large numbers of matched transistors, then reserve the optimal locations for the most critical devices.

18. Do not allow the NBL shadow to intersect the active gate area.
The NBL shadow should not fall across the active gate region of any precisely matched transistor. If the direction of the NBL shift is unknown, allow adequate overlap of NBL over the transistor on all sides. If the magnitude of the

NBL shift is also unknown, then overlap NBL over the active gate region by at least 150% of the maximum epi thickness.

19. Connect gate fingers using metal straps.
Connect the gate fingers of moderately and precisely matched transistors, using metal rather than poly. Minimally matched transistors can use a poly comb structure to simplify the connection of the gate electrodes.

20. Consider using NMOS transistors rather than PMOS transistors.
NMOS transistors generally match better than PMOS transistors. Whenever circuit considerations allow, consider using NMOS transistors rather than PMOS transistors.

12.5 SUMMARY

Many circuit designers think of MOS transistors primarily as building blocks for digital logic. While they are indispensable in this role, they also have many other important applications. Modern mixed-mode integrated circuits rely heavily upon MOS transistors for power switching and low-current analog functions.

MOS power transistors have revolutionized power switching. The new generation of high-efficiency switch-mode power supplies relies almost exclusively upon MOS power transistors. Similarly, virtually all low-voltage power distribution circuits use MOS transistors. Now that BiCMOS processes offer integrated power MOS transistors with specific on-resistances approaching those of discrete devices, it has become economically feasible to integrate many power switches into a single integrated circuit. The low on-resistances of the power switches also minimize power dissipation, allowing the use of compact surface-mount packages.

MOS transistors have also found many applications in analog signal processing circuitry. Although bipolar transistors remain entrenched in a few applications, the vast majority of analog circuitry can be implemented using MOS transistors. MOS circuits are usually smaller and consume less power than their bipolar counterparts. Even though most modern analog circuits are implemented in BiCMOS processes, the vast majority of the circuitry consists of MOS transistors. Circuit designers continue to develop new applications for MOS transistors, so future integrated circuits will probably incorporate an even larger percentage of MOS circuitry.

12.6 EXERCISES

Refer to Appendix C for layout rules and process specifications.

12.1. Suppose an extended-voltage transistor with a drain depletion width x_d equal to 10% of the pinched-off region width x_p can withstand a drain-to-source voltage of 10 V. What drain-to-source voltage could a similar device withstand if x_d were increased to 50% of x_p?

12.2. Suggest a structure for a self-aligned extended-voltage PMOS transistor. Draw a cross section of a representative transistor using this structure.

12.3. If the thin gate oxide of an extended-drain NMOS having no field-relief structure can withstand 10 V, then which of the following biasing conditions are allowable, and why?

a. Asymmetric NMOS, $V_{GS} = 6$ V, $V_{DS} = 10$ V.
b. Asymmetric NMOS, $V_{GS} = 7$ V, $V_{DS} = 16$ V.
c. Asymmetric NMOS, $V_{GS} = 3$ V, $V_{DS} = 16$ V.
d. Symmetric NMOS, $V_{GS} = -13$ V, $V_{DS} = -16$ V.
e. Symmetric NMOS, $V_{GS} = 20$ V, $V_{DS} = 0$ V.

12.4. Lay out asymmetric and symmetric extended-drain NMOS transistors, each having drawn dimensions of 2(15/10). The N-well drain geometry should abut the N moat source geometry beneath the gate as shown in Figure 12.6. The overlap of the poly gate over the N-well should equal exactly 3 μm. Include abutting backgate contacts for the asymmetric transistor. Why can't abutting backgate contacts be used for the symmetric transistor?

12.5. Compute the maximum theoretical $R_{DS(on)}$ for a 50000/2 NMOS power transistor operating at $5 < V_{GS} < 15$ V and $-40 < T_j < 150$°C. Assume that the device's threshold voltage equals 0.7 ± 0.2 V with a temperature coefficient of −2 mV/°C, and that its process transconductance equals 35 uA/V^2 ± 20% at 150°C.

12.6. Determine the specific on-resistance (in $\Omega \cdot \text{mm}^2$) for a power device having an $R_{DS(on)}$ of 165 mΩ and an area of 2.26 mm^2. Use this information to determine the area required for a 100 mΩ power transistor. Assume both $R_{DS(on)}$ values do not include bondwire or leadframe resistance.

12.7. Compute the ideal ratio B/L for a metal system consisting of a first layer of metal that is 7500 Å thick and a second layer of metal that is 14,000 Å thick. Lay out a 20,000/2 PMOS transistor, using this ratio and the analog BiCMOS layout rules in Appendix C. Divide the transistor into sufficient fingers to produce a roughly square aspect ratio. Fill the well with NBL and ring the outer edge of the well with deep-N+ to provide a backgate contact for the transistor. Include all necessary metallization.

12.8. Compute the approximate metallization resistance of the transistor in Exercise 12.7. Do not include the resistance of the metal-2 buses extending beyond the interdigitated region of the transistor.

12.9. Assume the source and drain leads of the transistor in Exercise 12.7 run 25 μm from the drawn edge of the well to the edge of their respective bondpads, and that each bondpad connects to a 600 μm-long 1mil-diameter gold bondwire. Calculate the total metallization resistance of the transistor, including bondwires. Assume that the resistance between the edge of the bondpad and the bondwire is negligible.

12.10. Lay out a minimum-size, annular, lateral DMOS transistor, using the analog BiCMOS rules in Appendix C, supplemented by the following rules for the DMOS layer:

1. DMOS width	5 μm
2. DMOS spacing to DMOS	4 μm
3. DMOS spacing to PMOAT	0 μm
4. POLY extends into DMOS	2 μm
5. POLY overhang of DMOS	4 μm
6. MOAT overlap of DMOS	2 μm
7. CONT extends into DMOS	2 μm

A DMOS to PMOAT spacing of 0 μm implies that the outer edge of the PSD plug should coincide with the inner edge of the annular DMOS ring. Include all necessary metallization.

12.11. If the length of the DMOS channel equals 1 μm and the inner edge of the channel coincides with the outer edge of the drawn DMOS geometry, then what is the drawn width of the transistor constructed in Exercise 12.9?

12.12. Lay out a standard bipolar epi-FET having drawn dimensions of 30/8. Assume that the base pinch plate must extend at least 2 μm into the isolation.

12.13. Construct a minimum-size circularly symmetric epi-FET. Include all necessary metallization. What are the drawn width and length of this device?

12.14. A cross-coupled NMOS differential pair of transistors, each having dimensions 100/10, has a three-sigma random mismatch of ±2.85 mV. Estimate the three-sigma random mismatch of a similar differential pair where the transistors each have dimensions of 1000/5.

12.15. Lay out a pair of differential NMOS transistors, each having dimensions of 1000/5, to obtain the best possible matching. The transistors may be divided into as many or as few segments as desired. Assume that backgate contacts are required only along the edges of the array. Include all necessary metallization, including the links connecting individual source/drain fingers and the links connecting the gate fingers.

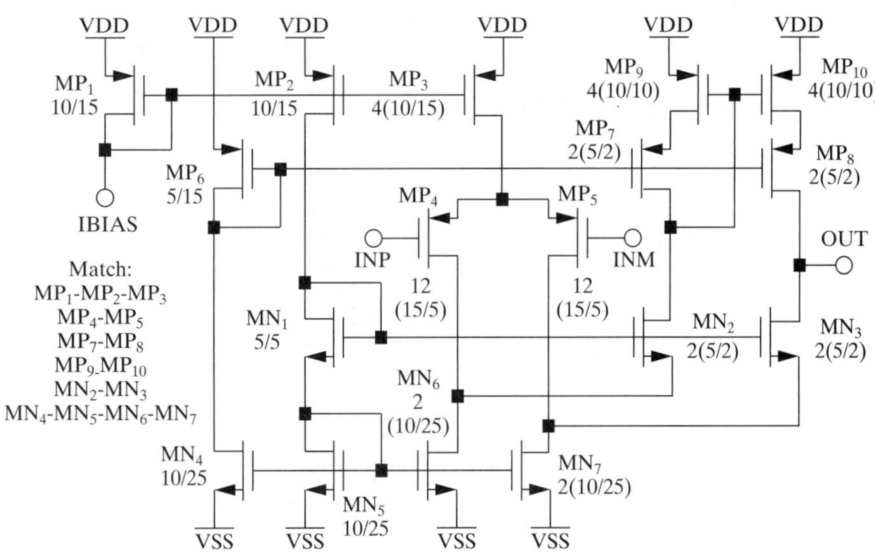

FIGURE 12.36 Folded-cascode MOS operational amplifier for Exercise 12.16.

12.16. Lay out the MOS operational amplifier shown in Figure 12.36 by following the recommendations for optimal matching. Use the poly-gate CMOS rules in Appendix C, and include all necessary backgate and substrate contacts. Assume that all PMOS transistors have backgates connecting to VDD.

12.17. Compare the matching of the following interdigitation patterns:

a. ${}_SA_DA_SB_D$ – ${}_DA_SA_D$ – ${}_DB_SA_DA_S$

b. ${}_DA_SA_DA_SB_DB_SA_DA_SA_D$

c. ${}_DB_SA_DA_SA_DA_SA_DA_SB_D$

Which pattern provides the best matching, and why?

12.18. What are the chiralities of each of the following interdigitation patterns?

a. ${}_DA_SB_DB_SA_D$

b. ${}_SA_D$ – ${}_DB_SB_DB_S$ – ${}_DA_S$

c. ${}_SA_DA_S$ – ${}_SB_D$ – ${}_DA_SA_D$

Which patterns exhibit orientation-dependent mismatches?.

13 Special Topics

The previous chapters have presented the details of constructing and matching resistors, capacitors, diodes, and transistors. Integrated circuits also contain a number of more specialized components, including merged devices, guard rings, tunnels, bondpads, and ESD protection devices.

Merged devices appear separate from one another in schematics, but they are combined in the layout. Mergers not only save space but in some cases also improve performance. The designer must weigh the benefits of mergers against the possibility of introducing unexpected interactions between merged devices.

Guard rings prevent minority carriers injected by one device from interfering with the operation of another device. Not only do guard rings prevent latchup, but they also block noise coupling that might otherwise interfere with the operation of low-power circuitry.

Tunnels are low-value resistors used as signal crossing points. Single-level-metal layouts almost always require tunnels. Multiple-level-metal layouts need not use tunnels, but they occasionally offer a convenient way to route leads to otherwise inaccessible areas of the die. Tunnels introduce parasitics that can degrade circuit performance, so each proposed tunnel must be carefully analyzed for potential problems.

Bondpads allow the connection of the integrated circuit to the external world. Most bondpads require electrostatic discharge (ESD) protection circuitry. *Trimpads* and *testpads* are accessible only during wafer-level probing, so they do not require ESD protection.

13.1 MERGED DEVICES

The largest spacings in standard bipolar are those associated with the isolation diffusion. Most circuits contain components whose tanks are connected to the same potential. Considerable space can be saved by placing these devices in common tanks. Figure 13.1A shows three minimum-geometry NPN transistors laid out side by side, while Figure 13.1B shows the same three transistors merged into a common tank. The merged devices require only about 70% of the area of the separate devices.

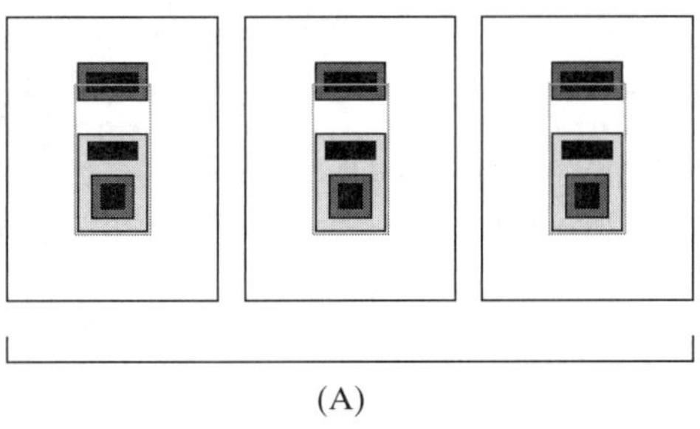

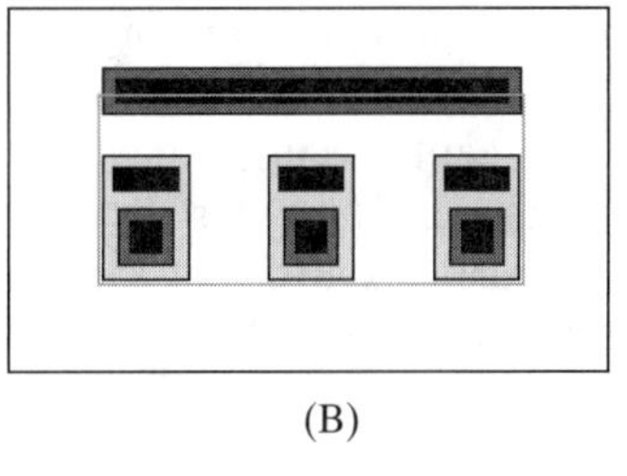

FIGURE 13.1 (A) Three separate NPN transistors and (B) the same transistors merged into one tank.

Additional area can be saved by reducing the size of the collector contact. If the three devices share a common base connection, then even more space can be conserved by merging the three emitters into a common base region.

The largest spacings in CMOS and BiCMOS processes are those associated with the isolated well (or wells). The merger of components into common wells can again save considerable die area. This is particularly true for designs containing large numbers of small components, such as minimum-size MOS transistors.

The merger of devices that should remain separate has caused many circuit malfunctions, some of which have proven remarkably difficult to diagnose and fix. Consequently, designers have become somewhat reluctant to merge devices, even when mergers would obviously result in significant area savings.

Most failures caused by merged devices are due to minority carrier injection, Ohmic debiasing, or capacitive coupling. If a designer understands these three mechanisms, then he or she can discern which mergers are safe and which should be avoided. Minority carrier injection has caused by far the most problems. Trouble can occur whenever a device injects minority carriers into a shared region such as a tank or a well. Some of the injected minority carriers transit to other devices, where they are collected by reverse-biased junctions. The flow of minority carriers between devices that should remain isolated causes leakage currents to appear at unexpected points in the circuit. These currents can cause circuit malfunctions ranging from subtle parametric shifts to catastrophic latchup. The following section discusses several device mergers known to exhibit serious flaws.

13.1.1. Flawed Device Mergers

Figure 13.2A shows a split-collector lateral PNP transistor. This structure consists of a pair of merged lateral PNP transistors sharing common emitter and base regions.

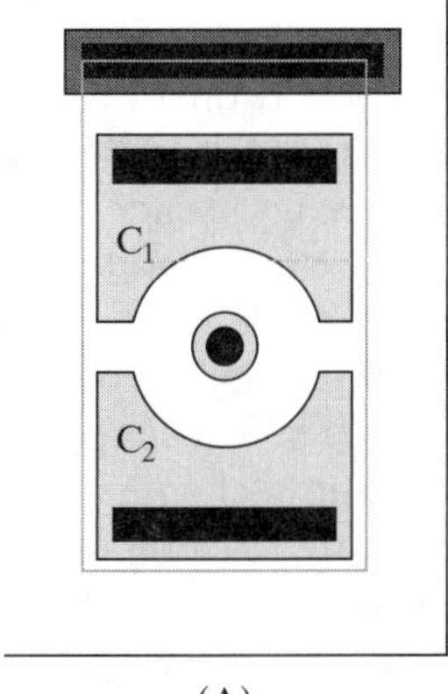

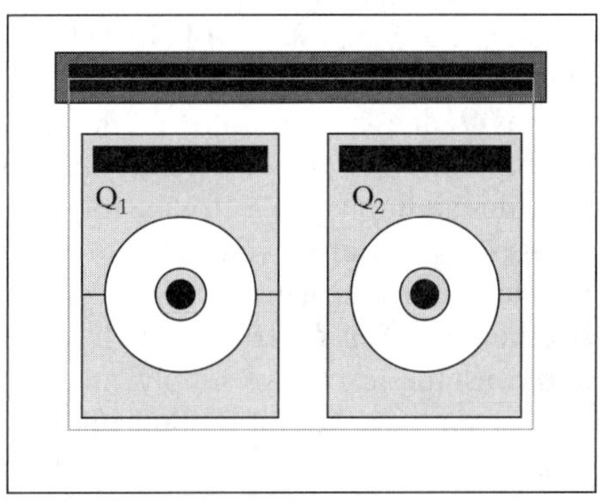

FIGURE 13.2 Merged lateral PNP structures susceptible to cross injection: (A) split-collector lateral PNP and (B) two lateral PNP transistors merged in a common tank.

Under normal operating conditions, collectors C_1 and C_2 both remain reverse biased. Holes flow radially from the shared emitter to the two collectors, and each collector intercepts about half of the total emitter current. Now suppose that collector C_1 saturates. The holes that should have been absorbed by C_1 are now reinjected from its surfaces. Most of these reinjected carriers flow to the isolation sidewalls and thence to the substrate, but some also flow from collector C_1 to collector C_2. This *cross injection* increases the current flowing out of C_2 and imbalances the apparent ratio between the two collectors.

Most designers know that the saturation of a split-collector PNP causes cross-injection, but many overlook the possibility of cross injection between separate lateral PNP transistors occupying a common tank. Figure 13.2B shows a tank containing two lateral PNP transistors, Q_1 and Q_2. Normally the collectors intercept most of the minority carriers injected by their respective emitters, so the two transistors remain effectively isolated from each other, even though both occupy the same tank. Now suppose that Q_1 saturates while Q_2 continues to operate in the normal active region. The holes injected by the emitter of Q_1 transit to its collector, but when this collector forward-biases, they are reinjected back into the tank. Since the collectors of Q_1 and Q_2 face each other across a relatively narrow gap, most of the carriers launched from Q_1 toward Q_2 reach its collector. Hence, when Q_1 saturates, about a quarter of its emitter current flows to Q_2. Cross injection occurs to some extent even if the lateral PNP transistors are not saturated. Some holes slip underneath the collector of a lateral transistor and emerge in the tank outside it. The percentage of carriers that escape in this manner increases as the collector-to-emitter voltage V_{CE} decreases. One should therefore avoid placing precisely matched lateral transistors in the same tank or well. Moderately matched lateral transistors can be merged if the devices operate at the same collector currents and collector-to-emitter voltages. The transistors must be arrayed symmetrically so that cross injection from one transistor to the other exactly counterbalances cross injection in the opposite direction.

The simplest way to prevent cross injection between lateral PNP transistors is to place each transistor in its own tank. However, this wastes so much space that designers have devised other methods of preventing (or at least minimizing) cross injection between merged devices. For example, two lateral PNP transistors can be separated from one another by a P-bar or an N-bar (Section 4.4.2). These bars block most, but not all, of the cross-injected minority carriers. The designer should consider what would happen to the circuit if, for example, 5% of the current injected by the saturating device were to reach adjacent devices. If this amount of current could cause a malfunction, then the devices require separate tanks.

Another example of minority carrier injection involves the merger of an NPN transistor, Q_2, driving a lateral PNP, Q_1 (Figure 13.3).[1] The collector of Q_2 is electrically common to the base of Q_1. This merged device will probably operate satisfactorily as long as Q_1 does not saturate. If it does, then holes reinjected by its collector will flow to the base of NPN Q_2. This additional base current drives Q_2 harder than before and provides additional collector current. The increased collector current feeds Q_1 and increases its emitter current. This situation is a classic example of SCR latchup. Once latchup has been triggered, it will continue until the power is interrupted. The die may overheat and self-destruct at high supply voltages; otherwise it simply malfunctions and consumes excessive supply current. The potential for latchup makes this merger risky, even if the lateral PNP never saturates during normal operation. If some transient condition saturates the lateral PNP, then the structure will latch up.

---

[1] C. Jones, "Bipolar Parasitics," unpublished manuscript, 1987, pp. 27–29.

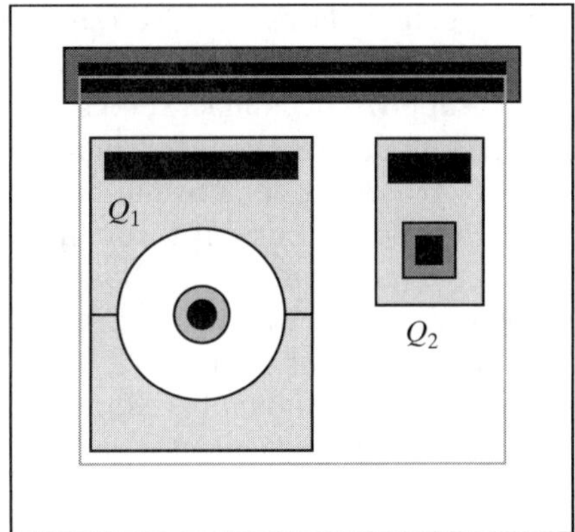

FIGURE 13.3 Example of a device merger prone to latchup due to minority-carrier injection.

P-bars and N-bars may not suffice to prevent device latchup. In order for latchup to occur, the product of the beta of the PNP, β_P, and the beta of the NPN, β_N, must exceed unity. The addition of a bar between the two transistors adds a third term, η_c, representing the fraction of minority carriers intercepted by the bar. The condition for latchup (Section 11.2.7) then becomes

$$\beta_N \beta_P (1 - \eta_c) > 1 \qquad [13.1]$$

The beta of a standard bipolar NPN transistor may equal 300 or more. The beta of the lateral PNP is somewhat lower, but usually exceeds 10. In order to prevent latchup with these betas, the efficiency of the bar must exceed 0.997, or 99.7%. N-bars will almost certainly fall short of this efficiency, and P-bars may not always achieve it. If there is the slightest chance that a lateral PNP may saturate, then it should not occupy the same tank as an NPN transistor. Identical concerns apply to the merger of analog BiCMOS lateral PNP and NPN transistors in a common well.

Figure 13.4 illustrates another pair of merged devices prone to latchup.[2] This example is particularly noteworthy because the latchup stems from the interaction of two separate mechanisms, namely, Ohmic debiasing and minority carrier injection.

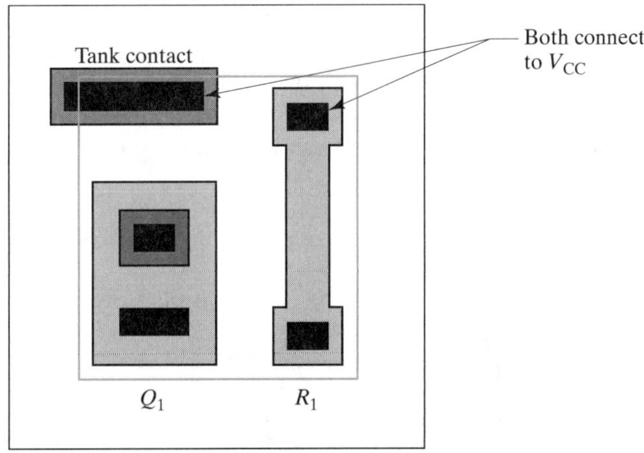

FIGURE 13.4 Another example of a device merger prone to latchup due to minority-carrier injection.

[2] W. F. Davis, *Layout Considerations*, unpublished manuscript, 1981, pp. 32–33.

This structure places an NPN transistor Q_1 in the same tank as a base resistor R_1. The NPN is configured as an emitter follower, and one end of the base resistor connects to the supply. One contact serves as the collector contact of Q_1 and the tank contact of R_1. Both the base-collector junction of Q_1 and the base-tank junction of R_1 remain reverse biased under normal conditions. This situation may change if Q_1 draws enough collector current through the shared tank contact. If the voltage drop between the tank contact and the intrinsic collector of Q_1 becomes large enough, then the positive end of R_1 will forward-bias into the tank. This is an example of *Ohmic debiasing.* Some of the minority carriers injected by R_1 reach the base of Q_1, where they provide additional base drive. Transistor Q_1 now pulls even more collector current. The additional collector current increases the Ohmic debiasing experienced by R_1, causing R_1 to inject additional holes into the tank. The resulting positive feedback causes the circuit to latch up. As with the previous example, latchup occurs because of the presence of an SCR, which in this case consists of base resistor R_1, the shared tank, and the base and emitter of Q_1.

Although the structure in Figure 13.4 contains an SCR, it will not latch up unless triggered by a voltage drop produced by Ohmic debiasing within the tank. The voltage drop required to trigger the SCR equals about 0.3 V at 150°C. (See Section 4.4.1.) If the NPN transistor conducts an average of 100 μA, then the tank resistance must equal 3 kΩ to produce 0.3 V of debiasing. The vertical resistance between the tank contact and the NBL can equal hundreds, if not thousands, of Ohms if the transistor lacks a deep-N+ sinker. The inclusion of even a minimum plug of deep-N+ reduces the tank resistance to no more than a few hundred Ohms. The structure in Figure 13.4 is therefore likely to latch up without deep-N+, but it is very unlikely to do so if a sinker is present.

Ohmic debiasing can also cause capacitive coupling of noise into sensitive nodes. Using the merger of Figure 13.4 as an example, suppose that the debiasing is insufficient to actually trigger latchup. Even so, the current flowing through Q_1 still causes some voltage drop within the tank. If the operation of Q_1 causes a rapid fluctuation of the collector current, then this will in turn generate a high-frequency ripple on the tank voltage. This signal can couple through the capacitance of the reverse-biased junction surrounding R_1. If R_1 is part of a sensitive circuit, then the noise injected by tank voltage fluctuations can cause problems. Designers should avoid merging noisy circuitry and sensitive circuitry in the same tank. Although some such mergers function satisfactorily, many do not.

Figure 13.5 shows another pair of problematic merged devices. This example merges an NPN transistor, Q_1, with a Schottky diode, D_1. The collector of Q_1 connects to the cathode of D_1 through the tank. Since Schottky diodes are majority-carrier devices, this might appear a safe merger. Unfortunately, this is not the case. Most Schottky diodes incorporate a field-relief guard ring consisting of a P-type diffusion.

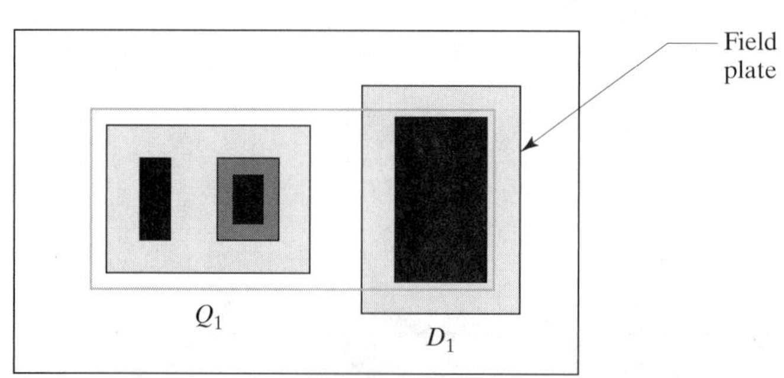

FIGURE 13.5 Another structure prone to minority-carrier cross injection, consisting of an NPN transistor, Q_1, and a Schottky diode, D_1.

This guard ring begins to inject minority carriers into the tank as soon as the voltage across the Schottky rises above the forward voltage of the guard-ring junction. The series resistance of small Schottky diodes can equal hundreds or thousands of Ohms, so their guard rings are easily debiased into conduction.

The structure in Figure 13.5 uses a field-plated Schottky diode instead of a guard-ringed Schottky. This eliminates the possibility of the guard ring forward-biasing into the tank, but low levels of cross injection can still occur due to the presence of the Schottky barrier itself. Although a rectifying Schottky junction conducts primarily by means of majority carriers, small numbers of minority carriers are also injected into the semiconductor side of the junction. The minority carriers injected by Schottky diode D_1 can travel to the base of Q_1, where they appear as additional base drive. This mechanism causes parametric shifts and could potentially trigger latchup.

The type of malfunction that occurs in the circuit in Figure 13.5 can potentially occur in an ordinary NPN transistor. If a heavily doped diffusion does not entirely enclose the collector contact, the portion of the contact touching the lightly doped epi forms a Schottky barrier. This barrier can inject minority carriers in much the same way as the Schottky diode in Figure 13.5. This problem normally occurs in structures that have been incorrectly laid out, but misalignments caused by poor photolithography have been known to produce the same result in structures that pass all applicable design rules.

13.1.2. Successful Device Mergers

This section presents two device mergers of the sort often encountered in standard bipolar layouts. There are countless possible mergers, and the examples given here provide only a general impression of what a skilled designer can achieve. Additional mergers can be discovered by examining the layout of almost any standard bipolar integrated circuit.

The *Darlington pair* in Figure 13.6A consists of a power NPN transistor, Q_1, and a smaller predrive transistor, Q_2, both sharing a common collector connection. Each

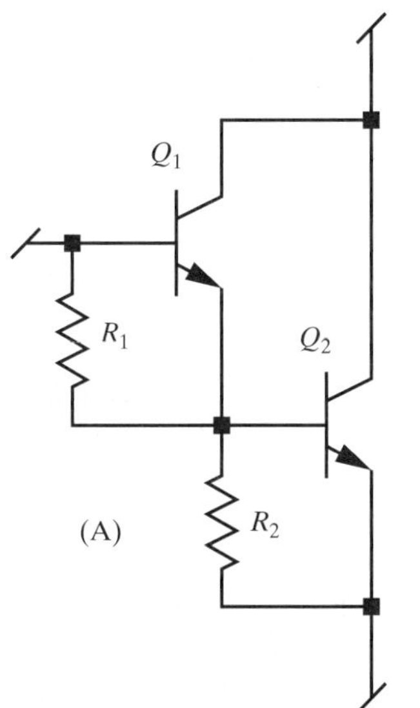

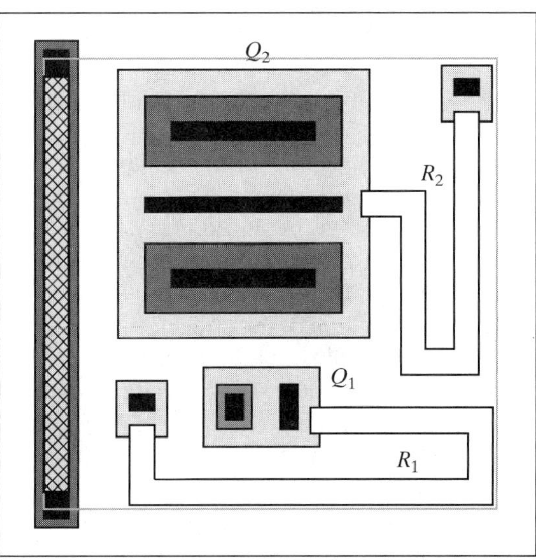

FIGURE 13.6 (A) Schematic and (B) layout of a merged Darlington pair (metal omitted for clarity).

transistor also has an associated base turnoff resistor. The layout of Figure 13.6B shows how all four of these components can occupy the same tank. The tank contact consists of a bar of deep-N+ placed along the left edge of the tank. The deep-N+ sinker does not encircle the power device, Q_2, because such an arrangement would consume additional die area. In general, only saturating NPN transistors and large power devices require full deep-N+ rings. Q_2 is unlikely to saturate because the extrinsic collector-to-emitter voltage V_{CE} cannot drop below the sum of the extrinsic V_{BE} of Q_2 and the extrinsic V_{SAT} of Q_1. The extrinsic V_{BE} of Q_2 probably approaches 1 V at high current levels, so the tank contact can debias by the better part of a volt before Q_2 begins to saturate. The deep-N+ bar in Figure 13.6B probably exhibits no more than 5 to 10 Ω of vertical resistance, so it can handle several hundred milliamps of current without debiasing enough to allow Q_2 to saturate. Q_1 can saturate, but the current flow through this device is small enough to limit substrate injection to manageable levels.

Even if Q_1 saturates, it will not interfere with the operation of the Darlington. When Q_1 saturates, it will be delivering as much current as possible into Q_2. Holes injected by Q_1 into the common tank will be collected by R_1, R_2, or Q_2. The holes collected by R_1 will either return to the base of Q_1 or will flow to the base of Q_2. The influx of additional base drive into either transistor causes no harm because both transistors are already conducting all the current they can. Holes collected by Q_2's base simply represent additional base drive for Q_2. Holes collected by R_2 will probably flow to the emitter of Q_2, there to combine with the much larger current flowing through this transistor. Again, the influx of extra current causes no harm. In conclusion, it makes little difference whether or not Q_1 saturates.

The base contact of each NPN transistor also serves as a contact head for its respective base turnoff resistor. This merger can save considerable area, but the HSR implant must be spaced far enough away from the emitter to prevent this implant from raising the doping concentration in the intrinsic base region of the NPN. This layout achieves the necessary separation without enlarging the transistor by running the HSR implant into the transistor base behind the base contact.

The devices in Figure 13.6 have been arranged to enable interconnection with one level of metal. The reader may wish to mentally trace the connections between the contacts. The collector lead enters the tank from the left and the emitter lead exits from the right. These leads can be as wide as desired. The lead connecting the emitter of Q_1 to the base of Q_2 passes between the tank contact and the body of Q_2.

Figure 13.7A shows another example of a circuit that can benefit from device mergers. Transistors Q_1 and Q_2 act as a differential pair. Three-fourths of the emitter current of Q_1 and Q_2 is shunted to ground, while one-fourth feeds a current mirror consisting of Q_5 and Q_6. The output of this circuit is taken through a substrate PNP emitter follower, Q_4. An identical substrate PNP, Q_3, balances the load on the circuit and eliminates a systematic offset otherwise created by the base current of Q_4.

The layout of this circuit requires no fewer than four tanks. Q_1 and Q_2 have separate base connections and therefore require individual tanks. Q_3 and Q_5 can reside in a common tank, as can Q_4 and Q_6. The larger collectors of Q_1 and Q_2 connect to ground through an extension of the collector into the surrounding isolation. This saves considerable area by eliminating the spacing between the collector and the isolation. The size of these transistors can be further reduced by aligning the smaller collector segments to allow them to fit within a narrower tank (Figure 13.7B). The narrower tank cannot contain enough NBL to fully floor the active region of the transistor, but NBL outdiffusion prevents any significant leakage of minority carriers to the substrate.

The mergers of Q_3–Q_5 and Q_4–Q_6 present more significant problems. Each of these structures places a substrate PNP in the same tank as an NPN. The substrate PNP injects holes that could trigger latchup if they reached the merged NPN transistors. The

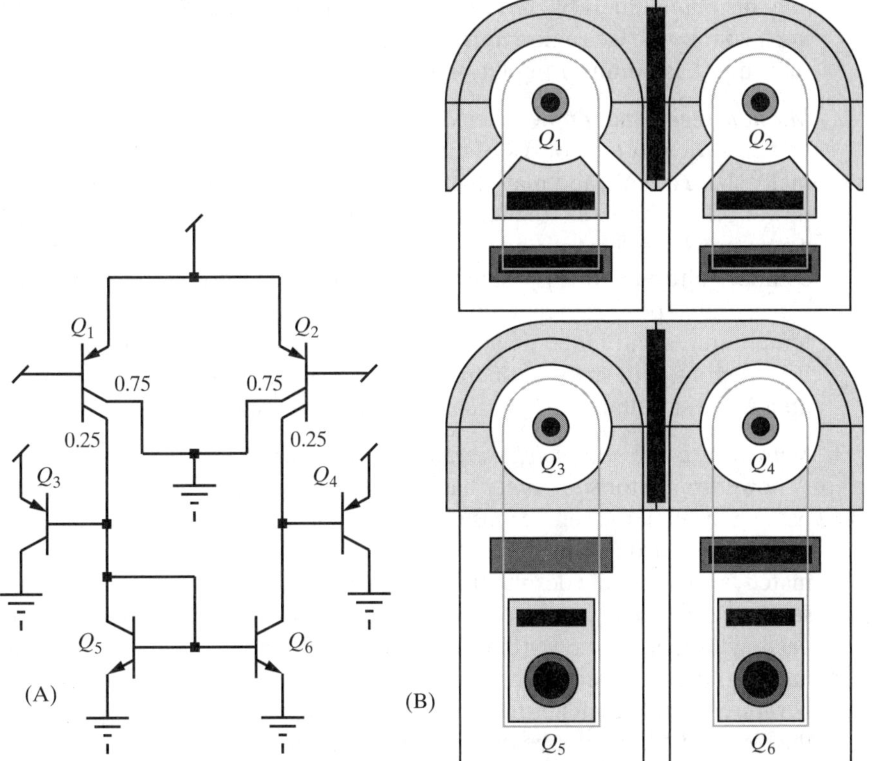

FIGURE 13.7 (A) Schematic and (B) layout of a merged operational amplifier input stage. The metallization has been omitted for clarity.

illustrated layout circumvents this problem by substituting lateral PNP's for the substrate PNP's normally used in this role. The lateral PNP collector functions as a P-bar. This collector also extends out into the isolation to save space. The P-bar may not entirely block minority carrier flow, but several standard bipolar designs have successfully used this layout. Latchup has not been observed, and any low level of hole collection experienced by Q_5 is also experienced by Q_6. Since Q_5 and Q_6 balance one another, the circuit inherently tolerates low levels of cross injection between Q_3–Q_5 and Q_4–Q_6.

The mergers of Q_3–Q_5 and Q_4–Q_6 are not quite identical, because Q_4–Q_6 requires a tank contact while Q_3–Q_5 does not. Both mergers incorporate identical strips of emitter to ensure matching, but only Q_4–Q_6 includes a tank contact. The contact and its associated metallization have little, if any, impact on matching, so Q_3–Q_5 omits these.

13.1.3. Low-Risk Merged Devices

Some device mergers are easy to perform and pose few risks. These mergers should be used whenever possible because the benefits they provide greatly outweigh their disadvantages. Examples of low-risk mergers include the following:

1. *Multiple segments of a single matched device.*
   Matched devices often consist of multiple segments connected in series or parallel. These segments can frequently share a tank or well, and sometimes they can share other diffusions as well. For example, a matched NPN transistor can consist of several separate emitters occupying a common base region inside a common tank. The compactness of merged layouts makes them less sensitive to gradients than conventional layouts. One must still exercize care when constructing these merged devices. Segments merged in a common tank

may become vulnerable to tank modulation. Multiple NPN emitters occupying a common base region must reside far enough from one another to ensure that their base-emitter depletion regions do not touch.

2. *Multiple segments of power devices.*
Power devices often consist of multiple segments, or *fingers.* These fingers normally share a tank, and may share other diffusions as well. For example, a power NPN transistor frequently consists of multiple emitter fingers occupying a common base region inside a common tank. Merged power devices are more compact, although this is not always a desirable feature. If the device dissipates considerable power, then a more dispersed structure may actually reduce peak temperatures within the device. Some power bipolar transistors interdigitate base regions with strips of deep-N+ to spread power dissipation over a larger area while simultaneously reducing collector resistance.

3. *Sense transistors and their associated power transistors.*
A power transistor sometimes has an associated sense transistor. The current passing through the sense transistor is smaller than, but proportional to, the current passing through the power transistor. The sense and power transistor must match fairly precisely despite the presence of large thermal gradients. Ideally, the sense device should consist of two equal segments located on an axis of symmetry passing through the power device. Each segment should reside approximately halfway between the center of the power device and its periphery so that its temperature roughly matches the average temperature of the whole power device. If the sense device cannot be divided into segments, then it should reside on the periphery of the power device on one of its axes of symmetry.

4. *Schottky clamps and their associated NPN transistors.*
Schottky-clamped NPN transistors are usually constructed as merged devices. The merged Schottky clamp can use the same deep-N+ sinker as the transistor, and if necessary it can also use an extension of the NPN base region as a guard ring.

5. *NPN Darlington transistors.*
Darlington transistors can use a layout similar to that in Figure 13.6. Most Darlington transistors are power devices that require custom layouts. Only a little additional time and effort are required to incorporate the predrive transistor and turnoff resistors in the same tank.

6. *Base turnoff resistors for NPN transistors.*
NPN transistors often require base turnoff resistors. If these resistors are P-type diffused devices, then they can occupy the same tanks as their associated NPN transistors. If the base turnoff resistor connects to substrate potential, then one can simply run the resistor out into the isolation. This technique saves the space required for one resistor head, but a substrate contact should reside near the resistor to minimize debiasing.

13.1.4. Medium-Risk Merged Devices

Another class of device mergers is easy to perform and offers substantial area savings, but these mergers are not without an element of risk. The risks involved are understood, and they can usually be avoided without much difficulty. Examples of moderate-risk device mergers include the following:

1. *MOS transistors in common wells.*
Designers routinely merge MOS transistors into common wells to save space. This practice has become so widespread that digital designers frequently take

it for granted. Actually, such mergers pose a certain amount of risk because cross injection can occur between the merged source/drain regions, whether these reside in a common well or in the epi. Any problems that occur usually stem from output transistors whose source/drain regions connect to pins other than supply or ground. Externally induced transients on such pins can forward-bias source/drain regions into backgates. All output transistors require their own minority carrier guard rings (Section 13.2). If possible, output transistors should not be merged with other transistors that do not connect to the same pin. For example, an output PMOS transistor in an N-well CMOS process should occupy its own well. On the other hand, the two PMOS transistors of a NAND gate both connect to the same output and can therefore occupy the same well.

When constructing merged transistors, the designer should always provide as much backgate contact as possible to minimize debiasing should a merged transistor forward-bias. If the process includes a suitable buried layer, then the well should contain as large an area of the buried layer as possible. Well contacts become more important as the size of the well and the number of transistors it contains increase. Very large MOS transistors often require integrated backgate contacts (Section 11.2.7).

Some circuits contain MOS transistors that regularly forward-bias into their backgate regions during normal operation. For example, some charge pumps contain devices that forward-bias during startup. These devices do not necessarily connect to pins and are not always easily identified. The circuit designer should clearly identify all such devices so that the layout designer can protect them by using guard rings and separate wells.

2. *Diffused resistors in common tanks.*
Diffused resistors are frequently merged in common tanks to save space. This practice allows the construction of compact resistor arrays that are less vulnerable to stress and thermal mismatch. Cross injection cannot occur between the merged resistors as long as none of them forward-bias into the tank. Problems can occur if any of the resistors connect to a pin. Such resistors should occupy their own tanks, and they may also require guard rings if the process is prone to latchup or if the transients occur during normal operation. Some circuits operate one or more resistors under biasing conditions that might cause minority carrier generation. The circuit designer should clearly mark each such resistor so that it can receive its own tank, along with any necessary guard rings. The layout designer should also be wary of merging noise-sensitive resistors with components carrying high-frequency signals, because of the potential for capacitive coupling. If in doubt, use separate tanks.

3. *Lateral PNP transistors.*
Lateral PNP transistors sharing the same base connection can occupy the same tank. Many bipolar designs make extensive use of lateral PNP mergers. As long as none of the merged transistors saturate, their collectors act as P-bars isolating them from one another. Minority carrier cross injection can occur if any of the transistors saturates. P-bars and N-bars (Section 4.4.2) can at least partially block cross injection between adjacent lateral PNP transistors, but the safest solution consists of placing the saturating transistors in their own tanks.

4. *Split-collector lateral PNP transistors.*
A split-collector lateral PNP is really a type of merged lateral PNP. A single split-collector transistor can perform the role of several ordinary lateral transistors

while consuming much less die area. As long as none of the collectors saturates, few holes escape between the segments of a split-collector device. The saturation of any of the split collectors causes the currents flowing through the remaining collectors to increase. No way exists to block cross injection in split-collector transistors, short of replacing the offending split-collector devices with separate transistors.

5. *Zener diodes.*
   Emitter-base Zener diodes can be merged with other components in a common tank as long as the tank voltage always equals or exceeds the voltage on the Zener's cathode. This condition ensures that the parasitic NPN transistor does not conduct. Series-connected Zener diodes can also occupy a common tank biased to a potential equal to or greater than that of the cathode end of the Zener string.

13.1.5. Devising New Merged Devices

Any imaginative designer will find many additional opportunities to merge devices. Before implementing a proposed device merger, determine whether it can pass the following three tests:

1. *Can any of the merged devices inject minority carriers into the shared tank or well?*
   If not, then the merged devices are safe from cross injection and latchup. Possible sources of minority carriers include saturating bipolar transistors, forward-biased Schottky diodes, and diffusions connecting to an external pin. Designers should be particularly wary of merging NPN transistors with other devices that can potentially inject minority carriers into the tank, because the resulting PNPN structure may latch. Potential sources of minority carriers either should reside in their own tanks or should be guarded by P-bars or N-bars, unless the designer can show that cross injection will not upset the operation of the circuit.
2. *Can any of the merged devices pull substantial current through the tank or well contact?*
   If so, then these devices may cause debiasing. Those that pull relatively small amounts of current (up to a few milliamps) rarely cause objectionable debiasing as long as the tank or well contact contains a plug of deep-N+. Higher current devices require more extensive deep-N+ regions to prevent debiasing.
3. *Can noise coupling upset the circuit?*
   If the merged tank or well contains both noisy devices and noise-sensitive devices, then capacitive coupling between these can degrade circuit performance. The potential for noise coupling is particularly great if the noisy device pulls significant current through the tank contact.

13.1.6. The Role of Merged Devices in Analog BiCMOS

Standard bipolar designs typically consist of a relatively small number of devices, each custom crafted to fit available space. Mergers between devices become extremely attractive under these circumstances. Designers who are already used to crafting custom layouts for each device have few qualms about merging two such devices to save a little area. Although the savings generated by each merger are relatively small, they can add up to savings of 10% or more of total die area.

Analog BiCMOS designers are much more reluctant to use mergers. Most analog BiCMOS layouts contain so many components that time considerations preclude custom crafting of every device. Various software tools have been developed to accelerate the speed of laying out complicated circuits. These tools, which include parameterized cells and device generators, seldom have the flexibility needed to create device mergers of the sort seen in standard bipolar. Circuit designers have also become accustomed to using standardized device layouts to ensure that their device models accurately reflect the performance of silicon.

Despite these objections, analog BiCMOS designers can still benefit from the judicious use of mergers. Examples include mergers between matched bipolar transistors that produce more compact structures, and the embedment of sense transistors within power transistors to obtain better thermal coupling. The benefits of these mergers often outweigh the disadvantages of the additional time required to construct the necessary devices and the uncertainty they introduce into device modeling.

BiCMOS designs also make extensive use of mergers between CMOS transistors, including the placement of devices in a common well and the sharing of a single source/drain finger between two separate devices. These mergers not only save space, but also can improve circuit performance through enhanced matching and reduced parasitic capacitance. These types of mergers are easily and quickly implemented using parameterized cells and device generators.

The analog BiCMOS designer must remain alert to the potential for interactions between merged devices, even those as seemingly innocuous as a pair of MOS transistors merged in a common well. Minority carrier injection, Ohmic debiasing, and capacitive coupling do not restrict themselves to standard bipolar designs. The principles discussed in this section apply to all integrated circuit processes, without exception.

13.2 GUARD RINGS

Of all the many types of failures that plague integrated circuits, none is so frustrating and so elusive as latchup. Devices that operate properly in one circuit latch up the moment they are inserted into another. Sometimes a device operates properly for hundreds or thousands of hours before it latches. Simulation rarely uncovers latchup problems, and neither do most forms of testing.

The most frequent causes of device latchup are external transients that pull device pins above supply or below ground. Common sources of such transients include low-level ESD events; momentary power interruptions; inductive kick back from relays, motors, and solenoids; and inductive spiking of rapidly switched signals. Proper board-level design minimizes, but does not eliminate, these transients. Circuit designers must ensure that their designs can withstand at least moderate levels of transient injection without latching up or otherwise malfunctioning.

Power supply pins and substrate connections rarely trigger latchup, but any other pin (including grounds not connected to substrate) can cause problems. The designer should trace every lead from such a pin back through the circuit to determine whether or not it connects to any diffusions. Each diffusion connecting directly to the pin can inject minority carriers when the pin flies above supply or below ground. Diffusions connecting to pins through deposited resistors still pose a concern if the series resistance is less than about 50 kΩ. Larger value deposited resistors reduce injected currents so much that they no longer pose any significant threat.

Latchup can be suppressed by enclosing each vulnerable diffusion (or device) in a suitable minority carrier guard ring. Multiple diffusions connecting to a common pin can share a common guard ring. ESD devices residing around the periphery of the die can often share a common guard ring separating the core of the die from the ESD devices and bondpads. Many older standard bipolar designs omitted guard rings from some or all pins, but new designs should not follow this practice because it sometimes results in costly redesigns.

13.2.1. Standard Bipolar Electron Guard Rings

Any tank connecting to a device pin can inject electrons into the substrate. Standard bipolar does not include the layers necessary to construct an *electron-blocking guard ring* (EBGR). This process does, however, support the construction of an *electron-collecting guard ring* (ECGR). The structure in Figure 13.8A is arguably the best electron guard ring that can be constructed in standard bipolar. It consists of a strip of deep-N+ residing in an N-tank augmented by both NBL and emitter.[3] This combination of diffusions forms the deepest possible guard ring and therefore collects the largest possible fraction of electrons. The presence of deep-N+ also helps prevent Ohmic debiasing. This guard ring would ideally connect to the highest available supply voltage to drive the depletion region as deeply as possible into the substrate. This style of guard ring also functions if it is connected to ground, but grounded guard rings are more susceptible to debiasing. Grounded guard rings are sometimes used to minimize power dissipation caused by minority carrier injection, which sometimes becomes a concern in high-current designs. If a grounded guard ring is used to minimize power dissipation, it can be supplemented by a second guard ring connected to the supply and placed outside of the grounded guard ring. This secondary guard ring will provide protection if the grounded guard ring saturates.

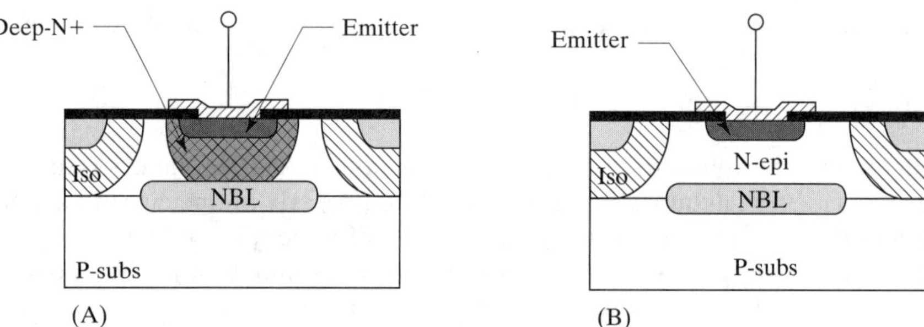

FIGURE 13.8 Electron-collecting guard rings for standard bipolar: (A) preferred structure and (B) alternate structure.

One can sometimes take advantage of adjacent tanks that connect to the supply. If these tanks are strategically situated between the point of minority carrier injection and adjacent sensitive circuitry, then they become very effective guard rings. All tanks used for this purpose should contain as much NBL as possible and should use deep-N+ sinkers to minimize debiasing. The efficiency of an electron-collecting guard ring also increases if it is placed next to the source of injected carriers to take advantage of the proximity effect.

The guard ring in Figure 13.8B should only be used if deep-N+ is not available. The vertical resistance of the epi layer separating the NBL from the emitter diffusion makes this guard ring extremely vulnerable to Ohmic debiasing. This structure is still marginally effective as long as it connects to a power supply, but it is virtually useless when connected to ground.

Electron guard rings constructed in standard bipolar are only marginally effective, yet they consume vast amounts of die area. Most designers omit these guard

[3] A similar guard ring for a BiCMOS process is presented in E. Bayer, W. Bucksch, K. Scoones, K. Wagensohner, J. Erdeljac, and L. Hutter, "A 1.0 μm Linear BiCMOS Technology with Power DMOS Capability," *Proc. Bipolar/BiCMOS Circuits and Technology Meeting,* 1995, pp. 137–141.

rings to save space and rely instead on large spacings and a few strategically placed hole guard rings to prevent latchup. These measures usually suffice for linear circuits such as operational amplifiers and voltage regulators. Devices that switch inductive loads are another matter entirely, as these loads can generate extremely energetic transients during normal operation. Even if these transients do not cause latchup, they can still inject noise into sensitive circuitry. High-frequency MOSFET gate drivers can also experience severe transients caused by resonance in the gate lead. The output circuitry of MOSFET gate drivers and inductive load drivers must be carefully shielded by electron guard rings to minimize noise coupling and latchup sensitivity.

Electron guard rings become much more effective if the process incorporates a P+ substrate. The P+/P− interface generates an electric field that traps most of the injected electrons in the P− epi. Those few that do penetrate into the P+ substrate quickly recombine. The P+ substrate makes deep guard rings such as the one in Figure 13.8A extremely effective, particularly if they are biased to a high-enough potential to drive a depletion region down to meet the P+ substrate. Section 4.4.2 discusses the theory of operation of minority carrier guard rings in further detail, and also presents several more specialized structures which might prove useful in certain circumstances. The structures presented in the current chapter should, however, prove adequate for most purposes.

13.2.2. Standard Bipolar Hole Guard Rings

Any P-type region can inject holes into a tank. Hole guard rings can prevent these carriers from flowing to adjacent P-type regions or to the sidewalls of the tank. Two types of hole guard rings exist: the *hole-collecting guard ring* (HCGR) and the *hole-blocking guard ring* (HBGR). Figure 13.9A shows a typical hole-collecting guard ring deployed to prevent holes from reaching the sidewalls of a tank. The presence of NBL prevents the holes from flowing down to the substrate and instead forces them to flow laterally. The guard ring consists of a reverse-biased base diffusion surrounding the point of injection. This diffusion acts as the collector of a lateral PNP transistor. Any holes reaching the depletion region surrounding the guard ring are drawn into it. Hole-collecting guard rings are normally grounded to maximize the reverse bias between the tank and the guard ring. This not only drives the depletion region deeper into the tank but also minimizes the effects of Ohmic debiasing within the guard ring itself. Grounded hole guard rings tie to the same potential as the isolation system, so they can be merged to save space. Examples of such merged guard rings include the P-bar in Figure 4.28 and the grounded collectors of transistors Q_3 and Q_4 in Figure 13.7. A hole-collecting guard ring can also be tied to the tank potential, but this reduces its effectiveness and does not save any appreciable space.

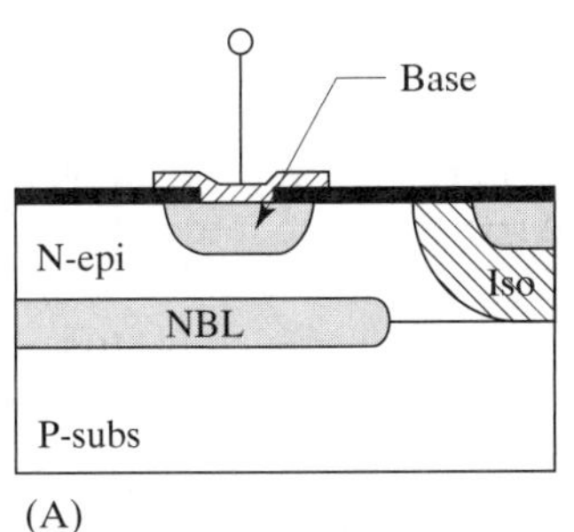

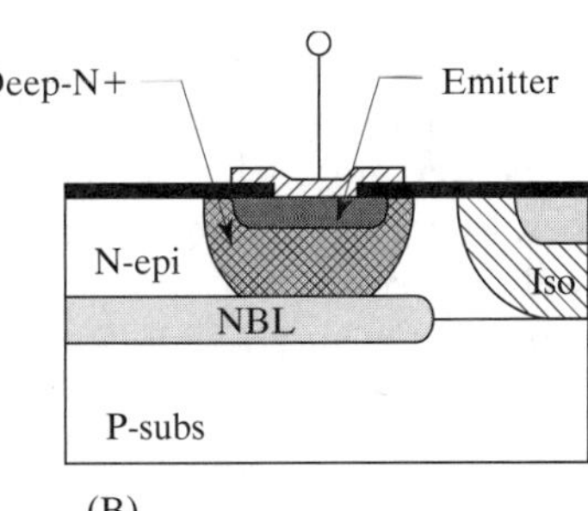

FIGURE 13.9 Hole guard rings for standard bipolar: (A) hole-collecting guard ring and (B) hole-blocking guard ring.

Figure 13.9B shows a typical example of a hole-blocking guard ring.[4] This type of guard ring surrounds the point of injection with heavily doped N-type regions. The N+/N− interface generates an electric field that acts as a barrier to the passage of holes. Most holes overcoming this barrier recombine before they can traverse the N+ region. The electron current required to sustain recombination flows through contacts to the N+ region. The structure in Figure 13.9B relies on NBL to block the downward travel of holes, and deep-N+ to block their lateral movement. In order to be fully effective, a hole-blocking guard ring must contain no gaps or holes. The only practical way to achieve this goal consists of entirely encircling the injector with a ring of deep-N+. Partial hole-blocking rings such as the N-bar in Figure 4.28 may allow a substantial fraction of the holes to escape around the gaps at either end. In some processes, these gaps can be eliminated by extending the deep-N+ bar into the isolation on either side of the tank. Most processes do not allow this configuration because of leakage across the isolation/deep-N+ junction. Both hole-collecting and hole-blocking guard rings can provide efficiencies of 95% or better in typical standard bipolar processes. Hole-collecting guard rings placed inside hole-blocking guard rings can produce efficiencies in excess of 99%.

Standard bipolar designs use relatively few hole guard rings because this process rarely requires them. Standard bipolar designs seldom experience latchup due to hole injection into the substrate, because the deep-P+ isolation and the large spacing between components both help to reduce the beta product of the parasitic SCR (Section 11.2.7). Hole injection into the substrate becomes a problem only if it overwhelms the capability of the substrate contact system. This is unlikely to occur because the grid of P+ isolation diffusion helps to magnify the effective area of substrate contacts, while the P− substrate helps to limit the maximum injected current and to contain substrate debiasing within relatively limited regions of the die. Hole guard rings are usually employed only to prevent cross injection between merged components (Section 13.2.1). P-type regions connecting to external pins that are neither power supplies nor substrate ground are usually isolated by placing them in their own tanks. This practice requires about the same amount of space as the construction of hole guard rings and requires less effort.

13.2.3. Guard Rings in CMOS and BiCMOS Designs

CMOS designs are more prone to latchup than standard bipolar. This vulnerability results in part from the smaller dimensions of modern CMOS and BiCMOS processes and in part from differences between isolation systems. CMOS-derived processes usually substitute a lightly doped epitaxial layer for the vertical P+ isolation of standard bipolar. The light doping increases the gain of the lateral bipolar transistor formed across the isolation and makes it more probable that minority carrier injection will trigger SCR action. The light doping of the P-epi also makes it more difficult to extract substrate current. Most of these processes rely on the presence of a P+ substrate to reduce their vulnerability to latchup through the substrate, but scrupulous care must be taken to block lateral conduction by using guard rings.

Newer, more advanced processes are even more vulnerable than their predecessors because of the continual reduction in lateral spacings. The introduction of retrograde wells to reduce well resistance has provided substantial benefits, but modern submicron CMOS and BiCMOS processes remain extremely vulnerable to latchup.

---

4 *Ibid.*

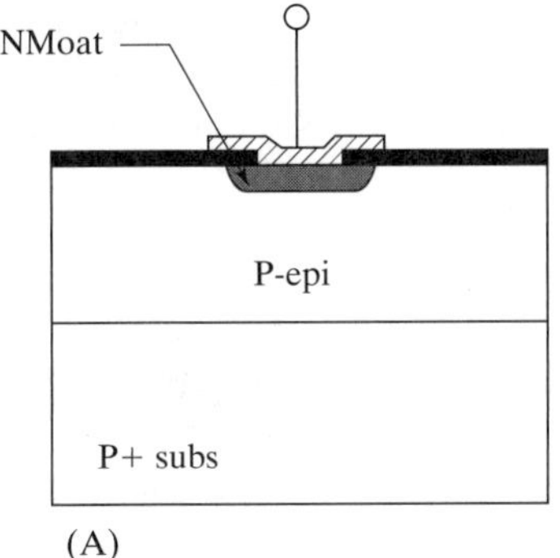

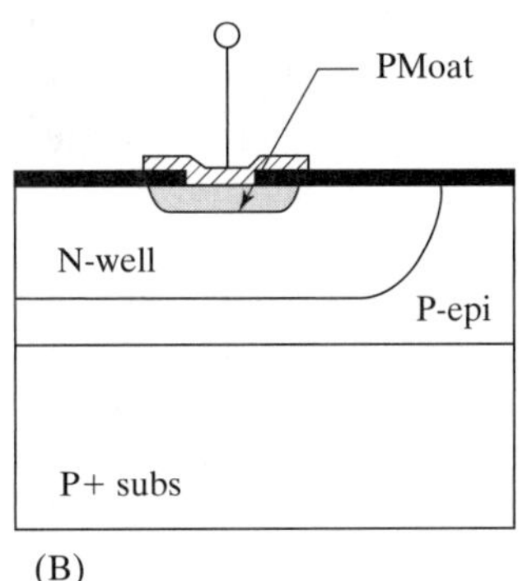

FIGURE 13.10 Minority-carrier guard rings for N-well CMOS: (A) electron-collecting guard ring and (B) hole-collecting guard ring.

Figure 13.10A shows an electron-collecting guard ring implemented in a CMOS process. This structure consists of an NMoat ring placed in the P-epi surrounding the source of injected electrons. The NSD implant is relatively shallow, so it can intercept only a fraction of the carriers. This type of guard ring relies on the P+ substrate beneath the wells to prevent minority carriers from bypassing the guard ring by burrowing through the substrate. Unfortunately, the presence of an electric field across the P+/P− interface tends to repel electrons from the substrate and to channel them laterally toward adjacent wells. This phenomenon makes it difficult to construct truly effective barriers against minority carrier injection into the substrate. Connecting the NMoat ring to a power supply rather than to ground helps only marginally, since the added depth of the depletion region does not penetrate more than a small fraction of the epitaxial layer. In low-voltage CMOS processes where the NMoat resides in a P-well, the guard ring should connect to substrate potential rather than to a power supply. The increased surface doping of the low-voltage P-well reduces the width of the NSD/P-well depletion region and causes the electric field across it to increase. High electric-field intensities can trigger avalanche multiplication within the depletion region. The resulting debiasing actually improves the collection efficiency of the guard ring, but it may cause problems elsewhere on the die. The addition of an N-well geometry to the electron-collecting guard ring of Figure 13.10A increases its depth and therefore improves its collection efficiency. Unfortunately, most N-well diffusions are so lightly doped that they cannot collect enough current to stop latchup.

Figure 13.10B shows a hole-collecting guard ring implemented in a CMOS process. This guard ring consists of a ring of PMoat placed in the N-well around the source of injected holes. This type of guard ring is generally not very effective because most of the holes flow down to the substrate rather than laterally to the guard ring. Increasing the width of the guard ring does little to improve its effectiveness. Retrograde wells may greatly improve the efficiency of hole–collecting guard rings. In effect, the heavily doped lower portion of a retrograde well acts as a buried layer. The N+/N− interface generates an electric field that tends to confine holes within the N-well. The holes thus flow laterally to the guard ring rather than vertically to the substrate.

The minority carrier guard rings that can be constructed in a CMOS process usually have very limited effectiveness. The two types of guard rings tend to reinforce one another, so the best design practice consists of using both electron and hole collecting guard rings around every device that might inject minority carriers. Since CMOS logic does not inject any substantial level of minority carriers during normal operation, digital devices can meet latchup requirements if every device connecting to an output pin receives a guard ring. The designer should examine each pin that neither connects to a power supply nor to substrate potential. Each source/drain region connecting to such a pin requires a guard ring. PMOS transistors require hole-collecting guard rings even when placed in their own wells. NMOS transistors require electron-collecting guard rings. A combination of guard rings and backgate contacts should suppress most forms of latchup, but they may prove inadequate for handling the severe minority carrier injection problems associated with inductive kickback and resonance. Analog designers must also consider the possibility that nodes internal to the circuit could cause minority carrier injection and latchup. Examples of such situations include capacitors connected as positive feedback elements and nodes associated with charge pumps.

Analog BiCMOS processes normally include NBL and deep-N+. The presence of these layers allows the construction of deep electron-collecting guard rings similar to the one in Figure 13.8A. These guard rings are especially effective on designs using a P+ substrate because the built-in potential of the P-epi/substrate interface helps confine the electrons within the epi. A deep-N+ guard ring on a thin-epi P+ process may collect 90% or more of the electrons injected into the epi.[5]

Newer processes with shallow, heavily doped N-well regions may not provide sufficient built-in potential between the N-well and the NBL to contain holes. This problem, which might be termed *NBL permeability*, has been observed on several low-voltage BiCMOS processes.

The efficiency of hole guard rings suffers if the NBL cannot efficiently block hole flow to the substrate. The addition of a hole blocking guard ring may actually increase substrate injection through a permeable NBL.[6] This seemingly paradoxical behavior probably results from a reduction in the effective volume of the N-well. The hole-blocking guard ring repels holes from the portion of the well it occupies, concentrating them within the remaining volume of the well. The higher concentration of holes near the NBL/N-well interface increases the injection rate of carriers into the substrate. This *reduction in volume effect* should not affect hole-collecting guard rings, but the presence of a permeable NBL still reduces their collection efficiency.

Analog BiCMOS designs also exhibit excessive substrate resistance. Even if the design uses a P+ substrate, the presence of a lightly doped P-epi makes it difficult to establish a low-resistance substrate contact. Even relatively low levels of substrate injection can produce substantial substrate debiasing. Substrate debiasing can be prevented by blocking minority carriers before they can reach the substrate through the use of hole guard rings. All high-current saturating NPN transistors should incorporate such guard rings to prevent substrate debiasing and noise coupling.

Analog BiCMOS designs sometimes use a P− substrate to avoid the necessity of growing two epitaxial layers. Designs constructed on a P− substrate are even more susceptible to latchup because electron guard rings no longer benefit from the presence of an electron barrier at the P−/P+ interface. Many designs can still achieve satisfactory levels of immunity to transient-induced latchup, providing that every potential source of minority carrier injection is surrounded by a suitable guard ring. Even the most conservatively designed guard rings may prove unable to handle the severe minority carrier injection problems associated with inductive kickback and resonance. Such designs may require the use of a P+ substrate despite the additional cost associated with the second epitaxial deposition.

Dielectrically isolated processes can also experience latchup due to low-voltage components constructed in a common tank, such as often occurs in digital logic. The placement of an isolation ring around the devices that would otherwise inject into a common well or field region will usually suffice to prevent latchup. However, current

[5] R. R. Troutman, "Epitaxial Layer Enhancement of N-Well Guard Rings for CMOS Circuits," *IEEE Electron Device Letters*, Vol. EDL-4, #12, 1983, pp. 438–440.

[6] N. Gibson, unpublished report, 1998.

may flow through the isolated devices and cause minority carrier injection from devices not directly connected to pins. In such cases, multiple devices may require isolation rings, or one may need to insert deposited resistors to minimize the magnitude of the current flowing through the isolated devices.

13.3 SINGLE-LEVEL INTERCONNECTION

Most modern processes offer at least two levels of metallization. Since leads can freely cross one another, the placement of components becomes constrained only by matching and packing. Routing almost never presents a problem as long as the designer leaves a little space between components. Given time, almost any designer can compress the wasted space out of such a layout to produce a reasonably densely packed design.

Interconnection becomes much more difficult if the process offers only one level of metallization. The lack of second metal makes it difficult to cross leads. Although low-value resistors can be inserted to create crossing points, these *tunnels* consume die area and add resistance and capacitance that degrades the performance of the circuit. A properly arranged layout contains surprisingly few tunnels. The components in such a layout are arranged to minimize the number of crossing points. Leads often route across resistors or between the terminals of transistors, and sometimes they even tunnel through tanks or base diffusions.

Single-level interconnection requires far greater skill and ingenuity than multilevel interconnection. The designer must anticipate possible blockages between components and mentally shuffle them to clear a path. A move that clears one blockage often creates others. Skilled designers have a sort of "geometric intuition" that aids them in placing components and routing leads. This intuition seems largely an innate talent and not a learned skill. There are, however, a number of specific skills and techniques that can help any designer better cope with single-level-metal designs. Although these skills may not seem to have any application in modern multilevel-metal processes, many designs dedicate the upper metal layers for power routing, electrostatic shielding, or optical shielding. A skilled designer must therefore understand how to route designs with a minimum number of layers of interconnection.

13.3.1. Mock Layouts and Stick Diagrams

The greatest challenge of single-level routing lies in properly arranging the components to minimize the number of tunnels required. The presence of matched components often complicates this task so that even skilled designers have to try several arrangements before finding a suitable one. These trial arrangements usually take the form of rough sketches, or *mock layouts,* similar to that in Figure 13.11. The transistors appear in this sketch as rectangles with emitter, base, and collector marked. The resistors appear as strips with connections to either end. Dummies and resistor tanks do not appear in the sketch. The tank contacts are marked "TC." Merged devices occupying a common tank are shown abutting one another, as in the case of Q_3 and Q_4. Although crude, this sketch illustrates all of the important features of the proposed layout.

This particular layout contains a large number of matched devices. In order to obtain the best possible matching, each set of these components has been arranged symmetrically around one axis of the layout as advocated in Section 7.2.10. The axis of symmetry passes horizontally through the middle of the sketch.

Resistors R_3 and R_4 consist of 160 Ω/□ base material and have values of 621 Ω and 4 kΩ, respectively. These resistors are not in simple integer ratio ($R_4/R_3 = 6.441$).

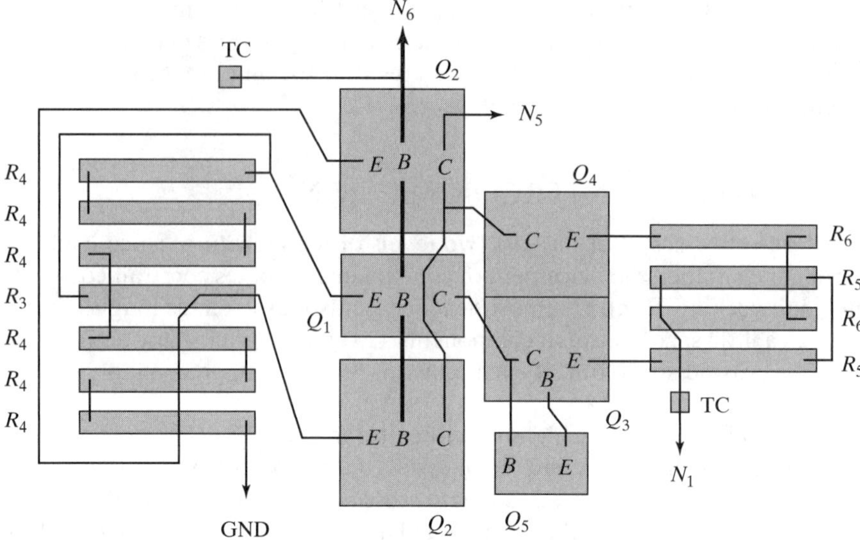

FIGURE 13.11 Mock layout for a portion of the circuit shown in Figure 14.2. The following components must match one another as accurately as possible: R_3–R_4, Q_1–Q_2, Q_3–Q_4, and R_5–R_6.

This complicates, but does not prohibit, the construction of a sectioned and interdigitated array. Suppose that R_3 is taken as the unit resistor for the array. R_4 then requires a minimum of seven segments. Unfortunately, the centroid of a one-segment resistor cannot perfectly align to the centroid of a seven-segment resistor. In order to achieve a true common-centroid layout, R_4 must consist of eight partial segments. This arrangement is not particularly compact because all of the segments are relatively short. (R_3 contains 3.88 squares.) A better arrangement consists of six segments of 666.7 Ω each for R_4 and a partial segment for R_3. The sliding contact in R_3 also allows the resistor ratio to be tweaked. R_3 occupies the center of the array to ensure that the centroids align (Section 7.2.10). The six segments of R_4 have been interconnected to cancel thermoelectrics (Section 7.2.11). Resistors R_5 and R_6 each consist of 18.75 squares of 160 Ω/□ base diffusion. The mock layout shows that each resistor consists of two segments of 9.375 squares that are interdigitated to form a compact array.

Transistors Q_1 and Q_2 form a 6:1 ratioed pair. The layout of such transistors normally involves splitting the larger transistor into two halves placed on either side of the smaller transistor (Figure 9.23A). Transistors Q_3 and Q_4 are matched, minimum-size lateral PNP transistors. These transistors can reside in a common tank because they share the same base connection. They are placed side-by-side to improve matching and to simplify interconnection. P-bars and N-bars are unnecessary because neither transistor saturates in normal operation.

Although this circuit contains several crossing points, none requires a tunnel. The lead interconnecting the emitters of Q_2 routes through the resistor array R_3–R_4. The collector lead could follow the same route, but this requires separating the resistor segments. Instead, transistors Q_1 and Q_2 have been stretched to allow Q_2's collector lead to route between the base and collector of Q_1. This configuration actually requires little or no elongation of the transistor tanks because the presence of deep-N+ in Q_1 and Q_2 already necessitates a large base-to-collector spacing.

The mock layout in Figure 13.11 has not been drawn to scale, but the rectangles representing the components have roughly the same proportions as the components themselves. Sometimes, designers carry this type of sketch one step further by using paper plots of the actual components. All of the components are plotted to the same scale, typically either 100:1 or 250:1. The individual components are cut out and shuffled about on a large sheet of paper until a suitable arrangement appears, and then the components are glued down and the interconnections are marked in pencil or ink. Cut-and-paste mock layouts (*paper dolls*) are especially useful for designing tight-packed layouts, since all of the dimensions of the components are to scale. One can carry out much the same activity by simply placing devices (or rectangles representing them) on the layout and then shuffling them about.

CMOS designers sometimes use another type of mock layout called a *stick diagram*. Although stick diagrams were originally intended to portray digital logic cells, they can represent analog circuitry. Figure 13.12 shows a schematic and a stick diagram of a CMOS NAND gate. The thick, black horizontal lines represent PMoat and NMoat regions. The NMoat usually lies at the bottom of the diagram and the PMoat at the top. The thick, gray vertical lines represent poly traces. A transistor forms wherever poly crosses either PMoat or NMoat. Contacts are represented by X-marks and metal leads by thin, black lines. Stick diagrams of analog circuits usually show NMoat and PMoat regions in different colors to aid in distinguishing them from one another. The names of nodes and devices may both appear on the stick diagram, and additional notations may be added to identify resistors and capacitors.

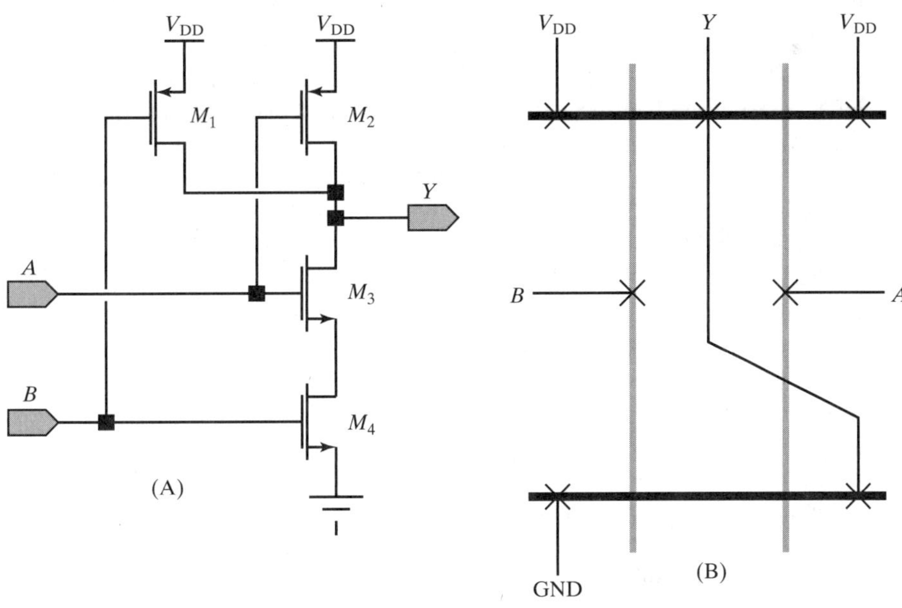

FIGURE 13.12 (A) Schematic and (B) stick diagram of a CMOS NAND gate.

13.3.2. Techniques for Crossing Leads

No matter how hard one tries to avoid them, most single-level metal layouts still require crossing leads. The following rules summarize the techniques available for crossing leads using only one level of metal. These techniques were originally developed for standard bipolar designs, but they are also applicable to multiple-level metallization designs where the upper metal layers are required for power routing, electrostatic shielding, or some other purpose.

1. *Cross leads over resistors.*
   A lead routed across a resistor provides a crossing point without consuming additional area, but not every lead can safely cross every resistor. Some resistors

require field plating that restricts or even prevents lead crossings. Other resistors are susceptible to noise coupling from overlying leads. Lightly doped materials such as 2 kΩ/□ HSR may also experience voltage modulation effects. Variations in hydrogenation make it advisable to avoid routing leads over precisely matched poly resistors.

2. *Rearrange the terminals of a device.*
   Crossing points can often be eliminated by altering the arrangement of device terminals. For example, the CEB layout for an NPN transistor places the emitter terminal between the collector and base, while the alternate CBE layout places the base terminal between the collector and emitter (Figure 8.14). The CBE layout exhibits slightly more collector resistance than the CEB layout, but the difference rarely affects circuit operation.

3. *Stretch devices to allow leads to pass through them.*
   Most types of devices can stretch to accommodate one or more leads between their terminals. Figure 8.15 shows three examples of stretched NPN transistors, and Figure 5.14 shows an example of a stretched HSR resistor. Stretched devices usually possess more parasitic resistance and capacitance than unstretched devices, so their use sometimes affects circuit operation. If one of a group of matched devices employs a stretched layout, then so should all the others.

4. *Connect signals through merged devices.*
   Certain types of devices lend themselves to use as tunnels. For example, the NPN transistor in Figure 8.15C contains a stretched base region with two contacts. This component actually merges an NPN transistor and a base tunnel into the same tank. A similar type of merger uses multiple tank contacts rather than multiple base contacts. These types of merged tunnels insert parasitic resistances and capacitances that can affect device operation. If large currents flow through the tunnel, the resulting debiasing may also affect circuit operation.

5. *Insert tunnels.*
   Tunnels, or *cross unders,* are low-value resistors incorporated into the layout to allow leads to cross one another. Various types of tunnels exist, but all share similar disadvantages. They not only consume die area, but they also insert parasitic resistance and capacitance into the tunneled lead. The insertion of a tunnel into a high-current lead can cause excessive voltage drops and power dissipation. Tunnels can also upset matching by introducing voltage drops where they cannot be tolerated. Since the placement of tunnels affects circuit operation, the circuit designer must ultimately approve or reject each potential tunnel. When all of the tunnels have been placed, the circuit designer should add their resistances and capacitances to the circuit and resimulate to see if any critical parameters have shifted.

6. *Rearrange the bondpads.*
   If high-current leads must cross one another to reach their respective bondpads, consider rearranging the bondpads to eliminate the crossing point. Sometimes one bondpad arrangement may lend itself to interconnection much more readily than others.

The layout designer usually requires guidance from the circuit designer to determine what types of stretches and tunnels are allowable. One simple and effective means of communicating this information consists of an annotated schematic prepared by the circuit designer. This diagram requires only a few minutes to prepare, yet it can save many hours of layout effort. Table 13.1 describes a simple annotation

Category	Precautions	Marking
Power leads	No tunnels allowed; leads must equal or exceed a certain width.	Highlight in **red;** mark width over lead.
Noisy leads	Do not cross sensitive devices.	Highlight in **yellow.**
Sensitive leads	Do not tunnel. Do not place substrate contacts in sensitive ground leads.	Highlight in **green.**
Sensitive devices	Do not cross noisy leads over sensitive devices.	Highlight in **green.**

TABLE 13.1 A simple schematic annotation scheme.

scheme, using different colors to highlight components and signals. The annotated schematic should also include lists of matched components, guard rings, and other special requirements that might influence the routing of the leads.

13.3.3. Types of Tunnels

Tunnels can be constructed from any diffusion having a relatively low sheet resistance. In standard bipolar, the candidates include base, emitter, deep-N+, and NBL. CMOS and BiCMOS processes often use gate poly jumpers instead of tunnels. The base diffusion usually has a sheet resistance of 100 to 200 Ω/□, while the other three materials usually have sheet resistances of about 10 Ω/□. Of these, only the base diffusion can occupy a common tank with other components. Most standard bipolar designs contain a number of merged base tunnels and a few stand-alone tunnels of other types.

All tunnels add series resistance. In the case of base tunnels, this resistance usually equals a few hundred Ohms. Although this may not seem like much resistance, it is sufficient to throw off device matching and to cause certain types of circuits to completely malfunction. A typical tunnel also exhibits a few hundred femtofarads of parasitic junction capacitance. Some high-speed circuits contain nodes that would be seriously slowed by even this small amount of capacitance. Any attempt to reduce the series resistance of the tunnel by widening also increases its shunt capacitance. Larger tunnels also become more vulnerable to junction leakage and to minority carrier collection.

The sheet resistance of the emitter diffusion is an order of magnitude lower than that of base. The simplest sort of emitter tunnel simply consists of a strip of emitter diffusion placed in a tank (Figure 13.13). The tank-substrate junction provides the necessary isolation between the signal and the underlying substrate. The tank requires no contact other than that provided by the tunnel itself. The addition of NBL does not significantly reduce the tunnel resistance or improve latchup immunity. NBL actually increases the parasitic shunt capacitance, so emitter tunnels generally omit it.

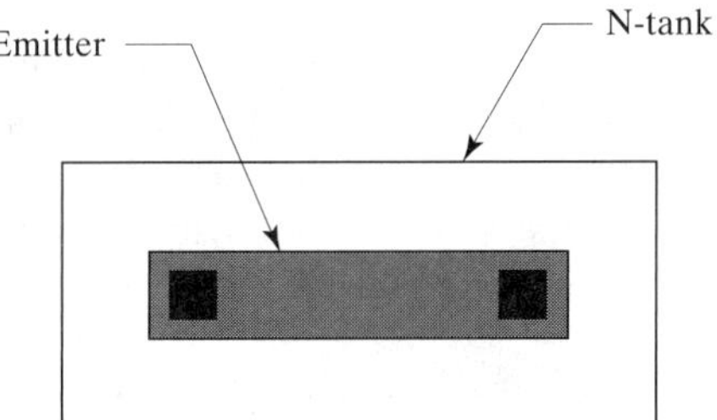

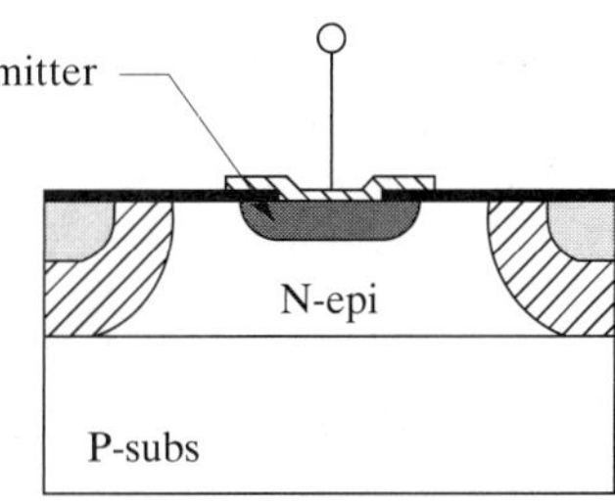

FIGURE 13.13 Layout and cross section of a conventional emitter tunnel.

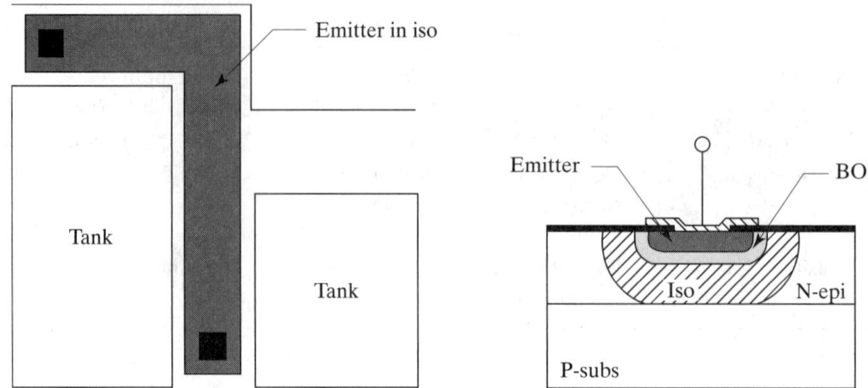

FIGURE 13.14 Layout and cross section of an emitter-in-iso tunnel.

The emitter-in-iso tunnel in Figure 13.14 saves considerable area by eliminating a tank geometry. The emitter diffusion can counterdope the isolation, but the breakdown of the resulting N+/P+ junction may equal only a few volts. Junctions with breakdown voltages this low tend to leak, but emitter-in-iso tunnels can still be used to route the substrate return line around the die.[7] Processes with emitter-iso breakdown voltages of 6 V or higher can often employ emitter-in-iso tunnels to route other signals, but the capacitance of the emitter-iso junction is relatively large (typically about 1.5 FF/μm^2). Some circuits have taken advantage of this high capacitance to fabricate emitter-iso junction capacitors. These capacitors can occupy unused areas of the isolation, allowing the construction of large junction capacitors with little or no increase in die area.

Some applications require a resistance that is lower than emitter alone can provide. A combination of all available N-type regions (N-epi, NBL, deep-N+, and emitter) will produce a slightly lower sheet resistance (Figure 13.15). A stack of this sort usually has a sheet resistance of about 5 Ω/□. NBL provides little benefit without deep-N+, so designs that do not use deep-N+ cannot make effective use of stacked tunnels.

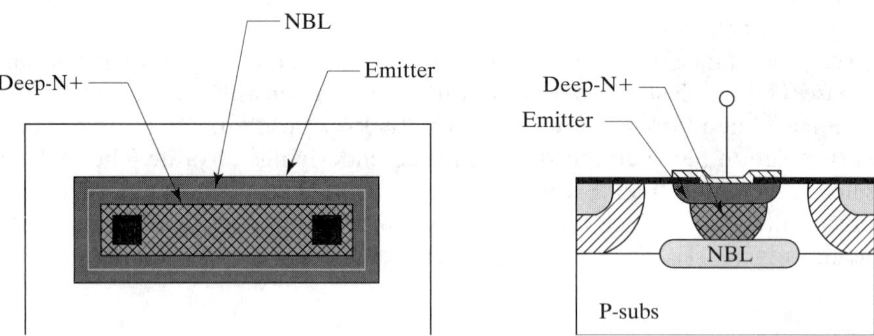

FIGURE 13.15 Layout and cross section of a low-resistance, stacked tunnel containing emitter, deep-N+, and NBL.

Figure 13.16 shows an NBL tunnel. This type of tunnel bridges between two adjacent tanks, eliminating the need for a tank contact in one. The strip of isolation between the two tanks functions as a highly effective P-bar that prevents cross injection from one tank to the other. One must employ caution when employing NBL tunnels so as not to exceed the NBL/isolation breakdown voltage.

---

[7] Davis, pp. 16–17.

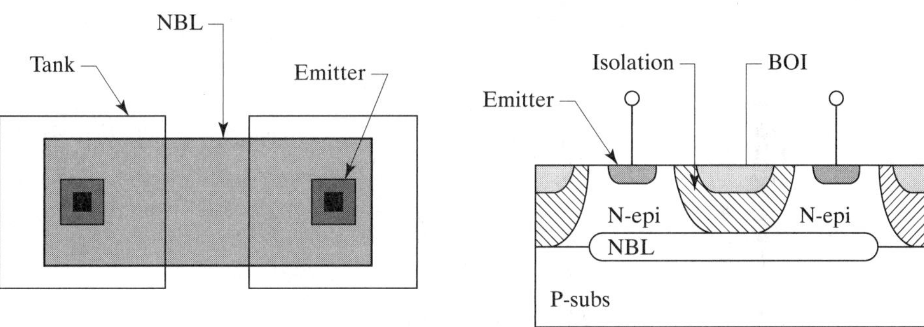

FIGURE 13.16 Layout and cross section of an NBL tunnel.

13.4 CONSTRUCTING THE PADRING

The *padring* of an integrated circuit consists of scribe streets, pads, ESD structures, and guard rings. Each of these plays a vital role in determining the success or failure of the design. Many circuits have failed because of inadequate ESD protection, misplaced bondpads, or missing guard rings. The following sections provide guidance that should help the layout designer avoid most of these mistakes.

13.4.1. Scribe Streets and Alignment Markers

Scribe streets must surround the die to provide room for the passage of the sawblade used to separate the dice. The saw consumes a strip of silicon about 25 μm wide, but the scribe streets must be three or four times wider to provide room for misalignment. Oxide and nitride tend to crack during sawing, and metal clogs the sawblade; therefore, most scribe streets consist of bare silicon. The edges of the die abutting the scribe street are often fitted with special structures called *scribe seals* to prevent contaminants from seeping underneath the exposed edges of the protective overcoat (Section 4.2.2). Additional structures may reside within the scribe street itself. Some fabs place *alignment markers* within the streets. These markers are used to align photomasks to previous steps of the process and are destroyed during sawing. Sometimes arrays of test devices are also placed in the scribe streets. These devices can be used to evaluate the performance of the wafer before it is sawn and assembled. The test devices also provide a means of characterizing large numbers of devices for statistical device modeling. The test devices can be destroyed during sawing because they already will have been tested and so will have served their purpose.

Most wafer fabs specify the scribe streets required for their process. Sometimes the scribe street occupies a separate database prepared by the fab and sent to the mask shop independently of the main design. Alternatively, the scribe street structures may be provided to the layout designer to place around the edges of the main layout. Regardless of whether they appear in the layout, the scribe seals abut the die on all four sides. Since these seals usually incorporate substrate contacts, they can form a useful addition to the substrate contact system. A thin strip of metal placed around the edges of the pad ring will make contact to the scribe seal metallization. In addition to providing substrate contacts, the scribe seal metallization also provides a convenient method of routing the substrate potential around the periphery of the die. The width of the metallization in the scribe seal adds to the width of metal in the padring, producing a relatively wide lead that can conduct significant current without debiasing or electromigration failure (Figure 13.17). Because of its convenient placement around the periphery of the die and the relatively low return resistance to the

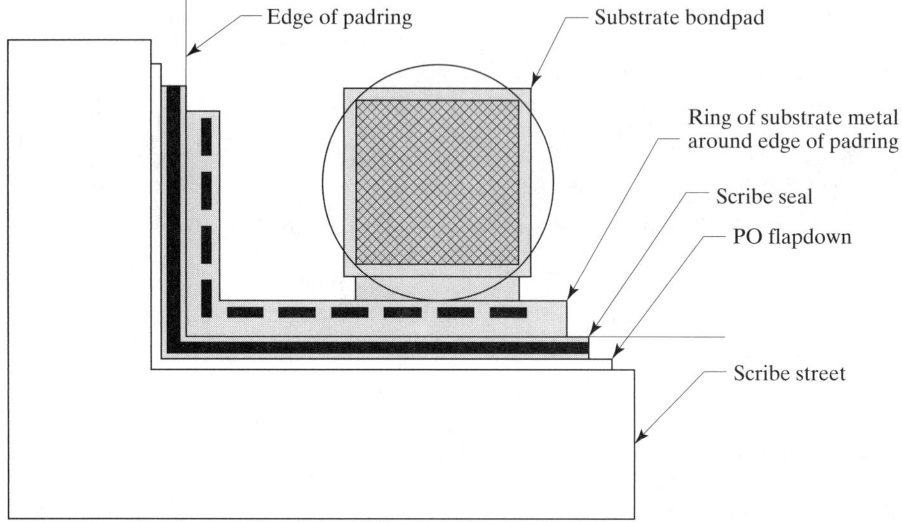

FIGURE 13.17 Diagram illustrating the relationship between the scribe street and the padring.

substrate bondpad, the substrate metallization often forms the return path for the ESD structures (Section 13.5).

The layout designer must seek the fab's guidance to determine the proper scribeline selection and placement. Some fabs may impose additional requirements, including limitations on the size and aspect ratio of the die. For example, some older steppers only accept dice whose dimensions are an integer number of mils. The efficient use of photolithographic equipment may also dictate that the die dimensions fall within a certain range of values. These issues should be resolved during die floorplanning so the designer can maximize the profitability of the design. These issues become much more difficult to resolve as the layout nears completion, so do not wait until the last moment to obtain the necessary information!

13.4.2. Bondpads, Trimpads, and Testpads

Most integrated circuits connect to the external world through bondwires. These wires consist of either gold or aluminum, and they range in diameter from 20–250 μm (0.8–10.0 mils). The most common form of bonding uses gold wires approximately 25 μm (1mil) in diameter attached to the die by means of ball bonds. The ball bonding process uses a hydrogen flame to create a tiny gold ball on the end of the bondwire. A capillary tube presses this ball down against the exposed aluminum metal with enough force to cause the two metals to alloy together (Section 2.7.1). The bonding process deforms the soft gold ball into a pancakelike structure (Figure 13.18). The actual area of gold-aluminum alloying is usually about the same diameter as the original bondwire, but the metal pad on which the ballbond rests must be two to three times larger to account for the misalignments that inevitably occur during automated bonding. Each ballbond thus requires an exposed metal pad several mils across. These specialized structures are called *bondpads*.

The simplest conceivable bondpad consists of a square of metal placed underneath a matching opening in the protective overcoat (PO). The metal must overlap the opening sufficiently to seal the die against the ingress of mobile ions (Section 4.2.2), even if overetching and misalignment occur. If the process offers more than one level of metal, then the bondpad typically includes plates of each metal placed coincident to one another. The interlevel oxides should be removed from the area of the bondpad

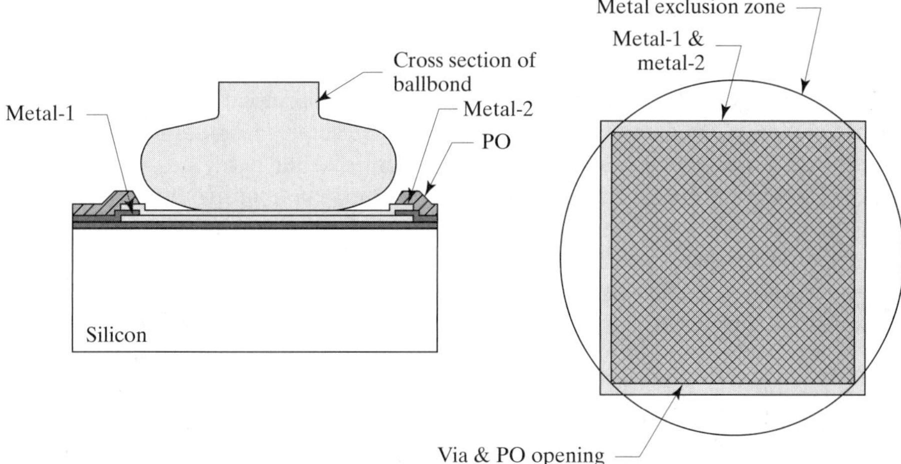

FIGURE 13.18 Cross section and layout of a double-level-metal bondpad intended for ballbonding.

opening so the ballbond lands on the stacked metal layers. Figure 13.18 shows a typical double-level-metal bondpad constructed according to these principles. The bondpad consists of a square of metal-2 placed over an identical square of metal-1. These metal plates overlap a PO opening and a via, both of which also coincide to one another. Layout rules generally prohibit the placement of any devices beneath bondpads, because the high pressures generated during bonding can cause stress-induced failures in the devices.

Dice are usually bonded using the smallest possible diameter of wire, since this enables the use of the smallest bondpads. A 25 μm-diameter gold bondwire packaged in plastic can conduct approximately one amp of continuous current (Section 14.3.3). Higher currents require either larger diameter wires or multiple bondwires connected in parallel. Every pin of a typical package can accommodate two (or possibly three) bondwires. Some surface-mount packages are so small that they can accommodate only one bondwire per pin, and these same packages are usually so thin that they can use only the finest wire diameters. The designer should verify that the package can handle the required number and diameter of bondwires during the earliest stages of floorplanning.

Small-diameter bondwires have relatively large resistances. The resistance of a bondwire can be estimated using the equation

$$R_w \cong \frac{\omega L}{D^2} \qquad [13.2]$$

where R_w is the resistance of the bondwire in Ohms, L is its length in μm, and D is its diameter in μm. The constant of proportionality ω equals approximately 27.9 mΩ · μm for gold and 35.6 mΩ · μm for aluminum.[8] Bondwires are typically about 100 μm long, so a typical 25 μm-diameter gold bondwire exhibits a resistance of 30 mΩ. Equation 13.1 does not consider the resistance of the leadframe, nor the resistance of the bond contact, each of which may add a few additional milliohms of resistance. Gold bondwires are usually limited to a maximum of 50 μm (2 mils) in diameter, but much larger aluminum bondwires are available. Although it is technically possible to bond a die by using different types or diameters of wire, this requires a

[8] These values are only approximations based on the bulk resistances of gold and aluminum. The actual resistance of a bondwire is affected by impurities and work hardening.

corresponding number of passes through the bonding equipment. The additional time and expense are rarely justifiable unless the design requires the use of extremely large-diameter wire.

Historically, gold ballbonds have required square bondpad openings about three times the diameter of the wire. The increasing precision of modern bonding equipment has enabled many assembly sites to accept smaller bondpads. The layout designer should obtain the current guidelines from the assembly site for the specific type and diameter of wire that will be used to bond the die. Aluminum wire must be wedge bonded rather than ballbonded, and this usually requires an elongated bondpad placed at a specific angle relative to the fingers of the leadframe. The rules for aluminum wirebonding can become quite complex, and the designer should seek guidance from the assembly site before attempting to lay out a design employing it.

The locations of the bondpads must simultaneously satisfy several conflicting requirements. The bondpads cannot lie too close to one another, or the capillary tube will damage one bond while placing the next. The bondwires must not pass too close to adjacent pads lest the capillary damage the wires while placing subsequent bonds. Long bondwires can short to one another or to adjacent bondpads due to a phenomenon called *wiresweep.* The injection-molding process forces molten plastic over the die, and the viscous drag of the plastic on the bondwires causes them to move. Larger diameter wires are more rigid and can better resist wiresweep, allowing them to span larger distances. Wiresweep also makes it inadvisable to cross one bondwire over another. The best bonding arrangements consist of a ring of pads placed around the periphery of the die in locations that allow the shortest and most direct wirebonds. Parts packaged in ceramic or metal can ignore the limitations imposed on plastic packages due to wiresweep, but they must still follow the spacing rules required to prevent capillary damage.

It is often difficult to find a suitable bonding arrangement for a die having a large number of bonds. Some assembly sites provide software tools that can evaluate a proposed bonding arrangement to ensure manufacturability. Others require that any potential bonding arrangement pass through a review process in order to obtain production approval. The layout designer should always check to make sure that the bondpad arrangement meets the approval of the assembly site before beginning the top-level layout. If this is not done and the finished layout does not meet the assembly site's requirements, substantial time and effort will be required to correct the problems.

The placement of bondpads also restricts the routing of adjacent metal leads. Misalignment or excessive bonding force may cause the bond to press against the protective overcoat adjacent to the bondpad opening. The resulting stress can crack the protective overcoat and can even damage underlying leads. Many assembly sites state that metal leads that do not connect to the bondpad must not pass within a certain distance of it. For ballbonds, this requirement usually takes the form of a circular exclusion zone centered on the bondpad. Many designers mark this exclusion zone with a circle placed on a special drawing layer. The layout in Figure 13.18 shows an example of one of these so-called *bondpad circles.* The assembly site will usually specify the dimensions of the exclusion zones. If no guidelines are available, assume that the bondpad circle passes through the four vertices of a square bondpad opening of minimum dimensions. Wedge bonding also requires exclusion zones, but the dimensions of these zones depend on the placement of the pads in relation to the leadframe.

Historically, many designers placed tanks (or wells) underneath all bondpads and probepads. These tanks were generally left unconnected. They were intended to protect the die against shorts caused by the probe needles scratching through the bondpad metallization and field oxide during wafer-level testing. If such shorts occurred, they would connect the bondpad to a tank rather than to the substrate. This would—theoretically—prevent the device from failing. The placement of unconnected tanks under bondpads is actually a very questionable practice. If the bondpad shorts to the tank, then the tank may inject electrons into the substrate. This means that the tank requires an electron-collecting guard ring, which wastes considerable space. Most modern designs do not place tanks or wells under pads unless they form part of some adjacent device whose tank or well connects to the pad.

Some assembly sites also require that the bondpad for pin #1 be visually distinct from all of the others. This requirement originally arose from the limitations of early machine vision systems used in automated bonding equipment. Even though most modern machines no longer require a distinct pad #1, it still provides a convenient visual reference point for operators who must inspect the mounted dice. A variety of different techniques have been used to mark pad #1, the simplest of which notches the four corners of the protective overcoat (PO) opening (Figure 13.19A). These notches are usually about 10 μm deep. If possible, the metal pattern should also contain notches corresponding to those of the PO opening on at least two corners of the bondpad. Another technique marks pad #1 with an octagonal PO opening (Figure 13.19B), while a third employs a circular opening (Figure 13.19C). The requirements of the assembly site may dictate a choice among these options. Otherwise, the designer should probably follow the conventions established by previous designs.

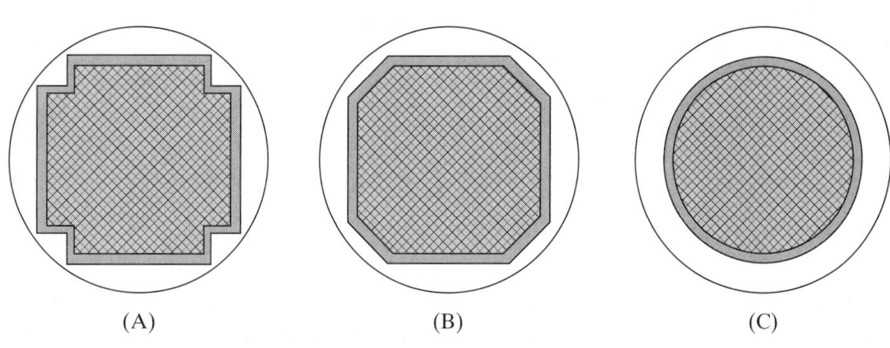

FIGURE 13.19 Three unique styles of bondpads sometimes used to identify pin #1.

Since trimpads and testpads allow access only for probe needles, they need not follow all of the requirements that apply to bondpads. The size of trimpads and testpads depends on the diameter of the probe needles and on the alignment tolerances of the probing equipment. These requirements usually prove somewhat less restrictive than those associated with bonding, so trimpads and testpads are usually smaller and more closely spaced than bondpads. The relatively high currents associated with trimming sometimes necessitate the use of larger diameter needles, so trimpads may require larger PO openings than do testpads. Both types of pads are customarily placed around the periphery of the die to simplify the design of the probe card. Additional testpads for engineering evaluation are sometimes placed in the interior of the die, but these are usually removed before the design reaches production, to minimize the ingress of contaminants.

Probing does not generate the same level of mechanical stresses as bonding, so many assembly sites allow the placement of trimpads and testpads over active circuitry. This concession virtually eliminates the area penalty associated with the use of small numbers of fuses or Zener zaps. Trimpads placed over active area consist of a square of top-level metal and an associated PO opening. They require neither vias nor lower level metal, so they can reside over any portion of the circuit free of top-level metal. One should employ caution when placing matched components partially or entirely under trimpads due to the possibility of hydrogenation-induced mismatches (Section 7.2.8). Some processes use a very thick layer of top-level metallization that provides enough mechanical compliance to dissipate the mechanical stresses that are induced during bonding. These processes may also allow the placement of bondpads over active area. This practice can result in substantially more compact designs, but only if the metallization system can support the resulting stresses.

13.5 ESD STRUCTURES

In addition to bondpads, the padring also contains structures intended to safely dissipate the energy of ESD events. These *ESD structures* must reside near their respective bondpads to minimize lead resistances and inductances that might otherwise interfere with their operation. The metallization resistance placed in series with an ESD device should not exceed 2–3 Ω. The scribe seal metallization can provide the necessary low-resistance path without consuming much additional die area. In order to take full advantage of the scribe seal, the ESD structures must either reside between the bondpads and the scribe seal, or between adjacent bondpads.

The effects of ESD vary depending on the type of component involved.[9] PN junctions are usually destroyed by overheating. Since considerable energy is required to melt even a few cubic microns of silicon, diffused junctions are generally quite robust. An avalanching junction dissipates most of its heat within its depletion region. Since lightly doped junctions have wider depletion regions, they can dissipate more energy than can heavily doped junctions. A lightly doped junction also has additional series resistance that dissipates some of the ESD energy. Large junctions are more robust than small ones because they contain a correspondingly greater volume of silicon within their depletion regions. The collector-base and collector-substrate junctions of standard bipolar processes are so large and so lightly doped that they can withstand most ESD transients without damage. Base-emitter junctions are more vulnerable because of their smaller dimensions and heavier doping. The base-emitter junctions of NPN transistors are also susceptible to avalanche-induced beta degradation (Section 4.3.3). The shallow, heavily doped junctions used in CMOS processes are more easily damaged than the deep, lightly doped junctions of standard bipolar. CMOS transistors with silicided source/drain regions (*clad moats*) are particularly fragile because of the lack of ballasting and the presence of silicide immediately adjacent to the depletion region. High-voltage CMOS devices with square SOA characteristics (Section 12.2.1) are often extremely rugged, and if they are sufficiently large, they can act as their own ESD devices. (Such devices are said to be *self-protecting*.) Transistors with electrical SOA limitations, on the other hand, will self-destruct if ESD devices do not clamp applied voltages so as to prevent avalanche.

Thin insulating films, such as those used in MOS transistors and deposited capacitors, are extremely fragile. High voltages will rupture these films within nanoseconds. Even if the insulating film does not rupture, it may suffer degradation that causes it to fail during normal operation due to time-dependent dielectric breakdown (TDDB). These delayed failures are very difficult to detect during high-speed automated testing. The only proven way to prevent such failures is to limit the voltage

---

[9] A. Amerasekera, W. van den Abeelen, L. van Roozendaal, M. Hannemann, and P. Schofield, "ESD Failure Modes: Characteristics, Mechanisms and Process Influences," *IEEE Trans. on Electron Devices*, Vol. 39, #2, 1992, pp. 430–436.

across the dielectric to safe values. Deposited resistors can also suffer dielectric breakdown through the thick-field oxide or the interlevel oxide (ILO), but hundreds of volts are required to rupture these layers. Any diffusion connected to a bondpad will avalanche long before the field oxide or the ILO ruptures, and will therefore protect them.

Early bipolar integrated circuits rarely incorporated any intentional ESD protection, but they generally withstood the rigors of ordinary handling because their junctions were sufficiently robust to absorb and dissipate low-level ESD strikes without damage. A few parts undoubtedly suffered damage during handling, but most of these failures were erroneously attributed to processing defects or infant mortality. CMOS integrated circuits proved much more fragile. Large numbers of early devices were destroyed through gate oxide rupture during normal handling. Once the mechanisms responsible for these failures were identified, designers began to recognize that even bipolar circuits were potentially vulnerable. A variety of protective structures were proposed, some of which worked and some of which did not. The reasons for success and failure were poorly understood, so ESD protection gained a reputation of being a "black art." This reputation is largely undeserved because ESD devices obey the same principles that govern other components. The following sections examine several types of ESD devices often used to protect analog integrated circuits. The strengths and weaknesses of each device are explained in terms of their structures and electrical properties. Using this information, the layout designer can construct ESD structures for a variety of applications.

13.5.1. Zener Clamp

The simplest ESD device consists of a Zener diode connected between the bondpad and the substrate return line (Figure 13.20A). Possible choices include the emitter-base Zener of standard bipolar and the NSD/P-epi and PSD/N-well Zeners of analog CMOS. An ideal Zener diode would impose a positive clamp voltage equal to its reverse breakdown voltage and a negative clamp voltage equal to its forward drop. Most Zener diodes contain enough internal series resistance to make the clamp voltages much larger than these ideal values. A minimum-size emitter-base Zener has from 100 to 300 Ω of internal series resistance, and NSD/P-epi and PSD/N-well diodes have even more. These resistances actually increase the robustness of the Zener by spreading the ESD energy over a larger volume of silicon, but in so doing they cause the bondpad voltage to rise above the theoretical clamp voltage by some tens of volts. This consideration severely limits the usefulness of Zener diodes as ESD structures.

The source/drain regions of NMOS and PMOS transistors can sometimes protect themselves against ESD damage. Consider the case of a large NMOS transistor whose drain connects to a pin. Negative ESD transients forward-bias the NSD/P-epi junction. Most of the resulting voltage drop occurs within the P-epi.

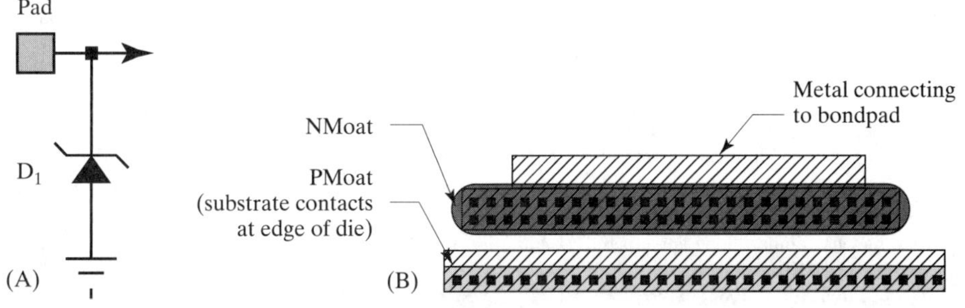

FIGURE 13.20 (A) Schematic diagram and (B) layout of the simple Zener clamp ESD circuit.

Positive ESD transients avalanche the NSD/P-epi junction, dumping energy into its depletion region. Since the depletion region contains less silicon than the P-epi, NMOS transistors are more vulnerable to positive transients than to negative ones. A similar analysis shows that PMOS transistors are most susceptible to negative transients. If a pin connects to both PMOS and NMOS transistors, then both conduct some portion of the ESD pulse and the more fragile device will determine the circuit's survival or failure. Large transistors are more robust than small ones because they can dissipate energy within a larger volume of silicon. A large number of small transistors usually provides the same degree of protection as a single large transistor. A 10V nonsilicided, single-diffused drain MOS device with a drawn drain width of 500–1000 μm will usually withstand 2 kV HBM, 200 V MM and 500 V CDM. Since PMOS and NMOS transistors avalanche under different biasing conditions, a large NMOS will not necessarily protect a small PMOS, or *vice versa.* If both PMOS and NMOS transistors connect to a pin, then the total PMOS drain area and the total NMOS drain area must both suffice to independently withstand the ESD strike.

Low-voltage CMOS processes are much more vulnerable to ESD damage than previous-generation, higher voltage processes, due in part to the extreme shallowness of the low-voltage source/drain diffusions and in part to the higher backgate doping required to prevent punchthough. Both of these factors reduce the volume of silicon in the depletion regions. Clad-moat devices sometimes exhibit localized breakdown due to their lack of source/drain ballasting. Once localized breakdown occurs, ESD performance no longer improves with increasing device area. The ESD performance of clad-moat devices can be improved by removing silicide from the perimeter of the source/drain implants in order to provide a small amount of ballasting. Robustness can be further improved by increasing the overlap of the (unsilicided) source/drain diffusions over their respective contacts on any diffusions that might avalanche due to ESD.[10] Most researchers attribute the increased robustness to ballasting, but some studies suggest that minority carrier injection from the source/drain contacts may also play a role.[11]

Highervoltage transistors that use extended drains may require larger device sizes to ensure self-protection due to restrictions in their SOA capability. Devices that exhibit electrical SOA limitations generally cannot self-protect regardless of their dimensions. The device sizes required for self-protection must be determined empirically by testing an array of devices of varying sizes.

If the area of the source/drain regions connected to a pad is insufficient to ensure self-protection, then a dedicated ESD protection device can be connected to the pad. In an N-well CMOS process, an NSD/P-epi diode (often called a *thick-oxide device* in the literature) usually offers the best protection for a given die area. Figure 13.20B shows a typical layout for such a diode.[12] The elongated NMoat region is placed alongside a strip of substrate contacts forming part of the scribe seal. The close proximity of the substrate contacts minimizes the series resistance of the device. The aspect ratio of this device allows it to be placed between a bondpad and the scribe seal metallization, or between adjacent bondpads. The corners of the NMoat diffusion are filleted to prevent premature avalanche breakdown. The radius of these fillets should

---

[10] T. L. Polgreen and A. Chatterjee, "Improving the ESD Failure Threshold of Silicided n-MOS Output Transistors by Ensuring Uniform Current Flow," *IEEE Trans. on Electron Devices*, Vol. 39, #2, 1992, pp. 379–388.

[11] T. J. Maloney, "Contact Injection: A Major Cause of ESD Failure in Integrated Circuits," *EOS/ESD Symposium Proc. EOS-8*, 1986, pp. 166–172.

[12] A somewhat similar diode appears in R. J. Antinone, P. A. Young, D. D. Wilson, W. E. Echols, M. G. Rossi, W. J. Orvis, G. H. Khanaka, and J. H. Yee, *Electrical Overstress Protection for Electronic Devices* (Park Ridge, NJ: Noyes Publications, 1986), p. 19.

equal or exceed the junction depth of the diffusion. The overlap of the NMoat region over its contacts should exceed the minimum layout dimension by 1 to 2 μm to provide additional ballasting. In clad-moat processes, the silicide block mask should block silicide for a distance of at least 1 to 2 μm from the drawn junction. The diode should contain at least 500 μm^2 of NMoat, and it should be surrounded by an electron-collecting guard ring and by as many substrate contacts as area permits. This device provides a reasonable degree of protection for NMOS and PMOS source/drain regions that are not large enough to protect themselves. In some cases, a low-value series resistor may be required to ensure that the ESD current flows through the Zener clamp rather than through the protected device.

13.5.2. Two-Stage Zener Clamps

Even a large protection Zener has an internal series resistance well in excess of 10 Ω. A 2 kV HBM strike produces peak currents of about 1.3A, which in turn produce voltage drops of tens of volts across the series resistance of the Zener. CDM strikes can produce even higher currents and larger voltages. These ESD-induced transients will damage or destroy a thin gate oxide. Although the Zener cannot protect the gate dielectric by itself, it can reduce the peak voltage of the ESD transient from hundreds or thousands of volts to mere tens of volts. A second protection structure connected in series with the first can provide enough additional clamping to protect the thin gate oxide. The schematic diagram in Figure 13.21A shows the conceptual arrangement of the resulting *two-stage ESD clamp.* Zener diode D_1 clamps the pad voltage to a maximum of perhaps 100 V. A second Zener, D_2, connects to the pad through a series limiting resistor, R_1. The presence of R_1 limits the current flow through D_2, enabling this second Zener to limit the voltage across the gate oxide to safe levels. In order for the circuit to function properly, the resistance of R_1 should equal at least 10 times the series resistance of D_2. A relatively small Zener diode may exhibit several hundred Ohms of internal series resistance, so R_1 typically equals several kilohms. The inclusion of this resistance limits the slew rate of the gate voltage, but this is often desirable since excessively large transient currents can damage gate dielectrics. R_1 also adds time delays of a few nanoseconds that may interfere with certain high-speed applications.

Figure 13.21B shows one possible layout for the series limiting resistor R_1 and the secondary protection Zener D_2. Resistor R_1 consists of a wide strip of lightly doped polysilicon with multiple small contacts on either end. The relatively large physical dimensions of this resistor help to ensure that it can successfully dissipate the energy dumped into it during an ESD transient. Diffused resistors are more robust than poly resistors because they dissipate a portion of their energy through a distributed avalanche mechanism, so many authors suggest using them instead of

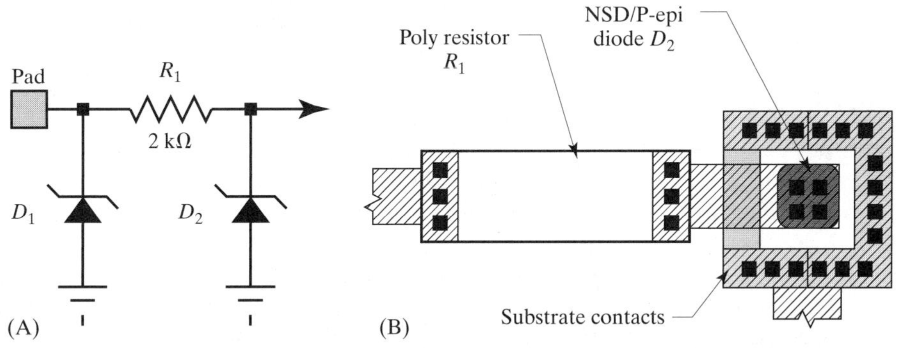

FIGURE 13.21 (A) Schematic diagram and (B) partial layout of a two-stage Zener clamp. The layout of primary protection diode D_1 is identical to that in Figure 13.20B.

poly resistors.[13] Poly resistors will survive 2 kV HBM and 200 V MM, provided that their resistance equals at least several hundred Ohms, they are at least 5 to 8 μm wide, and each end of the resistor includes at least six to eight minimum contacts. Fewer contacts are required for higher value resistors. Figure 13.21 illustrates a 2 kΩ poly resistor with three contacts at each end. Regardless of the type of resistor chosen, it should not include any bends, because current focuses near the inner corner of the bend, producing a hot spot that will fail before the remainder of the resistor. Zener diode D_2 consists of a relatively small plug of NMoat placed inside a concentric ring of substrate contacts. These substrate contacts not only minimize the series resistance of the Zener but also help minimize substrate debiasing near the secondary protection device. If these substrate contacts were omitted, or if they were located farther away from the secondary Zener, then the large transient currents flowing through the primary Zener D_1 could induce tens of volts of substrate debiasing near D_2. This debiasing would add to D_2's clamp voltage, potentially resulting in the destruction of the gate oxide that the structure is intended to protect. If possible, the secondary Zener should be placed 50 to 100 μm away from the primary one to further reduce the effects of substrate debiasing. One common arrangement places D_1 between the bondpad and the scribe R_1 alongside the bondpad, and D_2 on the inner side of the bondpad. Both D_1 and D_2 should be enclosed within an electron-collecting guard ring. The charged device model generates such extreme currents that metal debiasing may render the secondary clamps ineffectual unless they reside near the devices whose gates they protect (Section 13.5.8).

Two-stage ESD structures similar to that in Figure 13.21 have successfully protected MOS gates on many moderate-voltage CMOS processes. Since this type of ESD structure contains too much series resistance to allow its use on anything other than a high-impedance input terminal, it is often called an *input ESD device.* A similar two-stage ESD circuit can be constructed for some types of low-impedance applications, such as for the protection of the outputs of relatively small CMOS logic gates. This alternative type of structure uses a primary Zener diode D_1 identical to that employed in the input ESD circuit. The series resistance R_1 is reduced to 50 to 500 Ω, and the source/drain diffusions of the output MOS transistors serve as the secondary Zener diode, D_2. Although the source/drain junctions avalanche, the presence of the series resistance limits the current to safe levels. Larger output transistors can employ proportionally smaller series resistors. This type of ESD structure is sometimes called an *output ESD device.* These devices have successfully protected even minimum-area source/drain implants.

Sometimes a circuit includes both source/drain diffusions and gate electrodes connected to the same bondpad. A combination of input and output ESD circuitry will successfully protect such a circuit. A single primary protection device connects from the bondpad to the substrate return line. Two separate limiting resistors are required: a low-value one for the source/drain implants and a much higher value one for the gate electrodes. The source/drain implants serve as their own secondary protection, but the gate electrodes require the addition of a secondary Zener.

13.5.3. Buffered Zener Clamp

The availability of bipolar transistors allows the construction of very robust ESD circuits. Figure 13.22A shows a *buffered Zener clamp* that uses an NPN transistor to reduce the effective series resistance of a Zener diode. Emitter-base Zener D_1 provides

[13] A. R. Pelella and H. Domingos, "A Design Methodology for ESD Protection Networks," *EOS/ESD Symposium Proc. EOS-7,* 1985, pp. 24–40.

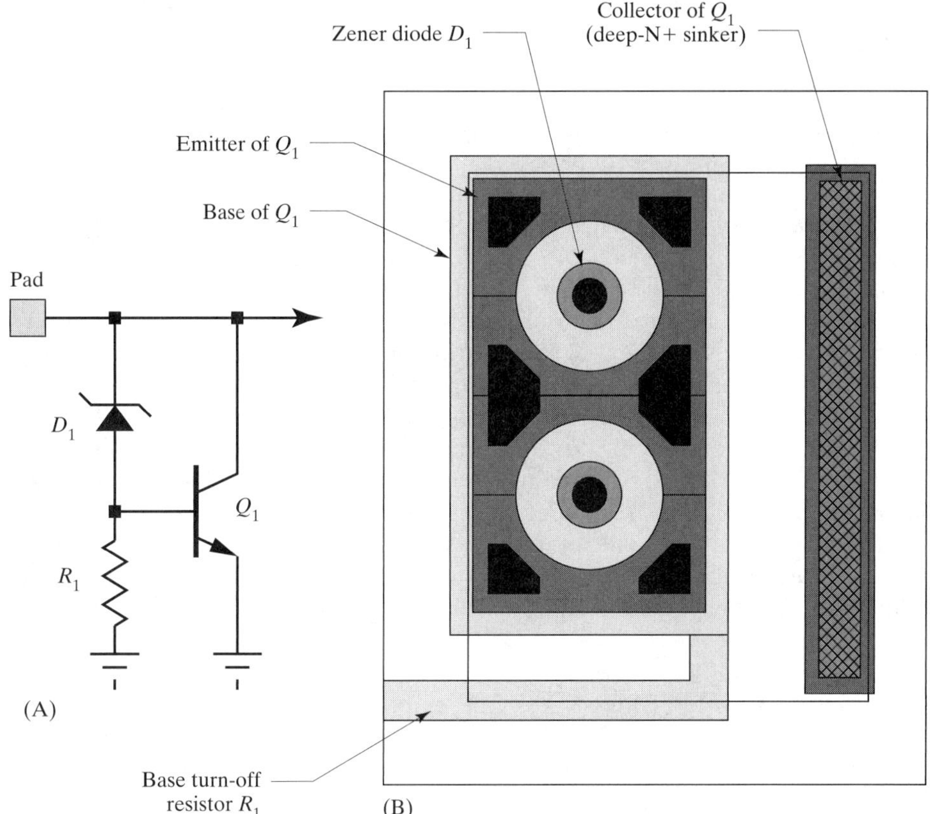

FIGURE 13.22 Schematic and layout of a buffered Zener clamp (metallization omitted for clarity).[14]

base drive to a much larger NPN transistor Q_1. This transistor multiplies the current through the Zener by its own effective beta. The positive clamp voltage of this structure equals the sum of an emitter-base breakdown and a diode drop. Assuming a standard bipolar V_{EBO} of 6.8 V, the positive clamp voltage lies near 8 V. The collector-substrate junction of Q_1 clamps negative ESD transients to one diode drop (plus substrate debiasing).

The positive clamp voltage of the buffered Zener remains roughly constant as long as the voltage across the collector resistance of Q_1 does not exceed about 7 V. If necessary, the clamp voltage can be increased by the inclusion of a second Zener diode, or by the addition of one or more diode-connected transistors in series with the Zener. The maximum clamp voltage this structure can safely support equals the $V_{CEO(sus)}$ of the NPN transistor. If one attempts to obtain a higher clamp voltage, the NPN transistor will avalanche and snap back to $V_{CEO(sus)}$. This type of snapback characteristic forms the basis of the V_{CES} clamp discussed in the next section.

The buffered Zener clamp dissipates most of its energy in its large base-collector depletion region. An NPN transistor with an emitter area of 300 to 600 μm^2 will usually provide 2 kV HBM and 200 V MM protection. Larger NPN transistors can provide protection against proportionately higher ESD voltages. The ultimate limits of this structure are probably determined more by the metallization and the bondwires than by the ability of the silicon to absorb ESD energy.

---

[14] M. Corsi, R. Nimmo, and F. Fattori, "ESD protection of BiCMOS Integrated Circuits which need to operate in the Harsh Environment of Automotive or Industrial" *(sic), EOS/ESD Symposium Proc. EOS-15,* 1993, pp. 209–213.

All of the components of the buffered Zener clamp can occupy a common tank. In the structure in Figure 13.22B, the emitter of the power transistor Q_1 consists of a series of annular emitter geometries. Inside each hole are smaller plugs of emitter diffusion that form the cathodes of Zener diode D_1. Both the emitters of Q_1 and the cathodes of D_1 are enclosed in a common base region, a portion of which extends out into the isolation to form base resistor R_1. All of these merged devices reside within a common tank having a single shared deep-N+ sinker. The buffered Zener clamp operates as follows: When the cathodes of Zener diode D_1 avalanche into the shared base region, they inject holes into it. These holes forward-bias the base-emitter junction of Q_1 and cause the large NPN transistor to conduct. The ESD transient only lasts for a few hundred nanoseconds, which is not enough time for a hot spot to form and collapse. Since thermal runaway cannot occur, the shape of the emitter diffusion can be tailored to improve performance in other ways. The annular shape of the emitters of Q_1 ensures rapid and even turn-on of all portions of Q_1's emitter. Base resistor R_1 holds Q_1 off during normal operation. R_1 should have a relatively low value (such as 1 kΩ) to avoid transient disruptions due to capacitive coupling across the collector-base junction of Q_1.

Although Figure 13.22B illustrates a buffered Zener clamp for a standard bipolar process, these structures are actually better suited to the protection of analog BiCMOS circuitry, because the smaller spacings of the latter process reduce the size of the structure to the point where it can reside in the pad ring. This structure has successfully protected the gate oxide of a 20 V analog BiCMOS process against 2 kV HBM and 200 V MM ESD strikes. The extremely low series resistance of the buffered Zener often eliminates the need for a secondary breakdown ESD device for gate oxides with rupture voltages of 20 V or more. Lower voltage gate oxides usually require a secondary protection structure similar to that in Figure 13.21A. The buffered Zener clamp can be adapted to higher voltage applications by inserting additional Zener diodes or diode-connected transistors in series with the anode of Zener D_1. These additional devices can also be merged into a common tank with the other portions of the ESD circuit.

13.5.4. V_{CES} Clamp

Figure 13.23A shows an ESD circuit that uses the collector-base breakdown of an NPN transistor to clamp positive ESD transients. The initial breakdown voltage of

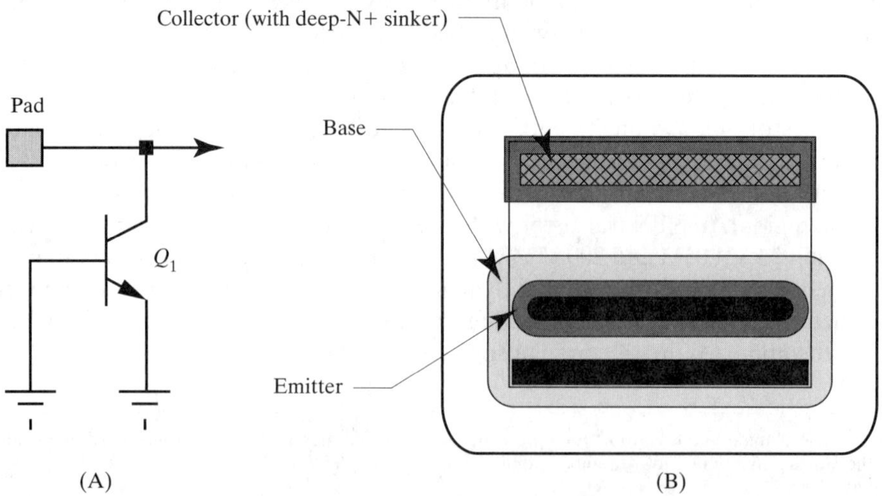

FIGURE 13.23 (A) Schematic diagram of the V_{CES} clamp and (B) a layout of a suitable NPN transistor.

this circuit equals the V_{CES} rating of transistor Q_1. Once conduction has begun, it does not cease until the voltage across the transistor drops below $V_{CEO(sus)}$. These two thresholds are sometimes called the *trigger voltage* (or *strike voltage*) and the *sustain voltage.* A typical 40 V standard bipolar transistor has a nominal trigger voltage of about 65 V and a nominal sustain voltage of about 45 V. The snapback from the higher trigger voltage to the lower sustain voltage decreases the voltage drop across the NPN and helps reduce the energy dissipated in the transistor. Despite the relatively high breakdown voltage of this structure, it is easily capable of withstanding 2 kV HBM and 200 V MM ESD strikes. An emitter area of about 300 to 500 μm^2 provides this level of protection in standard bipolar. Larger emitter areas provide proportionately higher levels of ESD protection. This structure has successfully served as the primary protection device for a 20 V analog BiCMOS gate oxide against 2 kV HBM and 200 V MM ESD events.[15]

ESD devices with snapback characteristics cannot safely protect low-impedance pins operating at or beyond their sustain voltage. If a transient triggers snapback, and the external circuit can supply enough current to sustain conduction, then the ESD device will continue to conduct indefinitely. The resulting power dissipation quickly overheats and destroys the integrated circuit. If the external circuitry cannot provide enough current to sustain conduction, then the ESD device can protect a pin even if it operates at a voltage in excess of the device's sustain rating. This type of application should not be contemplated unless the designer has full characterization data for the ESD device and can confidently state that the application will never deliver enough current to sustain conduction.

Figure 13.23B shows a typical layout of an NPN transistor used as a V_{CES} clamp. The filleted corners on the base diffusion raise the strike voltage of the device slightly, and also help make conduction slightly more uniform. Many designers also fillet other diffusions as shown in the illustration, but these fillets make no substantial difference in the operation of the transistor as a V_{CES} clamp. The transistor can be made more rugged by increasing the overlap of the base and emitter diffusions over their respective contacts by 1 to 2 μm. This precaution is particularly useful if the process does not include silicide or refractory barrier metal, since pure aluminum contacts are far more vulnerable to alloying failures than other types of contacts.

13.5.5. V_{ECS} Clamp

If the emitter and collector terminals of a bipolar transistor are swapped, the device will continue to operate as a bipolar transistor. When such a transistor is biased into conduction, it is said to operate in the *reverse active,* or *inverse active,* mode. The collector-base junction forward-biases and injects minority carriers into the base, which are collected by the emitter-base junction. An NPN transistor operated in reverse active mode has a very low beta because the substitution of the lightly doped collector for the heavily doped emitter drastically reduces emitter-injection efficiency. The heavily doped base-emitter junction also avalanches at a much lower voltage than the lightly doped collector-base junction. Because of this reduction in breakdown voltage, a transistor operated in reverse active mode makes an excellent low-voltage ESD device. Suppose the transistor in Figure 13.23B is used as an ESD clamp with its emitter connected to a bondpad and its base and collector connected to ground (Figure 13.24A). The trigger voltage of the V_{ECS} *clamp* equals the V_{EBO} of the NPN

---

[15] J. Z. Chen, X. Y. Zhang, A. Amerasekera, and T. Vrotsos, "Design and Layout of a High ESD Performance NPN Structure for Submicron BiCMOS/Bipolar Circuits," *International Reliability Physics Symposium,* 1996, pp. 227–232.

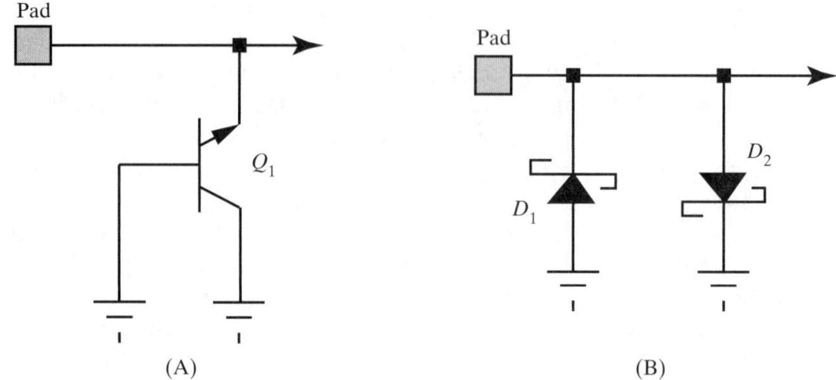

FIGURE 13.24 Schematic diagrams of (A) the V_{ECS} clamp and (B) the antiparallel-diode clamp.

transistor, and its sustain voltage equals approximately 60 to 80% of this voltage. Since analog BiCMOS NPN transistors typically have a V_{EBO} of 8 to 10 V, these devices can serve as primary protection devices for pins that do not operate at more than 5 V. This type of ESD device has a very low series resistance due to the presence of the NBL and the deep-N+ sinker. Devices with emitter areas of less than 600 μm^2 have successfully withstood 2 kV HBM and 200 V MM ESD strikes, and somewhat larger emitter areas have protected circuits to 10 kV HBM. These devices present a low-impedance path to negative ESD transients as well as positive ESD transients, a benefit that few other ESD devices offer. The V_{ECS} clamp does not require electron-collecting guard rings because its tank does not connect to a pin. This structure does contain a parasitic substrate PNP transistor that can inject large majority-carrier currents into the substrate, so the clamp should be surrounded by as many substrate contacts as possible to minimize debiasing in the surrounding substrate.

The V_{ECS} structure is an extremely useful device for protecting low-voltage pins. Higher voltage ESD circuits can be created by stacking V_{ECS} clamps in series, but this arrangement increases both the area required for the device and its series resistance. A buffered Zener or a V_{CES} clamp will probably provide better performance than stacked V_{ECS} clamps.

13.5.6. Antiparallel Diode Clamps

Many integrated circuits require multiple ground pins. Sometimes all of the ground pins connect to the substrate, but some are frequently separated to minimize noise coupling and substrate injection. Those ground pins that do not connect to substrate require some form of ESD protection. The most common configuration consists of a pair of back-to-back (or *antiparallel*) diodes. One can employ diode-connected NPN transistors, diode-connected substrate PNP transistors, or Schottky diodes (Figure 13.24B). Since the voltage drop across these devices is relatively small, they experience much less internal heating than do other types of ESD devices, and therefore they can be made somewhat smaller. Schottky diodes with areas of several thousand square microns will generally provide protection against 2 kV HBM and 200 V MM ESD strikes, and diode-connected transistors can be made even smaller than this. All of these structures require enclosure within electron-collecting guard rings.

13.5.7. Grounded-Gate NMOS Clamps

Many early CMOS ESD circuits used NMOS transistors as lateral NPN transistors. The NSD diffusions of the NMOS transistor act as the collector and emitter of the NPN, while the P-epi acts as the base. If one of the two NMoat regions connects to the bondpad and the other to the substrate return line, then the bipolar transistor will act as a V_{CES} clamp. Early ESD devices constructed along these lines also included a gate electrode on top of a thick-field oxide region separating the two NSD diffusions. Some designers connected this gate electrode to the bondpad, while others connected it to the substrate return. This metal electrode actually had little effect upon the transistor regardless of its connection because the thick-field threshold exceeded the NSD/P-epi breakdown voltage. This structure was popularly called the *thick-field transistor*.

The thick-field transistor often provided very poor ESD protection. Its shallow, heavily doped junctions frequently melted during ESD transients, causing device leakage if not outright failure. The snapback characteristics of the thick-field device also caused many problems. The sustain voltage of this structure usually equaled about 60% of the NSD/P-epi breakdown voltage, a value that was often less than the maximum operating voltage. Transients on the pin triggered the thick-field transistor into conduction while power was still applied to the device, causing catastrophic failure.

A number of more successful variants on the thick-field transistor have been developed. One consists of nothing more than a conventional NMOS transistor whose gate and source connect to ground, and whose drain connects to the pin (Figure 13.25A). This structure is called a *grounded-gate NMOS* (GGNMOS). An important feature of this device is an additional 1–2 μm of space between the source and drain contacts and the polysilicon gate to provide ballasting. Clad-moat processes require the use of a silicide block mask to obtain the necessary ballasting resistance between contacts and channel. Properly ballasted GGNMOS structures often prove surprisingly robust despite their shallow source/drain diffusions. Some grounded-gate devices use a channel length so short that the device breaks down by punchthrough rather than by avalanche.[16] This type of device exhibits less snapback than the classical GGNMOS, but its breakdown voltage may prove rather variable due to its extremely short channel length.

Another variation on the grounded-gate NMOS utilizes a capacitor connected from the pad to the gate of the NMOS, and a turn-off resistor connected from the gate to ground (Figure 13.25B). This structure is commonly known as the *gate-coupled NMOS* (GCNMOS).[17] The rapid rise of voltage during an ESD event couples energy

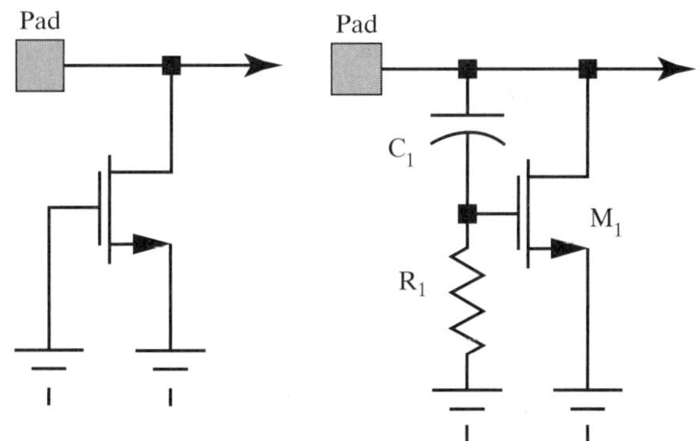

FIGURE 13.25 Two grounded-gate NMOS clamps: (A) thin-oxide GGNMOS, (B) gate-clamped NMOS.

[16] J. K. Keller, "Protection of MOS Integrated Circuits from Destruction by Electrostatic Discharge," *EOS/ESD Symposium Proc.*, 1980, pp. 73–80.

[17] C. Duvvury and C. Diaz, "Dynamic Gate Coupling of NMOS for Efficient Output ESD Protection," *International Reliability Physics Symp.*, 1992, pp. 141–150.

across capacitor C_1, turning on NMOS transistor M_1. This process reduces the peak voltage required to trigger conduction in the transistor and ensures relatively uniform conduction through all portions of the device. Many GCNMOS structures have been proposed, including ones in which the coupling capacitance consists of nothing more than the drain-to-gate overlap capacitance of an extended-drain MOS transistor.

All GCNMOS structures share a common defect, in that they can be triggered into conduction by rapid transients in the application circuit. Power buses experience such transients whenever someone connects a live power supply to the circuit. This type of situation, called a *hot-plug event*, can easily destroy an integrated circuit with a GCNMOS structure connected to a power supply pin. Attempts to make GCNMOS structures immune to such transients by reducing the value of the turn-off resistor R_1 rarely succeed, as the slew rates encountered in hot-plug events are similar to those generated by ESD transients. The GCNMOS structure, and all similar structures triggered by the rate of rise of a signal, should never protect low-impedance pins that might conceivably experience transients in the application.

GGNMOS and GCNMOS protection structures generally require electron-collecting guard rings to prevent minority carrier injection into the substrate. A large area of substrate contacts should also reside close to the protection structure.

13.5.8. CDM Clamps

The machine model has long been criticized as unrepresentative of actual handling conditions. The *charged device model* (CDM) was developed to address these complaints. The CDM test uses a contact discharge of the package capacitance to develop the ESD strike. This situation presents a quite different set of stresses than either the machine model or the human-body model. The package capacitance seldom amounts to more than a few picofarads, and CDM discharges are customarily done at relatively low voltages, typically 500 V. The amount of energy delivered by a CDM strike is therefore much smaller than the energy delivered by either machine model or human-body model strikes. The energy of a CDM strike seldom poses much threat to properly designed junctions. However, the absence of any external limiting resistance or inductance ensures that large currents briefly flow during the CDM event. These currents can cause significant resistive drops within primary ESD devices, metallization, and diffusions. Even though these transient stresses only last a few nanoseconds, they can easily destroy gate oxides. Many designs that successfully pass 2 kV human-body model and 200 V machine model tests now fail the new 500 V CDM requirements.

New protection structures called *CDM clamps* have been developed to protect gate oxides against CDM transients. A CDM clamp is a specialized version of a secondary protection structure placed near the gate oxide that it protects. The close coupling between the CDM clamp and the protected device prevents parasitic resistances and inductances from generating unwanted voltage drops that add to the clamp voltage.[18] Typical rules stipulate that CDM clamps should reside within 500 μm of the devices they protect, and that they should connect to ground or power as close to the protected devices as possible.

CDM clamps resemble ordinary secondary protection devices in that they contain a series limiting resistor and a clamp device. The resistor, which typically consists of polysilicon, usually has a value of between 0.5 and 2 kilohms. The clamp device may consist of either a small Zener diode or a small grounded-gate NMOS

[18] L. R. Avery, "ESD Protection Structures to Survive the Charged Device Model (CDM)," *EOS/ESD Symposium Proc.*, 1987, pp. 186–191.

transistor. Neither structure sees significant resistive heating because the transient lasts only a few tens of nanoseconds, and the series limiting resistor prevents any large current from flowing through the structure.

Most CDM clamps include two secondary protection devices, one connected to ground, and one connected to the supply. Both clamps often consist of nothing more than a small GGNMOS structure (Figure 13.26A). However, a sneak current can pass through the device connected between the pin and the power supply if the pin voltage exceeds the supply voltage. This situation sometimes occurs in designs that have multiple power supplies, or in interface circuitry. If this current path could cause problems, then one can place two back-to-back GGNMOS structures between the power supply and the pin (Figure 13.26B). This configuration is sometimes called a *self-protecting CDM clamp*.

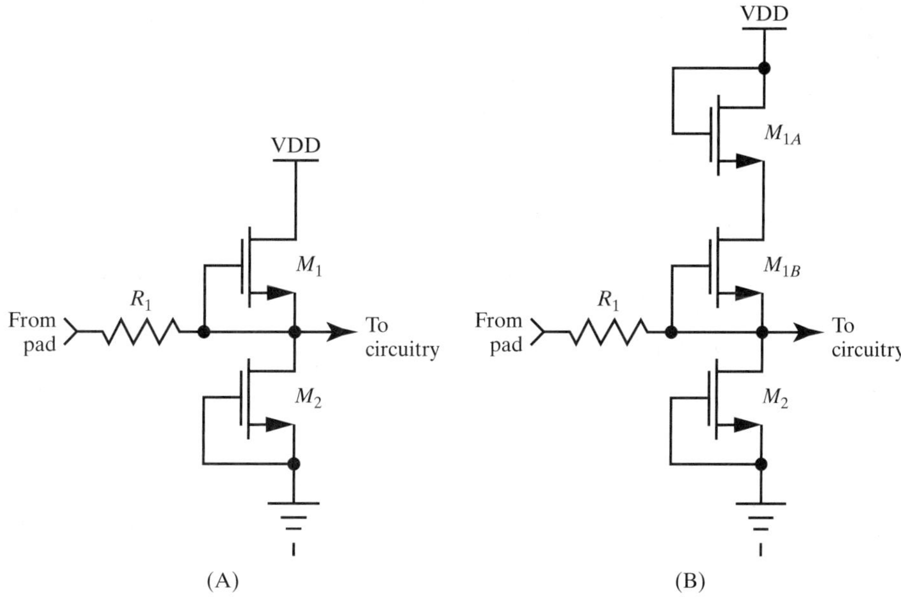

FIGURE 13.26 CDM clamps using grounded-gate NMOS transistors: (A) basic structure, and (B) self-protecting structure.

CDM clamps generally require one or more guard rings. In the case of the GGNMOS structures of Figure 13.26, a single electron-collecting guard ring placed around both clamps will suffice to provide complete protection. Alternatively, one can increase the value of the series limiting resistor to prevent any significant current from flowing into the substrate. A resistance of 50 kΩ will typically suffice for this purpose.

13.5.9. Lateral SCR Clamps

Many processes use some variant of the *silicon-controlled rectifier* (SCR) to provide ESD protection.[19] An SCR is a four-layer semiconductor device that exhibits strong snapback characteristics that arise from positive feedback between the NPN transistor and PNP transistor inherent in the PNPN structure. A *lateral SCR* consists of a PMOS transistor immediately adjacent to an NMOS transistor. Figure 13.27A shows the layout, cross section, and equivalent electrical circuit of this structure. The

19 L. R. Avery, "Using SCR's as Transient Protection Structures in Integrated Circuits," *EOS/ESD Symposium Proc.*, 1983, pp. 177–180.

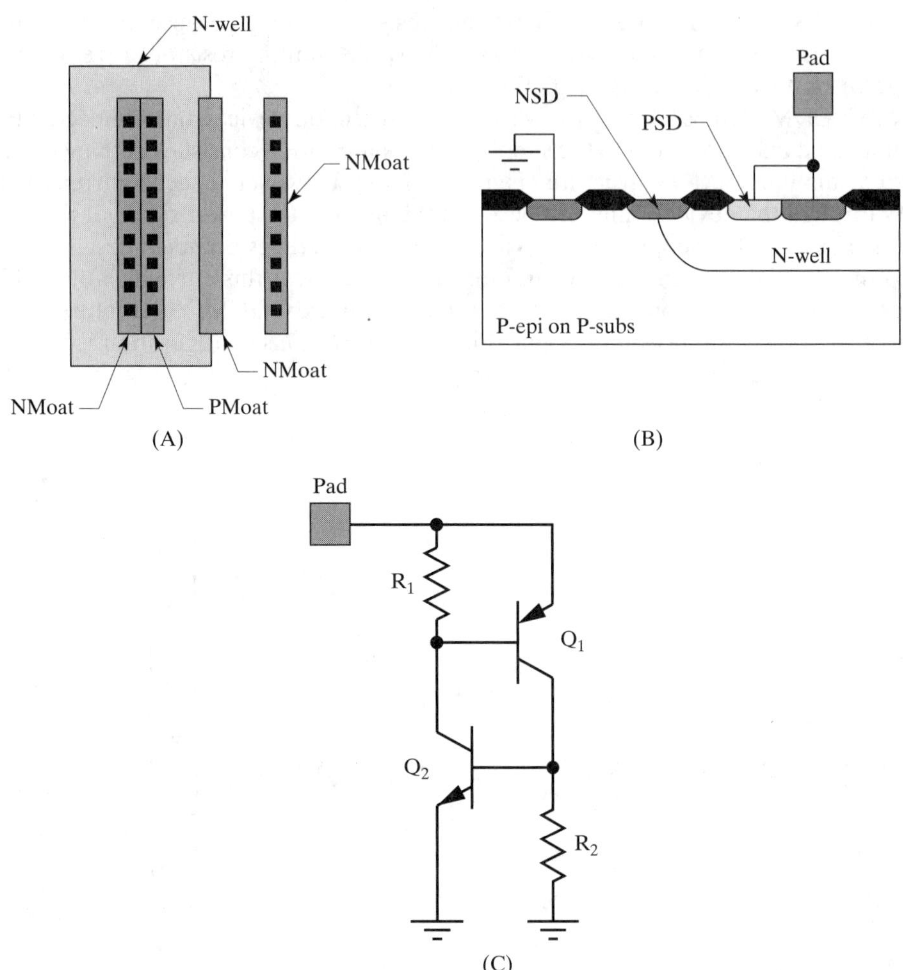

FIGURE 13.27 Lateral SCR clamp: (A) layout, (B) cross section, and (C) equivalent electrical circuit.[20]

emitter of PNP transistor Q_1 consists of a PSD diffusion placed inside an N-well. The well functions as the base of the PNP, and the surrounding P-epi functions as its collector. The N-well also serves as the collector of NPN transistor Q_2. The P-epi functions as the base of this transistor, and an NSD diffusion acts as its emitter. R_1 represents the resistance of the N-well, while R_2 represents the resistance of the P-epi and the substrate. This structure is almost identical to that responsible for CMOS latchup, but an additional strip of NMoat has been placed along the edge of the N-well to reduce the breakdown voltage of the well/epi junction. The avalanche of this junction triggers the device into conduction.

The SCR is triggered into conduction by the collector-base avalanche of either Q_1 or Q_2. Suppose Q_2 avalanches first. Carriers injected into the base of Q_2 cause it to conduct. Q_2 now pulls current from the base of Q_1, causing it to turn on and provide additional base drive for Q_2. Each of the two transistors now provides base drive to the other, and conduction continues until the input voltage drops so low

---

[20] C. Duvvury and R. Rountree, "A Synthesis of ESD Input Protection Scheme," *EOS/ESD Symposium Proc.*, 1991, pp. 88–97.

that R_1 and R_2 can extract more current than the transistors can supply. If R_1 and R_2 are both relatively large, then the SCR may have a sustain voltage of less than 2 V. Smaller values of resistance will produce higher sustain voltages, but the relationship between resistance and sustain voltage is difficult to predict. In practice, a multitude of SCR structures are constructed and measured to determine which structure gives the desired strike and sustain voltages.

SCR clamps are extremely robust structures. During a human-body-model strike, the low sustain voltage of the SCR forces the external 1.5 kΩ resistor to dissipate the vast majority of the energy. CDM strikes do not contain enough energy to imperil the junctions of a typical SCR structure. The machine model poses the most strenuous test of an SCR clamp; but many of these structures, especially those incorporating ballasting into the NSD and PSD contacts, have withstood severe machine model testing. SCR structures often provide much smaller ESD solutions than the other available options. They are especially favored for protecting CMOS processes that cannot construct V_{CES} or V_{ECS} clamps.

SCR clamps must never be used to protect pins connected to low-resistance circuitry operating at or above the sustain voltage. Momentary transients can trigger the SCR, and continued conduction will eventually overheat and destroy the integrated circuit. The spacings of the four elements of the lateral SCR can be increased to ensure that the sustain voltage lies above the operating voltage. Any increase in sustain voltage will also increase the amount of energy dissipated within the SCR structure. Devices with higher sustain voltages thus require larger areas to safely dissipate the energy of the ESD strike.

The trigger voltage of the lateral SCR clamp is often too high to protect low-voltage CMOS circuits. Rate-triggered SCR clamps have been developed that include capacitors connected from the pad to the base of Q_2, or from the base of Q_1 to ground. Rapidly slewing transients generated by an ESD strike cause these capacitors to trigger the SCR before its strike voltage has been reached. Rate-triggered SCR clamps can provide excellent protection, but like all rate-fired structures, they should never be employed on pins that may experience transients during normal operation.

13.5.10. Selecting ESD Structures

Pins connected directly to the substrate, or connected only to relatively robust diffusions, can usually survive without the addition of dedicated ESD structures. Most other pins require some form of ESD protection. The following guidelines offer some specific advice for several commonly encountered situations:

1. *Pins connecting to base or emitter diffusions.*
   The relatively low sheet resistances of base and emitter diffusions render them vulnerable to ESD damage. Larger diffusions may spread the energy over sufficient area to protect themselves, but localized heating often damages smaller diffusions. The minimum diffusion area capable of self-protection depends on process parameters and testing conditions, but it is probably safe to say that a 500 μm^2 160 Ω/□ base diffusion will survive 2 kV HBM and 200 V MM. Smaller diffusions should include some form of ESD clamp that avalanches before the base diffusion, such as a V_{CES} clamp or a V_{ECS} clamp. Series limiting resistors are rarely necessary because either the diffusion or the region enclosing it is usually quite resistive.

2. *Pins connecting to the emitters of NPN transistors.*
   The emitters of vertical NPN transistors are vulnerable to avalanche-induced beta degradation. If possible, the circuit should be designed to eliminate any

direct connection between an emitter and a bondpad other than substrate ground. Otherwise, an ESD clamp device must be connected to the bondpad and a series resistance of several hundred Ohms placed between the bondpad and the emitter. The circuit designer must consider the impact of this resistance on circuit operation. The emitters of power NPN transistors sometimes operate at substrate potential, but return through a separate pin. In this case, an antiparallel diode clamp will provide adequate protection without requiring the insertion of any series resistance. Poly-emitter NPN transistors must never be allowed to avalanche, so the ESD circuit must be supplemented by clamp diodes and current limiting resistors to ensure the safety of such devices.

3. *Pins connecting to CMOS gates.*
   CMOS gate dielectrics are so fragile that they usually require some form of two-stage ESD protection. The primary protection device need only limit the voltage at the pad to a few hundred volts. The secondary ESD protection device should clamp the gate voltage to no more than 75% of the oxide rupture voltage. If the secondary ESD device returns through the substrate, then its clamp voltage must include any substrate debiasing generated either by itself or by the primary device. The series limiting resistor between the primary and secondary devices should have a resistance several times larger than that of the secondary protection structure. The resistor may consist either of a diffusion or of polysilicon, but poly resistors should be at least 5 to 8 μm wide and should contain at least six or eight contacts at either end to help prevent excessive localized heating. Resistors used in ESD devices should not contain any bends, as these generate localized hot spots that may fail before the remainder of the resistor. Zeners used as secondary protection devices may require series limiting resistors of several kilohms. The secondary protection device and limiting resistor can sometimes be omitted if the primary protection device can clamp the voltage at the pad to approximately 75% of the gate oxide rupture voltage. The high currents generated by machine-model testing make this very difficult to achieve, but V_{ECS} clamps have successfully protected a 20 V gate oxide against 2 kV HBM and 200 V MM. CDM testing will almost certainly require secondary protection, and these devices frequently have to reside near the device to be protected in order to prevent substrate debiasing from developing excessive voltage drops. Some low-voltage CMOS processes have oxide rupture voltages below the trigger voltages of conventional avalanche-triggered ESD structures, in which case rate-triggered devices or SCRs must be used.

4. *Pins connecting to moat regions.*
   Some types of moat regions will protect themselves against ESD, while others will not. Silicided moats almost always require some form of additional ESD protection, as do moats with breakdown voltages of less than 5 to 8 V. Nonsilicided moats of transistors with breakdown voltages of 10 V or more will probably protect themselves against 2 kV HBM and 200 V MM transients, provided that the total drawn area of each type of moat diffusion exceeds 500 μm^2. A large NSD diffusion will not necessarily protect a small PSD diffusion, or *vice versa.* The exact moat areas required to provide self-protection vary depending on processing parameters and testing conditions. Small moat regions, particularly clad ones, generally require some form of additional ESD protection. A single-stage ESD circuit will suffice if this structure can clamp the voltage at the bondpad to less than the avalanche voltage of the moat diffusions. V_{ECS} clamps and buffered Zener clamps can sometimes provide this level of protection, but Zener clamps usually have too much internal series resistance. A series limiting

resistance of a few hundred Ohms enables the use of a Zener clamp as a protection device for small moat regions. Large silicided moats often exhibit localized breakdown due to lack of ballasting. Consider using a silicide block mask to remove the silicide from the periphery of moat regions connected to bondpads. The unsilicided moat periphery slightly increases the resistance of the transistor, but one can compensate by increasing the size of the device.

5. *Pins connecting both to moat regions and to CMOS gates.*
   The moats may serve as a primary protection device if they are sufficiently large; otherwise a primary protection device must be connected to the bondpad. Small moats, or ones made especially vulnerable by silicidation, may require a series limiting resistor of 50 to 200 Ω. Unless the primary protection device has a very low series resistance, it cannot protect the gates without the addition of a secondary protection device. A resistor of several hundred Ohms to several kilohms should be connected between the pad and the gates, and a suitable secondary protection structure should be placed after this resistor. This structure now has separate conduction paths for gates (which require large series resistances) and moats (which do not). CDM structures may also be required.

6. *Pins connecting only to polysilicon.*
   The voltages generated during human-body model testing are sufficient to rupture the thick-field oxide and interlevel oxide surrounding polysilicon resistors and leads. If a bondpad does not directly connect to any diffusion, then the voltages across the oxide surrounding the polysilicon may rise to destructive levels. An N-well geometry placed beneath the bondpad and connected to it by means of a ring of NMoat contacts encircling the bondpad will provide adequate protection while consuming very little die area. This structure can inject electrons into the substrate, and thus it generally requires the addition of an electron-collecting guard ring.

7. *Pins connecting to capacitors.*
   Thin oxide or nitride dielectrics require the same type of protection as gate dielectrics. Junction capacitors usually contain a thin, heavily doped diffusion that requires protection similar to an emitter region.

8. *Pins connecting to Schottky diodes.*
   Field-plated Schottky diodes should not operate in avalanche breakdown because their depletion regions are very thin and are located immediately adjacent to a silicide layer. Large Schottky diodes can be protected by adding a field-relief guard ring that avalanches before the Schottky contact. Smaller Schottky diodes may be protected by a large area of moat diffusion forming part of another device connected to the same pin. If no suitable moat region exists, then a field-relief guard ring and a series resistance of a few hundred Ohms should provide adequate protection.

9. *Bondpads operating at substrate potential, but not connected to substrate.*
   These bondpads are usually ground returns isolated from substrate to minimize noise coupling. ESD protection is not required if these pads are bonded to the same pin as the substrate pad. Otherwise, an antiparallel diode clamp connected between the pad and the substrate return will provide sufficient protection for most applications.

10. *Multiple bondpads connecting to the same pin through multiple bondwires.*
    Many dice use multiple bondwires attached to a common pin. If two or more bondpads connect to the same pin through separate bondwires, then only one

of these pads requires a primary ESD device. Series limiting resistors and secondary protection devices must be placed on every bondpad requiring them, as secondary protection placed on one bondpad cannot protect circuitry connected to another bondpad.

11. *Test pads and probe pads.*
Test pads and probe pads normally do not require ESD protection because they are encapsulated within the package and therefore do not experience ESD transients.

When placing ESD structures, always remember to include any necessary guard rings and substrate contacts. Of the types of ESD structures just discussed, only the V_{ECS} clamp does not require guard rings. The guard rings should be placed in the padring during its construction. They require so much room that they are often very difficult to add later.

13.6 EXERCISES

Refer to Appendix C for layout rules and process specifications.

13.1. Lay out three minimum-size standard bipolar NPN transistors without deep-N+ sinkers. Place the transistors side by side as shown in Figure 13.1A, and measure the area of a rectangle enclosing all three devices. Now lay out a merged device similar to that in Figure 13.1B. Assume that the area consumed by the merged device equals the area of its tank. What is the ratio of the area of the merged device to the area of the three separate devices?

13.2. Describe the risks posed by each of the following mergers:

a. Two HSR resistors placed in the same tank, one of which connects to a bondpad.
b. A base resistor merged in the same tank as a Schottky diode.
c. A lateral PNP transistor merged with an NPN transistor.
d. A junction capacitor merged with a Darlington NPN transistor pair.
e. Two substrate PNP transistors merged in the same tank.

13.3. Propose measures that will minimize the risks associated with each of the mergers in Exercise 13.2.

13.4. Lay out a merged Darlington NPN transistor capable of conducting 100 mA. Use standard bipolar layout rules and assume a maximum emitter current density of 8 $\mu A/\mu m^2$. Use a wide-emitter, narrow-contact structure with an emitter overlap of contact of 6 μm for the power device, and connect a 5 kΩ base turnoff resistor between its emitter and collector. Size the predrive transistor, assuming that the power transistor has a minimum beta of 20 at 100 mA. Assume that the predrive transistor does not require a minimum-width base turnoff resistor. Include all necessary metallization.

13.5. Assume that the Darlington transistor in Exercise 13.4 must operate at voltages exceeding the thick-field threshold of the process. Modify the layout to include all necessary field plates and channel stops.

13.6. Lay out the totem pole driver circuit in Figure 13.28, using standard bipolar layout rules. Use wide-emitter, narrow-contact transistors with an emitter overlap of contact of 6 μm for Q_1, Q_2, and D_1. The power supply voltage VCC can exceed the thick-field threshold of the process, and both OUT and PGND may go below substrate potential during switching transients. The leads connecting to VCC, OUT, and PGND must have widths of at least 15 μm. Include all necessary metallization and label all leads and devices.

13.7. Modify the MOS operational amplifier in Exercise 12.16 to provide latchup protection, assuming that only the output pin (OUT) connects directly to a bondpad. Make the circuit as robust as possible without using either deep-N+ or NBL.

13.8. (A) Draw a stick diagram of the flip-flop of Figure 13.29. Use a single VDD bus across the top of the cell and a single GND bus across the bottom of the cell. All

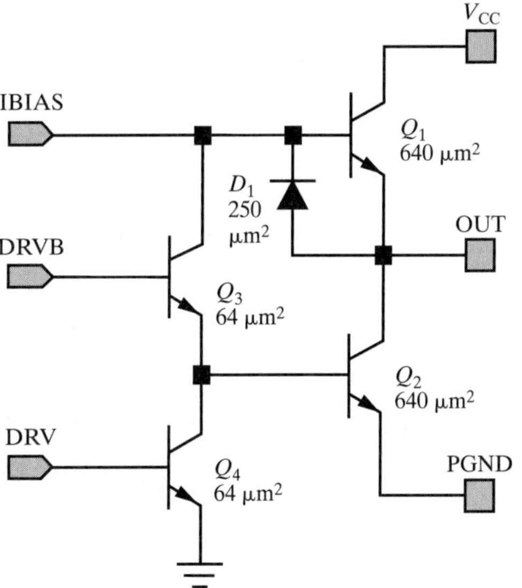

FIGURE 13.28 Schematic of totem pole output stage. All dimensions are emitter areas.

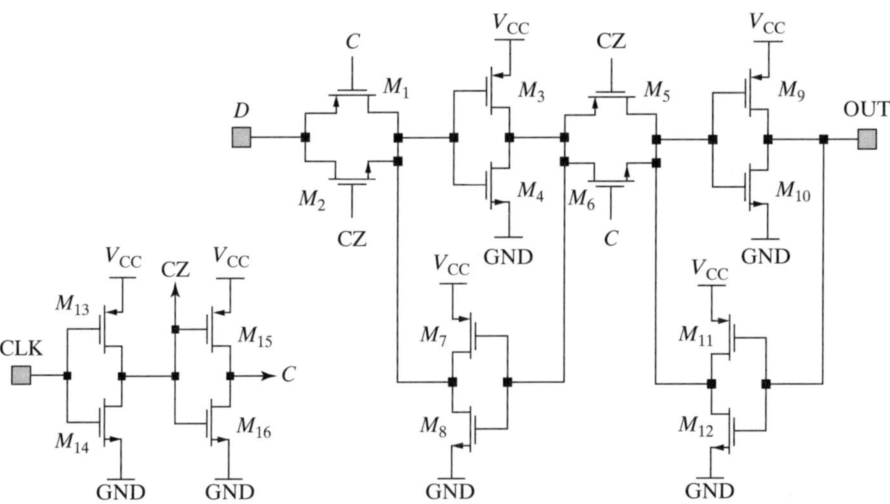

FIGURE 13.29 Schematic of *D*-type flip-flop.

NMOS transistors are 4/3 and all PMOS transistors are 7/3. Do not use any metal-2 within the cell. Poly can be used to route gate leads, and short poly jumpers can be placed in source and drain leads if necessary. (B) Following the stick diagram as closely as possible, lay out the flip-flop using the CMOS rules in Appendix C. Label all leads and devices.

13.9. Construct a padring for an analog CMOS layout. The die must be exactly 2750 μm wide by 2150 μm high, including scribe. The lower left-hand corner of the die must reside at the origin (0, 0). Denote the extents of the die, using a rectangle on a layer called BOUNDARY. The scribe streets lie along the left side and bottom of the die, and each is exactly 110 μm wide. Denote the locations for the scribe by drawing two rectangles on the BOUNDARY layer. Draw substrate metal coincident with the edge of the padring, as shown in Figure 13.17. This metal should include a 12 μm strip of metal-1 and a 12 μm strip of metal-2 that exactly coincide with one another. Place PMOAT underneath the substrate metallization. Provide contacts between the substrate metallization and the PMOAT, and vias between the two layers of substrate metallization.

13.10. Construct bondpads for the padring in Exercise 13.9. Assume that the pads require a square nitride opening 75 μm across, and that they must contain both metal-1 and metal-2. The two metal geometries should exactly coincide. Place a single, large via 75 μm across coincident to the nitride opening. Denote the metal exclusion zone by using a circle with a diameter of 90 μm centered on the bondpad opening. Place eight of these bondpads in the padring of Exercise 13.9 in the following locations: top left, top center (2), top right, bottom left, bottom center (2), and bottom right. The pad at the bottom right connects to pin #1. Choose a suitable way of denoting this pad, and modify the layout accordingly. Number the pads in counterclockwise order and label each.

13.11. Construct a Zener clamp similar to that in Figure 13.20, using the CMOS layout rules in Appendix C. The NMoat region should have a total area of at least 650 μm^2, and should contain two rows of contacts. Overlap the NMoat over contact by 3 μm and fillet both ends of the diode. Place two of these Zener clamps in the layout in Exercise 13.10 to protect pins 3 and 5. These clamps should reside between the respective pads and the substrate metallization. Include all necessary guard rings.

13.12. Construct a two-stage Zener clamp, using the diode of Exercise 13.11 as a primary protection device. The series resistor should equal 1 kΩ and have a width of 8 μm. Place at least eight contacts on either end of the resistor. The secondary protection structure requires an NMoat area of 100 μm^2. Overlap the NMoat over the contacts by at least 3 μm and encircle the device with a ring of substrate contacts. All metallization within the ESD structure should have a width of at least 6 μm, and if vias are used in any of these leads, use a minimum of eight vias. Place two of these clamps in the layout of Exercise 13.11 so that they protect pins 1 and 7. Include all necessary guard rings.

13.13. Construct a V_{ECS} clamp, using the analog BiCMOS layout rules in Appendix C. Assume that the clamp requires an emitter area of 350 μm^2. Instead of a single contact for the emitter, use two rows of minimum contacts. Overlap the emitter over the contact by 4 μm and fillet both ends of the emitter. The deep-N+ sinker should completely encircle the clamp to form a hole-blocking guard ring. Compare the area of this structure to the area consumed by all of the components of the two-stage Zener clamp in Exercise 13.12 (excluding guard ring).

13.14. Lay out a thick-field transistor for use as an ESD structure in a CMOS design. The source and drain consist of strips of NMoat, each 20 μm long and just wide enough to contain two rows of minimum contacts. Overlap NMoat over the contacts by 2 μm and include fillets on both source and drain. Separate the two moat regions by 6 μm and place a metal-1 plate over the region between the source and drain to act as a "gate." Connect this gate to the source. Compare the area consumed by this clamp with the area consumed by the Zener clamp in Exercise 13.11.

14 Assembling the Die

The first step in laying out an integrated circuit is estimating the die area. Layout designers should not rely on preliminary area estimates, as these are seldom accurate. The area of each circuit block, or *cell,* should be computed separately. The total die area equals the sum of the areas of all the cells plus the area required for wiring, bondpads, scribe seals, and scribe streets. As area estimates tend toward optimism, the prudent designer always includes a generous safety margin.

Once the dimensions of the die and the areas of all of the cells have been determined, a floorplan should be constructed. A good floorplan includes an outline of the die, placements of all pads, and the sizes and locations of all the major cells. The initial floorplan often requires revision as the layout progresses.

The completed floorplan serves as a template for constructing the individual cells. Each cell requires a multitude of resistors, capacitors, transistors, and diodes. When these components have been laid out, the designer must arrange them to optimize matching, packing, and ease of interconnection. The components are then connected to form a cell, and the cells are connected to form the completed die. Previous chapters have examined the design of individual components and their placement relative to one another. This chapter examines the process of planning a die layout, constructing a die floorplan, and interconnecting the finished cells to form the complete die.

14.1 DIE PLANNING

The layout of an integrated circuit requires considerable planning and forethought. An experienced designer knows what tasks must be accomplished and in what order; the layout progresses smoothly and all of the components fit into their assigned places. A novice attempting the same feat soon discovers that it is not as easy as it seems. Days or weeks of effort often come to naught because of unforeseen complications. Most of these difficulties are usually due to incorrect die area estimates, misplaced components, and inadequate wiring channels. The cautious designer can avoid most of these problems by spending a few hours planning the layout.

Using the information gathered during the planning phase, one can estimate the total die area and the cost of manufacture. Assuming that the design appears profitable, a floorplan can be developed showing the size, shape, and location of each cell. This floorplan forms the basis for the top-level layout of the die. Floorplans become particularly valuable on larger designs where many people must simultaneously create portions of the layout.

14.1.1. Cell Area Estimation

The first phase of the planning effort consists of compiling a list of all the cells used in the design. If detailed schematics are available, then this task amounts to little more than listing the cells found in the top-level schematic. If no schematics exist, then the circuit designer must prepare a list based on a careful examination of the specifications. The list should only contain cells appearing in the top-level schematic, and should exclude cells occupying the lower levels of the schematic hierarchy. The top-level cells may number as few as three or four, or as many as thirty or forty, depending on the scale of the design. In a few designs, especially those of great complexity, the floor plan must include the first sublevel of the hierarchy, to provide sufficient detail. The list should also include any power devices that require specific locations due to routing or matching considerations.

The designer must now estimate the area required by each cell. Some cells may have already been laid out for previous designs, in which case accurate area estimates are easily obtained by measurement. If a previous design contains a similar cell, then its layout may provide a close approximation of the area required by the new cell. If no previous layout exists, then the area of the cell must be computed from the areas of the individual components. The following sections explain how to rapidly estimate the size of the areas required by various types of components. These estimates are, of necessity, somewhat imprecise, but planners are generally allowed a margin of error of at least ±20%. Area estimates are usually given in either square millimeters (mm^2) or in thousands of square mils ($kmil^2$), where 1 $kmil^2 = 0.645$ mm^2.

Resistors

The area, A, required to construct one or more resistors can be estimated from the formula

$$A \cong \frac{1.2\,RW_r(W_r + S_r)}{R_s} \qquad [14.1]$$

where R equals the desired resistance, R_s is the applicable sheet resistance, W_r is the width of the resistor, and S_r is the spacing between adjacent resistor stripes. The factor of 1.2 helps account for the space consumed by dummy resistors, contact heads, and less-than-ideal layouts. For example, 122 kΩ of 2 kΩ/□ HSR with a width of 6 μm and a spacing of 12 μm will consume an estimated 7900 μm^2 of die area. Resistors of different widths or materials must be computed separately.

Capacitors

The area required by capacitors depends on the capacitance per unit area of the dielectric. For finger-style junction capacitors, an average capacitance per unit area can be computed using an existing capacitor as a guideline. Estimates based on oxide thickness underestimate the size of capacitors because they do not include contacts and isolation spacings. For example, suppose that a 50pF finger-style junction capacitor has a measured area of 27,500 μm^2. This capacitor has an average capacitance per unit area of 1.8 fF/μm^2.

Vertical Bipolar Transistors

The area of vertical NPN and substrate PNP transistors must be computed separately, but the same principles apply to both types of devices. The area required for a minimum-emitter device equals the area consumed by its tank; this is best measured from the layout of an existing device. The device area will not scale linearly with emitter area because the emitter forms only a small part of the transistor. It is usually not worth the effort to obtain exact area values for small transistors. One can safely assume that transistors with emitter areas of two to five times the minimum require 150% of the minimum device area. Larger transistors should be roughly sketched out, and their area estimated on this basis. Figure 14.1 shows a sketch of a wide-emitter, narrow-contact transistor. On the basis of the dimensions indicated, this device would consume an area of 38,800 μm^2 and would contain 4,000 μm^2 of emitter.

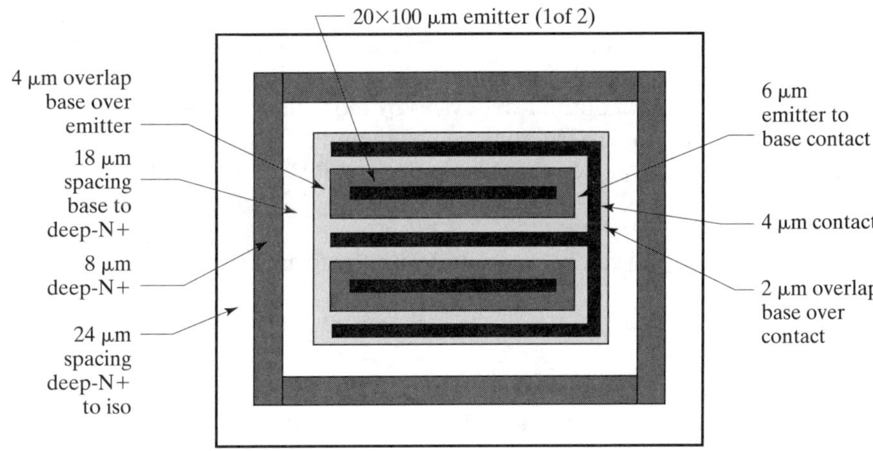

FIGURE 14.1 Sketch of a wide-emitter, narrow-contact transistor used for estimating its area.

Lateral PNP Transistors

The area required by a minimum lateral PNP transistor can be obtained by measuring the tank area of an existing device. Larger transistors are normally constructed either by placing multiple copies of a single device in a common tank or by stretching the transistor along one axis. In either case, the device area scales approximately linearly with collector size. Split-collector transistors require additional area, so count each such device as 150% of the minimum device area.

MOS Transistors

The area, *A*, required for a finger-style MOS transistor can be approximated by

$$A \cong 1.3\, W_g(L_g + S_{gg}) \qquad [14.2]$$

where W_g equals the width of the gate, L_g equals the length of the gate, and S_{gg} equals the spacing between adjacent gate stripes of a multiple-finger transistor. The factor of 1.3 helps account for the space consumed by the terminations on either end of the transistor array, well spacings, and less-than-ideal packing. This formula usually underestimates the area required by small transistors, particularly if they require guard rings or separate wells.

MOS Power Transistors

MOS power transistors are usually specified in terms of their on-resistance, $R_{ds(on)}$. Area estimations based on device models or SPICE simulations do not properly account for metallization resistance. Estimates based on measured specific on-resistance R_{sp} give much better results. The area required to obtain a desired $R_{ds(on)}$ equals

$$A \cong \frac{R_{sp}}{R_{ds(on)} - R_p} \qquad [14.3]$$

The variable R_p accounts for the resistance of the package, including bondwires and leadframe. The bondwires contribute the largest component of the package resistance. A typical 25 μm-diameter gold bondwire contributes about 25 to 50 mΩ (Section 13.4.2). Larger diameter bondwires or multiple bondwires placed in parallel can greatly reduce this resistance.

The accuracy of Equation 14.3 depends on how closely the proposed transistors resemble the test devices. The proposed transistor should have the same gate length, and the R_{sp} and $R_{ds(on)}$ figures should be measured at the same gate-to-source voltage. Because R_{sp} varies with device area, the area of the proposed transistors should not differ from the area of the test devices by more than a factor of five. Further, the dimensions of the finger structure and the metallization pattern of the proposed transistor and the test device should closely resemble one another.

Computing Cell Area

The area of a cell A_{cell} can be estimated with the formula

$$A_{cell} = P_f \Sigma A \qquad [14.4]$$

where ΣA equals the sum of the areas of all of the individual components. The packing factor, P_f, accounts for the area consumed by isolation and device interconnection as well as the area wasted by imperfect packing. Standard bipolar designs employing single-level metal typically have packing factors of 1.5 to 3.0. Values toward the lower end of this range represent well-packed designs using custom-crafted devices and extensive device mergers. The values toward the upper end of this range represent designs using standardized components and few or no device mergers. Standard bipolar designs using double-level metal require less area and will typically have packing factors of 1.5 to 2.0. Double-level-metal analog CMOS or BiCMOS designs using standardized components usually achieve packing factors of 1.4–1.8. Triple-level metal provides little improvement unless the cell contains an extraordinary amount of interconnection or dense logic circuitry.

Suppose that the bandgap circuit in Figure 14.2 will be laid out in a standard bipolar process using single-level metallization. Three of the transistors require individual sketches: Q_1, Q_2, and Q_{10}. The first two form the ratioed pair of the Brokaw bandgap cell, while Q_{10} is a small power device. The area of the other devices can be estimated by the procedures discussed. Table 14.1 shows the results of this calculation.

14.1.2. Die Area Estimation

Three factors contribute to the overall die area: the circuitry it contains, the ring of pads around its periphery, and the scribe streets separating it from adjacent dice. The circuitry resides in the middle of the die, forming the *core*. The pads lie around the periphery of the die, forming the *padring*. Ideally, both the core and the padring should contain no wasted space. Practical designs almost never meet this goal. In a

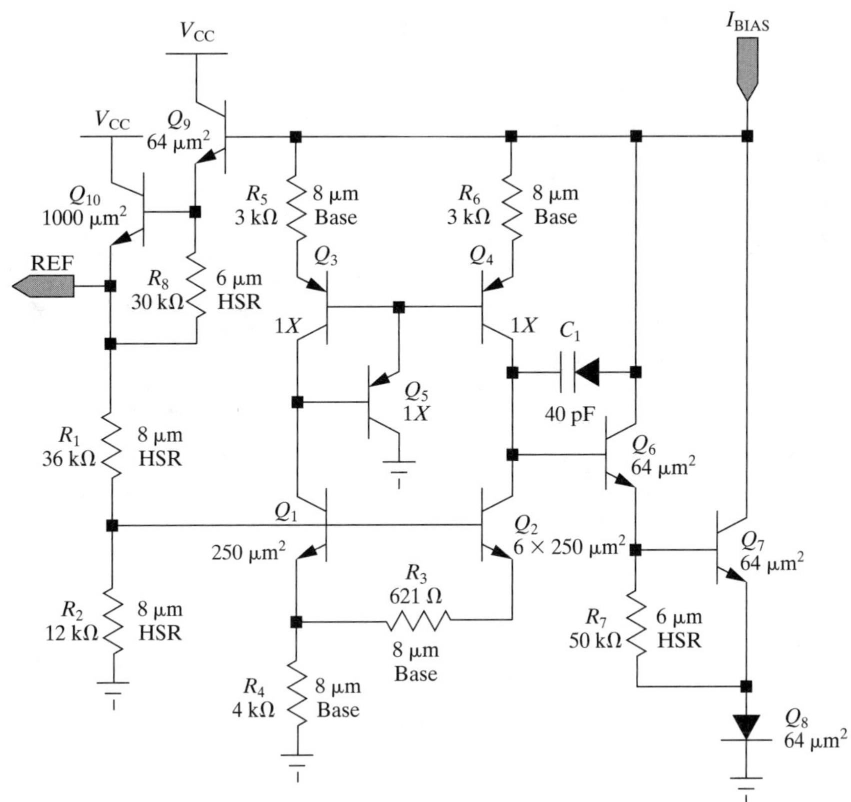

FIGURE 14.2 Schematic of the sample circuit block, a simple bandgap regulator constructed in a standard bipolar process.

TABLE 14.1 Estimated area for the simple bandgap regulator.

Device	Amount	Area
8 µm 160 Ω/□ base resistance	10.621 kΩ	14,200 µm²
8 µm 2 kΩ/□ HSR resistance	48.0 kΩ	4,600 µm²
6 µm 2 kΩ/□ HSR resistance	80.0 kΩ	5,200 µm²
Junction capacitance, 1.8 fF/µm²	40 pF	22,200 µm²
Minimum NPN transistors @ 2,200 µm²	4	8,800 µm²
Minimum PNP transistors @ 4,100 µm²	3	12,300 µm²
Bandgap NPN transistors @ 3,100 µm²	7	21,700 µm²
Output NPN transistors @ 6,600 µm²	1	6,600 µm²
Total area of components		95,400 µm²
Estimated cell area (P_f = 2)		0.19 mm²

core-limited design, the core packs tightly into the space inside the padring, but there are not enough pads to fill the ring (Figure 14.3A). The gaps between the pads are often used for ESD structures and trim circuitry. A *pad-limited* design has so many pads that the space remaining inside the padring exceeds the area required by the core (Figure 14.3B). One can sometimes place a second ring of bondpads inside the first, offsetting the inner bondpads so that they fall between the outer bondpads. This type of layout should be carefully reviewed by the assembly/test site to ensure that it is manufacturable. The estimation procedure must determine whether the design is core-limited or pad-limited before a final area estimate is possible.

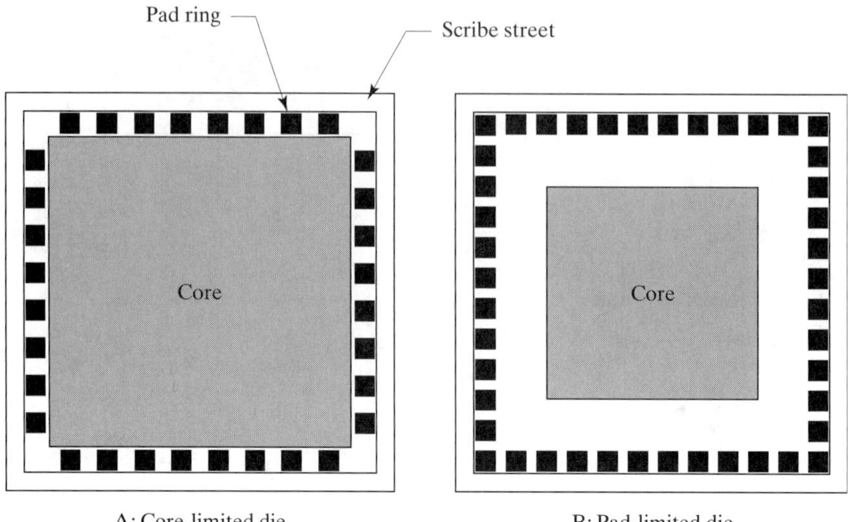

FIGURE 14.3 Comparison of core-limited and pad-limited dice.

The first step in computing the estimated die area consists of computing the core area A_c

$$A_c = R_f P_f \Sigma A_{\text{cell}} + P_f \Sigma A_{\text{pwr}} \qquad [14.5]$$

where ΣA_{cell} represents the sum of the areas of all the individual cells, and ΣA_{pwr} represents the sum of the areas of all the power devices not contained in any cell. The core does not include bondpads, trimpads, ESD devices, scribe seals, or scribe streets. The *routing factor* R_f accounts for the area consumed by top-level wiring. Typical routing factors for a design with several hundred top-level signals are 1.3 to 1.5 for single-level metal, 1.2 to 1.3 for double-level metal, and 1.1 to 1.2 for triple-level metal. Designs making extensive use of poly routing fall somewhere between these values. For example, a design routed on metal-1 and poly-1 might achieve a routing factor of 1.3, as compared with a routing factor of 1.4 for metal-1 alone. The *packing factor* P_f represents the areas of wasted space between cells. A die containing 20 or 30 moderate-sized cells should achieve a packing factor of 1.1 to 1.2. Layouts incorporating very large or very odd-shaped cells often have much higher packing factors. Conversely, hand-optimized designs can achieve a packing factor close to one. However, hand optimization requires additional time and effort, particularly for larger designs.

The following formula computes the estimated area of the entire die A_{die} based upon a square aspect ratio:

$$A_{\text{die}} = \left(\sqrt{A_c} + 2W_{\text{pr}} + W_s\right)^2 \qquad [14.6]$$

The design rules should specify the scribe width W_s, typically 75 to 125 μm. The width of the padring W_{pr} usually equals about 130% of the width of a bondpad. The design rules generally specify the minimum allowed dimensions of the bondpads. For ball-bonded gold wires, the minimum bondpad width equals between two and three times the diameter of the wire. According to these guidelines, a 25 μm-diameter gold wire requires a 75 μm bondpad and a 100 μm padring. These approximations are adequate for preliminary area estimates, but the final area estimate should be made only after the padring has been constructed (Section 13.4).

The foregoing computations assume that the padring contains enough space to contain all of the required pads. The minimum die perimeter P_{min} required to place the pads approximately equals

$$P_{\text{min}} \cong (N_p + 4)(W_p + S_p) + 4W_s \qquad [14.7]$$

where N_p equals the total number of pads, W_p equals the width of a pad, W_s equals the width of the scribe, and S_p equals the minimum spacing allowed between adjacent pads (measured between the facing edges of adjacent pads). This formula assumes that the pads can be placed relatively close to the corners of the die and that circuitry cannot reside underneath the bondpads. If the layout rules specify a minimum allowed distance from the corner of the die to the edges of the nearest bondpads, then add eight times this distance to the estimate of P_{min} computed previously.

The *perimeter utilization factor P* represents the fraction of available pad perimeter used by the pads. Assuming that the die has a square aspect ratio,

$$P = \frac{P_{\text{min}}}{4\sqrt{A_{\text{die}}}} \qquad [14.8]$$

If P is less than one, then the design is core-limited, and the estimated die area is probably correct. The following formula will determine the approximate number of ESD devices N_e that will fit between the pads:

$$N_e \cong \frac{N_p(W_p + S_e)}{P(W_p + S_p)} \qquad [14.9]$$

Here, S_e equals the minimum spacing between pads required to fit one ESD structure. If N_e exceeds the number of ESD devices used by the design, then they should all fit into the padring. If there are too many ESD devices, then add the area of the devices that will not fit into the padring to the area of the core and repeat the calculations using Equations 14.5 to 14.8.

If the perimeter utilization factor P exceeds one, then the design is pad-limited and the die area must increase. The total die area A_{die} for a pad-limited die with a square aspect ratio equals

$$A_{\text{die}} = \frac{P_{\text{min}}^2}{16} \qquad [14.10]$$

The pad-limited die will have wasted space equal to the difference between the area estimates of Equations 14.6 and 14.10. It is possible to slightly increase the amount of usable die periphery by elongating the die, but reasonable aspect ratios rarely provide enough additional perimeter to transform a pad-limited die into a core-limited one.

14.1.3. Gross Profit Margin

Managers and marketers use die-area estimates to determine the profitability of a design. The figure of merit most often used for this purpose is the *gross profit margin* (GPM), defined as the percentage of the sales price remaining after manufacturing costs have been subtracted. The procedure used to determine GPM is worth examining, as it provides some insight into the economics of integrated circuit manufacture.

The first step consists of computing the number of dice obtainable from one wafer N_d:

$$N_d = \frac{\eta \pi d^2}{4A_d} \quad [14.11]$$

In this equation, d represents the diameter of the wafer in millimeters and A_d represents the area of the die in square millimeters. The *wafer utilization factor* η represents the fraction of the wafer's surface covered by potentially usable dice. Some of the dice around the edges will be incomplete, either because they extend beyond the edge of the wafer or because they fall outside the field of exposure. Wafer utilization factors range from 0.7 to 0.9, depending on die size and photolithography techniques. As an example of the use of this equation, consider a 10 mm^2 die constructed on an 8″ (200 mm) wafer with a wafer utilization factor of 0.85. This wafer will yield 2670 dice.

The cost of a functional die C_d can be determined by using the formula

$$C_d = \frac{C_w}{N_d Y_p} \quad [14.12]$$

where C_w represents the cost of one wafer, including probing and sawing, and Y_p represents the *probe yield*, defined as the fraction of the potentially usable dice that pass wafer probe. Probe yields depend on the area of the die, the complexity of the process, and the robustness of the design. Probe yields of analog integrated circuits usually range from 0.8 to 0.95, although both lower and higher figures are possible. Continuing the previous example, suppose that each 8″ wafer costs $750 and the probe yield equals 90%. Assuming 2670 potential dice per wafer, each good die costs 31¢.

The total cost C_t of an integrated circuit equals

$$C_t = \frac{C_d + C_a}{Y_a} \quad [14.13]$$

where C_d represents the die cost and C_a the assembly cost (including packaging, symbolization, final testing, storage, and shipping). The *assembly yield* Y_a usually exceeds 0.95 because the vast majority of the defective dice have already been rejected during wafer probe. Continuing the previous example, if each good die costs 31¢, assembly costs 15¢, and the assembly yield equals 0.95, the cost of the finished integrated circuit equals 48¢.

The *gross profit margin* (GPM) is computed from the total cost C_t and the sales price S:

$$\text{GPM} = \frac{S - C_t}{S} \cdot 100\% \quad [14.14]$$

If an integrated circuit costs 48¢ to produce and sells for $1.00, then it will have a GPM of 52%. The gross profit margin must cover all the costs associated with running a large company, including sales and distribution, engineering, research and development, fixed overhead, and administrative costs. A GPM of at least 50% is desirable.

14.2 FLOORPLANNING

The final phase of the planning process consists of creating a sketch of the layout, called a *floorplan*, showing the placement of the bondpads and the locations and shapes of all the cells. During the layout, the floorplan serves as a guide for constructing the padring. If any cell requires significantly more or less space than allocated, the floorplan should be revised accordingly. Once most or all of the cells have been completed, the floorplan can be used as a template for assembling the top level layout.

The information required to construct a floorplan includes area estimates for each cell as well as an area estimate for the whole die. The designer must also obtain a complete listing of all pads and the order of their placement. Table 14.2 shows a sample worksheet containing the information needed to produce a floorplan of a small analog integrated circuit.

TABLE 14.2 Sample floorplanning worksheet.

Device:	Dual operational amplifier
Process:	Standard bipolar, double-level metal
Dimensions in:	Microns
Package type:	8-pin DIP
Die area estimate:	1.33 mm^2 ($P_f = 1$, $R_f = 1.2$)
Bondpad width:	75 μm
Padring width:	100 μm estimated
Scribe width:	75 μm

Circuit Block	Area	Dedicated Pins	Shared Pins
AMP1	0.32 mm^2	IN1+, IN1−, OUT1	V+, V−
AMP2	0.32 mm^2	IN2+, IN2−, OUT2	V+, V−
BIAS	0.13 mm^2	None	V+, V−

Note: AMP1 and AMP2 are identical to each other.

Pin #	Pin Name	Function of Pin
1	OUT1	Output of first amplifier
2	IN1−	Inverting input of first amplifier
3	IN1+	Noninverting input of first amplifier
4	VEE	Negative supply (connect to substrate)
5	IN2+	Noninverting input of second amplifier
6	IN2−	Inverting input of second amplifier
7	OUT2	Output of second amplifier
8	VCC	Power supply

Having obtained the necessary information, the next step consists of sketching out the padring. Assuming a square die aspect ratio, a 1.33 mm^2 die will have dimensions of 1153 × 1153 μm. This distance must be rounded off to the nearest increment allowed by the stepper, as determined by the photomask vendor. Many older steppers require the die size to equal an integer number of mils. A die that is 1153 μm on a side would become 46 mils (1168.4 μm) on a side. Modern steppers use metric dimensions instead of the older English units.

Once the dimensions of the die have been determined, a leadframe must be chosen. Figure 14.4 shows a drawing of a typical 8-pin *dual in-line package* (DIP) leadframe. The large square tab in the middle of the drawing is called the *mount pad*. The die must be slightly smaller than the mount pad to allow for misalignment. Given an allowance of 125 μm per side, an 1.5 × 2.0 mm mount pad can accommodate a die with maximum dimensions of 1.375 × 1.875 mm. This leadframe can easily accommodate a 1.153 mm-square die. Given a choice of several leadframes, choose the smallest one that will accommodate the die. Excessively large leadframes may reduce the assembly yield because of wiresweep and sag.

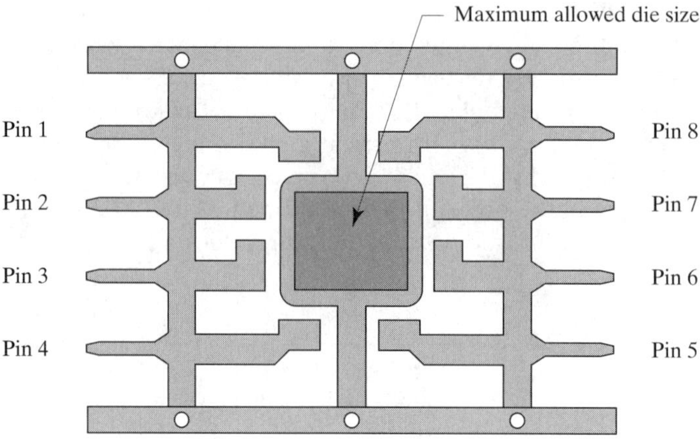

FIGURE 14.4 Sample leadframe drawing for an 8-pin DIP package.

The leadframe in Figure 14.4 illustrates a common arrangement in which the lead fingers actually encircle the mount pad. Thus, while pin one emerges from the top left of the package, the bondwire attaches to its lead finger directly above the mount pad. The bondpads should be placed to obtain the shortest and most direct bondwire routings. This precaution not only minimizes the chance of wires shorting to one another but also reduces the amount of gold wire required to bond the die.

The floorplan should also show the location of the scribe streets. Some processes require the scribe street on the bottom and left sides, others on the top and right sides, and still others place half-width scribe streets on all four sides of the die. All of these arrangements result in the same pattern of scribe streets on the wafer, but each requires a slightly different layout. For the purposes of this example, we will assume that the scribe occupies the top and right sides of the die and that the origin of the layout resides at the lower left corner.

The floorplan sketch includes a rectangle marking the extents of the die and a second smaller rectangle delimiting the area reserved for the core (Figure 14.5). A strip along the top and right-hand sides of the die shows where the scribe fits. The individual cells are represented by rectangles having the appropriate areas. Since this design contains two identical amplifiers, it makes sense to use mirror-image placements of a single amplifier cell. This not only saves layout effort but also ensures that the two amplifiers will have similar electrical characteristics. The placement of the amplifiers must allow wires to route to the appropriate pins. AMP1 connects to pins 2, 3, and 4, while AMP2 connects to pins 6, 7, and 8. AMP1 should therefore reside on the left side of the die and AMP2 on the right. The BIAS block fits into the middle between the two amplifiers. This arrangement produces rather elongated circuit blocks, but there is no reason why this should cause problems. The die's aspect ratio remains square, and the elongation of the amplifier blocks actually helps improve matching by allowing the sensitive input circuitry to be placed far away from the power devices. The shape of the bias circuit is somewhat awkward, but not unworkable.

The die area estimate reserved 20% of the core area for routing. This space has been incorporated into the floorplan in the form of two narrow strips running vertically across the entire die. The actual placement of the leads will be determined later; these strips merely reserve room for them.

The next step consists of placing the bondpads. This requires that the floorplan be superimposed on a copy of the leadframe drawing. The bondpads should initially sit adjacent to their respective leadframe fingers (Figure 14.6A). This arrangement results in the shortest bondwires and the largest separation between the wires, but it

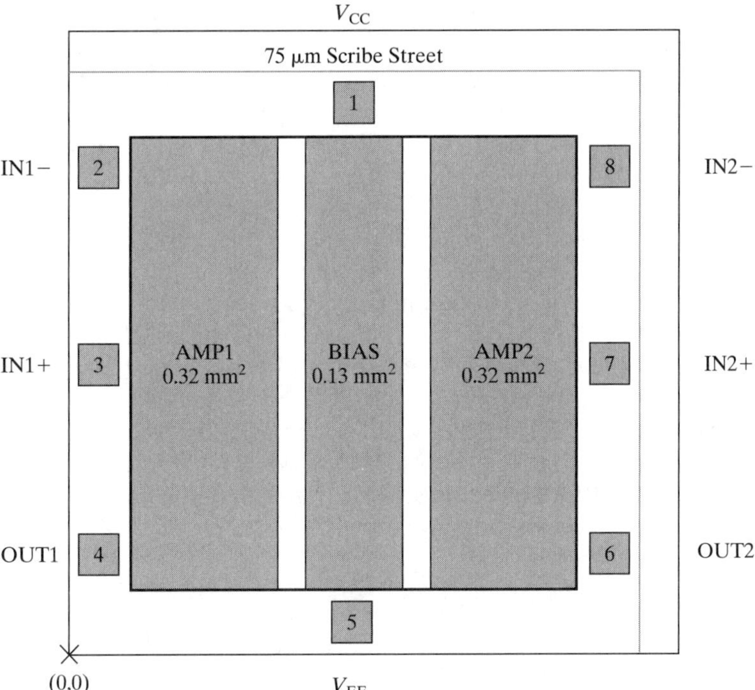

FIGURE 14.5 Floorplan of the eight-pin dual op-amp die.

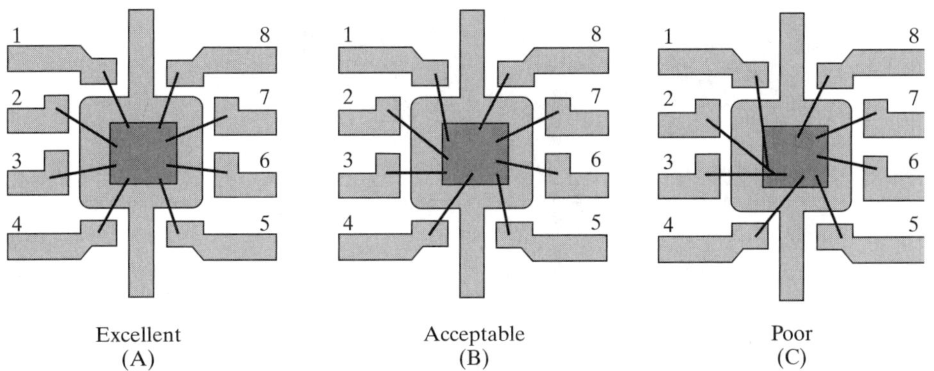

FIGURE 14.6 Three possible placements of bondpads for an 8-pin leadframe.

does not necessarily provide the best interconnection between pads and circuit blocks. The pads can move slightly to accommodate layout, but if they are moved too far the design becomes unmanufacturable. In order to facilitate automated bonding, the bondwires should not intersect, nor should they even closely approach one another. Figure 14.6B shows an acceptable bonding arrangement, while Figure 14.6C shows an unacceptable one. This arrangement causes the wire for pin #2 to pass too close to the ball bond for pin #1. Similarly, the wire for pin #3 comes too close to the ball bond for pin #2. These faults make it almost impossible to bond the die without damaging some of the wires. Once the bondpad arrangement has been tentatively decided, the designer should forward a bonding diagram to the assembly site to allow them to check for potential bonding problems.

The floorplan in Figure 14.5 uses the pad arrangement shown in Figure 14.6B, and places the three input/output pads of AMP2 directly opposite the three corresponding input/output pads of AMP1. This symmetric placement helps simplify the connection of these pads to their respective amplifiers. The two power pads move to the top and bottom of the die. These locations not only help minimize interference between adjacent bondwires but also help to ensure that the power leads can run to all three blocks.

If the design incorporates high-current circuitry, then the designer should check the routing of high-current leads. Electromigration sets a lower limit on the width of a high-current lead, but metal resistance often forces the use of much wider leads. All high-current leads should be kept as short as possible to minimize unnecessary metal resistance. The anticipated locations of each high-current lead should be marked on the floorplan, along with the equivalent DC current that they must conduct. The resulting diagram will show whether any awkward or unnecessarily long leads exist.

Figure 14.7 shows a diagram of the dual op-amp, highlighting the locations of the high-current leads. The VCC lead routes directly across the top of the bias cell. This is an acceptable arrangement for a double-level-metal layout because the lead can be routed in metal-2 and the circuitry in metal-1. The lead routing does place a limitation on the layout of the BIAS block, which otherwise might not have been apparent. If the BIAS block must use metal-2, then the power lead can slide laterally into either of the two routing channels.

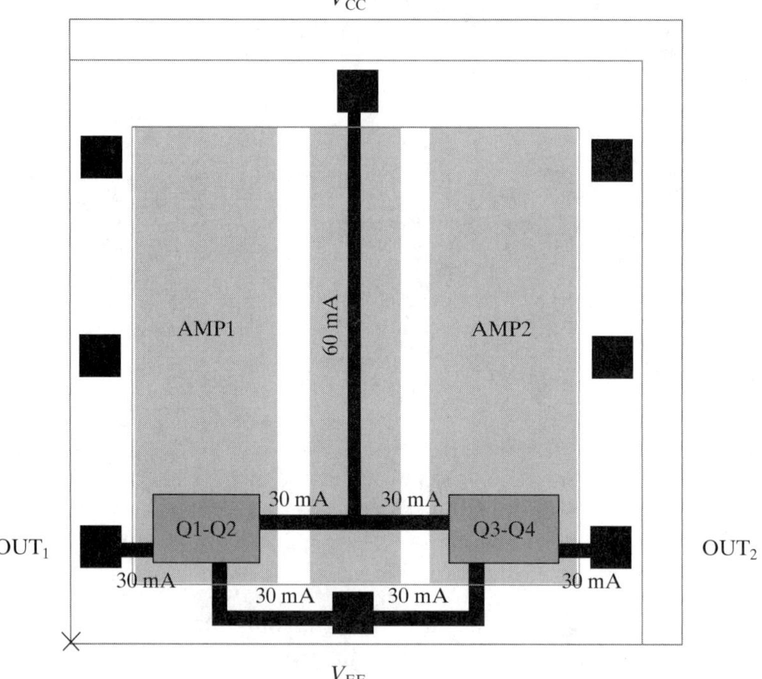

FIGURE 14.7 Floorplan of dual op-amp showing the routing of leads to transistors *Q*1 to *Q*4.

Although the floorplan in Figure 14.7 explicitly shows the VEE leads connecting to each amplifier, in practice these leads form part of the scribe seal metallization. Most dice arrange their power and ground return signals to take advantage of the scribe seal metallization as much as possible, and this design is no exception. An additional width of metal placed around the periphery of the die abutting the scribe seal allows the metal to carry additional current. Substrate contacts placed under this metallization augment those in the scribe seal and form a significant portion of the substrate contact system of the die.

As previously mentioned, electromigration determines the minimum width of high-current leads. The minimum allowed lead width W_{min} (in microns) equals

$$W_{\text{min}} = \frac{10^{12} I_{\text{DC}}}{J_{\text{max}} t} \qquad [14.15]$$

where I_{DC} is the equivalent DC current in amps, J_{max} equals the maximum allowed current density in A/cm^2, and t is the thickness of the metal in Angstroms (Å). The values for J_{max} and t depend on the process and operating conditions (Section 14.3.3). If the lead passes over oxide steps, then the minimum thickness of the lead at the crossing point should be used for electromigration calculations.

For example, suppose $J_{\text{max}} = 5 \cdot 10^5$ A/cm^2 and $t = 8$ kÅ. Given these values, Equation 12.14 indicates that a lead carrying 60mA requires a width of 15 μm. The sheet resistance R_s for a metal lead can be computed with the formula

$$R_s = \frac{10^8 \rho}{t} \qquad [14.16]$$

where ρ is the resistivity of the metal in Ω · cm and t is the thickness of the metal in Angstroms (Å). The resistivity of aluminum containing 0.5% copper and 2% silicon equals about 2.8 μΩ · cm. Refractory barrier metals are much more resistive than aluminum alloys, so the sheet resistance of the metallization should be computed on the basis of the thickness of the aluminum and not on the thickness of the full metallization. The thinning of the metal system over oxide steps can be ignored when computing lead resistance. An 8 kÅ aluminum-copper-silicon metal system has a sheet resistance of about 35 mΩ/□. The 15 μm VCC lead in Figure 14.7 runs about 1000 μm and contains about 67 squares of metal that have a total resistance of 2.3 Ω and generate a total voltage drop of about 140 mV. Since both amplifiers connect to this lead, voltage drops generated by one amplifier can interfere with the performance of the other amplifier, a problem called *crosstalk.* Crosstalk can be reduced by making the lead wider, but a better solution is to run two separate power leads, one for each amplifier. This is an example of a *Kelvin connection.* (See Section 14.3.2.)

Sometimes vias must be placed in large power leads. The vias not only increase the resistance of the lead but also limit the amount of current that it can conduct before electromigration causes the vias to fail. A via can carry approximately the same current as a lead whose width equals the width of the via perpendicular to the direction of current flow. Consider the case of a 4 μm-wide via through metal that can carry 4 mA/μm. This via could conduct approximately 16 mA. Typical vias have a resistance of about 0.1 Ω each. A 1-amp lead would therefore require a minimum of 63 vias, which together would add about 2 mΩ of resistance. An array of vias that large would consume a significant amount of area, and their presence would have to be considered while constructing the floorplan.

The floorplan for the dual op-amp has now been completed. Although this example was fairly simple, the same principles apply to larger and more complex dice. The arrangement of the circuit blocks becomes more difficult, and often begins to resemble the construction of a jigsaw puzzle (Figure 14.8). The placement and width of routing channels also become much more critical. The careless placement of a block can easily constrict a routing channel. The resulting *choke points* can needlessly complicate routing, especially if they are not found until the routing has begun. The layout of Figure 14.8 contains two obvious choke points.

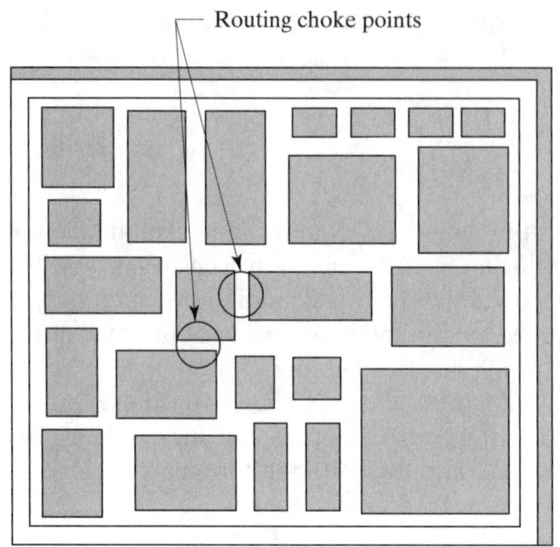

FIGURE 14.8 Floorplan of a complex die showing two potential choke points in the routing.

14.3 TOP-LEVEL INTERCONNECTION

Most analog and mixed-signal layouts actually benefit from manual interconnection. Autorouting software can interconnect blocks very quickly, but a human designer better understands such factors as wiring resistance, electromigration, noise coupling, and heat distribution. A skilled layout designer also takes advantage of every opportunity to insert substrate contacts, bypass capacitors, probe points, and test pads. Even if an autorouter is used, the layout designer must still verify that each analog signal has been properly routed. This process of verification (and the inevitable corrections) may require as much time as manual interconnection for a small analog design.

Some designers advocate running most of the wires across or through the circuit blocks themselves, a technique called *maze routing.* Others advocate running the wires alongside the individual blocks, a technique known as *channel routing.* Maze routing can save 5 to 10% of the initial die area, but it often requires two or three times as long to complete. Most modern designs use channel routing because it is quicker to implement and easier to modify.

14.3.1. Principles of Channel Routing

Channel routing requires at least two levels of interconnection. These consist, at a minimum, of one level of metal and one level of polysilicon. Unsilicided gate poly usually has a sheet resistance of 20 to 50 Ω/□, while silicided poly can routinely achieve sheet resistances of less than 5 Ω/□. Many signals can tolerate the insertion of short poly jumpers, especially if these are silicided. On the other hand, most signals cannot tolerate the resistance of long poly runs. This consideration makes it difficult to produce a compact wiring arrangement by using only one level of metal, so most modern designs use at least two. Multiple metal layers reduce the need for poly routing, but the use of limited amounts of poly can still help clear congested wiring channels. As long as the designer carefully chooses which signals to route in poly, the presence of long polysilicon traces has little or no impact on circuit performance.

In a double-level-metal design, channel routing allocates one layer of metallization for vertical routing and a second for horizontal routing. It matters little which

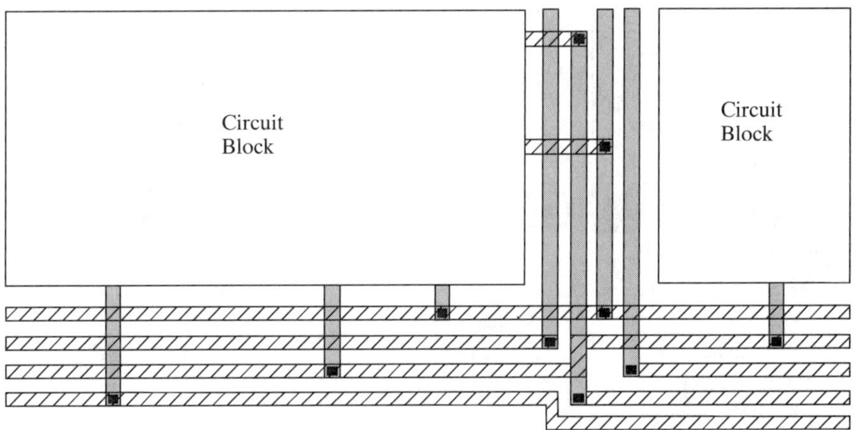

FIGURE 14.9 A portion of a channel-routed layout.

layer runs horizontally and which runs vertically. Figure 14.9 shows portions of two routing channels and an intersection between them. As this example shows, leads can exit from either routing channel at any point by jumping from one metal layer to the other. This orderly arrangement can be maintained only as long as all of the leads reside on the designated layers. If each signal is routed on whichever metal layer seems convenient at the time, then the design soon becomes a tangle of unwieldy (and unnecessary) metal jumpers.

The leads on successive layers should run at right angles to one another. For example, metal-3 leads should run at right angles to metal-2 leads. Similarly, poly should run at right angles to metal-1. Thus, if poly runs horizontally, metal-1 should run vertically, metal-2 horizontally, and metal-3 vertically. This arrangement minimizes the need for jumpers and results in the best utilization of channel space.

Metal systems are often specified in terms of their *metal pitch.* The metal pitch P_m equals the sum of the minimum drawn metal width W_m and the minimum drawn metal spacing S_m:

$$P_m = W_m + S_m \qquad [14.17]$$

For example, a process that can fabricate 2 μm leads spaced 1.5 μm apart has a metal pitch of 3.5 μm. The width of a wiring channel can be determined using the formula

$$W_c = NP_m + S_m \qquad [14.18]$$

where W_c equals the width of a wiring channel that can just accommodate N minimum-width leads. Continuing the previous example, a six-lead wiring channel would require a width of 22.5 μm. The width W_c includes the space on either side of the channel between the leads and adjacent metallization.

In order to ensure optimal packing, the vias should not require enlarged metal heads. This requirement will be satisfied if the following inequality holds:

$$W_m \geq W_v + 2O_{mv} \qquad [14.19]$$

Here, W_v equals the minimum width of a via and O_{mv} equals the minimum overlap of metal over via. If this inequality does not hold, then W_m should be increased. For example, suppose a process is capable of fabricating a 2 μm metal lead, but it requires 1.5 μm vias and a metal overlap of via of 0.5 μm. The width of

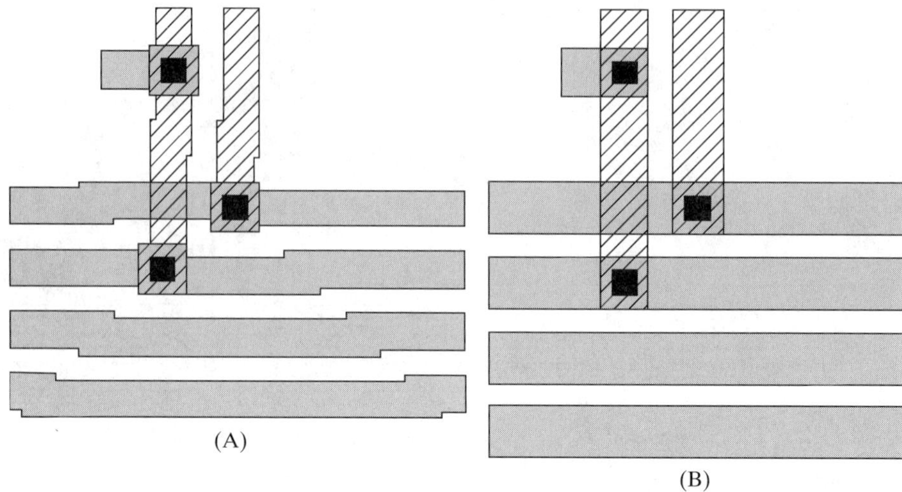

FIGURE 14.10 (A) A layout that requires enlarged via heads (B) becomes much simpler when the lead width is slightly increased.

the metal head required to contain the via is 2.5 μm, which is 0.5 μm larger than the specified metal width. In order to maintain proper packing in the wiring channels, the designer should increase the width of the metal leads in the channel to 2.5 μm. The increased lead width wastes a little area, but it greatly simplifies wiring. Figure 14.10A shows a layout employing enlarged via heads, while Figure 14.10B shows the same layout reworked using wider leads. Despite the presence of a little wasted space, the layout of Figure 14.10B is obviously preferable to that of Figure 14.10A.

In most designs, two or three routing channels carry a large percentage of the signals. These *primary routing channels* usually intersect somewhere near the center of the die. These intersections may easily become choke points unless the channels are made quite wide. As a conservative rule, each primary routing channel should contain space for about 20% of the top-level signals. A design containing 100 top-level signals requires room for about 20 leads in each of its primary routing channels. The width of the primary wiring channels can decrease as they approach the edges of the die. Similarly, the feeder channels branching off the primary channels can be made proportionately narrower. The resulting pattern of routing channels resembles the path of watercourses across a plain. Even the narrowest channels should always allow capacity for 3 to 5 signals, as it is very difficult to guess exactly where leads need to route, and it is very difficult to increase channel widths once the top-level interconnection has commenced.

Poly routing can reduce the width of the primary routing channels by perhaps 30%. This reduction in width assumes that about a third of the leads are routable in poly. The use of poly routing under both horizontal and vertical routing channels can lead to gridlock in the vicinity of major intersections. This type of gridlock is best avoided by running poly leads in only one direction, preferably in the same direction as metal-2. If possible, one should align the largest (or longest) primary routing channel so that it contains poly.

14.3.2. Special Routing Techniques

Certain leads require special consideration during routing. Some leads are particularly sensitive to voltage drops or noise coupling, while others require wider metal to minimize voltage drops and to prevent electromigration. This section discusses some of the techniques used to handle these concerns.

Kelvin Connections

The metallization resistance, although small, is not always negligible. Consider an amplifier circuit containing a pair of matched bipolar transistors whose emitters connect to a ground return line carrying 100 μA (Figure 14.11A). The two emitter leads do not tap into the ground return at the same point, so the ground current generates a small voltage differential between them. Suppose the ground return lead between points *A* and *B* contains 10 squares of 30 mΩ/□ metal. A 100 μA current flowing through 0.3 Ω develops a voltage of 30 μV. Any variation in ground current produces a corresponding variation in this voltage drop, which the circuit amplifies. If the ground current fluctuates by ±10% and the amplifier has 80 dB of voltage gain, then the output will fluctuate by 0.3 V! This is clearly an unacceptable situation.

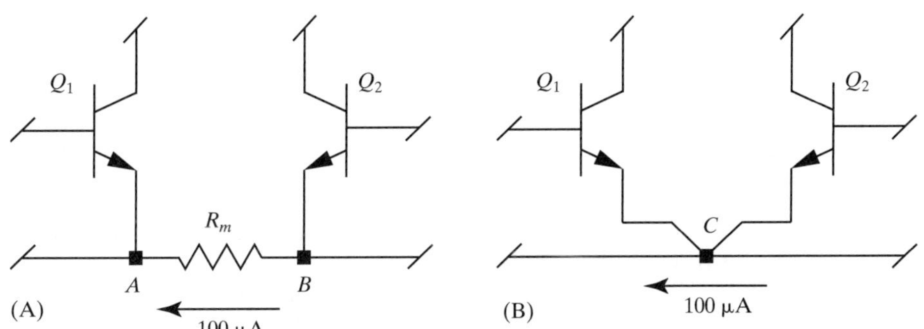

FIGURE 14.11 (A) The effects of voltage drops in a ground return line (B) can be eliminated through the use of a Kelvin connection.

The magnitude of the voltage drop in the ground lead depends on the distance between points A and B. Figure 14.11B shows the same circuit with one modification: both emitter leads now return to a common point *C*. Since the leads return to the same point, the ground current cannot generate any differential between them. Ground drops elsewhere along the lead cause the voltage on both emitters to vary in unison, but most circuits exhibit a high degree of immunity to such *common-mode variations.*

The common point *C* is called a *star node* or a *Kelvin connection.* These points are often represented on schematics by leads entering a solder dot at 45° angles, as in Figures 14.11 and 14.12. Kelvin connections have many other applications. Figure 14.12A shows a pair of Kelvin connections that allow accurate sensing of the voltage developed across a metal resistor. Leads F_1 and F_2 carry a large current through resistor R_1, while leads S_1 and S_2 connect the resistor to a sensing circuit. F_1 and F_2 are called *force leads,* while S_1 and S_2 are called *sense leads.* Figure 5.16 shows sample layouts of metal resistors with force and sense leads. The sense leads carry very little current to minimize the voltage drops occurring in them. In high-precision applications, the currents flowing through the sense leads are carefully balanced, and the layouts of the leads are adjusted so that each contains the same number of squares. This precaution ensures that any voltage drops that do occur in the sense leads are common-mode variations.

Figure 14.12B shows another application for Kelvin connections. A voltage regulator, VR_1, must provide a large amount of current to an external circuit. Voltage drops inevitably occur in power lead F_1 and in ground lead F_2. A separate set of sense leads, S_1 and S_2, is employed to sense the voltage across the load at Kelvin connections K_1 and K_2. This arrangement ensures that voltage drops in the force leads do not degrade the regulation of the circuit. The manufacturers of commercial power supplies call this arrangement *remote sensing.* Integrated voltage regulators

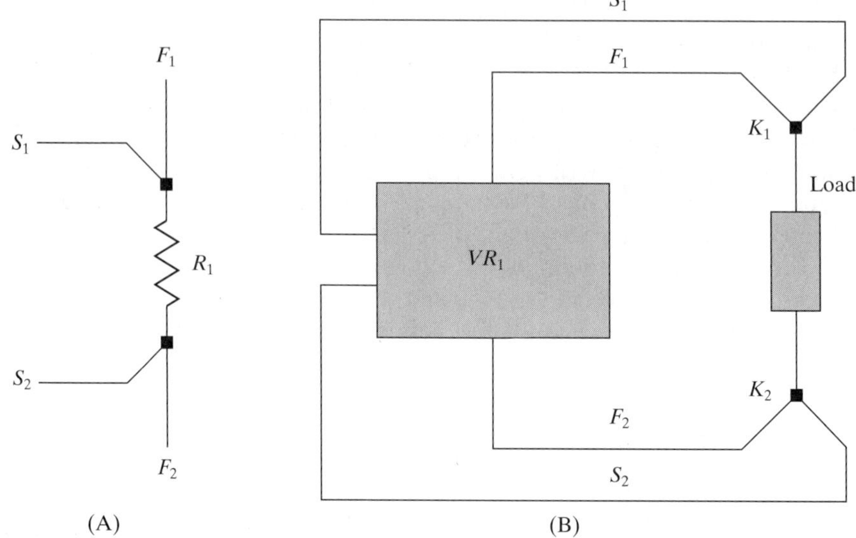

FIGURE 14.12 Additional applications of Kelvin connections: (A) metal sense resistor and (B) remote sensing.

sometimes support remote sensing by providing separate force and sense pins. If the package does not provide enough pins to accommodate separate force and sense connections, two bondwires can be connected to the same pin to eliminate voltage drops in the IC metallization and the bondwires. Even if the package cannot accommodate double bonding, the designer can still run separate force and sense leads to the output and ground bondpads.

Noisy Signals and Sensitive Signals

Broadly speaking, electrical noise consists of any unintended or undesired signal. All electrical devices generate some noise, but the level of this *device noise* is so low that it affects only very sensitive circuits. Most of the noise problems encountered in integrated circuits are caused by capacitive coupling of signals from one circuit node to another. A capacitor appears whenever leads cross or run alongside one another. Although these capacitors are very small, the amount of energy coupled through them increases with frequency. At sufficiently high frequencies, even intersections between minimum-width leads can couple objectionable amounts of noise energy from one circuit to another.

In practice, capacitively coupled noise becomes a concern only if one or more signals contain appreciable energy at frequencies in excess of 1 MHz. Most analog signals have relatively little high-frequency content, but digital signals are quite another matter. The crisp transitions so prized by digital designers generate high-frequency harmonics extending well beyond 1 GHz. Each digital signal generates a burst of noise every time it switches states. All digital signals that experience state transitions during normal circuit operation should be considered potential noise sources. All power switching transistors should be considered noise sources, as should signals connecting to pins seeing rapid transients during normal operation.

The mere existence of any number of noisy signals is harmless. Problems arise only if the integrated circuit also contains one or more signals that are unusually sensitive to capacitively coupled noise. Analog signals are far more noise-sensitive than digital signals, but not all analog signals are equally sensitive. The most sensitive nodes are those that carry very low-level signals at high impedance levels. For example, the input of an amplifier is far more noise sensitive than its output because any signal present at the input is amplified by the gain of the amplifier. Furthermore, the output of an amplifier is usually low-impedance, while its inputs are often very high-impedance. The following types of signals are among the most noise sensitive:

- Inputs to high-gain amplifiers and precision comparators
- Inputs to analog-to-digital converters (ADCs)
- Outputs of precision voltage references
- Analog ground lines to high-precision circuitry
- Precision high-value resistor networks
- Very low-level signals, regardless of impedance
- Very low-current circuitry of any sort

Most layout designers do not possess the knowledge and experience required to correctly identify all of the noisy and sensitive nodes in a complicated analog circuit. Instead, the circuit designer must identify these signals, preferably by marking them on the schematic in some distinctive manner. Once the layout designer understands which signals are noisy and which are sensitive, the signals can be routed appropriately.

Noisy signals should not run on top of sensitive signals, or *vice versa.* If a crossing must occur, the area of intersection should be minimized. The circuit designer should examine each crossing to determine if it requires electrostatic shielding (Section 7.2.12). The usual method of constructing such shielding in double-level-metal designs is to run one signal in poly and the other in metal-2 (Figure 14.13A). A plate of metal-1 interposed between the two signals acts as an electrostatic shield. The shield should connect to a quiet low-impedance node such as an analog ground line. The shielding plate should extend beyond the area of intersection by 2 to 3 μm to block fringing fields.

Noisy signals should not run adjacent to sensitive signals. If such an arrangement seems unavoidable, then the layout designer should run another signal between the two (Figure 14.13B). The shield lead may consist of some other relatively low-noise, low-impedance signal, such as the output of a digital logic gate that seldom changes states. Alternatively, it may consist of an extra ground line or supply line added to the layout specifically to shield the noisy signal from the sensitive one. In order for the shield lead to perform its function, it must connect to a fairly low-impedance node in the circuit. Supply and ground lines satisfy this requirement, as will the output of digital logic gates.

Whenever possible, the designer should place noisy circuitry as far away from sensitive circuitry as possible. This usually requires that noisy circuitry occupy one portion of the die and sensitive circuitry another. This type of arrangement often proves beneficial from other standpoints because it separates sensitive circuitry from large power devices.

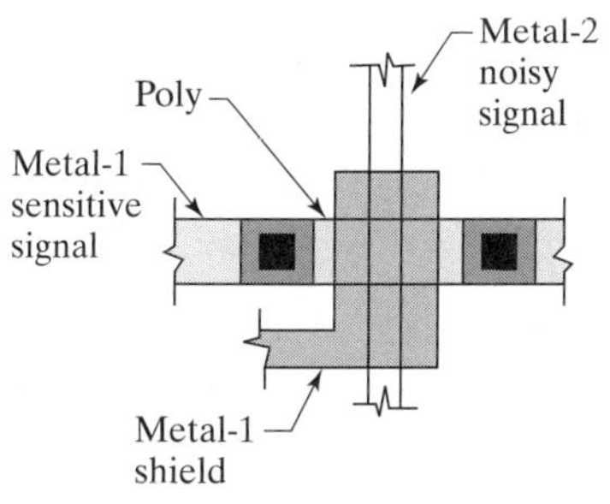

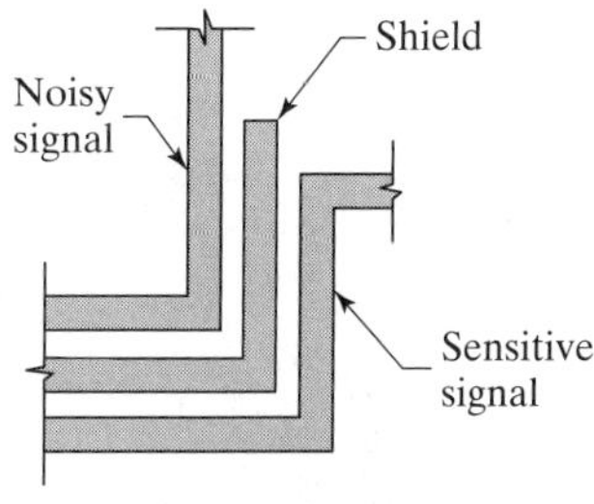

FIGURE 14.13 Electrostatic shielding techniques: (A) shield placed beneath a noisy signal and above a sensitive signal at their point of intersection, and (B) a shield line run between a noisy signal and a sensitive signal.

Sensitive signals should never run farther than necessary. The shorter these signals, the fewer opportunities they provide for noise coupling. A layout designer should always try to place analog circuits so that the sensitive signals running between them do not pass through other circuit blocks. Although this is not always possible, the fewer long sensitive leads a design contains, the easier it is to prevent capacitive coupling problems.

Noise may also couple from one circuit to another through the substrate. This problem can be minimized to a certain degree by proper grounding. Most designs use separate grounds for noisy circuitry and sensitive analog components. The question then becomes whether to connect the substrate to the noisy ground or to the quiet one. Connecting the substrate to the quiet ground would couple noise from the substrate into the most sensitive analog circuitry. This approach seems clearly undesirable. On the other hand, connecting the substrate to the noisy ground would ensure that the substrate itself can inject the maximum amount of noise into delicate analog circuitry above it. Neither approach seems optimal. For those designs in which substrate noise coupling is a serious concern, the designer should consider routing three separate grounds: one for noisy circuitry, one for sensitive analog circuitry, and a third for the substrate contacts. The substrate ground can connect to one of the other grounds at the bondpad, or better yet, at the lead finger through a separate bondpad and bondwire.

For particularly sensitive circuitry, one may wish to consider isolating components from the substrate. PMOS transistors, for example, are isolated by the well that acts as their backgate. NMOS transistors can sometimes be isolated by placing them within a section of isolated P-epi floored by NBL and surrounded by a ring of either deep-N+ or N-well. By connecting the wells of the PMOS transistors and the isolated P-epi of the NMOS transistor to quiet analog voltage rails, one can obtain nearly complete isolation of sensitive analog CMOS circuitry from the underlying substrate. Dielectric isolation can provide similar benefits, provided that one connects the substrate to a quiet ground return.

14.3.3. Electromigration

As discussed in Section 4.1.2, electromigration limits the current density that can safely flow through metallization. If the current density becomes too large, the pressure of carrier collisions on the aluminum metal atoms causes a slow displacement of the metal. This eventually results in voiding or lateral extrusions (*whiskers*) that can short adjacent signals.[1]

Electromigration obeys a relationship that was first discovered by J. R. Black and is therefore called *Black's Law*:[2]

$$\mathrm{MTF} = \frac{1}{AJ^2} e^{\frac{E_a}{kT_j}} \qquad [14.20]$$

Here, MTF is the mean time to metallization failure, in hours, A is a rate constant having units of $\mathrm{cm^4/A^2/hr}$, J is the current density passing through the lead in $\mathrm{A/cm^2}$, E_a is the activation energy in electron volts (eV), k is Boltzmann's constant ($8.6 \cdot 10^{-5}$ eV/K), and T_j is the absolute junction temperature in Kelvin.

[1] J. R. Black, "Electromigration—A Brief Survey and Some Recent Results," *IEEE Trans. Electron Devices,* Vol. ED-16, #4, 1969, pp. 338–348.

[2] J. R. Black, "Mass Transport of Aluminum by Momentum Exchange with Conducting Electrons," *Proc. 1967 Ann. Symp. on Reliability Physics,* IEEE Cat 7-15C58, 1967.

Since the rate of electromigration varies exponentially with temperature, high-temperature testing can accelerate the failure process to allow experimental determination of constants A and E_a. On the basis of such testing, one can derive a maximum current density, J_{max}, corresponding to any desired operating life. As one might expect, J_{max} is a strong function of temperature. A typical value for copper-doped aluminum is $5 \cdot 10^5$ A/cm² at 85°C. This value is widely accepted throughout the industry

$$I_{max} = 10^{-9} J_{max} W t \qquad [14.21]$$

where I_{max} is the maximum current (in mA) that the lead can safely carry, J_{max} is the maximum allowed current density in A/cm², W is the width of the lead in microns, and t is the thickness of the metallization in Angstroms (Å). A factor of 10^{-9} is inserted in Equation 14.21 to balance the units. As an example of the use of this equation, suppose a 10 kÅ lead is constructed of copper-doped aluminum having $J_{max} = 5 \cdot 10^5$ A/cm². If this lead is 10 μm wide, then it can conduct no more than 50 mA.

Metal leads passing over oxide steps often experience less-than-perfect step coverage. Most processes state a minimum guaranteed step coverage in terms of a percentage of normal metal thickness. The carrying capacity of any lead crossing an oxide step must be derated by the percentage step coverage. Standard bipolar processes typically quote a 50% step coverage figure. Thus, if the 10 μm lead discussed previously crosses an oxide step, then its current carrying capacity would drop to 25 mA.

Most parts are assumed to experience junction temperatures of 85°C or less for the majority of their operating life, even if the parts are rated for higher junction temperatures. Certain types of devices, particularly those that experience substantial differences between junction and ambient temperatures, may actually operate at high temperatures for long periods of time. If such conditions are anticipated, then the current carrying capacity of the leads must be multiplied by a derating factor, D, equal to

$$D = e^{\frac{E_a}{2k}\left(\frac{1}{T_j} - \frac{1}{T_o}\right)} \qquad [14.22]$$

where T_o is the junction temperature at which I_{max} was computed in Kelvin, T_j is the anticipated maximum operating temperature in Kelvin, E_a is the activation energy in electron volts, and k is Boltzmann's constant. A typical activation energy for electromigration-induced voiding in pure aluminum is 0.5 eV, while values for copper-doped aluminum[3] are closer to 0.7 eV. Suppose we wish to compute the de rating factor for 125°C (398K), given an original current density calculation performed at 85°C. Assuming $E_a = 0.7$ eV, the derating factor D equals 0.32. Therefore, a lead that can safely carry 25 mA at 85°C can safely carry only 32% of this current, or 8 mA, at 125°C.

If a lead does not carry a constant current, but rather a time-varying current, then its current-handling capability is increased. One type of time-varying current, consisting of short pulses repeated at frequent intervals, often occurs in digital logic and MOS gate drivers. The derating factor D for pulsed-current operation equals[4]

$$D \cong \frac{1}{d^2} \qquad [14.23]$$

---

3 H. V. Schreiber, "Activation Energies for the Different Electromigration Mechanisms in Aluminum," *Solid-State Elect.*, Vol. 24, 1981, pp. 583–589.

4 J. S. Suehle and H. A. Schafft, "Current Density Dependence of Electromigration t_{50} Enhancement Due to Pulsed Operation," *Proc. International Reliability Physics Symposium*, 1990, pp. 106–110.

where d is the duty cycle of the signal and D is the derating factor. For example, a lead conducting current only 50% of the time has a duty cycle of 0.5 and a derating factor of 4. If the lead can safely carry 25 mA DC, then it can handle 50 mA pulses at a duty cycle of 0.5. The derating factor computed in Equation 14.23 assumes that the pulsed current flows in only one direction. The rate of AC electromigration is somewhat slower than the rate of DC electromigration because some of the displacement occurring during one phase reverses during the other phase. The magnitude of the derating factor for AC operation is difficult to determine because it varies with the exact waveforms involved, but a conservative estimate is $D = 1.5$. Thus, if a lead could handle 50 mA unipolar pulses, then it could handle 75 mA bipolar pulses.

Another common type of time-varying signal consists of a sinusoid, or sine wave. A lead can handle a sinusoidal current with a peak value equal to about three times its DC rating. Thus, if a lead can handle 25 mA DC, then it can handle a sinusoidal current with a peak value of 75 mA (which equals 150 mA peak to peak or 106 mA root mean square).

Bondwires also have limited current-carrying capability. If too much current flows through a wire, its internal temperature may rise to the point where it fails, either gradually through electromigration, or suddenly through fusing or burning. Military specification Mil-M-38510 sets a maximum allowed current, I_{max}, (in mA) equal to[5]

$$I_{max} = kd^{3/2} \qquad [14.24]$$

where d is the bondwire diameter in mils and k is a constant equal to approximately 480 mA·mil$^{2/3}$ for aluminum and 650 mA·mil$^{2/3}$ for gold. This law is derived from the classical three-halves power law of radiative processes.[6] While it applies to long bondwires suspended in air, it does not apply to bondwires embedded in plastic or other encapsulating materials. These substances act as thermal insulators and prevent radiative heat transfer. The temperature of the wire is therefore determined solely by conduction of heat through the wire to its endpoints and through the encapsulation. Assuming that the wire is relatively long and that the encapsulation is a good thermal insulator, then the temperature of the wire depends only on the amount of power dissipated and not on its surface area or volume. Under these conditions, the current-handling capability of a bondwire scales linearly with its diameter. Large-diameter bondwires are placed under a disadvantage when they are encapsulated because they do not gain the benefit of their larger surface area.

In practice, bondwires that are less than about 100 mil long conduct enough heat to their ends to produce a noticeable increase in their current-carrying capacity.[7] For this reason, most designers use slightly larger current values than Mil-M-38510 suggests. Gold wires encapsulated in plastic are often scaled using the rule of "one amp per mil of diameter," and aluminum wires are usually rated to carry about half the current of an equivalent gold wire. Aluminum wires are not allowed to carry as much current as gold wires because they are more vulnerable to electromigration and high-temperature corrosion.

---

5 "Maximum Current in Wires," *Semiconductor Reliability News,* Vol. I, #10, 1989, p. 9.

6 W. H. Preece, "On the Heating Effects of Electric Currents," *Proc. Roy. Soc.(London),* April 1884, December 1887, April 1888.

7 From bondwire fusing data given in "Fusing Currents of Bond Wires," *Semiconductor Reliability News,* Vol. VIII, #12, 1996, p. 8. See also B. Krabbenborg, "High Current Bond Design Rules Based on Bond Pad Degradation and Fusing of the Wire," *Microelectronics Reliability,* Vol. 39, 1999, pp. 77–88.

14.3.4. Minimizing Stress Effects

The coefficients of thermal expansion of packaging materials rarely match that of silicon, and encapsulation usually takes place at an elevated temperature (Section 7.2.9). The stresses that accumulate as the die cools become permanently frozen into the finished part. On larger dice, or on those mounted using solder or gold eutectic, the stress levels at the corners of the die may become severe enough to damage the die.[8] Common forms of damage include sheared bondwires, broken metal traces, and delamination of the protective overcoat from the underlying metallization.

Some assembly sites prohibit the placement of circuitry in the stress-prone regions around the corners of the die. The prohibited regions usually take the form of *stress triangles* extending some 100–250 μm from each corner of the die (Figure 14.14). The designer should not place circuit components, leads, or bondpads in these triangles. Test structures can still reside in the stress triangles, as can identification markings and alignment structures.

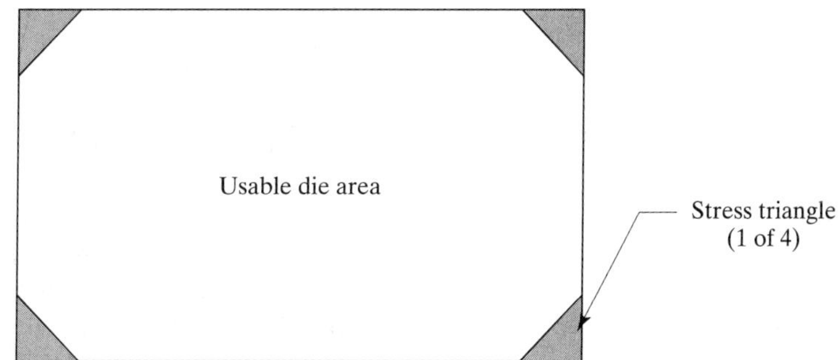

FIGURE 14.14 Layout of a die showing the placement of stress triangles.

Stress triangles are usually added to large dice, but not to small ones. The benefits of adding stress triangles to a large die outweigh the relatively small percentage of die area they consume. The stress levels on a small die are lower, yet the stress triangles consume a larger fraction of the available area. At some point, the triangles become more burdensome than they are worth. Most dice under 5 mm^2 do not incorporate stress triangles, and those under 10 mm^2 often leave them off unless the die experiences unusually stressful conditions, such as solder mounting or gold eutectic bonding.

Certain precautions should always be taken when placing components in the corners of a die. Matched components should never occupy the corners of a die (Section 7.2.10). If possible, one should also avoid placing bondpads in the corners of a die to minimize the possibility of sheared bonds. This prohibition does not apply to trimpads and testpads since neither of these receives bondwires.

Leads should not make right angles near a corner of the die. Stresses concentrate on the outside vertex of such a lead, and this can, in turn, lead to delamination and cracking of the protective overcoat. The designer should insert a short 45° segment into each such lead to help distribute the stresses more evenly (Figure 14.15A). This precaution helps prevent delamination and subsequent damage to the metal system.

---

[8] J. R. Dale and R. C. Oldfield, "Mechanical Stresses Likely to be Encountered in the Manufacture and Use of Plastically Encapsulated Devices," *Microelectronics and Reliability,* Vol. 16, 1977, pp. 255–258.

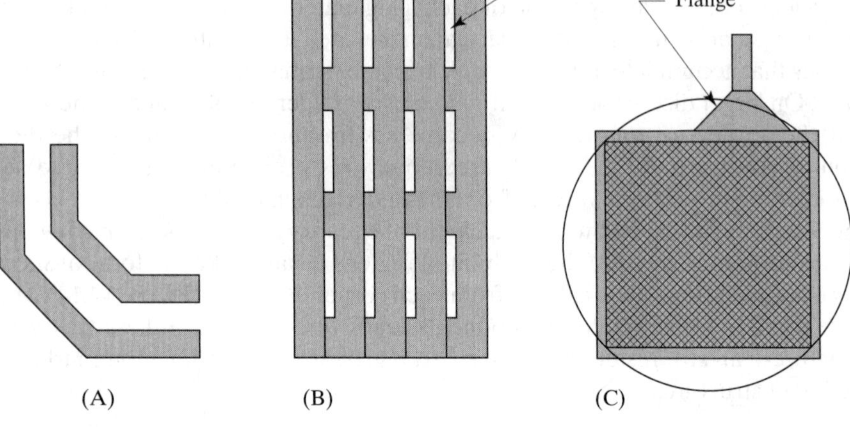

FIGURE 14.15 Various stress-reduction techniques: (A) 45° bends in leads, (B) slots in a wide metal-2 lead, and (C) flanged bondpad.

The presence of large areas of top-level metal near the corners of a die can also cause delamination. The planarization techniques that improve the step coverage of upper level metal also eliminate the irregularities that help lock the protective overcoat to the die. The addition of an array of narrow slots in the metal helps restore these irregularities and consequently discourages delamination. The slots should be oriented in the same direction that the current flows through the lead so that they do not greatly reduce its effective cross-sectional area (Figure 14.15B). Leads less than 25 μm wide generally do not require slots. An array of vias placed between two levels of metal also produces surface irregularities that will lock the protective overcoat in much the same way that the addition of slots will. Thus, arrays of vias do not require the addition of slots.

The bonding process also produces high levels of stress. Although this stress only lasts for an instant, it can still damage adjacent metal traces or components. Most processes therefore define a circular exclusion zone around each ball bond and a rectangular or trapezoidal exclusion zone around each wedge bond (Section 13.4.2). Leads that do not connect to the bondpad should not enter this zone, and active circuitry should also reside outside it. Many designers also add a triangular or rectangular *flange* of metal to each lead entering a bondpad (Figure 14.15C). This helps ensure that the stresses associated with bonding do not sever the lead from the bondpad. Modern bonding equipment has minimized the need for these flanges by more accurately controlling the forces applied during bonding, but the flanges consume so little area that it seems unreasonable to eliminate them.

14.4 CONCLUSION

A layout designer must understand not only how to design individual components but also how to interconnect these components to form a complete die. This chapter has touched upon the skills required to predict the size and structure of the die, and those required to actually assemble it. These are perhaps the most difficult of all of the skills the layout designer must learn. They partake of the essential character of all analog layout, in that they are largely arts and not sciences. As such, they require a certain degree of intuition and sound judgment that can only be learned through experience. The information presented here represents only a foundation, and the designer must continue to learn through actually practicing the art of layout.

14.5 EXERCISES

Refer to Appendix C for layout rules and process specifications.

14.1. Estimate the area of the following components in a standard bipolar process. (show all computations and state all assumptions used in the estimations):
a. 250 kΩ of 8 μm-wide HSR resistance.
b. A minimum-area lateral PNP transistor.
c. 100 pF of junction capacitance (assume 1.8 fF/μm^2).
d. A power NPN transistor with an emitter area of 1200 μm^2.
e. A bondpad for 30 μm-diameter ballbonded gold wire.

14.2. Estimate the total area of a circuit laid out in standard bipolar, using single-level metallization containing 410 kΩ of 6 μm HSR resistance, 55 kΩ of 8 μm base resistance, 50 pF of junction capacitance (at 1.8 fF/μm^2), eleven minimum NPN transistors, seven minimum lateral PNP transistors, two split-collector lateral PNP transistors, one 4X lateral PNP, and one power NPN transistor requiring an emitter area of 660 μm^2. Justify your choice of packing factor.

14.3. Compute the core area of a design containing six cells having the following areas: 0.35, 0.27, 0.21, 0.18, 0.10, and 0.08 mm^2. The design also includes two power transistors that each consume 0.77 mm^2. The design will be laid out in double-level metal and contains about 50 top-level signals. Justify your choices of routing and packing factors.

14.4. Compute the estimated die area for each of the following designs:
a. Core area of 10.1 mm^2 using 23 pads.
b. Core area of 1.49 mm^2 using 42 pads.
c. Core area of 8.8 mm^2 using 11 pads.

Assume that all designs require a bondpad width of 75 μm, a spacing between adjacent bondpads of 65 μm, and a scribe width of 75 μm. State whether each design is core-limited or pad-limited.

14.5. Using the parameters specified in Exercise 14.4, estimate the number of bondpads that can be packed around the periphery of a square die having a total area of 3.9 mm^2. How many additional pads could be obtained if the die had an aspect ratio of 1.5:1?

14.6. A device with a die area of 7.6 mm^2 is fabricated on 150 mm^2 wafers. Assume a wafer utilization factor of 0.85.
a. Approximately how many dice will each wafer yield?
b. If each wafer costs $650 and yields 77% functional dice, what is the cost of a single functional die?
c. If packaging and final testing cost 11.5¢ per device and 97% of devices pass the final test, what is the cost of a completed integrated circuit?
d. If the device is sold for 83¢, what is the GPM?
e. What would be a more reasonable sale price, and why?

14.7. Suppose an improved testing technique increases the probe yield of the device in Exercise 14.6 to 92%. This technique adds 4¢ to the cost of each die probed. Is it worthwhile to implement the new technique? Why or why not?

14.8. A certain integrated circuit contains four operational amplifiers, each with a cell area of 0.41 mm^2. The circuit also contains a bias circuit that consumes 0.15 mm^2. This circuit requires 14 bondpads: VCC, VEE, IN1+, IN1−, OUT1, IN2+, IN2−, OUT2, IN3+, IN3−, OUT3, IN4+, IN4−, and OUT4. VCC and VEE are the power and ground pins, respectively. The remaining pins are the inputs and outputs of each of the four amplifiers. The package has 14 pins, with pin #1 bonded to the top center of the die and the remaining pins arranged in counterclockwise order.
a. Suggest a reasonable pinout for this device.
b. Draw a die floorplan showing the placement of each of the five cells of this design.
c. Each amplifier requires high-current leads to VCC, VEE and its output pin. Assuming that the design uses only a single level of metallization, draw a diagram illustrating a suitable routing pattern.

14.9. A voltage reference circuit contains a reference cell occupying 0.13 mm^2 and a small power transistor requiring 0.16 mm^2.

a. Assuming the die uses single-level metallization and a hand-packed layout, estimate the core area of the die.
b. Estimate the complete die area, assuming a 110 μm-wide scribe and a total of three bondpads, each requiring a total area of 9000 μm (pad plus ESD protection).
c. Draw a die floorplan for this device. The power transistor must connect to the VIN and VOUT pads, which are at the top right and bottom right of the die, respectively. The reference cell must connect to the GND pin, which must lie somewhere along the left side of the die.
d. Indicate the preferred location for critical matched devices.

14.10. An octal buffer circuit contains eight buffers, each of which requires an area of 0.09 mm^2. Each buffer has an input and an output, and all require connections to VCC and GND.

a. Assuming that the die uses single-level metallization and must be assembled quickly, estimate the core area of the die.
b. Estimate the total die area, assuming a 110 μm-wide scribe and a total of 18 pads, each requiring 9000 μm of area.
c. Draw a die floorplan for this device. Pin #9 must be GND and pin #18 must be PWR; suggest a reasonable arrangement for the eight inputs and eight outputs. Assume pad #1 must occupy the top left corner of the die and the remaining pads are arranged in counterclockwise order.
d. Each buffer requires high-current connections to VCC, GND, and its output pin. Draw a diagram of a routing pattern that will connect all eight cells without requiring any tunnels in power leads.

14.11. Calculate the width required to construct a wiring channel capable of holding 12 leads, using the CMOS layout rules of Appendix C.

14.12. Suppose the output lead of a buffer carries a digital signal with a duty cycle of 50%. When the output is on, it sources 360 mA of current, and when it is off it sinks negligible current.

a. If the metallization consists of 10 kÅ of copper-doped aluminum capable of carrying a constant current of $5 \cdot 10^5$ A/cm^2 at 85°C, how wide must the output lead be made in order for it to withstand electromigration?
b. How wide must the lead be made to safely cross an oxide step that may induce 30% metallization thinning?
c. How wide must the lead be made if the device will operate continuously at a junction temperature of 125°C? Assume an activation energy of 0.7 eV.

14.13. The power output of an amplifier circuit conducts an average current of 3.1A under worst-case conditions.

a. Assuming the part is packaged in plastic, how many 30 μm-diameter gold bondwires are required to safely conduct this current?
b. How many 50 μm-diameter aluminum wires would be required to conduct the same current?

Appendix A

Table of Acronyms Used in the Text

AC	Alternating Current
A/D	Analog to Digital
ARC	Anti-Reflective Coating
BCLDD	Buried-Channel Lightly Doped Drain
BiCMOS	Bipolar and Complementary Metal-Oxide-Semiconductor
BiFET	Bipolar and junction Field-Effect Transistor
BJT	Bipolar Junction Transistor
BOI	Base Over Isolation
BOX	Buried OXide
BPSG	BoroPhosphoSilicate Glass
CDI	Collector Diffused Isolation
CDM	Charged Device Model
CMOS	Complementary Metal-Oxide-Semiconductor
CMP	Chemical-Mechanical Polish
CTE	Coefficient of Thermal Expansion
CVD	Chemical Vapor Deposited
D/A	Digital to Analog
DC	Direct Current
DCML	Differential Current-Mode Logic
DDD	Double-Diffused Drain
DI	Dielectric Isolation
DIP	Dual In-line Package
DLM	Double-Level Metal
DMOS	Double-Diffused Metal-Oxide-Semiconductor

DRAM	Dynamic Random-Access Memory
DSW	Direct Step on Wafer
ECGR	Electron Collecting Guard Ring
ECL	Emitter-Coupled Logic
EEPROM	Electrically Erasable Programmable Read-Only Memory
EOS	Electrical OverStress
EPROM	Erasable Programmable Read-Only Memory
ESD	ElectroStatic Discharge
FAMOS	Floating-Gate Avalanche-injection Metal-Oxide-Semiconductor*
FBSOA	Forward-Biased Safe Operating Area
FET	Field-Effect Transistor
FLOTOX	FLOating-Gate Tunneling OXide
GCNMOS	Gate-Coupled N-channel Metal-Oxide-Semiconductor
GGNMOS	Grounded-Gate N-channel Metal-Oxide-Semiconductor
GIDL	Gate-Induced Drain Leakage
GOI	Gate Oxide Integrity
GPM	Gross Profit Margin
HBGR	Hole Blocking Guard Ring
HBM	Human-Body Model
HBT	Heterojunction Bipolar Transistor
HCGR	Hole Collecting Guard Ring
HF	Hydrofluoric acid (chemical formula)
HSD	High-Side Drive
HSR	High-Sheet Resistor
IC	Integrated Circuit
ILO	InterLevel Oxide
IPTAT	Current Proportional to Absolute Temperature
JFET	Junction Field-Effect Transistor
JI	Junction Isolation
LDD	Lightly Doped Drain
LDMOS	Lateral Double-Diffused Metal-Oxide-Semiconductor
LDO	Low DropOut
LED	Light-Emitting Diode
LOCOS	LOCal Oxidation of Silicon
LPCVD	Low-Pressure Chemical Vapor Deposition
LSD	Low-Side Drive
LSTTL	Low-Power Schottky-clamped Transistor-Transistor Logic
MLO	MultiLevel Oxide
MM	Machine Model

*This term is a trademark of Intel Corporation.

MOS	Metal-Oxide-Semiconductor
MOSFET	Metal-Oxide-Semiconductor Field-Effect Transistor
NBL	N-Type Buried Layer
NBTI	Negative Bias Temperature Instability
NMOS	N-Channel Metal-Oxide-Semiconductor
NSD	N-Type Source/Drain
NVM	Non Volatile Memory
ONO	Oxide-Nitride-Oxide
OR	Oxide Removal
OTP	One-Time Programmable
PBL	P-Type Buried Layer
PBTI	Positive Bias Temperature Instability
PG	Pattern Generation
PMOS	P-Channel Metal-Oxide-Semiconductor
PO	Protective Overcoat
PPM	Parts per Million
PROM	Programmable Read-Only Memory
PSD	P-Type Source/Drain
PSG	PhosphoSilicate Glass
Q	Quality factor
RESURF	REduced SURface Field
RF	Radio Frequency
RIE	Reactive Ion Etch(ing)
ROM	Read-Only Memory
SCL	Space Charge Layer
SCR	Silicon Controlled Rectifier
SDD	Single-Diffused Drain (or Single-Doped Drain)
SI	*Systéme Internationale* (the metric system)
SILC	Stress-Induced Leakage Current
SIMOX	Separation by IMplanted OXygen
SLM	Single-Level Metal
SOA	Safe Operating Area
SOG	Spin-On Glass
SOI	Silicon On Insulator
SOIC	Small-Outline Integrated Circuit
SOS	Silicon on Sapphire
SPICE	Simulation Program with Integrated Circuit Emphasis
SSA	Super Self-Aligned
STI	Shallow Trench Isolation
TCR	Temperature Coefficient of Resistivity
TDDB	Time-Dependent Dielectric Breakdown

TEOS	TetraEthOxySilane
TTL	Transistor-Transistor Logic
UHVCVD	Ultra-High Vacuum Chemical Vapor Deposition
UV	UltraViolet
VDMOS	Vertical Double-Diffused Metal-Oxide-Semiconductor
VLSI	Very Large-Scale Integration
VPTAT	Voltage Proportional to Absolute Temperature

Appendix

B The Miller Indices of a Cubic Crystal

A *crystal* consists of an orderly arrangement of atoms or molecules extending indefinitely in all directions. This arrangement is periodic in the sense that the same patterns of atoms or molecules reappear at regular intervals along certain axes. One can imagine the crystal consisting of a large number of submicroscopic building blocks called *unit cells.* In the case of crystals belonging to the cubic system, the unit cells are tiny cubes that stack in rows and columns to create a rectilinear crystal lattice (Figure B.1). As the illustration suggests, the resulting crystal is not always cubic in shape. In every case, the underlying periodic structure of the crystal remains invariant regardless of its external form or dimensions.

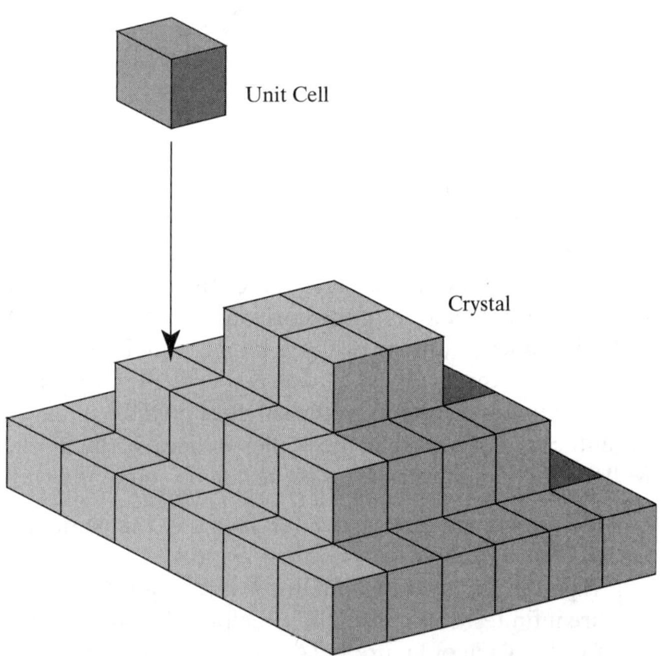

FIGURE B.1 Relationship between a cubic crystal and its unit cell.

A silicon ingot is a single cubic crystal, and each wafer cut from the ingot constitutes a slice cut from this crystal. The properties of the wafer depend on the angle of the cut relative to the axes of the crystal lattice. A set of numbers called *Miller indices* are useful for labeling various cuts of silicon and for labeling directions on the surface of a wafer. This appendix discusses the system of Miller indices used for cubic crystals, including those of silicon and germanium.

A plane intersecting a cubic crystal can be assigned a set of three Miller indices that together specify its orientation relative to the crystal axes. In order to compute these indices, one must first identify the three crystal axes. These lie orthogonal to one another and correspond to the X, Y, and Z axes of the Cartesian coordinate system. The cubic unit cells stack in neat rows and columns aligning to these three axes. The location of any point can be given in terms of multiples of the width of the unit cell along each of the three crystal axes. A plane intersecting the lattice can now be described in terms of its X, Y, and Z intercepts. For example, the intercepts of the plane of Figure B.2A are $X = 1$, $Y = 3$, and $Z = 3$. Similarly, the intercepts of the plane of Figure B.2B are $X = 3$, $Y = 2$, and $Z = 2$.

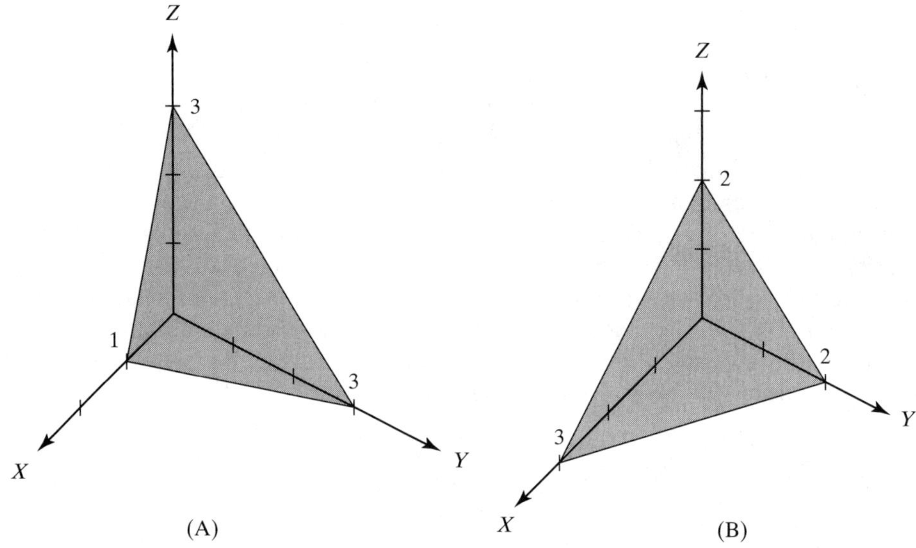

FIGURE B.2 Examples of two planes that intersect a cubic lattice. The tick marks represent multiples of the unit cell dimension.

A crystal plane may lie parallel to one or more of the crystal axes, in which case its intercepts with those planes lie at infinity. For example, the intercepts of a horizontal plane are $X = \infty$, $Y = \infty$, and $Z = 1$. Miller indices avoid infinite entries by using the reciprocals of the intercepts rather than the intercepts themselves. For example, the reciprocal intercepts of the horizontal plane are $X = 0$, $Y = 0$, and $Z = 1$. A set of Miller indices consists of three integer numbers corresponding to the reciprocal intercepts for the X, Y, and Z axes, respectively. These three reciprocal intercepts are always expressed in terms of the smallest possible integer values and are enclosed in parentheses. For example, the Miller indices of the horizontal plane are (001). The Miller indices of any arbitrary plane can be computed with the following rules:

1. Determine the X, Y, and Z intercepts of the plane. The intercepts of the plane of Figure B.2A are $X = 1$, $Y = 3$, and $Z = 3$.
2. Take the reciprocals of the three intercepts. If one or more of the intercepts are infinite, assume that their reciprocals equal zero. The reciprocal intercepts for the plane of Figure B.2A are $X = 1$, $Y = 1/3$, and $Z = 1/3$.

3. Multiply the reciprocal intercepts by their lowest common denominator to obtain three integers. The lowest common denominator for the reciprocal intercepts of the plane of Figure B.2A is three, so the new reciprocal intercepts are $X = 3, Y = 1$, and $Z = 1$.
4. Enclose the resulting reciprocal intercepts in parentheses to form the Miller indices. The Miller indices for the plane of Figure B.2A are (311). Those for the plane of Figure B.2B are (233). If one or more of the numbers happens to be negative, then a bar is placed over this number in the Miller indices.

Planes whose Miller indices are permutations of one another are said to be *equivalent.* For example, the (001), (010), and (100) planes are all equivalent. This does not mean that these are all one and the same plane; in fact, these three planes lie at right angles to one another. Rather, it means that all three of these planes have identical crystallographic properties, which also implies that they have identical chemical, mechanical, and electrical properties. Equivalent planes are denoted by a trio of Miller indices enclosed in braces. Thus, the set of equivalent planes {100} includes the (001), (010), and (100) planes. Remember that Miller indices enclosed in braces refer to a set of planes and not to any one specific plane. There is no such thing as a {100} wafer—the surface of the wafer might be a (001) plane or a (010) plane or a (100) plane, but it cannot be all three at the same time!

Miller indices can also be used to describe directions relative to the crystal lattice. The line passing perpendicularly through the plane (ABC) has the Miller indices [ABC]. The Miller indices for the *X*, *Y*, and *Z* axes are [100], [010], and [001], respectively. Directions whose Miller indices are permutations of one another are also said to be equivalent. For example, the directions [100], [010], and [001] are equivalent to one another. Equivalent directions are denoted by a trio of Miller indices placed in angle brackets. For example, the set of equivalent directions <100> includes the [100], [010], and [001] directions. Since an axis actually aligns to two direction vectors—that point in opposite directions—the Miller indices for an axis should be enclosed in angle brackets.

Appendix C

Sample Layout Rules

The rules given in this appendix were developed for use with the problems presented in the text. The numerical values given here do not correspond to any single process, but they are broadly representative of industry practices during the late 1980s. No attempt has been made to bring the rules in line with the current state of the art because this would greatly increase their complexity without providing any additional insight. For much the same reasons, many seldom-used rules have been omitted entirely or have been relegated to individual problem descriptions.

SECTION C.1 STANDARD BIPOLAR RULES

The standard bipolar process presented here is a 30 V process using a single P+ isolation. The isolation spacings for the process are significantly wider than the illustrations in the text might suggest. Modern processes generally use up–down isolation to reduce these spacings, bringing them more in line with the illustrations. Table C.1 lists key electrical parameters of this process.

The baseline process uses eight coding layers: NBL, TANK, DEEPN, BASE, EMIT, CONT, METAL1, and POR. The TANK and POR (protective overcoat removal) layers code for openings in the isolation diffusion and protective overcoat, respectively. Base over isolation (BOI) is automatically generated from the geometries coded on the TANK layer. Two process extensions are described: one producing 2 kΩ/□ HSR resistors and one generating Schottky diodes. Table C.2 lists the baseline layout rules, while Tables C.3 and C.4 list the rules for the HSR and Schottky contact process extensions, respectively. All of the rules are given in microns, assuming a coding grid of 2 μm. Section C.3 explains the syntax used to describe the layout rules.

TABLE C.1 Standard bipolar parametric specifications.

Parameter	Min	Mean	Max	Units
NPN beta	100	200	300	
NPN V_{EBO}	6.4	6.8	7.2	V
NPN V_{CBO}	40			V
NPN V_{CEO}	30			V
Base sheet	130	160	190	Ω/□
Emitter sheet	5	7	10	Ω/□
Pinched base sheet	1.5	3	4.5	kΩ/□
HSR sheet	1.6	2	2.4	kΩ/□
Thick-field threshold	35			V

TABLE C.2 Standard bipolar baseline rules.

Rule	Value
1. NBL width	8 μm
2. TANK width	8 μm
3. TANK spacing to TANK	6 μm
4. TANK overlap NBL	22 μm
5. DEEPN width	8 μm
6. TANK overlap DEEPN	24 μm
7. BASE width	6 μm
8. BASE spacing to DEEPN	18 μm
9. BASE spacing to BASE	14 μm
10. TANK overlap BASE	22 μm
11. EMIT width	6 μm
12. BASE spacing to EMIT	12 μm
13. EMIT spacing to EMIT	6 μm
14. TANK overlap EMIT	18 μm
15. BASE overlap EMIT	4 μm
16. CONT width	4 μm
17. EMIT spacing to CONT	6 μm
18. CONT spacing to CONT	4 μm
19. BASE overlap CONT	2 μm
20. EMIT overlap CONT	2 μm
21. METAL1 width	6 μm
22. METAL1 spacing to METAL1	4 μm
23. METAL1 overlap CONT	2 μm
24. POR width	10 μm
25. POR spacing to POR	10 μm
26. METAL1 overlap POR	4 μm

TABLE C.3 Standard bipolar HSR extension rules.

Rule	Value
27. HSR width	6 μm
28. HSR spacing to DEEPN	16 μm
29. HSR spacing to BASE	14 μm
30. HSR spacing to EMIT	10 μm
31. HSR spacing to CONT	4 μm
32. HSR spacing to HSR	12 μm
33. BASE overhang HSR	2 μm
34. HSR extends into BASE	2 μm
35. TANK overlap HSR	20 μm

TABLE C.4 Standard bipolar Schottky extension rules.

36. SCONT spacing DEEPN	12 μm
37. SCONT spacing to BASE	6 μm
38. SCONT spacing to EMIT	4 μm
39. SCONT spacing to HSR	4 μm
40. SCONT spacing to CONT	4 μm
41. SCONT spacing to SCONT	4 μm
42. SCONT overlap CONT	2 μm
43. METAL1 overlap SCONT	2 μm

SECTION C.2 POLYSILICON-GATE CMOS RULES

This section describes a 10 V, N-well, poly-gate CMOS process. The LDD NMOS has a minimum allowed channel length of 4 μm, while the SDD PMOS allows channel lengths as short as 3 μm. Both transistors use the same N-type gate poly, and a single boron threshold adjust sets both threshold voltages simultaneously. A combination of boron and phosphorus channel stops ensure that both the NMOS and the PMOS thick-field thresholds lie safely above the operating voltage of the process. CMOS latchup is minimized by the use of a P+ substrate and by optional availability of NBL as part of the analog BiCMOS process extension. This process cannot fabricate Schottky diodes because it uses titanium silicide to minimize contact resistance. Table C.5 lists key electrical parameters of this process.

TABLE C.5 Polysilicon-gate CMOS parametric specifications.

Parameter	**Min**	**Mean**	**Max**	**Units**
NMOS V_t	0.5	0.7	0.9	V
NMOS k	50	70	90	$\mu A/V^2$
NMOS V_{DS}	10			V
NMOS V_{GS}	12			V
PMOS V_t	−0.9	−0.7	−0.5	V
PMOS k	17	25	33	$\mu A/V^2$
PMOS V_{DS}	12			V
PMOS V_{GS}	12			V
Thick-field thresholds	15			V
Poly-1 sheet	20	30	40	Ω/□
Poly-2 sheet (w/PSD)	160	200	240	Ω/□
Base sheet	400	500	600	Ω/□
Gate oxide capacitance	0.85	0.95	1.05	$fF/\mu m^2$
Poly-poly capacitance	1.3	1.5	1.7	$fF/\mu m^2$
NPN beta	40	80	120	
NPN V_{EBO}	7	8	9	V
NPN V_{CBO}	15			V
NPN V_{CEO}	12			V

The baseline process as described in Table C.6 uses 11 masks: NWELL, MOAT, NSD, PSD, CHST, POLY1, CONT, METAL1, VIA, METAL2, and POR. These 11 masks are normally coded by using nine drawing layers: NWELL, NMOAT, PMOAT, POLY1, CONT, METAL1, VIA, METAL2, and POR. The NMOAT drawing layer simultaneously produces geometries on the MOAT and NSD masks. The PMOAT drawing layer simultaneously produces geometries on the MOAT and PSD masks. The information for the CHST mask is obtained from the NWELL and MOAT coding layers. The geometries on the POR mask represent openings in the protective

TABLE C.6 Poly-gate CMOS baseline rules.

Rule	Value
1. NWELL width	5.0 μm
2. NWELL spacing to NWELL	15.0 μm
3. NMOAT width	3.0 μm
4. NMOAT spacing to NWELL	9.5 μm
5. NMOAT spacing to NMOAT	5.5 μm
6. NWELL overlap NMOAT	1.0 μm
7. PMOAT width	3.0 μm
8. PMOAT spacing to NWELL	7.0 μm
9. PMOAT spacing to NMOAT *(note 1)*	4.0 μm
10. PMOAT spacing to PMOAT	5.5 μm
11. NWELL overlap PMOAT	2.0 μm
12. POLY1 width	2.0 μm
13. POLY1 spacing to NMOAT	2.0 μm
14. POLY1 spacing to PMOAT	2.0 μm
15. POLY1 spacing to POLY1	2.0 μm
16. POLY1 overhang NMOAT	1.0 μm
17. POLY1 overhang PMOAT	1.0 μm
18. NMOAT overhang POLY1	4.0 μm
19. PMOAT overhang POLY1	4.0 μm
20. CONT width	1.0 μm exactly
21. CONT spacing to POLY1	2.0 μm
22. CONT spacing to CONT	2.0 μm
23. NMOAT overlap CONT	1.0 μm
24. PMOAT overlap CONT	1.0 μm
25. POLY1 overlap CONT	1.0 μm
26. METAL1 width	2.0 μm
27. METAL1 spacing to METAL1	2.0 μm
28. METAL1 overlap CONT	1.0 μm
29. VIA width *(note 2)*	1.0 μm exactly
30. VIA spacing to CONT *(note 3)*	2.0 μm
31. VIA spacing to VIA	2.0 μm
32. METAL1 overlap VIA	1.0 μm
33. METAL2 width	2.0 μm
34. METAL2 spacing to METAL2	2.0 μm
35. METAL2 overlap VIA	1.0 μm
36. POR width	4.0 μm
37. POR spacing to POR	4.0 μm
38. METAL2 overlap POR	2.0 μm

Notes: [1] PMOAT allowed to abut NMOAT if the two are connected by METAL1.
[2] Except for bondpads.
[3] VIA must not touch CONT.

overcoat. All of these layout rules are given in microns and presume a coding grid of 0.5 μm. Section C.3 explains the syntax used to describe the layout rules.

This process supports two extensions, the first of which adds a second layer of poly-silicon to the process, using the POLY2 mask. This poly is deposited in a near-intrinsic state and is doped with PSD, using the PSD drawing layer to produce poly resistors. A thin oxide-nitride-oxide dielectric deposited between the two polysilicon layers provides poly-poly capacitors. Table C.7 lists all of the rules associated with this extension.

TABLE C.7 Poly-2 extension rules.

Rule	Value
39. POLY2 width	2.0 μm
40. POLY2 spacing to POLY2	2.0 μm
41. NSD spacing to POLY2	2.0 μm
42. PSD spacing to POLY2	2.0 μm
43. POLY2 overlap CONT	1.0 μm
44. POLY1 overlap POLY2	1.5 μm
45. NSD overlap POLY2	1.5 μm
46. PSD overlap POLY2	1.5 μm
47. PSD spacing to NMOAT	2.0 μm
48. NSD spacing to PMOAT	2.0 μm

The second process extension adds analog BiCMOS functionality, using three additional masks: NBL, DEEPN, and BASE. Data coded on the BASE drawing layer automatically generates geometries on both the MOAT and the BASE masks. The addition of an N-buried layer forces the process to incorporate a second epitaxial layer for compatibility with a P+ substrate. This process extension allows the creation of vertical NPN transistors, lateral PNP transistors, and substrate PNP transistors. Table C.8 lists the layout rules required to construct analog BiCMOS structures.

TABLE C.8 BiCMOS extension rules.

Rule	Value
47. NBL width	5.0 μm
48. NBL spacing to NBL	19.0 μm
49. NWELL spacing to NBL	16.5 μm
50. NWELL overlap NBL	5.0 μm
51. DEEPN width	5.0 μm
52. DEEPN spacing to NMOAT	6.0 μm
53. DEEPN spacing to PMOAT	7.5 μm
54. DEEPN spacing to DEEPN	9.0 μm
55. NWELL overlap DEEPN	2.0 μm
56. BASE width	3.0 μm
57. BASE spacing to NMOAT	4.5 μm
58. BASE spacing to PMOAT	6.0 μm
59. BASE spacing to DEEPN	8.0 μm
60. BASE spacing to BASE	6.5 μm
61. BASE overlap NSD	1.5 μm
62. PSD overlap BASE	1.0 μm
63. NWELL overlap BASE	3.0 μm
64. NSD spacing to CONT	2.5 μm
65. NSD spacing to PSD	3.0 μm
66. NSD spacing to PSD	3.5 μm
67. NSD overlap CONT	1.0 μm
68. PSD overlap CONT	1.0 μm
69. BASE overlap CONT	1.0 μm

SECTION C.3 LAYOUT RULE SYNTAX

The layout rules listed in this appendix follow a format that is similar (but not identical) to that of Chameleon, a layout verification program developed at Texas Instruments beginning in 1976 and currently supported and marketed by K2 Technologies. This notation resembles plain English and is therefore particularly easy for novices to understand. Those employing other verification programs should have little difficulty translating the rules into the appropriate format, particularly since all of the rules fall into one of five elementary categories: width, spacing, overlap, overhang, and extent into. The next sections explain each of these types of rules.

Width

A width check verifies that all dimensions of every geometry on a given layer equal or exceed a minimum feature size. The syntax for a width check is

LAYER1 width   *N* μm

where *N* is the minimum allowed dimension of all geometries on LAYER1. In order for the geometry of Figure C.1 to pass the above width check, dimensions A to D must all equal or exceed the minimum width, *N*. Occasionally the dimension will be followed by the notation "exactly." In such cases, the width of every figure on the indicated layer must equal exactly *N* microns, neither more nor less.

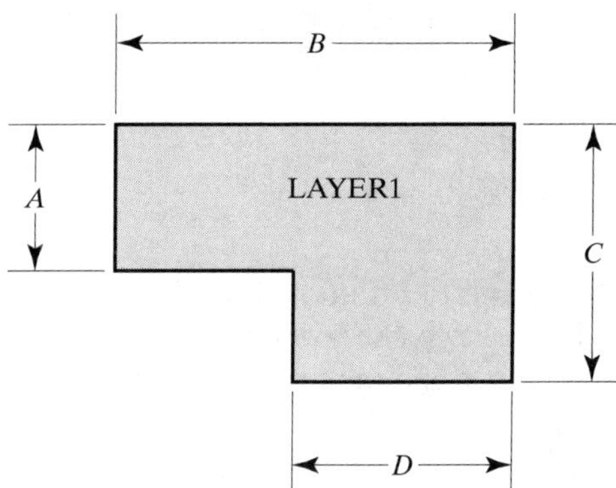

FIGURE C.1 Dimensions checked by "LAYER1 width."

Spacing

A spacing check verifies that all geometries on the specified layers maintain a minimum separation from one another. The syntax for a spacing check is

LAYER1 spacing to LAYER2   *N* μm

This check determines the minimum distance from each geometry on LAYER1 to each geometry on LAYER2. A violation occurs if any of these distances is less than *N*. Elements that touch or overlap do not violate the spacing. A spacing check may also be applied to a single layer, using the syntax

LAYER1 spacing to LAYER1   *N* μm

In this case, the check applies not only to the separation between any two geometries on LAYER1, but also to separate portions of a single geometry, such as the turns of a serpentine resistor. (For an example, see dimension C in Figure C.2.)

Overlap

An overlap check only applies to geometries on a first layer that are partially or wholly enclosed by geometries on a second layer (Figure C.3). The syntax of an overlap check is

LAYER1 overlap LAYER2   *N* μm

Wherever a geometry on LAYER1 encloses a geometry on LAYER2, the amount by which the former geometry overlaps the latter must equal or exceed *N*. If

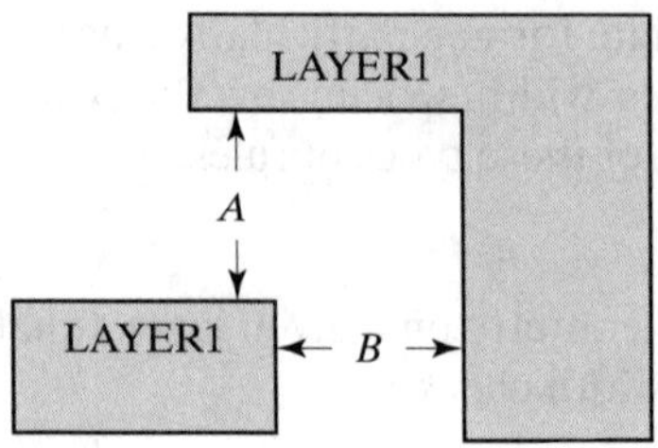

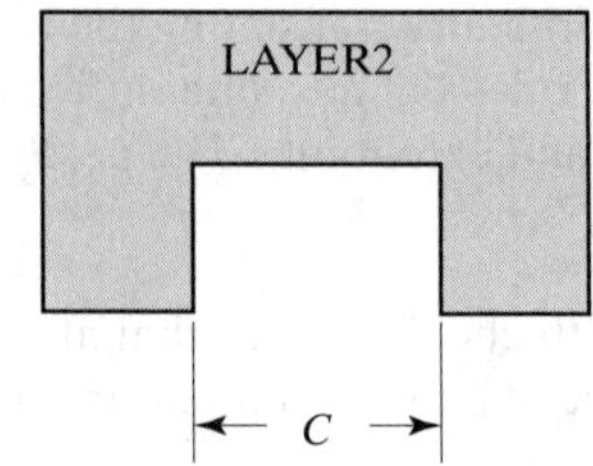

FIGURE C.2 Dimensions checked by "LAYER1 spacing to LAYER1."

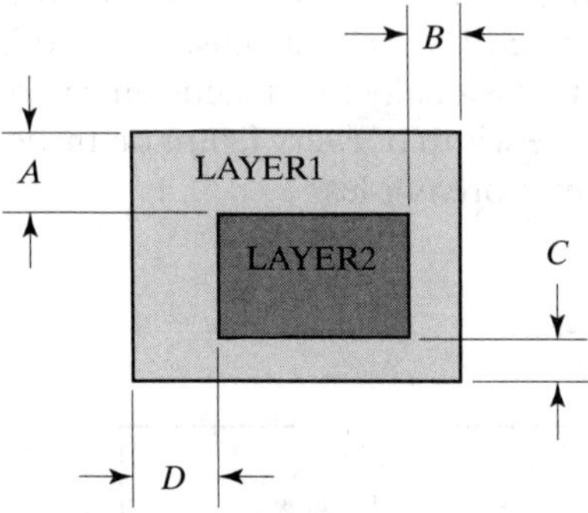

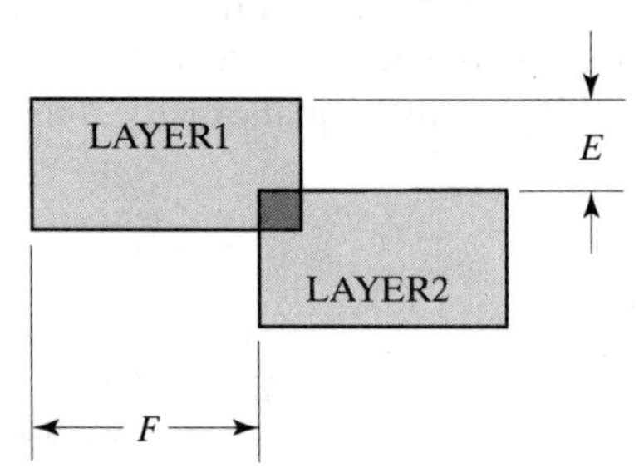

FIGURE C.3 Dimensions checked by "LAYER1 overlap LAYER2."

the geometries do not completely overlap, the sides that do not overlap automatically pass the check. Figure C.3 shows examples of fully and partially overlapped geometries and the dimensions checked in each case.

Overhang

An overhang check applies only to cases where a geometry on one layer partially overlaps a geometry on another (Figure C.4). The syntax of an overhang check is

LAYER1 overhang LAYER2 N μm

Wherever a geometry on LAYER1 partially overlaps a geometry on LAYER2, the amount by which the former geometry extends beyond the latter must equal or exceed N. Cases where the former geometry is fully enclosed by the latter automatically pass the check.

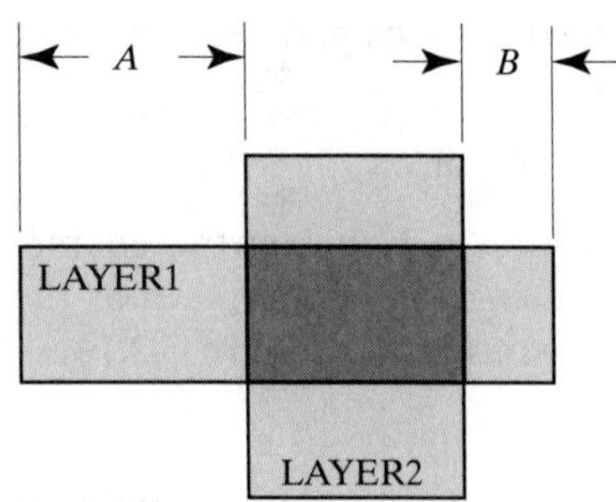

FIGURE C.4 Dimensions checked by "LAYER1 overhang LAYER2."

Extent Into

An extent into check applies only to cases where a geometry on one layer partially overlaps a geometry on another (Figure C.5). The syntax of an extent into check is

LAYER1 extends into LAYER2 N μm

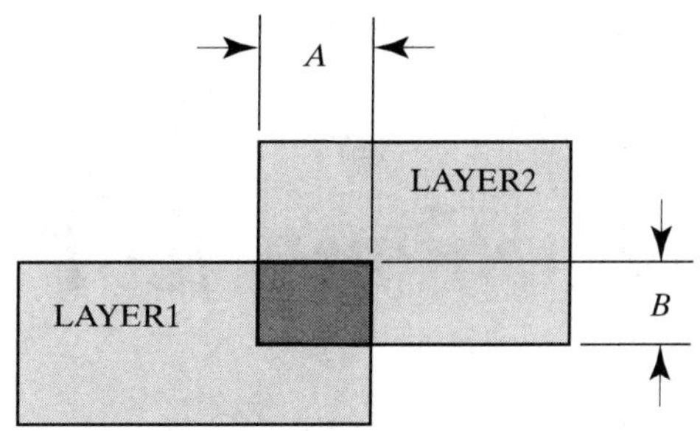

FIGURE C.5 Dimensions checked by "LAYER1 extends into LAYER2."

Wherever a geometry on LAYER1 partially overlaps a geometry on LAYER2, the amount by which the former geometry extends into the latter must equal or exceed N. Cases where the former geometry is fully enclosed by the latter automatically pass the check.

Appendix

D Mathematical Derivations

Equations 11.10A–B, 12.7–12.9, 12.14, and 12.15 have been derived specifically for this text. This appendix contains the details of these derivations, which were deemed too complicated for inclusion within the body of the text.

EQUATIONS 11.10A–B

In a circularly symmetric device, current flows radially from source to drain. The length of the channel therefore equals the difference between the radii of the outer and inner edges of the channel, or $L = 1/2(B - A)$. The width of the channel W can be determined from the Shichman-Hodges equation for the linear region. Rearranging this equation in terms of L gives

$$L = k'\frac{W}{I_D}\left(V_{\text{gst}} - \frac{V_{\text{DS}}}{2}\right)V_{\text{DS}} \qquad \text{[D.6]}$$

Differentiating to find dL/dV_{DS} gives

$$\frac{dL}{dV_{\text{DS}}} = k'\frac{W}{I_D}(V_{\text{gst}} - V_{\text{DS}}) \qquad \text{[D.7]}$$

For any infinitesimal length of channel, dL, the corresponding width, W, equals $2\pi L$, where $L = 0$ denotes the center of the annular structure. Separating terms and integrating gives

$$\int_0^{V_{\text{DS}}}(V_{\text{gst}} - V_{DS})\,dV_{DS} = \frac{I_D}{k'}\int_{A/2}^{B/2}\frac{dL}{2\pi L} \qquad \text{[D.8]}$$

The limits of these integrals assume that the source is at the inner edge of the channel ($L = A/2$) and the drain is at the outer edge of the channel ($L = B/2$).

Integrating gives

$$V_{gst}V_{DS} - \frac{V_{DS}}{2} = \frac{I_D}{2\pi k'}\ln(B/A) \quad \text{[D.9]}$$

Gathering terms,

$$I_D = \frac{2\pi k'}{\ln(B/A)}\left(V_{gst} - \frac{V_{DS}}{2}\right)V_{DS} \quad \text{[D.10]}$$

This equation is analogous to the Shichman-Hodges equation for the linear region with a *W/L* ratio equal to

$$\frac{W}{L} = \frac{2\pi}{\ln(B/A)} \quad \text{[D.11]}$$

Substituting the previously determined value of L gives

$$W = \frac{\pi(B - A)}{\ln(B/A)} \quad \text{[D.12]}$$

EQUATION 12.7

Let a source/drain finger consist of a uniform rectangular strip having width W, length L, and sheet resistance, R_s. Let variable x define position along the length of the finger. Assume that an equal amount of current flows into the finger at every point along its length, so the current $I(x)$ increases linearly from $x = 0$ to $x = L$. Let I_{max} equal the current $I(x)$ at $x = L$. The voltage drop V from $x = 0$ to $x = L$ equals

$$V = \int_0^L \frac{R_s I(x)}{W}dx \quad \text{[D.13]}$$

Since $I(x) = I_{max}x/L$, this reduces to

$$V = \frac{R_s I_{max}}{WL}\int_0^L x\,dx = \frac{R_s I_{max} L}{2W} \quad \text{[D.14]}$$

$$V = \frac{R_s L}{2W}I_{max} \quad \text{[D.15]}$$

The resistance of any source/drain finger may be divided into three components: (1) the portion of the finger consisting only of metal-1, (2) the portion of the finger consisting of a sandwich of metal-1 and metal-2, and (3) the portion of the finger under a plate of metal-2. If we assume that the metal-2 buses are equipotential across their width, then the resistance of the finger consists of only components (1) and (2). The voltage drop V_1 across component (1) equals

$$V_1 = \frac{R_{s1}B}{2W}I_1 \quad \text{[D.16]}$$

where I_1 equals the current flowing through component (1) of a single source/drain finger. Since component (1) has a length of B, while the full length of one finger equals L, and since there are N_D pairs of source/drain fingers, the current I_1 equals

$$I_1 = \frac{B}{LN_D}I_D \quad \text{[D.17]}$$

where I_D equals the total drain current of all fingers. Substituting equation D.17 into equation D.16 yields

$$V_1 = \frac{R_{s1}B^2}{2WLN_D} I_D \qquad \text{[D.18]}$$

The voltage drop V_2 across component (2) can likewise be computed as

$$R_2 = \frac{R_{s12}A}{2W} I_2 + \frac{R_{s12}A}{W} I_1 \qquad \text{[D.19]}$$

where I_2 equals the total current entering the finger within component (2) and I_1 equals the current flowing into component (2) from component (1). I_2 can be computed in a similar manner to that used to determine I_1. Substituting I_1 and I_2 into equation D.19 gives

$$V_2 = \frac{R_{s12}A}{2W}\left(\frac{A}{LN_D}\right) I_D + \frac{R_{s12}A}{W}\left(\frac{B}{LN_D}\right) I_D \qquad \text{[D.20]}$$

$$V_2 = \frac{R_{s12}A}{2W}\left(\frac{A + 2B}{LN_D}\right) I_D \qquad \text{[D.21]}$$

But since $L \equiv A + 2B$, it follows that

$$R_2 = \frac{R_{s12}A}{2WN_D} I_D \qquad \text{[D.22]}$$

The total voltage drop across the fingers of the transistor equals twice the sum of V_1 and V_2 because there are both source and drain fingers involved. To this must be added the voltage drop V_B across the metal-2 buses:

$$V_B = \frac{HR_{s2}}{2B} I_D \qquad \text{[D.23]}$$

The total voltage drop V_M therefore equals

$$V_M = \left(\frac{R_{s1}B^2}{WLN_D} + \frac{R_{s12}A}{WN_D} + \frac{HR_{s2}}{2B}\right) I_D \qquad \text{[D.24]}$$

Since the metallization resistance $R_M \equiv V_M/I_D$, it follows that

$$R_M \frac{R_{s1}B^2}{WLN_D} + \frac{R_{s12}A}{WN_D} + \frac{HR_{s2}}{2B} \qquad \text{[D.25]}$$

EQUATIONS 12.8 AND 12.9

To determine the optimum value of B, take $\partial R_M/\partial B$ and set it equal to zero. This determines an inflection point in the function $R_M(B)$. Begin by replacing A by $(L - 2B)$ and differentiating:

$$\frac{\partial R_M}{\partial B} = \frac{2BR_{s1}}{WN_DL} - \frac{2R_{s12}}{WN_D} \qquad \text{[D.26]}$$

Setting this equal to zero, we obtain

$$\frac{BR_{s1}}{L} = R_{s12} \tag{D.27}$$

$$\frac{B}{L} = \frac{R_{s12}}{R_{s1}} \tag{D.28}$$

Assuming both metal layers have resistivity ρ, $R_{S1} = \rho/t_1$, where t_1 is the thickness of metal-1; and $R_{S12} = \rho/(t_1 + t_2)$, where t_2 is the thickness of metal-2. Substituting these relationships into equation D.28 gives

$$\frac{B}{L} = \frac{t_1}{t_1 + t_2} \tag{D.29}$$

EQUATION 12.14

This equation is derived from the Shichman-Hodges equation for saturation. Let transistor M_1 have drain current I_{D1}, transconductance k_1, and effective gate voltage V_{gst1}. Let transistor M_2 have drain current I_{D2}, transconductance k_2, and effective gate voltage V_{gst2}. Since $I_{D1} \equiv I_{D2}$, we have

$$k_1 V_{gst1}^2 = k_2 V_{gst2}^2 \tag{D.30}$$

Rearranging gives

$$\frac{k_1}{k_2} = \left(\frac{V_{gst2}}{V_{gst1}}\right)^2 \tag{D.31}$$

Let $\Delta V_{gs} \equiv V_{gs1} - V_{gs2}$; then $V_{gs2} = V_{gs1} - \Delta V_{gs}$. Let $\Delta V_t \equiv V_{t1} - V_{t2}$; then $V_{t2} = V_{t1} - \Delta V_t$. Substituting these relationships into equation D.31 yields

$$\frac{k_1}{k_2} = \left(\frac{V_{gst2} - V_{t2}}{V_{gst1}}\right) = \left(\frac{V_{gs1} - \Delta V_{gs} - V_{t1} + \Delta V_t}{V_{gst1}}\right)^2 \tag{D.32}$$

$$\frac{k_1}{k_2} = \left(1 + \frac{\Delta V_t - \Delta V_{gs}}{V_{gst1}}\right)^2 \tag{D.33}$$

$$\sqrt{\frac{k_1}{k_2}} = 1 + \frac{\Delta V_t - \Delta V_{gs}}{V_{gst1}} \tag{D.34}$$

$$\left(\sqrt{\frac{k_1}{k_2}} - 1\right)V_{gst1} = \Delta V_t - \Delta V_{gs} \tag{D.35}$$

Collecting ΔV_{gs} yields

$$\Delta V_{gs} = \Delta V_t - V_{gst1}\left(\sqrt{\frac{k_1}{k_2}} - 1\right) \tag{D.36}$$

Let $\Delta k \equiv k_1 - k_2$; then $k_1 = k_2 + \Delta k$. Substituting this into equation D.36 gives

$$\Delta V_{gs} = \Delta V_t - V_{gst1}\left(\sqrt{1 + \frac{\Delta k}{k_2}} - 1\right) \tag{D.37}$$

By the binominal expansion,

$$\sqrt{1+x} = \sum_{n=0}^{\infty} \binom{1/2}{n} x^n = 1 + \frac{x}{2} - \frac{x^2}{8} + \frac{3x^3}{48} + \cdots \qquad \text{[D.38]}$$

If x is sufficiently small, then only the first two terms are significant. Applying this expansion to equation D.37 gives

$$\Delta V_{gs} \cong \Delta V_t - V_{gst1} \frac{\Delta k}{2k_2} \qquad \text{[D.39]}$$

EQUATION 12.15

This equation is derived from the Shichman-Hodges equation for saturation. Let transistor M_1 have drain current I_{D1}, transconductance k_1, and effective gate voltage V_{gst1}. Let transistor M_2 have drain current I_{D2}, transconductance k_2, and effective gate voltage V_{gst2}. The ratio of the two drain currents I_{D2}/I_{D1} equals

$$\frac{I_{D2}}{I_{D1}} = \frac{k_2}{k_1} \left(\frac{V_{gst2}}{V_{gst1}} \right)^2 \qquad \text{[D.40]}$$

Let $\Delta V_t \equiv V_{t1} - V_{t2}$; then $V_{t2} = V_{t1} - \Delta V_t$. Substituting this into equation D.40 gives

$$\frac{I_{D2}}{I_{D1}} = \frac{k_2}{k_1} \left(\frac{V_{gs2} - V_{t2}}{V_{gst1}} \right)^2 = \frac{k_2}{k_1} \left(\frac{V_{gs2} - V_{t1} + \Delta Vt}{V_{gst1}} \right)^2 \qquad \text{[D.41]}$$

But since $V_{gs1} \equiv V_{gs2}$, it follows that

$$\frac{I_{D2}}{I_{D1}} = \frac{k_2}{k_1} \left(\frac{V_{gs1} - V_{t1} + \Delta V_t}{V_{gst1}} \right)^2 = \frac{k_2}{k_1} \left(\frac{V_{gst1} + \Delta V_t}{V_{gst1}} \right)^2 \qquad \text{[D.42]}$$

Expanding yields

$$\frac{I_{D2}}{I_{D1}} = \frac{k_2}{k_1} \left(\frac{V_{gst1}^2 + 2\Delta V_t V_{gst1} + \Delta V_t^2}{V_{gst1}^2} \right) \qquad \text{[D.43]}$$

So long as $\Delta V_t << V_{gst1}$, we have

$$\frac{I_{D2}}{I_{D1}} \cong \frac{k_2}{k_1} \left(1 + \frac{2\Delta V_t}{V_{gst1}} \right) \qquad \text{[D.44]}$$

Appendix

E Sources of Layout Editor Software

This textbook does not discuss layout software, because these programs constantly evolve and any information given here would soon become obsolete. The following are several sources of layout editors and documentation:

- J. P. Uyemura, *Physical Design of CMOS Integrated Circuits Using L-EDIT™* (Boston: PWS Publishing Company, 1995). This text provides an excellent introduction to layout editing software, using the L-EDIT™ package developed and sold by Tanner Research. The text also includes a student version of the program on a floppy disk.
- Tanner Research, located at **http://www.tanner.com**. Tanner Research sells a professional version of L-EDIT that provides a much more capable editing environment than the student version included with *Physical Design of CMOS Integrated Circuits Using L-EDIT™*.
- Cadence Design Systems, located at **http://www.cadence.com**. Cadence sells Virtuoso, which is one of the most widely used layout editors in the industry.
- Mentor Graphics, located at **http://www.mentorg.com**. Mentor sells IC Graph, another widely used layout editor.

Index

A

Acceptors, 8
Accumulation, biasing mode, 239
Acronyms, 607–610
Adaptive RESURF, 510
Adjusted transistors, 454
Aging, 246
Aligner, 42
Alloy-42, 75, 274
Aluminum:
  metallization, 137
  resistance of, 71
Aluminum-copper-silicon (Al-Cu-Si) metal system, 196
Aluminum fuses, 217
Aluminum-gate processes, 447
Analog BiCMOS, 114–130
  available devices, 121–125
  essential features, 115–116
  fabrication sequence, 116–121
    base implant, 118
    channel stop implants, 119
    epitaxial growth, 117
    inverse moat, 118
    LOCOS processing and dummy gate oxidation, 119
    metallization and protective overcoat (PO), 120–121
    N-buried layer, 116–117
    N-well diffusion and deep-N+, 117–118
    polysilicon deposition and pattern, 120
    process comparison, 121
    source/drain implants, 120
    starting material, 116
    threshold adjust, 119–120
  NPN transistors, 121–123
  PNP transistors, 123–125
  process extensions, 125–130
    advanced metal systems, 126
    dielectric isolation, 126–130
  resistors, 125
Analog BiCMOS bipolar transistors, 347–349
Analog circuits, and hot carrier degradation, 483
Analog CMOS processes, 341
Anisotropic nature, of relative ion etching, 47
Annealing, 56
Annular transistors, 462–464, 518
Anode(s), 13–14, 232, 410
  Schottky diodes, 17
Antenna effect, 133, 141–143, 182
  areal antenna ratio, 141
  effects, 141–142
  peripheral antenna ratio, 141
  preventative measures, 142–143
Antimony, 7, 52, 58–59
Antiparallel diode clamps, 570
Antireflective coating (ARC), 268
Aqua regia, 67
Areal antenna ratio, 141
Areal capacitance, 231
Areal variations, 257
Arrayed-emitter transistor, 337
Arsenic, 7, 52
Ashing, 43
Asperities, 228
Assembly, semiconductors, 73–77
  Alloy-42, 75
  ball, 76–77
  ball bonding, 76–77
  bondpads, 76
  capillary, 76
  direct-step-on-wafer (DSW) processing, 74
  direct write on wafer (DWW), 74
  dual-in-line package (DIP), 75
  gold preform, 75
  lead frame, 75
  mount and bond, 74–75
  mount pad, 75
  packaging, 77
  probepads, 76
  process control structures, 73
  stitch bond, 77
  test dice, 73–74
  testpads, 76
  wafer-level testing (wafer probing), 74
  wedge bonding, 76
Assembly/test site, 73
Assembly yield, 588
Asymmetric extended-drain transistors:
  NMOS, 487
  PMOS, 488
Asymmetric junction field-effect transistors (JFETs), 34
Asymmetric metal-oxide-semiconductor (MOS) transistors, 27
Avalanche breakdown, bipolar junction transistors (BJTs), 308–310
Avalanche diodes, 19, 409. *See also* Zener diodes
Avalanche energy rating, 494
Avalanche-induced beta degradation, 153–154, 182
  avoiding, 353
  effects, 153–154
  preventative measures, 154

Avalanche multiplication, 18
Avalanche multiplication factor, 338

B

Backgate contacts, CMOS transistors, 464–467
Backgate doping, 435–436
Backside contact, 181
Ball, 76–77
Ball bonding, 76–77
Ballasting:
  base-side, 313–314
  emitter, 363, 365
Bamboo poly, 209
Bandgap energy, 4
Bandgap reference, 394
Bandgap voltage, 394
Bare dice, 74–75
Barium strontium titanate, 228
Base-collector capacitance, 349–350
Base-emitter junction capacitors, 237–239
BASE layer, 614–616
BASE mask, 618
Base over isolation (BOI), 159, 614
Base pinch resistors, 191, 202
Base punchthrough, 25
Base push-out, 355–356, 367
Base resistance, and bipolar transistors, 350–351
Base resistors, 90–91, 199
Base-side ballasting, 313–314
Base transit time, 351
Base tunnels, 555
Base turnoff resistors, for NPN transistors, 542
Bent-gate transistors, 505
Berkeley *k*, 434
Beta, 23–24
Beta multiplication, 309
Beta rolloff, 308
Bias, 256
BICMOS small-signal bipolar transistors, 341–358
  analog BiCMOS bipolar transistors, 347–349
  CMOS PNP transistors, 341–344
  fast bipolar transistors, 349–351
  oxide-isolated transistors, 354–356
  polysilicon-emitter transistors, 351–353
  shallow-well transistors, 345–346
  silicon-germanium transistors, 356–358
Binary weighted resistor segments, 218
Bipolar junction transistors (BJTs), 21–25, 35, 306–359, 361–381
  applications of, 360–405
  avalanche breakdown, 308–310
  base, 21
  base-emitter junction of, 307
  base punchthrough, 25
  behavior of, 24
  beta, 23–24
  beta rolloff, 308
  bipolar transistor matching rules, 396–402
  categories of, 358
  collector, 21
  collector-to-base breakdown, 25
  current gain, 23
  cutoff, 21–22
  Early effect, 25
  emitter, 21
  emitter injection efficiency, 23
  first appearance of, 1
  forward active region, 22
  Gummel number, 23
  high-level injection, 24
  high transconductance of, 360–361
  I-V characteristics, 24–25
  intrafinger debiasing, 363–364
  layout of, 358
  matching bipolar transistors, 381–395
  matching rules, 396–402
  operation, 306–320
  power bipolar transistors, 361–381
    failure mechanisms of NPN power transistors, 361–368
  reverse active region, 22
  saturation region, 24–25
  secondary breakdown, 311–312
  super-beta transistors, 25
  thermal runaway, 310
  transconductance, 360–361
Bipolar transistors, 491
  parasitics of, 318–320
Bird's beak, 49–50, 100
Bits, 467
BJTs, *See* Bipolar junction transistors (BJTs)
Black's Law, 600
Blanket process, 41
Blocking voltage:
  metal-oxide-semiconductor (MOS) transistors, 483
  rating, 105
Blown, defined, 216
Bond forcing, excessive, 560
Bond over active circuitry (BOAC), 73
Bondpad circles, 560
Bondpads, 76, 534, 558–562
Bondwires, 560, 602
Boron, 8, 52
Boron-doped silicon, 8
Boron suckup, 48
Borophosphosilicate glass (BPSG), 144
BOUNDARY layer, 579
BOX, *See* Buried oxide (BOX)
Breakdown voltage, 409
Buffered Zener clamps, 566–568
Built-in potential, 12
Buried channel, 455
Buried-channel lightly doped drain (BCLDD), 486

Buried layers:
  and epitaxy, 58–59
  ion implantation, 56
Buried oxide (BOX), 60–61, 148, 180, 346
Buried Zeners, 152, 412–415
Bytes, 467

C

Cadmium sulfide, 9
Capacitance:
  areal, 231
  base-collector, 349–350
  defined, 226
  dielectric, 229
  distributed, 197
  junction, 349–350
  mismatch coefficient of, 258
  peripheral, 231
  zero-bias, 230
Capacitor arrays, 302
Capacitor matching, rules for, 300–303
Capacitors, 226–253
  base-emitter junction, 237–239
  capacitance, 226–246
  comb, 231–232
  defined, 226
  dielectric, 227
  dielectric constant, 227
  dielectric strength, 227–228
  electrodes, 227
  fringing, 229–230
  gate oxide, 229
  high-permittivity, 246
  inductance, 246–251
  integrated, 252
  junction, 230–233, 252, 300
  lateral flux, 245
  matching, 254–305
  MOS, 229, 239–241
  ONO, 229
  oxide, 229
  parallel-plate, 227
  parasitics, 235–237
  plate, 231
  poly-poly, 229, 241–243
  random variation, 258
  relative permittivity, 227
  sandwich (stacked) capacitor, 240
  stack, 243–245
  standard bipolar, 92–93
    junction, 92
  thin-film, 229, 252
  unit, 261, 279, 300
  variability, 232–235
    process variation, 232–233
    voltage modulation and temperature variation, 233–235
Capillary, 76
Carbon, 4
Carrier lifetime, 6
Carrier saturation velocity, 311
Carriers, 5, 7–8
  majority, 8
  minority, 8
Cathedral PNP, 329
Cathode(s), 14, 232, 410
  Schottky diodes, 17
CB-shorted diode, 88
CDM clamps, 136, 572–573
  self-protecting, 573
Cell area estimation, 582–584
  capacitors, 582
  computing cell area, 584
  lateral PNP transistors, 583
  MOS power transistors, 584
  MOS transistors, 583
  resistors, 582
  vertical bipolar transistors, 583
Cells, 581
  unit, 611
Centroidal symmetry, principle of, 276
Centroids, 521
  defined, 276
Ceramic capacitors, 246
Chained implants, 56
Chameleon, 618
Chamfers, defined, 338
Channel routing, 594
Channel stop implants, 100–101, 160
  and CMOS transistors, 452–453
Channel stoppers, 160
Channels, 26, 57
Charge spreading, 156–157, 295
  defined, 157
  and parasitic channels, 156–164
    effects, 156–158
    preventative measures, 159–164
Charge storage capacitors, 295
Charged device model (CDM), 135
  development of, 572
Chemical mechanical polishing (CMP), 61, 70
Chemical vapor etching, 46–47
Chirality, of a transistor, 524
Choke points, 593
Christmas-tree device, 372–373
CHST mask, 616–617
Chstop (coding layer), 488–489
Circular loop inductors, 247
Clad gate, 439
Clad moats, 112, 426, 439, 562
  technology, 211
Clad poly, 67
  process, 209–210

Clamping diode, 314–315
Clamps:
  antiparallel diode, 570
  buffered Zener, 566–568
  CDM, 136, 572–573
    self-protecting, 573
  grounded-gate NMOS, 570–571
  lateral SCR, 573–575
  rate-triggered SCR, 575
  Schottky, 313
  self-protecting CDM, 573
  two-stage Zener, 565–566
  V_{CES}, 568–570
  Zener, 563–568
    buffered, 566–568
    two-stage, 565–566
Cleavage planes, 39
CMOS latchup, 98, 170–171
CMOS transistors:
  constructing, 446–467
    backgate contacts, 464–467
    channel stop implants, 452–453
    coding the MOS transistor, 447–449
    N-well and P-well processes, 449–452
    scaling the transistor, 456–459
    threshold adjust implants, 453–455
    variant structures, 459–464
Coding, 448
Coding layers, 448
Coincidence, rule of, 280
Collector-diffused isolation (CDI), 115, 450
  NPN transistor, 345–346
Collector efficiency, 316
Collector-to-base breakdown, 25
Comb capacitors, 231–232
Common-centroid layout, 277–281
  and thermoelectrics, 287
Common-mode variations, 597
Compactness, rule of, 280
Complementary metal-oxide-semiconductor (CMOS), 306, 431
Complementary MOS (CMOS) circuits, 97
Composite dielectrics, 300–302
  compared to oxide dielectrics, 300
Compound semiconductors, 9
Conduction, 4
Conductivity, 186
  modulation, 195, 290
Constant-field scaling, 456–458
Constant-voltage scaling, 456
CONT layer, 614–616
CONT mask, 616–617
Contact OR, 86
Contact potential, 12, 287
  Schottky barriers, 16
Contact resistance, 189
Contact spiking, 62
Contacts, 85
  backside, 181
  Ohmic, 19–21
  over active gate, 518
Contamination, 143–147
  dry corrosion, 144–145
  mobile ion, 144
Control gate, floating-gate transistors, 470
Copper metallization, 71–73
  dual damascene copper, 71
  power copper process, 71–72
Core-limited design, 585
Counterdoping, 8
Covalent bonding, 3
Critical current density, 365, 367
Cross-couple arrayed capacitors, 302
Cross-coupled pair, 281
Cross-coupling, 302
Cross injection, 178–179, 535
Crossing leads, techniques for, 553–555
Crosstalk, 593
Crossunders, 94, 201
Cruciform-emitter transistors, 373–374
Crystal, 611
Crystal growth, 38–39
Crystalline sodium chloride, 3
Cubic crystal, Miller indices of, 611–613
Current-controlled device, 307
Current crowding, 249, 325
Current gain, 23
Current hogging, 313
Czochralski process, 38–39
Czochralski silicon, 58, 140

D

Dangling bonds, 150, 353, 437
Darlington pair, 539–540
Debiasing:
  emitter, 363
  intrafinger, 363–364, 364
  Ohmic, 537–538, 545
  substrate, 165–169, 182
Deep N-well, 345
Deep-P+ diffusion, 377–378
Deep-P+ isolation, 115
  diffusion, 81
DEEPN layer, 614–616
DEEPN mask, 618
Deglazing, 53
Dendrites, 136
Depletion-mode NMOS, MOS transistors, 28
Depletion-mode transistors, 434
Depletion regions, 11–13
  thickness of, 13
Device doping, and hot carrier injection (HCI), 149–150

Device matching, rules for, 295–303
Device physics, 1–36
Device transconductance, 432–434
Diamond crystals, 4
Diborane, 53
Die area estimation, 584–587
Die planning, 581–588
  cell area estimation, 582–584
  die area estimation, 584–587
  gross profit margin (GPM), 587–588
Dielectric breakdown, 133, 138–141, 182
  effects, 138
  preventative measures, 139–141
Dielectric capacitance, 229
Dielectric constant, 227
Dielectric, defined, 227
Dielectric isolation, 60–62
Dielectric polarization, 146
Dielectric relaxation, 294–295, 300
Dielectric strength, 227–228
Differential pair (diff pair), 387
Differential trimming, 218–219
Diffused dummies, 297–298
Diffused resistors, 299
Diffusion, 9–11
  defined, 9
  deposition (predeposition), 51
  doping profile, 54
  drive/drive-in, 51
  emitter push, 54
  and etch effects, 516–519
  limitations of, 53–54
  NBL push, 54
  outdiffusion, 54
  oxidization-enhanced, 55
Diffusion currents, 10
Diffusion interactions, 268–270
Diffusion penetration of polysilicon, 517–518
Diode-connected transistors, 88, 406–409
Diode rectification, 14
Diodes, 13, 406–429
  avalanche diodes, 19, 409
  in BiCMOS processes, 422–425
    BiCMOS Schottky diodes, 423–425
  CB-shorted diode, 88
  clamping diode, 314–315
  in CMOS processes, 422–425
    CMOS junction diodes, 422–423
    CMOS Schottky diodes, 423–425
  forward-biased diodes, 14–16
  history of, 406
  light-emitting diode (LED), 6
  matching, 425–428
    matching PN junction diodes, 425–426
    matching Schottky diodes, 428
    matching Zener diodes, 426–428
  P-type Schottky diodes, 18
  palladium silicide Schottky diodes, 419
  platinum silicide Schottky diodes, 419
  PN diodes, 13–16
  Schottky, 16–18, 94, 415–420, 454, 538–539
  in standard bipolar, 406–421
    diode-connected transistors, 406–409
    power diodes, 420–422
    Schottky diodes, 415–420
    Zener diodes, 409–415
  surface Zener diodes, 410–412
Direct electron tunneling, 138
Direct write on wafer (DWW), 74
Discrete ceramic capacitors, 246
Dishing, 70
Dispersion, rule of, 280
Distributed backgate contacts, 466
Distributed capacitance, 197
DMOS, *See* Double-diffused MOS (DMOS)
Dogbone resistor, 190–191
Donors, 7
Dopant-enhanced oxidation, 48–49
Doped semiconductors, 7
Double-diffused drain (DDD), 485
Double-diffused JFETs, 478
Double-diffused MOS (DMOS), 482, 505–511
  compared to the DDD transistor, 505
  DMOS NPN, 510–511
  lateral DMOS (LDMOS) transistor, 506–508
  RESURF transistors, 508–510
Double-level metal, 94
Double polysilicon self-aligned transistors, 354–355
Downbond, 181
Drawing layers, 448
Drawn dimensions, 457
Drawn length, 105, 187, 190, 448
Drawn width, 105, 187–188, 190, 448–449
Drift, 9–11
  defined, 9–10
  long-term, 273, 293
Drift current, 10
Drift region, 484
Dry corrosion, 144–145, 182
  effects, 144–145
  preventative measures, 145
Dry etching, 46
Dry oxides, 44–45
Dual-collector transistors, 318
Dual damascene copper, 71–72
Dual-doped poly CMOS transistors, 455
Dual in-line package (DIP), 75, 589
Dumbbell resistor, 190
Dummy gate oxidation, 50, 101–102
Dummy resistors, 266–267
Dummy tiles, 521
Dynamic antisaturation circuit, 379

E

Early effect, 25, 395, 402
Early voltage, 395, 402
Eddy currents, 248–249
Edge effects, 514
Effective gate voltage, 432
Effective length, 448
Effective mobility, 433, 522
Effective width, 449
Electrical overstress (EOS), 133–143
  antenna effect, 141–143
  dielectric breakdown, 133, 138–141
  electromigration, 133, 136–138
  electrostatic discharge (ESD), 134–136
Electrical SOA, 493–496
Electrically erasable programmable read-only memories (EEPROM), 468
Electrically programmable read-only memory (EPROM), 221
Electrodes, 227
Electromigration, 133, 136–138, 182, 600–602
  effects, 136–137
  preventative measures, 137–138
Electron-blocking guard rings (EBGRs), 546
Electron-collecting guard rings (ECGRs), 173, 546
Electron guard rings, 546–547
Electron-hole pairs, formation of, 6
Electron shading, 141
Electrostatic discharge (ESD), 133, 134–136, 182, 430 *See also* ESD structures
  effects, 135
  preventative measures, 135–136
  self-protecting circuitry, 136
Electrostatic interactions, 288–295
  charge spreading, 292–293
  dielectric polarization, 293–294
  dielectric relaxation, 294–295
  voltage modulation, 288–292
Electrostatic shielding, 290
Electrostatically shield matched capacitors, 302
Electrothermal SOA, 493, 496–497
Elongated-emitter lateral PNP, 336
EMIT layer, 614–616
Emitter ballasting, 363, 365
Emitter-base Zener diodes, 544
Emitter-coupled pair, 387
Emitter current focusing, 311
Emitter debiasing, 363
Emitter degeneration, 384–386
Emitter-in-iso buried Zener, 413
Emitter injection efficiency, 23
Emitter punchthrough, 64
Emitter push, 54
Emitter resistors, 91, 201–202
Emitter saturation current, 307
Emitter tunnels, 555–556
Enhancement-mode MOS transistor, 434
Enhancement-mode NMOS (enhancement NMOS), 28
EOS, *See* Electrical overstress (EOS)
Epi-base transistors, 347
Epi-FETs, 476–478
Epi pinch resistors, 205–206, 476
Epitaxy, 57–59
  buried layers, 58
  liquid-phase, 57
  low-pressure chemical vapor deposited (LPCVD), 57–58
  NBL shadow, 59
  pattern shift, 59
EPROM trims, 221–222
Erasable programmable read-only memory (EPROM), 221, 468
ESD, *See* Electrostatic discharge (ESD)
ESD structures, 561–578
  antiparallel diode clamps, 570
  CDM clamps, 572–573
  grounded-gate NMOS clamps, 570–571
  lateral SCR clamps, 573–575
  selecting, 575–578
  V_{CES} clamp, 568–570
  Zener clamps, 563–568
    buffered, 566–568
    two-stage, 565–566
Etch-and-regrowth technique, 490
Etch effects, and diffusion, 516–519
Etch guards, 266
Etch rate variations, 265–267
  dummy resistors, 266–267
  poly-poly capacitors, 266–267
Etching, 188
Excess minority carrier concentration, 11
Extended-base NPN transistors, 347
Extended drain, 486–487
Extended-drain, high-voltage transistors, 113–114
Extended-voltage transistors, 482–491
  channel length modulation can be minimized and punchthrough averted either by reducing the operating, 483
  extended-drain transistors, 486–489
  LDD and DDD transistors, 483–486
  multiple gate oxides, 489–491
  punchthrough, 483
Extrinsic drain, 484
Extrinsic semiconductors, 6–9
  technology, 9

F

Failure mechanisms, 133–184
  contamination, 143–147
  electrical overstress (EOS), 133–143
  of NPN power transistors, 362–368
    emitter debiasing, 362–363

hot spot, 364
Kirk effect, 366
secondary breakdown, 364–365
thermal runaway, 364–365
parasitics, 164–182
surface effects, 148–164
FAMOS transistor, 469–472
Farad (F), 226–227
Faraday shielding, 290
Fast BiCMOS processes, 341
Fast bipolar transistors, 349–351
Fast surface states, 437
Ferromagnetic materials, 247
FETs, *See* Field-effect transistors (FETs)
Field-effect transistors (FETs), 25, 430–481 *See also* Complementary metal-oxide-semiconductor (CMOS); Junction field-effect transistors (JFETs); Metal-oxide-semiconductor (MOS) transistors
floating-gate transistors, 467–474
history of, 430
junction field-effect transistors (JFETs), 430, 474–479
metal-oxide-semiconductor field-effect transistors (MOSFETs), 430
Field oxides, 45, 50
Field plates, 152–153, 333–334, 511
Field plating, 160–162, 299
Field regions, 99, 446
Field relief guard ring, 417
Field-relief structure, 114, 487
Filamentation, 495
Fill metal, 70, 521
Filler-induced stresses, 272
Filleted corners, 230
Fingers, 523, 542, 545
Fixed oxide charge, 150, 436
Flange, 604
Flanging, 161
Flash memories, 468
Flats, 39–40
Flawed device mergers, 535–539
Floating-gate avalanche-injection metal-oxide-semiconductor (FAMOS) transistor, 469–472
Floating-gate transistors, 431, 467–474
accumulation of charges within the oxide, 472
control gate, 470
defined, 468
Fowler-Nordheim tunneling, 471
operation principles, 469–472
single-poly EEPROM memory, 472–473
Floating-gate tunneling oxide (FOTOX) transistors, 471–472
Floorplan, 581
defined, 588–589
Floorplanning, 588–594
Force leads, 597
Forming gas, 330
Forward active current gain, 306
Forward active region, 22
Forward-bias safe operating area (FBSOA), 311–312
Forward-biased diodes, 14–16
behavior of, 14–15
current through, 15–16
Schottky diodes, 17
FOTOX transistor, *See* Floating-gate tunneling oxide (FOTOX) transistors
Four-terminal NPN, 318
Fowler-Nordheim tunneling, 139
and floating-gate transistors, 471
Fringing, 229–230, 302
Fuse trim schemes, 219
Fused leadframe, 181, 376
Fused silica, 43
Fuses, 216–219
aluminum, 217
polysilicon, 217

G

Gallium arsenide, 9
Gate-coupled NMOS (GCNMOS), 571–572
Gate delay, 457
Gate dielectric, 431–432
Gate doping block mask, 210
Gate-induced drain leakage (GIDL), 445–446
Gate oxide capacitors, 229
Gate oxide integrity (GOI), 140
General matching rules, 296–297
Germanium, 4, 140
Giant isotope effect, 150
Gigabytes (GB), 467
Glass, 53
Gold ballbonds, 560
Gold preform, 75, 273
Gradients, 521
Grains, 136
Greatest common factor, 279
Gross profit margin (GPM), 587–588
Grounded-gate NMOS clamps, 570–572
Grounded-gate NMOS (GGNMOS), 571
Grounded guard rings, 546
Grown-junction transistors, 50–51
Guard rings, 160, 172–175, 442–443, 534, 545–550
in CMOS and BiCMOS designs, 548–551
electron, 546–547
electron-collecting (ECGRs), 173
grounded, 546
hole-blocking (HBGRs), 176–177
hole-collecting (HCGRs), 173
minority-carrier-collecting, 172–173
multiple, 177
standard bipolar electron guard rings, 546–547
standard bipolar hole guard rings, 547–548
Gummel number, 23, 340, 348, 351

H
H-emitter transistor, 374
Heat sink, 284
Henry (H), 246
Heterojunction bipolar transistors (HBTs), 357
High-level injection, 24, 366
High-permittivity capacitors, 246
High-quality oxide dielectrics, 295
High-sheet implant, 95
High-sheet poly resistors, 211
High-sheet-resistance (HSR) implants, 202–203
High-sheet resistors, 94–95, 202–205
High-side drive (HSD) layout, 508
High-voltage bipolar transistors, 337–340
Hillocks, 136
Holding, 444
Hole-blocking guard ring (HBGR), 176–177, 547
Hole-collecting guard ring (HCGR), 173, 421, 547–548
Holes, 5
Hot carrier injection (HCI), 148–151, 182, 445
  and decrease in threshold voltages, 149–150
  effects, 148–150
  MOS transistors, 31–32
  preventative measures, 150–151
Hot carriers, 10, 387, 442
Hot-dog transistor, 336
Hot hole degradation, 106
Hot-plug event, 572
Hot spot, 310, 364
Human body model (HBM), 134
Hydrogen compensation, 152, 270
Hydrogenation, 270–271

I
IC Graph (Mentor Graphics), 627
III-V compound semiconductors, 9
Impact ionization, 18
Implant energy, 56
Indirect-bandgap semiconductors, 6
Inductance, 246–252
  defined, 246
Inductors:
  circular loop, 247
  construction, 250–252
  defined, 226, 246
  integrating, guidelines for, 251–252
  parasitics, 248–250
    current crowding, 249
    eddy currents, 248–249
    skin effect, 249
  planar, 247–248
Ingots, 39
Insulated-gate field effect transistor (IGFET), 29
Integrated capacitors, categories of, 252
Integrated transformers, 248
Integrating inductors, guidelines for, 251–252
Interconnection:
  parasitics, 261–263
  single-level, 551–556
  top-level, 594–604
Interconnection parasitics, 261–263
Interdigitated arrays, 277–278
Interdigitated backgate contacts, 466
Interdigitated-emitter transistor, 369–371
Interdigitation pattern (or weave) ABBA, 278
Interlevel oxide (ILO), 110, 197, 563
International System of Units (SI), 186, 226, 246
Intrafinger debiasing, 363–364
Intrinsic collector current, 313
Intrinsic semiconductors, 6–7
Inverse moat mask, 99
Inversion, 26
  strong, biasing mode, 239
Ion implantation, 28, 55–57
  buried layer, 56
  chained implants, 56
  implant dose, 56
  implant energy, 56
Ionic bonding, 2–3
Ions, 3
Isobaric contour plot, 275–276
Isobars, 275–276
Isolation mask, 82
Isothermal contour plot, 284
Isotherms, 285
Isotropic nature, of wet etching, 47

J
JFETS, *See* Junction field-effect transistors (JFETs)
Junction capacitance, 349–350
Junction capacitors, 92, 230–233, 252, 300
  forward-biasing of, 239
  series resistance of, 238
  value of, 239
Junction field-effect transistors (JFETs), 32–34, 430, 474–479
  annular N-JFET, 478–479
  asymmetric, 34
  backgate, 32
  body, 32
  channel, 32
  channel length modulation, 33–34
  depletion regions, 32–33
  double-diffused JFETs, 478
  drain, 32–33
  epi-FETs, 476–478
  gate, 32
  history of, 474
  layout, 476–479
  linear region, 32
  modeling, 474–475
  N-well JFET, 477–478
  P-channel JFETs, 478

pinchoff voltage, 475
saturation, 33
saturation current, 475–476
source, 32
symmetric, 34
turnoff voltage, 33–34
Junction isolation (JI), 81–82
Junction leakage, 443–444
magnitude of, 444
Junction temperature, 283
Junction-to-ambient thermal impedance, 283
Junction-to-case thermal impedance, 284
Junctions, *See* PN junctions

K
K2 Technologies, 618
Kelvin connections, 206–207, 593, 597–598
Kilobytes (KB), 467
Kirk effect, 366–368, 495–496
Kooi effect, 50, 101

L
L-EDIT (Tanner Research), 627
Laser trims, 222–223
Latchup, 165, 171
Lateral autodoping, 59, 117
Lateral DMOS (LDMOS) transistor, 506–508
Lateral flux capacitors, 245
Lateral NPN transistors, 344
Lateral outdiffusion, 452–453
Lateral PNP transistors, 89–90, 348–349, 377, 543
betas, 90
constructed in analog BiCMOS, 349
saturation in, 315–316
self-alignment of, 90
Lateral SCR clamps, 573–575
Lateral transistors, matching rules, 400–402
Layout editor software, sources of, 627
Layout rule syntax, 618–621
extent info, 620–621
overhang, 620
overlap, 619–620
spacing, 619
width, 618
Layout rules (sample), 614–621
Leakage, 16
Leakage current, 443–444
Leaker, 142
Least significant bit (LSB), 218
Light-emitting diode (LED), 6
Lightly doped drain (LDD) transistors, 112–113, 150, 485
Linear region, 432
MOS transistors, 30
Linewidth control, 191
Liquid-phase epitaxy, 57
Local oxidation of silicon (LOCOS), 49–50
LOCOS processing and dummy gate oxidation, 100–101
Long-tailed pair, 387
Long-term drifts, 273, 293
Look-ahead trimming, 219
Low-dropout (LDO) voltage regulators, 377
Low-level injection, 366
Low-pressure chemical vapor deposited (LPCVD) epitaxy, 57–58
Low-risk merged devices, 541–545
Low-side drive (LSD) layout, 508

M
Machine model (MM), 134–135
Magnetically coupled, use of term, 247
Majority-carrier device, 17
Majority carriers, 8
Mask layers, 448
Matched devices, 254
Matched resistors, 191, 193
Matching coëfficient, 257
Matching lateral transistors, rules for, 400–402
Matching, MOS transistors, 511–527
Matching PN junction diodes, 425–426
Matching Schottky diodes, 428
Matching vertical transistors, rules for, 397–399
Matching Zener diodes, 426–428
Mathematical derivations, 622–626
Maze routing, 594
Mean, 255
Meander resistors, 189
Mechanical stress and package shift, 271–273
Megabytes (MB), 467
Memories, 467
flash, 468
Merged bipolar-CMOS (BiCMOS) processes, 80
Merged devices, 534–545
flawed device mergers, 535–539
low-risk merged devices, 541–545
role in analog BiCMOS, 544–545
successful device mergers, 539–541
Metal-oxide-semiconductor field-effect transistors (MOSFETs), 430, *See also* Field-effect transistors (FETs)
modeling, 431–438
device transconductance, 432–434
threshold voltage, 434–438
operation, 431–446
parasitics of, 438–446
breakdown mechanisms, 440–442
CMOS latchup, 442–443
leakage mechanisms, 443–446
Metal-oxide-semiconductor (MOS) transistors, 25–32, 35
accumulation, 26
applications of, 482–533
asymmetric, 27
backgate, 26–27

backgate doping, 27–28
backgate effect, 31
behavior of, 30
blocking voltage, 483
body, 26
body effect, 31
channel, 26
channel length modulation, 31
common-centroid layout of, 523–527
depletion-mode NMOS (depletion NMOS), 28
double-diffused MOS (DMOS), 505–511
drain, 26–27
enhancement-mode NMOS (enhancement NMOS), 28
extended-voltage transistors, 482–491
gate, 26
gate dielectric, 26
gate voltage, 26–27
hot carrier injection, 31–32
I-V characteristics, 29–30
inversion, 26
linear region, 30
matching, 511–527
channel length modulation, 515
contacts over active gate, 518
diffusion and etch effects, 516–520
diffusion penetration of polysilicon, 517–518
diffusions near the channel, 518–519
and fill metal, 521
gate area, 513–514
gate oxide thickness, 514–515
geometric effects, 513–516
hydrogenation, 520–521
minimal matching, 528
moderate matching, 528
orientation, 515–516
oxide thickness gradients, 522
PMOS versus NMOS transistors, 519–520
polysilicon etch rate variations, 516–517
precise matching, 528
rules for, 528–531
stress gradients, 522
thermal gradients, 522–523
thermal/stress effects, 521–523
MOS capacitor, 26
N-channel MOS transistors, 27
ohmic region, 30
operating voltage, 483
P-channel MOS (PMOS) transistors, 27
pinched-off channel, 30–31
power MOS transistors, 491–511
saturation region, 30
source, 26–27
subthreshold conduction, 31
symbols, 28–29
symmetric, 27
threshold adjust implant, 28
threshold voltage, 26, 27–28
triode region, 30
V_t adjust implant, 28
Metal pitch, 595
Metal resistors, 206–208
METAL1 layer, 614–616
METAL1 mask, 616–617
METAL2 mask, 616–617
Metallic bonding, 2
Metallization, 62–73
copper, 71–73
deposition and removal of aluminum, 63–65
interlevel oxide (ILO), 69–70
multilevel oxide (MLO), 69
planarization, 69–70
protective overcoat (PO), 63, 69–71
refractory barrier metal, 65–67
silicidation, 67–68
single-level-metal (SLM) interconnection system, 62–63
spin-on glass (SOG), 69–70
Metallurgical-grade polysilicon, 37
Metallurgical junction, 11, 268
Metals, 1–2
Metric system, 186
Mil, 86
Miller effect, 97, 350
Miller indices, 40
of cubic crystal, 611–613
Minimal matching, 296, 396
MOS transistors, 528
Minority-carrier-collecting guard rings, 172–173
Minority-carrier device, 16
Minority-carrier injection, 165, 169–180, 182, 444–445, 537, 545
effects, 169–170
preventative measures:
cross-injection, 178–180, 182
substrate injection, 171–178
Minority-carrier lifetime, 349
Minority carriers, 8
Mismatch coefficient:
of capacitance, 258
of resistance, 259
Mismatches:
causes of, 257–295
diffusion interactions, 268–270
electrostatic interactions, 288–295
etch rate variations, 265–267
hydrogenation, 270–271
interconnection parasitics, 261–263
mechanical stress and package shift, 271–273
pattern shift, 263–265
photolithographic effects, 267–268
process biases, 260–261
random variation, 257–260

stress gradients, 273–283
temperature gradients and thermoelectrics, 283–288
device matching, rules for, 295–303
measuring, 254–257
random, 256
six-sigma, 256–257
standard deviation of the, 255
systematic, 256, 257
three-sigma, 256–257
Mixed-signal integrated circuits, 114
Moat-2 geometry, 490–491
MOAT mask, 618
Moat regions, 50, 99
Mobile ion contamination, 144
effects, 145
preventative measures, 146–147
Mobile ions, 182, 293
defined, 145
Mobile oxide charge, 436
Mobility, of holes, 5
Mock layouts, 551–553
Moderate matching, 296, 396
MOS transistors, 528
Modulation:
conductivity, 195
tank, 195
voltage, 193
Molybdenum headers, 274
Monocrystalline silicon, piezoresistivity of, 275
MOS capacitors, 26, 229, 239–241
MOS transistors, *See* Metal-oxide-semiconductor (MOS) transistors
MOSFETS, *See* Metal-oxide-semiconductor field-effect transistors (MOSFETs)
Mount pad, 75, 589
Multilevel oxide (MLO), 103
Multiple gate oxides, 489–491

N

N-bars, 179–180, 537, 543
N-buried layer (NBL), 58–59, 263
and analog BiCMOS, 116–117
N-channel MOS transistors, 27
N-epi, 449–450
N-source/drain (NSD) implant, 447
N-type channel stop implant, 100
N-type gate poly (NPoly) mask, 455–456
N-type semiconductors, 7–8
N-well, 98, 449
processes, 449–452
N-well BiCMOS processes, 451
N-well CMOS processes, 99
N-well resistors, 211–212
Native thresholds, 453
Native transistors, 454
Natural thresholds, 453
Natural transistors, 454
NMOS, 105–106
PMOS, 106–107
Natural V_t mask, 454
NatVT mask, 454
NBL layer, 614–615
NBL mask, 618
NBL permeability, 550
NBL push, 54
NBL shadow, 59, 263, 263–265
NBL tunnels, 556–557
Negative bias temperature instability, 154–156, 182
effects, 155
preventative measures, 155–156
Negative resist, 42
Nesting, 126
Neutral base region, 23
Nichrome, 212, 222
NMoat, 105, 208, 447, 553
NMOS transistors, 104–106
PMOS vs., 519–520
symmetric extended-drain, 487–488
Noble silicides, 416
Nodes, 141–142
Noisy signals, 598–599
Nonlinearity, resistors, 193–195
Nonmetals, 1–2
Nonvolatile memory (NVM), 467
NPN Darlington transistors, 542
NPN power transistors:
failure mechanisms of:
emitter debiasing, 362–363
hot spot, 364
Kirk effect, 366–368
secondary breakdown, 364–365
thermal runaway, 364–365
NPN transistors, 9
standard bipolar, 86–88
NSD and PSD resistors, 211
NSD mask, 616–617
NSD/P-epi leakers, 143
NWELL mask, 616

O

Offset voltage, 381
Ohmic contacts, 19–21
Ohmic debiasing, 537–538, 545
Ohmic region, MOS transistors, 30
Ohms, 186
Ohm's law, 10
Ohms per square, 90
One-dimensional array, 280–281
One-time programmable (OTP) EPROM, 222
1 1/2-level metal, 110
ONO capacitors, 229

Operating voltage:
- metal-oxide-semiconductor (MOS) transistors, 483
- rating, 105

Optical shrink, 457–458
Outdiffusion, 54, 188, 350
Overvoltage stress testing (OVST), 140–141
Oxide capacitors, 229
Oxide-isolated transistors, 354–356
Oxide sidewall spacer, 112, 484–485
Oxide step, 48
Oxide thickness gradients, 522
Oxides, 43–50
- boron suckup, 48
- buried oxide (BOX), 60–61
- chemical vapor etching, 46–47
- dry, 44
- dry etching, 46
- effects of growth and removal, 47–49
- field, 45, 50
- growth and deposition, 44–45
- local oxidation of silicon (LOCOS), 49–50
- pad, 98, 99
- phosphorus plow, 48
- plasma etching, 46–47
- reactive ion etching (RIE), 46–47
- removal, 45–47
- wet, 45
- wet etching, 46

Oxidization-enhanced diffusion, 55
Oxygen precipitates, 140

P

P-bars, 178–180, 537, 543
P-buried layer (PBL), 93
P-channel MOS (PMOS) transistors, 27
P-epi, 449–452
P-source/drain (PSD) implant, 447
P-type channel stop implant, 100
P-type gate poly (PPoly) mask, 455–456
P-type Schottky diodes, 18
P- type semiconductors, 8
P-type V_t adjust (PVT), 455
P-well, 449
- defined, 449
- processes, 449–452

P-well CMOS process, 99
Package shifts, 272–273
Packing factor, 586
Pad-limited design, 585
Pad oxides, 98, 99
Padring, 584
- bondpads, 76, 534, 558–562
- constructing, 557–562
- scribe streets and alignment markers, 557–558
- testpads, 534, 558–562
- trimpads, 217, 534, 558–562

Palladium silicide Schottky diodes, 419
Paper dolls, 553
Parallel-plate capacitors, 227
Parasitic channels:
- and charge spreading, 156–164, 182
  - effects, 156–158
  - preventative measures, 159–164

Parasitic components, 164, 197
Parasitic PNP, 313
Parasitic transistors, 452
Parasitics, 164–182
- capacitors, 235–237
- inductors, 248–250
  - current crowding, 249
  - eddy currents, 248–249
  - skin effect, 249
- minority-carrier injection, 169–180
- of MOS transistors, 438–446
  - breakdown mechanisms, 440–442
  - CMOS latchup, 442–443
  - leakage mechanisms, 443–446
- resistors, 197–200
- substrate debiasing, 165–169
- substrate influence, 180–182

Pattern distortion, 82, 263–264
Pattern shift, 59, 263–265
- magnitude of, 264

Pattern washout, 264
Patterning, 43
Pellicles, 43
Perimeter utilization factor, 587
Periodic table, 1–2
Peripheral antenna ratio, 141
Peripheral capacitance, 231
Peripheral variations, 257
Permeability of free space, 247
Permittivity of free space, 433
Phosphine, 53
Phosphors, 6
Phosphorus, 7, 52
- and hydrogen compensation, 271

Phosphorus-doped silicon, 7
Phosphorus oxychloride ($POCl_3$), 52
Phosphorus pileup, 48, 54
Phosphorus plow, 48
Phosphosilicate-doped glass (PSG) layer, 86
Phosphosilicate glass (PSG), 293
Photolithographic effects, mismatches, 267–268
Photolithography, 41–43, 188
- patterning, 43
- pellicles, 43
- photomasks, 42–43
- photoresists, 41–42
- reticles, 42–43
- stepped working plate, 43

Photomasks, 42–43, 268

Photoresists, 41–42
Piezoresistivity, 274–275
Pilling-Bedworth ratio, 47
Pinch resistors, 91–92
Pinched base, 92
Pinched-off channel, 30–31
Pinchoff, 30
Pinchoff voltage, 205
Pinholes, 228
Planar inductors, 247–248
Planar junction, 19
Planar process, 51
Planarization, 69–70
  resist etch-back, 70
Plasma etching, 46–47
Plate capacitors, 231
Platinum silicide Schottky diodes, 419
PMoat, 105, 179, 208, 447, 553
PMOS thick-field threshold, 154
PN diodes, 13–16
PN junctions, 9, 11–21
  depletion regions, 11–13
  Ohmic contacts, 19–21
  PN diodes, 13–16
  Schottky diodes, 16–18
  summary of behavior of, 13
  Zener diodes, 18–19
PNP transistors, standard bipolar, 88–89
Pocket implants, 486
Poly dummies, 297
Poly-emitter transistors, 154
Poly-poly capacitors, 229, 241–243
  etch rate variations, 266–267
Poly resistors, 195, 208–211, 265, 299
Poly routing, and primary routing channels, 596
POLY1 mask, 616–617
POLY2 mask, 617
Polycide, 300
Polycrystalline, 37
Polycrystalline silicon, 275
Polysilicon deposition, 59
Polysilicon emitter (poly emitter), defined, 352
Polysilicon-emitter transistors, 351–353
  advantages of, 353
Polysilicon etch rate variations, 516–517
Polysilicon fuses, 217
Polysilicon-gate CMOS, 96–114
  available devices, 104–109
    capacitors, 109
    NMOS transistors, 104–106
    PMOS transistors, 106–107
    resistors, 107–109
    substrate PNP transistors, 107
  capacitors, 109
  essential features, 97–98
  fabrication sequence, 98–104
    channel stop implants, 100
    contacts, 103
    epitaxial growth, 98
    inverse moat, 98–99
    LOCOS processing and dummy gate oxidation, 100–101
    metallization, 103
    N-well diffusion, 98–99
    polysilicon deposition and patterning, 102
    protective overcoat (PO), 103–104
    source/drain implants, 102–103
    starting material, 98
    threshold adjust, 101–102
  PMOS transistors, 106–107
  process extensions, 109–114
    double-level metal, 110
    extended-drain, high-voltage transistors, 113–114
    lightly doped drain (LDD) transistors, 112–113
    shallow trench isolation (STI), 110–111
    silicidation, 111–112
  resistors, 107–109
  substrate PNP transistors, 107
POR layer, 614–615
POR mask, 616–617
Ports, 461
Positive bias temperature instability (PTBI), 155
Positive resist, 42
Postpackage trimming, 273
Power BiCMOS processes, 341
Power bipolar transistors, 361–381
  failure mechanisms of NPN power transistors, 361–368, 362–368
    emitter debiasing, 362–363
    hot spot, 364
    Kirk effect, 366
    secondary breakdown, 364–365
    thermal runaway, 364–365
  layout of power NPN transistors, 368–376
    Christmas-tree device, 372–373
    cruciform-emitter transistor, 373–374
    interdigitated-emitter transistor, 369–371
    power transistor layout in analog BiCMOS, 374–376
    selecting a power transistor layout, 376
    wide-emitter narrow-contact transistor, 371–372
  matching bipolar transistors, 381–386
    emitter degeneration, 384–386
    filler-induced stress, 393–395
    NBL shadow, 386–387
    random variations, 382–384
    stress gradients, 391–393
    systematic mismatch, 395–396
    thermal gradients, 387–391
  power PNP transistors, 376–378
  saturation detection and limiting, 378–381
Power copper process, 71–72
Power-delay product, 457

Power diodes, 420–422
Power MOS transistors, 491–511
compared to bipolar transistors, 491
conventional, 498
computation of metallization resistance, 500–501
diagonal device, 500
nonconventional structures, 503–505
rectangular device, 499–500
electrical SOA, 493–496
electrothermal SOA, 493, 496–497
MOS safe operating area (SOA), 492–493
rapid transient overload, 497–498
Power packaging, 273
Power PNP transistors, 376–378
Power transistor layout:
in analog BiCMOS, 374–376
selecting, 376
Power transistors, defined, 491
Precise matching, 296, 396–397
MOS transistors, 528
Prels, 461
Primary routing channels, and poly routing, 596
Probe card, 74
Probe pads, 577
Probe yield, 588
Process biases, 260–261
and the length of a resistor, 260
Process extensions, 93–96
polysilicon-gate CMOS:
double-level metal, 94
Schottky diodes, 94
up-down isolation, 93
Process plasma-induced damage, 141
Process transconductance, 433
Programmable read-only memory (PROM), 467–468
Protective overcoat (PO), 69–71
Proteins, 74
Proximity effect, 167
PSD mask, 616–617
PSD/N-well leakers, 143
Punchthrough, 440–441, 483
stoppers, 486
stops, 441

Q

Quality factor (Q), 249–250
Quartz, 37, 42
Quasisaturation, 321
Quatrefoil, 427

R

Radio-frequency (RF) integrated circuits, 246
Random mismatches, 256
Random variation, 257–260
capacitors, 258
resistors, 258–259
Rapid transient overloads, 497–498
Rate-triggered SCR clamps, 575
Ratioed bipolar transistors, 360–361
Ratioed Schottky diodes, 428
Reactive ion etching (RIE), 46–47
Reactive ions, defined, 47
Read-only memory (ROM), 467
Recombination centers, 6
Rectifiers, 14
Reduced surface field (RESURF) transistors, *See* RESURF transistors
Reduction in volume effect, 550
Reflow, 65
Refractory barrier metal, and electromigration, 137
Refractory silicides, 416
Relative permeability, 247
Relative permittivity, 227
Remote sensing, 597–598
Resist etch-back planarization, 70
Resistance, 185
contact, 189
mismatch coefficient of, 259
sheet, 187
Resistivity, 186
Resistor matching, rules for, 296–300
Resistor variability:
contact resistance, 196
nonlinearity, 193–195
process variation, 191–192
temperature variation, 192–193
Resistors, 185–225
adjusting resistor values, 213–223
base, 199
base pinch, 191, 202
contact resistance, 196
diffused, 213
dogbone, 190–191
dumbbell, 190
emitter, 201–202
epi pinch, 205–206
high-sheet, 202–205
laid out as cross-coupled pairs, 281
layout, 187–190
matched, 191, 193
matching, 254–305
meander, 189
metal, 206–208
N-well, 211–212
nonlinearity, 193–195
NSD and PSD, 211
parasitics, 197–200
poly, 195, 208–211
random variation, 258–259
resistivity, 191-196
and sheet resistance, 185–187
serpentine, 189, 191, 299

standard bipolar, 90–92
base, 90–91
emitter, 91
high-sheet, 94–95
pinch, 91–92
thin-film, 212–213
trimming, 216–223
EPROM trims, 221–222
fuses, 216–219
laser trims, 222–223
Zener zaps, 219–220
tweaking, 213–216
metal options, 215–216
sliding contacts, 214–215
trombone slide, 215
uses of, 185
variability, 191–196
contact resistance, 196
nonlinearity, 193–195
process variation, 191–192
temperature variation, 192–193
RESURF technology, 482
RESURF transistors, 150, 508–510
adaptive RESURF, 510
early high-voltage devices, 510
lateral-vertical depletion region interaction, 509
Reticles, 42–43
Retrograde well, 177, 441, 449–450
Reverse active region, 22
Reverse-biased diode, behavior of, 14
Reverse-biased junction, 312–313
Reverse-biased Schottky diode, 17
Reverse breakdown, 18
Reverse conduction, 16
Reverse recovery time, 312
Rough sketches, 551–553
Routing factor, 586
Rules of common-centroid layout, 280

S

Sacrificial gate oxide, 101–102
Sample layout rules, 614–621
layout rule syntax, 618–621
extent info, 620–621
overhang, 620
overlap, 619–620
spacing, 619
width, 618
polysilicon-gate CMOS rules, 616–618
standard bipolar rules, 614–616
Sampling, 444
Sandwich (stacked) capacitors, 240
Saturation region, 24–25, 432
MOS transistors, 30
Saturation voltage, 318
Scaling factor S, 456
Schottky barriers, 16, 94
Ohmic conduction, 20
Schottky-clamped NPN, 315
Schottky clamps, 313
Schottky contact mask, 416
Schottky diodes, 16–18, 94, 415–420, 454, 538–539
anode, 17
and base-collector junctions, 349–350
cathode, 17
current-voltage characteristics, 17
P-type, 18
ratioed, 428
reverse-biased, 17
Scribe seals, 147–148, 557
Scribe streets, 557–558
Secondary breakdown, 311–312, 364–365
Secondary breakdown voltage, 365
Secondary collector, 378–379
Seebeck coefficient, 287
Seebeck effect, 287
Seebeck voltage, 20–21
Selective gate shrinks, scaling laws for, 458
Self-aligned implanted subcollector (SIC), 355
Self-aligned structures, 56–57
Self-protecting, 562
Self-protecting CDM clamps, 573
Self-protecting devices, 562
Semiconductor-grade polysilicon, 38
Semiconductors, 1–11, 37–79
assembly, 73–77. *See also* Silicon manufacture
Alloy-42, 75
ball, 76–77
ball bonding, 76–77
bondpads, 76
capillary, 76
direct-step-on-wafer (DSW) processing, 74
direct write on wafer (DWW), 74
dual-in-line package (DIP), 75
gold preform, 75
lead frame, 75
mount and bond, 74–75
mount pad, 75
packaging, 77
probepads, 76
process control structures, 73
stitch bond, 77
test dice, 73–74
testpads, 76
wafer-level testing (wafer probing), 74
wedge bonding, 76
capillary, 76
compound, 9
covalent bonding, 3
defined, 2
diffusion, 50–55
diffusion and drift, 9–11

direct-bandgap, 6
direct-step-on-wafer (DSW) processing, 74
direct write on wafer (DWW), 74
dual-in-line package (DIP), 75
extrinsic, 6–9
generation and Recombination, 4–6
gold preform, 75
history of, 37
III-V compound semiconductors, 9
impediments to manufacturing large quantities of, 37
indirect-bandgap, 6
ion implantation, 55–57
lead frame, 75
metallization, 62–73
N-type, 7–8
oxide growth and removal, 43–50
P-type, 8
photolithography, 41–43
patterning, 43
pellicles, 43
photoresists, 41–42
reticles, 42–43
stepped working plate, 43
silicon deposition and etching, 57–62
silicon manufacture, 37–40
Semisimple figure, 333
Sense leads, 597
Sense transistors, 542
Sensitive signals, 598–599
Serpentine resistors, 189, 191, 299
Serpentine transistors, 461–462
Shadow effect, 172
Shallow N-well, 345
Shallow P-well, 345, 423
Shallow trench isolation (STI), 61, 110–111
and oxide-isolated transistors, 354
polysilicon-gate CMOS, 110–111
Shallow-well transistors, 345–346
Sheet resistance, 90, 187
Shells, 2
Shichman-Hodges equations, 432, 444, 448, 492
Short-emitter effect, 342
Short resistor segments, 298
Shrinking a transistor, 457–458
Shrunk dimensions, 457
Sichrome, 212, 222
SiGe, 356
Silica, 37
Silicidation, 111–112
Silicide block mask, 210, 343
Silicided polysilicon, 300
Silicides, 18
Silicon, 4
crystal structure of, 39–40
defined, 37
Silicon-controlled rectifier (SCR), 171, 573–575
Silicon crystal, 3–4
Silicon deposition and etching, 57–62
dielectric isolation, 60–62
epitaxy, 57–59
polysilicon deposition, 59
Silicon dioxide (SiO_2), 43–44
Silicon-germanium transistors, 356–358
Silicon ingot, 612
Silicon manufacture, 37–40, 39
(100) plane, 40–41
(111) plane, 40–41
cleavage planes, 39
crystal growth, 38–39
Czochralski process, 38–39
flats, 39–40
ingots, 39
metallurgical-grade polysilicon, 37
Miller indices, 40
polycrystalline, 37
quartz, 37
semiconductor-grade polysilicon, 38
silica, 37
silicon:
crystal structure of, 39–40
defined, 37
unit cells, 39–40
wafers, 39
Silicon nitride, 70
Silicon-on-insulator (SOI), 60
Silicon-on-sapphire (SOS), 60
SIMOX (separation by implanted oxygen) process, 61–62
Single-diffused drain (SDD) transistors, 485–486
Single-level interconnection, 551–556
mock layouts, 551–553
stick diagrams, 553
Single-level metal (SLM), 94
Single-poly EEPROM memory, 472–474
Single-well processes, 449–450
Singly doped drain (SDD), 112
Sinker, 83–84
Sintering, 64
Six-sigma mismatch, 256–257
Skin effect, 249
Sliding contacts, 214–215
Sliding heads, 215
Slow surface states, 437
Small-signal devices, 491
Small-signal transistors, 320
Snapback, 309
Soakage, 246, 294–295
Sodium line, 86
Sodium metal, 2
Soft breakdown, 410
Solenoid, 247
Solid-state device, defined, 1

Space charge layer, 12–13
Specific on-resistance, 498
SPICE, 186, 434
Spin-on glass (SOG), 45, 53
  metallization, 69–70
Split-collector lateral PNP transistors, 333, 543–544
Split collector transistors, 334–335
  examples of, 334
Split field plates, 294
Sputtering, 65
Square SOA characteristic, 494–495
Squares, 187
Stack capacitors, 243–245
Staged oxidation, 490
Standard bipolar, 81–96
  base implant, 84
  base-over isolation (BOI), 84
  capacitors, 92–93
    junction, 92
  contact, 85
  deep-N+ diffusion (sinker), 83–84
  diodes in:
    diode-connected transistors, 406–409
    power diodes, 420–422
    Schottky diodes, 415–420
    Zener diodes, 409–415
  electron guard rings, 546–547
  emitter diffusion, 84–85
  emitter pilot, 85
  essential features, 81–82
  fabrication sequence, 82–86
    epitaxial growth, 83
    N-buried layer, 82–83
    starting material, 82
  hole guard rings, 547–548
  isolation diffusion, 83
  lateral PNP transistors, 330–337
  metallization, 85–86
  NPN transistors, 86–88, 320–326
  phosphosilicate-doped glass (PSG) layer, 86
  PNP transistors, 88–89
  polysilicon-gate CMOS, process extensions, 93–96
  protective overcoat (PO), 86–87
  resistors, 90–92
    base, 90–91
    emitter, 91
    high-sheet, 94–95
    pinch, 91–92
  small-signal transistors, 320–340
    high-voltage bipolar transistors, 337–340
    standard bipolar lateral PNP transistors, 330–337
    standard bipolar NPN transistors, 320–326
    standard bipolar substrate PNP transistor, 326–330
    super-beta NPN transistors, 340–341
  substrate PNP transistor, 326–330
  thick emitter oxide, 85
  thin emitter oxide, 85
Star node, 597
Stepped working plate, 43
Steppers, 43
Stepping, 43
Stick diagrams, 553
Stitch bond, 77
Straggle, defined, 56
Stress gradients, 273–283, 393, 522
  common-centroid layout, 277–281
  defined, 276
  gradients and centroids, 275–277
  location and orientation, 281–283
  piezoresistivity, 274–275
Stress-induced leakage current (SILC), 139, 445, 472
Stress triangles, 603
Strong inversion, biasing mode, 239
Subcircuit models, 197
Substrate debiasing, 165–169, 182
  dielectrically isolated substrates, 169
  effects, 166–167
  heavily doped substrates, 168
  lightly doped substrates with heavily doped isolation, 168
  lightly doped substrates with lightly doped isolation, 169
  preventative measures, 167–169
Substrate influence, 180–182
  effects, 180
  preventative measures, 180–181
Substrate PNP, 88–90
Substrate terminal, 312
Subsurface Zeners, 412
Subthreshold conduction, 437, 444
  MOS transistors, 31
Super-beta transistors, 25, 96, 340–341
Surface effects, 148–164
  avalanche-induced beta degradation, 153–154
  hot carrier injection (HCI), 148–151
  negative bias temperature instability, 154–156
  parasitic channels and charge spreading, 156–164
  Zener walkout, 151–153
Surface scattering, 433
Surface state charge (Q_{ss}), 45, 437
Surface states, 45
Surface Zener diodes, 410–412
Sustain voltage, 440, 494
Symmetric extended-drain transistors:
  NMOs, 487–488
  PMOS, 488
Symmetric junction field-effect transistors (JFETs), 34
Symmetric metal-oxide-semiconductor (MOS) transistors, 27
Symmetry, rule of, 280
Systematic mismatches, 256, 257
Système Internationale (SI), *See* International System of Units (SI)

T

Tail, diffusions, 268
TANK layer, 614–615
Tank modulation, 195, 288–289, 299
Tanks, 81
Tantalum capacitors, 246
Temperature coefficient of resistivity (TCR), 192–193
Temperature gradients and thermoelectrics, 283–288
Test dice, 73–74
Testpads, 76, 534, 558–562, 577
  size of, 561
Tetraethoxysilane (TEOS), 45
Thermal feedback, 387
Thermal gradients, 285–287, 522–523
Thermal impedance, 283
Thermal runaway, 310, 364–365
Thermal voltage, 307
Thermoelectric effect, 20, 287
Thermoelectric potential, 287
Thick-field oxide, 563
Thick-field threshold, 82
Thick-field transistors, 571
Thin-film capacitors, 229, 252
Thin-film interference, 45
Thin-film resistors, 212–213
Thin-film tantalum capacitor, 246
Three-sigma mismatches, 256–257
Three-terminal NPN, 318
Threshold adjust implant mask, 454
Threshold adjust implants, and CMOS transistors, 453–455
Threshold voltage, 26, 432
  of a MOS transistor, 434–438
Tiling, 281
Time-dependent dielectric breakdown (TDDB), 139, 433, 441, 562–563
Tombstone PNP, 329
Top-level interconnection, 594–604
  channel routing, principles of, 594–596
  electromigration, 600–602
  special routing techniques, 596–600
  stress effects, minimizing, 603–604
Transconductance, *See* Field-effect transistors (FETs)
Transformers, 248
Transistors, 406–409
  adjusted, 454
  annular, 462–464, 518
  bent-gate, 505
  BICMOS small-signal bipolar, 341–358
  bipolar junction transistors (BJTs), 21–25, 35, 306–359, 361–381
  depletion-mode, 434
  diode-connected, 87
  double-diffused MOS (DMOS), 505, 508–510
  double polysilicon self-aligned, 354–355
  dual-collector, 318
  dual-doped poly CMOS, 455
  enhancement-mode MOS, 434
  epi-base, 347
  extended-base NPN, 347
  extended-drain, high-voltage, 113–114
  FAMOS, 469–472
  fast bipolar, 349–351
  field-effect transistors (FETs), 25, 430–481
  floating-gate, 431, 467–474
  floating-gate tunneling oxide (FOTOX), 471–472
  grown-junction, 50–51
  heterojunction bipolar (HJBTs), 357
  high-voltage bipolar, 337–340
  hot-dog, 336
  insulated-gate field effect (IGFET), 29
  interdigitated-emitter, 369–371
  junction field-effect (JFETs), 32–34, 430, 474–479
  lateral NPN, 344
  lateral PNP, 348–349, 377, 543
  lightly doped drain (LDD), 112–113, 150
  metal-oxide-semiconductor (MOS), 430–46
    applications of, 482–533
    matching, 511–527
  native, 454
  natural:
    NMOS, 105–106
    PMOS, 106–107
  NPN, 121–123
  oxide-isolated, 354–356
  P-channel MOS (PMOS), 27
  parasitic, 452
  PNP, 123–125
  polysilicon-emitter, 351–353
  power bipolar, 361–381
  power MOS, 491–511
  RESURF, 150, 508–510
  serpentine, 461–462
  shallow-well, 345–346
  sharing common source or drain connections, 459–460
  shrinking, 457–458
  single-diffused drain (SDD), 485–486
  small-signal transistors, 320
  split collector transistors, 334–335
    examples of, 334
    lateral PNP transistors, 543–544
  standard bipolar small-signal, 320–340
  super-beta, 25, 96
    NPN, 340–341
  symmetric extended-drain:
    NMOS, 487–488
    PMOS, 488
  symmetric junction field-effect (JFETs), 34
  symmetric metal-oxide-semiconductor (MOS), 27
  thick-field, 571
  vertical, 313, 397–399, 583
  waffle, 505
  walled-emitter, 354–355

washed-emitter, 350
wide-emitter narrow-contact, 371–372
Transmission gate, 443
Trap-assisted tunneling, 139
Traps, 6
Trench isolation, and latchup immunity, 177–178
Trigger voltage, 440
Trimmers, 213
Trimming, 468
differential, 218–219
look-ahead, 219
Trimming resistors, 216–223
EPROM trims, 221–222
fuses, 216–219
laser trims, 222–223
Zener zaps, 219–220
Trimpads, 217, 534, 558–562
size of, 561
Triode region, 432
MOS transistors, 30
Trombone slide, 215
Tubs, 81
Tunneling, 18–19, 138–139
direct electron, 138
Fowler-Nordheim, 139
trap-assisted, 139
Zener diodes, 18–19
Tunneling oxide, 471
Tunnels, 94, 201, 534, 551
base, 555
emitter, 555–556
NBL, 556–557
types of, 555–556
Turn, 247
Turnoff time, 312
Tweaking resistors, 213–216
metal options, 215–216
sliding contacts, 214–215
trombone slide, 215
Tweaks, 213
Twin-well process, 449
Two-dimensional array, 281
Two-stage Zener clamps, 565–566

U

Ultrahigh-vacuum chemical vapor deposition (UHVCVD), 356
Unconnected dummies, 266
Unit capacitors, 261, 279, 300–301
Unit cells, 39–40, 611
Up-down isolation, 93

V

Vacuum tubes, 1
Valence, 2
Valence electrons, 2
Variant structures, CMOS transistors, 459–464
ΔV_{BE} generator, 207
V_{CES} clamp, 568–570
Velocity saturation, 11
Vertical transistors, 313, 583
matching rules, 397–399
Vertilat PNP, 329
VIA mask, 616–617
Virtuoso (Cadence Design Systems), 627
Volatile memory, 467
Voltage:
bandgap, 394
blocking, 483
rating, 105
breakdown, 409
Early, 395, 402
offset, 381
operating:
MOS transistors, 483
rating, 105
pinchoff, 205
saturation, 318
Seebeck, 20–21
sustain, 440, 494
thermal, 307
threshold, 26, 432, 434–438
trigger, 440
Voltage modulation, 193
Voltmeter, 12

W

Wafer boat, 44
Wafer bonding, 61–62
Wafer cleaving, 62
Wafer-level testing (wafer probing), 74
Wafer utilization factor, 588
Wafers, 39
Waffle transistors, 504–505
defects, 505
Walled-emitter transistors, 354–355
Washed emitter, defined, 350
Washed-emitter transistor, 350
Wedge bonding, 76
Wells, 449
Wet etching, 46
Wet oxides, 45
Whiskers, 600
Wide-emitter narrow-contact transistor, 371–372
Width bias, 188
Wiresweep, 560

Z

Zapping, 220
Zener breakdown, 19

Zener clamps, 563–568
  buffered, 566–568
  two-stage, 565–566
Zener diodes, 18–19, 409–415, 544
  avalanche multiplication, 18
  buried Zeners, 412–415
  impact ionization, 18
  reverse breakdown, 18
  surface, 410–412
  tunneling, 18–19
Zener walkback, 152
Zener walkout, 151–153, 182, 411
  buried Zeners, 152
  defined, 151
  effects, 151–152
  preventative measures, 152–153
Zener zaps (zap Zeners), 209, 219–220, 468, 561–562
Zero bias, 14
Zero-bias capacitance, 230
Zero-biased Schottky diode, 17